Chemical Modification of Biological Polymers

SERIES EDITOR

Roger L. Lundblad
Lundblad Biotechnology
Chapel Hill, North Carolina, U.S.A.

PUBLISHED TITLES

Approaches to the Conformational Analysis of Biopharmaceuticals
Roger L. Lundblad

Application of Solution Protein Chemistry to Biotechnology
Roger L. Lundblad

Approaches to the Conformational Analysis of Biopharmaceuticals
Roger L. Lundblad

Development and Application of Biomarkers
Roger L. Lundblad

Chemical Modification of Biological Polymers
Roger L. Lundblad

Chemical Modification of Biological Polymers

Roger L. Lundblad

PROTEIN SCIENCE SERIES

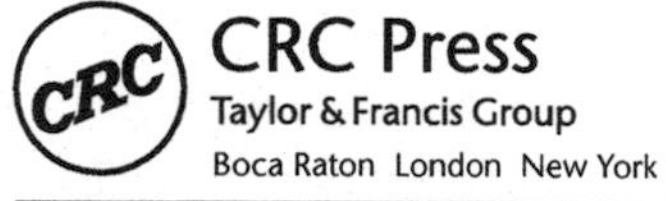

CRC Press is an imprint of the
Taylor & Francis Group, an informa business

CRC Press
Taylor & Francis Group
6000 Broken Sound Parkway NW, Suite 300
Boca Raton, FL 33487-2742

First issued in paperback 2019

ISBN-13: 978-1-4398-4898-2 (hbk)
ISBN-13: 978-0-367-38241-4 (pbk)

Visit the Taylor & Francis Web site at
http://www.taylorandfrancis.com

and the CRC Press Web site at
http://www.crcpress.com

This book is dedicated to
Dr. Christine Vogel Sapan and other students who
have become colleagues over time and
provided continued inspiration through
insightful and penetrating questions.

Contents

Preface

This work is intended to provide a comprehensive review of the chemical modification of biopolymers including proteins, nucleic acids, and polysaccharides. That said, I clearly understand that I have missed considerable information. This has become painfully apparent as I used multiple information retrieval systems. An article that might be found by one system is totally missed by other systems. This is compounded by my own personal retrieval system, which is based on some 45+ years of working in protein chemistry. So, apologies to those investigators whom I have missed; I would appreciate receiving notice of omitted materials. The explosion in current literature has compounded the problem as has what appears to be a total breakdown in any effort to standardize abbreviations and acronyms.

I have tried to document the development and use of reagents rather than focusing exclusively on current use. In doing this, I have taken the liberty of including some personal observations about some studies, most notably those in the laboratories of Stanford Moore and William Stein at the Rockefeller Institute (now Rockefeller University).

Perusal of any contents of current biochemistry journal, even those with protein or proteomics in the title, will show that the chemical modification of biopolymers is not a "hot" topic. However, I still felt that the material in this book should be placed into a format that can be more easily retrieved in today's electronic environment. That said, I am mightily suspicious of the current electronic environment (see *The Shallows* by Nicholas Carr). Regardless of format, I hope that this information will be of value to current investigators.

Roger L. Lundblad
Chapel Hill, North Carolina

Acknowledgments

I am indebted to the usual suspects, including the long-suffering and patient Barbara Norwitz and equally patient and even longer-suffering Jill Jurgensen for their help in bringing this material to print. I am also indebted to Professor Bryce Plapp at the University of Iowa for his continued and somewhat inexplicable patience with the thermodynamically challenged.

Author

Roger L. Lundblad is a native of San Francisco, California. He received his undergraduate education at Pacific Lutheran University and his PhD in biochemistry at the University of Washington. After postdoctoral work in the laboratories of Stanford Moore and William Stein at The Rockefeller University, he joined the faculty of the University of North Carolina at Chapel Hill. He joined the Hyland Division of Baxter Healthcare in 1990. Currently, Dr. Lundblad works as an independent consultant at Chapel Hill, North Carolina, and writes on biotechnological issues. He is an adjunct professor of pathology at the University of North Carolina at Chapel Hill.

1 Functional Groups in Biopolymers and Factors Influencing Reactivity

The goal of this chapter is to introduce some basic concepts of organic chemistry, which provide the basis for the reactivity of the several functional groups on biological polymers. While the polymers are diverse, the various functional groups are similar and common in organic chemistry, and, thus, there are some general concepts that apply to proteins, nucleic acids, and polysaccharides; reactivity of alkyl hydroxyl groups is one example as such would include serine or threonine hydroxyl in proteins, the ring hydroxyl groups in ribose/deoxyribose in nucleic acids, and ring hydroxyls in polysaccharides. Likewise, amino groups are present on proteins, nucleic acids, and polysaccharides.

Reactivity of individual functional groups is influenced by, for example, ionization with sulfhydryl groups or the formation of a neutral species with amino groups or carboxyl groups. As will be shown in the following chapters, thiol groups are essentially unreactive, while the thiolate anion is the reactive species. The protonated amino group is essentially unreactive; release of the proton to produce the neutral amine is associated with reactivity. While model compounds can be a guide to ionization, local electrostatic factors have a profound effect. This latter consideration will be mentioned several times in the following text in an attempt to underscore the importance of this concept.

Most chemical modification reactions of biological polymers are S_{N^2} reactions (substitution, nucleophilic, bimolecular) of second order, although there are examples of S_{N^1} (substitution, nucleophilic, unimolecular) reactions and free-radical-mediated reactions. There are a few examples of elimination reactions such as the formation of dehydroalanine from serine or cysteine.[1–4] The author cannot understate the importance of considering natural biopolymers such as polysaccharides, nucleic acids, and proteins as organic polymers and not macromolecules endowed with vitalistic properties. It is granted that the author, having worked with blood coagulation proteins years ago, might be a little sensitive to this issue.

The biological polymers can be composed of diverse monomer units as is the case with proteins, substantially less diverse with nucleic acids, and occasionally homopolymers with polysaccharides. Additional diversity is added through the modification of monomer units as with the various posttranslational modifications of proteins and, for example, sulfation of polysaccharides. It is not possible to be as inclusive of material as the author would like and the reader is referred to other reviews on this topic.[5–22] In addition, there are several volumes of *Methods in Enzymology*,[23–33] which are extremely useful as well as other review articles.[34–37] The reader is strongly advised to consult this early literature, even if it means actually going to the library, as much original art has been lost in passing from review to review. Extreme caution should be used in statements starting with "It is widely known..." or "The reaction was performed by the method of...".

Reaction of functional groups in biopolymers depends (mostly)* on the nucleophilicity of the specific functional group. So, let me start with a question from a work in 1987, "...What is a nucleophile?"[38] Consideration of various concepts suggests that, with biopolymers, the definition is based on the kinetic data for a substitution or displacement reaction[39]; a more practical definition may be the possession of a pair of electrons that can form a new bond with another molecule.[40] The kinetic data may yield conclusions that are based on extrinsic conditions, such as solvent, as well as intrinsic nucleophilicity since intrinsic nucleophilicity can be enhanced.[38] Understanding of intrinsic nucleophilicity can be challenged,[41] prompting study of gas phase reactions to avoid solvent issues.[42] An electrophile can accept a pair of electrons. Nucleophiles can be considered as "bases" and electrophiles as "acids."[43,44] The concept of hard and soft bases and acids[43] provides insight into intrinsic nucleophilicity.[44] For example, a sulfur center nucleophile, which is a "soft" nucleophile, reacts more rapidly with an alkylating agent (soft electrophile) than does an oxygen center nucleophile, while an acylating agent is a harder electrophile and the advantage of the sulfur center nucleophile is reduced.[44] Hard metal ions such as Mg^{2+} prefer binding to oxygen, while soft metal ions such as Cu^{2+} prefer sulfur.[45]

Solvation has a potent effect on nucleophilic substitution reactions.[44] For example, the order of reactivity of halides in the displacement of tosylate from *n*-hexyl tosylate in MeOH (I > Br > Cl) is reversed in dimethylsulfoxide.[46] Landini and Maia[47] observed that the second-order rate constant for the S_{N^2} substitution reaction of *n*-hexyl methane sulfonate by various inorganic anions does vary with solvent. While there was little difference in the rate constant for iodide in MeOH and DMSO (5.8×10^4 M^{-1} s^{-1} versus 5.3×10^4 M^{-1} s^{-1}), the value for chloride was $0.9 \times 10 M^{-1}$ s^{-1} in MeOH and $36 \times 10 M^{-1}$ s^{-1} in DMSO. Abrams has examined the effect of solvent on nucleophilic substitution and suggests that the major effect is on the transition state.[48] Solvent environment/local electrostatic potential also influences functional group pK_a, and, hence, nucleophilicity and the effect of local environment on the reactivity of functional groups is discussed in more detail in the following.

Most of the work on the chemical modification of biological polymers used proteins; as such, much of the chemistry in the literature is derived from work on proteins. Nonetheless, the basic chemistry is of value for a given functional group, such as an amino group, irrespective of polymer type. It might be useful to introduce the concept of selective chemical modification versus nonselective chemical modification. Selective chemical modification is described as modification of a given functional group in a biological polymer such as the modification of cysteine residues in proteins with maleimides,[49] the modification of adenine nucleobases with diethylpyrocarbonate,[50] the grafting of Lucifer yellow VS dyes onto chitosan chains and

* Reaction rate can also be enhanced by increasing local concentration as with the use of affinity reagents.

linking with glutaraldehyde,[51] and the selective introduction of functional groups onto silicon nitride/silicon oxide surfaces and subsequent modification with glutaraldehyde to provide sites for immobilzation.[52] A somewhat different approach is taken by Yalpani[53] with respect to polysaccharides where nitration (cellulose nitrate) or acetylation (cellulose acetate) are considered nonselective modifications, while esterification (formation of tosylate) at the primary position (position 6) is a selective modification (favored 200 to 1 over modification at the 3rd position in cellulose). Fixation of tissues with formaldehyde or glutaraldehyde is another example of nonspecific chemical modification albeit on a macroscale as is the browning reaction in cooking. Cohen[54] discusses selectivity versus reactivity with respect to carbonyl halides. Here, the less reactive the halide, the more selective the reaction and thus the discrimination between various functional groups; thus, a fluoride derivative is less reactive than a chloride derivative. There is further discussion of this concept in Chapter 3.

Proteins are the most complex of the biological polymers considered in the current work. Excluding posttranslational modification, there are 20 naturally occurring amino acids found in proteins (18 L-amino acids, 1 imino acid, and glycine); posttranslational modifications include glycosylation, phosphorylation, methylation, acetylation, hydroxylation, sulfation, and the attachment of C-terminal GPI anchors.[55,56] The individual amino acids vary in nucleophilic character with some that have aliphatic side chains such as leucine and isoleucine are considered essentially unreactive except for free radical insertion, while others vary considerably in reactivity as, for example, with serine and cysteine. Seven of the 20 amino acids (lysine, histidine, arginine, tyrosine, tryptophan, aspartic acid, and glutamic acid) have functional side chains that are subjected to facile modification; serine and threonine can be modified with chemical reagents but with more difficulty (unless as with serine residues at enzyme active sites). Five of these residues (usually) carry a charge at physiological pH and modification of three residues, lysine, aspartic acid, and glutamic acid, can change the charge and properties of a protein.[57–59] Modification of arginine can also change the charge but is not pursued as frequently. In addition, modification can be accomplished at the C-terminal carboxyl and amino-terminal amino group. The acid dissociation constants for "typical" amino acid functional groups are presented in Table 1.1. Some more recent data[60–62] has also been included in Table 1.1 and deserves comment. In particular, note the difference in the values for the sulfhydryl group of cysteine where the "older" value is 10.46, while the more recent value is 6.8 ± 2.7.[62] The later value[62] is an average value for a cysteine residue in a protein with a range from 2.5 to 11.1; the value for cysteine in an alanine pentapeptide is 8.6. The ionization of a specific functional group in a protein is influenced by intrinsic pK_a of the specific functional group and the effect of the local electrostatic potential.[63–66] The intrinsic pK_a is the pK_a of the functional group transferred from bulk solution into a protein with no interaction with other functional groups in that protein. In the case of cysteine mentioned earlier, the low pK_a value is usually associated with a residue involved in catalytic function.[67]

Proteins vary considerably in composition. Globular proteins are differentiated from outer membrane proteins[68] and from connective tissue proteins such as elastin and collagen; elastin and collagen contain disproportionate amounts of protein and

TABLE 1.1
Dissociation of Ionizable Groups in Proteins

Potential Nucleophile	pK_a[a]	pK_a[b]
γ-Carboxyl (glutamic acid)	4.25	4.2 ± 0.9
β-Carboxyl (aspartic acid)	3.65	3.5 ± 1.2
α-Carboxyl (isoleucine)	2.36	3.3 ± 0.8
Sulfhydryl (cysteine)	10.46	6.8 ± 2.7
α-Amino (isoleucine)	9.68	7.7 ± 0.5
Phenolic hydroxyl (tyrosine)	10.13	10.3 ± 1.2
ε-Amino (lysine)	10.79	10.5 ± 1.1
Imidazole (histidine)	6.00	6.6 ± 1.0
Guanidino (arginine)	12.48	—
Serine (hydroxyl)		13.06 ± 0.5[c]

[a] Taken from Mooz, E.D., Data on the naturally occurring amino acids, in *Practical Handbook of Biochemistry and Molecular Biology*, G.D. Fasman (ed.), CRC Press, Boca Raton, FL, 1989. Also see Dawson, R.M.C., Elliott, D.C., Elliott, W.H., and Jones, K.M., *Data for Biochemical Research*, Oxford University Press, Oxford, U.K., 1969.

[b] Taken from Grimsley, G.R., A summary of the measured p*K* values of the ionizable groups in folded proteins, *Protein Sci.* 18, 247–251, 2009.

[c] Determined for *N*-acetylserineamide. See Bruice, T.C., Fife, T.H., Bruno, J.J., and Brandon, N.E., Hydroxyl group catalysis. II. The reactivity of the hydroxyl group of serine. The nucleophilicity of alcohols and the ease of hydrolysis of their acetyl esters as related to their pKa, *Biochemistry* 1, 7–12, 1962.

glycine as well as unique residues such as hydroxyproline and hydroxylysine.[69,70] It is a bit difficult to make generalizations about the amino composition of proteins but some amino acids such as histidine, tryptophan, and methionine are usually present at low concentrations, while alanine and valine are present at higher concentrations.[68,71] Accessibility of amino acid residues to solvent is a variable[72,73] and efforts are made to use compositional data to predict solution behavior based on residue exposure.[74–78] I would be remiss if I did not acknowledge the contributions[79,80] of the late Fred Richards to the concept of surface and buried residues in proteins. The concept of surface and buried residues can be ascribed to early work by Fred Richards at Yale University. Arthur Lesk has written an excellent book[81] on protein structure, which provides a lucid summary for accessible and buried surface area. Miller and coworkers[82] evaluated the solvent-accessible surface residues in 46 monomer proteins. The majority of exposed surface area is provided by hydrophobic amino acids (58%) with lesser contribution from polar (24%) and charged (19%) amino acids; interior residues (buried) are 58% hydrophobic and 39% polar but only 4% charged. There is asymmetric distribution of accessible residues[83] consistent with the existence of hydrophobic residues at domain interface regions[83,84] and the existence of anion-binding exosites[85] important for regulatory protease function.[86]

Functional group availability is also a factor in the modification of nucleic acids, for example, by formaldehyde.[87] Formaldehyde does not react with hydrogen-bonded exocyclic amino groups in nucleobases in DNA and RNA[88–91] and is therefore a probe of nucleic acid structure.[92–98] Formaldehyde distinguishes between single-stranded and doubled-stranded nucleic acids.[95,99] Formaldehyde does denature nucleic acids.[92,100] Reaction with formaldehyde in supramolecular complexes can reveal information about DNA–protein interactions[101,102] and can be reversible.[101–103] However, such interactions can be missed because of their rapid nature.[104] The reaction of formaldehyde is used to identify regions in DNA where hydrogen bonding is in equilibrium ("breathing").[92,105] Reaction with formaldehyde has a long history in the preparation of vaccines,[106,107] toxoids,[108] and allergoids.[109] Formaldehyde (formalin) has a long history of use for "fixing" tissue prior to clinical analysis.[110–113]

Nucleic acids (oligonucleotides and polynucleotides) are biological heteropolymers. While proteins are heteropolymers composed of 20 or more individual monomer units (see above), nucleic acids have fewer monomer units. The monomer unit of a nucleic acid is referred to as a nucleotide*; a nucleotide is composed of a phosphoryl group covalently bound to either the 3′ or 5′ hydroxyl of a ribose or deoxyribose moiety coupled to a nitrogenous base via glycosidic linkage. The 2′-hydroxyl group does have a role in transesterification reactions in intron splicing, hammerhead ribozyme, and other RNA cleavage reactions[114] and can be modified by selected electrophiles,[115] although it should be noted that the 2′-hydroxyl group has a high pK_a (ca. 12.50).[116] The acid–base properties of a nucleic acid reside in nitrogenous bases that are referred to as nucleobases, a combination word formed with nucleotide and base. The nucleobases in RNA are adenine, guanine, cytosine, and uracil; the nucleobases in DNA are adenine, guanine, cytosine, and thymine. Table 1.2 provides a partial listing on ionizable groups in nucleic acid and derivative forms. The pK_a values for nucleobases in ribozymes have been suggested to be modulated by metal ions,[117] raising the low pK_a values on the nucleobases to a physiological range,[118] although other explanations have been provided.[119] Nucleoside pK_a values are perturbed toward neutrality in RNA and DNA.[120] There are some more recent studies[121,122] on the ionization of nucleobases in ribozymes as well as an active-site labeling study.[123] It is possible to selectively modify a specific base in a polynucleotide using the concept of complementarity addressing (addressed; sequence-specific) modification[124,125] where a reactive group such as an haloalkyl function[126] is attached to an oligonucleotide sequence "specific" for binding to the target DNA sequence.[127] The chemical modification of nucleic acids is discussed in detail in Chapter 6. Current interest in the chemical modification of nucleic acids is directed at the use of footprinting to determine site of a nucleic acid–protein interaction[128] and the formation of DNA adducts.[129–131] The approach to chemical modification of nucleic acids in the current work focuses on the reaction of chemical reagents with nucleic acids and precursor nucleobases. Selective 2′-hydroxyl acylation analyzed by primer extension

* The term nucleotide was originally used to define the phosphoryl derivative of a nucleoside within an RNA or DNA molecule. The term has a broader definition today in describing any phosphorylated derivative of a nucleoside with a nucleoside defined as glycoside consisting of ribose or deoxyribose in glycosidic linkage with a heterocyclic nitrogenous base.

TABLE 1.2
Dissociation Constants for Ionizable Groups in Nucleic Acids[a]

Functional Group	pK_a
Adenine amino group	4.15,[b] 4.12[c]
Adenine imino group (N7)[d]	9.80,[b] 9.67[c]
Adenosine-5′-monophosphate (phosphoric acid)	6.40[e]
8-Hydroxyadenosine (N1)	2.9[f]
8-Hydroxyadenosine (N7)	8.7[f]
Guanine amino group	3.3[b]
Guanosine N7	2.11[g]
Guanine imino group (N1)	9.2[b]
7-Methylguanosine N1	7.01[g]
Uracil imino group	9.5[b]
Uracil-6-carboxylic acid (orotic acid) carboxyl group	2.07[e]
Uracil-6-carboxylic acid imino group	9.45[e]
Uracil-5-carboxlic acid (isoorotic acid) carboxyl group	4.16[e]
Uracil-5-carboxylic acid imino group	8.89
Cytosine amino group	4.60[b]
Cytosine imino group	12.16[b]
Pyrimidine (1,3-diazine; metadiazine)	1.30[b]
Purine pK_1 (N7)	2.52[b]
Purine pK_2 (N1)	8.90[b]

[a] The serious reader is directed to some of the more classic studies on nucleic acids including Chargaff, E. and Davidson, J.N. (eds.), *The Nucleic Acids: Chemistry and Biology*, Academic Press, New York, 1955; Saenger, W., *Principles of Nucleic Acid Structure*, Springer-Verlag, New York, 1984; Neidle, S. (ed.), *Oxford Handbook of Nucleic Acid Structure*, Oxford University Press, Oxford, U.K., 1999; Neidle, S., *Principles of Nucleic Acid Structure*, Elsevier, Amsterdam, the Netherlands, 2008.

[b] Bendich, A., Chemistry of purines and pyrimidines, in *The Nucleic Acids. Chemistry and Biology*, E. Chargaff and J.N. Davidson (eds.), Academic Press, New York, Chapter 3, pp. 81–136, 1955. The data was obtained by spectrophotometry and/or titration.

[c] At 30°C (Lewis, S. and Tann, N.W., Reactions of nucleic acids and their components. Part II. Thermodynamic constants of adenine, *J. Chem. Soc.* 1466–1467, 1962) using pH titration with a glass electrode.

[d] More recent studies on the acid–base properties of nucleobases have used NMR technology (Kampf, G., Kopinos, L.E., Griesser, R. et al., Comparison of acid-base properties of purine derivatives in aqueous solution. Determination of intrinsic proton affinities of various basic sites, *J. Chem. Soc. Perkin Trans.* 2, 1320–1327, 2002) as a reflection of the power of NMR for the study of acid–base chemistry at the molecular level (Jameson, R.F., Hunter, G., and Kiss, T., A ^{1}H nuclear magnetic resonance study of the deprotonation of L-Dopa and adrenaline, *J. Chem. Soc. Perkin Trans.* 2, 1105–1110, 1980).

TABLE 1.2 (continued)
Dissociation Constants for Ionizable Groups in Nucleic Acids[a]

[e] pH titration with KOH (Tucci, E.R., Doody, E., and Li, N.G., Acid dissociation constants and complex formation constants of several pyrimidine derivatives, *J. Phys. Chem.* 65, 1570–1574, 1961).

[f] ^{15}N NMR spectroscopy (Cho, B.P. and Evans, F.E., Structure of oxidatively damaged nucleic acid adducts. 3. Tautomerism, ionization and protonation of 8-hydroxyadenosine studied by ^{15}N NMR spectroscopy, *Nucleic Acids Res.* 19, 1041–1047, 1991). A pK_a value of 4.19 for the N1 hydrogen of adenine was determined by calorimetry (Zimmer, S. and Biltonen, R., The thermodynamics of proton dissociation of adenine, *J. Solution Chem.* 1, 291–298, 1972).

[g] Potentiometric and ^1H NMR (Kampf, G., Kapinos, L.E., Griesser, R. et al., Comparison of the acid-base properties of purine derivatives in aqueous solution. Determination of intrinsic proton affinities of various basic sites, *J. Chem. Soc. Perkin Trans.* 2, 1320–1327, 2002).

(SHAPE) is a method of chemical modification of the 2′-hydroxyl on the ribose of RNA with acylating agents such as benzoyl cyanide[132] to study RNA structure. The term "chemical modification of nucleic acids" is used in chemogenetics[133,134] and the preparation of chemical modified siRNAs[135,136] where a chemical modified nucleobase is incorporated into nucleic acid.

The monomer composition of polysaccharides is usually less complex than either nucleic acids or proteins, although modification of basic monomer units by, for example, sulfation or acetylation can introduce complexity. In a simple polysaccharide such as starch, only hydroxyl functions are available. Modified polysaccharides such as hyaluronic acid, heparan sulfate, and heparin contain carboxyl groups, amino groups, and sulfonic acid groups, which are subjected to chemical modification. Carbohydrates are subject to nucleophilic modification by S_{N^1} or S_{N^2} mechanisms.[137] Substitution at anomeric carbons takes place by S_{N^1} mechanisms, while substitution at primary or secondary carbons uses S_{N^2} mechanisms. There are significant stereochemical effects in displacement reactions at primary and secondary carbons and the electron-rich oxygen tends to repel nucleophiles. The C_2 is the least reactive, C_3 and C_4 equally reactive, while reaction at C_6 is the easiest. Carbohydrates are susceptible to oxidation; for example, oxidation by periodate of *cis*-diols generates two carbonyl groups.[138] Halogens and hypohalites (sodium hypochlorite) oxidize aldoses to aldonic acids.[139] As a practical note, a reducing sugar such as glucose contains an aldehyde function which can be oxidized to a carboxylic acid while sucrose does not contain an aldehyde function and is thus not a reducing sugar. The modification of carbohydrates is discussed in greater detail in Chapter 7.

While we think of biological polymers as being "special" as compared to the various commercial plastic polymers such as polyacrylate and polyethylene, proteins, polynucleotides, and polysaccharides are nevertheless polymers. As such, proteins in

TABLE 1.3
Solvent Effects on Apparent pK_a Values for Amino Acids and Related Compounds[a]

	ΔpK_a		
Functional Group	**86% EtOH**	**65% EtOH**	**20% Dioxane**
CH_3COOH	+2.24	+1.19	+0.37
Alanine—COOH	+1.79	+1.19	+0.23
Alanine—αNH_3^+	+0.13	+0.30	−0.05
Lys-COOH	+1.73	+1.05	+0.10
Lys-αNH_3^+	+0.18	+0.05	−0.15
Lys-εNH_3^+	+0.14	0.00	−0.20
Arg-COOH	+1.12	+1.32	+0.11
Arg-αNH_3^+	−0.01	+0.36	−0.10
Arg-guanidino NH_3^+	+0.25	+1.52	+0.55

[a] See Frohliger, J.O., Gartska, R.A., Irwin, H.H., and Steward, O.W., Determination of ionization constants of monobasic acids in ethanol-water solvents by direct potentiometry, *Anal. Chem.* 40, 1400–1411, 1963; Frohliger, J.O., Dziedzic, J.E., and Steward, O.W., Simplified spectrophotometric determination of acid dissociation constants, *Anal. Chem.* 42, 1189–1191, 1970.

particular can be converted into plastics.*[,140–143] One of the earliest protein plastics was derived from fibrinogen.[144,145] Polysaccharides are also plasticized[146–149] but this author could not find a report of plasticized polynucleotides. As with conventional plastic polymers, the properties of a protein plastic are derived, in part, from the nature of the plasticizer used.

The reactivity of any given functional group is the local microenvironment. For example, consider the effect of the addition of an organic solvent, ethyl alcohol, on the pK_a of acetic acid. In 100% H_2O, acetic acid has a pK_a of 4.70. The addition of 80% ethyl alcohol results in an increase of the pK_a to 6.9. In 100% ethyl alcohol the pK_a of acetic acid is 10.3 (Table 1.3). These are particularly important in considering the reactivity of nucleophilic groups such as amino, cysteine, carboxyl groups, and the phenolic hydroxyl group. In the case of the primary amines present in protein, these functional groups are essentially unreactive except in the free base form. In other words, the proton present at neutral pH must be removed from the ε-amino group of lysine before this functional group can function as an effective nucleophile. In the cases of amines, the pK_a is lowered with the addition of organic solvent[150] showing the preference for an uncharged species (see Table 1.3). Considering the importance of this information, it is surprising that there are not more studies in this area. Some 70 years ago, Richardson[151] concluded that lowering the dielectric

* Not to be confused with bio plastic or bioplastic, which appears to be a marketing term for plastic derived from biomass. Our colleagues in marketing appear to feel that petroleum-based products are not derived from organic sources.

constant decreases the acidity (increases the pK_a) of carboxylic acids with little effect on the dissociation of protonated amino groups. These observations were confirmed by Duggan and Schmidt.[152] The increase in the pK_a of carboxyl groups in organic solvents has a favorable effect on transpeptidation reactions[153,154] where the carboxyl groups are required to be protonated. While it may be a bit of an oversimplification, it is useful to understand that an uncharged group is favored in a hydrophobic environment so the pK_a of an acid is increased, while the pK_a for dissociation of a conjugate acid such as the ammonium form of the ε-amino group of lysine would be decreased. The reader is directed to a study by García-Moreno and coworkers[155] where the valine at position 66 in staphylococcal nuclease (a "buried" residue) was replaced with a lysine; the pK_a of the lysine residue in the engineered protein (V66K) was ≤6.38. This study used the changes in the ionization constant of a "buried" residue from the value in water as a means to estimate the effective dielectric constant. These investigators also provide a listing of residues in other proteins with perturbed pK_a values. The values for functional groups at catalytic sites in both nucleic acids and proteins are also perturbed.[156] The reader is also directed to other studies on perturbation of the pK_a values for functional groups in proteins.[157–161] Finally, while it is possible to make a generalization such as it is generally accepted that the pK_a for a buried lysine residue decreases, while the pK_a value for dicarboxylic acid residue increases, there are exceptions where a "buried" lysine at position 38 in staphylococcal nuclease has a normal or slightly elevated pK_a, while aspartic acid or glutamic acid at the same position have the expected elevated pK_a values (7.0 and 7.2, respectively).[162]

Other factors that can influence the pK_a of a functional group include hydrogen bonding with an adjacent functional group, the direct electrostatic effect of the presence of a charged group in the immediate vicinity of a potential nucleophile, and direct steric effects on the availability of a given functional group (see earlier discussion of differential reaction of carbon atoms in hexoses). The role of hydrogen bonding in functional group reactivity was mentioned earlier with nucleic acids. There are other examples of the effect of hydrogen bonding on functional group reactivity[163–165] and the reader is directed to an excellent article by Taylor and Kennard for a general discussion of hydrogen bond geometry[166] and the earlier referenced discussion by Glusker.[45] The effect of neighboring group on function group reactivity is related to hydrogen bonding and the "buried" effect described earlier. The reader is directed to several studies that address this issue.[167–169]

Another excellent example of the effect of a neighboring group on the reaction of a specific amino acid residue is provided by the comparison of the rates of modification of the active-site cysteinyl residue by chloroacetic acid and chloroacetamide in papain.[170,171] A rigorous evaluation of the effect of pH and ionic strength on the reaction of papain with chloroacetic acid and chloroacetamide demonstrated the importance of a neighboring imidazolium group in enhancing the rate of reaction at low pH. Similar results had been reported earlier by Gerwin[172] for the essential cysteine residues in streptococcal proteinase. The essence of the experimental observations is that the plot of the pH dependence of the second-order rate constant for the reaction with chloroacetic acid is bell shaped with an optimum at about pH 6.0, while that of chloroacetamide is S-shaped approaching maximal rate of reaction at pH 10.0.

Gerwin demonstrated that the reaction of chloroacetic acid and chloroacetamide with reduced glutathione did not demonstrate this difference in pH dependence. Other examples of the effect of neighboring functional groups provided by the study of the pH dependence of the reaction of 2,4-dinitrophenyl acetate with a lysine residue at the active site of phosphonoacetaldehyde hydrolase demonstrate a decrease in the pK_a value to 9.3 as a result of positively charged environment[173] because of proximity to the amino terminal and the effect of remote sites on the reactivity of histidine residues in ribonuclease A.[174] It would be remiss not to mention the seminal observations of Schmidt and Westheimer[175] on the reaction of 2,4-dinitrophenyl propionate with the active-site lysine of acetoacetate decarboxylase demonstrating a pK_a of 5.9 for this residue. These examples clearly demonstrate the effect of electrostatic effects in the reactivity of amino acid residues in proteins. The effect of metal on hydroxyl group reactivity in ribozymes provides another example of the influence of electrostatic effects.[117–119]

Partitioning between bulk solution and local microenvironment contributes to chemical reactivity. This partitioning can cause a "selective" increase (or decrease) in reagent concentration in the vicinity of a potentially reactive species. The most clearly understood example of this is the process of affinity labeling[176]; the conceptually similar process of complementarity addressing nucleic acids has been mentioned earlier.[127] Another consideration is the partitioning of a reagent such as tetranitromethane between the polar, aqueous environment and the interior of the protein, which is nonpolar (hydrophobic). Tetranitromethane is an *organic* compound and, in principle, can react equally well with exposed and "buried" tyrosyl residues.[177] Skov and coworkers[178] modified horse heart cytochrome *c* with tetranitromethane (fourfold molar excess over tyrosine). Two of the four tyrosine residues, Tyr48 and Tyr67, were modified. These residues have reduced exposure to solvent. In more recent work,[179] Battghyány and coworkers observed that peroxynitrite readily modified Tyr97 and Tyr74l under more rigorous conditions, and that all four tyrosine residues were modified by peroxynitrite with dinitration and trinitration. Tyr48 was the least susceptible to modification with peroxynitrite. These investigators also studied modification with tetranitromethane (10-fold molar excess over tyrosine) and obtained modification comparable to that obtained with peroxynitrite; Tyr67 was most susceptible to nitration with tetranitromethane. The reader is also directed to a study by Hnízda[180] and coworkers on microenvironmental influences on the reactivity of lysine and histidine residues in proteins (lysozyme and human serum albumin). They concluded that while a modification is an indication of surface accessibility, other factors also contributed to reactivity. Again, a reminder that while much of our information has been obtained from proteins, the concepts are equally valid for nucleic acids and polysaccharides.

The author together with Dr. Claudia Noyes wrote a book on the chemical modification of proteins.[5] At that point in time, there was considerable interest in the use of solution chemistry to study protein structure and function. This early work on protein chemistry established the concept of functional group reactivity at enzyme active sites, the role of protein functional groups in binding sites including exosites, as well as the participation of functional groups in protein–protein interaction. Current state-of-the-art studies in these areas has moved to the use of analytical techniques such as nuclear magnetic resonance (NMR), electron spin resonance (ESR), mass spectrometry, and crystallographic analysis tools of structural biology. Thus, it could be argued that the use of solution protein chemistry to study functional group reactivity in biological polymers is bit archaic. However, the author has been in this area for what seems like a short period of time but yet is many years and is cynical to the extent to realize that what goes around, comes around. Chemical modification can be used to advantage in the characterization of conformational changes in biopharmaceuticals.[181]

Establishing the stoichiometry of modification is a relatively straightforward process. First, the molar quantity of modified residues is established by analysis. This could be by spectrophotometry as, for example, with the trinitrophenylation of primary amino groups, the nitration of tyrosine with tetranitromethane, or the alkylation of tryptophan with 2-hydroxy-5-nitrobenzyl bromide or by amino acid analysis to determine either the loss of a residue as, for example, in photooxidation of histidine and the oxidation of the indole ring of tryptophan with *N*-bromosuccinimide or the appearance of a modified residue such as with *S*-carboxymethylcysteine or N^1- or N^3-carboxymethylhistidine. In the situation where spectral change or radiolabel incorporation is used to establish stoichiometry, analysis must be performed to determine that there is no reaction with another amino acid. For example, the extent of oxidation of tryptophan by *N*-bromosuccinimide can be determined spectrophotometrically, but amino acid analysis or mass spectrometric analysis was *required* to determine if modification has also occurred with another amino acid such as histidine or methionine. Mass spectrometry has largely eclipsed the use of these classical techniques for characterization of biopolymer modification.[182–188] Other techniques such as Raman spectroscopy,[189] near-infrared spectrophotometry,[190] neutron scattering,[191] and small-angle x-ray scattering (SAXS)[192] have also proved useful.

It is clear that the evolution of mass spectrometry over the past two decades from an esoteric, specialized laboratory resource to a technique that is as common in the protein chemistry as amino acid analysis has provided another tool for the evaluation of protein structure after chemical modification.[119–130]

The reaction pattern of a given reagent with free amino acids or amino acid derivatives does not necessarily provide the basis for reaction with such amino acid residues in protein nor would, for that matter, the reaction of a nucleic base predict modification of a nucleobase in a nucleic acid. Furthermore, the reaction pattern of a given reagent with one protein cannot necessarily be extrapolated to all proteins. The results of a chemical modification can be markedly affected by reaction conditions (e.g., pH, temperature, solvent and/or buffer used, degree of illumination, etc.). Establishment of stoichiometry does not necessarily mean that this modification has occurred at a unique residue (unique in terms of position in the linear peptide chain—not necessarily unique with respect to reactivity). It is, of course, useful if there is a change in biological activity (catalysis, substrate binding, ion binding, etc.), which occurs concomitant with the chemical modification. Ideally, one would like to establish a direct relationship (i.e., 0.5 mol mol^{-1} of protein with 50% activity modification; 1.0 mol mol^{-1} of protein with 100% activity modification). More frequently, there is the situation where there are several moles of a given residue modified per mole of protein but there is reason to suspect stoichiometric chemical modification. In some of these situations it is possible to fractionate the protein into uniquely modified species. An early study[193] on the modification of the lysine groups of insulin with acetyl-*N*-hydroxysuccinimide as a function of pH provides an excellent example, which remains of current value. The separation of carboxymethyl-His12-pancreatic ribonuclease from carboxymethyl-His119-pancreatic ribonuclease is a classic example of this type of a situation.[194] More recently, it has been possible to separate various derivatives of lysozyme obtained from the modification of carboxyl groups.[195] Frequently, however, while there is good evidence that multiple modified species are obtained as a result of the reaction, apparently, it is not possible to separate uniquely modified species. In the reaction of tetranitromethane with thrombin,[196] apparent stoichiometry of inactivation was obtained with equivalent modification of two separate tyrosine residues (Tyr71 and Tyr85 in the B chain) and it was not possible to separate these derivatives.

Assessing stoichiometry of modification from the functional consequences of such modification with any degree of comfort is a far more difficult proposition. First, there must be a clear, unambiguous signal that can be effectively measured. In a situation where there are clearly multiple sites of reaction, which can be distinguished by analytical techniques, the approach advanced by Ray and Koshland is useful.[197] This analysis is based on establishing a relationship between the rate of loss of biological activity and the rate of modification of a single residue. A similar approach advanced by Tsou[198–200] is based on establishing a relationship between the number of residues modified and the change in biological activity. Horiike and McCormick[201] have explored the approach of relating changes in activity to the extent of chemical modification. These investigators state that the original concepts that form the basis of this approach are sound, but that extrapolation from a plot of activity remaining versus residues modified is not necessarily sound. Such extrapolation is only valid if the "nonessential" residues react much slower (rate at least 10 times slower). Given a situation where all residues within a given group are equally reactive toward the reagent in question, the number of essential residues obtained from such a plot is correct only when the total number of residues is equal to the number of essential residues, which is, in turn, equal to 1. However, it is important to emphasize that this approach is useful when there is a difference in the rate of reaction of an *essential* residue or residues and all other residues in that class as in the example of the modification of histidyl residues with diethylpyrocarbonate in lactate dehydrogenase[202,203] and pyridoxamine-5′-phosphate oxidase.[204] A major advantage in relating changes in "activity" to a specific chemical modification is being able to demonstrate that the reversal of modification is directly associated with the reversal of the change(s) in biological activity. Demonstrating that the "effects" of a specific chemical modification are reversible lends support *against* the argument that such "effects" are a result of irreversible and "nonspecific" conformational change. The issue is complicated when there is more than one residue modified in the course of the chemical reaction. Whether the residues are like or unlike amino acids, it still is difficult to assign the functional consequences to the modification of a single residue. The mathematical approaches described earlier provide an approach to this specific situation. Another complication arises when there is "incomplete" inactivation at the completion of the chemical modification,[196,205–207] which is discussed in more detail in the following.

A key difference between the use of site-specific chemical modification and site-specific mutagenesis to study protein structure and function is the ability to measure the rate of reaction of a specific amino acid residue or residues. The reader is referred to a review by Rakitzis[208] for a discussion of the kinetics of protein chemical modification. There has been continuing use of this approach during the 20 years since the publication of this article.[209–214] An example of the use of reaction rate is provided from the study of the modification of an aminopeptidase by diethylpyrocarbonate.[215] It was demonstrated that the reaction of the aminopeptidase with diethylpyrocarbonate resulted in the modification of histidine residues. A difference of the reactivity of the two histidine residues modified by diethylpyrocarbonate in the presence and absence of calcium ions permitted the identification of one of the two histidine residues as critical for the binding of calcium ions. Careful analysis of the effect of pH on the reaction rate in the presence and absence of calcium ions allowed the assignment of pK_a value to the two residues.

The functional characterization of the modified protein can provide a significant challenge. The discussion of this problem is biased toward the study of enzymes but the same general considerations are valid for receptors, protein ligands, structural proteins, and carrier proteins such as hemoglobin and transferrin. The functional characterization is relatively straightforward when activity is totally abolished such as that which occurs when the active-site histidine in a serine protease is modified with a peptide chloromethylketone.[216] A more difficult problem is encountered with a modified protein with fractional activity.[196,207–209] The most critical aspect in the characterization of the modified protein is the method used to determine activity. The rigorous determination of binding constants and kinetic constants is absolutely essential; the reporting of percent change in activity is clearly inadequate. The reader is directed to several classic works in this area[217,218] as well as more recent expositions in this area.[219–223] For the reader who, like the author, is somewhat challenged by physical biochemistry, consideration of some more basic information[224–226] will be useful. Finally, the reader is directed to an excellent review by Plapp.[227] While the discussion is directed toward the use of site-specific mutagenesis for the study of enzymes, much of the content is equally applicable to the characterization of chemically modified proteins. Particular consideration should be given to the section on kinetics with emphasis on the importance of *V/K* (catalytic efficiency) for evaluation of the effect of a modification on catalytic activity and the discussion on the importance of understanding that K_m is not necessarily a measure of affinity. Evaluation contribution of individual residues to the overall catalytic process is also discussed. The reader is referred to another review article for consideration of this latter issue.[228] This type of analysis would markedly increase the value of studies where several different reagents are used for the chemical modification of a protein.

Characterization of a partially modified biological polymer is also of importance in the characterization of biopharmaceuticals. Frequently, there is some activity lost in the transition from active pharmaceutical ingredient to final drug product. It is critical to understand the change in activity. For example, does the loss of 10% of the activity mean the loss of 10% of product or is there 100% of product with 90% activity? The careful use of enzyme kinetics and binding assays can resolve these issues.

REFERENCES

1. Bartone, N.A., Bentley, J.D., and MacLaren, J.A., Determination of dehydroalanine residues in proteins and peptides—An improved method, *J. Protein Chem.* 10, 603–607, 1991.
2. Galonde, A.K., Trent, J.O., and Spatola, A.F., Understanding base-assisted desulfurization using a variety of disulfide-bridged peptides, *Biopolymers* 71, 7296–7300, 2001.
3. Wang, H., Zhang, J., and Xian, M., Facile formation of dehydroalanine from *S*-nitrosocysteines, *J. Am. Chem. Soc.* 131, 13238–13239, 2009.
4. Thakur, S.S. and Balaram, P., Characterization of alkali induced formation of lanthionine, trisulfides, and tetrasulfides from peptide disulfides using negative ion mass spectrometry, *J. Am. Soc. Mass Spectrom.* 20, 783–791, 2009.
5. Lundblad, R.L. and Noyes, C.M., *Chemical Reagents for Protein Modification*, CRC Press, Boca Raton, FL, 1984.
6. Lundblad, R.L., *Chemical Reagents for Protein Modification*, 2nd edn., CRC Press, Boca Raton, FL, 1991.
7. Walker, J.M. (ed.), *Methods in Molecular Biology, Vol. 1, Proteins*, Humana Press, Clifton, NJ, 1984.
8. Shivley, J.E. (ed.), *Methods of Protein Microcharacterization. A Practical Handbook*, Humana Press, Clifton, NJ, 1986.
9. Oxender, D.L. and Fox, C.F. (eds.), *Protein Engineering*, Alan R. Liss, New York, 1987.
10. Fasman, G.D., *Practical Handbook of Biochemistry and Molecular Biology*, CRC Press, Boca Raton, FL, 1989.
11. Lundblad, R.L. and MacDonald, F., *Handbook of Biochemistry and Molecular Biology*, 4th edn., CRC Press, Boca Raton, FL, 2010.
12. Walker, J.M. (ed.), *Methods in Molecular Biology, Vol. 3, New Protein Techniques*, Humana Press, Clifton, NJ, 1988.
13. Lundblad, R.L., *Techniques in Protein Modification*, CRC Press, Boca Raton, FL, 1994.
14. Means, G.E., Zhang, H., and Le, M., The chemistry of protein functional groups, in *Proteins. A Comprehensive Treatise*, G. Allen (ed.), JAI Press, Stamford, CT, Vol. 2, Chapter 2, 1999.
15. DeSantis, G. and Jones, J.B., Chemical modification of enzymes for enhanced functionality, *Curr. Opin. Biotechnol.* 10, 324, 1999.
16. Stark, G.R., Recent developments in chemical modification and sequential degradation of proteins, *Adv. Prot. Chem.* 24, 261, 1970.
17. Offord, R.E., Chemical approaches to protein engineering, in *Protein Engineering. A Practical Approach*, M.J.E. Sternberg and R. Wetzel (eds.), IRL Press at Oxford University Press, Oxford, U.K., Chapter 10, 1992.
18. Li-Chan, E.C.Y., Methods to monitor process-induced changes in food proteins, in *Process-Induced Changes in Food*, F. Shahadi and C.T. Ho (eds.), Plenum Press, New York, Chapter 2, 1998.
19. Davis, B.G., Chemical modification of biocatalysts, *Curr. Opin. Biotechnol.* 14, 379, 2003.
20. Basle, E., Joubert, N., and Pucheault, M., Protein chemical modification on endogenous amino acids, *Chem. Biol.* 17, 213–227, 2010.
21. Price, N.C. (ed.), *Protein LabFax*, Bios Scientific Publishers, Academic Press, Oxford, U.K., 1996.
22. Engel, P.C. (ed.), *Enzymology LabFax*, Bios Scientific Publishers, Academic Press, Oxford, U.K., 1996.

23. Hirs, C.H.W. (ed.), Enzyme structure, *Methods Enzymol.* 11, 1967.
24. Hirs, C.H.W. and Timasheff, S.N. (eds.), Enzyme structure, *Methods Enzymol.* 25, 1972.
25. Hirs, C.H.W. and Timasheff, S.N. (eds.), Enzyme structure, *Methods Enzymol.* 47, 1976.
26. Hirs, C.H.W. and Timasheff, S.N. (eds.), Enzyme structure, *Methods Enzymol.* 91, 1983.
27. Packer, L. (ed.), Oxygen radicals in biological systems, *Methods Enzymol.* 233, 1994.
28. Lee, Y.C. and Lee, R.T., Neoglycoconjugates, Part A. Synthesis, *Methods Enzymol.* 242, 1994.
29. Spears, K., Biochemical spectroscopy, *Methods Enzymol.* 246, 1995.
30. Purich, D., Enzyme kinetics, *Methods Enzymol.* 249, 1995.
31. Packer, L., Biothiols, Part A. Monothiols and dithiols, protein thiols, and thiyl radicals, *Methods Enzymol.* 251, 1995.
32. Packer, L., Nitric oxide, Part A. Sources and detection of NO; No synthase, *Methods Enzymol.* 268, 1996.
33. Marriott, G. and Parker, I. (eds.), Biophotonics, Part A, *Methods Enzymol.* 360, 2003.
34. Bell, N.M. and Micklefield, J., Chemical modification of oligonucleotides for therapeutic, bioanalytical and other applications, *Chembiochem* 10, 2691–2703, 2009.
35. Mäe, M., Andaloussi, S.E., Lehto, T., and Langel, U., Chemically modified cell-penetrating peptides for the delivery of nucleic acids, *Expert Opin. Drug Deliv.* 6, 1195–1205, 2009.
36. Canalle, L.A., Löwik, D.W., and van Hest, J.C., Polypeptide-polymer bioconjugates, *Chem. Soc. Rev.* 39, 329–353, 2010.
37. Stenius, P. and Andresen, M., Chemical modification of viruses and virus-like particles, *Curr. Topics Microbiol. Immunol.* 327, 1–21, 2009.
38. Harris, J.M. and McManus, S.P. (eds.), *Nucleophilicity*, American Chemical Society, Washington, DC, 1987.
39. Harris, J.M. and McManus, S.P., Introduction to nucleophilic reactivity, in *Nucleophilicity*, J.M. Harris and S.P. McManus (eds.), American Chemical Society, Washington, DC, Chapter 1, pp. 1–20, 1987.
40. Bunnett, J.F., Nucleophilic reactivity, *Annu. Rev. Phys. Chem.* 14, 271–290, 1963.
41. Brauman, J.I., Dodd, J.A., and Han, C.C., Intrinsic nucleophilicity, in *Nucleophilicity*, J.M. Harris and S.P. McManus (eds.), American Chemical Society, Washington, DC, Chapter 2, pp. 23–33, 1987.
42. Chen, X. and Brauman, J.I., Hydrogen bonding lowers intrinsic nucleophilicity of solvated nucleophiles, *J. Am. Chem. Soc.* 130, 15038–15046, 2008.
43. Pearson, R.G., Hard and soft bases and acids, *J. Am. Chem. Soc.* 85, 3533–3539, 1963.
44. Miller, B., *Advanced Organic Chemistry. Reactions and Mechanisms*, 2nd edn., Prentice-Hall, Upper Saddle River, NJ, Chapter 13, 2004.
45. Glusker, J.P., The binding of ions to proteins, in *Protein*, G. Allen (ed.), JAI Press, London, U.K., Vol. 2, Chapter 4, 1999.
46. Fuchs, R.L. and Cole, J.L., Transition state enthalpies of transfer in the reaction of nucleophiles with α-hexyl tosylate, *J. Am. Chem. Soc.* 95, 3194–3197, 1973.
47. Landini, D. and Maia, A., Anion nucleophilicity in ionic liquids: A comparison with traditional molecular solvents of different polarity, *Tetrahedron Lett.* 46, 3961–3963, 2005.
48. Abraham, M.H., Solvent effect on reaction rate, *Pure Appl. Chem.* 57, 1055–1064, 1985.
49. Smith, M.E.B., Schumacher, F.F., Ryan, C.P. et al., Protein modification, bioconjugation, and disulfide bridging using bromomaleimides, *J. Am. Chem. Soc.* 132, 1960–1965, 2010.
50. Furlong, J.D. and Lilly, D.M.J., Highly selective chemical modification of cruciform loops by diethyl pyrocarbonate, *Nucleic Acids Res.* 14, 3995–4007, 1986.
51. Cruz, J., Kawasaki, M., and Gorski, W., Electrode coatings based on chitosan scaffolds, *Anal. Chem.* 72, 680–686, 2000.
52. Banuls, M.J., Gonzalez-Pedro, V., Barrios, C.A. et al., Selective chemical modification of silicon nitride/silicon oxide nanostructures to develop label-free biosensors, *Biosens. Bioelectron.* 25, 1460–1465, 2010.

53. Yalpani, M., Chemistry of polysaccharide modification and degradation, in *Carbohydrate Structures, Syntheses and Dynamics*, P. Finch (ed.), Kluwer, Dordrecht, the Netherlands, Chapter 8, pp. 294–318, 1999.
54. Cohen, S., Biological reactions of carbonyl halides, in *The Chemistry of Acyl Halides*, S. Patai (ed.), Wiley Interscience, London, U.K., Chapter 10, pp. 313–348, 1972.
55. Walsh, C.T., Garneau-Tsodikova, S., and Gatto, G.J., Jr., Protein posttranslational modifications: The chemistry of proteome diversifications, *Angew. Chem. Int. Ed.* 44, 7342–7372, 2005.
56. Patthy, L., *Protein Evolution*, 2nd edn., Blackwell, Oxford, U.K., 2008.
57. Cassini, G., Illy, S., and Pileni, M.P., Chemically modified proteins solubilized in AOT reverse micelles. Influence of protein charges on intermicellar interactions, *Chem. Phys. Lett.* 221, 205–212, 1994.
58. Franco, T.T., Andrews, A.T., and Asenjo, J.A., Conservative chemical modification of proteins to study the effects of a single protein property on partitioning in aqueous two-phase systems, *Biotechnol. Bioeng.* 49, 290–299, 1996.
59. Zschörnig, O., Paasche, G., Thieme, C. et al., Modulation of lysozyme charge influences interaction with phospholipid vesicles, *Colloids Surf. B Biointerfaces* 42, 69–73, 2005.
60. Thurkill, R.L., Grimsley, G.R., Scholtz, J.M., and Pace, C.N., pK values of the ionizable groups of proteins, *Protein Sci.* 15, 1214–1218, 2006.
61. Pace, C.N., Grimsley, G.R., and Scholtz, J.M., Protein ionizable groups: pK values and their contribution to protein stability and solubility, *J. Biol. Chem.* 284, 13285–13289, 2009.
62. Grimsley, G.R., Scholtz, J.M., and Pace, C.N., A summary of the measured pK values of the ionizable groups in folded proteins, *Protein Sci.* 28, 247–251, 2009.
63. Drummond, C.J., Grieser, F., and Healy, T.W., Acid-base equilibria in aqueous micellar solutions, *J. Chem. Soc. Faraday Trans. I* 85, 521–535, 1989.
64. Allewell, N.M., Oberoi, H., Hariharan, M., and LiCata, V.J., Electrostatic effects of in proteins: Experimental and computational approaches, in *Proteins*, G. Allen (ed.), JAI Press, London, U.K., Vol. 2, 1997.
65. Juffer, A.H., Argos, P., and Vogel, H.J., Calculating acid-dissociation constants of proteins using the boundary element method, *J. Phys. Chem. B* 101, 7664–7673, 1997.
66. Davies, M.N., Toseland, C.P., Moss, D.P., and Flower, D.R., Benchmarking pK_a prediction, *BMC Biochem.* 7, 18, 2006.
67. Nelson, N.J., Day, A.E., Zeng, B.B. et al., Isotope-coded, iodoacetamide-based reagent to determine individual cysteine pK_a values by matrix-assisted laser desorption/ionization time-of-flight mass spectrometry, *Anal. Biochem.* 375, 187–195, 2008.
68. Gromiha, M.M. and Suwa, M., A simple statistical method for discriminating outer membrane proteins with better accuracy, *Bioinformatics* 21, 961–968, 1995.
69. Piez, K.A. and Gross, J., The amino acid composition and morphology of some invertebrate and vertebrate collagens, *Biochim. Biophys. Acta* 34, 24–39, 1959.
70. Maestro, M.M., Turnay, J., Olmo, N. et al., Biochemical and mechanical behavior of ostrich pericardium as a new biomaterial, *Acta Biomater.* 2, 213–219, 2006.
71. Nunn, B.L. and Keil, R.G., Size distribution and amino acid chemistry of base-extractable proteins from Washington coast sediment, *Biogeochemistry* 75, 177–200, 2005.
72. Gromiha, M.M. and Ahmad, S., Role of solvent accessibility in structure based drug design, *Curr. Comput Aided Drug Des.* 1, 223–235, 2005.
73. Bernadó, P., Blackledge, M., and Sancho, J., Sequence-specific solvent accessibilities of protein residues in unfolded protein ensembles, *Biophys. J.* 91, 4536–4543, 2006.
74. Ponnuswamy, P.K., Muthasamy, R., and Manavalan, P., Amino acid composition and thermal stability of proteins, *Int. J. Biol. Macromol.* 4, 186–190, 1982.
75. Gekko, K. and Hasegawa, Y., Compressibility-structure relationship of globular proteins, *Biochemistry* 25, 6563–6571, 1986.

76. Gromiha, M.M., Obatake, M., and Sarai, A., Important amino acid properties for enhanced thermostability from mesophilic to thermophilic proteins, *Biophys. Chem.* 82, 51–67, 1999.
77. Wang, C.H. and Damodaran, S., Thermal gelation of globular proteins: Weight-average molecular weight dependence of gel strength, *J. Agric. Food Chem.* 38, 1157–1164, 1990.
78. Lienqueo, M.E., Mahn, A., Navarro, G. et al., New approaches for predicting protein retention time in hydrophobic interaction chromatography, *J. Mol. Recogn.* 19, 260–269, 2006.
79. Richards, R.M. and Richmond, T., Solvents, interfaces and protein structure, *Ciba Found. Symp.* 60, 23–45, 1977.
80. Richards, F.M., Areas, volumes, packing and protein structure, *Annu. Rev. Biophys. Bioeng.* 6, 151–176, 1977.
81. Lesk, A.M., *Introduction to Protein Architecture*, Oxford University Press, Oxford, U.K., 2004.
82. Miller, S., Janin, J., Lesk, A.M., and Clothia, C., Interior and surface of monomeric proteins, *J. Mol. Biol.* 196, 641–656, 1987.
83. Argos, P., An investigation of protein and domain interfaces, *Protein Eng.* 2, 101–113, 1988.
84. Keskin, O., Ma, B., and Nussinov, R., Hot regions in protein-protein interactions: The organization and contribution of structurally conserved hot spot residues, *J. Mol. Biol.* 345, 1281–1294, 2005.
85. Sheehan, J.P. and Sadler, J.E., Molecular mapping of the heparin-binding exosite of thrombin, *Proc. Nat. Acad. Sci. USA* 91, 5518–5522, 1994.
86. Wang, M., Zajicek, J., Geiger, J.H. et al., Solution structure of the complex of VEK-30 and plasminogen kringle 2, *J. Struct. Biol.* 169, 349–359, 2010.
87. Fraenkel-Conrat, H., Reaction of nucleic acids with formaldehyde, *Biochim. Biophys. Acta* 15, 307–309, 1954.
88. Grossman, L., Stollar, D., and Herrington, K., Some aspects of the structure of heat-denatured and ultraviolet-inactivated deoxyribonucleic acid, *J. Chim. Phys. Chim. Biol.* 58, 1078–1089, 1961.
89. Fasman, G.D., Lindblow, C., and Grossman, I., The helical conformation of polycytidylic acid. Studies on the forces involved, *Biochemistry* 3, 1015–1021, 1964.
90. Feldman, M.Y., Reactions of nucleic acids and nucleoproteins with formaldehyde, *Prog. Nucl. Acids Res. Mol. Biol.* 13, 1–49, 1973.
91. McGhee, J.D. and von Hippel, P.H., Formaldehyde as a probe of DNA structure I. Reactions with exocyclic amino groups of DNA bases, *Biochemistry* 14, 1281–1296, 1975.
92. Stevens, C.L., Chay, T.R., and Loga, S., Rupture of base pairing in double-stranded poly(riboadenylic acid)-poly(cytidylic acid) by formaldehyde: Medium chain length, *Biochemistry* 16, 3727–3739, 1977.
93. Jacob, R.J., Lebowitz, J., and Kleinschmidt, A.K., Locating interrupted hydrogen bonding in the secondary structure of PM2 circular DNA by comparative denaturation mapping, *J. Virol.* 13, 1176–1185, 1974.
94. McGhee, J.D. and von Hippel, P.H., Formaldehyde as a probe of DNA structure. 3. Equilibrium denaturation of DNA and synthetic polynucleotides, *Biochemistry* 16, 3267–3276, 1977.
95. Landy, A., Ross, W., and Foeller, C., Generation of DNA fragments by enzymatic cleavage at sites sensitive to denaturation, *Biochim. Biophys. Acta* 299, 264–272, 1973.
96. Sarkar, N.K. and Dounce, A.L., A spectroscopic study of the reaction of formaldehyde with deoxyribonucleic and ribonucleic acids, *Biochim. Biophys. Acta* 29, 160–169, 1961.
97. Stollar, D. and Grossman, L., The reaction of formaldehyde with denatured DNA: Spectrophotometric, immunologic, and enzymic studies, *J. Mol. Biol.* 4, 31–38, 1962.
98. Sarkar, N.K. and Dounce, A.L., A spectroscopic study of the reaction of formaldehyde with deoxyribonucleic and ribonucleic acids, *Biochim. Biophys. Acta* 49, 160–169, 1961.

99. Boedtker, H., Reaction of ribonucleic acid with formaldehyde. I. Optical absorbance studies, *Biochemistry* 6, 2718–2727, 1967.
100. Stevens, C.L., Destabilization of secondary structure in polyadenylic acid by formaldehyde, *Biopolymers* 13, 1515–1533, 1974.
101. Jackson, V., Studies on histone organization in the nucleosome using formaldehyde as a reversible cross-linking agent, *Cell* 15, 945–954, 1978.
102. Perez-Romero, P. and Imperiale, M.J., Assaying protein-DNA interactions in vivo and in vitro using chromatin immunoprecipitation and electrophoretic mobility shift assays, *Methods Mol. Med.* 131, 123–139, 2007.
103. Niranjanakumari, S., Lasda, E., Brazas, R., and Garcia-Blanco, M.A., Reversible cross-linking combined with immunoprecipitation to study RNA-protein interactions *in vivo*, *Methods* 26, 182–190, 2002.
104. Schmiedeberg, L., Skene, P., Deaton, A., and Bird, A., A temporal threshold for formaldehyde crosslinking and fixation, *PLoS One* 4, e4636, 2009.
105. Shikama, K. and Miura, K.I., Equilibrium studies on the formaldehyde reaction with native DNA, *Eur. J. Biochem.* 63, 39–46, 1976.
106. Salk, J.E. and Gori, J.B., A review of theoretical, experimental, and preclinical considerations in the use of formaldehyde for the inactivation of polio virus, *Ann. N. Y. Acad. Sci.* 83, 609–637, 1960.
107. Fine, D.L., Jenkins, E., Martin, S.S. et al., A multisystem approach for development and evaluation of inactivated vaccines for Venezuelan equine encephalitis virus (VEEV), *J. Virol. Methods* 163, 424–432, 2010.
108. Keller, J.E., Characterization of new formalin-detoxified botulinum neurotoxin toxoids, *Clin. Vaccine Immunol.* 15, 1374–1379, 2008.
109. Salgado, J., Casadevall, G., Puigneró, V., and Queralt, J., Characterization of allergoids from ovalalbumin *in vitro* and *in vivo*, *Immunobiology* 196, 375–386, 1996.
110. Hood, B.L., Conrads, T.P., and Veenstra, T.D., Mass spectrometric analysis of formalin-fixed paraffin-embedded tissue: unlocking the proteome within, *Proteomics* 6, 4106–4114, 2006.
111. Farragher, S.M., Tanney, A., Kennedy, R.D., and Paul Harkin, D., RNA expression analysis from formalin fixed paraffin embedded tissues, *Histochem. Cell Biol.* 130, 435–445, 2008.
112. Berg, D., Hipp, S., Malinowsky, K. et al., Molecular profiling of signalling pathways in formalin-fixed and paraffin-embedded cancer tissues, *Eur. J. Cancer* 46, 47–55, 2010.
113. Grizzle, W.E., Special symposium: Fixation and tissue processing models, *Biotech. Histochem.* 84, 185–193, 2009.
114. Lilly, D.M., Structure, folding and catalysis of the small nucleolytic ribozymes, *Curr. Opin. Struct. Biol.* 9, 330–338, 1999.
115. Wilkinson, K.A., Vasa, S.M., Deigan, K.E. et al., Influence of nucleotide identity on ribose 2′-hydroxyl reactivity in RNA, *RNA* 15, 1314–1321, 2009.
116. Acharya, S., Földesi, A., and Chattopadhyaya, J., The pK_a of the internucleotidic 2′-hydroxyl group in diribonucleoside (3′–5′) monophosphates, *J. Org. Chem.* 68, 1906–1910, 2003.
117. Smith, M.D., Mehdizadeh, R., Olive, J.E., and Collins, R.A., The ionic environment determines ribozyme cleavage rate by modulation of nucleobase pK_a, *RNA* 14, 1942–1949, 2008.
118. Lee, T.S., Giambasu, G.M., Sosa, C.P. et al., Threshold occupancy and specific cation binding modes in the hammerhead ribozyme active site are required for active conformation, *J. Mol. Biol.* 388, 195–206, 2009.
119. Guo, M., Spitale, R.C., Volpini, R. et al., Direct Raman measurement of an elevated base pK_a in the active site of a small ribozyme in a precatalytic conformation, *J. Am. Chem. Soc.* 131, 12908–12909, 2009.

120. Moody, E.M., Brown, T.S., and Bevilacqua, P.C., Simple method for determining nucleobase pK_a values by indirect labeling of a pK_a of neutrality in dsDNA, *J. Am. Chem. Soc.* 126, 10200–10201, 2004.
121. Tang, C.L., Alexov, E., Pyle, E.M., and Honig, B., Calculation of pKas in RNA: On the structural origins and functional roles of protonated nucleotides, *J. Mol. Biol.* 366, 1475–1496, 2007.
122. Suydam, I.T. and Strobel, S.A., Fluorine substituted adenosines as probes of nucleobase protonation in functional RNAs, *J. Am. Chem. Soc.* 130, 13639–13648, 2009.
123. Thomas, J.M. and Perrin, D.M., Active site labeling of G8 in the hairpin ribozyme: Implications for structure and mechanism, *J. Am. Chem. Soc.* 128, 16540–16545, 2006.
124. Knorre, D.G. and Vlassov, V.V., Complementary-addressed (sequence-specific) modification of nucleic acids, *Prog. Nucleic Acid Res. Mol. Biol.* 32, 291–320, 1985.
125. Knorre, D.G., Vlassov, V.V., and Zarytova, V.F., Reactive oligonucleotide derivatives and sequence-specific modification of nucleic acids, *Biochimie* 67, 785–789, 1985.
126. Knorre, D.G., Zarytova, V.G., Karpova, G.G., and Stephanovich, L.E., Specific modification of nucleic acids inside the cell with alkylating derivatives of oligonucleotides ethylated at internucleotide phosphates, *Nucleic Acid Symp. Ser.* 9, 195–198, 1981.
127. Federova, O.S., Adeenah-Zadah, A., and Knoore, D.G., Cooperative interactions in the tandem of oligonucleotide derivatives arranged at complementary target. Quantitative estimates and contribution of the target secondary structure, *FEBS Lett.* 369, 287–289, 1995.
128. Manfield, I.W. and Stockley, P.G., Methylation interference footprinting of DNA-protein complexes, *Methods Mol. Biol.* 543, 105–120, 2009.
129. Richardson, F.C. and Richardson, K.K., Sequence-dependent formation of alkyl DNA adducts: A review of methods, results, and biological correlates, *Mutat. Res.* 233, 127–138, 1990.
130. Shrivastav, N. and Essigmann, J.M., Chemical biology of mutagenesis and DNA repair: Cellular responses to DNA alkylation, *Carcinogenesis* 31, 59–70, 2010.
131. Gates, K.S., An overview of chemical processes that damage cellular DNA: Spontaneous hydrolysis, alkylation, and reactions with radicals, *Chem. Rev. Toxicol.* 22, 1747–1760, 2009.
132. Mortimer, S.A. and Weeks, K.M., Time-resolved RNA SHAPE chemistry: Quantitative RNA structure analysis in one-second snapshots and at single-nucleotide resolution, *Nat. Protoc.* 4, 1413–1421, 2009.
133. Sigler, P.B., Chemogenetics, *Nat. Struct. Biol.* 1, 3–4, 1994.
134. Strobel, S.A., Ribozyme chemogenetics, *Biopolymers* 48, 65–81, 1998.
135. Behlke, M.A., Chemical modification of siRNAs for *in vivo* use, *Oligonucleotides* 18, 305–319, 2008.
136. Ge, Q., Dallas, A., Ilves, H. et al., Effects of chemical modification on the potency, serum stability, and immunostimulatory properties of short siRNAs, *RNA* 16, 118–130, 2010.
137. Binkley, R.W., Nucleophilic substitution reactions, in *Modern Carbohydrate Chemistry*, Marcel Dekker, New York, Chapter 9, 1988.
138. Binkley, R.W., Oxidation, in *Modern Carbohydrate Chemistry*, Marcel Dekker, New York, Chapter 10, 1988.
139. Binkley, R.W., Unprotected sugars, in *Modern Carbohydrate Chemistry*, Marcel Dekker, New York, Chapter 7, 1988.
140. Lefèvre, T., Subirade, M., and Pézolet, M., Molecular description of the formation and structure of plasticized globular protein films, *Biomacromolecules* 6, 3209–3219, 2005.
141. Althamneh, A.I., Griffin, M., Whaley, M., and Barone, J.R., Conformational changes and molecular mobility in plasticized proteins, *Biomacromolecules* 9, 3181–3187, 2008.
142. Pruneda, E., Peralta-Hernández, J.M., Esquivel, K. et al., Water vapor permeability, mechanical properties and antioxidant effect of Mexican oregano-soy based edible films, *J. Food Sci.* 73, C488–C493, 2008.

143. Gonzalez-Gutierrez, J., Partal, P., Garcia-Morales, M., and Gallegos, C., Development of highly-transparent protein/starch-based bioplastics, *Bioresour. Technol.* 101, 2007–2013, 2010.
144. Ferry, J.D. and Morrison, P.R., Chemical, clinical, and immunological studies on the products of human plasma fractionation. XVI. Fibrin clots, fibrin films, and fibrinogen plastics, *J. Clin. Invest.* 23, 566–572, 1944.
145. Ferry, J.D., Modified fibrinogen plastics, U.S. Patent 2,385,802.
146. Shi, Y.C., Two- and multi-step annealing of cereal starches in relation to gelatinization, *J. Agric. Food Chem.* 56, 1097–1104, 2008.
147. Hagesaether, E. and Sande, S.A., Effect of pectin type and plasticizer on *in vitro* mucoadhesion of free films, *Pharm. Dev. Technol.* 13, 105–114, 2008.
148. Ma, X., Jian, R., Chang, P.R. et al., Fabrication and characterization of citric acid-modified starch nanoparticles/plasticized-starch composites, *Biomacromolecules* 9, 3314–3320, 2008.
149. Bajdik, J., Marciello, M., Caramella, C. et al., Evaluation of surface and microstructure of differently plasticized chitosan films, *J. Pharm. Biomed. Anal.* 49, 655–659, 2008.
150. Crowhurst, L., Llewellyn Lancaster, N., Pérez Arlandis, J.M., and Welton, T., Manipulating solute nucleophilicity with room temperature ionic liquids, *J. Am. Chem. Soc.* 126, 11549–11555, 2004.
151. Richardson, G.M., The principle of formaldehyde, alcohol, and acetone titrations. With a discussion of the proof and implication of the zwitterionic conception, *Proc. Rov. Soc. B* (London) 115, 121–141, 1934.
152. Duggan, E.L. and Schmidt, C.L.A., The dissociation of certain amino acids in dioxane-water mixtures, *Arch. Biochem.* 1, 453–471, 1943.
153. Canova-Davis, E. and Carpenter, F.H., Semisynthesis of insulin: Specific activation of arginine carboxyl group of the B chain of desoctapeptide—(B23–36)-insulin (Bovine), *Biochemistry* 20, 7053–7058, 1981.
154. Canova-Davis, E., Kessler, T.J., and Ling, V.T., Transpeptidation during the analytical proteolysis of proteins, *Anal. Biochem.* 196, 39–45, 1991.
155. García-Moreno, B., Dwyer, J.J., Gittis, A.G. et al., Experimental measurement of the effective dielectric in the hydrophobic core of a protein, *Biophys. Chem.* 64, 211–224, 1997.
156. Harris, T.K. and Turner, G.J., Structural basis of perturbed pK_a values of catalytic groups in enzyme active sites, *IUBMB Life* 53, 85–98, 2002.
157. Mehler, E.L., Fuxreiter, M., Simon, L., and Garcia-Moreno, E.B., The role of hydrophobic microenvironments in modulating pK_a shifts in proteins, *Proteins Struct. Funct. Genetics* 48, 283–292, 2002.
158. Edgecomb, S.P. and Murphy, K.P., Variability of the pK_a of histidine side-chains correlates with burial within proteins, *Proteins Struct. Funct. Genetics* 49, 1–6, 2002.
159. Kim, J., Mao, J., and Gunner, M.R., Are acidic and basic groups in buried proteins predicted to be ionized, *J. Mol. Biol.* 348, 1283–1298, 2005.
160. Thurkill, R.L., Grimsley, G.R., Scholtz, M., and Pace, C.N., Hydrogen bonding markedly reduces the pK of buried carboxyl groups in proteins, *J. Mol. Biol.* 362, 594–604, 2006.
161. Castanada, C.A., Fitch, C.A., Majumdar, A. et al., Molecular determinants of the pK(a) values of Asp and Glu residues in staphylococcal nuclease, *Proteins Struct. Funct. Bioinformatics* 77, 570–588, 2009.
162. Harms, M.M., Castanada, C.A., Schlessman, J.L. et al., The pK(a) values of acidic and basic residues buried at the same internal location in a proteins are governed by different factors, *J. Mol. Biol.* 389, 34–47, 2009.
163. Sadekov, I.D., Minkin, V.I., and Lutskii, A.E., English translation of original Russion, *Upspekhi Khimi*, 39, 380–411, 1970.

164. Szedja, W., Effects of internal hydrogen bonding on the reactivity of the hydroxyl-group in esterification under phase transfer conditions, *J. Chem. Soc. Chem. Commun.* 5, 215–216, 1981.
165. Yamauchi, K., Hosokawa, T., and Kinoshita, M., Possible effect of hydrogen bonding on methylation of pyrimidine and pyridone nucleosides, *J. Chem. Soc. Perkin Trans.* 1, 13–15, 1989.
166. Taylor, R. and Kennard, O., Hydrogen-bond geometry in organic crystals, *Acc. Chem. Res.* 17, 320–326, 1984.
167. Oda, Y., Yamazaki, T., Nagayama, K. et al., Individual ionization constants of all the carboxyl groups in ribonuclease HI from *Escherichia coli* determined by NMR, *Biochemistry* 33, 5275–5284, 1994.
168. Dyson, H.J., Jeng, M.R., Tennant, L.L. et al., Effects of buried charged groups on cysteine thiol ionization and reactivity in *Escherichia coli* thioredoxin: Structural and functional characterization of mutants of Asp 26 and Lys 57, *Biochemistry* 36, 2622–2636, 1997.
169. Lindman, S., Linse, S., Mulder, F.A. et al., pK(a) values for side-chain carboxyl groups of a PGH1 variant explain salt and pH-dependent stability, *Biophys. J.* 92, 257–266, 2007.
170. Chaiken, I.M. and Smith, E.L., Reaction of chloroacetamide with the sulfhydryl groups of papain, *J. Biol. Chem.* 244, 5087, 1969.
171. Chaiken, I.M. and Smith, E.L., Reaction of the sulfhydryl group of papain with chloroacetic acid, *J. Biol. Chem.* 244, 5095, 1969.
172. Gerwin, B.I., Properties of the single sulfhydryl group of streptococcal proteinase. A comparison of the rates of alkylation by chloroacetic acid and chloroacetamide, *J. Biol. Chem.* 242, 451, 1967.
173. Zhang, G., Mazurkie, A.S., Dunaway-Mariano, D., and Allen, K.N., Kinetic evidence for a substrate-induced fit in phosphonoacetaldehyde hydrolase catalysis, *Biochemistry* 41, 13370, 2002.
174. Fisher, B.M., Schultz, L.W., and Raines, R.T., Coulombic effects of remote subsites on the active site of ribonuclease A, *Biochemistry* 37, 17386, 1998.
175. Schmidt, D.E., Jr. and Westheimer, F.H., pK of the lysine amino group at the active site of acetoacetate decarboxylase, *Biochemistry* 10, 1249, 1971.
176. Plapp, B.V., Application of affinity labeling for studying structure and function in enzymes, *Methods Enzymol.* 87, 469, 1982.
177. Myers, B.H. and Glazer, A.N., Spectroscopic studies of the exposure of tyrosine residues in proteins with special references to the subtilisins, *J. Biol. Chem.* 26, 412, 1971.
178. Skov, K., Hofmann, T., and Williams, G.R., The nitration of cytochrome c, *Can. J. Biochem.* 47, 750, 1969.
179. Battghyány, C., Souza, J.M., Durán, R. et al., Time course and site (s) of cytochrome c tyrosine nitration by peroxynitrite, *Biochemistry* 44, 8038–8046, 2005.
180. Hnízda, A., Šantrůček, J., Šanda, M. et al., Reactivity of histidine and lysine side-chains with diethylpyrocarbonate—A method to identify surface exposed residues in proteins, *J. Biochem. Biophys. Methods* 70, 1091–1097, 2008.
181. Lundblad, R.L., *Approaches to Conformational Analysis of Biopharmaceuticals*, CRC/Taylor & Francis, Boca Raton, FL, 2010.
182. Bennett, K.L., Smith, S.V., Lambrecht, R.M. et al., Rapid characterization of chemically-modified proteins by electrospray mass spectrometry, *Bioconjug. Chem.* 7, 16–22, 1996.
183. Fligge, T.A., Kast, J., Burns, K., and Przybylski, M., Direct monitoring of protein-chemical reactions by utilizing nanoelectrospray mass spectrometry, *J. Am. Soc. Mass Spectrom.* 10, 112–118, 1999.
184. Bennett, K.L., Kussman, M., Björk, P. et al., Chemical cross-linking with thiol-cleavable reagents combined with differential mass spectrometric peptide mapping—A novel approach to assess intermolecular protein contacts, *Protein Sci.* 9, 1503–1518, 2000.

185. Fenaille, F., Guy, P.A., and Tabet, J.C., Study of protein modification by 4-hydroxy-2-nonenol and other short chain aldehydes analyzed by electrospray ionization tandem mass spectrometry, *J. Am. Soc. Mass Spectrom.* 14, 215–226, 2003.
186. Glish, G.L. and Vachet, R.W., The basics of mass spectrometry in the twenty-first century, *Nat. Rev. Drug Discov.* 2, 140–150, 2003.
187. Turner, K.B., Yi-Brunozzi, H.Y., Brinson, R.G. et al., SHAMS: Combining chemical modification of RNA with mass spectrometry to examine polypurine tract-containing RNA/DNA hybrids, *RNA* 15, 1605–1613, 2009.
188. Simpson, D.M. and Benyon, R.J., Acetone precipitation of proteins and the modification of peptides, *J. Proteome Res.* 9, 444–450, 2010.
189. Zhao, Y., Ma, C.Y., Yuen, S.-N., and Phillips, D.L., Study of succinylated food proteins by Raman spectroscopy, *J. Food Agric. Chem.* 52, 1815–1823, 2004.
190. Peydecastaing, J., Bras, J., Vaca-Garcia, C. et al., NIR study of chemically modified cellulosic biopolymers, *Mol. Cryst. Liq. Cryst.* 448, 717–724, 2006.
191. Dahmani, M.R., Ramzi, M., Rochas, C., and Guenet, J.M., Thermoreversible gelation in aqueous binary solvents of chemically modified agaroses, *Int. J. Biol. Macromol.* 3, 147–153, 2003.
192. Egli, M. and Pallan, P.S., The many twists and turns of DNA: template, telomere, tool, and target, *Curr. Opin. Struct. Biol.* 20, 262–275, 2010.
193. Lindsay, D.G. and Shall, S., The acetylation of insulin, *Biochem. J.* 121, 737–745, 1971.
194. Crestfield, A.M., Stein, W.H., and Moore, S., Alkylation and identification of the histidine residues at the active site of ribonuclease, *J. Biol. Chem.* 238, 2418, 1963.
195. Yamada, A., Imoto, T., Fujita, K., Okasaki, K., and Motomura, M., Selective modification of aspartic acid-101 in lysozyme by carbodiimide reaction, *Biochemistry* 20, 4836, 1981.
196. Lundblad, R.L., Noyes, C.M., Featherstone, G.L., Harrison, J.H., and Jenzano, J.W., The reaction of bovine alpha-thrombin with tetranitromethane. Characterization of the modified protein, *J. Biol. Chem.* 263, 3729, 1988.
197. Ray, W.J., Jr. and Koshland, D.E., Jr., A method for characterizing the type and numbers of groups involved in enzyme action, *J. Biol. Chem.* 236, 1973, 1961.
198. Tsou, C.L., Relation between modification of functional groups of proteins and their biological activity. I. A graphical method for the determination of the number and type of essential groups, *Sci. Sin.* 11, 1535, 1962.
199. Tsou, C.L., Kinetics of substrate reaction during irreversible modification of enzyme activity, *Adv. Enzymol.* 61, 381, 1988.
200. Zhou, J.M., Liu, C., and Tsou, C.L., Kinetics of trypsin inhibition by its specific inhibitors, *Biochemistry* 28, 1070, 1989.
201. Horiike, K. and McCormick, D.B., Correlations between biological activity and the number of functional groups chemically modified, *J. Theor. Biol.* 79, 403, 1979.
202. Holbrook, J.J. and Ingram, V.A., Ionic properties of an essential histidine residue in pig heart lactate dehydrogenase, *Biochem. J.* 131, 729, 1973.
203. Bloxham, D.P., The chemical reactivity of the histidine-195 residue in lactate dehydrogenase thiomethylated at the cysteine-165 residue, *Biochem. J.* 193, 93, 1981.
204. Horiike, K., Tsuge, H., and McCormick, D.B., Evidence for an essential histidyl residue at the active site of pyridoxamine (pyridoxine)-5′-phosphate oxidase from rabbit liver, *J. Biol. Chem.* 254, 6638, 1979.
205. Levy, H.M., Leber, P.D., and Ryan, E.M., Inactivation of myosin by 2,4-dinitrophenol and protection by adenosine triphosphate and other phosphate compounds, *J. Biol. Chem.* 238, 3654, 1963.
206. Grouselle, M. and Pudlis, J., Chemical studies on yeast hexokinase. Specific modification of a single tyrosyl residue with 1-ethyl-3-(3-dimethylaminopropyl) carbodiimide, *Eur. J. Biochem.* 74, 471, 1977.

207. Mäkinen, K.K. et al., Chemical modification of *Aeromonas* aminopeptidase. Evidence for the involvement of tyrosyl and carboxyl groups in the activity of the enzyme, *Eur. J. Biochem.* 128, 257, 1982.
208. Rakitzis, E.T., Kinetics of protein modification reactions, *Biochem. J.* 217, 341, 1984.
209. Rakitzis, E.T., Kinetic analysis of regeneration by dilution of a covalently modified protein, *Biochem. J.* 268, 669, 1990.
210. Page, M.G.P., The reaction of cephalosporins with penicillin-binding protein 1bγ from *Escherichia coli, Biochim. Biophys. Acta* 1205, 1994.
211. Dubus, A., Normark, S., Kania, M., and Page, M.G.P., Role of asparagine 152 in catalysis of β-lactam hydrolysis by *Escherichia coli* AmpC β-lactamase studied by site-directed mutagenesis, *Biochemistry* 34, 7757, 1995.
212. Yang, S.J. et al., Involvement of tyrosine residue in the inhibition of plant vacuolar H^+-pyrophosphatase by tetranitromethane, *Biochim. Biophys. Acta* 1294, 89, 1996.
213. Chu, C.L. et al., Inhibition of plant vacuolar H^+-ATPase by diethylpyrocarbonate, *Biochim. Biophys. Acta* 1506, 12, 2001.
214. Hsiao, Y.Y. et al., Diethylpyrocarbonate inhibition of vacuolar H^+-pyrophosphatase possibly involves a histidine residue, *J. Protein Chem.* 21, 51, 2002.
215. Yang, S.H., Wu, C.H., and Lin, W.Y., Chemical modification of aminopeptidase isolated from Pronase, *Biochem. J.* 302, 595, 1994.
216. Coggins, J.R., Kray, W., and Shaw, E., Affinity labelling of proteinases with tryptic specificity by peptides with C-terminal lysine chloromethyl ketone, *Biochem. J.* 137, 579–585, 1974.
217. Dixon, M. and Webb, E.C., *Enzymes*, 3rd edn. Academic Press, New York, 1979.
218. Siegal, I.H., *Enzyme Kinetics. Behavior and Analysis of Rapid Equilibrium and Steady-State Enzyme Systems*, Wiley-Interscience, New York, 1975.
219. Purich, D.L. (ed.), *Contemporary Enzyme Kinetics and Mechanism*, Academic Press, New York, 1983.
220. Northrup, D.B., Rethinking fundamentals of enzyme action, *Adv. Enzymol.* 73, 25, 1999.
221. Wang, J. et al., A graphical method of analyzing pH dependence of enzyme activity, *Biochim. Biophys. Acta* 1435, 177, 1999.
222. Ragin, O., Gruaz-Guyon, A., and Barbet, J., Equilibrium expert: An add-in to Microsoft Excel for multiple binding equilibrium simulations and parameter estimations, *Anal. Biochem.* 310, 1–14, 2002.
223. Liao, F. et al., Kinetic substrate quantification by fitting the enzyme reaction curve to the integrated Michaelis-Menten equation, *Anal. Bioanal. Chem.* 375, 756, 2003.
224. Brey, W.S., *Physical Chemistry and Its Biological Applications*, Academic Press, New York, 1978.
225. Sheehan, D., *Physical Biochemistry: Principles and Applications*, John Wiley & Sons, Ltd., Chichester, U.K., 2000.
226. Price, N.C. et al., *Physical Chemistry for Biochemists*, 3rd edn., Oxford University Press, Oxford, U.K., 2001.
227. Plapp, B.V., Site-directed mutagenesis: A tool for studying enzyme catalysis, *Methods Enzymol.* 249, 91, 1995.
228. Peracchi, A., Enzyme catalysis: Removing chemically "essential" residues by site-directed mutagenesis, *Trends Biochem. Sci.* 26, 497, 2001.

2 Modification of Amino/Amidino Groups in Proteins

Histidine (imidazole nitrogen), lysine (ε-amino), arginine (guanidino), and the α-amino group of N-terminal amino acids are nitrogen nucleophiles in proteins. This chapter discusses the modification of lysine, α-amino groups, and arginine, while histidine is discussed with tryptophan in Chapter 4. The modification of these functional groups is dependent on nucleophilicity, which is a product of the micro-environment as discussed in Chapter 1, and the pK_a values of functional groups is an approximate (but not perfect) measure of nucleophilicity assessed by reactivity.[1–6] The reader is directed to a more thorough consideration of this issue in Chapter 1. The discussion will start with α-amino groups, then the ε-amino groups, and conclude with a discussion of the guanidino function of arginine.

α-AMINO GROUPS (N-TERMINAL AMINO GROUPS)

The average pK_a for the α-amino group in protein is 7.7 (see Table 1.1) with a range of 6.8–9.1.[7] The availability of the α-amino group for modification is variable and in some cases the α-amino group is blocked.[8–10] Cyanate (see below) reacts with α-amino groups and ε-amino groups to yield carbamyl derivatives with α-amino groups being somewhat more reactive.[11] Also, as noted in the following, cyanate derived from the dismutation of urea can block amino-terminal groups. It is possible to use the difference in pK_a values between the α-amino group and the ε-amino group to allow preferential modification of the α-amino group at lower pH. Stark[12] suggested that α-amino groups will react 100 times faster than ε-amino groups. Selective modification at the N-terminal amino acid can also be obtained with isothiocyanate derivatives[13] and by using lysine-deficient peptides.[14] The selective modification of N-terminal serine or threonine by periodate[15–18] is discussed in detail in Chapter 3. It is possible to selectively modify the α-amino groups of proteins by chemical transamination with glyoxylate at a slightly acid pH.[19,20] This modification has been applied to *Euglena* cytochrome C-552. This reaction was performed in 2.0 M sodium acetate, 0.10 M acetic acid, 0.005 M nickel sulfate, and 0.2 M sodium glyoxylate and resulted in the complete loss of the amino-terminal residue. Snake venom phospholipase A_2 has been subjected to chemical transamination.[20] This reaction was performed in 2.0 M sodium acetate, 0.4 M acetic acid, 0.010 M cupric ions, and 0.1 M glyoxylic acid, pH 5.5. The various modification reactions for amino-terminal amino acids are shown in Figure 2.1. The ketoacyl function resulting from transamination

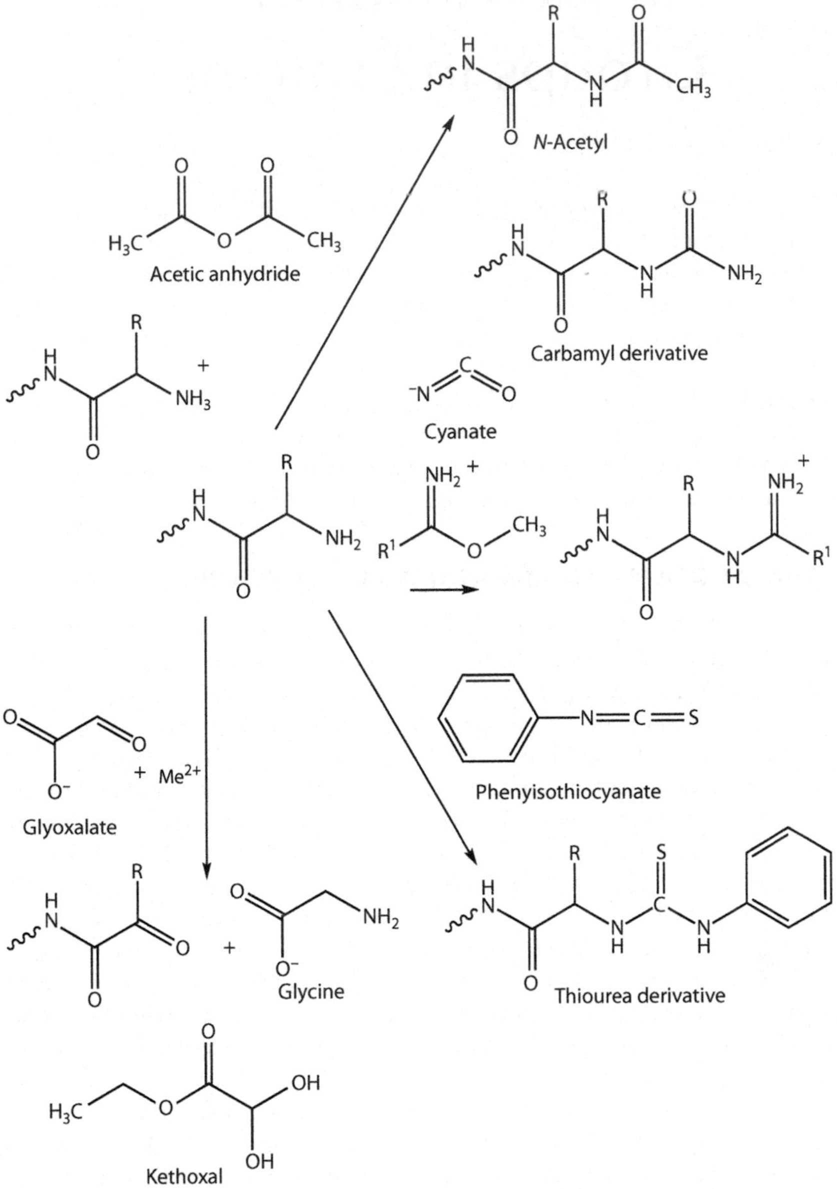

N-Acetyl

Acetic anhydride

Carbamyl derivative

Cyanate

Phenyisothiocyanate

Glyoxalate

Glycine

Thiourea derivative

Kethoxal

FIGURE 2.1 (See caption on facing page.)

reaction is expected to be highly reactive as a structurally related compound, kethoxal (Figure 2.1), is used as a probe for ribosomes where it reacts with guanine[21–23] and has been recently shown to react with arginine.[24] Nitrous acid deamination of the amino-terminal isoleucine of chymotrypsin results in inactivation.[25] Similar results were observed earlier with trypsin.[26] A complex group of products can be obtained with the nitrous acid deamination reaction stemming from the initial formation of a diazo group, which decomposes to a carbanion,[27] and nitrous acid also modifies other amino acids including lysine, tyrosine, and tryptophan.[26,28] N-terminal amino groups appear to be modified more rapidly than the ε-amino groups of lysine. Nitrous acid treatment also results in the deamination of nucleobases[29] and is used for the limited hydrolysis of sulfated glucosaminoglycans such as heparin and heparan sulfate.[30] It is of interest that the specificity of hydrolysis depends on the pH; at pH 1.5, cleavage in sulfated glycosamine depends on the distribution of *N*-sulfated glucosamine, while at pH 4.5, cleavage occurs at unsubstituted amino sugars, and *N*-sulfated glucosamine glycosidic linkages are not cleaved.[31]

Nitrous acid has been demonstrated to have variable results depending on the protein and conditions. Philpot and Small[32] treated pepsin with nitrous acid and suggested that inactivation (50%) was associated with modification of tyrosine residues, although there was also modification of tryptophan. Hoffman subsequently suggested[33] that while the amino-terminal isoleucine residue was modified rapidly with nitrous acid, the 40% inactivation observed was due to the modification of tryptophan. It should be mentioned that, unlike serine proteases,[34]

FIGURE 2.1 Modification of α-amino group in proteins. Shown are five approaches for the modification of amino groups in proteins. Acetic anhydride is promiscuous with modification also occurring at the ε-amino group of lysine (Chen, J., Smith, D.L., and Griep, M.A., The role of the 6 lysines and the terminal amine of *Escherichia coli* single-strand binding protein in its binding of single-stranded DNA, *Protein Sci.* 7, 1781–1788, 1998; Schepmoes, A.A., Zhang, Q., Petriis, B.O. et al., *N*-Terminal enrichment: Developing a protocol to detect specific proteolytic fragments, *J. Biomol. Tech.* 20, 263–265, 2009). Cyanate also reacts with the ε-amino group of lysine as well as other functional groups (Stark, G.R., Reactions of cyanate with functional groups of proteins. IV. Inertness of aliphatic hydroxyl groups. Formation of carbamyl- and acylhydantoins, *Biochemistry* 4, 2363–2367, 1965; Stark, G.R., Modification of proteins with cyanate, *Methods Enzymol.* 25, 579–584, 1972). Nitrous acid results in oxidative deamination of the α-amino groups (Kurosky, A. and Hofmann, T., Kinetics of the reaction of nitrous-acid with model compounds and proteins, and conformational state of N-terminal groups in chymotrypsin family, *Can. J. Biochem.* 50, 1282–1296, 1972). The α-amino groups of protein can be preferentially modified with organic isothiocyanates at neutral pH, while modification of ε-amino groups occurs at more alkaline pH (Rana, T.M. and Meares, C.F., N-terminal modification of immunoglobulin polypeptide chains tagged with isothiocyanato chelates, *Bioconjug. Chem.* 1, 357–362, 1990); specifically shown in this figure is the modification with phenylisothiocyanate, which serves as the basis for the Edman degradation (Edman, P., Phenylthiohydantoins in protein analysis, *Ann. N. Y. Acad. Sci.* 88, 602–610, 1960). The final reaction shows the transamination reaction involving glyoxalate and divalent cation (Dixon, H.B.F. and Fields, R., Specific modification of NH_2 terminal residues by transamination, *Methods Enzymol.* 25, 409–419, 1972).

there is no evidence to suggest the involvement of the amino-terminal residues in maintaining the integrity of the active site of pepsin.

The chemistry of nitrous acid is intertwined with the chemistry of nitric oxide.[35] Indeed Furgott mentioned the reversible dismutation of nitrous acid to form nitric oxide as a seminal observation in the identification of endothelium-derived relaxing factor (EDRF).[36,37] Despite the large interest in compounds such as nitric oxide and peroxynitrite, understanding the chemistry of nitrous acid is challenging. Granted that there is little current use of nitrous acid in solution protein chemistry, there is use in carbohydrate chemistry and nucleic acid chemistry. There is also use of nitrous acid to remove free amino acids and amino sugars from complex mixtures prior to analytical procedures.[38,39]

The chemistry of nitrous acid is complex (Figure 2.2). Nitrous acid is formed *in situ* from the acidification of a nitrite salt (most frequently sodium nitrite). Philpot and Small[32] have shown that concentration of nitrous acid in a solution of sodium nitrite in sodium acetate decreases by 50% in going from pH 3.7 to 4.2 to less than 10% at pH 5.2. Nitrous acid is unstable (Figure 2.2) decomposing to form nitrate and nitric oxide.[40,41]

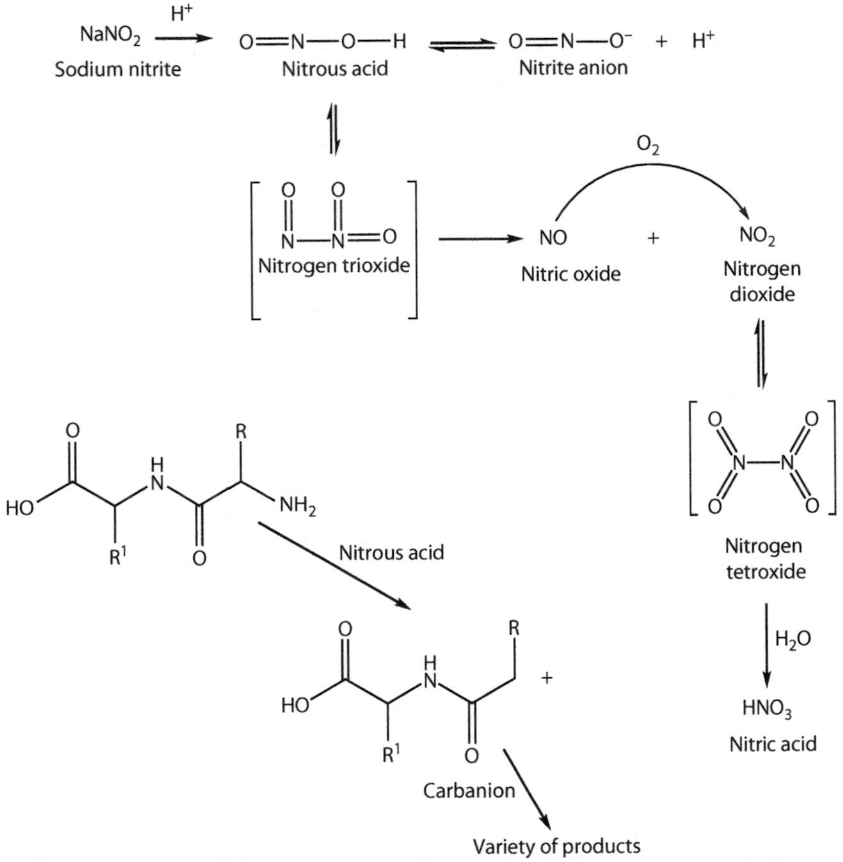

FIGURE 2.2 The reaction of nitrous acid for the modification of α-amino acids in proteins. Shown is one of the processes for the production of various nitrogen oxides and the deamination of the α-amino group of a peptide.

The nitrate anion has been demonstrated to be stable at alkaline pH.[42] Daiber and coworkers[43] observed that the freezing of nitrite in sodium phosphate solutions results in the nitration of tyrosine and the formation of *S*-nitrosocysteine. It is suggested that the nitration and nitrosylation are a result of the formation of nitrous acid from nitrite, reflecting the change in the pH of sodium phosphate solutions with change in temperature.[44] Pikel-Cleland and coworkers[45] report that there can be as much as a 3 pH unit decrease in sodium phosphate buffer on freezing. Butler and Ridd[46] suggest that the formation of nitric acid from nitrous acid involves dinitrogen oxide (nitrogen trioxide, nitrous anhydride, N_2O_3) as an intermediate. Aga and Hughes[47] review the preparation and purification of nitric oxide gas and emphasize the ease of oxidation to nitrogen dioxide (NO_2) which dimerizes to form dinitrogen tetroxide (N_2O_4) which is an effective nitrosation reagent. In the presence of water, nitric oxide and dioxygen readily forms nitrogen trioxide (nitrous anhydride) which, in addition to forming nitric acid as described above, is also suggested by Kurosky and Hofmann[48] to be the active species in the reaction of nitrous acid with amino acid and proteins.

There is some support for the use of 1-fluoro-2,4-dinitrobenzene (see below) to selectively modify α-amino groups at lower pH (7.0).[49,50] While not perfect, there is a sense that there will be preferential reaction of FDNB at α-amino groups in proteins.[51]

I would be remiss not to mention inteins where an N-terminal cysteine residue may be selectively modified.[52,53] This technology is most useful for the preparation of semisynthetic proteins. There is one report where this technology was used for the attachment of a protein to a surface.[54] In addition, this technology can be used to label the C-terminus as well.[55]

The average pK_a for the epsilon amino group of lysine is 10.79 but there is a large variation in the pK_a values. Schmidt and Westheimer[56] obtained a pK_a of 5.9 for the amino group at the active site of acetoacetate decarboxylase b using the pH dependence of acylation by 2,4-dinitrophenyl propionate (Figure 2.3); as noted by the authors, it is some 4 pK_a units less than that of an "ordinary" ε-amino group of lysine. Subsequent information leads to the suggestion that the pK_a of Lys115, the residue modified by 2,4-dinitrophenylpropionate, is influenced by proximity to Lys116. While this is an elegant explanation, the solving of the crystal structure revealed that Lys115 was, in fact, in a hydrophobic pocket.[57] Not withstanding the difficulties in the interpretation of such data,[58] it is clear that the pH dependence of this reaction is consistent with the presence of a more reactive amino group at lower pH[59] and, hence, a decrease in the pK_a (see Table 2.1). It is thought that a "buried" lysine residue tends to have a lower pK_a as a neutral species is favored in an apolar environment,[60] although there is an effect of local environment[61,62] including buried water.[63] However, while a lysine may lower pK_a value from being located in an apolar environment, this may not translate into enhanced reactivity. Approximately 15%–20% of the lysine residues in bovine serum albumin are not reactive to modification with organic acid anhydrides[64,65] such as acetic anhydride or succinic anhydride except at very high reagent concentration[66] and are considered buried. Similar results are obtained with ovalbumin in reaction with succinic anhydride and acetic anhydride[67] where 25% of the lysine residues are accessible to modification with a high molar excess (1000-fold excess) of reagent. The great majority of studies on the effect of conformation on lysine group reactivity have been performed with acetic anhydride with lesser

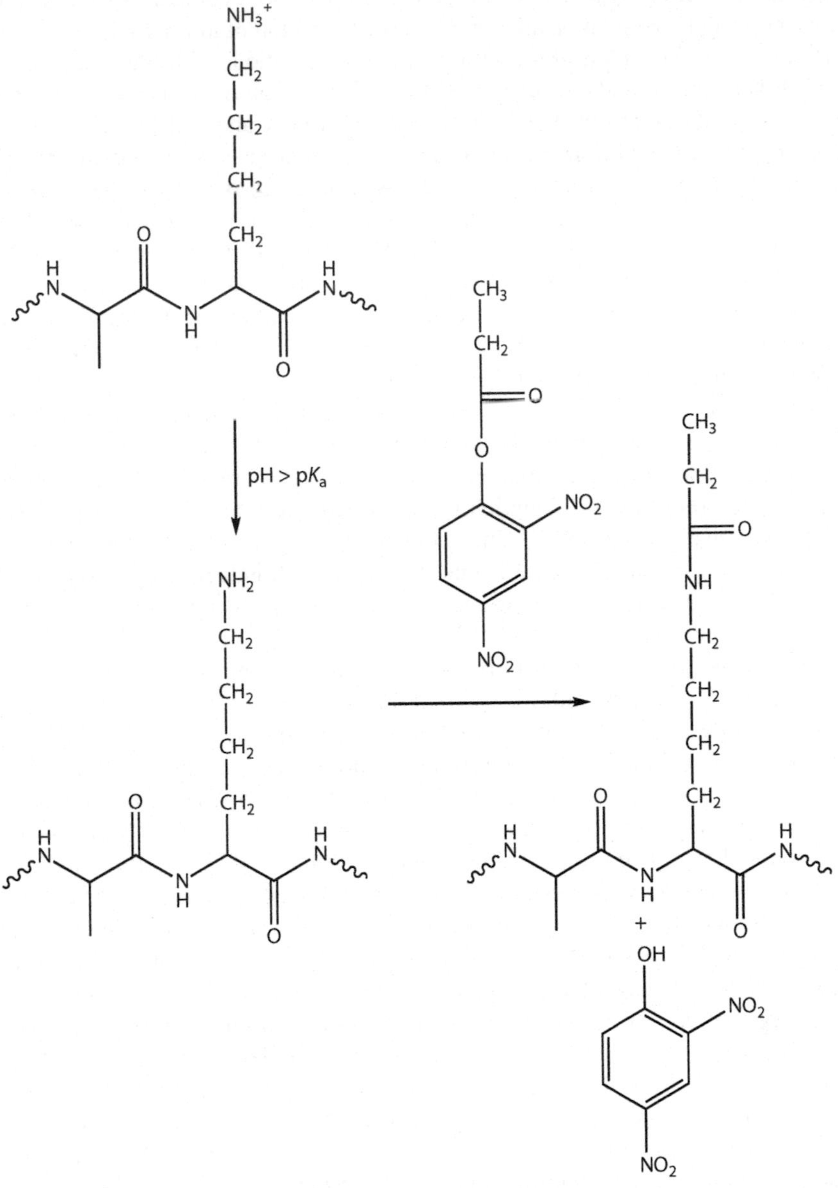

FIGURE 2.3 The reaction of *p*-nitrophenyl propionate with a lysine residue. It is this reaction that was used to establish a low pK_a value for the lysine residue at the active site of acetoacetate decarboxylase (Schmidt, D.E., Jr. and Westheimer, F.H., PK of the lysine amino group at the active site of acetoacetate decarboxylase, *Biochemistry* 10, 1249–1253, 1971).

TABLE 2.1
Some pK_a Values for Lysine in Proteins as Determined by Chemical Modification

Protein and Reagent	pK_a	Reference
Acetoacetate decarboxylase/2,4-dinitrophenyl propionate. The concept was advanced that the pK_a of Lys115 (the residue modified by *p*-nitrophenylacetate) was influenced by the proximity of Lys116. The concept was extended by a number of investigators. Highberger and coworkers (Highberger, L.A., Gerlt, J.A., and Kenyon, G.L., Mechanism of the reaction catalyzed by acetoacetate decarboxylase Importance of lysine 116 in determining the pK(a) of active-site lysine 115, *Biochemistry* 35, 41–46, 1996). These investigators substituted Lys115 with cysteine (K115C) or asparagine (K115N) in acetoacetate decarboxylase with complete loss of activity. However, the crystal structure (Gerlt, J.A., Acetoacetate decarboxylase: Hydrophobics, not electrostatics, *Nat. Chem. Biol.* 5, 454–455, 2009) shows Lys115 in a hydrophobic pocket, which could also explain the decreased pK_a.	5.9	[1]
Lys41 in bovine pancreatic ribonuclease. The studies with FDNB suggested the pK_a of 8.8; the reactivity at Lys41 was reduced in the presence of 8.0 M urea. A pK_a value for Lys41 of 9.11 was obtained from reaction with 4-sulfonoxy-2-nitrofluorobenzene.	8.8 (FDNB) 9.11 (SFNB)	[2] [3]
Melittin modification with TNBS at two lysine residues (Lys21 and Lys23). Reaction at amino-terminal glycine was 20 times slower.	Lys21 6.5 Lys 23 8.6	[4]
Hydroxymethylbilane synthase (porphobilogen deaminase). PLP used for the modification of lysine residues. Inactivation studies (with several lysine residues modified) suggested a pK_a of 6.7. Subsequent work showed that two lysine residues (Lys55 and Lys59) modified under those reaction conditions (Miller, A.D., Packman, L.C., Hart, G.J. et al., Evidence that pyridoxal phosphate modification of lysine residues (lys55 and lys59) causes inactivation of hydroxymethylbilane synthase (porphobilinogen deaminase), *Biochem. J.* 262, 119–124, 1989). The modification of these two residues appears to be mutually exclusive in that modification of one residue precludes modification at the second site. Further work (Haedener, A., Alefounder, P.R., Hart, G.J. et al., Biosynthesis of porphyrins and related macrocycles. Part 36. Investigations of putative active-site lysine residues in hydroxymethylbilane synthase. Preparation and characterization of mutants in which (a) lys55, (b) lys59, and both lys55 and lys59 have been replaced by glutamine, *Biochem. J.* 271, 487–491, 1990) showed that while both Lys55 and Lys59 are near the enzyme active site of hydroxymethylbilane synthase, neither is essential for activity.	6.7	[5]
Pig liver phosphomelavonate kinase; modification with PLP	8.15 @ 24°C 7.9 @ 31°C	[6]

(continued)

TABLE 2.1 (continued)
Some pK_a Values for Lysine in Proteins as Determined by Chemical Modification

Protein and Reagent	pK_a	Reference
(Na^+ + K^+)-dependent ATPase modification with acetic anhydride or TNBS	7.5–8.0	[7]
Avian liver phosphoenolpyruvate carboxykinase modification with PLP	8.1	[8]
Sarcosine oxidase (*Corynebacterium* sp. U-96). Two lysine residues are modified by reaction with iodoacetamide. Modification of lysine residues with iodoacetamide is unusual but not unknown (Parker, R.C., Stanley, S., and Kristol, D.S., Reaction of iodoacetamide with polylysine, α-acetyllysine, and ε-acetyllysine, *Int. J. Biochem.* 6, 863–866, 1975).	6.7 8.5	[9]
Fructose-1,6-bisphosphate aldolase modification with PLP. Modification occurs at Lys107; the low pK_a of Lys107 is thought to reflect local electrostatic environment including interaction with Lys146 (St-Jean, M., Blonski, C., and Sygusch, J., Charge stabilization and entropy reduction of central lysine residues in fructose-bisphosphate aldolase, *Biochemistry* 48, 4528–4537, 2009)	8	[10]

1. Schmidt, D.E. Jr. and Westheimer, F.H., p*K* of the lysine amino at the active sites of acetoacetate decarboxylase, *Biochemistry* 10, 1249–1253, 1971.
2. Murdock, A.L., Grist, K.L., and Hirs, C.H.W., On the dinitrophenylation of bovine pancreatic ribonuclease A. Kinetics of the reaction in water and 8.0M urea, *Arch. Biochem. Biophys.* 114, 375–390, 1966.
3. Carty, R.P. and Hirs, C.H.W., Modification of bovine pancreatic ribonuclease A with 4-sulfonoxy-2-nitrofluorobenzene. Effect of pH and temperature on the rate of reaction, *J. Biol. Chem.* 243, 5254–5265, 1968.
4. Quay, S.C. and Tronson, L.P., Conformational studies of aqueous melittin. Determination of ionization constants of lysine-21 and lysine-23 by reactivity toward 2,4,6-trinitrobenzene sulfonate, *Biochemistry* 22, 700–707, 1983.
5. Hart, G.J., Leeper, F.J., and Battersby, A.R., Modification of hydroxymethylbilane synthase (porphobilinogen deaminase) by pyridoxal-5′-phosphate. Demonstration of an essential lysine residue, *Biochem. J.* 222, 93–102, 1984.
6. Bazaes, S., Beytía, E., Jabalquinto, A.M. et al., Pig liver phosphomelavonate kinase 2. Participation of cysteinyl and lysyl groups in catalysis, *Biochemistry* 19, 2305–2310, 1980.
7. Robinson, J.D. and Flashner, M.S., Modification of the (Na^+ + K^+)-dependent ATPase by acetic anhydride and trinitrobenzene sulfonate: Specific changes in enzymatic properties, *Arch. Biochem. Biophys.* 196, 350–362, 1979.
8. Guidinger, P.F. and Nowak, T., An active-site lysine in avian liver phosphoenolpyruvate carboxykinase, *Biochemistry* 30, 8851–8861, 1991.
9. Mukouyama, E.B., Oguchi, M., Kodera, Y. et al., Low p*Ka* lysine residues at the active site of sarcosine oxidase from *Corynebacterium* sp. U-96, *Biochem. Biophys. Res. Commun.* 320, 846–851, 2004.
10. Dax, C., Coincon, M., Syngusch, J., and Blonski, C., Hydroxynaphthaldehyde phosphate derivatives as potent covalent Schiff base inhibitors of fructose-1,6-biphosphate aldolase, *Biochemistry* 44, 5430–5443, 2005.

use of succinic anhydride and citraconic anhydride. There are several studies with other reagents that are of interest in considering the reactivity of buried functional groups. Wells[68] showed that the "burial" of lysine residues in phospholipase A_2 from *Crotalus adamateus* venom in the presence of calcium ions blocks reaction with diethylpyrocarbonate (DEPC); this is a "side reaction" when using DEPC to modify histidine but can be differentiated by being an irreversible reaction as compared to the reversibility of monosubstitution at histidine (see Chapter 4). Mita and coworkers[69] showed that 93% of the lysine residues could be modified by *O*-methylisourea in calf thymus histone H1 with the same rate constant ($1.8 \times 10^{-5}\ s^{-1}$; pH 10, 4°C), while the remaining lysine residues are modified only in the presence of 6.0M guanidine hydrochloride. *S*-Methylthioacetimidate was used by Running and Reilly[70] to categorize lysine residues in ribosomal proteins, which are either buried or in contact with ribosomal RNA. In summary, while a buried lysine residue may have a decreased pK_a, this does not translate into increased reactivity with electrophiles.

The majority of lysine residues in a protein are located on the surface of a protein and while there are examples of equivalent reactivity as mentioned earlier for modification of a histone with *O*-methylisourea,[70] there are many examples of major differences in reactivity for surface lysine residues. In the work by Schmidt and Westheimer[56] cited earlier, the reactivity of the active-site ε-amino group of acetoacetate decarboxylase with 2,4-dinitrophenylproprionate was used to determine a value of 5.9 for the pK_a of this group. This is an example of using functional group reactivity to measure the dissociation constant with the assumption, in the case of the ε-amino group of lysine, that only the unprotonated form reacts with the reagent (Figure 2.1). This is a well-cited study but there were earlier observations on low pK_a values for ε-amino groups in proteins. Murdock and coworkers[71] estimated the pK_a of Lys41 in bovine pancreatic ribonuclease at 8.8 from reaction with fluorodinitrobenzene (FDNB); the rate of reaction was reduced in the presence of 8.0M urea emphasizing the importance of conformation in the observed enhanced reactivity at this site. The overall rate of reaction of RNase with FDNB was increased in the presence of 8.0M urea reflecting the increased solubility of the reagent in this solvent. Carty and Hirs[72] confirmed the decreased pK_a/increased reactivity of Lys41 in RNase with 4-sulfonoxy-2-nitrofluorobenzene. Hart and coworkers[73] established a pK_a of 6.7 for a lysine residue in hydroxymethylbilane synthase using reaction with pyridoxal-5′-phosphate (PLP). Quay and Tronson[74] used reaction with 2,4,6-trinitrobenzenesulfonate (TNBS) to establish a pK_a of 6.5 for Lys21 and 8.6 for Lys23 in melittin.

It is important to recognize the importance of the acylation of the ε-amino groups of lysine as a posttranslational modification. The acetylation of the ε-amino groups by acetyltransferase using acetyl-CoA as a donor is a well-known posttranslational modification (Figure 2.4)[75–77] as is the modification with ubiquitin[78] and small ubiquitin-like modifier (SUMO) proteins.[79] The nonenzymatic acetylation of proteins by acetyl-CoA has been reported (Figure 2.4).[80] Kashiwagi and coworkers[81] reported the nonenzymatic acetylation of the C-2 amino group of sphinganine by acetyl-CoA. The acetylation of lysine by aspirin (acetylsalicylic acid) (Figure 2.4) has been reported for a variety of proteins including albumin[82,83] and hemoglobin.[84] Transglutaminase is an enzyme that catalyzes a modification of lysine generally associated with cross-linking,[85] which can be used to attach probes.[86–88]

The modification of the ε-amino group of lysine in proteins has proved useful for the identification of amino residues important for biological function, a measure of conformation and ligand interaction, and as "handles" for attachment for matrices. A list of reagents commonly used for the modification of amino groups in proteins is presented in Table 2.2. A more extensive discussion of some specific reagents used for the modification of ε-amino group of lysine in proteins is given in the following.

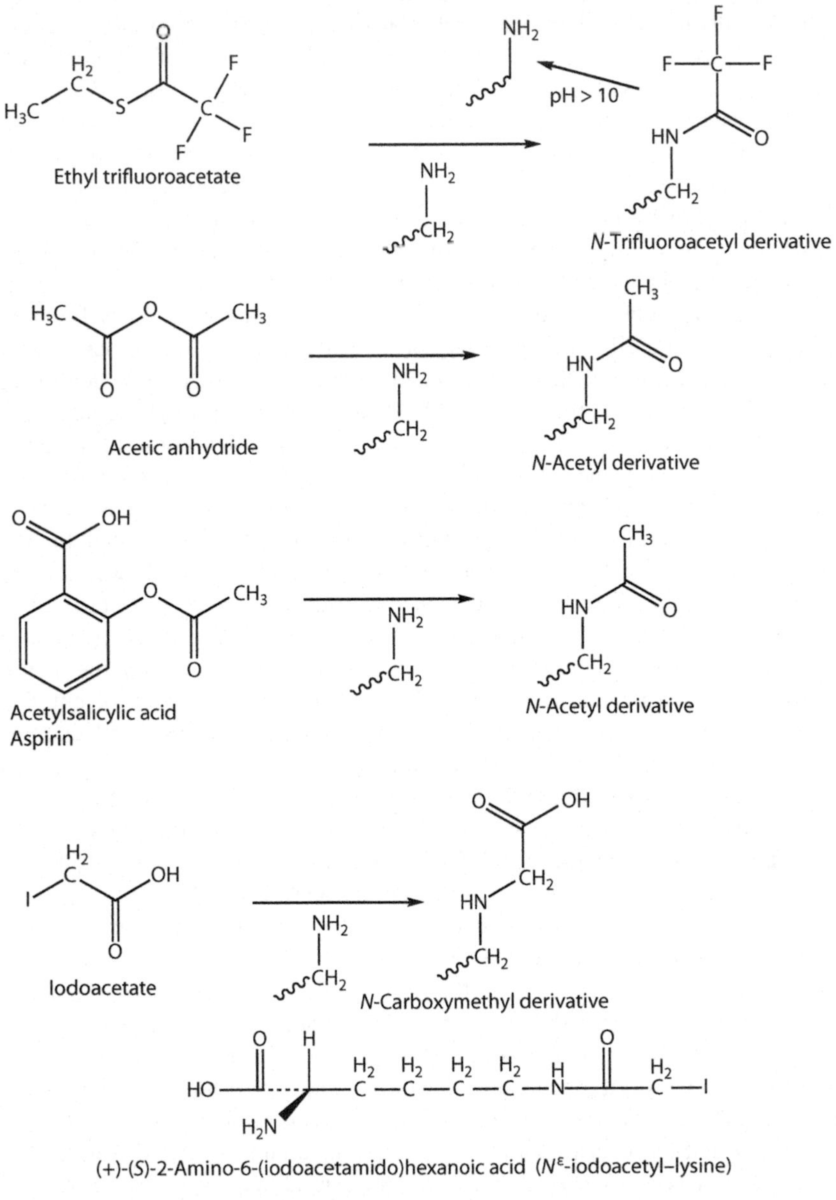

Ethyl trifluoroacetate

pH > 10

N-Trifluoroacetyl derivative

Acetic anhydride

N-Acetyl derivative

Acetylsalicylic acid
Aspirin

N-Acetyl derivative

Iodoacetate

N-Carboxymethyl derivative

(+)-(*S*)-2-Amino-6-(iodoacetamido)hexanoic acid (N^{ε}-iodoacetyl–lysine)

FIGURE 2.4 (See caption on facing page.)

Acylation of amino groups in proteins with acid anhydrides (Figure 2.5) or less frequently with acid chlorides is an extensively used modification procedure. Modification most often occurs at the ε-amino group of lysine and somewhat less frequently at α-amino groups, and amino sugars.[89] Reaction can also occur at other nucleophilic functional groups including sulfhydryl, phenolic hydroxyl, histidine imidazole nitrogens, and at aspartic or glutamic via mixed acid anhydride formation. Most of these latter modifications are either exceedingly transient or labile under conditions (mild base) where *N*-acyl groups are stable. While the majority of acylation reactions use acetic anhydride (Figure 2.4) acylation can be performed with a variety of acid anhydrides including citraconic anhydride,[90,91] maleic anhydride,[92–94] succinic anhydride,[95–97] 3-hydroxyphthalic anhydride,[98] trimellitic anhydride,[98] methyltetrahydrophthalic anhydride,[99,100] cis-aconitic anhydride,[96,97] fatty acid anhydrides,[101] hexahydrophthalic anhydride,[100,102] and phthalic anhydride (Figure 2.5).[103]

Succinic anhydride (Figure 2.6) has also proved useful in the modification of lysine.[104] Modification of lysine residues with succinic anhydride results in charge reversal. Reaction with succinic anhydride frequently results in the dissociation of multimeric proteins and has also been used to "solubilize" insoluble proteins. Meighen and coworkers[105] have produced a "variant" form of bacterial luciferase through reaction with succinic anhydride. The succinylated protein retained the dimeric subunit structure of the native enzyme. By complementation experiments involving the mixing/hybridization of the modified and native enzyme, it was determined that succinylation of bacterial luciferase resulted in the inactivation of the β-subunit without markedly affecting the function of the α-subunit. Shetty and Rao[106] studied the reaction of succinic anhydride with arachin. In this study, reaction of the protein was performed in 0.1 M sodium phosphate, pH 7.8 with the pH maintained over the course of the reaction by the addition of 2.0 M NaOH. The extent of modification was determined by reaction of the unmodified primary amino groups on the protein with trinitrobenzenesulfonic acid (see below). With a 200:1 molar excess of succinic anhydride, 82% of the available amino groups were succinylated with concomitant dissociation of the subunits of

FIGURE 2.4 Some examples of the reaction of several small acyl organics with lysine residues in proteins. A variety of organic derivatives of the "acetyl" group can be used to acetylate amino groups in proteins. Acetylation also occurs *in vivo* (Soppa, J., Protein acetylation in archaea, bacteria, and eukaryotes, *Archaea* pii:820681, 2010; Sadoul, K., Wang, J., Diagouraga, B., and Khochbin, S., The tale of protein lysine acetylation in the cytoplasm, *J. Biomed. Biotechnol.* 2011, 970382, 2011). Acetic anhydride is the most frequently used reagent to acetylate lysine residues in proteins. The acetylation of lysine is also accomplished with aspirin [2-(acetyloxy) benzoic acid; acetylsalicylic acid] (Liyasova, M.S., Schopfer, L.M., and Lockridge, O., Reaction of human albumin with aspirin in vitro: Mass spectrometric identification of acetylated lysines 199, 402, 519, and 545, *Biochem. Pharmacol.* 79, 784–791, 2010). Acetylation with acetyl coenzyme A is an enzyme-catalyzed reaction that can result in autoacetylation (Stavropoulos, P., Nagy, V., Blobel, G., and Hoelz, A., Molecular basis for the autoregulation of the protein acetyl transferase Rtt109, *Proc. Natl. Acad. Sci. USA* 105, 12236–12241, 2008; Wu, H.Y., Huang, F.Y., Chang, Y.C. et al., Strategy for determination of in vitro protein acetylation sites by using isotope-labeled acetyl coenzyme A and liquid chromatography-mass spectrometry, *Anal. Chem.* 80, 6178–6189, 2008).

TABLE 2.2
Reagents for the Modification of ε-Amino Groups in Proteins

Reagent	Other Sites Modified[a]	References
Acetic anhydride	α-Amino groups, tyrosine, histidine, cysteine, and carboxyl groups	[1–13]
Methyl acetyl phosphate	α-Amino groups	[14–21]
PLP		[22–29]
Reductive alkylation		[30–40]
2,4,6-Trinitrobenzenesulfonic acid	α-Amino groups	[40–50]
NHS	α-Amino groups; hydroxyl groups	[51–56]
Imidoesters	α-Amino groups	[57–58]
Iminothiolane (Traut's reagent)		[59–63]

[a] Specificity is determined by reaction conditions (e.g., pH, buffer ion) and the protein being modified.

1. Shen, S. and Strobel, H.W., Role of lysine and arginine residues of cytochrome P450 in the interaction between cytochrome P4502B1 and NADPH-cytochrome P450 reductase, *Arch. Biochem. Biophys.* 304, 257, 1993.
2. Suckau, D., Mak, M., and Przybylski, M., Protein surface topology-probing by selective chemical modification and mass spectrometric peptide mapping, *Proc. Natl. Acad. Sci. USA* 89, 5630–5634, 1992.
3. Cervenansky, C., Engstrom, A., and Karlsson, E., Study of structure-activity relationship of fasciculin by acetylation of amino groups, *Biochim. Biophys. Acta* 1199, 1–5, 1994.
4. Ohguro, H. et al., Topographical study of arrestin using differential chemical modifications and hydrogen deuterium exchange, *Protein Sci.* 3, 2428–2434, 1994.
5. Zappacosta, F. et al., Surface topology of Minibody by selective chemical modification and mass spectrometry, *Protein Sci.* 6, 1901–1909, 1997.
6. Hochleitner, E.O. et al., Characterization of a discontinuous epitope of the human immunodeficiency virus (HIV) core protein p24 by epitope excision and differential chemical modification followed by mass spectrometric peptide mapping analysis, *Protein Sci.* 9, 487–496, 2000.
7. Hlavica, P., Schulze, J., and Lewis, D.F.V., Functional interaction of cytochrome P450 with its redox partners: A critical assessment and update of the topology of predicted contact regions, *J. Inorg. Biochem.* 96, 279–297, 2003.
8. Calvete, J.J. et al., Characterisation of the conformational and quaternary structure-dependent heparin-binding region of bovine seminar plasma protein PDC-109, *FEBS Lett.* 444, 260–264, 1999.
9. Smith, C.M. et al., Mass spectrometric quantification of acetylation at specific lysines within the amino-terminal tail of histone H4, *Anal. Biochem.* 316, 23–33, 2003.
10. Yadev, S.P., Brew, K., and Puett, D., Holoprotein formation of human chorionic gonadotropin: Differential labeling with acetic anhydride, *Mol. Endocrinol.* 8, 1547–1558, 1994.
11. Gao, J. et al., Determination of the effective charge of a protein in solution by capillary electrophoresis, *Proc. Natl. Acad. Sci. USA* 91, 12027–12030, 1994.
12. Gao, J. et al., Using capillary electrophoresis to follow the acetylation of the amino groups of insulin and to estimate their basicities, *Anal. Chem.* 68, 3093–3100, 1995.
13. Taralp, A. and Kaplan, H., Chemical modification of lysophilized proteins in non-aqueous environments, *J. Prot. Chem.* 16, 183–193, 1997.
14. Kluger, R. and Tsui, W.-C., Methyl acetyl phosphate. A small anionic acetylating agent, *J. Org. Chem.* 45, 2723, 1980.
15. Ueno, H., Pospischil, M.A., Manning, J.M., and Kluger, R., Site-specific modification of hemoglobin by methyl acetyl phosphate, *Arch. Biochem. Biophys.* 244, 795, 1986.

TABLE 2.2 (continued)
Reagents for the Modification of ε-Amino Groups in Proteins

16. Ueno, H., Pospischil, M.A., and Manning, J.M., Methyl acetyl phosphate as a covalent probe for anion-binding sites in human and bovine hemoglobins, *J. Biol. Chem.* 264, 12344, 1989.
17. Ueno, H. and Manning, J.M., The functional, oxygen-linked chloride binding sites of hemoglobin are contiguous within a channel in the central cavity, *J. Prot. Chem.* 11, 177, 1992.
18. Ueno, H., Popowicz, A.M., and Manning, J.M., Random chemical modification of the oxygen-linked chloride-binding sites of hemoglobin: Those in the central dyan axis may influence the transition between deoxy- and oxy-hemoglobin, *J. Prot. Chem.* 12, 561, 1993.
19. Manning, J.M., Preparation of hemoglobin derivatives selectively or randomly modified at amino groups, *Meth. Enzymol.* 231, 225–246, 1994.
20. Kataoka, K. et al., Identification of active site lysyl residues of phenylalanine dehydrogenases by chemical modification with methyl acetylphosphate combined with site-directed mutagenesis, *J. Biochem.* 116, 1370–1376, 1994.
21. Xu, A.S., Labotka, R.J., and London, R.E., Acetylation by human hemoglobin by methyl acetylphosphate. Evidence of broad region-selectivity revealed by NMR studies, *J. Biol. Chem.* 274, 26629–26632, 1999.
22. Lo Bello, M. et al., Chemical modification of human placental glutathione transferase by pyridoxal-5′-phosphate, *Biochim. Biophys. Acta* 1121, 167, 1992.
23. Paine, L.J. et al., The identification of a lysine residue reactive to pyridoxal-5-phosphate in the glycerol dehydrogenase from the thermophile *Bacillus stearothermophilus, Biochim. Biophys. Acta* 1202, 235, 1993.
24. Basu, S., Basu, A., and Modak, M.J., Pyridoxal 5′-phosphate mediated inactivation of *Escherichia coli* DNA polymerase I: Identification of lysine-635 as an essential residue for the processive mode of DNA synthesis, *Biochemistry* 27, 6710, 1988.
25. Hale, J.A. and Maloney, P.C., Pyridoxal-5-phosphate inhibition of substrate selectivity mutants of UhpT, the sugar 6-phosphate carrier of *Escherichia coli, J. Bacteriol.* 184, 3756–3758, 2002.
26. Celestina, F. and Suryananarayana, T., Biochemical characterization and helix stabilizing properties of HSNP-C' from the thermoacidophilic archeon *Sulfolobus acidocaldarius*, *Biochem. Biophys. Res. Commun.* 267, 614–618, 2000.
27. Talarico, T.L., Guise, K.J., and Stacey, C.J., Chemical characterization of pyridoxylated hemoglobin polyoxyethylene conjugate, *Biochim. Biophys. Acta* 1476, 53–65, 2000.
28. Lapko, A.G. and Ruckpaul, K., Discrimination between conformational states of mitochondrial cytochrome P-450scc by selective modification with pyridoxal-5-phosphate, *Biochemistry (Moscow)* 63, 568–572, 1998.
29. Anderberg, S.J., Topological disposition of lysine 943 in native Na^{+}/K(+)-transporter ATPase, *Biochemistry* 34, 9508–9516, 1995.
30. Brown, E.M. et al., Accessibility and mobility of lysine residues in β-lactoglobulin, *Biochemistry* 27, 5601, 1988.
31. Fujita, Y. and Noda, Y., Effect of alkylation with different sized substituents on thermal stability of lysozyme, *Int. J. Pept. Protein Res.* 40, 103, 1992.
32. Rypniewski, W.R., Holden, H.M., and Rayment, I., Structural consequences of reductive methylation of lysine residues in hen egg white lysozyme: An X-ray analysis at 1.8 Å resolution, *Biochemistry* 32, 9851, 1993.

(*continued*)

TABLE 2.2 (continued)
Reagents for the Modification of ε-Amino Groups in Proteins

33. Kim, T.K. and Burgess, D.J., Pharmacokinetic characterization of ^{14}C-vascular endothelial growth factor controlled release microspheres using a rat model, *J. Pharm. Pharmacol.* 54, 897–905, 2002.
34. Kinesetter, O. et al., Mono-N-terminal poly(ethylene glycol)-protein conjugates, *Adv. Drug Deliver. Rev.* 54, 477–485, 2002.
35. Yammani, R.R., Seetharam, S., and Seetharam, B., Identification and characterization of two distinct ligand binding regions of cubilin, *J. Biol. Chem.* 276, 44777–44788, 2001.
36. Zhang, M., Thulin, E., and Vogel, H.J., Reductive methylation and pKa determination of the lysine side chains in calbindin D9k. *J. Prot. Chem.* 13, 527–535, 1994.
37. Means, G.E. and Feeney, F.E., Reductive alkylation of proteins, *Anal. Biochem.* 224, 1–16, 1995.
38. Rayment, I., Reductive alkylation of lysine residues to alter crystallization properties of proteins, *Methods Enzymol.* 276, 171–179, 1997.
39. Kurinov, I.V. et al., X-ray crystallographic analysis of pokeweed antiviral protein-II after reductive methylation of lysine residues, *Biochem. Biophys. Res. Commun.* 275, 549–552, 2000.
40. Iwata, K., Eyles, S.J., and Lee, J.P., Exposing asymmetry between monomers in Alzheimer's amyloid fibrils via reductive alkylation of lysine residues, *J. Am. Chem. Soc.* 123, 6728–6729, 2001.
41. Coffee, C.J. et al., Identification of the sites of modification of bovine liver glutamate dehydrogenase reacted with trinitrobenzenesulfonate, *Biochemistry* 10, 3516, 1971.
42. Means, G.E., Congdon, W.I., and Bender, M.L., Reactions of 2,4,6-trinitrobenzenesulfonate ion with amines and hydroxide ion, *Biochemistry* 11, 3564–3571, 1972.
43. Kurono, Y. et al., Kinetic-study on rapid reaction of trinitrobenzenesulfonate with human serum albumin, *J. Pharm. Sci.* 70, 1297–1298, 1981.
44. Cahalan, M.D. and Pappone, P.A., Chemical modification of potassium channel gating in frog myelinated nerve by trinitrobenzene sulfonic acid, *J. Physiol.* 342, 119–143, 1983.
45. Whitson, P.A., Burgumn, A.A., and Matthews, K.S., Trinitrobenzenesulfonate modification of the lysine residues in lactose repressor protein, *Biochemistry* 23, 6046, 1984.
46. Sams, C.F. and Matthews, K.S., Diethyl pyrocarbonate reaction with the lactose repressor protein affects both inducer and DNA binding, *Biochemistry* 27, 2277, 1988.
47. Haniu, M. et al., Structure-function relationship of NAD(P)H: Quinone reductase: Characterization of NH_2-terminal blocking group and essential tyrosine and lysine residues, *Biochemistry* 27, 6877, 1988.
48. Xia, C. et al., Chemical modification of GSH transferase P1–1 confirms the presence of Arg-13, Lys-44 and one carboxylate group in the GSH-binding domain of the active site, *Biochem. J.* 293, 357, 1993.
49. Cayot, P. and Tainturier, G., The quantitation of protein amino groups by the trinitrobenzenesulfonic acid method: A reexamination, *Anal. Biochem.* 249, 184–200, 1997.
50. Cayot, P., Roullier, L., and Tainturier, G., Electrochemical modifications of proteins. 1. Glycitolation, *J. Agr. Food Chem.* 47, 1915–1923, 1999.
51. Zeng, Q., Reuther, R., Oxsher, J., and Wang, Q., Characterization of horse spleen apoferritin reactive lysines by MALDI-TOF mass spectrometry combined with enzymatic digestion, *Bioorg. Chem.* 36, 255–260, 2008.
52. Mädler, S., Bich, C., Touboul, D., and Zenobi, R., Chemical cross-linking with NHS esters: A systematic study on amino acid reactivities, *J. Mass Spectrom.* 44, 694–706, 2009.
53. Dreger, M., Leung, B.W., Brownless, G.G., and Deng, T., A quantitative strategy to detect changes in accessibility of protein regions to chemical modification on heterodimerization, *Protein Sci.* 18, 1448–1458, 2009.

TABLE 2.2 (continued)
Reagents for the Modification of ε-Amino Groups in Proteins

54. Frost, S.H., Jensen, H., and Lindegren, S., In vitro evaluation of avidin antibody pretargeting using ^{211}At-labeled and biotinylated poly-L-lysine as effector molecule, *Cancer* 116 (4 Suppl), 1001–1010, 2010.
55. Ori, A., Free, P., Courty, J. et al., Identification of heparin-binding sites in proteins by selective labeling, *Mol. Cell. Proteomics* 8, 2256–2265. 2009.
56. Mentinova, M. and McLuckey, S.A., Covalent modification of gaseous peptide ions with N-hydroxysuccinimide ester reagent ions, *J. Am. Chem. Soc.* 132, 18248–18257, 2010.
57. Gonzalez de Peredo, A., Saint-Pierre, C., Latour, J.M. et al., Conformational changes of the ferric uptake regulation protein upon metal activation and DNA binding; first evidence of structural homologies with the diphtheria toxin repressor, *J. Mol. Biol.* 310, 83–91, 2001.
58. Running, W.E. and Reilly, J.P., Ribosomal proteins of *Deinococcus radiodurans*: Their solvent accessibility and reactivity, *J. Proteome Res.* 8, 1228–1246, 2009.
59. Miyata, K., Kakizawa, Y., Nishiyama, N. et al., Block catiomer polyplexes with regulated densities of charge and disulfide cross-linking directed to enhance gene expression, *J. Am. Chem. Soc.* 126, 2355–2361, 2004.
60. Nacharaju, P., Boctor, F.N., Manjula, B.N., and Acharya, S.A., Surface decoration of red blood cells with maleimidophenyl-polyethylene glycol facilitated by thiolation with iminothiolane: An approach to mask A, B, and D antigens to generate universal red blood cells, *Transfusion* 45, 374–383, 2005.
61. Fowler, J.M., Stuart, M.C., and Wong, D.K., Self-assembled layer of thiolated protein G as an immunosensor scaffold, *Anal. Chem.* 79, 350–354, 2007.
62. Rees, J.N., Florang, V.R., Eckert, L.L., and Doorn, J.A., Protein reactivity of 3,4- dihydroxyphenylacetaldehyde, a toxic dopamine metabolite, is dependent on both the aldehyde and the catechol, *Chem. Res. Toxicol.* 22, 1256–1263, 2009.
63. Lim, C.H., Cho, H.M., Choo, J. et al., Fluorescence-based peptide screening using ligand peptides directly conjugated to a thiolated glass surface, *Biomed. Microdevices* 11, 663–669, 2009.

this protein. The reaction of chymotrypsinogen with succinic anhydride has been studied.[107] In these experiments, the reaction was performed under ambient conditions in 0.05 M sodium phosphate, pH 7.5. During the course of the reaction, the pH was maintained at 7.5 by the addition of 1.0 M NaOH. Chymotrypsinogen (1 g) was dissolved in the sodium phosphate buffer, and 50 mg of succinic anhydride was added over a 30-min period. Under these conditions, 8 of the 14 lysine residues were modified. A related reaction involves the trimesylation of amino groups in proteins.[108] This reaction involves the modification of the protein with di(trimethysilyethyl)trimesic acid. Removal of the blocking groups results in an extremely polar derivative. The procedure is suggested to have value in the solubilization of membrane proteins.

Citraconic anhydride (Figure 2.2) has proved useful since the modification of lysine residues with this reagent is a reversible reaction. Reaction conditions for the modification of lysine residues in proteins are similar to those described earlier for other carboxylic acid anhydrides. Atassi and Habeeb[109] have discussed the use of this reagent in some detail. As an example, the reaction of egg white lysozyme with citraconic anhydride has been studied.[110] With multiple additions of reagent, all primary amino groups were modified at pH 8.2 (the pH of the reaction mixture was maintained with a pH-stat). The product of the reaction was heterogeneous as judged by polyacrylamide gel electrophoresis. All citraconyl groups could be removed by treatment with 1.0 M hydroxylamine, pH 10.0. This treatment also resulted in an electrophoretically homogeneous species. Complete removal of the citraconyl groups could also be achieved by incubation at pH 4.2 for 3 h at 40°C. Reaction with citraconic anhydride has been used to dissociate nucleoprotein complexes.[111] Modification of

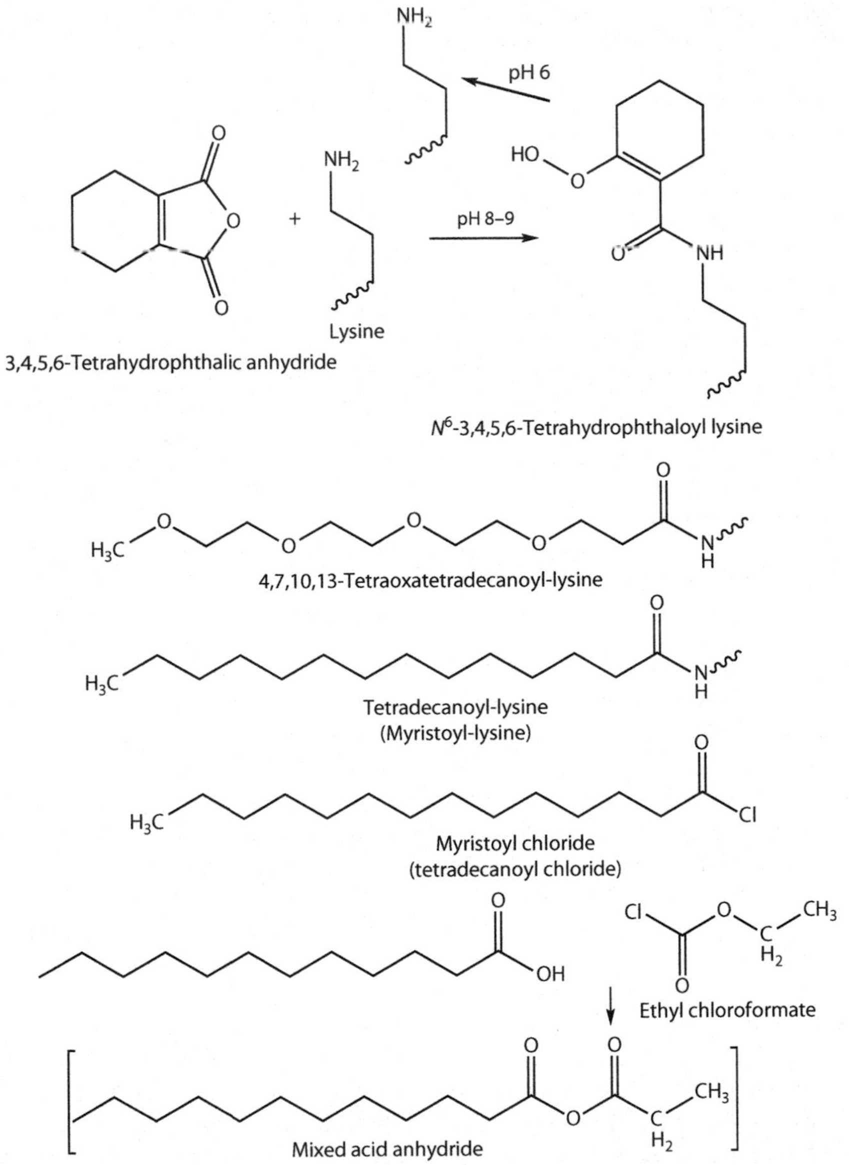

FIGURE 2.5 (See caption on facing page.)

the lysine residues with citraconic anhydride (pH 8.0–9.0 maintained with pH-stat) resulted in a marked change in the charge relationship between the α-amino groups of lysine and the phosphate backbone of the nucleic acid, allowing subsequent separation of protein from nucleic acid. The citraconyl groups were subsequently removed from this protein by incubation at pH 3.0–4.0 at 30°C for 3 h. Ramanathan and colleagues[112] modified lysine residues in riboflavin-binding protein with citraconic anhydride resulting in the loss of ability to bind riboflavin and immunogenicity; both binding capacity and immunogenicity were recovered with removal of the citraconyl groups (0.1 M glycine hydrochloride, pH 2.5 for 6 h at 37°C).

Urabe and coworkers[113] prepared various mixed carboxylic acid anhydrides of tetradecanoic acid and oxa derivatives, which varied in their "hydrophobicity." This represented an attempt to change the surface properties of the enzyme molecule, in this case, thermolysin. The carboxylic acid anhydrides were formed *in situ* from the corresponding acid and ethylchloroformate in dioxane with triethylamine. The modification reaction was performed in 0.013 M barbital, 0.013 M $CaCl_2$, pH 8.5, containing 39% (v/v) dioxane and was terminated with neutral hydroxylamine, which also served to remove *O*-acyl derivatives. The extent of reaction was determined by titration with trinitrobenzenesulfonic acid. Derivatives obtained with tetradecanoic acid and 4-oxatetradecanoic acid were insoluble. Derivatives obtained with both 4,7,10-trioxatertradecanoic acid and 4,7,10,13-tetraoxotetradecanoic acid had approximately seven amino groups modified per mole of enzyme, showed little if any loss in either proteinase or esterase activity, and possessed enhanced thermal stability.

Pool and Thompson[101,114] selectively attached long chain (C_{12}–C_{16}) fatty acids to the amino-terminal amino acid of chemical modified (guanidated) bovine pancreatic trypsin inhibitor (BPTI). The guanidated protein was obtained by reaction of the BPTI bound to trypsin, which protected Lys15 at the inhibitor active site.

FIGURE 2.5 The use of large acyl groups to modify lysine residues in proteins. The top of the figure shows the modification of a lysine residue with 3,4,5,6-tetrahydrophthalic anhydride (Gibbons, I. and Schachman, H.K., A method for the separation of hybrids of chromatographically identical oligomeric proteins. Use of 3,4,5,6-tetrahydrophthaloyl groups as a reversible "chromatographic handle," *Biochemistry* 15, 52–60, 1976; Wearne, S.J., Factor Xa cleavage of fusion proteins. Elimination of non-specific cleavage by reversible acylation, *FEBS Lett.* 263, 23–26, 1990; Palacián, E., Gonzalez, P.J., Piñero, M. et al., Dicarboxylic acid anhydrides as dissociating agents of protein-containing structures, *Mol. Cell. Biochem.* 97, 101–111, 1990). The center part of the figure shows the oxyfatty acid acyl groups used to modify proteins (Urabe, I., Yamamoto, M., Yamada, Y., and Okada, H., Effect of hydrophobicity of acyl groups on the activity and stability of acylated thermolysin, *Biochim. Biophys. Acta* 524, 435–441, 1978). Myristyl chloride (tetradecanoyl chloride) is used to modify the amino terminal of guanidated BPTI (Pool, C.T. and Thompson, T.E., Chain length and temperature dependence of the reversible association of model acylated proteins with lipid bilayers, *Biochemistry* 37, 10246–10255, 1998). The bottom of the figure shows the *in situ* development of acid anhydrides using ethyl chloroformate for the modification of proteins (Paul, B. and Korytnyk, W., Alanine derivatives with reactive groups, *J. Pharm. Sci.* 67, 642–645, 1978; Lévai, F., Liu, C.M., Tse, M.M., and Lin, E.T., Pre-column fluorescence derivatization using leucine-coumarnylamide for HPLC determination of mono- and dicarboxylic acids in plasma, *Acta. Physiol. Hung.* 83, 39–46, 1995).

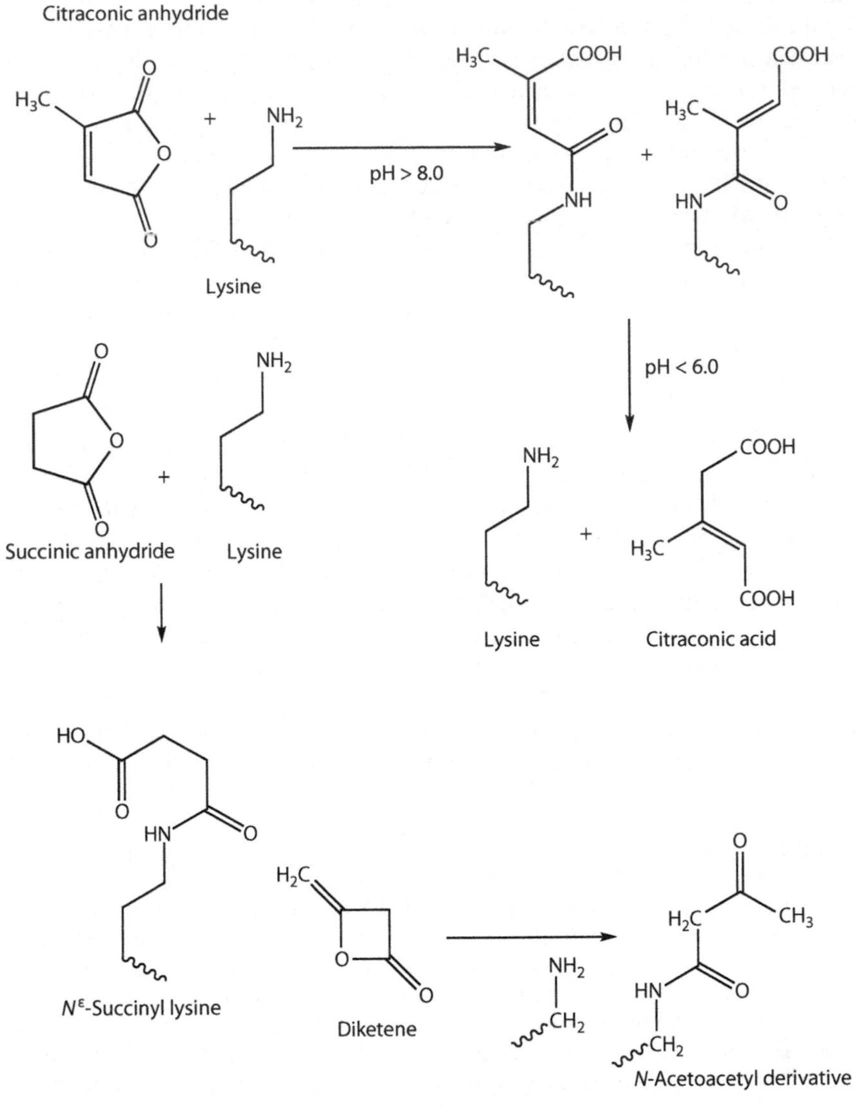

FIGURE 2.6 Modification of lysine with succinic anhydride, citraconic anhydride and ketene. Shown is the reaction of succinic anhydride or citraconic anhydride with lysine. The reaction of citraconic anhydride with lysine is reversible (Palacián, E., González, P.J., Piñeiro, M., and Hernández, F., Dicarboxylic acid anhydrides as dissociating agents of protein-containing structures, *Mol. Cell. Biochem.* 97, 101–111, 1990). Unlike succinic anhydride, ketene introduces the acetoacetyl group, which is neutral but similar in shape to succinic anhydride (Jones, A., Covinsky, K.E., and Sweeny, S.A., Effects of amino group modification in discoidal apolipoprotein A-1-egg phosphatidylcholine-cholesterol complexes on their reactions with lecithin: Cholesterol acyltransferase, *Biochemistry* 24, 3508–3513, 1985).

The inhibitor–enzyme complex is then modified with the fatty acid anhydrides (Figure 2.5) to selectively acylate the α-amino group. Fatty acid chlorides can also be used for the acylation reaction.[115]

3,4,5,6-Tetrahhydrophthalate (Figure 2.5) can be used to reverse modify amino groups in proteins.[116] The use of this reagent enabled the separation of hybrids of chromatographically identical oligomeric proteins. Subsequent use of this reagent[117] eliminated the nonspecific cleavage of fusion proteins. Howlett and Wardrop[118] were able to dissociate the components of human erythrocyte membrane by the use of 3,4,5,6-tetrahydrophthalic anhydride. The reaction was performed in 0.02 M Tricine, pH 8.5. The 3,4,5,6-tetrahydrophthalic anhydride was introduced into the reaction mixture as a dioxane solution (a maximum of 0.10 mL/5 mL reaction mixture). The pH was maintained at 8.0–9.0 with 1.0 M NaOH. The reaction was considered complete when no further change in pH was observed. The extent of modification was determined by titration with trinitrobenzenesulfonic acid. The reaction could be reversed by incubation for 24–48 h at ambient temperature following the addition of an equal volume of 0.1 M potassium phosphate, pH 5.4 (the final pH of the reaction mixture was 6.0). The use of dicarboxylic acid anhydrides as dissociating agents of multi-protein structures has been reviewed by Palacián and coworkers.[119]

Modification of amino groups with acetic anhydride (Figure 2.4) is one of the oldest[120] and still most extensively used approach for the chemical modification of proteins. Early studies used acetic anhydride in saturated sodium acetate.[121] This reaction is somewhat self-limiting in that the pH of the reaction decreases as a result of the increase in acetic acid concentration. Performance of the reaction in a pH-stat avoids this issue. However, in the absence of the acetate formed in the acetic anhydride/sodium acetate system, there is extensive formation of *O*-acetyl tyrosine. It was suggested[121] that reaction occurred with groups on the surface of the protein since lysine is a charged amino acid residue. As with other reagents, the issue of "accessibility" is a question more of microenvironment and reaction rate rather than an "all-or-none" phenomenon. The use of acetic anhydride as a conformational probe is discussed in greater detail in the following.

Acetylation has been used to study calcitonin[122] and a bacterial cytochrome.[123] Acetic and maleic anhydrides have been used to study porcine pancreatic elastase.[124] In these studies, the reaction was carried out in a pH-stat or in various buffers. Inactivation of elastase at neutral pH (7.6, pH-stat) with acetic anhydride was reversible and activity recovered when acetic anhydride was no longer added to the reaction mixture. At pH 9.5, reactivation was complete within 1 min. It was established that reaction occurred at both lysyl and tyrosyl residues. It is relatively easy to differentiate between the two sites of modification, since *O*-acyl tyrosyl residues are unstable at pH ≥ 9.0. The inactivation by acetic anhydride was irreversible at pH 10.5 or at 9.0–9.5 in 2 and 4 M urea. The amino-terminal valine was not available for modification at pH 7.4 but could be modified at pH 11.0. Modification of this residue could be achieved in the presence of urea (4.0 M) at a lower pH (9.0). It was suggested that the reversible inactivation was likely due to the transient acetylation of a histidine residue (see Chapter 4) and that the irreversible inactivation was a reflection of the modification of the N-terminal valine residue.

Change in protein charge resulting from the modification of ε-amino groups of lysine in proteins has been mentioned earlier. The change in protein charge can result in changes in electrophoretic or chromatographic mobility. In a study that, to my surprise, has been largely ignored, Smit[125] used native electrophoresis to study the modification of human interleukin-3 with acetic anhydride. Changes in electrophoretic mobility correlated well with the amino group modification as measured by reaction with trinitrobenzenesulfonic acid.[126] This study is also quite useful in that it provides a study of the effect of pH on the modification of amino groups with acetic anhydride. Human interleukin-3 has eight amino groups for potential modification. Approximately two groups were available for modification at pH 5.0 (acetate), three at pH 6.0 (MES), and approximately eight at pH 9.0 (borate). Furth and Hope[127] have observed a change in electrophoretic mobility of bovine neurophysin II upon nitration (TNM) using starch gel electrophoresis. Meighen and Schachman[128] showed that the succinylation of aldolase resulted in little change in the sedimentation coefficient but great change in electrophoretic mobility as positive charges are converted to negative charges. Thomas and Moss[129] reported changes on the electrophoretic mobility of human placental alkaline phosphatase on modification with acetic anhydride or succinic anhydride. Zschomig and coworkers[130] observed that succinylation of lysozyme resulted in a decrease in the isoelectric point, while modification with carbodiimide increased the isoelectric point. These studies were performed under native conditions (without denaturation) on matrices including starch gel and polyacrylamide. While SDS should eliminate these changes, the binding of SDS to proteins is complex (see below); as an example, Armstrong and Roman[131] showed the anomalous behavior of proteins with a high content of acidic acid residues. Garcia-Ortega and coworkers[132] observed that a very acidic protein (isoelectric point of 2.8–3.3), ribonuclease U2, did not bind to SDS.

The use of acetic anhydride to modify amino groups in proteins evolved into a much more sophisticated approach over the last decade with advances in analytical technology. In particular, mass spectrometry has become extremely useful for the study of chemical modification in proteins.[133–136] A number of studies on the use of acetic anhydride have appeared in the past 10 years and are listed in Table 2.1. The following sections describe certain of these studies in some detail. Suckau and colleagues[137] used modification with acetic anhydride to study the surface topology of hen egg white lysozyme. Modification was performed in 0.5 M NH_4CO_3, pH 7.0 [maintained by the addition of NH_4OH (30% NH_c)] with a 10–10,000 molar excess (to amino groups) of acetic anhydride for 30 min at 20°C. Analysis by mass spectrometry before and after tryptic hydrolysis permitted the identification of modified residues and the assignment of relative reactivity of the individual amino groups. This is an excellent paper, which has been extensively cited since its publication in 1992 and continues to be of interest to investigators.[138] Palczewski and colleagues[139] used reaction with acetic anhydride to study conformational changes in arrestin as well as the association of arrestin with P-Rho. The effect of light/dark cycles on reactivity with acetic anhydride was evaluated as well as that on the interaction of arrestin with P-Rho. The initial modification was performed with low levels (1–10 mM) of deuterated acetic anhydride at 0°C in 100 mM sodium borate, pH 8.5. This was followed by modification with higher concentrations of acetic anhydride (20 mM) in

100 mM sodium borate, pH 9.0, containing 6.0 M guanidine hydrochloride. The ratio of deuterated to protiated modification permitted the identification of residues "protected" from modification by interaction with P-Rho as well three lysine residues that were more reactive as a result of the interaction. A similar approach was used by Scaloni and coworkers[140] to study the interaction of thyroid transcription factor 1 homeodomain with DNA. A 1–10-fold molar excess of acetic anhydride was added to free or DNA-complexed thyroid transcription factor 1 homeodomain in 20 mM Tris-HCl-75 mM KCl, pH 7.5, at 25°C for 10 min. The acetylated samples were subjected to cyanogen bromide cleavage in 70% formic acid at room temperature or 18 h in the dark. The fractions were separated by HPLC and subjected to analysis by mass spectrometry. D'Ambrosio and coworkers[141] have also used this differential modification approach to study the dimeric structure of porcine aminoacylase 1. Reaction with acetic anhydride was performed in 10 mM NH_4CO_3, 1 mM DTT, and 1 mM $ZnCl_2$, pH 7.5, at 25°C for 10 min using a 100–5000-fold molar excess of reagent. Modification occurred readily at the amino terminus and at 8 of the 17 lysine residues.

Calvete and coworkers[142] used a clever approach to identify the heparin-binding domain of bovine seminal plasma protein PDC-109. The PDC-109 protein was bound to heparin–agarose in 16.6 mM Tris, 50 mM NaCl, 1.6 mM EDTA, and 0.025% NaN_3, pH 7.4. After washing the column to remove protein not bound to the matrix, the column was recycled at room temperature with the application buffer containing acetic anhydride (25–1600-fold molar excess over protein lysine). A similar experiment was performed with 1,2-cyclohexanedione to study arginine modification. Six basic residues were protected from modification by binding to the heparin matrix.

Zappacosta and coworkers[143] measured amino group reactivity for the conformational analysis of minibody by reaction with a low concentration of acetic anhydride. The modification reaction was performed in 50 mM NH_4CO_c, pH 7.5, at 25°C for 10 min with a twofold molar excess (to amino groups) of acetic anhydride. Two lysine residues were highly reactive, one lysine and the α-amino group were less reactive, and one lysine residue was not modified. Minibody is described as a group of small, engineered antibody-like proteins, which have potential for both therapeutic and diagnostic use.[144–147] The study by Zappacosta and coworkers[143] was particularly useful in the study of conformation as the solubility of the minibody precluded nuclear magnetic resonance (NMR) or crystallographic analysis. Tomer and coworkers[148] used the reaction with acetic anhydride to define a discontinuous epitope in human immunodeficiency virus core protein p24. The protein was bound to the immobilized antibody and digested with endoprotease lys-C. Modification was then accomplished with acetic anhydride (10,000-fold molar excess of reagent, pH 7.8, 50 mM NH_4CO_c, 20 min) followed by a 100,000-fold excess of the hexadeuterated reagents under the same conditions; pH was maintained by the addition of 10% NH_4OH.

Taralp and Kaplan[149] examined the reaction of acetic anhydride with lyophilized α-chymotrypsin *in vacuo*. α-Chymotrypsin was lyophilized from an unbuffered solution at pH 9.0 in one chamber in a reaction vessel. ^{3}H-Acetic anhydride was added to another compartment in the reaction vessel. The reaction vessel was evacuated and placed in an oven at 75°C. Several reaction vessels were used and removed at various time intervals for analysis. The proteins were then modified with

^{14}C-acetic anhydride and the ratio of ^{3}H to ^{14}C was used to determine the extent of modification. While complete modification of amino groups is achieved at pH 9.0 in aqueous solution, in the nonaqueous system, only 25% of the ε-amino groups and 90% of the α-groups were modified. It also appeared that mixed anhydrides formed with carboxyl groups on the protein surface.

Smith and coworkers[150] used the reaction with acetic anhydride to determine the extent of posttranslational acetylation in histone H4. Histone H4 was modified with deuterated acetic anhydride (dried histone samples were suspected in deuterated glacial acid acetic and deuterated acetic anhydride added; the reaction mixture was allowed to stand for 6h at ambient temperature). The modified samples were subjected to mass spectrometric analysis. The extent of endogenous acetylation was determined from the ration of protiated to deuterated fragmentation ions.

Competitive labeling (trace labeling) is a technique for determining the ionization state or constant and intrinsic reactivity of individual amino groups in a protein.[151] The method is based on the hypothesis that the individual amino groups will compete for a trace amount of radiolabeled reagent (the reagent is selected on the basis of nonselective reactivity with amino groups; with most studies, acetic anhydride has been the reagent of choice). The extent of radiolabel incorporation into the protein at a given site will then be a function of the pK_a, microenvironment, and inherent nucleophilicity of that particular amino group.[151] After the reaction with the radiolabeled reagent is complete, the protein is denatured, and complete modification at each amino group is achieved by the addition of an excess of unlabeled reagent. A reproducible digestion method (i.e., tryptic or chymotryptic hydrolysis) is used to obtain peptides from the completely modified protein. The peptides are separated by a chromatographic technique, and the extent of radiolabel at each site is determined. The extent of radiolabel incorporation at a given site is a function of the reactivity of that individual amino group under the reaction conditions used at the radiolabel step. An alternative approach[152,153] involves a "trace" labeling step with tritiated acetic anhydride followed by complete modification with unlabeled acetic anhydride under denaturing conditions. This modified protein is then mixed with a preparation of the same protein, which has been uniformly labeled with the ^{14}C-labeled acetic anhydride. Digestion and separation of peptide is performed by conventional techniques (see above), and the extent of radiolabeling is determined. The ratio of ^{3}H to ^{14}C in peptides containing amino groups is an indication of functional group reactivity. This method is somewhat more sensitive than the original method. Reductive methylation (see below) has also been used for the study of protein conformation.[154] Although this is a laborious technique, the data obtained are excellent and provide considerable insight into the solution structure of proteins. This technique has been used for the study of troponin-T,[155] troponin-C,[156] troponin-I,[157] calmodulin,[158–160] and tropomyosin.[161] In particular, studies[160,161] that have used this technique to assess conformational change in solution have been particularly rewarding. The reader is also referred to a review on the use of chemical modification for the study of protein therapeutics.[162]

Trifluoroacetylated derivatives (Figure 2.4) have been used in the study of protein conformation as this derivative can introduce ^{19}F as an NMR probe. The trifluoroacetyl group can be introduced via the anhydride[163] or via ethylthiotrifluoroacetate.[164–166]

Ethylthiotrifluoroacetate was used to modify cytochrome *c* in 0.14M sodium phosphate, pH 8.0.[167,168] The pH was maintained at 8.0 using a pH-stat. Singly substituted derivatives of cytochrome *c* can be separated by chromatography on anion-exchange resin (Bio Rex 70) and carboxymethyl cellulose. It is critical to avoid lyophilization during the preparation of the various derivatives. These derivatives have been subjected to further investigation[169,170] including NMR.[171] Recent advances in the NMR spectroscopy of proteins[172] have somewhat diminished the value of the necessity of an extrinsic probe. Trifluoroacetylation was originally introduced for use in protein structure determination as peptide bonds with a carboxyl group contributed by trifluoroacetylated lysine is not cleaved by trypsin and the trifluoroacetyl group can be removed by treatment with 1.0M piperidine (0°C).[173]

Mahley and coworkers have prepared the acetoacetyl derivatives of lipoproteins by reaction with diketene in 0.3 borate, pH 8.5 (Figure 2.4).[174,175] The modification of tyrosyl and seryl residues also can occur under these conditions, but the *O*-acetoacetyl groups can be removed by dialysis against a mild alkaline buffer such as bicarbonate. The modification at lysyl residues can be reversed by 0.5M hydroxylamine, pH 7.0, at 37°C. A 0.06M solution of diketene was prepared by taking 50μL diketene into 10mL of 0.1M sodium borate, pH 8.5. The extent of modification was determined by subsequent titration with FDNB. The effect of the modification of lysine residues on the *in vivo* clearance of lipoproteins in rats has been investigated.[175] The modification of lysine residues with diketene results in charge neutralization as opposed to charge reversal as observed with citraconic anhydride (Figure 2.6) or charge maintenance with reductive methylation.[176] Brubaker and colleagues[177] have used acetoacetylation to evaluate the role of positively charged lysine residues in apolipoprotein A-1 function of cholesterol efflux. Diketene is also of similar bulk to succinic anhydride or citraconic anhydride (Figure 2.6).

Lysine residues can be modified by reaction with α-ketoalkyl halides (Figure 2.4) such as iodoacetic acid.[178] Acylation can occur at pH > 7.0, but the rate of reaction is much slower than reaction with cysteinyl residues. Both the mono- and disubstituted derivatives have been reported. The monosubstituted derivative migrates close to methionine on amino acid analysis, while the disubstituted derivative migrates near aspartic acid. It should be noted that the reaction with α-ketoalkyl halides is not considered particularly useful for the modification of primary amino groups. Heinrikson[179] did report preferential modification of Lys41 with bromoacetate in pancreatic ribonuclease. It should be noted that reaction with α-ketoalkyl halides such as iodoacetate can be a possible side reaction occurring during the reduction and carboxymethylation of proteins. N^{ε}-Carboxymethyl lysine is a degradation product of fructoselysine in glycated proteins[180] and is a major immunological epitope in glycated proteins.[181] There is an age-dependent accumulation of N^{ε}-carboxymethyl lysine.[182] N^{ε}-carboxymethyl lysine formation has been reported in collagen as a result from reaction with glucuronic acid.[183] N^{ε}-carboxymethyl lysine can be synthesized by reaction with glyoxylic acid following by $NaCNBH_3$.[182] The synthesis of (+)-(S)-2-amino-5-(iodoacetamido) petanoic acid and (+)-(S)-2-amino-6-(iodoacetamido) hexanoic acid has been reported[184] as has their use as active-site directed inhibitors of ornithine decarboxylase and extrahepatic arginase.[185]

Both fluoronitrobenzene and FDNB (Figure 2.7) have been of considerable value in protein chemistry since Sanger and Tuppy's work on the structure of insulin.[186] Welches and Baldwin[187] have examined the reaction of bacterial luciferase with FDNB; activity was lost at the rate of $157\,M^{-1}\,min^{-1}$ (pH 7.0; 0.05 M phosphate). N-terminal amino groups, lysine, histidine, tyrosine, and cystine residues are modified with FDNB. In order to assess the significance of reaction at primary amino groups, the cysteinyl residues were "blocked" with methyl methanethiosulfonate

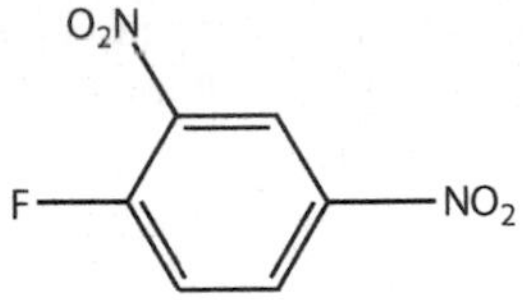

Fluorodinitrobenzene (FDNB)
1-Fluoro-2,4-dinitrobenzene

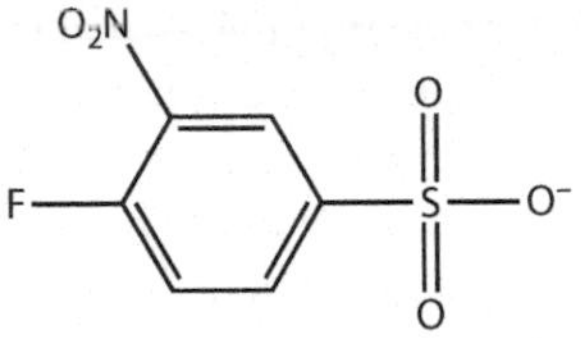

4-Sulfonyl-2-nitrofluorobenzene

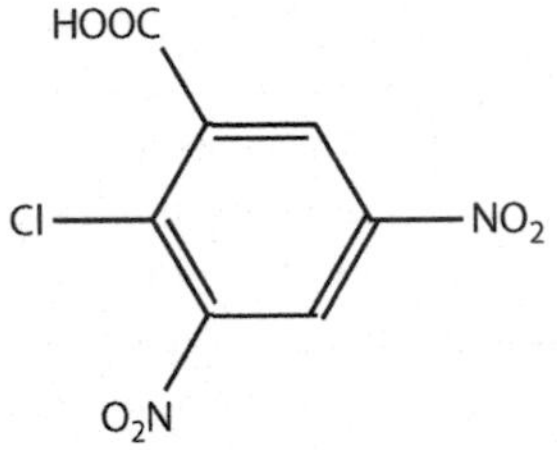

2-Carboxy-4,6-dinitrochlorobenzene

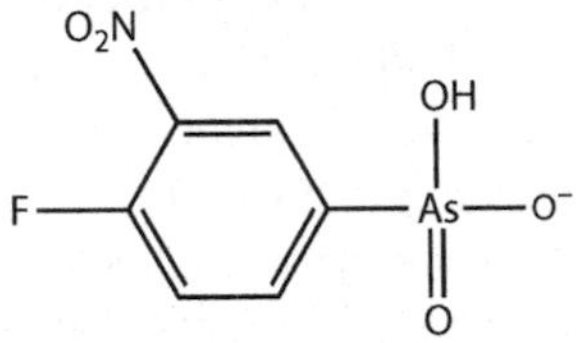

4-Arsono-2-nitrofluorobenzene

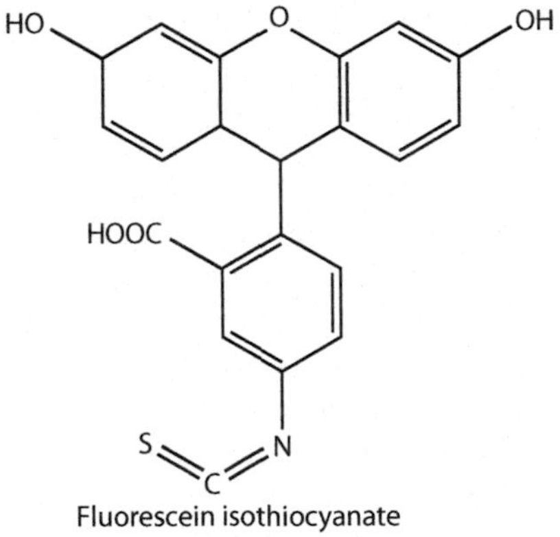

Fluorescein isothiocyanate

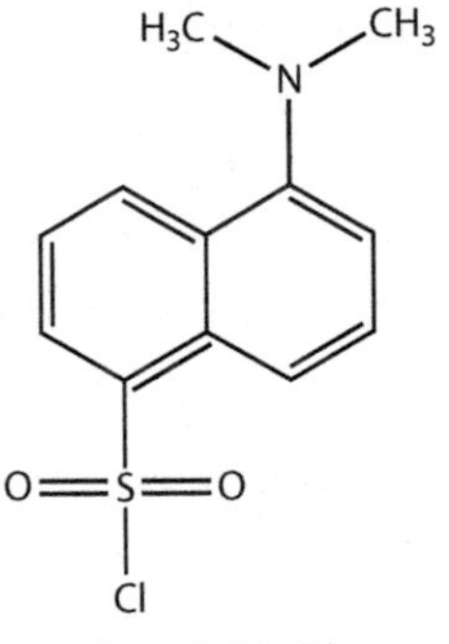

Dansyl chloride

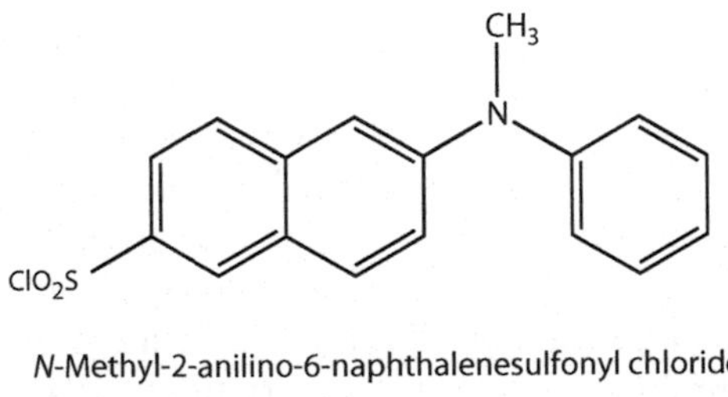

N-Methyl-2-anilino-6-naphthalenesulfonyl chloride

FIGURE 2.7 (See caption on facing page.)

(MMTS) prior to modification with FDNB. Reaction of luciferase with MMTS resulted in greater than 95% loss of catalytic activity, which was completely reversed with 2-mercaptoethanol. The small amount of residual activity present after treatment with MMTS is further reduced on treatment with 2,4-dinitrofluorobenzene. The loss of luciferase activity observed with FDNB was not reversed by 2-mercaptoethanol; Shatiel[188] has shown that the modification of tyrosine, cysteine, and histidine by FDNB is reversed with 2-mercaptoethanol, while the modification of amino groups with FDNB is stable to thiolytic cleavage. The combination of the results obtained with the MMTS-modified enzyme and the lack of activity recovery with 2-mercaptoethanol suggests a role for an amino group on the β-subunit in addition to the active-site cysteine on the α-subunit in activity of luciferase. Quantitative analysis was not performed, but qualitative analysis suggested that the modification occurred at the α-amino group of methionine on the α- and/or β-subunits. A combination of site-specific mutagenesis and site-specific chemical modification with FDNB was used to study lysine residues in angiogenin.[189] Limited modification of angiogenin with FDNB yielded two dinitrophenylated products, which could be separated by C^{18}-HPLC. One derivative was modified by Lys50 with slightly decreased ribonuclease activity, while the other derivative was modified by Lys60 with a greater loss of enzymatic activity. Angiogenin is a 124 residue protein having endoribonuclease activity[190] with significant homology with bovine pancreatic ribonuclease including the residues involved in catalytic activity. Lys41 is readily (and selectively) modified by FDNB in pancreatic ribonuclease[71] and it is surprising that the homologous residue Lys40 is modified readily in angiogenin by FDNB; mutagenesis at this site does demonstrate the importance of this residue for angiogenin activity. Olah and colleagues used FDNB to label proteins isolated from the 30S complex of the *Escherichia coli* ribosome.[191] The modified proteins were isolated by reverse-phase HPLC and, after characterization, assembled into reconstituted *E. coli* ribosomes and used to identify the site of incorporation. Lysine residues not involved in imido-cross-linking of proteins can be identified by FDNB as this reagent does not react with imidoesters[192] and to identify cross-linked peptides obtained by proteolysis after reductive methylation of ε-amino groups[193] at lower pH (see above). FDNB was used to define the chemical nature of epitopes in riboflavin-binding protein and the relationship to riboflavin binding activity.[112] Modification with FDNB resulted in a loss of

FIGURE 2.7 Various aromatic chromophores and fluorophores used for the modification of primary amino groups in proteins. Note is taken of 4-sulfonyl-2-nitrofluorobenzene, which is referred to as 4-sulfonyloxy-2-nitrofluorobenzene (Carty, R.P. and Hirs, C.H.W., Modification of bovine pancreatic ribonuclease A with 4-sulfonyloxy-2-nitrofluorobenzene. Isolation and identification of modified proteins, *J. Biol. Chem.* 243, 5244–5253, 1968). The term sulfonyloxy is used to refer to a linkage via the sulfur-oxygen and another functional group such as the nitrogen in N-sulfonyloxy carbamates used in synthetic organic chemistry (Donohoe, T.J., Chughtai, M.J., Klauber, D.J. et al., N-Sulfonyloxy carbamates as reoxidants for the tethered aminohydroxylation reaction, *J. Am. Chem. Soc.* 128, 2514–2515, 2006). The sulfonyl and arsano derivatives were developed to improve the solubility and specificity of the parent 2,4-fluorodinitrobenzene (Hummel, C.F., Gerber, B.R., and Carty, R.P., Chemical modification of ribonuclease A with 4-arsono-2-nitrofluorobenzene, *Intl. J. Pept. Protein Res.* 24, 1–13, 1984).

antigenicity and binding activity as did modification with citraconic anhydride or succinic anhydride; modification with methyl acetamidate, which does not result in a change in charge, did not result in the loss of binding activity or antigenicity. Recent use of FDNB has focused on it role as a hapten in sensitization reactions[194–196] and for determining "reactive" or available lysine[197] in food products.[198–201] Torbatinejad and coworkers[201] compared FDNB and guanidation to measure reactive lysine in food products. The formation of Maillard products in food processing is not uncommon[202] and the various lysine derivatives are not available for nutrition; FDNB measures available or reactive lysine that is of nutritional value; it has, however, been suggested that only a bioassay measures nutritionally valuable lysine in food products.[203]

Several other aromatic halides have been developed for the modification of nucleophiles in proteins. Carty and Hirs[72] developed the use of 4-sulfonyloxy-2-nitrofluorobenzene (Figure 2.7) for the modification of amino groups in pancreatic ribonuclease. Use of this compound, developed a 4-fluoro-3-nitrobenzene sulfonic acid by Zahn and Lebkucher,[204] did not present the solubility and reactivity problems posed by 2,4-dinitrofluorobenzene. The lysine residue at position 41 in pancreatic ribonuclease is the site of major substitution, which is a reflection of the lower pK_a for the ε-amino group of this residue.[205] It was possible to qualitatively determine the classes of amino groups in ribonuclease; these were the α-amino group, nine "normal" amino groups, and Lys41. The reactivity of Lys41 was influenced by neighboring functional groups and this effect was lost at $pH > 11$ or on thermal denaturation of the protein. Hummel and colleagues[206] subsequently reported on the use of 4-arsono-2-nitrofluorobenzene (Figure 2.7) as a reagent for introducing a spectral probe into RNase at Lys41. This is an interesting reagent but seems only to have been used by this research group in one subsequent publication.[207] The 4-arsono-4-nitrophenyl chromophore was an useful probe responding to the binding of 3′-cytidylic acid; there is also evidence that the arsonate dianion bound close to the active-site histidine residues[206] as bromoacetate reacts primarily with His119 at rate 45 times slower than the native enzyme. It was thought that 4-arsono-4-nitrophenyl chromophore binds at the anion binding site of RNase, thus resembling an affinity label. However, there are several products from this reaction; modification at Lys41 is 45% of the reaction product with 11% modification at the amino terminal with 25% modification at both sites.[207]

The reaction of 2-carboxy-4,6-dinitrochlorobenzene (Figure 2.7) with proteins has been explored.[208,209] This reagent reacts with amino, sulfhydryl, and amino groups. This reagent has also been used for the modification of specific lysine residues in cytochrome *c*.[210,211] The modification reaction (approximately sixfold molar excess of reagent) was performed in 0.2 M sodium bicarbonate, pH 9.0, at ambient temperature for 24 h. The extent of modification was determined as described by Brautigan and coworkers.[208] The maximum absorbance of derivatives formed with various model alkylamines such as aminobutyric acid or lysine peptides derived from cytochrome *c* was 436 nm with an extinction coefficient of $6.9 \times 10^3\ M^{-1}\ cm^{-1}$. The spectral analysis of modified cytochrome *c* is complicated by absorbance by the Soret band at 416 nm requiring measurement at 450 nm (extinction coefficient of $6.2 \times 10^3\ M^{-1}\ cm^{-1}$). As the modification by the carboxy derivative results in a change in net charge of the modified protein, it is possible to use chromatographic fractionation (sulfoethyl-cellulose) to obtain six fractions with differing lysine group modification.

Fluorescent probes can be used to determine residue accessibility and to evaluate protein conformation. The reaction of 1-dimethylaminonaphthalene-5-sulfonyl chloride (dansyl chloride) (Figure 2.7) has been useful both in the structural analysis and amino group modification with proteins. Reaction of trypsin with dansyl chloride in 0.1 M phosphate, pH 8.0 results in the modification of the amino-terminal isoleucine and one lysine residue.[212] Reaction of dansyl chloride with phosphoenolpyruvate carboxylase has been used to introduce a fluorescent probe into this protein.[213] A somewhat specific modification of one of the eight lysine residues was achieved. The extent of modification was determined by spectral analysis at 355 nm using an extinction coefficient of $3.4 \times 10^3\ M^{-1}\ cm^{-1}$. Hiratsuka and Uchida[214] examined the reaction of *N*-methyl-2-anilino-6-naphthalenesulfonyl chloride with lysyl residues in cardiac myosin. There was a difference in the nature of the reaction in the presence and absence of a divalent cation. *N*-Methyl-2-anilino-6-naphthalenesulfonyl chloride has been suggested for use as a fluorescent probe for hydrophobic regions of protein molecules.[215–217] The extent of incorporation of the *N*-methyl-2-anilino-6-naphthalenesulfonyl (Figure 2.7) moiety into protein can be determined by spectral analysis at 327 nm ($\varepsilon = 2.0 \times 10^4\ M^{-1}\ cm^{-1}$).[215,216] The hydrazide derivative of *N*-methyl-2-anilino-6-naphthalenesulfonic acid has been used for the measurement of carbonyl compounds.[217] Modification of protein amino groups with isothiocyanate derivatives of various dyes has proved to be an effective means of introducing structural probes into proteins at specific sites.[218] Eosin isothiocyanate has been used to modify the lysyl residues in phosphoenol-pyruvate carboxylase.[213] The reagent was dissolved in dimethylsulfoxide/50 mM HEPES, pH 8.0 (50:50) immediately prior to use and added to the protein (in 50 mM HEPES, pH 8.0). The modified derivatives were used to determine the spatial proximity of the modified lysine residues using resonance energy transfer. Fluorescein isothiocyanate (Figure 2.7) has been used to modify cytochrome P-450 (reaction performed in 30 mM Tris, pH 8.0, containing 0.1% Tween 80, 2 h at 0°C in the dark),[219] actin (2 mM borate, pH 8.5, 3 h at ambient temperature then at 4°C for 16 h),[220] and ricin (pH 8.1, 6°C for 4 h).[221] The study on actin[220] required modification of a reactive cysteine residues with *N*-ethylmaleimide prior to reaction with fluorescein isothiocyanate. The extent of modification with fluorescein isothiocyanate can be determined by spectroscopy using an extinction coefficient of $80 \times 10^3\ M^{-1}\ cm^{-1}$ at 495 nm (1% SDS with 0.1 M NaOH)[219] or $74.5 \times 10^3\ M^{-1}\ cm^{-1}$ (0.1 M Tris, pH 8.0).[220] Antibodies labeled with fluorescein have been used as targeted phototoxic agents.[222] In this approach, the fluorescein moiety is iodinated, resulting in a photodynamic sensitizer. Fluorescent probes are very useful probes of polymer microenvironment and the reader is directed to studies on the effect of microenvironment on the fluorescence of arylaminophthalenesulfonates.[223,224]

Cyanate reacts with several functional groups in proteins (Figure 2.8) including primary amines and was a useful reagent in protein structural analysis.[225] Stark and coworkers[226] pursued the observation that ribonuclease was inactivated by urea in a time-dependent reaction. It was established that this inactivation was a reflection of the content of cyanate in the urea preparation. The formation of cyanate from urea (Figure 2.8) was reported as early as 1897 by Walker and Kay.[227] It should be noted that the thrust of Walker and Kay work[227] was directed at the effect of alcohol on the formation of urea from cyanate and ammonia. Subsequent work by Fawsitt[228]

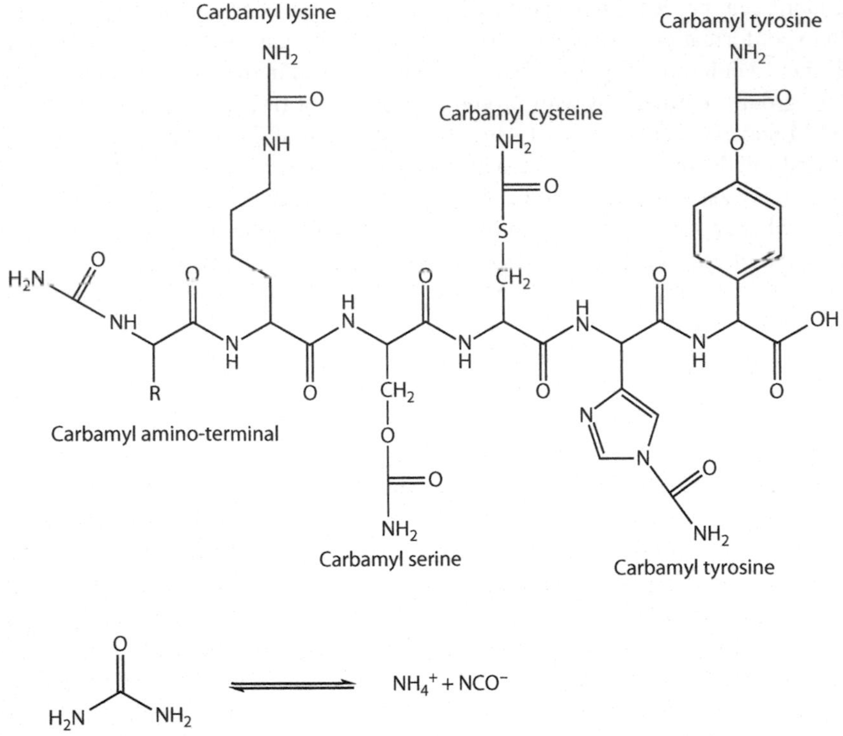

FIGURE 2.8 Products of the reaction of cyanate (or alkylisocyanates) with functional groups in proteins. Cyanate can react with several functional groups in proteins as shown but only with the products obtained with primary amino groups (either α-amino groups of N-terminal amino acids in proteins or the ε-amino group of lysine). (See Stark, G.R., Modification of proteins with cyanate, *Method Enzymol.* 25, 579–584, 1972.)

reported that the rate of formation of cyanate from urea was enhanced in acidic solution but was independent of acid concentration. Shaw and Bordeaux[229] suggested that the enhancement of cyanate formation from urea was not due an effect on the decomposition of urea, which is a first-order reaction[230] but rather on the reverse reaction, the formation of urea from cyanate and ammonia. Hagei and coworkers[231] reported a systematic study on the formation of cyanate from urea in solution suggesting that cyanate formation is primarily a product of urea concentration and, differing from earlier observations, cyanate formation is enhanced at increased pH. These investigators emphasize that the rate of cyanate formation is slow and dependent on temperature. The reaction of proteins with cyanate derived from urea has become less of a problem reflecting high purity of reagents and is only reported infrequently.[232–238] The rate of reaction of cyanate with amino acids is proportional to the pK_a of the amino group.[239] At pH 7.0 or below, the ε-amino group of lysine is some 100-fold less reactive compared to the α-amino group of amino acids.[239–241] The reaction with sulfhydryl groups is much faster than that observed for amino groups.[242] The carbamyl derivative of histidine is quite unstable as is the corresponding derivative of cysteine.[242–246] Concern should be shown to reactions at residues other than amines. For example, the reaction of chymotrypsin with cyanate results in loss of catalytic activity associated with the carbamylation of the active-site serine residue.[247] The reaction with an aliphatic hydroxyl group is unusual[248] and formation of carbamoyl-serine at the active site of chymotrypsin was subsequently confirmed by crystallographic analysis.[249] Modification of the active serine of chymotrypsin by alkyl isocyanates such as butyl isocyanate has been reported.[250]

Manning and coworkers[251–254] established that the modification of sickle cell hemoglobin with cyanate increased the oxygen affinity of this protein. It has been established that the amino-terminal value of hemoglobin is more reactive to cyanate in deoxygenated blood than in partially deoxygenated blood. At pH 7.4, the amino-terminal valyl residues of oxyhemoglobin S are carbamylated 50–100 times faster than lysyl residues[254] consistent with the other observations about enhanced reactivity of α-amino groups. The same laboratory has examined the carbamylation of α- and β-chains in some detail.[253] With the deoxy protein, the ratio of radiolabel from ^{14}C-cyanate on the α-chain as compared to the β-chain is 1.7:1.0, while it is 1:1 with the oxy protein. The carbamylation of the amino-terminal valine residues of hemoglobin is approximately 2.5-fold greater in partially deoxygenated media as compared to fully oxygenated media. Thus, it would appear that the reactivity of the amino-terminal valine is a sensitive index of conformational change.[254] It is also of interest that removal of Arg141 (α) with carboxypeptidase B abolishes the enhancement of carbamylation observed with the removal of oxygen from hemoglobin. There was some interest in the use of cyanate[255] in the treatment of sickle cell trait as carbamylation prevents gelation of sickle cell hemoglobin (hemoglobin S) but such efforts are compromised by toxicity.

Weisgraber and coworkers[174] used carbamylation to explore the role of lysyl residues in the binding of plasma lipoprotein to fibroblasts. The reaction was performed in 0.3 M sodium borate, pH 8.0. The extent of modification was determined in two ways. In the first, the modified protein was subjected to acid hydrolysis. The amount of homocitrulline, the product of the reaction of the ε-amino group of lysine with cyanate, was considered equivalent to the number of lysine residues modified. However, homocitrulline is partially degraded on acid hydrolysis to produce lysine (17%–30%). In order to obviate this difficulty, these investigators removed a portion of the modified protein and measured the free or unmodified lysine by reaction with FDNB under denaturing conditions with 2,4-dinitrifluorobenzene. The number of lysine residues modified was therefore the sum of free lysine and homocitrulline obtained on amino acid analysis following acid hydrolysis. This is similar to the use of FDNB for measurement of lysine modification in food proteins as described earlier. Carbamylation (potassium cyanate) has been used to prepare allergoids.[256]

Homocitrulline can be used as a biomarker for cyanate formation. Kraus and coworkers[257] observed homocitrulline formation in leukocyte proteins in patients with end-stage renal disease as a result of cyanate formation from urea. Metwalli and coworkers[258] reported the formation of homocitrulline from cyanate derived from urea in heated milk. Wang and coworkers[259] demonstrated an alternative pathway for protein carbamylation in inflammation, which used myeloperoxidase-catalyzed oxidation of thiocyanate. The carbamylation of lipoprotein by myeloperoxidase is suggested to be of importance in atherosclerotic vascular disease.[260]

The reaction of cyanate with lysine residues in bovine pancreatic deoxyribonuclease A was compared with other reagents by Plapp and coworkers.[261] The modification with cyanate is performed at 37°C in 1.0M triethanolamine hydrochloride, pH 8.0. The extent of modification was determined by analysis for homocitrulline following acid hydrolysis. A time course of hydrolysis was utilized to provide for the accurate determination of homocitrulline, since this amino acid slowly decomposes to form lysine during acid hydrolysis (see above). Modification of amino groups with cyanate or trinitrobenzene sulfonate results in neutral derivatives of lysine, while reaction with *O*-methylisourea or methyl picolinimidate retains the positive charge. Reaction with cyanate in the absence of calcium ions resulted initially in the modification of the α-amino group and seven ε-amino groups and loss of activity, which could be recovered by the addition of calcium ions; modification of the remaining two lysine residues resulted in the loss of activity, which could not be recovered with calcium ions. Similar results were obtained with trinitrophenylation, while modification with either *O*-methylisourea or methyl picolinimidate did not result in the loss of activity.

The reaction of imidoesters with the primary amino groups of proteins (Figure 2.9) has the advantage of maintaining the net charge of the modified protein[262] as does the reaction of proteins with *O*-methylisourea (Figure 2.10).[263] The pK_a for the product amidines from the reaction of imidoesters with proteins is equal to or greater than 12.5 (pK_a of lysine is 10.79) so modification by either amidination or guanidation can increase protein basicity.[264] In addition to the modification of amino groups, imidoesters are used for cross-linking proteins and can study subunit association,[265–268] protein–protein interaction,[269] and the association of protein with other biological macromolecules such as nucleic acids.[270,271]

The use of imidoesters for the modification of amino groups in proteins was introduced by Hunter and Ludwig[272] in 1962. Ludwig[273] as well as others[274] have reported on the reversibility of this reaction (Figure 2.11). Browne and Kent[275] suggested that *N*-alkyl imidates were the preferential product in a reaction between ethyl acetimidate and *n*-propylamine at pH 8.0, which will react with ammonia to form the expected amidine or with water to yield the free amine, while the amidine products were the preferential product at pH 10. The reaction between ethyl acetimidate and either *n*-propylamine or ε-caproic acid was quite rapid ($t_{1/2}$ approximately 3–7 min). Subsequently Plapp[276] studied the reaction of ethyl acetimidate with liver alcohol dehydrogenase (LADH) and showed that while the rate of formation of amidine and *N*-alkyl imidate was equal at neutral pH, the rate of amidine derivative formation was much larger at pH 8 and above. This does not preclude the possibility that amidine product in LADH proceeds through an *N*-alkyl intermediate. Browne and Kent[275] also observed that the imidoesters were less stable at acid pH than at alkaline pH (for ethyl acetimidate, $t_{1/2}$ approximately 5 min at pH 8 at 35°C; approximately 15 min at pH 10). In the aforementioned studies with LADH,[276] Plapp observed that most of the amino groups in a protein could be modified with four hourly additions of ethyl acetimidate or methyl acetimidate in a solution with a pH between 8 and 11. In other studies, Jastrzebska and coworkers[277] noted that reagent stability was a complicating issue in the use of dimethyl suberimidate for cross-linking pericardium tissue for use in replacement surgery. Cross-linkage can occur with monofunctional imidates such as ethyl acetimidate (Figure 2.11).[278–280] Some functionalized alkyl imidates have

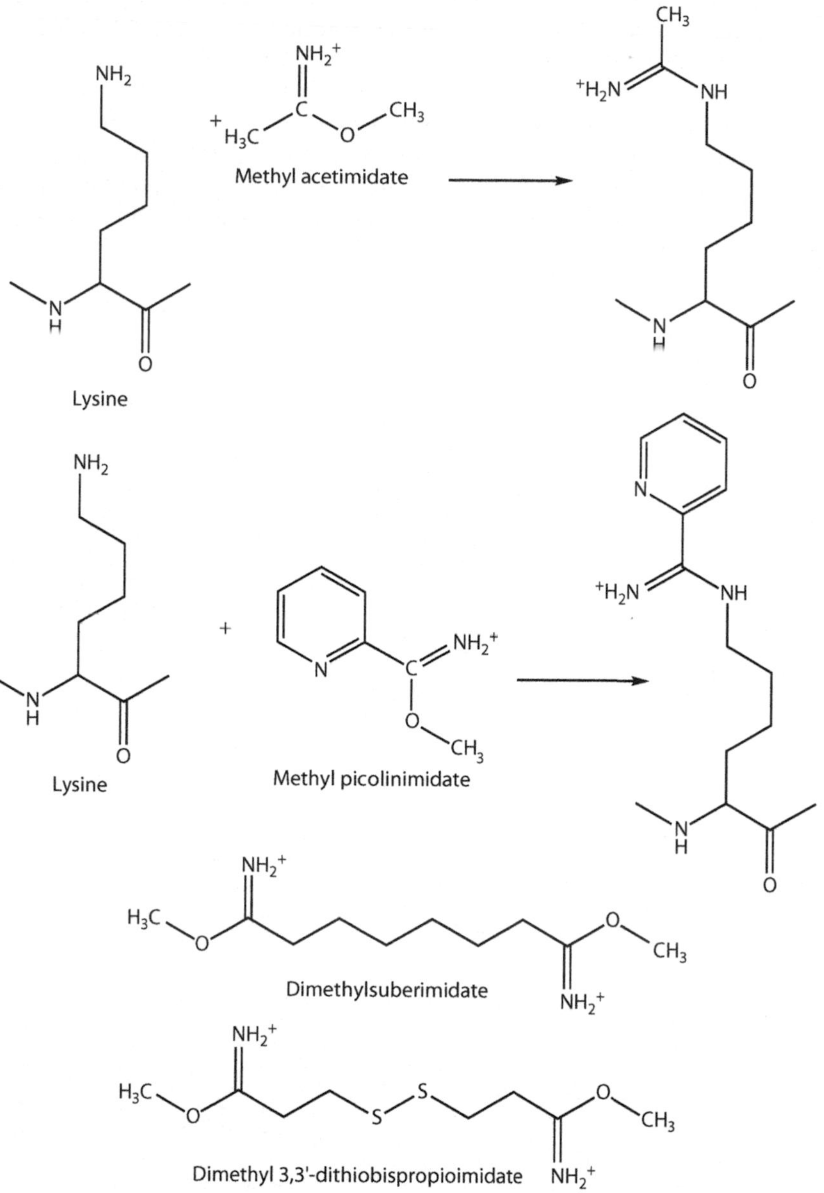

FIGURE 2.9 The reaction of imidoesters with proteins. Shown is the reaction of several imidoesters with the amino groups in proteins. Modification with imidoesters preserves the positive charge on proteins (Drews, G. and Rack, M., Modification of sodium and gating currents by amino group specific cross-linking and monofunctional reagents, *Biophys. J.* 54, 383–391, 1988). Methyl picolimidate can be used as a probe for functional groups in proteins (Perham, R.N., The reactivity of functional groups as a probe for investigating the topography of tobacco mosaic virus. The use of mutants with additional lysine residues in the coat protein, *Biochem. J.* 131, 119–126, 1973).

been developed including ethyl chloroacetimidate[281] and *S*-methylthioacetimidate (Figure 2.10).[282] Zoltobrocki and cowokers[283] reported on a variety of imidoesters including long-chain (C_6) and aromatic (e.g., pyridyl) derivatives (Figure 2.10). Ethyl chloroacetimidate[281] can also react with histidine and sulfhydryl groups in proteins and has been suggested as a cross-linking agent; ethyl chloroacetimidate has only been used sparingly since its description in 1972. Zoltobrocki and coworkers[283] used this reagent to modify LADH but did not observe any cross-linking. *S*-Methylthioacetimidate[282] will react at amino groups at low pH (pH 5.0) and has

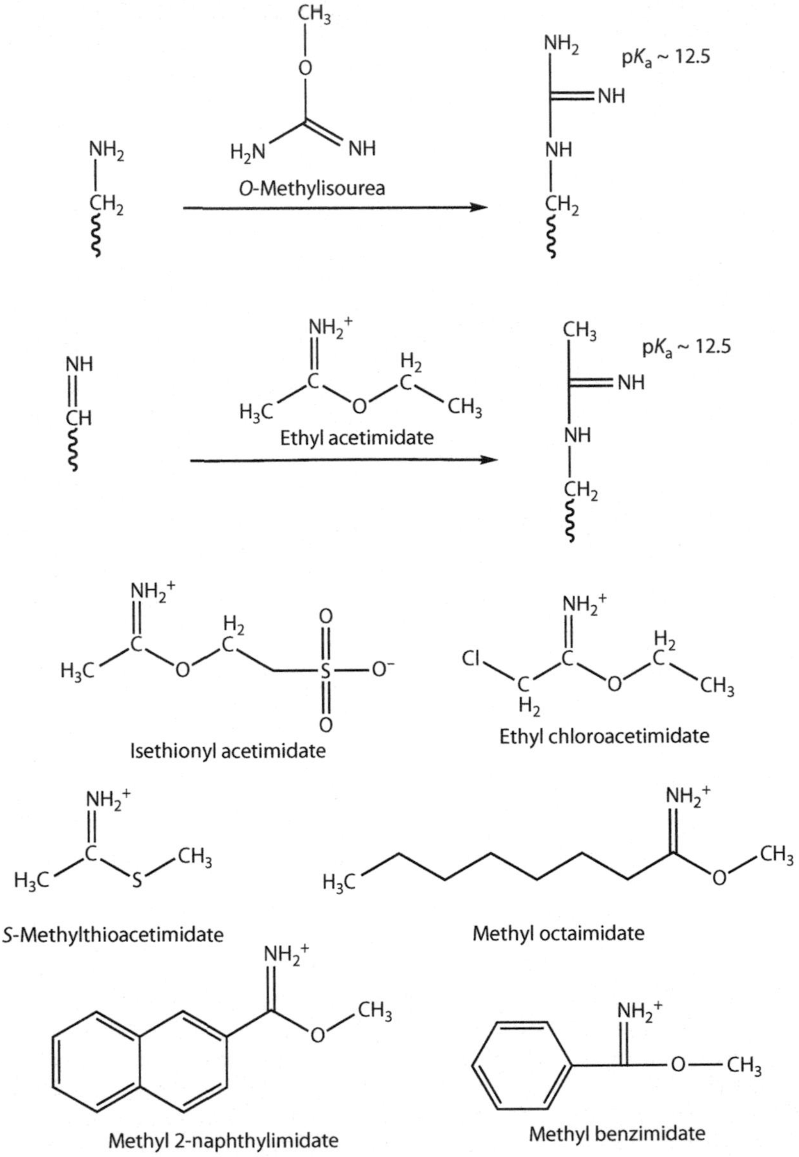

O-Methylisourea

pK_a ~ 12.5

Ethyl acetimidate

pK_a ~ 12.5

Isethionyl acetimidate

Ethyl chloroacetimidate

S-Methylthioacetimidate

Methyl octaimidate

Methyl 2-naphthylimidate

Methyl benzimidate

FIGURE 2.10 (See caption on facing page.)

been used for the mass spectrophotometric identification of proteins[284] and surface mapping of proteins.[285] Specific examples where *S*-methylthioacetimidate has been used for surface mapping include ubiquitin,[285] hemoglobin,[285] bovine carbonic anhydrase II,[285] and ribosomal proteins.[70,286,287]

Isothionyl acetimidate[288] was introduced as a membrane-impermeable derivative,[289] which can also be used to modify amino lipids such as phosphatidylethanolamine.[290]

The reaction of imidoesters with amino groups in proteins results in the formation of amidines (Figure 2.9). Amidines are not stable to acid hydrolysis[262] so amino acid analysis should not be used to measure the extent of modification in proteins. However, Ohara and Takahashi,[291] used amino acid analysis after hydrolysis to measure lysine modification by dimethylsuberimidate. A mixture of bislysylsubermidate and monolysylsuberimidate are obtained. These investigators obtained a first order rate constant of 3.4×10^{-3} h^{-1} for the hydrolysis of the amidine bond under standard conditions of acid hydrolysis for amino acid analysis. The majority of investigators measure free lysine after modification by reaction with FNDB (see above) or trinitrobenzenesulfonic acid (see below). The seminal work by Dutton and coworkers[292] used the ninhydrin reaction of determine the extent of amino group modification. Kim and coworkers[293] found that picolinimidation improved the quality of mass spectrometry signals. If the imidoester contains a chromophore as is the case with methyl picolinimidate, modification can be measured by spectrophotometry.[284]

Modification of protein amino groups with imidoesters has not been used extensively for protein modification. Despite this lack of use, there are several studies on the use of imidoesters for the modification of proteins, which should be discussed in some detail.

Sekiguchi and colleagues[294] modified glutamine synthetase from *Bacillus stearothermophilus* with ethyl acetamidate. Modification of approximately 26 (25.5) of

---

FIGURE 2.10 The reaction of *O*-methylisourea with proteins; comparison with amidination. Both amidation and guanidation of a protein result in products that retain the charge of the parent protein molecule (see Hunter, M.J. and Ludwig, M.L., The reaction of imidoesters with proteins and related small molecules, *J. Am. Chem. Soc.* 84, 3491–3504, 1962). As with other reactive groups, the imidoester can be used to deliver a variety of substituent groups. Isethionyl acetimidate is a membrane impermeable reagent comparable in size to ethyl acetimidate (Varela, I., Alvarez, J.F., Clemente, R. et al., Asymmetric distribution of the phosphatidylinositol-linked phospho-oligosaccharides that mimic insulin action in the plasma membrane, *Eur. J. Biochem.* 188, 213–218, 1990). Ethyl chloroacetimidate is a derivative that can serve as a cross-linking agent, which can react with thiols and histidine residues in proteins (Glomucki, M. and Diopoh, J., New protein reagents I. Ethyl chloroacetimidate in properties and its reaction with ribonuclease, *Biochim. Biophys. Acta* 263, 213–219, 1972). S-Methylthioacetimidate is a derivative that can be used to modify amino groups at low pH (pH 5.0) (Thumm, M., Hoenes, J., and Pfeiderer, G., S-Methylthioacetimidate is a new reagent for the amidination of proteins at low pH, *Biochim. Biophys. Acta* 923, 263–267, 1987). Methyl octaimidate, methyl 2-naphthylamidate, and methyl 2-pyridylimidate are derivatives developed for the modification of LADH (Zoltobrocki, M., Kim, J.C., and Plapp, B.V., Activity of liver alcohol dehydrogenase with various substituents on the amino groups, *Biochemistry* 13, 899–703, 1974). Methyl benzimidate was developed as one of the several hydrophobic imidoesters (Ampon, K. and Means, G.E., Immobilization of proteins on organic polymer beads, *Biotechnol. Bioeng.* 32, 689–697, 1988).

the 27 lysine residues (500-fold molar excess ethyl acetamidate to subunit) as determined by assay with trinitrobenzenesulfonic acid (see below) resulted in loss of 30% of activity (succinylation of 12 lysine residues resulted in total loss of activity). No change in CD spectra was observed on modification while there was a change in the quality of binding of TNS (2-*p*-toluidinylnaphthalene-6-sulfonic acid) as measured

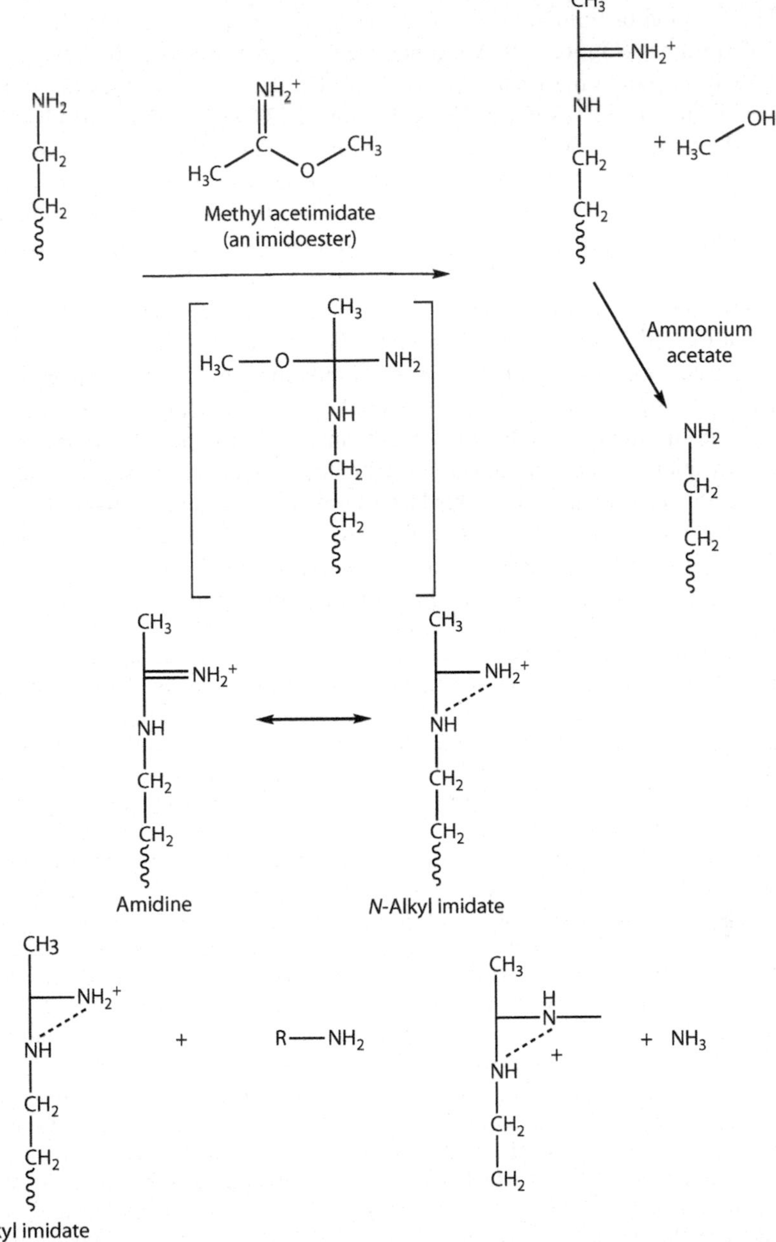

FIGURE 2.11 (See caption on facing page.)

by fluorescence. The modified enzyme demonstrated increased thermal stability. SDS-PAGE showed a small amount of material in the modified enzyme preparation migrating more slowly suggesting some intermolecular cross-linkage. As the modification reaction was performed at pH 9.5 (sodium phosphate), cross-linkage is not unexpected. It is noted that Cupo and coworkers observed increased stability of several proteins after guanidation with *O*-methylisourea.[295]

Plapp[296–298] observed that the activity of LADH could be enhanced by modification with cyanate, ethyl acetamidate, or methyl picolinimidate (Figure 2.10). This study was somewhat unique in that modification of the enzyme resulted in enhanced catalytic activity reflecting more rapid dissociation of the enzyme–coenzyme complex. It should be noted that the derivatized lysine (ε-picolinimidyllysine) reverts to lysine (60% yield) under the normal conditions of acid hydrolysis; an accurate value for ε-picolinimidyllysine could be obtained by taking several hydrolysis time points for analysis and extrapolating to zero time. Plapp and Kim[299] subsequently reported a method for the determination of ε-acetimidyllysine in proteins by acid hydrolysis with extrapolation (first order) to zero time. This approach is used for a variety of analytes where acid hydrolysis is required but the analyte is labile.[300–304] Plapp and coworkers[261] subsequently compared the reaction of methyl picolinimidate with other reagents for the modification of lysine residues in bovine pancreatic DNase in studies described earlier.

Ampon and Means[305] used modification of proteins with hydrophobic imidoesters as method for immobilizing these proteins onto organic polymer beads. A variety of hydrophobic imidoesters were evaluated including methyl benzimidate and methyl 3-phenylpropionimidate. Subsequent work involved the use of hydrophobic imidoesters for the immobilization of lipase.[306,307]

The modification of proteins and other biological polymers with 2-iminothiolane (originally described as methyl-4-mercaptobutyrimidate) (Figure 2.12) provides a method for introducing a sulfhydryl group into proteins by substitution at lysine residues. Perham and Thomas[308] had previously synthesized methyl-3-mercaptopropionimidate (Figure 2.12) for use as providing a site in tobacco mosaic virus (TMV) for the attachment of a heavy metal such as mercury for x-ray diffraction analysis. Richard Perham and Fred Richards had demonstrated that a single lysine residue in TMV was available for modification with methyl picolinimidate.[309] The introduction of a free sulfhydryl group provided the basis for use as cross-linking agent by Traut and colleagues[310–313] and is described by some investigators as Traut's

FIGURE 2.11 Mechanism and reversibility of the reaction of alkyl imidoesters with proteins. The predominant product of the reaction of imidoesters (methyl acetimidate as an example) in amidine, which can be reversed as shown in the presence of ammonium acetate (Ludwig, M.L., Reversible blocking of protein amino groups by the acetimidyl group, *J. Am. Chem. Soc.* 84, 4160–4162, 1962). The existence of the intermediate and formation of the *N*-alkyl imidate as a reaction product with protein amines and imidoesterase has been demonstrated. (Browne, D.T. and Kent, S.B.H., Formation of non-amidine products in the reaction of primary amines with imido esters, *Biochem. Biophys. Res. Commun.* 67, 126–132, 1975). The N-alkyl imidate provides the basis for the cross-linking reactions observed with alkyl acetimidates (Sweader, K.J., Crosslinking and modification of Na, K-ATPase by ethyl acetimidate, *Biochem. Biophys. Res. Commun.* 78, 962–969, 1978).

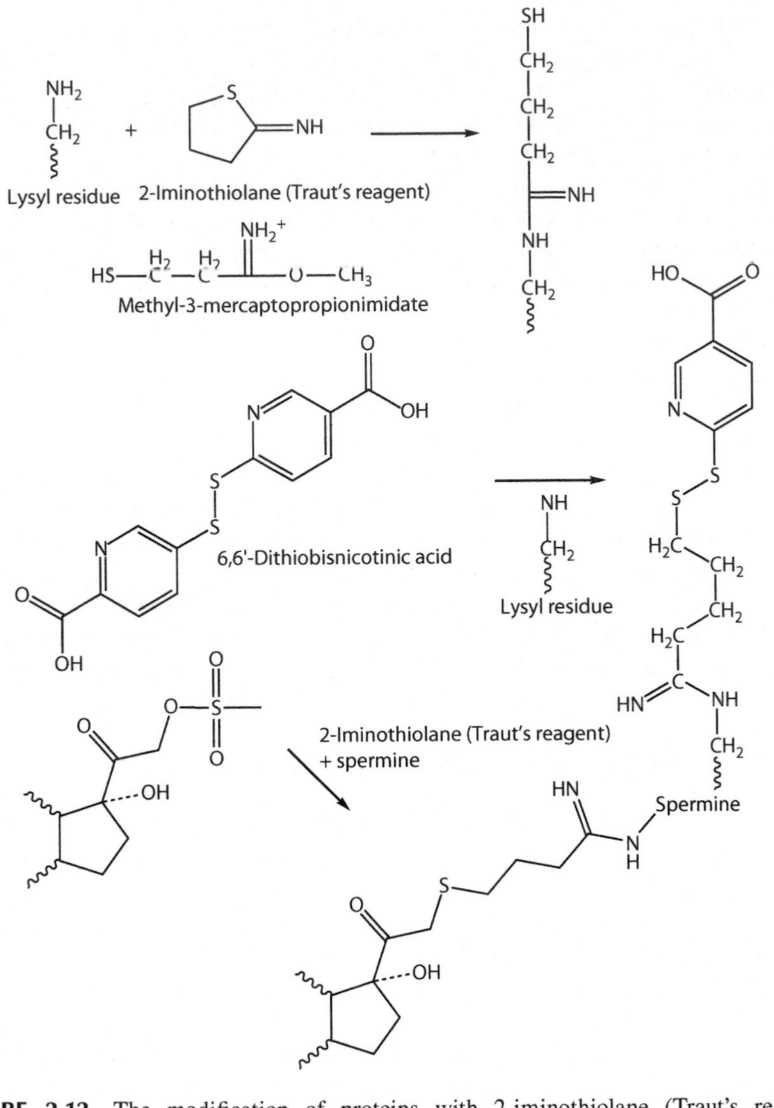

FIGURE 2.12 The modification of proteins with 2-iminothiolane (Traut's reagent). 2-Iminothiolane (known earlier as methyl-4-mercaptobutyrimidate) is used to introduce a sulfhydryl group into proteins by modification of amino groups. The sulfhydryl group can serve as a site for attachment of a variety of compounds and for cross-linking proteins (Traut, R.R., Bollen, A., Sun, T.-T. et al., Methyl-4-mercaptobutyrimidate as a cleavable cross-linking reagent and its application to the *Escherichia coli* 30S ribosome, *Biochemistry* 12, 3266–3273, 1973). One example is the coupling of 6,6′-dithiobisnicotinic acid to lysyl residue in a protein (Meunier, L., Bourgerie, S., Mayer, R. et al., *Bioconjug. Chem.* 10, 206–212, 1999), while the other describes the formation of a bioconjugate between glucocorticoids and spermine (Gruneich, J.A. and Diamond, S.L., Synthesis and structure-activity relationships of a series of increasingly hydrophobic cationic steroid lipofection reagents, *J. Gene Med.* 9, 381–391, 2007).

reagent.[314,315] 2-Iminothiolane is not subject to stability issues associated with other imidoesters.[310] 2-Iminothiolane has been used to couple steroids to spermine to prepare hydrophobic cationic steroid lipofection reagents[316] and subsequently to prepare a glucocorticoid-polyethyleneimine conjugate for nuclear gene delivery.[317,318]

A variety of carbonyl compounds including aldehyde and ketones react with amino groups in proteins (Figure 2.12). PLP is used for site-specific labeling of amino groups, while formaldehyde is used in reductive methylation of proteins. Formaldehyde and glutaraldehyde are also used in tissue fixation. The reaction of proteins with acetone is less common. Carbonyl compounds react with primary amines via formation of a Schiff base; this reaction is reversible but the product can be stabilized by reduction with sodium borohydride or sodium cyanoborohydride. Schiff bases can also form from natural products such as 4-hydroxy-2-nonenal.[319]

A number of investigators have used PLP (vitamin B_6) (Figure 2.13) to modify lysyl residues in proteins. PLP is the cofactor form of vitamin B_6 and plays an important role in biological catalysis.[320] PLP can react with both the ε-amino groups of lysine and the amino-terminal α-amino function in a protein. In general, PLP is far more reactive than pyridoxal because of hemiacetal formation in pyridoxal (Figure 2.12)[321,322] making the aldehyde group unavailable for reaction.[322] Taguchi and collegues[323] show a marked difference in the rate of reaction of aminoguanidine with PLP or pyridoxal. The reaction of PLP with aminoguanidine is 60% complete after 1 h (pH 7.0, 37°C), while there is approximately 5% product formation at 5 h with pyridoxal.

Shapiro and coworkers investigated the reaction of PLP with rabbit muscle aldolase.[324] The initial reaction produced a species with an absorbance maximum at 430–435 nm reflecting the protonated Schiff base form of the PLP–protein complex. After reduction with sodium borohydride, the absorbance maximum was at 325 nm, which is characteristic of the reduced Schiff base. This early study is quite useful for the spectral studies, the difference in reactivity of PLP and pyridoxal, and the reversibility of the initial Schiff base intermediate; inactivation was irreversible on reduction of the Schiff base complex but could be reversed by sulfhydryl compounds such as cysteine or cystamine.

Schnackerz and Noltmann[325] compared the reaction of PLP and other aldehydes in reaction with rabbit muscle phosphoglucose at pH 8.0. PLP (0.19 mM) resulted in 82% inactivation, while the following results were obtained with other aldehydes: pyridoxal (8.4 mM), 16% inactivation; acetaldehyde (75 mM), 75% inactivation; and acetone (75 mM), 31% inactivation. This last reaction is of interest as many investigators are unaware that acetone can react with amino groups in proteins. The reaction of acetone with primary amino groups has been known for some time[326] and is discussed in further detail in the following within the topic of reductive alkylation. The relative ineffectiveness of pyridoxal is like a reflection of hemiacetal formation as discussed earlier.

The reaction of ribulose 1,5-bisphosphate carboxylase/oxygenase with PLP has been studied by Paech and Tolbert.[327] PLP inactivated the enzyme with or without reduction with $NaBH_4$; the reaction demonstrated an optimum at pH 8.4. Spectral studies showed the formation of a species absorbing at 432 nm. As is characteristic for the Schiff base derivative, this peak disappears on reduction to yield a species

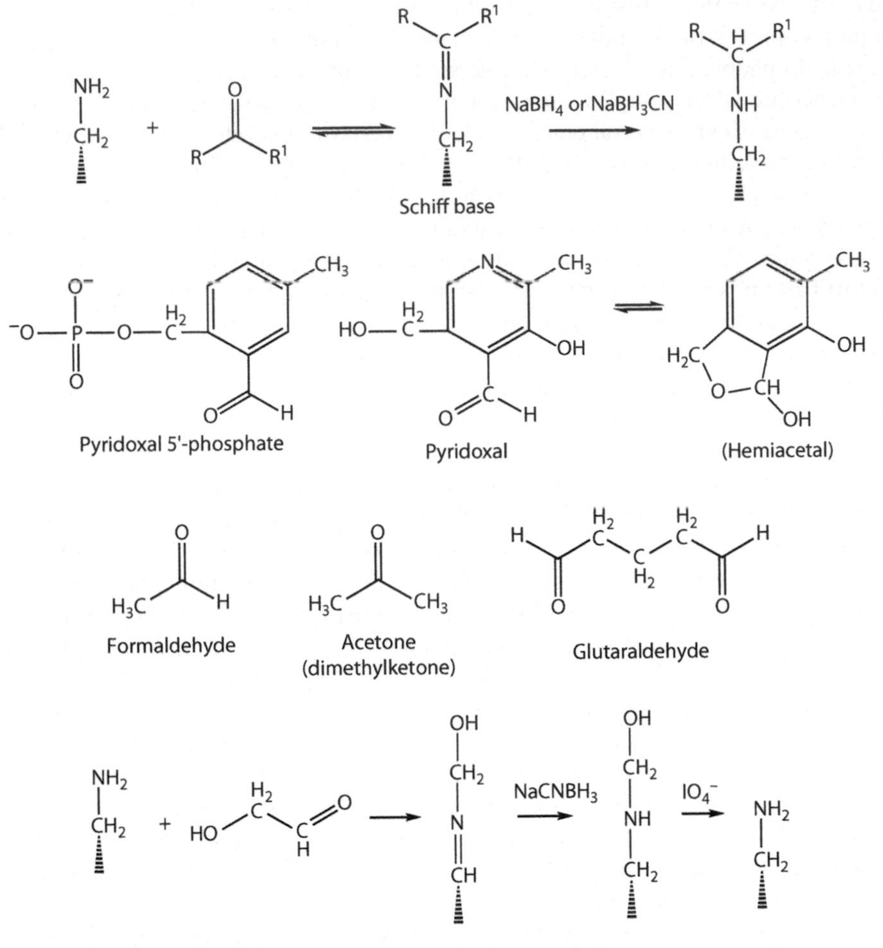

FIGURE 2.13 The reaction of carbonyl compounds with amino groups. Carbonyl compounds react in a reversible reaction with primary amines to form a Schiff base, which can be converted to a stable nitrogen–carbon bond by reduction with sodium borohydride or sodium cyanoborohydride. Shown also is the equilibrium between pyridoxal and the hemiacetal form (Metzler, D.E., Equilibria between pyridoxal and amino acids and their imines, *J. Am. Chem. Soc.* 79, 485–490, 1957). Reductive alkylation is used to describe the modification of primary amines in proteins with alkyl aldehydes such as formaldehyde and acetaldehyde (Means, G.E. and Feeney, R.E., Reductive alkylation of proteins, *Anal. Biochem.* 224, 1–16, 1995). The bottom reaction shows the reversible reductive alkylation of a protein by glycolaldehyde (Geoghagen, K.F., Ybarra, D.M., and Feeney, R.E., Reversible reductive alkylation of amino groups in proteins, *Biochemistry* 18, 5392–5399, 1979).

with an optimum at 325 nm ($\Delta\varepsilon = 4800\,M^{-1}\,cm^{-1}$). This supports the suggestion that the loss of activity observed on reaction with PLP is due to the formation of a Schiff base, which can be reduced with $NaBH_4$ to form a stable derivative.

PLP has seen limited use in the modification of proteins in the past several decades. Modification of proteins with PLP has been used to improve sensitivity of detection on blots using an antibody directed against PLP.[328] Western blotting has been used to identify PLP bound to phosphorylase in biopsies obtained from patients with McArdle's disease.[329–333] Cookson and coworkers[334] used a monoclonal antibody against PLP to analyze degradation products of phosphorylase. It was necessary to reduce the PLP peptide complex prior to SDS-PAGE as the PLP was dissociated by treatment with the detergent. Sodium cyanoborohydride was superior to sodium borohydride as treatment with sodium borohydride resulted in artifactual fragmentation of the protein samples. Modification of undigested phosphorylase with either sodium borohydride or sodium cyanoborohydride resulted in an apparent increase in the mass of the protein.

It should be emphasized that the modification of proteins with PLP is usually a reversible reaction unless the Schiff base intermediate is reduced. Cake and coworkers[335] have demonstrated that modification of activated hepatic glucocorticoid receptor with PLP obviated the binding of the receptor to DNA. Greatly reduced inhibition was seen with pyridoxamine-5′-phosphate, pyridoxamine, or pyridoxine. Inhibition could be reversed by gel filtration or treatment with dithiothreitol, while treatment with $NaBH_4$ resulted in irreversible inhibition of DNA binding. Nishigori and Toft[336–338] explored the reaction of PLP with the avian progesterone receptor. Reaction with PLP was performed in 0.02 M barbital, 10% (v/v) glycerol, 0.005 mM dithiothreitol, 0.010 M KCl, pH 8.0. The modification was stabilized by $NaBH_4$. It is of interest that these investigators noted that the modification was readily reversed in Tris buffer unless stabilized by $NaBH_4$. Peters and coworkers[339] reported on the inactivation of the ATPase activity in a bacterial coupling factor by reaction with PLP. The modification was performed in 0.050 M morpholinosulfonic acid, pH 7.5. The inhibition was readily reversed by dilution or by 0.01 M lysine and was, as expected, stabilized by $NaBH_4$. Gould and Engel[340] reported on the reaction of mouse testicular lactate dehydrogenase with PLP in 0.050 M sodium pyrophosphate, pH 8.7 at 25°C. This reaction resulted in the inactivation of the dehydrogenase activity. The inactivation was reversed by cysteine and stabilized by $NaBH_4$. These investigators reported that the observed absorption coefficient at 325 nm may be decreased as much as 50% with protein-bound PLP. Thus, estimation of the number of lysine residues modified using the absorption coefficient obtained with model compounds might provide only a minimum value.

Jones and Priest[341] have investigated the modification of apo-serine hydroxymethyltransferase with PLP and the subsequent use of the enzyme-bound pyridoxal phosphate as a structural probe. Cortijo and coworkers[342] have suggested the use of the ratio of absorbance at 415 and 335 nm of enzyme-bound PLP as an indication of the polarity of the medium. Binding of PLP to albumin is important for transport in blood. The binding of PLP to human albumin and bovine albumin has been studied by Fonda and coworkers[343] and there are marked differences between the two proteins. Human albumin has two classes of binding sites[343] while bovine albumin has two high-affinity sites and a larger group of lower-affinity sites.[344] There are also differences in binding modes as determined by spectral analysis.[343] The reader is also directed to a study by Haddad and coworkers on solvent effects on PLP.[345,346]

A substantial portion of the specificity of PLP in protein modification arises from electrostatic interaction(s) via the phosphate group with positively charged groups such as arginine or lysine in the protein. Amine compounds, such as lysine, have the potential to interfere in the reaction of PLP with proteins. Moldoon and Cidlowski[347] demonstrate that 0.1 M Tris, pH 7.4, markedly interfered with the modification of rat uterine estrogen receptor with PLP. These investigators also noted that, as in the other studies, 0.05 M lysine would block the modification reaction and could also reverse the modification if the Schiff base had not been reduced. Stock solutions of PLP were prepared in 0.01 M NaOH to avoid acid decomposition. The importance of local environmental factors in the specificity of modification by PLP is emphasized by Ohsawa and Gualerzi.[348] These investigators examined the modification of *E. coli* initiation factor by PLP in 0.020 M triethanolamine, 0.03 M KCl, pH 7.8. In the course of the studies, it was observed that PLP will not react with poly (AUG). These investigators also reported the preparation of N^6-pyridoxal lysine by reaction of PLP with polylysine in 0.01 M sodium phosphate, pH 7.2 at 37°C followed by reduction with $NaBH_4$. The reduction was terminated by the addition of acetic acid. Acid hydrolysis (6 N HCl, 110°C, 22 h) yielded N^6-pyridoxal-L-lysine. Bürger and Görisch[349] reported the inactivation of histidinol dehydrogenase upon reaction with PLP in 0.02 M Tris, pH 7.6. This modification could be reversed by dialysis unless the putative Schiff base was stabilized by reduction with NaH_4 (*n*-octyl alcohol added to prevent foaming). These investigators used a $\Delta\varepsilon$ for ε-amino pyridoxal lysine of 1×10^4 M^{-1} cm^{-1} at 325 nm.

While the aforementioned studies could be used to argue that PLP functions as an affinity label based on the binding of the anionic phosphate function, Valinger and colleagues[350] have argued that this is not the case in studies with 5′-deoxypyridoxal with bovine liver glutamate dehydrogenase. Other investigators have suggested that PLP can function as an affinity label.[351] Derivatives of PLP[352,353] have been developed as affinity labels for enzymes. It would appear that PLP can be used to label anion-binding sites in proteins with thrombin being a prime example.[354,355] It is clear that the presence of lysine residues in these anion-binding exosites is important in the modification of thrombin by PLP.[355]

Methyl acetyl phosphate is analogous to PLP. The reagent was originally developed as an affinity label for D-3-hydroxybutyrate dehydrogenase.[356] Manning and coworkers have examined the chemistry of the reaction of methyl acetyl phosphate with hemoglobin in some detail[357,358] It appears to be affinity label for the 2,3-diphosphoglycerate binding site.[357] More recent work with other proteins suggest that this reagent may be a useful probe for other anion binding sites in proteins.[358–360] Xu and coworkers[361] suggest the importance of methyl acetyl phosphate as a regio-selective reagent for modifying hemoglobin.

The modification of primary amines in proteins by reductive alkylation (Figure 2.13) has the advantage that the basic charge properties of the modified residue are preserved, resulting in only minor changes in tertiary structure.[362] Both monosubstituted and disubstituted derivatives can be prepared depending upon reaction conditions and the nature of the carbonyl compound. Reductive alkylation is similar to the reaction of PLP with proteins in that there is the initial reversible formation of a Schiff base. The technique is well established as discussed in the studies described in the following. Reductive methylation is likely the most common of the various reductive alkylation and is used for determination of lysine pK_a values in proteins and providing better NMR signals.[363–365] Reductive methylation has also proved useful for protein crystallization.[366–368] Reductive methylation with ^{13}C-enriched formaldehyde has been used to introduce an NMR probe for the study of protein conformation.[369]

A similar approach has been developed using deuterated acetone.[370] The use of [^{13}C]-formaldehyde in the reductive alkylation of ribonuclease has been reported.[371]

Reductive methylation of bovine trypsin reduces susceptibility to autolysis.[223] This reaction utilized formaldehyde/sodium borohydride in 0.2 M sodium borate, pH 9.2 in the cold. Unsubstituted amino groups were present after the reaction as demonstrated by titration with trinitrobenzenesulfonic acid. The amino-terminal isoleucine residue was not modified under these conditions. Subsequent work by Poncz and Dearborn[372] showed that peptide bonds with carboxyl group contributed by N^{ε}-dimethyllysine were resistant to trypsin. In related work, glycated trypsin has been shown to increased thermal stability.[373,374] Glycation has some properties similar to reductive alkylation and is discussed in greater detail in the following.

The introduction of sodium cyanoborohydride as a reducing agent for this reaction represented a real advance. Sodium cyanoborohydride is stable in aqueous solution at pH 7.0. Unlike sodium borohydride, which can reduce aldehydes and disulfide bonds, sodium cyanoborohydride only reduces the Schiff base formed in the initial process of reductive alkylation. The radiolabeling of proteins using ^{14}C-formaldehyde and sodium cyanoborohydride has been reported.[375] The modification was performed in 0.04 M phosphate, pH 7.0 at 25°C. The modification can be performed equally well at 0°C, but, as would be expected, it takes a longer period of time; there is no effect on the extent of the modification. In this regard, these authors estimated that the same extent of modification obtained in 1 h at 37°C could be achieved in 4–6 h at 25°C or 24 h at 0°C. Although the majority of experiments in this study were performed in phosphate buffer at pH 7.0, equivalent results can be obtained in Tris or HEPES buffer at pH 7.0. A greater extent of modification was observed with sodium cyanoborohydride at pH 7.0 than with sodium borohydride at pH 9.0. Jentoft and Dearborn have studied the use of sodium cyanoborohydride in some detail.[376] Sodium borohydride has the advantage of being able to reduce Schiff bases without reducing aldehydes. As with Dottavio-Martin and Ravel,[375] these investigators reported greater labeling efficiency with sodium cyanoborohydride than with sodium borohydride as well as the effectiveness of sodium cyanoborohydride at acidic pH. Both of these studies have been highly cited in subsequent studies on reductive methylation.

The effect of carbonyl compounds of different size on the extent of reductive alkylation has been examined by Fretheim and coworkers.[377] The extent of modification is more a reflection of the type of alkylating agent and reaction conditions than an intrinsic property of the protein under study. For example, nearly 100% disubstitution can be obtained with formaldehyde and approximately 35% disubstitution with *n*-butanol, while only monosubstitution can be obtained with acetone, cyclopentanone, cyclohexanone, and benzaldehyde. While most of the products of reductive alkylation retained solubility, the reaction products obtained with cyclohexanone and benzaldehyde tended to precipitate. Examination of the reductive alkylation of ovomucoid, lysozyme, and ovotransferrin with different aldehydes suggests that such modification occurs without major conformational change as judged by circular dichroism measurements.[378] The same study also examined the stability of the modified proteins by scanning differential calorimetry. The extensive modification of amino groups decreases thermal stability. The destabilization effect increases with increasing size (and hydrophobicity) of the modifying aldehyde. Subsequent work by Fujita and Noda[379] reported the reductive alkylation of lysozyme by aliphatic aldehydes of various chain lengths. Alkylation resulted in the loss of thermal stability; loss of thermal stability depended on both chain length and extent of modification. The use of hydroxyaldehyde had a smaller destabilizing effect than the corresponding aldehyde.[380] Wu and Means[381] have used reductive alkylation with a nonpolar aldehyde (dodecylaldehyde) to subsequently prepare insoluble proteins by binding of the modified protein to octyl-Sepharose.

Reductive alkylation has also been used to couple proteins to aldehyde matrices. Royer and coworkers[382] oxidized various carbohydrate matrices (dextran, Sepharose®, glass coated with glycerol silane) and coupled enzymes using sodium borohydride. This concept has been advanced by Gemeiner and Breier[383] who used either benzaldehyde derivatives of bead cellulose or bead cellulose oxidized with periodate with stabilization with either sodium borohydride or sodium cyanoborohydride; borohydride was more effective at pH 7 for benzaldehyde and at pH 9 for oxidized cellulose. However, sodium cyanoborohydride is more effective at or below pH 7 with oxidized cellulose. Lenders and coworkers[384] used sodium cyanoborohydride in coupling cellobiase to a dialdehyde dextran (periodate oxidized). Burteau and coworkers[385] also used cyanoborohydride-stabilized coupling for the immobilization (and stabilization) of penicillin amidase.

The reversible reductive alkylation of proteins by α-hydroxyl aldehydes or ketones (Figure 2.12) has been examined by Geoghagen and coworkers.[386] Both glycolaldehyde and acetol will react with the primary amino groups in proteins to yield derivatives, which can be cleaved with periodate under mild basic condition to yield the free amine. Acharya and Manjula[387,388] described the use of glyceraldehyde for the reversible reaction via the mechanism described earlier for glycoaldehyde. The periodate chemistry used in these reactions is the same as that described for the modification of N-terminal serine and threonine residues in Chapter 3.

Cyanide is a product of the reduction of Schiff bases by sodium borohydride, which inhibits the process of reductive alkylation. Jentoft and Dearborn[389] showed that the inhibition of reductive alkylation by cyanide can be blocked by nickel (II) or cobalt (III). This observation eliminates the previously observed necessity for recrystallization of sodium cyanoborohydride.[376] Other studies on the development of reagent alternative to sodium borohydride have been reported from other laboratories. Geoghegan and coworkers[390] compared sodium cyanoborohydride, dimethylamine borane, and trimethylamine borane with respect to effectiveness in reductive alkylation. Reduction at disulfide bonds was not observed with any of the three reagents. Dimethylamine borane was only slightly less effective than sodium cyanoborohydride, while trimethylamine borane was much less effective. This decrease in effectiveness in reductive alkylation is balanced by the absence of toxic byproducts such as cyanide evolving during the reaction. Quantitative reductive methylation (equal to or greater than one methyl group per lysyl residue) is achieved at 10 mM formaldehyde with dimethylamine borane and at 50 mM formaldehyde with trimethylamine borane. It should be noted that a similar extent of modification is obtained with 5 mM formaldehyde using sodium cyanoborohydride. In a subsequent study,[391] this laboratory reported the successful use of pyridine borane in the reductive alkylation of proteins.

Another class of α-hydroxy carbonyl compounds that reacts with protein to give interesting products are simple monosaccharides, which exist in solution in enol and keto forms. This process is referred to as glycation,[392,393] the nonenzymatic coupling of carbohydrate to protein. Wilson[394] showed that bovine pancreatic ribonuclease dimer would react with lactose in the presence of sodium cyanoborohydride to yield an active derivative that shows selectivity in uptake by the liver during *in vivo* experiments. The modification of ribonuclease dimer was performed in 0.2 M potassium phosphate, pH 7.4 (phosphate buffer was used to protect Lys41 from modification) at 37°C for 5 days with lactose and sodium cyanoborohydride. Under these conditions, 80% of the amino groups were modified. Bunn and Higgins[395] have explored the reaction of monosaccharides with protein amino groups in the presence of sodium cyanoborohydride in some detail. These investigators studied the reaction of hemoglobin with various monosaccharides in Krebs–Ringer phosphate buffer, pH 7.3. The extent of modification was determined using tritiated sodium cyanoborohydride. The rate of modification was demonstrated to be a direct function of the amount of each sugar in the carbonyl (or keto) form; as an example the first-order rate constant for the reaction of D-glucose (0.002% in carbonyl form) is 0.6×10^{-1} mM^{-1} h^{-1}, while for D-ribose (0.05% in carbonyl form), the first-order rate constant is 10×10^{-3} mM^{-1} h^{-1}. The nonenzymatic glycation of hemoglobin is important as an index of glucose in diabetes[243] and as the initial step in the Maillard reaction.[396–401]

Glyceraldehyde is a triose that has become important in the study of the glycation of proteins.[402–405] The reaction of glyceraldehyde with carbonmonoxyhemoglobin S was investigated by Acharya and Manning.[406] This reaction was performed with 0.010 M glyceraldehyde in phosphate-buffered saline, pH 7.4, and the resultant Schiff bases were stabilized by reduction with sodium borohydride. Using radiolabeled glyceraldehyde, these investigators were able to obtain support for the concept that there is selectivity in the reaction of sugar aldehydes with hemoglobin. The reaction product between glyceraldehyde and hemoglobin S did have stability properties without reduction that were not consistent with only Schiff base products. These investigators suggested that the glyceraldehyde-hemoglobin Schiff base could undergo an Amadori rearrangement to form a stable ketoamine adduct, which could be reduced with sodium borohydride to form a product identical to that obtained by direct reduction of the Schiff base (Figure 2.14). In a subsequent study, these investigators did show that the glyceraldehyde-hemoglobin S Schiff base could rearrange to form a ketamine via an Amadori rearrangement.[407] These investigators were able to use reaction with phenylhydrazine to detect the protein-bound ketamine adduct. The reaction of glyceraldehyde with hemoglobin S results in the modification of the amino-terminal valine in the β-chain forming a Schiff base that rearranges into an Amadori product as well as the modification of three lysine residues in the β-chain. Subsequent work from this work showed specificity in the reaction of glyceraldehyde with dipeptides and tripeptides.[408,409]

FDNB (see above) was introduced by Sanger[186,410] first for estimation of amino groups in insulin[186] and subsequently for amino acid sequence determination.[410] While a useful reagent, FDNB is not soluble in water (FDNB is usually used as an ethanolic solution) and thus not useful as more general reagent for the

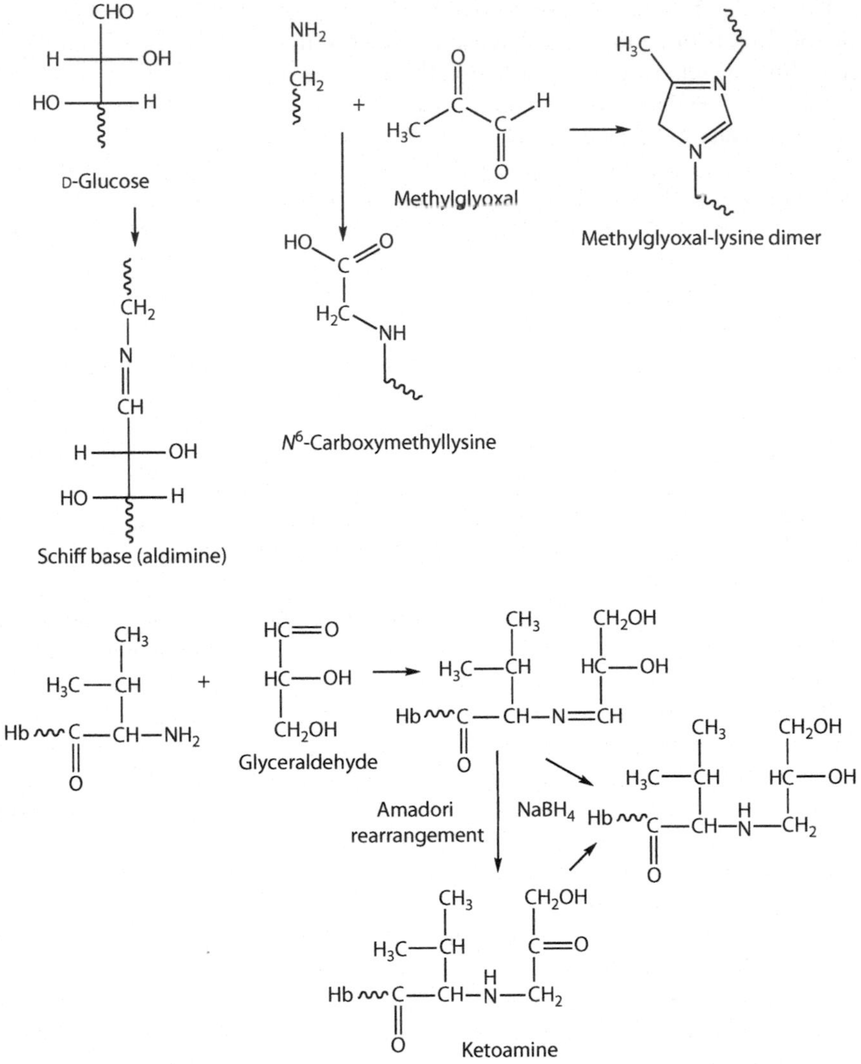

D-Glucose

Methylglyoxal

Methylglyoxal-lysine dimer

Schiff base (aldimine)

N^6-Carboxymethyllysine

Glyceraldehyde

Amadori rearrangement

$NaBH_4$

Ketoamine

FIGURE 2.14 The glycation of proteins. The reaction of methylglyoxal is shown as an example of the cross-linkage reaction, which can occur in advanced glycation end products (AGE). The reaction of glucose with amino groups to form a Schiff base is shown. This intermediate may go on to form complex AGE products (Watkins, N.G., Thorpe, S.H., and Baynes, J.W., Glycation of amino groups in protein. Studies of the specificity of modification of RNase by glucose, *J. Biol. Chem.* 260, 10629–10636, 1985). Also shown is the reaction of glyceraldehyde with the amino-terminal valine in the β-chain of hemoglobin S resulting in an Amadori rearrangement (Acharya, A.S. and Manning, J.M., Reactivity of the amino groups of carbonmonoxyhemoglobin S with glyceraldehyde, *J. Biol. Chem.* 255, 1408–1412, 1980).

study of proteins in solution. Okuyama and Satake[411] characterized the reaction of 2,4,6-trinitrobenzenesulfonic acid (TNBS) with amino acids and peptides as a substitute for FDNB. 2,4,6-Trinitrobenzenesulfonic acid has the advantage of solubility and lack of reaction with tyrosine or the imidazole ring in proteins. In a subsequent paper, Okuyama and coworkers[412] characterized the spectral properties of the reaction products of TNBS with amino acids and developed a method for quantification of the products at 340 nm. The reaction of TNBS with amino groups (Figure 2.15) has a long history for the study of function and reactivity of amino groups in proteins.[413–415] Goldfarb[413] presented a study on the kinetics of the reaction of TNBS with amino groups using absorbance 335 nm for quantitation of the products. In a subsequent study, Goldfarb[414] studied the reaction of TNBS with the amino groups of human serum showing the presence of 19 reactive ε-amino groups (there are 58–62 potentially available ε-amino groups and one α-amino group in human serum albumin[415]). The ε-amino groups could be divided into three sets (14, 4, and 1) with different kinetic characteristics. It should be noted that human serum albumin tends to

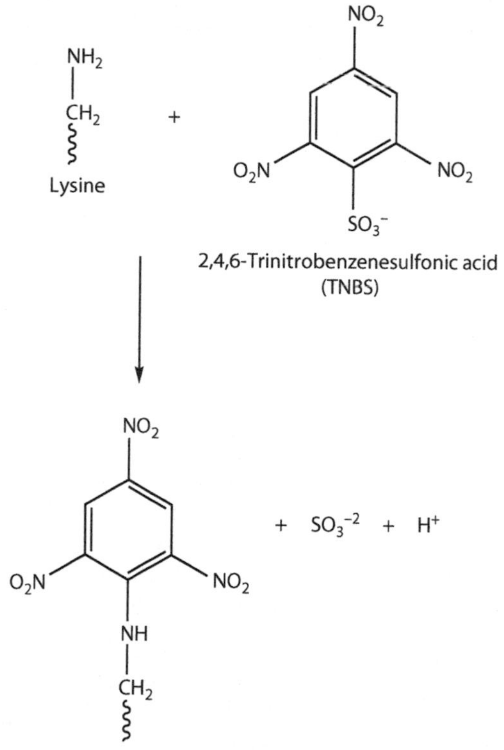

NH_2 | CH_2 Lysine + 2,4,6-Trinitrobenzenesulfonic acid (TNBS) → + SO_3^{-2} + H^+

FIGURE 2.15 The reaction of 2,4,6-trinitrobenzenesulfonic acid (TNBS) with amino groups. (See Fields, R., The rapid determination of amino groups with TNBS, *Methods Enzymol.* 25, 464–468, 1972; Cayot, P. and Tainturier, G., The quantification of protein amino groups by the trinitrobenzenesulfonic acid method: A reexamination, *Anal. Biochem.* 249, 184–200, 1987.)

be heterogeneous.[416] Habeeb[417] examined the reaction of TNBS with proteins in the presence or absence of denaturing agents using absorbance at 335 nm to follow the reaction. In studies with bovine serum albumin, the extent of modification with TNBS increased with increasing pH. Sodium dodecyl sulfate showed a small inhibition of reaction at pH 7.5, while at pH 9.5 there was a greatly reduced inhibition of reaction with TNBS; similar results were obtained with ovalbumin and human γ-globulin (Cohn Fraction II). Modification of bovine serum albumin with formaldehyde reduced the reaction with TNBS. Markus and coworkers[418] showed that the conformation change of human serum albumin in the presence of 6 M urea at pH 7.4 was prevented by the binding of SDS; this effect was reduced at alkaline pH. The binding of SDS to albumin (and presumably other proteins) is complex[419–421] and suggested that the hydrophobic portion of SDS binds to an apolar region, while polar head group binds to the positively charged ε-amino group of lysine.[417] Fields[419–421] examined the reaction of TNBS with amino groups noting the influence of the sulfite that is displaced in the reaction with amino groups (Figure 2.15). In the presence of an excess of sulfite, absorbance at 420 nm is a more sensitive index than absorbance in the 350–370 nm region (isosbestic point). However, absorbance at 420 nm is dependent upon the ability of the reaction product to form a complex with sulfite. The derivatives of α- and ε-amino groups have similar spectra with the α-amino derivatives having a slightly higher extinction coefficient at 420 nm ($\varepsilon = 2.20 \times 10^4\ M^{-1}\ cm^{-1}$) than ε-amino groups ($\varepsilon = 1.92 \times 10^4\ M^{-1}\ cm^{-1}$). Both of these derivatives have much higher extinction coefficients than the derivative obtained by reaction of TNBS with cysteinyl residues ($\varepsilon = 2.25 \times 10^3\ M^{-1}\ cm^{-1}$). The α-and ε-amino derivatives can be differentiated by their stability to acid or base hydrolysis. The α-amino derivatives are unstable to acid hydrolysis (8 h at 110°C) or base hydrolysis.[422] Cayot and Tainturier[423] have extended our understanding on the reaction of TNBS with proteins with particular emphasis on the study of glycated proteins. These investigators have carefully examined the conditions for the use of TNBS for the determination of amino groups in proteins. They observed that while raising the temperature of the reaction only slightly increases the rate of reaction with amino groups, the rate of the hydrolysis of the reagent is substantially increased. An increase of the pH of the reaction to 10 increases the rates of reaction establishing the importance of the nucleophilic characters of the amino group in this reaction. It is also observed that, as expected, accessibility of the functional groups is important for reaction with TNBS. They recommend reaction at pH 10 (0.1 M borate) with a 100-fold molar excess (to protein amino groups). Reaction is usually complete in 15 min and can be measured by absorbance at 420 nm. Amino acid concentration can also be determined with TNBS (A_{350}).[424]

Coffee and coworkers[425] have explored the reaction of trinitrobenzenesulfonic acid with bovine liver glutamate dehydrogenase. In these studies, the modification was performed in 0.04 M potassium phosphate, pH 8.0. Under these reaction conditions, the cysteinyl residues were not modified. The modification reaction could be followed as a function of time by absorbance at 367 nm (in the absence of NADH and ATP) or at 420 nm (in the presence of NADH and ATP). There was evidence to suggest that glutamate dehydrogenase could catalyze the conversion of TNBS to trinitrobenzene.[426] Subsequent work from another laboratory[427] demonstrated that there is also a nonenzymatic reaction between NADPH and TNBS resulting in the formation of trinitrobenzene.

The reaction of TNBS with simple amines and hydroxide ions has been studied in some detail by Means and coworkers.[428] The reaction of TNBS with hydroxide is first order with respect to both trinitrobenzenesulfonate and hydroxide ions. Reaction with amines was considered in some detail. In general, reactivity of trinitrobenzenesulfonate with amines increases with increasing basicity except that secondary amines and *t*-alkylamines are comparatively unreactive. The specific binding of trinitrobenzenesulfonate to proteins must be considered in the study of the reaction of this compound with proteins. Only amines with a $pK_a > 8.7$ follow a simple rate law. These investigators presented the following considerations regarding the reaction of trinitrobenzenesulfonic acid with proteins. Reactivity is a sensitive measure of the basicity of an amino group. Adjacent charged groups have an influence on the rate of reaction with an increase observed with a positively charged group and a decrease with a negatively charged group. Proximity to surface hydrophobic regions that can bind TNBS can increase the observed reactivity of a particular amino group. The reaction of TNBS with ammonium has also been investigated by Whitaker and coworkers.[429] This reaction was performed in tetraborate buffer and 1 mM sulfite. The rate of the reaction was determined by following the increase in absorbance at 420 nm. The rate of reaction with ammonium ($k = 0.128\,\text{min}^{-1}$) was slower than that with the average amine in a protein ($k = 0.907\,\text{min}^{-1}$ for enterotoxin). The reaction with ammonium does, however, provide a sensitive assay for ammonia (as low as 6 nmol) with a precision of 1%–2%.

Flügge and Heldt have explored the labeling of a specific membrane component with TNBS[430] and PLP.[431] The modification of the phosphate translocation protein in spinach chloroplasts with TNBS was performed in 0.050 M HEPES, 0.33 M sorbitol, 0.001 M $MgCl_2$, 0.001 M $MnCl_2$, and 0.002 M EDTA, pH 7.6, at 4°C for periods of time up to 15 min, at which point tritiated sodium borohydride was added to both terminate the reaction and radiolabel the trinitrophenyl derivatives.[432] It is possible to label components on the surface of membranes with TNBS, as the sulfonate moiety does not permit membrane penetration. The same is true for PLP.

The use of trinitrobenzenesulfonate in the selective modification of membrane surface components has been explored by Salem and coworkers.[433] This study involved the modification of intact cells with the TNBS (dissolved in methyl alcohol) diluted to a 1% methanolic solution. As mentioned previously, trinitrobenzene sulfonate does not pass across (or into) membranes, being more hydrophilic than, for example, FDNB.

Trinitrobenzenesulfonate can be used to identify ionization constants for amino groups in proteins. Hartman and coworkers[434,435] used TNBS to modify lysine residues in ribulosebiphosphate carboxylase/oxygenase. The reaction of TNBS with ribulosebiphosphate carboxylase/oxygenase resulted in inactivation of the spinach enzyme at Lys166, which exhibited a pK_a of 7.9.

N-Hydroxysuccinimide (NHS) esters were developed by Anderson and coworkers[436] for use in peptide synthesis and used first for protein modification by Blumberg and Vallee.[437] The use of NHS chemistry (Figure 2.16) is not extensively used for the modification of proteins but is used extensively for the cross-linkage of proteins, the

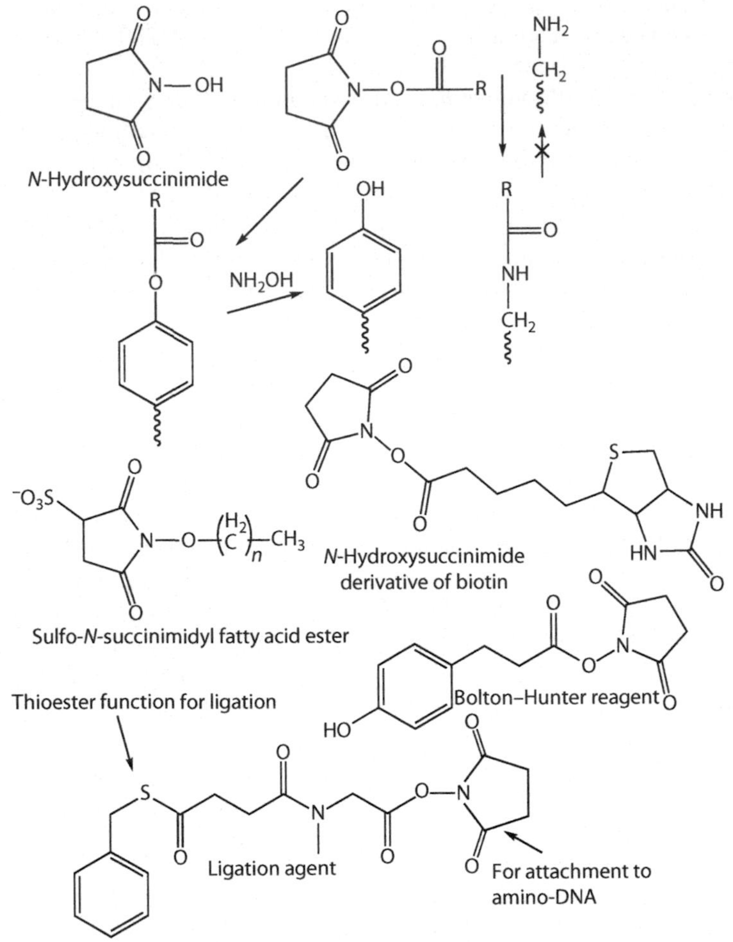

N-Hydroxysuccinimide

NH_2OH

N-Hydroxysuccinimide derivative of biotin

Sulfo-*N*-succinimidyl fatty acid ester

Bolton–Hunter reagent

Thioester function for ligation

Ligation agent

For attachment to amino-DNA

FIGURE 2.16 The reaction of N-hydroxysuccinimide with functional groups in proteins. Shown is the structure of NHS and the general structure of the active ester form. The reaction of an alkyl NHS ester with lysine or tyrosine. The product obtained from reaction with tyrosine can be reversed with hydroxylamine. (See Anderson, G.W., Zimmerman, J.E., and Callahan, F.M., The use of ester of *N*-hydroxysuccinimide in peptide synthesis, *J. Am. Chem. Soc.* 86, 1839–1842, 1964; Yem, A.W., Zurcher-Neely, H.A., Richard, K.A. et al., Biotinylation of reactive amino groups in native recombinant human interleukin-1β, *J. Biol. Chem.* 264, 17691–17697, 1989; Wetzl, B., Gruber, M., Oswald, B. et al., Set of fluorochromophores in the wavelength range from 450 to 700 nm and suitable for labeling proteins and amino-modified DNA, *J. Chromatog. A* 793, 83–92, 2003.) Shown also is a reagent used for coupling to DNA for subsequent ligation to a protein fragment to prepare a protein–DNA hybrid (Takeda, S., Tsukiji, S., Ueda, H., and Nagamune, T., Covalent split protein fragment-DNA hybrids generated through N-terminus-specific modification of proteins by oligonucleotides, *Org. Biomol. Chem.* 6, 2187–2194, 2008) as a well as sulfo-*N*-succinimidyl ester of fatty acid (Coort, S.L.M., Willerns, J., Coumans, W.A. et al., Sulfo-*N*-succinimidyl esters of long chain fatty acids specifically inhibit fatty acid translocase (FAT/CD36)-mediated cellular fatty acid uptake, *Mol. Cell. Biochem.* 239, 213–219, 2002).

preparation of protein conjugates, and for three-dimensional matrices such as hydrogels. The Bolton–Hunter reagent[438] for radiolabeling proteins is based on NHS chemistry and is used for pharmacokinetic studies.[439–443] Sulfo-*N*-hydroxysuccinidimidyl esters were developed by Staros[444] as reagents that were more soluble and membrane impermeant compared to NHS esters. The sulfo-NHS esters of fatty acids have been used to study the uptake and metabolism of long-chain fatty acids.[445,446] More recent studies[447] have demonstrated that sulfo-*N*-succinimidyl oleate is an inhibitor of the mitochondrial respiratory chain.

NHS esters have been used for a variety of purposes in the preparation of useful protein derivatives. There has not, however, been significant use of NHS esters in the study of protein structure function. Yem and coworkers[448] have used NHS chemistry to introduce biotin into recombinant interleukin-1-α. There were three primary sites of modification: the amino-terminal alanine, Lys93, and Lys94. There were few other single-modified products despite the presence of other lysine residues. Yem and coworkers did use the sulfo derivative in their studies. It was possible to separate the amino-terminal-modified product from the product modified at the lysine residues but it was not possible to separate the two lysine-modified species; the presence of singly modified species was inferred from isoelectric point measurement. Subsequent studies from this laboratory[449] showed modification of Lys103 with acrylodan without modification of Lys93 or Lys94. It is suggested that the hydrophobicity of the environment around Lys103 promotes interaction with the hydrophobic reagent acrylodan. Wilchek and Bayer[450] have reviewed the use of NHS derivatives of biotin for the labeling of proteins. Assays for protein biotinylation were also reviewed by Bayer and Wilchek in 1990.[451] Advances in analysis include commercial assays and mass spectrometry.[452] NHS-biotin has been used to label surface lysine residues in proteins.[453] Smith has developed a kinetic model for the modification of protein with NHS-biotin.[454] Fölsch has prepared the NHS ester of chloromercuriacetic acid.[455] NHS derivatives of agarose are used to prepare affinity matrices by providing a mechanism for the immobilization of proteins and other amino-containing molecules.[456] Kim and coworkers[457] have used NHS chemistry to immobilize human serum albumin on silica columns.

NHS esters will react with functional groups in proteins other than lysine including serine, threonine, and tyrosine as seen with NHS-based cross-linking agents.[458] Some of the earliest work on NHS esters involved the modification of a tyrosine residue in thermolysin with amino acid NHS esters.[437] This work resulted from the earlier observation of the "superactivation" of thermolysin with DEPC in the presence of β-phenylpropionyl-L-phenylalanine.[459] Similar results were obtained with a mixed acid anhydride of β-phenylpropionyl-L-phenylalanine leading to the studies with the amino acid NHS esters.[459] The effect was reversed with neutral hydroxylamine. Superactivation of other neutral proteases was observed with amino acid NHS esters and this effect was also reversed with hydroxylamine.[460] The reversibility with neutral hydroxylamine is consistent with the modification of tyrosine. Subsequent work by Blumberg[461] established Tyr110 in thermolysin as the site of modification with the amino acid NHS esters.

The recent work using NHS esters has focused on the preparation of conjugates and complex protein structures. Examples of conjugates include the preparation of single-chain vascular endothelial growth factor (ScVEGF) conjugates for positron emission tomography (PET) imaging of vascular endothelial growth factor (VEGF) receptors in angiogenic vasculature,[462] the pegylation of recombinant non-glycosylated erythropoietin,[463] and an adenovirus vector to a cell-penetrating peptide.[464] Teixeira and coworkers[465] used NHS chemistry to prepare a heparinized hydroxyapatite/collagen matrix. The reader is also directed to studies by Strehin and coworkers[466] on a "smart" hydrogel based on chondroitin sulfate and PEG and by Deng and coworkers on a collagen/glycopolymer hydrogel.[467] The use of NHS chemistry to mobilize proteins for affinity chromatography was mentioned earlier. NHS chemistry is also used for the immobilization of proteins for other solid phase assay applications.[468–471]

MODIFICATION OF ARGININE

The average pK_a for the guanidino group of arginine is 12.48 (Table 1.1) making it the most basic group in a protein. Arginine residues are usually involved in binding sites rather than catalytic sites.[472–480] Anion-binding sites are characterized by the presence of arginine and lysine.[481,482]

Present approaches to the site-specific modification of arginyl residues in proteins use three reagents: phenylglyoxal (and derivatives such as *p*-hydroxyphenylgloxal),[483] 2,3-butanedione,[484] and 1,2-cyclohexanedione.[485] The determination of the extent of arginine modification can be determined by amino acid analysis after acid hydrolysis, but conditions generally need to be modified to prevent loss of the arginine derivative.[485] While amino acid analysis is mentioned in several papers, details are not presented regarding the results. However, more frequently, such analysis is totally ignored in more recent publications. Many of the studies are qualitative in nature and such data is not required. Chromogenic reagents such as *p*-hydroxyphenylglyoxal and 4-hydroxy-3-nitrophenylglyoxal are described in the following. Mass spectrometry has been of value in identifying the modification of arginine in proteins.[486–492]

There is one method for the accurate determination of arginine, which is of both historical interest and current practical value. The Sakaguchi reaction (Figure 2.17) is based on the reaction of α-naphthol with guanidino compounds in the presence of hypohalite to yield a red product.[493] The Sakaguchi reaction[494,495] has been used, after acid hydrolysis, to determine the extent of arginine modification by 2,3-butanedione and 1,2-cyclohexanedione in rat liver microsomal stearoyl-coenzyme A desaturase[496] and the extent of arginine modification by 4-hydroxybenzil in gelatin.[497] This modification is part of a program for the development of novel protein-polymer grafts.[498] It is noted that the modification occurs under rigorous (at least for proteins) conditions (70% ethanolic NaOH at 23°C). It has also been used for the assay of arginine in food products.[499,500] It has been suggested[501] that the enzymatic assay for arginine is more valuable than the Sakaguchi reaction. Other applications of Sakaguchi chemistry have been in histochemical studies[502,503] and for the assay of guanidine compounds (guanidinoacetic acid) in screening for hepatic guanidinoacetate-methyl transferase activity.[504,505] There are two fundamental problems with the Sakaguchi reaction. The first is the short life of the reaction product, which can be addressed, in part, by the substitution of hypobromite for hypochlorite and the addition of urea to remove excess hypobromite.[506] Linearity is another issue[507] complicated by a lack of complete understanding of the reaction. A fluorometric method for the determination

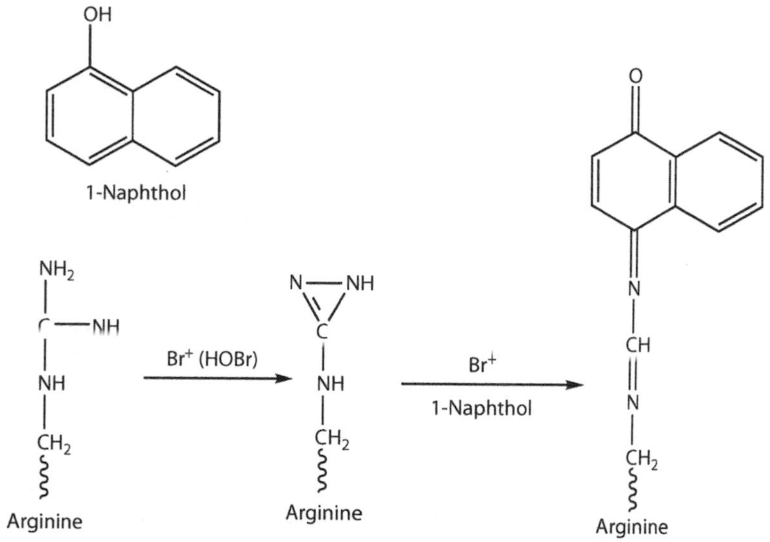

FIGURE 2.17 The Sakaguchi reaction for determination of arginine. The Sakaguchi reaction used reaction with hypobromite followed by addition of a chromogen such as 1-naphthol. The color product is not stable and a suggested structure is shown in this figure (Casdebaig, F., Mesnard, P., and Devaux, G., Mise au point. La réaction de sakaguchi, *Bull. Soc. Pharm. Bordeaux* 114, 82–94, 1976; Dupin, J.-P., Casadebaig, F., Pometan, J.-P. et al., La réaction de Sakaguchi, *Anns. Pharmaceut. Francoises* 37, 523–530, 1979).

of arginine using 9,10-phenanthrenequinone[508] has been described. This method is some 1000-fold more sensitive than the Sakaguchi reaction, but some concern remains concerning the absolute accuracy of the reagent for determination of arginine in peptide linkage. This reagent has also been used to study the modification of arginine in proteins by methylglyoxal,[509,510] glycation,[510,511] and oxidation.[510]

The use of phenylglyoxal (Figure 2.18) was developed by Takahashi[483] in 1968 and has since been applied to the study of the role of arginyl residues in proteins as shown in Table 2.1. As with other modifying agents, there has been increased use of phenylglyoxal to study complex systems.[512–531] It of no small interest that the use of chemical reagents started with their use in complex systems including whole animals.[531] Today's investigators have a considerable advantage over the workers cited in Webb's classic work[531] in having superb analytical tools such as mass spectrometry and immunochemistry for looking at these more complex systems. It is however, still challenging to consider these studies have being of value to solution protein chemistry is usually insufficient analytical information to unequivocally establish site and stoichiometry of modification. For example, many of these studies fail to recognize that phenylglyoxal, like glyoxal, will react with ε-amino groups at an appreciable rate.[483] Polymerization was noted in a sample incubated for 21 h. The amino-terminal lysine residue was rapidly modified under these conditions. The possible effect of light on the reaction of phenylglyoxal with arginine, as has been reported for 2,3-butanedione,[533–537] has not been studied. As established by Takahashi, the stoichiometry of the reaction involves the reaction of 2 mol of phenylglyoxal with

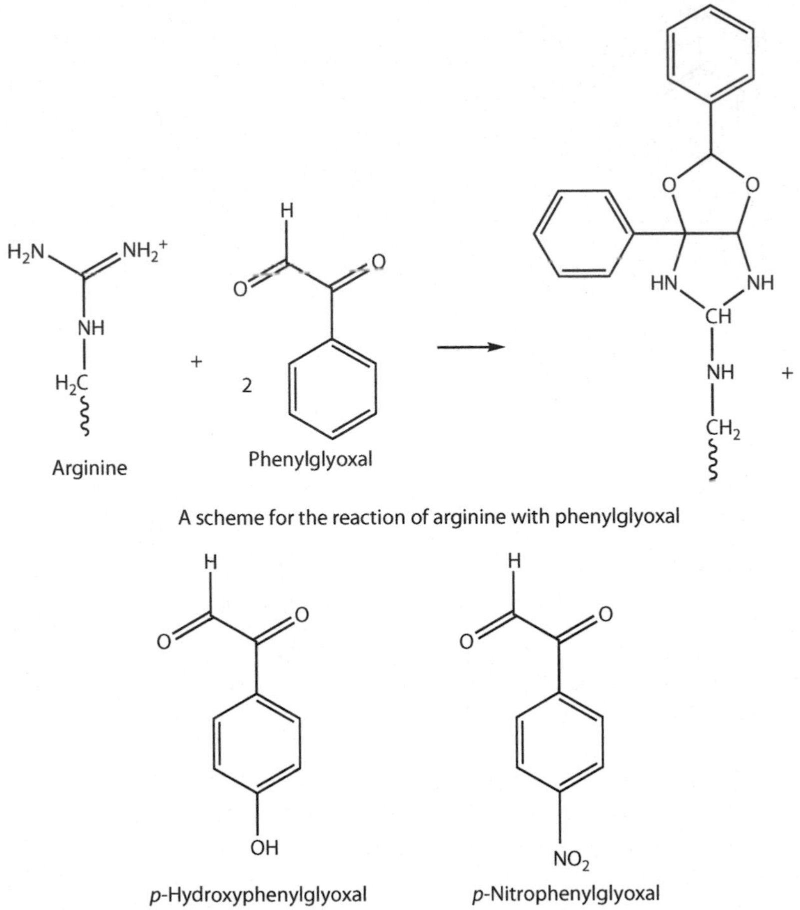

FIGURE 2.18 Phenylglyoxal and analogues for modification of arginine residues in proteins. (See Riordan, J.F., Arginyl residues and anion binding sites in proteins, *Mol. Cell. Biochem.* 26, 71–92, 1979.) For the chromophoric reagents, see Borders, C.L., Jr., Pearson, L.J., McLaughlin, A.E. et al., 4-Hydroxy-3-nitrophenylglyoxal. A chromophoric reagent for arginyl residues in proteins, *Biochim. Biophys. Acta* 568, 491–495, 1979.

1 mol of arginine. The [^{14}C]-labeled reagent can be easily prepared.[483,579] A facile modification of the original Riley and Gray[538] method that omits the vacuum distillation step has been reported by Schloss and coworkers.[539] The synthesis of phenyl [2-^{3}H]-glyoxal[540] has been reported.

Borders and coworkers[541] have reported the synthesis of a chromophoric derivative, 4-hydroxy-3-nitrophenylglyoxal (Figure 2.18). The adduct between 4-hydroxy-3-nitrophenylgyloxal and arginine absorbs light at 316 nm ($\varepsilon = 1.09 \times 10^4$ M^{-1} cm^{-1}). The derivative is unstable to acid hydrolysis (6 N HCl, 110°C, 24 h) but can be stabilized by the inclusion of thioglycolic acid. Fresh solutions of the reagent should be used to avoid problems associated with the decomposition of the reagent. This same group subsequently used this reagent to identify the reactive arginine in yeast

Cu, Zn superoxide dismutase.[542] The reaction of 4-hydroxy-3-nitrophenylglyoxal (50 mM BICINE, 100 mM $NaHCO_3$, pH 8.3) with yeast Cu, Zn superoxide dismutase is slower (0.57 M^{-1} min^{-1}) than that observed with phenylglyoxal (28 M^{-1} min^{-1}). A similar difference in the rate of reaction with the two reagents was observed with creatinine kinase.[541,543] 4-Hydroxy-3-nitrophenylglyoxal has been used to measure creatinine.[544] This reagent is used infrequently but is useful.[545–551]

p-Hydroxylphenylglyoxal (Figure 2.18) was developed by Yamasaki and colleagues[552] for the detection of available arginine residues in proteins. As with phenylglyoxal, it reacts with arginine under mild conditions (pH 7–9, 25°C, 30–60 min). The concentration of the resulting adduct (2:1 stoichiometry) with arginine can be determined at 340 nm ($\varepsilon = 1.83 \times 10^4$ M^{-1} cm^{-1}). The modification is slowly reversed under basic conditions. The general characteristics of the reaction of *p*-hydroxyphenylglyoxal are similar to those described for phenylglyoxal. Most studies on the modification of arginine in proteins have used multiple reagents but there are a few studies that have used *p*-hydroxyphenylglyoxal. These are presented in Table 2.2.

There have been several studies that have compared *p*-hydroxyphenylglyoxal and phenylglyoxal. It can be argued that *p*-hydroxyphenylglyoxal is more hydrophilic than phenylglyoxal. Béliveau and coworkers[553,554] showed that the rate of inactivation of phosphate transport by rat kidney brush border membranes was more rapid (k_{obsvd} = 0.052 s^{-1}) with phenylglyoxal than with *p*-hydroxyphenylglyoxal (k_{obsvd} = 0.012 s^{-1}). The rate of inactivation of glucose uptake by phenylglyoxal was slower than that for phosphate. Reaction order suggested that the inactivation is associated with a modification of a single arginine residue. Mukovyama and colleagues[555] have compared the modification of an L-phenylalanine oxidase from *Pseudomonas* sp. P.501 by phenylglyoxal and *p*-hydroxyphenylglyoxal. The rate for the phenylglyoxal modification of a single essential arginine residue was 10.6 M^{-1} min^{-1}, while the rate observed for the modification by 4-hydroxy-3-nitrophenylglyoxal was 15.1 M^{-1} min^{-1}. The most remarkable observation on differences between *p*-hydroxyphenylglyoxal and phenylglyoxal comes from studies by Linder and coworkers.[513] These investigators observed that treatment of mitochondria with phenylglyoxal (10 mM HEPES, 250 mM sucrose, 10 mM succinate, 100 μM EGTA, 3 μM rotenone, pH 8.0) results in the closing of the permeability pore, while reaction with *p*-hydroxyphenylglyoxal results in pore opening.

The reaction of arginine with phenylglyoxal is greatly accelerated in bicarbonate–carbonate buffer systems.[556] The reaction of methylglyoxal with arginine is also enhanced by bicarbonate, while a similar effect is not seen with either glyoxal or 2,3-butanedione. The molecular basis for this specific buffer effect is not clear at this time nor is it known whether reaction with α-amino functional groups occurs at a different rate than with other solvent systems used for this modification of arginine with phenylglyoxal. Yamasaki and coworkers[557] reported that *p*-nitrophenylglyoxal reacts with arginine in 0.17 sodium pyrophosphate, 0.15 M sodium ascorbate, pH 9.0 to yield a derivative, which absorbs at 475 nm. There is also the reaction with histidine (the imidazole ring is critical for this reaction in that the l-methyl derivative yielded a derivative that absorbed at 475 nm, while the 3-methyl derivative did not). Free sulfhydryl groups also yielded a product with absorbance at 475 nm, but its absorbance was only 3% of that of arginine. Branlant and coworkers[558] have used *p*-carboxyphenyl glyoxal in bicarbonate buffer at pH 8.0 to modify aldehyde reductase. A second-order rate constant of $26\,M^{-1}\,min^{-1}$ was observed in 80 mM bicarbonate and $2.9\,M^{-1}\,min^{-1}$ in 20 mM sodium phosphate, pH 7.0. Saturation kinetics was observed with this reagent under certain conditions suggesting that a binding event occurs prior to actual chemical modification. Phenylglyoxal also inactivated this enzyme with a second-order rate constant of $2.6\,M^{-1}\,min^{-1}$ in 20 mM sodium phosphate, pH 7.0. Inactivation is reversible on dialysis. Further work with *p*-carboxyphenylglyoxal has not been reported.

Eun[559] has examined the effect of borate on the reaction of arginine with phenylglyoxal and *p*-hydroxyphenylglyoxal. The base buffer of these studies was 0.1 M sodium pyrophosphate, pH 9.0. Spectroscopy was used to follow the rate of arginine modification. The rate of modification of either free arginine or *N*-acetyl-L-arginine with phenylglyoxal was 10–15 times faster than that of *p*-hydroxyphenylglyoxal in the base buffer system. The inclusion of sodium borate (10–50 mM) markedly increased the rate of the reaction (approximately 20-fold) of *p*-hydroxyphenylglyoxal with either arginine or *N*-acetyl-L-arginine, while there was only a slight enhancement of the phenylglyoxal reaction. In a related study,[547] the effect of phenylglyoxal on sodium channel gating in frog myelinated nerve was compared with that of *p*-hydroxyphenylglyoxal or *p*-nitrophenylglyoxal. Both *p*-hydroxyphenylglyoxal and *p*-nitrophenylglyoxal had less effect than phenylglyoxal in reduced sodium current. The results are discussed in terms of the differences in hydrophobicity of the reagents, but it is clear that the intrinsic difference in reagent effectiveness described by Eun may be responsible, in part, for the observed differences. Phenylglyoxal has been used to modify arginine residues in *Klebsialla aerogenes* urease.[560] Previous studies have shown Arg336 to be present at the enzyme active site. The R336Q variant shows greatly reduced enzyme activity (decreased k_{cat}, normal K_m). Modification of this variant with phenylglyoxal resulted in a further decrease of activity suggesting that the modification of non-active-site residues can eliminate activity.

2,3-Butanedione (Figure 2.19) is the second well-characterized reagent for the selective modification of arginyl residues in proteins. Yankeelov and coworkers introduced the use of this reagent.[484,561] There were problems with the specificity of the reaction and the time required for modification until the observation of Riordan[562] that borate had a significant effect on the nature of the reaction of 2,3-butanedione with arginyl residues in proteins. The ability of 2,3-butanedione to act as a photosensitizing agent for the destruction of amino acids and proteins in the presence of oxygen was emphasized in work by Fliss and Viswanatha.[533] As would be expected from consideration of early photooxidation work, tryptophan and histidine are lost most rapidly with methionine; cystine and tyrosine are lost at a much slower rate. Loss is not seen on irradiation in the absence of 2,3-butanedione. Azide (10 mM), a singlet oxygen scavenger, greatly reduces the rate of loss of amino acids. The absence of oxygen also greatly reduces the rate of loss of sensitive amino acids. These observations have been confirmed and extended by other laboratories.[534,535]

There are several recent studies on the use of 2,3-butanedione to modify proteins that are worthy of comment. Alkena and coworkers[563] have modified penicillin acylase (*E. coli*) with 2,3-butanedione (50 mM borate, pH 8.0, 25°C). The modified enzyme had decreased specificity (k_{cat}/K_m) for 2-nitro-5[(phenacetyl)amino]benzoic acid. Oligonucleotide-directed mutagenesis of two arginine residues yielded similar results,

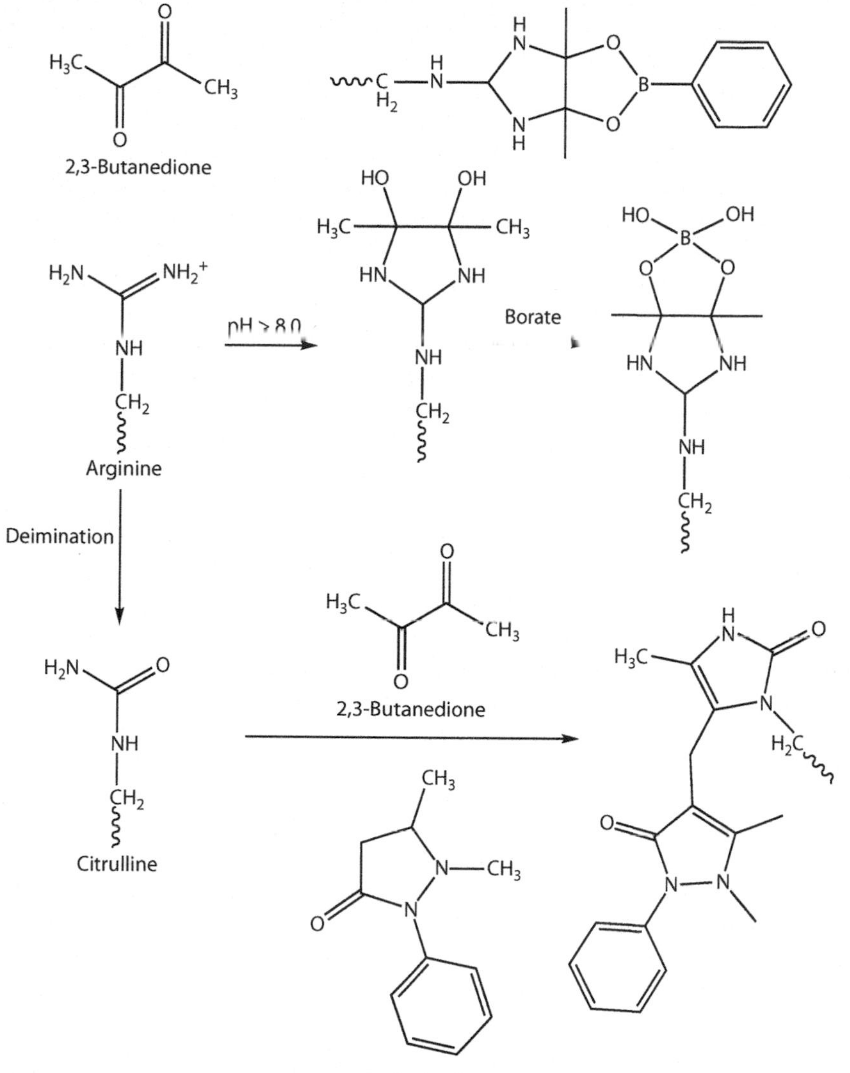

FIGURE 2.19 2,3-Butanedione modification of arginine residues in proteins. This is modification in the presence of borate (Riordan, J.F., Functional arginyl residues in carboxypeptidase A. Modification with butanedione, *Biochemistry* 12, 3915–3923, 1973). As shown, this reaction uses borate to stabilize the product of the reaction between 2,3-butanedione and arginine. It is also possible to use arylboronic acids to stabilize the product as shown with phenylboronic acid (Leitner, A. and Lindner, W., Probing of arginine residues in peptides and proteins using selective tagging and electrospray ionization mass spectrometry, *J. Mass Spectrom.* 38, 891–899, 2003). Also shown is the reaction of 2,3-butanedione with citrulline in the presence of antipyrine (Stensland, M., Holm, A., Kiehne, A., and Fleckenstein, B., Targeted analysis of protein citrullination using chemical modification and tandem mass spectrometry, *Rapid Commun. Mass Spectrom.* 23, 2754–2762, 2009).

although there is dependence on the replacement amino acid. The kinetic parameters of βR263K were similar to wild type and was inactivated by 2,3-butanedione. The βR263L variant showed a greater than 1000-fold decrease in k_{cat}/K_m. It is of interest that while phenylglyoxal also inhibited the enzyme, analysis was complicated by observation that phenylglyoxal was also a competitive inhibitor. Clark and Ensign[564] have studied the inactivation of 2-[(R)-2-hydroxypropylthio]ethanesulfonate dehydrogenase with 2,3-butanedione. The reaction was performed in 50 mM sodium borate, pH 9.0, at 25°C. In experiments for amino acid analysis of the modified enzyme, reaction was performed at 30°C for 30 min and then terminated by the addition of sodium borohydride to trap the adducts. The modification was specific for arginine with a second-order rate constant of 0.031 M^{-1} s^{-1} for the loss of enzyme activity. Leitner and Linder[488,565] have developed an approach for the general labeling of guanidino groups in proteins via reaction with 2,3-butanedione in the presence of an arylboronic acid (e.g., phenylboronic acid, see Figure 2.19) under alkaline conditions (pH 8–10). The sample is then subjected to electrospray ionization mass spectrometry without further processing. The same concept was used by these investigators for purification of peptides containing an arginine modified with 2,3-butanedione by using phenylboronic acid immobilized to a matrix.[566] This group extended the studies on phenylboronic acid to study of buried and available arginine residues in proteins[567] and for the identification of protein by peptide mapping.[568] This group has also studied the reaction of malondialdehyde with arginine.[569]

The use of 1,2-cyclohexanedione (Figure 2.20) under very basic conditions to modify arginyl residues was demonstrated in 1967.[570] However, it was not until Patthy and Smith[485] reported on the reaction of 1,2-cyclohexanedione in borate with arginyl residues in proteins that the use of this reagent became practical. These investigators reported that 1,2-cyclohexanedione reacted with arginyl residues in 0.2 M borate, pH 9.0. At alkaline pH, reaction of 1,2-cyclohexanedione with arginine forms N^5-(4-oxo-1,3-diazaspiro[4,4]non-2-yliodene)-L-ornithine (CHD-arginine), a reaction that cannot be reversed. Between pH 7.0 and 9.0 a compound is formed from arginine and 1,2-cyclohexanedione, N^7-N^8-(1,2-dihydroxycyclohex-1,2-ylene)-L-arginine (DHCH-arginine). This compound is stabilized by the presence of borate and is unstable in the presence of buffers such as Tris. This compound is readily converted back to free arginine in 0.5 M hydroxylamine, pH 7.0.

Patthy and Smith subsequently used this reagent to identify functional residues in bovine pancreatic ribonuclease A and egg white lysozyme.[571] Extent of modification of arginine residues in protein by 1,2-cyclohexanedione is generally assessed by amino acid analysis after acid hydrolysis. Under the conditions normally used for acid hydrolysis (6 N HCl, 110°C, 24 h), the borate-stabilized reaction product between arginine and 1,2-cyclohexanedione is unstable, and there is partial regeneration of arginine and the formation of unknown degradation products.[485] Acid hydrolysis in the presence of an excess of mercaptoacetic acid prevents the destruction of DHCD-arginine. Modification with cyclohexanedione is reversed by dialysis against mildly alkaline Tris buffer.[572] Ullah and Sethumadhaven[573] have demonstrated differences in the susceptibility of two phytases from *Aspergillus ficus* to modification of arginine. Phytase A was rapidly inactivated by either 1,2-cyclohexanedione (borate, pH 9.0) or phenylglyoxal ($NaHCO_3$, pH 7.5). Phytase B was resistant to inactivation by 1,2-cyclohexanedione and less susceptible than Phytase A to inactivation with

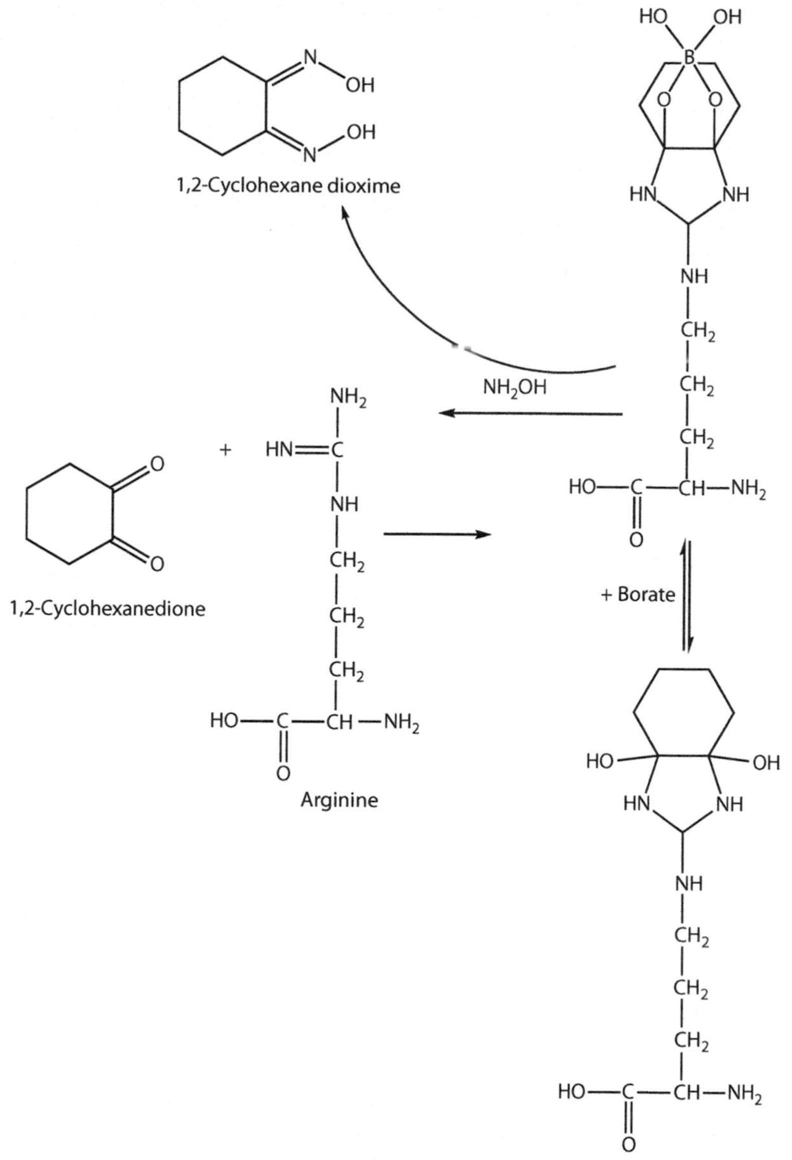

FIGURE 2.20 1,2-Cyclohexanedione modification of arginine. Shown is the reaction of 1,2-cyclohexanedione with arginine to form the DHCH-arginine derivative, which is stabilized by formation of a complex with borate at mild alkaline pH (Patthy, L. and Smith, E.L., Reversible modification of arginine residues. Application to sequence studies by restriction of tryptic hydrolysis to lysine residues, *J. Biol. Chem.* 250, 557–564, 1975). The modification is reversed by hydroxylamine resulting in the formation of the dioxime, which can be detected as the nickel complex (Patthy, L. and Smith, E.L., Identification of functional arginine residues in ribonuclease A and lysozyme, *J. Biol. Chem.* 250, 565–569, 1975). (For synthesis of 1,2-cyclohexane dioxime see Hach, C.C., Banks, C.V., and Diehl, H., 1,2-Cyclohexane dioxime, *Org. Synth. Coll.* 4, 229, 1963.)

phenylglyoxal. Calvete and colleagues[142] used a novel approach to the modification of arginine residues in bovine seminal plasma protein PDC-109. The protein was bound to a heparin–agarose column and the 1,2-cyclohexanedione (in 16 mM Tris-50 mM NaCl, 1.6 mM EDTA, 0.025% NaN_3, pH 7.4) circulated through the column overnight at room temperature. The modified protein was eluted with 1.0 M NaCl. Residues shielded from modification were presumed to be at the heparin-binding site.

Arginine is susceptible to oxidation and reaction with products of the oxidation or lipids (Figure 2.21) and carbohydrates to form a variety of products.[574–577]

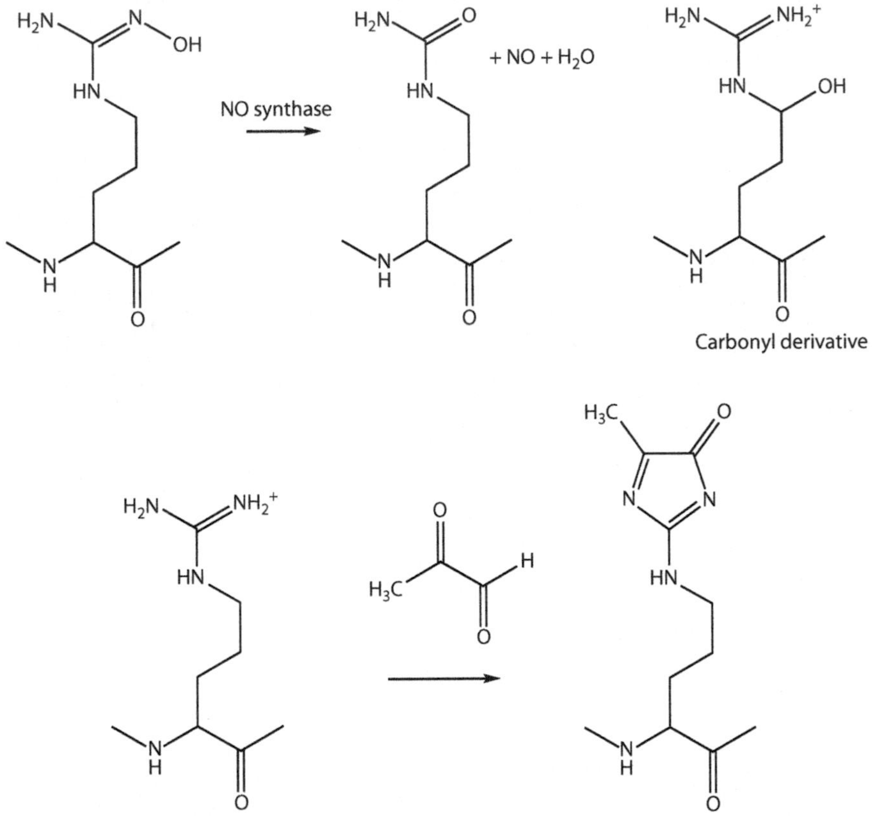

FIGURE 2.21 Oxidation of arginine and reaction of oxidation products with lysine. Shown are products from the enzymatic oxidation of arginine by nitric oxide synthetase (NO synthetase) (Mansuy, Y. and Boucher, J.-L., Oxidation of N-hydroxyguanidines by cytochromes P450 and NO-synthases and formation of nitric oxide, *Drug Metabol.* 3, 593–606, 2007) and nonenzymatic oxidation (Requena, J.R., Levine, R.L., and Stadtman, E.R., Recent advances in the analysis of oxidized proteins, *Amino Acids* 25, 221–226, 2003). N^{ω}-hydroxy-L-arginine is derived from arginine by NO synthase, while the carbonyl derivative is thought to be derived from metal-catalyzed oxidation. Also shown is the reaction of arginine with methylglyoxal form *N*δ-(1,5-dihydro-5-methyl-4-imidazolone-2-yl)orninthine (Lo, T.W., Westwood, M.E., McLellan, A.C. et al., Binding and modification of proteins by methylglyoxal under physiological conditions. A kinetic and mechanistic study with *N*α-acetylarginine, *N*α-cysteine, and *N*α-acetyllysine and bovine serum albumin, *J. Biol. Chem.* 269, 32299–32305, 1994).

An excellent review of the methods for the analysis of the various oxidation products is available.[575] Methylglyoxal is formed during aerobic glycolysis and reacts with arginine to form imidazolium adducts.[578–580] Ascorbic acid has been shown to react with N^{α}-acetyl-L-arginine to form N^{α}-acetyl-N^{δ}-[4-(1,2-dihydroxy-3-propyliden)-3-imidazolin-5-on-2-yl]-L-ornithine.[581] Arginine is also subject to methylation.[582,583]

Ninhydrin was developed as a reagent for the modification of arginine residues (Figure 2.22) in proteins by Takahashi.[584] The modification reaction was performed in 0.1 M morpholine acetate, pH 8.0, at 25°C in the dark. Modification of both arginine and lysine occurs under these reaction conditions. Specific modification of arginine residues may be accomplished by first modifying the lysine residues with citraconic anhydride. Verri and coworkers[526] observed that thiamin transport in renal brush border membranes was decreased by treatment with phenylglyoxal but not with ninhydrin. Earlier studies[585] on an enterokinase from kidney beans (*Phaseola*

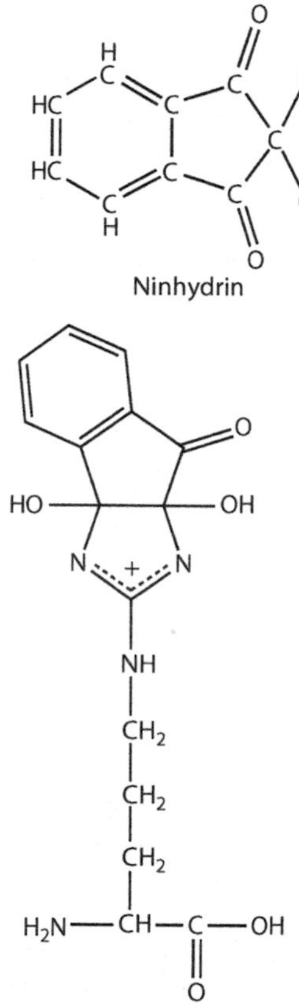

FIGURE 2.22 The reaction of ninhydrin with the guanidino group of arginine. Shown is the structure of ninhydrin and the proposed structure for the adduct of ninhydrin and the guanidino group of arginine (Takakashi, K., Specific modification of arginine residues in proteins with ninhydrin, *J. Biochem.* 80, 1173–1176, 1976; Chaplin, M.E., The use of ninhydrin as a reagent for the reversible modification of arginine residues in proteins, *Biochem. J.* 155, 457–459, 1976).

vulgaris) showed inactivation with either 1,2-cyclohexanedione or ninhydrin. Use of ninhydrin under basic conditions (0.4 M NaOH, alkaline ninhydrin) provides a specific method for the detection of arginine peptides.[586,587] For the purpose of historical accuracy, it is noted that Ruehmann reported on the reaction between ninhydrin (triketohydrindene hydrate) and guanidine in 1910[588]; this investigator also noted the relationship between ninhydrin and phenylglyoxal. As reported by Retinger,[589] ninhydrin is said to have been discovered by Aderhalden for use in his test for the presence of proteolytic enzymes in serum.[590–592]

Kethoxal (see Figure 2.1) was developed as an antiviral agent[593] and is used extensively for the modification of guanine residues in RNA.[594–596] Given structural similarity to phenylglyoxal and other reagents used for the modification of arginine, it is not surprising that kethoxal has been used for the modification of arginine in proteins.[22,24,597,598] However, the reaction is complex[598] and poorly characterized.

REFERENCES

1. Khan, M.N., Structure-reactivity relationships and the rate-limiting step in the reactions of methyl salicylate with primary amines, *Int. J. Chem. Kinet.* 19, 415–434, 1987.
2. Odiaka, T.I., Steric and electronic influences on the rate of addition of anilines to the tricarbonyl(cyclohexadienyl)iron(II) cation, *Inorg. Chim. Acta* 164, 143–147, 1989.
3. Bernasconi, C.F. and Stronach, M.W., Kinetic behavior of tetrachlorocyclopentadienyl-type anions. Deprotonation of 1,2,3,4-tetrachloro-1,3-cylcopentanediene and nucleophilic addition to 1,2,3,4-tetrachloro-6-phenylfulvene in 50% Me_2SO/50% water, *J. Am. Chem. Soc.* 112, 8446–8454, 1990.
4. King, J.F., Guo, Z.H., and Klassen, D.F., Nucleophilic attack vs. general base assisted hydrolysis in the reactions of acetic anhydride with primary and secondary amines. pH-yield studies in the recognition and assessment of the nucleophilic and general base behavior, *J. Org. Chem.* 59, 1095–1101, 1994.
5. Khan, M.N., Arlin, Z., George, A., and Walhab, I.A., Effects of mixed H_2O-CH_3CN solvents on the Bronsted coefficient for the intramolecular general base catalyzed cleavage of ionized phenyl salicylate in the presence of primary and secondary amines, *Int. J. Chem. Kinet.* 32, 153–164, 2000.
6. Garei, J. and Tawfik, D.S., Mechanism of hydrolysis and aminolysis of homocysteine thiolactone, *Chemistry* 12, 4144–4152, 2006.
7. Grimsley, G.R., Scholtz, J.M., and Pace, C.N., A summary of the measured p*K* values of the ionizable groups in folded proteins, *Protein Sci.* 18, 247–251, 2009.
8. Driessen, H.P., de Jong, W.W., Tesser, G.I., and Bloemendal, H., The mechanism of N-terminal acetylation of proteins, *CRC Crit. Rev. Biochem.* 18, 281–325, 1985.
9. Carr, S.A., Hemling, M.E., Bean, M.F., and Roberts, G.D., Integration of mass spectrometry in analytical biotechnology, *Anal. Chem.* 63, 2802–2824, 1991.
10. Perrier, J., Durand, A., Giardina, T., and Puigserver, A., Catabolism of intracellular N-terminal acetylated proteins: Involvement of acylpeptide hydrolase and acylase, *Biochimie* 87, 673–685, 2005.
11. Stevens, J.W. and Hascall, V.C., N-terminal carbamylation of the hyaluronic acid binding region and the link protein from the chondrosarcoma proteoglycan aggregate, *J. Biol. Chem.* 261, 15442–15449, 1986.
12. Stark, G.R., Reactions of cyanate with functional groups of proteins. III. Reactions with amino and carboxyl groups, *Biochemistry* 4, 1030–1036, 1968.

13. Huang, Z., Ni, C., Chu, Y. et al., Chemical modification of recombinant human keratinocyte growth factor 2 with polyethylene glycol improves bioavailability and reduces animal immunogenicity, *J. Biotechnol.* 142, 242–249, 2009.
14. Morishige, T., Yoshioka, Y., Inakura, H. et al., Creation of a lysine-deficient LIGHT mutant with the capacity for site-specific PEGylation and low affinity for a decoy receptor, *Biochem. Biophys. Res. Commun.* 393, 888–893, 2010.
15. Massen, J.A., Thielen, T.P., and Möller, W., Synthesis and application of two reagents for the introduction of sulfhydryl groups into proteins, *Eur. J. Biochem.* 134, 327–330, 1983.
16. Geoghegan, K.F., Emery, M.J., Martin, W.H. et al., Site-directed double fluorescent tagging of human renin and collagenase (MMP-1) substrate peptides using the periodate oxidation of N-terminal serine. An apparently general strategy for provision of energy-transfer substrates for proteases, *Bioconjug. Chem.* 4, 537–544, 1993.
17. Rose, K., Chen, J., Dragnovic, M. et al., New cyclization reaction at the amino terminus of peptides and proteins, *Bioconjug. Chem.* 10, 1038–1043, 1999.
18. Gaertner, H.F. and Offord, R.E., Site-specific attachment of functionalized poly (ethylene glycol) to the amino terminus of proteins, *Bioconjug. Chem.* 7, 38–44, 1996.
19. Dixon, H.B.F. and Fields, R., Specific modification of NH_2-terminal residues by transamination, *Methods Enzymol.* 25, 409–419, 1972.
20. Verheij, H.M., Egmond, M.R., and de Haas, G.H., Chemical modification of the α-amino group in snake venom phospholipases A_2. A comparison of the interaction of pancreatic and venom phospholipases with lipid-water interfaces, *Biochemistry* 20, 94, 1981.
21. Benkov, K. and Delihas, N., Analysis of ketoxal bound to ribosomal proteins from *Escherichia coli* 70S reacted ribosomes, *Biochem. Biophys. Res. Commun.* 60, 901–908, 1974.
22. Miura, K., Tsuda, S., Ueda, T. et al., Chemical modification of guanine residues of mouse 5S ribosomal RNA with kethoxal, *Biochim. Biophys. Acta* 739, 281–285, 1983.
23. Hogan, J.J., Gutell, R.R., and Noller, H.F., Probing the conformation of 18S rRNA in yeast 40S ribosomal subunits with kethoxal, *Biochemistry* 23, 3322–3330, 1984.
24. Akinsiku, O.T., Yu, E.T., and Fabris, D., Mass spectrophotometric investigation of protein alkylation by the RNA footprinting probe kethoxal, *J. Mass Spectrom.* 40, 1372–1381, 2005.
25. Dixon, J.W. and Hofmann, T., Reaction with nitrous acid of the "active site" N-terminal isoleucine in chymotrypsin and derivatives, *Can. J. Biochem.* 48, 671–681, 1970.
26. Scrimger, S.T. and Hofmann, T., The involvement of the amino-terminal amino acid in the activity of pancreatic proteases. II. The effects of nitrous acid on trypsin, *J. Biol. Chem.* 242, 2528–2533, 1967.
27. Ridd, J.F., Nitrosation, diazotization, and deamination, *Quart. Rev.* (London) 15, 418–441, 1961.
28. Bonnett, R. and Nicolaidou, P., Nitrosation and nitrosylation of haemoproteins and related compounds. Part 2. The reaction of nitrous acid with the side chains of α-acyl amino-acid esters, *J. Chem. Soc. Perkin Trans.* 1(8), 1969–1974, 1979.
29. Halliwell, B., Oxygen and nitrogen are pro-carcinogens. Damage to DNA by reactive oxygen, chlorine and nitrogen species: Measurement, mechanism and the effects of nutrition, *Mutat. Res.* 443, 37–52, 1999.
30. Edge, A.S. and Spiro, R.G., Structural elucidation of glycosaminoglycans through characterization of disaccharides obtained after fragmentation by hydrazine-nitrous acid treatment, *Arch. Biochem. Biophys.* 240, 560–572, 1985.
31. Conrad, H.E., Degradation of heparan sulfate by nitrous acid, in *Proteoglycan Protocols*, R.V. Iozzo (ed.), Humana Press, Totowa, NJ, Chapter 34, pp. 347–351, 2001.
32. Philpot, J.St.L. and Small, P.A., LXXI, The action of nitrous acid on pepsin, *Biochem. J.* 32, 542–551, 1938.

33. Hofmann, T., Amino-terminal groups in zymogen activation: Effect of nitrous acid on pepsin, *Can. J. Biochem.* 47, 1099–1101, 1969.
34. Kirchhofer, D., Lipari, M.T., and Santall, S., Utilizing the activation mechanism of serine proteases to engineer hepatocyte growth factor into a Met antagonist, *Proc. Natl. Acad. Sci. USA* 104, 5306–5311, 2007.
35. Martin, W., Smith, J.A., Lewis, M.J., and Henderson, A.H., Evidence that inhibitory factor extracted from bovine retractor penis is nitrite, whose acid-activated derivative is stabilized nitric oxide, *Br. J. Pharmacol.* 93, 579–586, 1988.
36. Aoki, N., Johnson, G., 3rd., and Lefer, A.M., Beneficial effects of two forms of NO administration in feline splanchnic artery occlusion shock, *Am. J. Physiol.* 258, G275–G281, 1990.
37. Furchgott, R., Preface, in *Nitric Oxide and Peripheral Nervous System*, N. Toda, S. Moncado, R. Furchgott, and E.A. Higgs (eds.), Portland Press, London, U.K., 2000.
38. Porter, W.H., Application of nitrous acid deamination of hexosamines to the simultaneous GLC determination of neutral and amino sugars in glycoproteins, *Anal. Biochem.* 63, 27–43, 1975.
39. Moulin, M., Deleu, C., Larher, F.R., and Bouchereau, A., High-performance liquid chromatography determination of pipecolic acid after precolumn ninhydrin derivatization using domestic microwave, *Anal. Biochem.* 308, 320–327, 2002.
40. Perry, D.L. and Phillips, S.L. (eds.), *Handbook of Inorganic Chemistry*, CRC Press, Boca Raton, FL, 1995.
41. Patnaik, P., *Handbook of Inorganic Chemicals*, McGraw-Hill, New York, 2003.
42. Muneta, P., Jasman, R., and Reid, L.M., Effect of freezing on nitrite stability in aqueous solutions, *J. Assoc. Off. Anal. Chem.* 70, 22–23, 1987.
43. Daiber, A., Bachschmid, M., Kavakli, C. et al., A new pitfall in detecting biological end products of nitric oxide - nitration, nitros(yl)ation and nitrite/nitrate artefacts during freezing, *Nitric Oxide* 9, 44–52, 2003.
44. Hill, J.P. and Buckley, P.D., The use of pH indicators to identify suitable environments for freezing samples in aqueous and mixed aqueous/nonaqueous solutions, *Anal. Biochem.* 192, 358–361, 1991.
45. Pikel-Cleland, K.A., Cleland, J., Anchordoquy, T.J., and Carpenter, J.F., Effect of glycine of pH changes and protein stability during freeze-thawing in phosphate buffer systems, *J. Pharm. Sci.* 91, 1969–1979, 2002.
46. Butler, A.R. and Ridd, J.H., Formation of nitric oxide from nitrous acid in ischemic tissue and skin, *Nitric Oxide* 10, 20–24, 2004.
47. Aga, R.G. and Hughes, M.N., The preparation and purification of NO gas and the use of NO releasers: The application of NO donors and other agents of nitrosative stress in biological systems, *Methods Enzymol.* 436, 35–48, 2008.
48. Kurosky, A. and Hofmann, T., Kinetics of the reaction of nitrous acid with model compounds and proteins, and the conformational state of N-terminal groups in the chymotrypsin family, *Can. J. Biochem.* 50, 1282–1296, 1972.
49. Neer, E.J. and Konigsberg, W., The characterization of modified human hemoglobin. II. Reaction with 1-fluoro-2,4-dinitrobenzene, *J. Biol. Chem.* 243, 1966–1970, 1968.
50. de Bruin, S.H. and Bucci, E., Reaction of 1-fluoro-2,4,-dinitrobenzene with the free α chains of human hemoglobin. Evaluation of the pK of the terminal amino group, *J. Biol. Chem.* 246, 5228–5233, 1971.
51. Yamada, H., Matsunaga, N., Domoto, H., and Imoto, T., Dinitrophenylation as a probe for the determination of environments of amino groups in proteins. Reactivities of individual amino groups in lysozyme, *J. Biochem.* 100, 233–241, 1986.
52. Lew, B., Mills, K.V., and Paulus, H., Characteristics of protein splicing in trans mediated by a semisynthetic split intein, *Biopolymers* 51, 355–362, 2000.
53. Muralidharan, V. and Muir, T.W., Protein ligation: An enabling technology for the biophysical analysis of proteins, *Nat. Methods* 3, 429–438, 2006.

54. Camarero, J.A., Kwon, Y., and Coleman, M.A., Chemoselective attachment of biologically active proteins to surfaces by expressed protein ligation and its application for "protein chip" fabrication, *J. Am. Chem. Soc.* 126, 14730–14731, 1994.
55. Volkmann, G. and Liu, X.Q., Protein C-terminal labeling and biotinylation using synthetic peptide and split-intein, *PLoS One* 4, e8381, 2009.
56. Schmidt, D.F., Jr. and Westheimer, F.H., p*K*a of the lysine amino group of acetoacetate decarboxylase, *Biochemistry* 10, 1249–1253, 1971.
57. Gerlt, J.A., Acetoacetate decarboxylase: Hydrophobics, not electrostatics, *Nat. Chem. Biol.* 5, 454–455, 2009.
58. Knowles, J.R., The intrinsic pKa values of functional groups in enzymes: Improper deductions from the pH-dependence of steady-state parameters, *CRC Crit. Rev. Biochem.* 4, 165–173, 1976.
59. Brocklehurst, K. and Dixon, H.B.F., The pH-dependence of second-order rate constants of enzymes modification may provide free-reactant p*K*a values, *Biochem. J.* 107, 859–862, 1977.
60. Stiles, W.E., Gittis, A.G., Lattman, E.E., and Shortle, D., In a staphylococcal nuclease mutant the side-chain of a lysine replacing valine 66 is fully buried in the hydrophobic core, *J. Mol. Biol.* 221, 7–14, 1991.
61. Kim, J., Mao, J., and Gunner, M.R., Are acidic and basic residues in buried proteins predicted to be ionized? *J. Mol. Biol.* 348, 1283–1298, 2005.
62. Milletti, F., Storchi, L., and Cruciani, G., Predicting protein p*K*(a) by environment similarity, *Proteins Struct. Funct. Bioinf.* 76, 484–495, 2009.
63. Fitch, C.A., Karp, D.A., Lee, K.K. et al., Experimental p*K*a values of buried residues: Analysis with continuum methods and role of water penetration, *Biophys. J.* 82, 3289–3304, 2002.
64. Jonas, A. and Weber, G., Partial modification of bovine serum albumin with dicarboxylic anhydrides. Physical properties of the modified species, *Biochemistry* 9, 4729–4735, 1970.
65. Tayyab, S. and Qasim, M.A., A correlation between changes in conformation and molecular properties of bovine serum albumin and succinylation, *J. Biochem.* 100, 1125–1136, 1986.
66. Muzaffar, M., Mir, K., Fazili, M., and Qasim, M.A., Chemical modification of buried lysine residues of bovine serum albumin and its influence on protein conformation and bilirubin binding, *Biochim. Biophys. Acta* 1119, 261–267, 1992.
67. Batra, P.P., Helix stability in succinylated and acetylated ovalbumin: Effect of high pH, urea and guanidine hydrochloride, *Biochim. Biophys. Acta* 1040, 102–108, 1990.
68. Wells, M.A., Effects of chemical modification on the activity of *Crotalus adamateus* phospholipase A_2. Evidence for an essential amino group, *Biochemistry* 12, 1086–1093, 1973.
69. Mita, K., Ichimura, S., Zama, M., and Hamana, K., Kinetics of chemical modification of arginine and lysine residues in calf thymus histone H1, *Biopolymers* 20, 1103–1112, 1981.
70. Running, W.E. and Reilly, J.P., Ribosomal proteins of *Deinococcus radiodurans*: Their solvent accessibility and reactivity, *J. Proteome Res.* 8, 1228–1246, 2009.
71. Murdock, A.L., Grist, K.L., and Hirs, C.H.W., On the dinitrophenylation of bovine pancreatic ribonuclease A, *Arch. Biochem. Biophys.* 114, 375–390, 1966.
72. Carty, R.P. and Hirs, C.H.W., Modification of bovine pancreatic ribonuclease A with 4-sulfonoxy-2-nitrofluorobenzene. Effect of pH and temperature on the rate of reaction, *J. Biol. Chem.* 243, 5254–5265, 1968.
73. Hart, G.J., Leeper, F.J., and Battersby, A.R., Modification of hydroxymethylbilane synthase (porphobilinogen deaminase) by pyridoxal-5′-phosphate. Demonstration of an essential lysine residue, *Biochem. J.* 222, 93–102, 1984.

74. Quay, S.C. and Tronson, L.P., Conformational studies of aqueous melittin: Determination of ionization constants of lysine-21 and lysine-23 by reactivity toward 2,4,6-trinitrobenzenesulfonate, *Biochemistry* 22, 700–707, 1983.
75. Mellert, H.S. and McMahon, S.B., Biochemical pathways that regulate acetyltransferase and deacetylase activity in mammalian cells, *Trends Biochem. Sci.* 34, 571–579, 2009.
76. Rodríguez-Navarro, S., Insights into SAGA function during gene expression, *EMBO Rep.* 10, 843–850, 2009.
77. Close, P., Creppe, C., Gillard, M. et al., The emerging role of lysine acetylation of non-nuclear proteins, *Cell. Mol. Life Sci.* 67, 1255–1264, 2010.
78. van Wijk, S.J. and Timmers, H.T., The family of ubiquitin-conjugating enzymes (E2s): Deciding between life and death of proteins, *FASEB J.* 24, 981–993, 2010.
79. Wilkinson, K.A. and Henley, J.M., Mechanisms, regulation and consequences of protein SUMOylation, *Biochem. J.* 428, 133–145, 2010.
80. d'Alayer, J., Expert-Bezançon, N., and Béguin, P., Time- and temperature-dependent acetylation of the chemokine RANTES produced in recombinant *Escherichia coli*, *Protein Expression Purif.* 55, 9–16, 2007.
81. Kashiwagi, M., Mikami, T., Chiba, M. et al., Occurrence of nonenzymatic N-acetylation of sphinganine with acetyl coenzyme A producing C_2H_2-ceramide and its inconvertibility to apoptotic C2-ceramide, *Biochem. Mol. Biol. Int.* 42, 1071–1080, 1997.
82. Walker, J.E., Lysine residue 199 of human serum albumin is modified by acetylsalicylic acid, *FEBS Lett.* 66, 173–175, 1976.
83. Liyasova, M.S., Schopfer, L.M., and Lockridge, O., Reaction of human albumin with aspirin in vitro: Mass spectrometric identification of acetylated lysines 199, 402, 519, and 545, *Biochem. Pharmacol.* 79, 784–791, 2010.
84. Xu, A.S., Macdonald, J.M., Labotka, R.J., and London, R.E., NMR study of the sites of human hemoglobin acetylated by aspirin, *Biochim. Biophys. Acta* 1432, 333–349, 1999.
85. Griffin, M., Casadio, R., and Bergamini, C.M., Transglutaminases: Nature's biological glues, *Biochem. J.* 368, 377–396, 2002.
86. Lin, C.W. and Ting, A.Y., Transglutaminase-catalyzed site-specific conjugation of small-molecule probes to proteins in vitro and on the surface of living cells, *J. Am. Chem. Soc.* 128, 4542–4543, 2006.
87. Mindt, T.L., Jungi, V., Wyss, S. et al., Modification of different IgG1 antibodies via glutamine and lysine using bacterial and human tissue transglutaminase, *Bioconjug. Chem.* 19, 271–278, 2008.
88. Murthy, S.N., Lukas, T.J., Jardetzky, T.S., and Lorand, L., Selectivity in the post-translational, transglutaminase-dependent acylation of lysine residues, *Biochemistry* 48, 2654–2660, 2009.
89. Kellam, B., De Bank, P.A., and Shakesheff, K.M., Chemical modification of mammalian cell surfaces, *Chem. Soc. Rev.* 32, 327–337, 2003.
90. Gould, A.R. and Norton, R.S., Chemical modification of cationic groups in the polypeptide cardiac stimula anthopleurin-A. *Toxicon* 33, 187–199, 1995.
91. Becker, L. et al., Identification of a critical lysine residue in apolipoprotein B-100 tat mediates noncovalent interaction with apolipoprotein(a), *J. Biol. Chem.* 276, 36155–36162, 2001.
92. Wink, M.R. et al., Effect of protein-modifying reagents on ecto-apyrase from rat brain, *Int. J. Biochem. Cell. Biol.* 32, 105–113, 2000.
93. Ehrhard, B. et al., Chemical modification of recombinant HIV-1 capsid protein p24 leads to the release of a hidden epitope prior to changes of the overall folding of the protein, *Biochemistry* 35, 9097–9105, 1996.
94. Paetzel, M. et al., Use of site-directed chemical modification to study an essential lysine in *Escherichia coli* leader peptidase, *J. Biol. Chem.* 272, 9994–10003, 1997.

95. Alcalde, M. et al., Succinylation of cyclodextrin glycosyltransferase from *Thermoanaerobacter* s501 enhances its transferase activity using starch as a donor, *J. Biotechnol.* 86, 71–80, 2001.
96. Swart, P.J. et al., Antiviral effects of milk proteins: Acylation results in polyanionic compounds with potent activity against human immunodeficiency virus types 1 and 2 *in vitro*, *AIDS Res. Hum. Retroviruses* 12, 769–775, 1996.
97. Swart, P.J. et al., Lactoferrin. Antiviral activity of lactoferrin, *Adv. Exp. Med. Biol.* 443, 205–213, 1998.
98. Neurath, A.R. et al., Blocking of CD4 cell receptors for the human immunodeficiency virus type 1 (HIV-1) by chemically modified milk proteins: Potential for AIDS prophylaxis, *J. Mol. Recognit.* 8, 204–216, 1995.
99. Jonsson, B.A. et al., Lysine adducts between methyltetrahydrophthalic anhydride and collagen in guinea pig lung, *Toxicol. Appl. Pharmacol.* 135, 156–167, 1995.
100. Lindh, C.H. and Jonsson, B.A., Human hemoglobin adducts following exposure to hexhydrophthalic anhydride and methylhexahydrophthalic anhydride, *Toxicol. Appl. Pharmacol.* 153, 152–160, 1998.
101. Pool, C.T. and Thompson, T.E., Methods for dual, site-specific derivatization of bovine pancreatic trypsin inhibitor: Trypsin protection of lysine-15 and attachment of fatty acids or hydrophobic peptides at the N-terminus, *Bioconjug. Chem.* 10, 221–230, 1999.
102. Kristiansson, M.H., Jonsson, B.A., and Lindh, C.H., Mass spectrometric characterization of human hemoglobin adducts formed *in vivo* by hexahydrophthalic anhydride, *Chem. Res. Toxicol.* 15, 562–569, 2002.
103. O'Brien, A.M., Smith, H.T., and O'Fagain, C., Effects of phthalic anhydride modification on horse radish peroxidase stability and activity, *Biotechnol. Bioeng.* 81, 233–240, 2003.
104. Klotz, I.M., Succinylation, *Methods Enzymol.* 11, 576, 1967.
105. Meighen, E.A., Nicolim, M.Z., and Hustings, J.W., Hybridization of bacterial luciferase with a variant produced by chemical modification, *Biochemistry* 10, 4062, 1971.
106. Shetty, K.J. and Rao, M.S.N., Effect of succinylation on the oligomeric structure of arachin, *Int. J. Pept. Protein Res.* 11, 305, 1978.
107. Shiao, D.D.F., Lumry, R., and Rajender, S., Modification of protein properties by change in charge. Succinylated chymotrypsinogen, *Eur. J. Biochem.* 29, 377, 1972.
108. Morton, R.C. and Gerber, G.E., Water solubilization of membrane proteins. Extensive derivatization with a novel polar derivatizing reagent, *J. Biol. Chem.* 263, 7989, 1988.
109. Atassi, M.Z. and Habeeb, A.F.S.A., Reaction of protein with citraconic anhydride, *Methods Enzymol.* 25, 546, 1972.
110. Habeeb, A.F.S.A. and Atassi, M.Z., Enzymic and immunochemical properties of lysozyme. Evaluation of several amino group reversible blocking reagents, *Biochemistry* 9, 4939, 1970.
111. Shetty, J.K. and Kinsella, J.E., Ready separation of proteins from nucleoprotein complexes by reversible modification of lysine residues, *Biochem. J.* 191, 269, 1980.
112. Ramanathan, L., Guyer, R.B., Buss, E.G., and Clagett, C.O., Chemical modifications of riboflavin-binding protein: Effects on function and antigenicity, *Mol. Immunol.* 17, 267–274, 1980.
113. Urabe, I., Yamamoto, M., Yamada, Y., and Okada, H., Effect of hydrophobicity of acyl groups on the activity and stability of acylated thermolysin, *Biochim. Biophys. Acta* 524, 435, 1978.
114. Pool, C.T. and Thompson, T.E., Chain length and temperature dependence of the reversible association of model acylated proteins with lipid bilayers, *Biochemistry* 37, 10246–10255, 1998.
115. Béven, L. et al., Ca^{2+}-myristoyl switch and membrane binding of chemical acylated neurocalcins, *Biochemistry* 40, 8152–8160, 2001.

116. Gibbons, I. and Schachman, H.K., A method for the separation of hybrids of chromatographically identical oligomeric proteins. Use of 3,4,5,6-tetrahydrophthaloyl groups as a reversible "chromatographic" handle, *Biochemistry* 15, 52–60, 1976.
117. Wearne, S.J., Factor Xa cleavage of fusion proteins. Elimination of non-specific cleavage by reversible acylation, *FEBS Lett.* 263, 23–26, 1990.
118. Howlett, G.J. and Wardrop, A.J., Dissociation and reconstitution of human erythrocyte membrane proteins using 3,4,5,6-tetrahydrophthalic anhydride, *Arch. Biochem. Biophys.* 188, 429, 1978.
119. Palacián, E., Gonzalez, P.J., Piñero, M., and Hernández, F., Dicarboxylic acid anhydrides as dissociating agents of protein-containing structures, *Mol. Cell. Biochem.* 97, 101–111, 1990.
120. Fraenkel-Conrat, H., Methods for investigating the essential groups for enzyme activity, *Methods Enzymol.* 4, 247–269, 1957.
121. Riordan, J.F. and Vallee, B.L., Acetylation, *Methods Enzymol.* 11, 565, 1967.
122. Merle, M. et al., Acylation of porcine and bovine calcitonin: Effects on hypocalcemic activity in the rat, *Biochem. Biophys. Res. Commun.* 79, 1071, 1977.
123. Aviram, I., The role of lysines in Euglena cytochrome C-552. Chemical modification studies, *Arch. Biochem. Biophys.* 181, 199, 1977.
124. Karibian, D. et al., On the reaction of acetic and maleic anhydrides with elastase. Evidence for a role of the NH2-terminal valine, *Biochemistry* 13, 2891, 1974.
125. Smit, V., Native electrophoresis to monitor chemical modification of human intgerleukin-3, *Electrophoresis* 15, 251–254, 1994.
126. Claasen, E., Kors, N., and Van, R.N., Influence of carriers on the development and localization of anti-trinitrophenyl antibody forming cells in the murine spleen, *Eur. J. Immunol.* 16, 271–276, 1986.
127. Furth, A.J. and Hope, D.B., Chemical modifications of the tyrosine residues inn bovine neurophysin-II, *Biochem. J.* 116, 545–553, 1970.
128. Meighen, E.A. and Schachman, H.K., Hybridization of native and chemically modified enzymes. I. Development of a general method and its application to the study of the subunit structure of aldolase, *Biochemistry* 9, 1163–1176, 1970.
129. Thomas, D.M. and Moss, D.W., Electrophoretic mobilities of native and chemically modified human placental alkaline phosphatase isoenzymes, *Biochem. J.* 122, 30P–31P, 1971.
130. Zschornig, O., Pasache, G., Cathrin, T. et al., Modulation of lysozyme charge influences interaction with phospholipid vesicles, *Colloid Surf. B* 42, 69–78, 2005.
131. Armstrong, D.J. and Roman, A., The anomalous electrophoretic behavior of the human papillovirus type 16 E7 protein is due to the high content of acidic amino acid residues, *Biochem. Biophys. Res. Commun.* 192, 1380–1387, 1993.
132. Garcia-Ortega, L., De los Rios, V., Martínez-Ruiz, A. et al., Anomalous electrophoretic behavior of a very acidic protein: Ribonuclease U2, *Electrophoresis* 26, 3407–3412, 2005.
133. Glish, G.L. and Vacher, R.W., The basics of mass spectrometry in the twenty-first century, *Nat. Rev. Drug Discov.* 2, 140–150, 2003.
134. Fligge, T.A. et al., Direct monitoring of protein-chemical reactions utilizing nanoelectrospray mass spectrometry, *J. Am. Soc. Mass Spectrom.* 10, 112–118, 1999.
135. Bennett, K.L. et al., Rapid characterization of chemically-modified proteins by electrospray mass spectrometry, *Bioconjug. Chem.* 7, 16–22, 1996.
136. Jahn, O. et al., The use of multiple ion chromatograms in on-line HPLC-MS for the characterization of post-translational and chemical modifications of proteins, *Int. J. Mass Spectrom.* 214, 37–51, 2002.
137. Suckau, D., Mak, M., and Przybylski, M., Protein surface topology probing by selective chemical modification and mass spectrometric peptide mapping, *Proc. Natl. Acad. Sci. USA* 89, 5630–5634, 1992.

138. Yeboah, F.K. et al., Effect of limited solid-state glycation on the conformation of lysozyme by ESI-MSMS peptide mapping and molecular modeling, *Bioconjug. Chem.* 15, 27–34, 2004.
139. Ohguro, H. et al., Topographic study of arrestin using differential chemical modification and hydrogen/deuterium exchange, *Protein Sci.* 3, 2428–2434, 1994.
140. Scaloni, A. et al., Topology of the thyroid transcription factor 1 homeodomain-DNA complex, *Biochemistry* 38, 64–72, 1999.
141. D'Ambrosio, C. et al., Probing the dimeric structure of porcine aminoacylase 1 by mass spectrometry and modeling procedures, *Biochemistry* 42, 4430–4443, 2003.
142. Calvete, J.J. et al., Characterisation of the conformation and quaternary structure-dependent heparin-binding region of bovine seminal plasma protein PDC-109, *FEBS Lett.* 444, 260–264, 1999.
143. Zappacosta, F., Ingallinella, P., Scaloni, A. et al., Surface topology of Minibody by selective chemical modification and mass spectrometry, *Protein Sci.* 6, 1901–1909, 1997.
144. Lepin, E.J., Leyton, J.V., Zhou, Y. et al., An affinity matured minibody for PET imaging of prostate stem antigen (PSCA)-expressing tumors, *Eur. J. Nucl. Med. Mol. Imaging* 37, 1529–1538, 2010.
145. Secchiero, P., Sblattero, D., Chiaruttini, C. et al., Selection and characterization of a novel agonistic human recombinant anti-TRAIL-R2 minibody with anti-leukemic activity, *Int. J. Immunopathol. Pharmacol.* 22, 73–83, 2009.
146. Leyton, J.V., Olafsen, T., Lepin, E.J. et al., Humanized radioiodinated minibody for imaging of prostate stem cell antigen-expressing tumors, *Clin. Cancer Res.* 14, 7488–7496, 2008.
147. Orita, T., Tsunoda, H., Yabuta, N. et al., A novel therapeutic approach for thrombocytopenia by minibody agonist of the thrombopoietin receptor, *Blood* 105, 562–566, 2005.
148. Hochleitner, E.O. et al., Characterization of a discontinuous epitope of the human immunodeficiency virus (HIV) core protein p24 by epitope excision and differential chemical modification followed by mass spectrometric peptide mapping analysis, *Protein Sci.* 9, 487–496, 2000.
149. Taralp, A. and Kaplan, H., Chemical modification of lyophilized proteins in nonaqueous environments, *J. Protein Chem.* 16, 183–193, 1997.
150. Smith, C.M. et al., Mass spectrometric quantification of acetylation at specific lysines within the amino-terminal tail of histone H4, *Anal. Biochem.* 316, 23–33, 2003.
151. Kaplan, H., Stevenson, K.J., and Hartley, B.S., Competitive labeling, a method for determining the reactivity of individual groups in proteins. The amino groups of porcine elastin, *Biochem. J.* 124, 289, 1971.
152. Bosshard, H.R., Koch, G.L.E., and Hartley, B.S., The aminoacyl tRNA synthetase-tRNA complex: Detection by differential labeling of lysine residues involved in complex formation, *J. Mol. Biol.* 119, 125, 1978.
153. Richardson, R.H. and Brew, K., Lactose synthase. An investigation of the interaction site of alpha-lactalbumin for galactosyltransferase by differential kinetic labeling, *J. Biol. Chem.* 255, 3377, 1980.
154. Rieder, R. and Bosshard, H.R., The cytochrome c oxidase binding site on cytochrome c. Differential chemical modification of lysine residues in free and oxidase-bound cytochrome c, *J. Biol. Chem.* 253, 6045, 1978.
155. Hitchcock, S.E., Zimmerman, C.J., and Smalley, C., Study of the structure of troponin-T by measuring the relative reactivities of lysines with acetic anhydride, *J. Mol. Biol.* 147, 125, 1981.
156. Hitchcock, S.E., Study of the structure of troponin-C by measuring the relative reactivities of lysines with acetic anhydride, *J. Mol. Biol.* 147, 153, 1981.
157. Hitchcock-De Gregori, S.E., Study of the structure of troponin-I by measuring the relative reactivities of lysine with acetic anhydride, *J. Biol. Chem.* 257, 7372, 1982.

158. Giedroc, D.P. et al., Differential trace labeling of calmodulin: Investigation of binding sites and conformational states by individual lysine reactivities. Effects of beta-endorphin, trifluoroperazine, and ethylene glycol bis(beta-aminoethyl ether)-N, N, N′ N-tetraacetic acid, *J. Biol. Chem.* 260, 13406, 1985.
159. Wei, Q. et al., Effects of interactions of with calcineurin of the reactivities of calmodulin lysines, *J. Biol. Chem.* 263, 19541, 1988.
160. Winkler, M.A. et al., Differential reactivities of lysines in calmodulin complexed to phosphatase, *J. Biol. Chem.* 262, 15466, 1987.
161. Hitchcock-De Gregori, S.E. et al., Lysine reactivities of tropomyosin complexed with troponin, *Arch. Biochem. Biophys.* 264, 410, 1988.
162. Lundblad, R.L., *Approaches to the Conformational Analysis of Biopharmaceuticals*, CRC Press, Boca Raton, FL, 2010.
163. Ohtani, E., Kusumi, T., and Kakisawa, H., An insight into the conformation of ptilomycalin A. The NMR properties of trifluoroacetylated spermidine analogues, *Tetrahedron Lett.* 33, 2525–2528, 1992.
164. Schallenberg, E.E. and Calvin, M., Ethyl trifluoroacetate as an acetylating agent with particular reference to peptide synthesis, *J. Am. Chem. Soc.* 77, 2779–2783, 1955.
165. Levy, D. and Paselk, R.A., The trifluoroacetylation of insulin, *Biochim. Biophys. Acta* 310, 398–405, 1955.
166. Paselk, R.A. and Levy, D., Fluorine nuclear magnetic resonance studies on trifluoroacetyl insulin derivatives. Effect of pH on conformation and aggregation, *Biochemistry* 13, 3340–3346, 1974.
167. Staudenmayer, N. et al., An enzyme kinetics and ^{19}F nuclear magnetic resonance study of selectively trifluoroacetylated cytochrome c derivatives, *Biochemistry* 15, 3198, 1976.
168. Smith, M.B. et al., Use of specific trifluoroacetylation of lysine residues in cytochrome c to study the reaction with cytochrome b5, cytochrome, c1 and cytochrome oxidase, *Biochim. Biophys. Acta* 592, 303, 1980.
169. Web, M., Stonehuerner, J., and Millett, F., The use of specific lysine modifications to locate the reaction site of cytochrome c with sulfite oxidase, *Biochim. Biophys. Acta* 593, 290, 1980.
170. Ahmed, A.J. and Millett, F., Use of specific lysine modifications to identify the site of reaction between cytochrome c and ferricyanide, *J. Biol. Chem.* 256, 1611, 1981.
171. Smith, M.B. and Millett, F., A ^{19}F nuclear magnetic resonance study of the interaction between cytochrome c and cytochrome c peroxidase, *Biochim. Biophys. Acta* 626, 64–72, 1980.
172. Lesage, A., Recent advances in solid-state NMR spectroscopy of spin I = 1/2 nuclei, *Phys. Chem. Chem. Phys.* 11, 6876–6891, 2009.
173. Goldberger, R.F., Trifluoroacetylation of ε-amino groups, *Methods Enzymol.* 11, 317–322, 1967.
174. Weisgraber, K.H., Innerarity, T.L., and Mahley, R.W., Role of the lysine residues of plasma lipoproteins in high affinity binding to cell surface receptors on human fibroblasts, *J. Biol. Chem.* 253, 9053, 1978.
175. Mahley, R.W. et al., Accelerated clearance of low-density and high-density lipoproteins and retarded clearance of E apoprotein-containing lipoproteins from the plasma of rats after modification of lysine residues, *Proc. Natl. Acad. Sci. USA* 76, 1746, 1979.
176. Jonas, A., Covinsky, K.E., and Sweeny, S.A., Effects of amino group modification in discoidal apolipoprotein A-1-egg phosphatidylcholine -cholesterol complex on their reactions with lecithin: Cholesterol acyltransferase, *Biochemistry* 24, 3508–3513, 1985.
177. Brubaker, G., Peng, D.-Q., Somerlot, B. et al., Apolipoprotein A-1 lysine modification: Effects on helical content, lipid binding and cholesterol acceptor activity, *Biochim. Biophys. Acta* 1761, 64–72, 2006.
178. Gurd, F.R.N., Carboxymethylation, *Methods Enzymol.* 11, 532, 1967.

179. Heinrikson, R.L., Alkylation of amino acid residues at the active site of ribonuclease, *J. Biol. Chem.* 241, 1393–1405, 1966.
180. Ahmed, M.U., Thorpe, S.R., and Baynes, J.W., Identification of N^{ε}-carboxymethyllysine as a degradation product of fructoselysine in glycated protein, *J. Biol. Chem.* 261, 4889–4894, 1986.
181. Ikeda, K., Higashi, T., Sano, H. et al., N^{ε}-(Carboxymethyl)lysine protein adduct is a major immunological epitope in proteins modified with advanced glycation end products of the Maillard reaction, *Biochemistry* 35, 8075–8083, 1996.
182. Dunn, J.A., McCance, D.R., Thorpe, S.R. et al., Age-dependent accumulation of N^{ε}-(carboxymethyl)lysine and N^{ε}-(carboxymethyl)hydroxylysine in human skin collagen, *Biochemistry* 30, 1205–1210, 1991.
183. Ehrlich, H., Hanke, T., Simon, P. et al., Carboxymethylation of the fibrillar collagen with respect to formation of hydroxyapatite, *J. Biomed. Mater. Res. Part B* 92, 542–551, 2010.
184. Seliskar, C.J. and Brand, J.L., Electronic spectra of 2-aminonaphthalene-6-sulfonate and related molecules. 2. Effects of solvent medium on absorption and fluorescence spectra, *J. Am. Chem. Soc.* 93, 5414–5420, 1971.
185. Gabellieri, E. and Stambini, G.B., ANS fluorescence detects widespread perturbation of protein tertiary structure, *Biophys. J.* 90, 3239–3245, 2006.
186. Sanger, F., The free amino groups of insulin, *Biochem. J.* 39, 507–515, 1945.
187. Welches, W.R. and Baldwin, T.O., Active center studies on bacterial luciferase: Modification of the enzyme with 2,4-dinitrofluorobenzene, *Biochemistry* 20, 512–517, 1981.
188. Shaltiel, S., Thiolysis of some dinitrophenyl derivatives of amino acids, *Biochem. Biophys. Res. Commun.* 29, 178–183, 1967.
189. Shapiro, R., Fox, E.A., and Riordan, J.F., Role of lysines in human angiogenin: Chemical modification and site-directed mutagenesis, *Biochemistry* 28, 1726–1732, 1989.
190. Li, W.M., Barnes, T., and Lee, C.H., Endoribonucleases—Enzymes gaining spotlight in mRNA metabolism, *FEBS J.* 277, 627–641, 2010.
191. Olah, T.V., Olson, H.M., Glitz, D.G., and Cooperman, B.S., Incorporation of single-dinitrophenylated proteins into the 30S subunit of *Escherichia coli* ribosomes b total reconstitution, *J. Biol. Chem.* 263, 4795–4800, 1988.
192. Kamra, A. and Gupta, M.N., A colorimetric procedure to monitor the extent of cross-linking of protein by bisimidoesters, *Biotechnol. Appl. Biochem.* 10, 287–293, 1988.
193. Chen, X., Yong, H., and Anderson, V.E., Protein cross-links: Universal isolation and characterization by isotopic derivatization and electrospray ionization mass spectrometry, *Anal. Biochem.* 273, 193–203, 1999.
194. Megherbi, R., Kiorpelidou, E., Foster, B. et al., Role of protein haptenization in triggering maturation events in the dendritic cell surrogate cell line THP-1, *Toxicol. Appl. Pharmacol.* 238, 120–132, 2009.
195. Kim, D., Kim, Y.-J., Seo, J.-N. et al., 2,4-Dinitrofluorobenzene modifies cellular proteins and induces macrophage inflammatory protein-2 gene expression via reactive oxygen species production in RAW 264.7 cells, *Immunol. Invest.* 38, 132–152, 2009.
196. Deepak, P., Kumar, S., Jr., Kishore, D., and Acharya, A., IL-13 from Th2-type cells suppresses induction of antigen-specific Th1 immunity in a T-cell lymphoma, *Int. Immunol.* 22, 53–63, 2010.
197. Hurrell, R.F., Carpenter, K.J., Sinclair, W.J. et al., Mechanisms of heat damage in proteins. 7. The significance of lysine-containing isopeptides and of lanthonine in heated proteins, *Br. J. Nutr.* 35, 383–395, 1976.
198. Carpenter, K.J., Steinke, F.H., Catignani, G.I. et al., The estimation of 'available lysine' in human foods by three chemical procedures, *Plant Foods Hum. Nutr.* 39, 129–135, 1989.

199. Pavlovic, S., Santos, R.C., and Gloria, M.B.A., Maillard reaction during the processing of 'doce de leite,' *J. Sci. Food Agric.* 66, 129–132, 1994.
200. Hermandez, A., Serrano, M.A., Munoz, M.M., and Castillo, G., Liquid chromatographic determination of total available free and intrachain lysine in various foods, *J. Chromatogr. Sci.* 39, 39–43, 2001.
201. Torbatinejad, N.M., Rutherford, S.M., and Moughan, P.J., Total and reactive lysine contents in selected cereal -based food products, *J. Agric. Food Chem.* 53, 4454–4458, 2005.
202. Pizzoferrato, L., Manzi, P., Vivanti, V. et al., Maillard reaction in milk-based foods: Nutritional consequences, *J. Food Prot.* 61, 235–239, 1998.
203. Moughan, P.J. and Rutherford, S.M., Available lysine in foods: A brief historical overview, *J. AOAC Int.* 91, 901–906, 2008.
204. Zahn, H. and Lebkucher, K.H., 4-Fluoro-3-nitrobenzene sulfonic acid as a reagent for amino end-group determination in proteins, *Biochem. Zeit.* 334, 133–148, 1961.
205. Hirs, C.H.W., Reactions with reactive aryl halides, *Methods Enzymol.* 11, 548–555, 1967.
206. Hummel, C.F., Gerber, B.R., Babich, A.M. et al., Modification of bovine pancreatic ribonuclease A with the site-specific reagent 4-arsono-2-nitrofluorobenzene-spectrophotometric titration of arsonophenyl ribonuclease A derivatives, *Biochemistry* 20, 4843–4852, 1981.
207. Hummel, C.F., Gerber, B.R., and Carty, R.P., Chemical modification of ribonuclease-A with 4-arsono-2-nitrofluorobenzene, *Int. J. Peptide Protein Res.* 24, 1–13, 1984.
208. Brautigan, D.L., Ferguson-Miller, S., and Margoliuash, E., Definition of cytochrome c binding domains by chemical modification. I. Reaction with 4-chloro-3,5-dinitrobenzoate and chromatographic separation of singly substituted derivatives, *J. Biol. Chem.* 253, 130, 1978.
209. Bello, J., Iijima, H., and Kartha, G., A new arylating agent, 2-carboxy-4,6-dinitrochlorobenzene. Reaction with model compounds and bovine pancreatic ribonuclease, *Int. J. Pept. Protein Res.* 14, 199, 1979.
210. Hall, J. et al., Role of specific lysine residues in the reaction of *Rhodobacter sphaeroides* cytochrome c2 with the cytochrome bc1 complex, *Biochemistry* 28, 2568, 1989.
211. Long, J.E. et al., Role of specific lysine residues in binding cytochrome c2 to the *Rhodobacter sphaeroides* reaction center in optimal orientation for rapid electron transfer, *Biochemistry* 28, 6970, 1989.
212. Franklin, J.G. and Leslie, J., Some enzymatic properties of trypsin after reaction with 1-dimethylaminonaphthalene-5-sulfonyl chloride, *Can. J. Biochem.* 49, 516, 1971.
213. Wagner, R., Podestá, F.E., González, D.H., and Andreo, C.S., Proximity between fluorescent probes attached to four essential lysyl residues in phosphoenolpyruvate carboxylase—A resonance energy transfer study, *Eur. J. Biochem.* 173, 561, 1988.
214. Hiratsuka, T. and Uchida, K., Lysyl residues of cardiac myosin accessible to labeling with a fluorescent reagent, N-methyl-2-anilino-6-naphthalenesulfonyl chloride, *J. Biochem.* 88, 1437, 1980.
215. Onondera, M., Shiokawa, H., and Takagi, T., Fluorescent probes for antibody active sites. I. Production of antibodies specific to the N-methyl-2-anilinonaphthalene-6-sulfonate group in rabbits and some fluorescent properties of the hapten bound to the antibodies, *J. Biochem.* 79, 195, 1976.
216. Cory, R.P., Becker, R.R., Rosenbluth, R., and Isenberg, I., Synthesis and fluorescent properties of some N-methyl-2-anilino-6-naphthalensulfonyl derivatives, *J. Am. Chem. Soc.* 90, 1643, 1968.
217. Haugland, R.P., *Molecular Probes. Handbook of Fluorescent Probes and Research Chemicals*, Molecular Probes, Inc., Eugene, OR, p. 37, 1989.
218. Zeitler, H.-J., *N*-Methyl-2-anilino-6-naphthalenesulfonyl hydrazine (2,6-mansyl hydrazine), *Anal. Biochem.* 88, 649–658, 1978.

219. Tuls, J., Geren, L., and Millett, F., Fluorescein isothiocyanate specifically modifies lysine 338 of cytochrome P-450scc and inhibits adrenodoxin binding, *J. Biol. Chem.* 264, 16421, 1989.
220. Miki, M., Interaction of Lys-61 labeled actin with myosin subfragment-1 and the regulatory proteins, *J. Biochem.* 106, 651–655, 1989.
221. Bellelli, A. et al., Binding and internalization of ricin labelled with fluorescein isothiocyanate, *Biochem. Biophys. Res. Commun.* 169, 602, 1990.
222. Devanathan, S., Dahl, T.A., Midden, W.R., and Neckers, D.C., Readily available fluorescein isothiocyanate-conjugated antibodies can be easily converted into targeted phototoxic agents for antibacterial, antiviral, and anticancer therapy, *Proc. Natl. Acad. Sci. USA* 87, 2980–2984, 1990.
223. Turner, D.C. and Brand, L., Quantitative estimation of protein binding site polarity. Fluorescence of N-arylaminonaphthalenesulfonates, *Biochemistry* 7, 3381, 1968.
224. Gasymov, O.K. and Glasgow, B.J., ANS fluorescence: Potential to augment the identification of the external binding sites of proteins, *Biochim. Biophys. Acta* 1774, 403–411, 2007.
225. Schreier, S.M., Steinkellner, H., Jirovetz, L. et al., S-Carbamoylation impairs the oxidant scavenging activity of cysteine: Its possible impact on increased LDL modification in uraemia, *Biochimie* 93, 772–777, 2011.
226. Stark, G.R., Stein, W.H., and Moore, S., Reaction of the cyanate present in aqueous urea with amino acids and proteins, *J. Biol. Chem.* 235, 3177–3181, 1960.
227. Walker, J. and Kay, S.A., Velocity of urea formation in aqueous alcohol, *Proc. Chem. Soc.* 177, 75–76, 1897.
228. Fawsitt, C.E., The decomposition of urea, *Zeit. Physiol. Chem.* 41, 601, 1902 (*J. Phys. Chem.* 7, 233, 1903).
229. Shaw, W.H.R. and Bordeaux, J.J., The decomposition of urea in aqueous media, *J. Am. Chem. Soc.* 77, 4729–4733, 1955.
230. Cohen-Adad, R., A study of the aqueous solution of urea at 25°C, *Bull. Soc. Chim.* 64–72, 1951.
231. Hagei, P., Gerding, J.J.T., Fieggen, W., and Bloemendal, H., Cyanate formation in solutions of urea. I. Calculation of cyanate concentrations at different temperatures and pH, *Biochim. Biophys. Acta* 243, 366–373, 1971.
232. Gerding, J.J.T., Koppes, A., Hagei, P., and Bloemendal, H., Cyanate formation in solutions of urea. II. Effect of urea on eye lens protein α-crystallin, *Biochim. Biophys. Acta* 243, 374–349, 1971.
233. Koepke, J.A., Akeson, A., and Pietruszko, R., Denaturation of horse-liver alcohol dehydrogenase in urea. Studies by gel filtration and electrophoresis, *Enzyme* 13, 177–187, 1972.
234. Francis, K.T., Thompson, R.W., and Krumdieck, C.L., Reaction of tetrahydrofolic acid with cyanate from urea solutions: formation of an inactive folate derivative, *Am. J. Clin. Nutr.* 30, 3028–3032, 1977.
235. Frantzen, F., Chromatographic and electrophoretic methods for modified hemoglobins, *J. Chromatogr. B* 699, 269–286, 1997.
236. Lippincott, J. and Apostol, I., Carbamylation of cysteine: A potential artifact in peptide mapping of hemoglobins in the presence of urea, *Anal. Biochem.* 267, 57–64, 1999.
237. Stark, G.R. and Smyth, D.G., The use of cyanate for the determination of NH_2-terminal residues in proteins, *J. Biol. Chem.* 238, 214, 1963.
238. Stark, G.R., Reactions of cyanate with functional groups of proteins. III. Reactions with amino and carboxyl groups, *Biochemistry* 4, 1030–1036, 1965.
239. Austad, T. and Engemyt, L.B., Nucleophilicity of the cyanate ion, *Acta Chem. Scand.* 25, 3535–3536, 1971.

240. Stark, G.R., Modification of proteins with cyanate, *Methods Enzymol.* 11, 590–594, 1967.
241. Stark, G.R., Modification of proteins with cyanate, *Methods Enzymol.* 25, 579–584, 1972.
242. Stark, G.R., On the reversible reaction of cyanate with sulfhydryl groups and the determination of NH_2-terminal cysteine and cystine in proteins, *J. Biol. Chem.* 239, 1411–1414, 1964.
243. Al-Rawi, H., Day, R.A., Farrar, C.R., and Williams, A., Elimination-addition mechanisms of acyl group: Transcarbamoylation in aminoalkylimidazoles carbamoylated on the heterocyclic nitrogen, *J. Chem. Soc. Perkins* 2, 1153–1159, 1979.
244. Anderson, P.M. and Carlson, J.D., Reversible reaction of cyanate with a reactive sulfhydryl group at the glutamine binding site at carbamoyl phosphate synthase, *Biochemistry* 14, 3688–3694, 1975.
245. Hu, J.J., Luke, A., Chellani, M. et al., pH-related effects of sodium cyanate on macromolecular synthesis and tumor cell division, *Biochem. Pharmacol.* 37, 2259–2266, 1988.
246. Hu, J.J., Dimaira, M.J., Zirvi, K.A. et al., Influence of pH on the modification of thiols by carbamoylating agents and effects on glutathione levels in normal and neoplastic cells, *Cancer Chemother. Pharmocol.* 24, 95–101, 1989.
247. Shaw, D.C., Stein, W.H., and Moore, S., Inactivation of chymotrypsin by cyanate, *J. Biol. Chem.* 239, PC671–PC673, 1964.
248. Stark, G.R., Reactions of cyanate with functional groups of proteins. IV. Inertness of aliphatic hydroxyl groups. Formation of carbamyl and acylhydantoins, *Biochemistry* 4, 2363–2367, 1965.
249. Robillard, G.T., Powers, J.C., and Wilcox, P.E., A chemical and crystallographic study of carbamoyl-chymotrypsin A, *Biochemistry* 11, 1773–1784, 1972.
250. Brown, W.E. and Wold, F., Alkyl isocyanates as active-site specific reagents for serine proteases. Identification of the active-site serine as the site of reaction, *Biochemistry* 12, 835–840, 1973.
251. Cerami, A. and Manning, J.M., Potassium cyanate as an inhibitor of the sickling of erythrocytes in vitro, *Proc. Natl. Acad. Sci. USA* 68, 1180, 1971.
252. Lee, C.K. and Manning, J.M., Kinetics of the carbamylation of the amino groups of sickle cell hemoglobin by cyanate, *J. Biol. Chem.* 248, 5861, 1973.
253. Njikam, N. et al., Carbamylation of the chains of hemoglobin S by cyanate in vitro and in vivo, *J. Biol. Chem.* 248, 8052, 1973.
254. Nigen, A.M., Bass, B.D., and Manning, J.M., Reactivity of cyanate with valine-1 (α) of hemoglobin. A probe of conformational change and anion binding, *J. Biol. Chem.* 251, 7638, 1976.
255. Balcerzak, S.P., Grever, M.R., Sing, D.E. et al., Preliminary studies of continuous extracorporeal carbamylation in the treatment of sickle cell anemia, *J. Lab. Clin. Med.* 100, 345–355, 1982.
256. Cirković, T., Gavrović-Jankulović, M., Prisić, S. et al., The influence of a residual group in the low-molecular-weight allergoids of *Artemisia vulgaris* pollen on their allergenicity, IgE- and IgG-binding properties, *Allergy* 57, 1013–1020, 2002.
257. Kraus, L.M., Elberger, A.J., Handorf, C.R. et al., Urea-derived cyanate forms ε-aminocarbamyl-lysine (homocitrulline) in leukocyte proteins in patients with end-stage renal disease on peritoneal dialysis, *J. Lab. Clin. Med.* 123, 882–891, 1994.
258. Metwalli, A.A., Lammers, W.L., and Boekel, M.A., Formation of homocitrulline during heating of milk, *J. Dairy Res.* 65, 579–589, 1998.
259. Wang, E., Nichols, S.J., Rodriguez, E.R. et al., Protein carbamylation links inflammation, smoking, uremia and atherogenesis, *Nat. Med.* 13, 1176–1184, 2007.
260. Sirpal, S., Myeloperoxidase-mediated lipoprotein carbamylation as a mechanistic pathway for atherosclerotic vascular disease, *Clin. Sci.* 116, 681–695, 2009.

261. Plapp, B.V., Moore, S., and Stein, W.H., Activity of bovine pancreatic deoxyribonuclease A with modified amino groups, *J. Biol. Chem.* 246, 939–945, 1971.
262. Ludwig, M.L. and Hunter, M.J., Amidination, *Methods Enzymol.* 11, 595–604, 1967.
263. Kimmel, J.R., Guanidation of proteins, *Methods Enzymol.* 11, 584–589, 1967.
264. Beardsley, R.L. and Reilly, J.P., Quantitation using enhanced signal tags: A technique for comparative proteomics, *J. Proteome Res.* 2, 15–21, 2003.
265. Charulatha, V. and Rajaram, A., Dimethyl 3,3′-dithiobispropionimidate: A novel cross-linking reagent for collagen, *J. Biomed. Mater. Res.* 54, 122–128, 2001.
266. Ogawa, H., Gomi, T., Takusagawa, F. et al., Evidence for a dimeric structure of rat liver serine dehydratase, *Int. J. Biochem. Cell. Biol.* 34, 533–543, 2002.
267. Kim, M.V., Seit-Nebi, A.S., Marston, S.B., and Gusev, N.B., Some properties of human small heat shock protein Hsp22 (H11 or HspB8), *Biochem. Biophys. Res. Commun.* 315, 796–801, 2004.
268. Azem, A., Tsfadia, Y., Hajouj, O. et al., Cross-linking with bifunctional reagents and its applications to the study of the molecular symmetry and the arrangement of subunits in hexameric protein oligomers, *Biochim. Biophys. Acta* 1804, 768–780, 2010.
269. Ouimet, P.M. and Kapoor, M., Nucleotide binding and hydrolysis properties of *Neurospora crassa* cytosolic molecular chaperones, Hsp70 and Hsp80, heat-inducible members of the eukaryotic stress-70 and stress-90 families, *Biochem. Cell. Biol.* 77, 89–99, 1999.
270. Tzima, E., Trotter, P.J., Orchard, M.A. et al., Annexin V relocated to the platelet cytoskeleton upon activation and binds to a specific isoform of actin, *Eur. J. Biochem.* 267, 4720–4730, 2000.
271. Dodson, M.S., Dimethyl suberimidate cross-linking of oligo(dT) to DNA-binding proteins, *Bioconjug. Chem.* 11, 876–879, 2000.
272. Hunter, M.J. and Ludwig, M.L., The reaction of imidoesters with proteins and related small molecules, *J. Am. Chem. Soc.* 84, 3491–3504, 1962.
273. Ludwig, M.L., Reversible blocking of protein amino groups by the acetimidyl group, *J. Am. Chem. Soc.* 84, 4160–4162, 1962.
274. Dimarchi, R.D., Garner, W.H., Wang, C.-C. et al., Characterization of the reaction of methylacetamidate with sperm whale myoglobin, *Biochemistry* 17, 2822–2829. 1978.
275. Browne, D.T. and Kent, S.B.H., Formation of non-amidine products in the reaction of primary amines with imido esters, *Biochem. Biophys. Res. Commun.* 67, 126–132, 1975.
276. Plapp, B.V., Imidoesters and the mechanism of liver alcohol dehydrogenase, in *Proteins: Structure and Function*, J.J. L'Italien (ed.), Plenum Press, New York, pp. 615–622, 1987.
277. Jastrzebska, M., Zelewska-Rejdaj, J., Wrzalik, R. et al., Dimethyl suberimidate cross-linked pericardium tissue: Raman spectroscopic and atomic force microscopy investigation, *J. Mol. Struct.* 744–747, 789–795, 2005.
278. Sweadner, K.J., Crosslinking and purification of Na, K-ATPase by ethyl acetimidate, *Biochem. Biophys. Res. Commun.* 78, 962–969, 1978.
279. Chao, T.L., Berenfeld, M.R., Gelbart, T., and Gabuzda, T.G., The effects of oxygen affinity and gelation of hemoglobin S crosslinked by reaction with methyl acetimidate, *Hemoglobin* 5, 47–72, 1981.
280. Richardson, V.B., Littlefield, L.G., and Colyer, S.P., Cytogenetic evaluations in human lymphocytes exposed to methyl acetimidate, a lysine-specific protein crosslinking agent, *Mutat. Res.* 180, 121–129, 1987.
281. Olomucki, M. and Diopoh, J., New protein reagents I. Ethyl chloroacetimidate, its properties and its reaction with ribonuclease, *Biochim. Biophys. Acta* 263, 213–219, 1972.
282. Thumm, M., Hoenes, J., and Pfeiderer, G., *S*-methylthioacetimidate is a new reagent for the amidination of proteins at low pH, *Biochim. Biophys. Acta* 923, 263–267, 1987.

283. Zoltobrocki, M., Kim, J.C., and Plapp, B.V., Activity of liver alcohol dehydrogenase with various substituents on the amino groups, *Biochemistry* 13, 899–903, 1974.
284. Beardsley, R.L. and Reilly, J.P., Fragmentation of amidinated peptide ions, *J. Am. Soc. Mass Spectrom.* 15, 158–167, 2004.
285. Janecki, D.J., Beardsley, R.L., and Reilly, J.P., Probing protein tertiary structure with amidination, *Anal. Chem.* 77, 7274–7281, 2005.
286. Lauber, M.A., Running, W.E., and Reilly, J.P., *B. subtilis* ribosomal proteins: Structural homology and post-translational modifications, *J. Proteome Res.* 8, 4193–4206, 2009.
287. Liu, X. and Reilly, J.P., Correlating the chemical modification of *Escherichia coli* ribosomal proteins with crystal structure data, *J. Proteome Res.* 8, 4466–4478, 2009.
288. Roseman, M., Litman, B.J., and Thompson, T.E., Transbilayer exchange of phosphatidylethanolamine for phosphatidyl choline and N-acetimidoylphosphatidylethanolamine in single-walled bilayer vesicles, *Biochemistry* 14, 4826–4830, 1975.
289. Lyles, D.S., McKinnon, K.P., and Parce, J.W., Labeling of the cytoplasmic domain of the influenza virus hemagglutinin with fluorescein reveals sites of interaction with membrane lipid bilayers, *Biochemistry* 24, 8121–8128, 1985.
290. Davis, T.N. and Cronan, J.E., Jr., An alkyl imidate labeling study of the organization of phospholipids and proteins in the lipid-containing bacteriophage PR4, *J. Biol. Chem.* 260, 663–671, 1985.
291. Ohara, O. and Takahashi, S., Chromatographic analysis of suberimidate-crosslinked lysine, *J. Liquid Chromatogr.* 7, 1665–1672, 1984.
292. Dutton, A., Adams, M., and Singer, S.J., Bifunctional imidoesters as cross-linking reagents, *Biochem. Biophys. Res. Commun.* 23, 730–739, 1966.
293. Kim, J.S., Kim, J., and Kim, H.J., Matrix-assisted laser desorption/ionization signal enhancement of peptides by picolinimidation of amino groups, *Rapid Commun. Mass Spectrom.* 22, 495–502, 2008.
294. Sekiguchi, T., Oshiro, S., Goingo, E.M., and Nosoh, Y., Chemical modification of ε-amino groups in glutamine synthetase from Bacillus stearothermophilus with ethyl acetamidate, *J. Biochem.* 85, 75–78, 1979.
295. Cupo, P., El-Deiry, W., Whitney, P.L., and Awad, W.M., Jr., Stabilization of proteins by guanidation, *J. Biol. Chem.* 255, 10828–10833, 1980.
296. Plapp, B.V., Brooks, R.L., and Shore, J.D., Horse liver alcohol dehydrogenase. Amino groups and rate limiting steps in catalysis, *J. Biol. Chem.* 248, 3470–3475, 1973.
297. Sekiguchi, T., Oshiro, S., Goingo, E.M., and Nosoh, Y., Chemical modification of ε-amino groups in glutamine synthetase from *Bacillus stearothermophilus* with ethyl acetamidate, *J. Biochem.* 85, 75–78, 1979.
298. Plapp, B.V., Enhancement of the activity of horse liver alcohol dehydrogenase by modification of amino groups at the active sites, *J. Biol. Chem.* 245, 1727–1735, 1970.
299. Plapp, B.V. and Kim, J.C., Determination of ε-acetimidyllysine in proteins, *Anal. Biochem.* 62, 291–294, 1974.
300. Moore, S. and Stein, W.H., Chromatographic determination of amino acids by use of automatic recording equipment, *Methods Enzymol.* 6, 819–831, 1963.
301. Sund, H. and Akeson, A., Amino acid composition of glutamate dehydrogenase, *Biochem. Zeit.* 340, 421–435, 1964.
302. Yokote, Y., Arai, K.M., and Akahane, K., Recovery of tryptophan from 25-minute acid hydrolysates of protein, *Anal. Biochem.* 152, 245–249, 1986.
303. Hauque, M.R. and Bradbury, J.H., Total cyanide determination of plants and foods using the picrate and acid hydrolysis methods, *Food Chem.* 77, 207–114, 2002.
304. Einbu, A. and Vårum, K.M., Characterization of chitin and its hydrolysis to GlcNAc and GlcN, *Biomacromoles* 9, 1870–1875, 2008.

305. Ampon, K. and Means, G.E., Immobilization of proteins on organic polymer beads, *Biotechnol. Bioeng.* 32, 689–697, 1988.
306. Basri, M., Ampon, K., Yunus, W.M.Z. et al., Amidination of lipase with hydrophobic imidoesters, *J. Am. Oil Chem. Soc.* 69, 579–583, 1992.
307. Ampon, K., Basri, M., Salleh, A.H. et al., Immobilization by adsorption of hydrophobic lipase derivatives to porous polymer beads for use in ester synthesis, *Biocatalysis* 10, 31–351, 1994.
308. Perham, R.N. and Thomas, J.O., Reaction of tobacco mosaic virus with a thiol-containing imidoester and a possible application to X-ray diffraction analysis, *J. Mol. Biol.* 62, 415–418, 1971.
309. Perham, R.N. and Richards, F.M., Reactivity and structural role of protein amino groups in tobacco mosaic virus, *J. Mol. Biol.* 33, 795–807, 1968.
310. Jue, R., Lambert, J.M., Pierce, L.R., and Traut, R.R., Addition of sulfhydryl groups to *Escherichia coli* ribosomes by protein modification with 2-iminothiolane (methyl 4-mercaptobutyrimidate), *Biochemistry* 17, 5399–5406, 1978.
311. Lambert, J.M., Jue, R., and Traut, R.R., Disulfide cross-linking of *Escherichia coli* ribosomal proteins with 2-iminothiolane (methyl 4-mercaptobutyrimidate): Evidence that the cross-linked protein pairs are formed in the intact ribosomal subunit, *Biochemistry* 17, 5406–5416, 1978.
312. Tolan, D.R. and Traut, R.R., Protein topography of the 40 S ribosomal subunit from rabbit reticulocytes shown by cross-linking with 2-iminothiolane, *J. Biol. Chem.* 256, 10129–10136, 1981.
313. Nag, B., Johnson, A.E., and Traut, R.R., Identification of the elongation factor Tu binding site on 70 S *E. coli* ribosomes by chemical crosslinking, *Indian J. Biochem. Biophys.* 32, 343–350, 1995.
314. Miyata, K., Kakizawa, Y., Nishiyama, N. et al., Block catiomer polyplexes with regulated densities of charge and disulfide cross-linking directed to enhance gene expression, *J. Am. Chem. Soc.* 126, 2355–2361, 2004.
315. Grintsevich, E.E., Galkin, V.E., Orlova, A. et al., Mapping of drebrin binding site of F-actin, *J. Mol. Biol.* 398, 542–554, 2010.
316. Ji, T., Muenker, M.C., Papineni, R.V. et al., Increased sensitivity in antigen detection with fluorescent latex nanosphere-IgG antibody conjugates, *Bioconjug. Chem.* 21, 427–435, 2010.
317. Gruneich, J.A. and Diamond, S.L., Synthesis and structure-activity relationships of a series of increasingly hydrophobic cationic steroid lipofection reagents, *J. Gene Med.* 9, 381–391, 2007.
318. Ma, K., Hu, M., and Qi, Y., Structure-transfection activity relationships with glucocorticoid-polyethyl-enime conjugate nuclear gene delivery system, *Biomaterials* 30, 3780–3789, 2009.
319. Rauniyar, N. and Prokai, L., Detection and identification of 4-hydroxy-2-nonenal Schiff-base adducts along with products of Michael addition using data-dependent neutral loss-driven MS3 acquisition: Method evaluation through an in vitro study on cytochrome c oxidase modifications, *Proteomics* 9, 5188–5193, 2009.
320. Dunathan, H.C., Stereochemical aspects of pyridoxal phosphate catalysis, *Adv. Enzymol.* 35, 79, 1971.
321. Metzler, D.E. and Snell, E.P., Spectra and ionization constants of the vitamin B_6 group and related 3-hydroxypyridine derivatives, *J. Am. Chem. Soc.* 77, 2431–2437, 1955.
322. Metzler, D.E., Equilibria between pyridoxal and amino acids and their imines, *J. Am. Chem. Soc.* 79, 485–490, 1957.
323. Taguchi, T., Sugiura, M., Hamada, Y., and Miwa, I., In vivo formation of a Schiff base of aminoguandine with pyridoxal phosphate, *Biochem. Pharmacol.* 55, 1667–1671, 1998.

324. Shapiro, S., Enser, M., Pugh, E., and Horecker, B.L., The effect of pyridoxal phosphate on rabbit muscle aldolase, *Arch. Biochem. Biophys.* 128, 554, 1968.
325. Schnackerz, K.D. and Noltmann, E.A., Pyridoxal-5′-phosphate as a site-specific protein reagent for a catalytically critical lysine residue in rabbit muscle phosphoglucose isomerase, *Biochemistry* 10, 4837, 1971.
326. Havran, R.T. and du Vigneaud, V., The structure of acetone-lysine vasopressin as established through its synthesis from the acetone derivative of S-benzyl-l-cysteinyl-l-tyrosine, *J. Am. Chem. Soc.* 91, 2696, 1969.
327. Paech, C. and Tolbert, N.E., Active site studies of ribulose-1,5-bisphosphate carboxylase/oxygenase with pyridoxal-5′-phosphate, *J. Biol. Chem.* 253, 7864–7873, 1978.
328. Kittler, J.M., Meisler, N.T., Viceps-Madorf, D. et al., A general immunochemical method for detecting proteins on blots, *Anal. Biochem.* 137, 210–216, 1984.
329. McConchie, S.M., Coakley, J., Edwards, R.H., and Benyon, R.J., Molecular heterogeneity in McArdle's disease, *Biochim. Biophys. Acta* 1096, 26–32, 1991.
330. Cookson, E.J., Flannery, A.V., Cidlowski, J.A., and Beynon, R.J., Immunological detection of degradation intermediates of skeletal-muscle glycogen phosphorylase *in vitro* and *in vivo*, *Biochem. J.* 288, 291–296, 1992.
331. Kent, A.B., Krebs, E.G., and Fischer, E.H., Properties of crystalline phosphorylase b, *J. Biol. Chem.* 232, 549, 1958.
332. Wimmer, M.J., Mo, T., Sawyers, D.L., and Harrison, J.H., Biphasic inactivation of porcine heart mitochondrial malate dehydrogenase by pyridoxal-5′-phosphate, *J. Biol. Chem.* 250, 710, 1975.
333. Bleile, D.M., Jameson, J.L., and Harrison, J.H., Inactivation of porcine heart cytoplasmic malate dehydrogenase by pyridoxal-5′-phosphate, *J. Biol. Chem.* 251, 6304–6307, 1976.
334. Cookson, E.J., Flannery, A.V., Cidlowski, J.A., and Beynon, R.J., Immunological detection of degradation intermediates of skeletal-muscle glycogen phosphorylase in vitro and in vivo, *Biochem. J.* 288, 291–296, 1992.
335. Cake, M.A., DiSorbo, D.M., and Litwack, G., Effect of pyridoxal phosphate on the DNA binding site of activated hepatic glucocorticoid receptor, *J. Biol. Chem.* 253, 4886, 1978.
336. Slebe, J.C. and Martinez-Carrion, M., Selective chemical modification and 19F NMR in the assignment of a pK value to the active site lysyl residue in aspartate transaminase, *J. Biol. Chem.* 253, 2093, 1978.
337. Nishigori, H. and Toft, D., Chemical modification of the avian progesterone receptor by pyridoxal-5′-phosphate, *J. Biol. Chem.* 254, 9155, 1979.
338. Sugiyama, Y. and Mukohata, Y., Modification of one lysine by pyridoxal phosphate completely inactivates chloroplast coupling factor 1 ATPase, *FEBS Lett.* 98, 276, 1979.
339. Peters, H., Risi, S., and Dose, K., Evidence for essential primary amino groups in a bacterial coupling factor F1 ATPase, *Biochem. Biophys. Res. Commun.* 97, 1215, 1980.
340. Gould, K.G. and Engel, P.C., Modification of mouse testicular lactate dehydrogenase by pyridoxal 5′-phosphate, *Biochem. J.* 191, 365, 1980.
341. Jones, C.W., III. and Priest, D.G., Interaction of pyridoxal-5′-phosphate with apo-serine hydroxymethyltransferase, *Biochim. Biophys. Acta* 526, 369, 1978.
342. Cortijo, M., Jimenez, J.S., and Lior, J., Criteria to recognize the structure and micropolarity of pyridoxal-5′-phosphate binding sites in protein, *Biochem. J.* 171, 497–500, 1978.
343. Fonda, M.L., Trauss, C., and Guempel, U.M., The binding of pyridoxal 5′-phosphate to human serum albumin, *Arch. Biochem. Biophys.* 288, 79–86, 1991.
344. Dempsey, W.B. and Christensen, H.N., The specific binding of pyridoxal 5′-phosphate to bovine plasma albumin, *J. Biol. Chem.* 237, 1113–1120, 1962.

345. Haddad, L.C., Thayer, W.S., and Jenkins, W.T., Solvent interactions with the active site of the pig heart cytosolic aspartate aminotransferase, *Arch. Biochem. Biophys.* 181, 66–72, 1977.
346. Sober, H.A., *Handbook of Biochemistry*, 2nd edn., The Chemical Rubber Company, Cleveland, OH, 1970.
347. Moldoon, T.G. and Cidlowski, J.A., Specific modification of rat uterine estrogen receptor by pyridoxal-5′-phosphate, *J. Biol. Chem.* 2, 55, 3100, 1980.
348. Ohsawa, H. and Gualerzi, C., Structure-function relationship in *Escherichia coli* inhibition factors. Identification of a lysine residue in the ribosomal binding site of initiation factor by site-specific chemical modification with pyridoxal phosphate, *J. Biol. Chem.* 256, 4905, 1981.
349. Bürger, E. and Görisch, H., Evidence for an essential lysine at the active site of l-histidinol: NAD+ oxidoreductase, a bifunctional dehydrogenase, *Eur. J. Biochem.* 118, 125, 1981.
350. Valinger, Z., Engel, P.C., and Metzler, D.E., Is pyridoxal 5′-phosphate an affinity label for phosphate-binding sites in proteins? The case of bovine glutamate dehydrogenase, *Biochem. J.* 294, 835–839, 1993.
351. Jaffe, M. and Bubis, J., Affinity labelling of the guanine nucleotide binding site of transducing by pyridoxal 5′-phosphate, *J. Protein Chem.* 21, 339–348, 1992.
352. Riva, F., Giarlosio, A., Voltatorri, C. et al., Affinity labeling reagent for pyridoxal phosphate dependent enzymes, *Biochem. Biophys. Res. Commun.* 66, 863–869, 1975.
353. Fukui, T. and Tanizawa, K., Synthesis and application of pyridoxal polyphosphoryl derivatives as active-site probes for nucleotide-binding enzymes, *Methods Enzymol.* 280, 41–50, 1997.
354. Griffith, M.J., Covalent modification of human α-thrombin with pyridoxal 5′-phosphate. Effect of phosphopyridoxylation on the interaction of thrombin of heparin, *J. Biol. Chem.* 254, 3401–3406, 1979.
355. Church, F.C., Pratt, C.W., Noyes, C.M. et al., Structural and functional properties of human α-thrombin and γ-thrombin. Identification of lysyl residues in α-thrombin that are critical for heparin and fibrin(ogen), *J. Biol. Chem.* 264, 18419–18425, 1989.
356. Kluger, R. and Tsui, W.-C., Methyl acetyl phosphate. A small anionic acetylating agent, *J. Org. Chem.* 45, 2723, 1980.
357. Ueno, H., Pospischil, M.A., Manning, J.M., and Kluger, R., Site-specific modification of hemoglobin by methyl acetyl phosphate, *Arch. Biochem. Biophys.* 244, 795, 1986.
358. Ueno, H., Pospischil, M.A., and Manning, J.M., Methyl acetyl phosphate as a covalent probe for anion-binding sites in human and bovine hemoglobins, *J. Biol. Chem.* 264, 12344, 1989.
359. Kataoka, K., Tanizawa, K., Fukui, T. et al., Identification of active site lysyl residues of phenylalanine dehydrogenase by chemical modification with methyl acetyl phosphate combined with site-directed mutagenesis, *J. Biochem.* 116, 1370–1376, 1994.
360. Raibekas, A.A., Bures, E.J., Siska, C.C. et al., Anion binding and controlled aggregation of human interleukin-1 receptor antagonist, *Biochemistry* 44, 9871–9879, 2005.
361. Xu, A.S.L., Labotka, R.J., and London, R.E., Acetylation of human hemoglobin by methyl acetylphosphate. Evidence of broad regio-selectivity revealed by NMR studies, *J. Biol. Chem.* 274, 26629–26632, 1999.
362. Means, G.E., Reductive alkylation of amino groups, *Methods Enzymol.* 47, 469, 1977.
363. Jentoft, J.E., Gerken, T.A., Jentoft, N., and Dearborn, D.G., ^{13}C-Methylated ribonuclease A. Carbon-13 NMR studies of the interaction of lysine 41 with active site ligands, *J. Biol. Chem.* 256, 231–236, 1981.
364. Zhang, M., Thulin, E., and Vogel, H.J., Reductive methylation and p*K*a determination of the lysine side chains in calbindin D9k, *J. Protein Chem.* 13, 527–535, 1994.

365. Abraham, S.J., Kobayashi, T., Solaro, J.R., and Gaponenko, V., Differences in lysine p*K*a values may be used to improve NMR signal dispersion in reductively methylated proteins, *J. Biomol. NMR* 43, 249–246, 2009.
366. Kim, Y., Quartey, P., Li, H. et al., Large-scale evaluation of protein reductive methylation for improving protein crystallization, *Nat. Methods* 5, 853–854, 2008.
367. Fan, Y. and Joachimiak, A., Enhanced crystal packing due to solvent reorganization through reductive methylation of lysine residues in oxidoreductases from *Streptococcus pneumonae*, *J. Struct. Funct. Genomics* 11, 101–111, 2010.
368. Sledz, P., Zheng, H., Murzyn, K. et al., New surface contacts formed upon reductive methylation: Improving the probability of protein crystallization, *Protein Sci.* 19, 1395–1404, 2010.
369. Dick, L.R., Geraldes, C.F.G.C., Sherry, A.D., Gray, C.W., and Gray, D.M., 13C NMR of methylated lysines of fd gene 5 protein: Evidence for a conformational change involving lysine 24 upon binding of a negatively charged lanthanide chelate, *Biochemistry* 28, 7896, 1989.
370. Brown, E.M., Pfeffer, P.E., Kumosinski, T.F., and Greenberg, R., Accessibility and mobility of lysine residues in β-lactoglobulin, *Biochemistry* 27, 5601, 1988.
371. Rice, R.H., Means, G.E., and Brown, W.D., Stabilization of bovine trypsin by reductive methylation, *Biochim. Biophys. Acta* 492, 316–321, 1977.
372. Poncz, L. and Dearborn, D.G., The resistance of tryptic hydrolysis to peptide-bonds adjacent to N^{ε}, *N*-dimethyllysine residues, *J. Biol. Chem.* 258, 1844–1850, 1983.
373. Pham, V.T., Ewing, E., Kaplan, H. et al., Glycation improves the thermostability of trypsin and chymotrypsin, *Biotechnol. Bioeng.* 101, 452–459, 2008.
374. Pham, V.T., Alosaar, I., Dyuhig, M.N., and Kaplan, H., Glass-immobilized glycated trypsin: A novel modified trypsin that is remarkably thermostable, *J. Mol. Catal. B-Enzym.* 58, 48–53, 2009.
375. Dottavio-Martin, D. and Ravel, J.M., Radiolabeling of proteins by reductive alkylation with [14C] formaldehyde and sodium cyanoborohydride, *Anal. Biochem.* 87, 562, 1978.
376. Jentoft, N. and Dearborn, D.G., Labeling of proteins by reductive methylation using sodium cyanoborohydride, *J. Biol. Chem.* 254, 4359–4365, 1979.
377. Fretheim, K., Iwai, S., and Feeney, R.F., Extensive modification of protein amino groups by reductive addition of different sized substituents, *Int. J. Pept. Protein Res.* 14, 451, 1979.
378. Fretheim, K., Edelandsdal, B., and Harbitz, O., Effect of alkylation with different size substituents on the conformation of ovomucoid, lysozyme and ovotransferrin, *Int. J. Pept. Protein Res.* 25, 601, 1985.
379. Fujita, Y. and Noda, Y., Effect of alkylation with different sized substituents on thermal stability of lysozyme, *Int. J. Pept. Protein Res.* 40, 103–109, 1992.
380. Fujita, Y., Hidaka, Y., and Noda, Y., Thermal stability of alkylated and hydroxyalkylated lysozymes, *Thermochim. Acta* 253, 177–125, 1995.
381. Wu, H.-L. and Means, G.E., Immobilization of proteins by reductive alkylation with hydrophobic aldehydes, *Biotechnol. Bioeng.* 23, 855, 1981.
382. Royer, G.P., Liberatore, F.A., and Green, G.M., Immobilization of enzymes on aldehydic matrices by reductive alkylation, *Biochem. Biophys. Res. Commun.* 64, 478–484, 1975.
383. Gemeiner, P. and Breier, A., Aldehydic derivatives of bead cellulose—Relationships between matrix structure and function in immobilization of enzymes catalyzing hydrolysis of high molecular substrates, *Biotechnol. Bioeng.* 24, 2573–2582, 1982.
384. Lenders, J.P., Germain, P., and Crichton, R.R., Immobilization of a soluble chemically thermostabilized enzyme, *Biotechnol. Bioeng.* 27, 572–578, 1985.
385. Burteau, N., Burton, S., and Crichton, R.R., Stabilization and immobilization of penicillin amidase, *FEBS Lett.* 258, 185–189, 1989.

386. Geoghegan, K.F., Ybarra, D.M., and Feeney, R.E., Reversible reductive alkylation of amino groups in proteins, *Biochemistry* 18, 5392–5399, 1979.
387. Acharya, A.S. and Manjula, B.N., Dihydroxypropylation of amino groups of proteins: Use of glyceraldehyde as a reversible agent for reductive alkylation, *Biochemistry* 26, 3524–3530, 1987.
388. Jentoft, J.E., Jentoft, N., Gerken, T.A., and Dearborn, D.G., 13C NMR studies of ribonuclease A methylated with [13C] formaldehyde, *J. Biol. Chem.* 254, 4366, 1979.
389. Jentoft, N. and Dearborn, D.G., Protein labeling by reductive methylation with sodium cyanoborohydride effect of cyanide and metal ions on the reaction, *Anal. Biochem.* 106, 186, 1980.
390. Geoghegan, K.F. et al., Alternative reducing agents for reductive methylation of amino groups in proteins, *Int. J. Pept. Protein Res.* 17, 345, 1981.
391. Cabacungan, J.C., Ahmed, A.J., and Feeney, R.E., Amino boranes as alternative reducing agents for reductive alkylation of proteins, *Anal. Biochem.* 124, 272, 1982.
392. Garlick, R.L., Mazer, J.S., Chylack, L.T., Jr. et al., Nonenzymatic glycation of human lens crystallin. Effect of aging and diabetes mellitus, *J. Clin. Invest.* 74, 1742–1749, 1984.
393. Watkins, N.G., Neglia-Fisher, C.I., Dyer, D.G. et al., Effect of phosphate on the kinetics and specificity of glycation of proteins, *J. Biol. Chem.* 262, 7207–7212, 1987.
394. Wilson, G., Effect of reductive lactosamination on the hepatic uptake of bovine pancreatic ribonuclease A dimer, *J. Biol. Chem.* 253, 2070, 1978.
395. Bunn, H.F. and Higgins, P.J., Reaction of monosaccharides with proteins: Possible evolutionary significance, *Science* 213, 222, 1981.
396. True, M.W., Circulating biomarkers of glycemia in diabetes management and implications for personalized medicine, *J. Diabetes Sci. Technol.* 3, 743–747, 2009.
397. Nowak-Wegrzyn, A. and Fiocchi, A., Rare, medium, or well done? The effect of heating and food matrix on food protein allergenicity, *Curr. Opin. Allergy Clin. Immunol.* 9, 234–237, 2009.
398. Zhang, Q., Ames, J.M., Smith, R.D. et al., A perspective on the Maillard reaction and the analysis of food glycation by mass spectrometry: Probing the pathogenesis of chronic disease, *J. Proteome Res.* 8, 754–769, 2009.
399. Maillard-Lefebvre, H., Boulanger, E., Daroux, M. et al., Soluble receptor for advanced glycation end products: A new biomarker in diagnosis and prognosis of chronic inflammatory diseases, *Rheumatology* 48, 1190–1196, 2009.
400. de Noni, I. and Pagani, M.A., Cooking properties and heat damage of dried pasta as influenced by raw material characteristics and processing conditions, *Crit. Rev. Food Sci. Nutr.* 50, 465–472, 2010.
401. Sato, T., Iwaki, M., Shimogaito, N. et al., TAGE (toxic AGEs) theory in diabetic complications, *Curr. Mol. Med.* 6, 351–358, 2006.
402. Uchida, K., Khor, O.T., Oya, T. et al., Protein modification by a Maillard reaction intermediate methylglyoxal. Immunochemical detection of fluorescent 5-methylimidazolone derivatives in vivo, *FEBS Lett.* 410, 313–318, 1997.
403. Syrový, I., Glycation of albumin: Reaction with glucose, fructose, galactose, ribose or glyceraldehyde measured using four methods, *J. Biochem. Biophys. Methods* 28, 115–121, 1994.
404. Morgan, P.E., Dean, R.T., and Davies, M.J., Inactivation of cellular enzymes by carbonyls and protein-bound glycation/glycoxidation products, *Arch. Biochem. Biophys.* 403, 259–269, 2002.
405. Seidler, N.W. and Yeargans, G.S., Effects of thermal denaturation on protein glycation, *Life Sci.* 70, 1789–1799, 2002.
406. Acharya, A.S. and Manning, J.M., Reactivity of the amino groups of carbonmonoxyhemoglobin S with glyceraldehyde, *J. Biol. Chem.* 255, 1406, 1980.

407. Acharya, A.S. and Manning, J.M., Amadori rearrangement of glyceraldehyde-hemoglobin Schiff base adducts. A new procedure for the determination of ketoamine adducts in proteins, *J. Biol. Chem.* 255, 7218, 1980.
408. Mori, N., Bai, Y., Ueno, H., and Manning, J.M., Sequence-dependent reactivity of model peptides with glyceraldehyde, *Carbohydr. Res.* 189, 49–63, 1989.
409. Bai, Y., Ueno, H., and Manning, J.M., Some factors that influence the nonenzymatic glycation of peptides and polypeptides by glyceraldehyde, *J. Protein Chem.* 8, 299–315, 1989.
410. Sanger, F. and Tuppy, H., The amino acid sequence in the phenylalanyl chain of insulin. I. The identification of lower peptides from partial hydrolysates, *Biochem. J.* 49, 463–481, 1951.
411. Okuyama, T. and Satake, K., The preparation and properties of 2,4,6-trinitrophenyl-amino acids and peptides, *J. Biochem.* 47, 454–460, 1947.
412. Sataki, K., Okuyama, T., Ohashi, M., and Shinoda, T., The spectrophotometric determination of amines, amino acids, and peptides with 2,4,6-trinitrobenzene-1-sulphonic acid, *J. Biochem.* 47, 654–660, 1960.
413. Goldfarb, A.R., A kinetic study of the reactions of amino acids and peptides with trinitrobenzenesulfonic acid, *Biochemistry* 5, 2570–2574, 1966.
414. Goldfarb, A.R., Heterogeneity of amino groups in proteins. I. Human serum albumin, *Biochemistry* 5, 2574–2578, 1966.
415. Saifer, A. and Palo, J., Amino acid composition of monomeric and polymeric human serum albumin, *Anal. Biochem.* 27, 1–14, 1969.
416. Girard, M., Bietlot, H.P., Mousseau, N. et al., Use of capillary electrophoresis for the characterization of human serum albumin heterogeneity, *Biomed. Chromotogr.* 12, 183–184, 1998.
417. Habeeb, A.F.S.A., Determination of free amino groups in proteins by trinitrobenzenesulfonic acid, *Anal. Biochem.* 14, 328–336, 1966.
418. Markus, G., Love, R.L., and Wissler, F.C., Mechanism of protection by anionic detergents against denaturation of serum albumin, *J. Biol. Chem.* 239, 3687–3693, 1964.
419. Aoki, K. and Hiramatsu, K., The interaction of bovine plasma albumin with cationic detergents at pH 9, *Anal. Biochem.* 60, 213–225, 1974.
420. Moriyama, Y., Watanabe, E., Kobayashi, K. et al., Secondary structural change of bovine serum albumin in thermal denaturation up 130ºC and protective of sodium dodecyl sulfate on the change, *J. Phys. Chem. B* 112, 16585–16589, 2008.
421. Fields, R., The rapid determination of amino groups with TNBS, *Methods Enzymol.* 25, 464, 1972.
422. Kotaki, A. and Satake, K., Acid and alkaline degradation of the TNP-amino acids and peptides, *J. Biochem.* 56, 299–307, 1964.
423. Cayot, P. and Tainturier, G., The quantification of protein amino groups by the trinitrobenzenesulfonic acid method: A reexamination, *Anal. Biochem.* 249, 184–200, 1997.
424. van Staden, J.F. and McCormack, T., Sequential-injection spectrophotometric determination of amino acids using 2,4,6-trinitrobenzenesulphonic acid, *Anal. Chim. Acta* 369, 163–170, 1998.
425. Coffee, C.J., Bradshaw, R.A., Goldin, B.R., and Frieden, C., Identification of the sites of modification of bovine liver glutamate dehydrogenase reacted with trinitrobenzenesulfonate, *Biochemistry* 10, 3516–3526, 1971.
426. Bates, D.J., Goldin, B.R., and Frieden, C., A new reaction of glutamate dehydrogenase: The enzyme-catalyzed formation of trinitrobenzene from TNBS in the presence of reduced coenzyme, *Biochem. Biophys. Res. Commun.* 39, 502–507, 1970.
427. Brown, A. and Fisher, H.F., A comparison of the glutamate dehydrogenase catalyzed oxidation of NADPH by trinitrobenzenesulfonate with the uncatalyzed reaction, *J. Am. Chem. Soc.* 98, 5682–5688, 1976.

428. Means, G.E., Congdon, W.I., and Bender, M.L., Reactions of 2,4,6-trinitrobenzenesulfonate ion with amines and hydroxide ion, *Biochemistry* 11, 3564, 1972.
429. Whitaker, J.R., Granum, P.E., and Aasen, G., Reaction of ammonia with trinitrobenzene sulfonic acid, *Anal. Biochem.* 108, 72, 1980.
430. Flügge, U.I. and Heldt, H.W., Specific labelling of the active site of the phosphate translocator in spinach chloroplasts by 2,4,6-trinitrobenzene sulfonate, *Biochem. Biophys. Res. Commun.* 84, 37, 1978.
431. Flügge, U.I. and Heldt, H.W., Specific labelling of a protein involved in phosphate transport of chloroplasts by pyridoxal-5′-phosphate, *FEBS Lett.* 82, 29, 1977.
432. Parrott, C.L. and Shifrin, S., A spectrophotometric study of the reaction of borohydride with trinitrophenyl derivatives of amino acids and proteins, *Biochim. Biophys. Acta* 491, 114–120, 1977.
433. Salem, N., Jr., Lauter, C.J., and Trams, E.G., Selective chemical modification of plasma membrane ectoenzymes, *Biochim. Biophys. Acta* 641, 366, 1981.
434. Hartman, F.C., Milanez, S., and Lee, K.H., Ionization constants of two active-site lysyl ε-amino groups of ribulosebiphosphate carboxylase/oxygenase, *J. Biol. Chem.* 260, 13968–13973, 1985.
435. Hartman, F.C., Milanez, S., and Lee, E.H., Use of trinitrobenzene sulfonate to determine the p*K*a values of two active-site lysines of ribulosebisophosphate carboxylase/oxygenase, in *Proteins: Structure and Function*, J.J. L'Italien (ed.), Plenum Press, New York, 1987.
436. Anderson, G.W., Zimmerman, J.E., and Callahan, F.M., The use of esters of N-hydroxysuccinimide in peptide synthesis, *J. Am. Chem. Soc.* 86, 1839–1842, 1984.
437. Blumberg, S. and Vallee, B.L., Superactivation of thermolysin by acylation with amino acid N-hydroxysuccinimide esters, *Biochemistry* 14, 2410–2419, 1975.
438. Boulton, A.E. and Hunter, W.M., The labelling of proteins to high specific radioactivities by conjugation to a ^{125}I-containing acylating agent, *Biochem. J.* 133, 529, 1973.
439. Song, S.H., Jung, K.H., Paik, J.Y. et al., Distribution and pharmacokinetic analysis of angiostatin radioiodine labeled with high stability, *Nucl. Med. Biol.* 32, 845–850, 2005.
440. Mougin-Degraef, M., Jestin, E., Bruel, D. et al., High-activity radio-iodine labeling of conventional and stealth liposomes, *J. Liposome Res.* 16, 91–102, 2006.
441. Patel, Z.S., Yamamoto, M., Ueda, H. et al., Biodegradable gelatin microparticles as delivery systems for the controlled release of bone morphogenetic protein-2, *Acta Biomater.* 4, 1126–1138, 2008.
442. Boado, R.J., Zhou, Q.H., Lu, J.X. et al., Pharmacokinetics and brain uptake of a genetically engineered bifunctional fusion antibody targeting the mouse transferrin receptor, *Mol. Pharm.* 7, 237–244, 2010.
443. Welp, A., Manz, B., and Peschke, E., Development and validation of a high throughput direct radioimmunoassay for the quantitative determination of serum and plasma melatonin (N-acetyl-5-methoxytryptamine) in mice, *J. Immunol. Methods* 358, 1–8, 2010.
444. Staros, J.V., N-Hydroxysulfosuccinimidyl active esters: Bis(N-hydroxysulfosuccinimide) esters of two dicarboxylic acids are hydrophilic, membrane-impermeant, protein cross-linkers, *Biochemistry* 21, 3950–3955, 1982.
445. Bonen, A., Luiken, J.J., Liu, S. et al., Palmitate transport and fatty acid transporters in red and white muscles, *Am. J. Physiol.* 275, E471–E478, 1998.
446. Coort, S.L., Willems, J., Coumans, W.A. et al., Sulfo-N-succinimidyl esters of long chain fatty acids specifically inhibit fatty acid translocase (FAS/CD36)-mediated cellular fatty acid uptake, *Mol. Cell. Biochem.* 239, 213–219, 2002.
447. Drahota, Z., Vrbacký, M., Nůsková, H. et al., Succinimidyl oleate, established inhibitor of CD36/FAT translocase inhibits complex III of mitochondrial respiratory chain, *Biochem. Biophys. Res. Commun.* 391, 1348–1351, 2010.
448. Yem, A.W., Zurcher-Neely, H.A., Richard, K.A. et al., Biotinylation of reactive amino groups in native recombinant human interleukin-1β, *J. Biol. Chem.* 264, 17691–17697, 1989.

449. Yem, A.W., Epps, D.E., Mathews, W.R. et al., Site-specific chemical modification of interleukin 1β by acrylodan at cysteine 8 and lysine 103, *J. Biol. Chem.* 267, 3122–3128, 1992.
450. Wilchek, M. and Bayer, E., Biotin-containing reagents, *Methods Enzymol.* 184, 123, 1990.
451. Bayer, E. and Wilchek, M., Protein biotinylation, *Methods Enzymol.* 184, 148, 1990.
452. Blazer, L.L. and Boyle, M.D., Use of protein chip mass spectrometry to monitor biotinylation reactions, *Appl. Microbiol. Biotechnol.* 74, 717–722, 2007.
453. Dreger, M., Leung, B.W., Brownlee, G.G., and Deng, T., A quantitative strategy to detect changes in accessibility of protein regions to chemical modification on heterodimerization, *Protein Sci.* 18, 1448–1458, 2009.
454. Smith, G.P., Kinetics of amine modification of proteins, *Bioconjug. Chem.* 17, 501–506, 2006.
455. Fölsch, G., N-hydroxysuccinimide ester of chloromercuriacetic acid, a new reagent for preparing mercury derivatives of amino acids, proteins and aminoacyl transfer ribonucleic acids, *Acta Chem. Scand.* 24, 1115–1117, 1970.
456. Cuatrecasas, P. and Parikh, I., Adsorbents for affinity chromatography. Use of N-hydroxysuccinimide esters of agarose, *Biochemistry* 11, 2291–2299, 1972.
457. Kim, H.S., Kye, Y.S., and Hage, D.S., Development and evaluation of N-hydroxysuccinimide-activated silica for immobilizing human serum albumin in liquid chromatography columns, *J. Chromatogr. A* 1049, 51–61, 2004.
458. Kalkof, S. and Sinz, A., Chances and pitfalls of chemical cross-linking with amino-reactive N-hydroxysuccinimide esters, *Anal. Bioanal. Chem.* 392, 305–312, 2008.
459. Blumberg, S., Holmquist, B., and Vallee, B.L., Reversible inactivation and superactivation by covalent modification of thermolysin, *Biochem. Biophys. Res. Commun.* 51, 987–992, 1974.
460. Holmquist, B., Blumberg, S., and Vallee, B.L., Superactivation of neutral proteases with N-hydroxysuccinimide esters, *Biochemistry* 15, 4675–4680, 1976.
461. Blumberg, S., Amino acid residue modified during superactivation of neutral protease: Tyrosine 110 of thermolysin, *Biochemistry* 18, 2815–2820, 1979.
462. Eder, M., Krivoshein, A.V., Backer, M. et al., ScVEGF-PEG-HBED-CC and ScVEGF-PEG-NOTA conjugates: Comparison of easy-to-label recombinant proteins for ^{68}Ga PET imaging of VEGF receptors in angiogenic vasculature, *Nucl. Med. Biol.* 37, 405–412, 2010.
463. Wang, Y.J., Hao, S.J., Liu, Y.D. et al., PEGylation markedly enhanced the in vivo potency of recombinant human non-glycosylated erythropoietin: A comparison with glycosylated erythropoietin, *J. Control. Release* 145, 306–313, 2010.
464. Kida, S., Eto, Y., Yoshioka, Y. et al., Evaluation of synthetic cell-penetrating peptides, Pro-rich peptide and octaargine derivatives, as adenovirus vector carriers, *Protein Pept. Lett.* 17, 164–167, 2010.
465. Teixeira, S., Yang, L., Dijkstra, P.J. et al., Heparinized hydroxyapatite/collagen three-dimensional scaffolds for tissue engineering, *J. Mater. Sci. Mater. Med.* 21, 2385–2392, 2010.
466. Strehin, I., Nahas, Z., Arora, K. et al., A versatile pH sensitive chondroitin sulfate-PEG tissue adhesive and hydrogel, *Biomaterials* 31, 2788–2797, 2010.
467. Deng, C., Li, F., Hackett, J.M. et al., Collagen and glycopolymer based hydrogel for potential corneal application, *Acta Biomater.* 6, 187–194, 2010.
468. Nelson, G.W., Perry, M., He, S.M. et al., Characterization of covalently bonded proteins on poly(methyl methacrylate) by X-ray photoelectron spectroscopy, *Colloids Surf. B* 78, 61–68, 2010.
469. Xu, Z., Du, W., Zhang, P. et al., Development of a protein biochip to identify 6 monoclonal antibodies against subtypes of recombinant human interferons, *Assay Drug. Dev. Technol.* 8, 212–218, 2010.

470. Kim, J., Cho, J., Seidler, P.M. et al., Investigations of chemical modifications of amino-terminated organic films on silicon substrates and controlled protein immobilization, *Langmuir* 26, 2599–2608, 2010.
471. El Khoury, G., Laurenceau, E., Chevolot, Y. et al., Development of miniaturized immunoassay: Influence of surface chemistry and comparison with enzyme-linked immunosorbent assay and western blot, *Anal. Biochem.* 400, 10–18, 2010.
472. Schumann, T., Grudzinska, J, Kuzman, D. et al., Binding-site mutations in the α1 subunit of the inhibitory glycine receptor convert the inhibitory metal ion Cu2+ into a positive modulatory, *Neuropharmacology* 56, 310–317, 2009.
473. Chiang, J.Y., Jordan, M.S., Horai, R. et al., Cbl enforces an SLP76-dependent signaling pathway for T cell differentiation, *J. Biol. Chem.* 284, 4429–4438, 2009.
474. Jiang, M., Chen, X., Wu, X.-H. et al., Catalytic mechanism of SHCHC synthase in the menaquinone biosynthesis of *Escherichia coli*: Identification and mutational analysis of the active site residues, *Biochemistry* 48, 6921–6931, 2009.
475. Bello, Z. and Grubmeyer, C., Roles for cationic residues at the quinolinic acid binding site of quinolinate phosphoribosyltransferase, *Biochemistry* 49, 1388–1395, 2010.
476. Suzuki, N., Hozumi, K., Urushibata, S. et al., Identification of α-dystroglycan binding sequences in the laminen α2 chain of LG4-5 module, *Matrix Biol.* 29, 143–151, 2010.
477. Ringpis, G.-E., Stagno, J., and Aphasizhev, R., Mechanism of U-insertion RNA editing in trypanosome mitochondria: Characterization of RET2 functional domains by mutational analysis, *J. Mol. Biol.* 399, 696–706, 2010.
478. Nyce, H.L., Stober, S.T., Abrams, C.F., and White, M.M., Mapping spatial relationships between residues in the ligand-binding domain of the 5-Ht3 receptor using a molecular ruler, *Biophys. J.* 98, 1847–1855, 2010.
479. Ryan, R.M., Kort, N.C., Sirivanta, T., and Vandenberg, R.J., The position of an arginine residue influences substrate affinity and K+ coupling in the human glutamate transporter, EAAT1, *J. Neurochem.* 114, 565–575, 2010.
480. Albury, M.S., Elliott, C., and Moore, A.L., Ubiquinol-binding site in the alternative oxidase: Mutagenesis reveals features important for substrate binding and inhibition, *Biochim. Biophys. Acta* 1797, 1933–1939, 2010.
481. Jamieson, K.V., Hubbard, S.R., and Meruelo, D., Structure-guided identification of a laminin binding site of the laminin receptor precursor, *J. Mol. Biol.* 405, 24–32, 2011.
482. Sivla, I., Sansone, S., and Brancaleon, L., An anionic porphyrins binds β-lactoglobulin A at a superficial site rich in lysine residues, *Protein J.* 28, 1–13, 2009.
483. Takahashi, K., The reaction of phenylglyoxal with arginine residues in proteins, *J. Biol. Chem.* 243, 6171, 1968.
484. Yankeelov, J.A., Jr., Mitchell, C.D., and Crawford, T.H., A simple trimerization of 2,3-butanedione yielding a selective reagent for the modification of arginine in proteins, *J. Am. Chem. Soc.* 90, 1664, 1968.
485. Patthy, L. and Smith, E.L., Reversible modification of arginine residues. Application to sequence studies by restriction of tryptic hydrolysis to lysine residues, *J. Biol. Chem.* 250, 557–564, 1975.
486. Hager-Braun, C. and Tomer, K.M., Characterization of the tertiary structure of soluble CD4 bound to glycosylated full-length HIV gp120 by chemical modification of arginine residues and mass spectrometric analysis, *Biochemistry* 41, 1759–1766, 2002.
487. Hager-Braun, C., Hochleitner, E.O., Gorny, M.K. et al., Characterization of a discontinuous epitope of the HIV envelope protein gp120 recognized by a human monoclonal antibody using chemical modification and mass spectrometric analysis, *J. Am. Soc. Mass Spectrom.* 21, 1687–1698, 2010.
488. Barnaby, O.S., Wa, C., Cerny, R.L. et al., Quantitative analysis of glycation sites on human serum albumin using 16O/18O-labeling and matrix-assisted laser desorption/ionization time-of-flight mass spectrometry, *Clin. Chim. Acta* 411, 1102–1110, 2010.

489. Hermansson, M., Artemenko, K., Ossipova, E. et al., MS analysis of rheumatoid arthritic synovial tissue identifies specific citrullination sites on fibrinogen, *Proteomics Clin. Appl.* 4, 411–518, 2010.
490. Barnaby, O.S., Cerny, R.L., Clarke, W., and Hage, D.G., Comparison of modification sites formed on human serum albumin at various stages of glycation, *Clin. Chim. Acta* 412, 277–285, 2011.
491. Blifernez, O., Wobbe, L., Niehaus, K., and Kruse, O., Protein arginine methylation modulates light-harvesting antenna translation in *Chlamydomonas reinhardtii*, *Plant J.* 65, 119–130, 2011.
492. Silva, A.M., Borralho, A.C., Pinho, S.A. et al., Cross-oxidation of angiotensin II by glycerophosphatidylcholine oxidation products, *Rapid Commun. Mass Spectrom.* 25, 1413–1421, 2011.
493. Greenstein, J.P. and Winitz, M., *Chemistry of the Amino Acids*, John Wiley & Sons, New York, Chapter 22, pp. 1841–1855, 1961.
494. Sakaguchi, S., Über eine neue farbenreaktion von protein und arginin, *J. Biochem.* 5, 25–31, 1925.
495. Izumi, Y., New Sakaguchi reaction, *Anal. Biochem.* 10, 218–226, 1965.
496. Enoch, H.G. and Strittmatter, P., Role of tyrosyl and arginyl residues in rat liver microsomal stearyl-coenzyme A desaturase, *Biochemistry* 17, 4927–4932, 1978.
497. Kaleem, K., Chertok, F., and Erhan, S., Protein-polymer grafts III A. Modification of amino groups Ib.i Modification of protein-bound arginine with 4-hydroxybenzil, *J. Biol. Phys.* 15, 71–74, 1987.
498. Kaleem, K., Chertok, F., and Erhan, S., Novel materials form protein-polymer grafts, *Nature* 325, 328–329, 1987.
499. Young, C.T., Automated colorimetric measurement of free arginine in peanuts as a means to evaluated maturity and flavor, *J. Agric. Food Chem.* 21, 556–558, 1973.
500. Wang, H., Liang, X-h., Zhao, R. et al., Spectrophotometric determination of arginine in grape juice using 8-hydroquinoline, *Agric. Sci. China* 7, 1210–1215, 2008.
501. Mira de Orduna, R., Quantitative determination of L-arginine by enzymatic end-point analysis, *J. Agric. Food Chem.* 49, 549–552, 2001.
502. Murakami, T., Kosaka, M., Sato, H. et al., The intensely positive charged perioneuronal net in the adult rat brain, with special reference to its reactions to oxine, chondroitinase ABC, hyaluonidase, and collagenase, *Arch. Histol. Cytol.* 64, 313–318, 2001.
503. Mahmoud, L.H. and el-Alfy, N.M., Electron microscopy and histochemical studies on four Egyptian helminthes eggs of medical importance, *J. Egypt. Soc. Paristol.* 33, 229–243, 2003.
504. Schulze, A. et al., Sakaguchi reaction: A useful method for screening guanidinoacetate-methyl transferase deficiency, *J. Inherit. Metab. Dis.* 19, 706, 1996.
505. Schulze, A. et al., Creatinine deficiency syndrome caused by guanidinoacetate methyl transferase deficiency: Diagnostic tools for a new inborn error of metabolism, *J. Pediatr.* 131, 626–631, 1997.
506. Weber, C.J., A modification of Sakaguchi's reaction for the determination of arginine, *J. Biol. Chem.* 86, 217–222, 1930.
507. Brand, E. and Kassell, B., Photometric determination of arginine, *J. Biol. Chem.* 145, 359–364, 1942.
508. Smith, R.E. and MacQuarrie, R., A sensitive fluorometric method for the determination of arginine using 9,10-phenanthrenequinone, *Anal. Biochem.* 90, 246, 1978.
509. Fan, X. et al., Methylglyoxal-bovine serum albumin stimulates tumor necrosis factor alpha secretion in RAW 264.7 cells though activation of mitogen-activating protein kinase, nuclear factor κB and intracellular reactive oxygen species formation, *Arch. Biochem. Biophys.* 409, 274–285, 2002.

510. Knott, H.H. et al., Glycation and glycoxidation of low-density lipoproteins by glucose and low-molecular mass aldehydes. Formation of modified and oxidized proteins, *Eur. J. Biochem.* 270, 3572–3582, 2003.
511. Miele, C. et al., Human glycated albumin affects glucose metabolism in L6 skeletal muscle cells by impairing insulin-induced insulin receptor substrate (IRS) signaling through a protein kinase C alpha-mediated mechanism, *J. Biol. Chem.* 278, 47376–47387, 2003.
512. Schepens, I. et al., The role of active site arginines of sorghum NADP-malate dehydrogenase in thioredoxin-dependent activation and activity, *J. Biol. Chem.* 275, 35792–35798, 2000.
513. Linder, M.D. et al., Ligand-selective modulation of the permeability transition pore by arginine modification. Opposing effects of *p*-hydroxyphenylglyoxal and phenylglyoxal, *J. Biol. Chem.* 277, 937–942, 2002.
514. Wu, X. et al., Alteration of substrate specificity through mutation of two arginine residues in the binding site of amadoriase II from *Aspergillus* sp. *Biochemistry* 41, 4453–4458, 2002.
515. Xu, G., Takamoto, K., and Chance, M.R., Radiolytic modification of basic amino acid residues in peptides: Probes for examining protein-protein interactions, *Anal. Chem.* 75, 6995–7007, 2003.
516. Leitner, A. and Linder, W., Probing of arginine residues in peptides and proteins using selective tagging and electrospray ionization mass spectrometry, *J. Mass Spectrom.* 38, 891–899, 2003.
517. Ericksson, O. et al., Inhibition of the mitochondrial cyclosporine A-sensitive permeability pores by the arginine reagent phenylglyoxal, *FEBS Lett.* 409, 361–364, 1997.
518. Ericksson, O., Fontaine, E., and Bernardi, P., Chemical modification of arginines by 2,3-butanedione and phenylglyoxal causes closure of the mitochondrial permeability transition pore, *J. Biol. Chem.* 273, 12664–12674, 1998.
519. Skysgaard, J.M., Modification of Cl^- transport in skeletal muscle of *Rana temporaria* with the arginine-binding reagent phenylglyoxal, *J. Physiol.* 510, 591–604, 1998.
520. Cook, L.J. et al., Effects of methylglyoxal on rat pancreatic beta-cells, *Biochem. Pharmacol.* 55, 1361–1367, 1999.
521. Winters, C.J. and Adreoli, T.E., Chloride channels is basolateral TAL membranes XVIII. Phenylglyoxal induces functional mcCIC-Ka activity in basolateral MTAL membranes, *J. Membr. Biol.* 195, 63–71, 2003.
522. Emmons, C., Transport characterization of the apical anion exchanger of rabbit cortical collecting duct beta-cells, *Am. J. Physiol.* 276, F635–F643, 1999.
523. Tamari, I. et al., Anion antiport mechanism is involved in the transport of lactic acid across intestinal epithelial brush-border membrane, *Biochim. Biophys. Acta* 1468, 285–292, 2000.
524. Raw, P.E. and Gray, J.C., The effect of amino acid-modifying reagents on chloroplast protein import and the formation of early import intermediates, *J. Exp. Bot.* 52, 57–66, 2001.
525. Crenastri, D. et al., Inhibitors of the Cl^-/HCO_3^- exchanger activate an anion channel with similar features in the epithelial cells of rabbit gall bladder: Analysis in the intact epithelium, *Pfuegers Arch.* 441, 456–466, 2001.
526. Verri, A. et al., Molecular characteristics of small intestinal and renal brush border thiamin transporters in rats, *Biochim. Biophys. Acta* 1558, 187–197, 2002.
527. Forster, I.E. et al., Modulation of renal type IIa Na^+/P_i cotransporter kinetics by the arginine modifier phenylglyoxal, *J. Membr. Biol.* 187, 85–96, 2002.
528. Hermann, A. et al., Involvement of amino acid side chains of membrane proteins in the binding of glutathione to pig cerebral cortical membranes, *Neurochem. Res.* 27, 389–394, 2002.
529. Castagna, M. et al., Inhibitors of the lepidopteran amino acid co-transporter KAAT1 by phenylglyoxal. Role of Arg76, *Insect Mol. Biol.* 11, 389–394, 2002.

530. Kucera, I., Inhibition by phenylglyoxal of nitrate transport in *Peroacoccus denitrificans*: A comparison with the effect of a protonophorous uncoupler, *Arch. Biochem. Biophys.* 409, 327–334, 2003.
531. Sacchi, V.F. et al., Glutamate 59 is critical for transport function of the amino acid cotransporter KAAT1, *Am. J. Physiol.* 285, C623–C632, 2003.
532. Webb, J.L., *Enzyme and Metabolic Inhibitors*, Academic Press, New York, 1963.
533. Fliss, H. and Viswanatha, T., 2,3-Butanedione as a photosensitizing agent: Application to α-amino acids and α-chymotrypsin, *Can. J. Biochem.* 57, 1267, 1979.
534. Gripon, J.-C. and Hofmann, T., Inactivation of aspartyl proteinases by butane-2,3-dione. Modification of tryptophan and tyrosine residues and evidence against reaction of arginine residues, *Biochem. J.* 193, 55, 1981.
535. Mäkinen, K.K., Mäkinen, P.-L., Wilkes, S.H., Bayliss, M.E., and Prescott, J.M., Photochemical inactivation of Aeromonas aminopeptidase by 2,3-butanedione, *J. Biol. Chem.* 257, 1765, 1982.
536. Williassen, N.P. and Little, C., Effect of 2,3-butanedione on human myeloperoxidase, *Int. J. Biochem.* 21, 755–759, 1989.
537. Inano, H., Ohba, H., and Tamaoki, B., Photochemical inactivation of human placental estradiol 17 β-dehydrogenase in the presence of 2,3-butanedione, *J. Steroid Biochem.* 19, 1617–1622, 1983.
538. Riley, H.A. and Gray, A.R., Phenylglyoxal, *in Organic Syntheses*, Collective Vol. 2, A.H. Blatt (ed.), John Wiley & Sons, New York, p. 509, 1943.
539. Schloss, J.V., Norton, I.L., Stringer, C.D., and Hartman, F.C., Inactivation of ribulosebisphosphate carboxylase by modification of arginyl residues with phenylglyoxal, *Biochemistry* 17, 5626, 1978.
540. Augustus, B.W. and Hutchinson, D.W., The synthesis of phenyl[2–3H]glyoxal, *Biochem. J.* 177, 377, 1979.
541. Borders, C.L., Jr., Pearson, L.J., McLaughlin, A.E., Gustafson, M.E., Vasiloff, J., An, F.Y., and Morgan, D.J., 4-Hydroxy-3-nitrophenylglyoxal. A chromophoric reagent for arginyl residues in proteins, *Biochim. Biophys. Acta* 568, 491, 1979.
542. Borders, C.L., Jr. and Johansen, J.T., Identification of Arg-143 as the essential arginine residue in yeast Cu, Zn superoxide dismutase by the use of a chromophoric arginine reagent. *Biochem. Biophys. Res. Commun.* 96, 1071–1078, 1980.
543. Borders, C.L., Jr. and Riordan, J.F., An essential arginyl residue at the nucleotide binding site of creatinine kinase, *Biochemistry* 14, 4699–4704, 1975.
544. Aminlari, M. and Vaseghi, T., A new colorimetric method for determination of creatinine phosphokinase, *Anal. Biochem.* 164, 397–404, 1987.
545. Wijnands, R.A., Müller, F., and Visser, A.J.W., Chemical modification of arginine residues in *p*-hydroxybenzoate hydroxylase from *Pseudomonas fluorescens*: A kinetic and fluorescence study, *Eur. J. Biochem.* 163, 535–544, 1987.
546. Julian, T. and Zaki, L., Studies on the inactivation of anion transport in human blood red cells by reversibly and irreversibly acting arginine-specific reagents, *J. Membr. Biol.* 102, 217–224, 1988.
547. Meves, H., Ruby, N., and Staempfil, R., The action of arginine-specific reagents on ionic and gating currents in frog myelinated nerve, *Biochim. Biophys. Acta* 943, 1–12, 1988.
548. Betakis, E., Fritzsch, G., and Zaki, L., Inhibition of anion transport in the human red blood cell membrane with *p*- and *m*-methoxyphenylglyoxal, *Biochim. Biophys. Acta* 1110, 75–80, 1992.
549. Boehm, R. and Zaki, L., Toward the localization of the essential arginine residues in the band 3 protein of human red blood cell membranes, *Biochim. Biophys. Acta* 2390, 238–242, 1996.
550. Zaki, L., Boehm, R., and Merckel, M., Chemical labeling of arginyl-residues involved in anion transport mediated by human band 3 protein and some aspects of its location in the peptide chain, *Cell. Mol. Biol.* 42, 1053–1063, 1996.

551. Belousova, L.V. and Muizhnek, E.L., Kinetics of chemical modification of arginine residues in mitochondrial creatine kinase from bovine heart: Evidence for negative cooperativity, *Biochemistry* (Moscow) 69, 455–461, 2004.
552. Yamasaki, R.B., Vega, A., and Feeney, R.E., Modification of available arginine residues in proteins by *p*-hydroxyphenylglyoxal, *Anal. Biochem.* 109, 32, 1980.
553. Béliveau, R., Bernier, M., Giroux, S., and Bates, D., Inhibition by phenylglyoxal of the sodium-coupled fluxes of glucose and phosphate in renal brush-border membrane, *Biochem. Cell. Biol.* 66, 1005–1012, 1988.
554. Strevey, J., Vachon V., Beaumier, B. et al., Characterization of essential arginine residues implicated in the renal transport of phosphate and glucose, *Biochim. Biophys. Acta* 1106, 110–116, 1992.
555. Mukouyama, E.B., Hirose, T., and Suzuki, H., Chemical modification of L-phenylalanine oxidase from *Pseudomonas* sp. 5012 by phenylglyoxal. Identification of a single arginine residue, *J. Biochem.* 123, 1097–1103, 1998.
556. Cheung, S.-T. and Fonda, M.L., Reaction of phenylglyoxal with arginine. The effect of buffers and pH, *Biochem. Biophys. Res. Commun.* 90, 940, 1979.
557. Yamasaki, R.B., Shimer, D.A., and Feeney, R.E., Colorimetric determination of arginine residues in proteins by p-nitrophenylglyoxal, *Anal. Biochem.* 111, 220, 1981.
558. Branlant, G., Tritsch, D., and Biellmann, J.-F., Evidence for the presence of anion-recognition sites in pig-liver aldehyde reductase. Modification by phenylglyoxal and p-carboxyphenyl glyoxal of an arginyl residue located close to the substrate-binding site, *Eur. J. Biochem.* 116, 505, 1981.
559. Eun, H.-M., Arginine modification by phenylglyoxal and (p-hydroxyphenyl)glyoxal: Reaction rates and intermediates, *Biochem. Int.* 17, 719, 1988.
560. Pearson, M.A. et al., Kinetic and structural characterization of urease active site variants, *Biochemistry* 39, 8575–8584, 2000.
561. Yankeelov, J.A., Jr., Modification of arginine in proteins by oligomers of 2,3-butanedione, *Biochemistry* 9, 2433, 1970.
562. Riordan, J.F., Functional arginyl residues in carboxypeptidase A. Modification with butanedione, *Biochemistry* 12, 3915–3923, 1973.
563. Alkena, W.B.L. et al., Role of αArg^{145} and βArg^{263} in the active site of penicillin acylase of *Escherichia coli*, *Biochem. J.* 365, 303–309, 2003.
564. Clark, D.D. and Ensign, S.A., Characterization of the 2-[(R)-2-hydroxypropyl]ethanesulfonate dehydrogenase from *Xanthobacter* strain PY2: Product inhibition, pH dependence of kinetic parameters, site-directed mutagenesis, rapid equilibrium inhibition, and chemical modification, *Biochemistry* 41, 2727–2740, 2002.
565. Leitner, A. and Lindner, W., Probing of arginine residues in peptides and proteins using selective tagging and electrospray ionization mass spectrometry, *J. Mass Spectrom.* 38, 891–899, 2003.
566. Foettinger, A., Leitner, A., and Linder, W., Solid-phase capture and release of arginine peptides by selective tagging and boronate affinity chromatagraphy, *J. Chromatogr. A.* 1079, 187–196, 2005.
567. Leitner, A. and Lindner, W., Functional probing of arginine residues in proteins using mass spectrometry and an arginine-specific covalent tagging concept, *Anal. Chem.* 77, 4481–4488, 2005.
568. Leitner, A., Amon, S., Rizzi, A., and Lindner, W., Use of the arginine-specific butanedione/phenylboronic acid tag for analysis of peptides and protein digests using matrix-assisted laser desorption/ionization mass spectrometry, *Rapid Commun. Mass Spectrom.* 21, 1321–1330, 2007.
569. Foettinger, A., Leitner, A., and Lindner, W., Derivativiszation of arginine residues with malonaldehyde for the analysis of peptides and protein digests by LC-ESI-MS/MS, *J. Mass Spectrom.* 41, 623–632, 2006.

570. Toi, K., Bynum, E., Norris, E., and Itano, H.A., Studies on the chemical modification of arginine. I. The reaction of 1,2-cyclohexanedione with arginine and arginyl residues of proteins, *J. Biol. Chem.* 242, 1036, 1967.
571. Patthy, L. and Smith, E.L., Identification of functional arginine residues in ribonuclease A and lysozyme, *J. Biol. Chem.* 250, 565, 1975.
572. Samy, T.S., Kappen, L.S., and Goldberg, I.H., Reversible modification of arginine residues in neocarzinostatin. Isolation of a biologically active 89-residue fragment from tryptic hydrolysate, *J. Biol. Chem.* 255, 3420–3426, 1980.
573. Ullah, A.H.J. and Sethumadhaven, K., Differences in the active site environment of *Aspergillus ficus* phytases, *Biochem. Biophys. Res. Commun.* 243, 458–462, 1998.
574. Baynes, J.W. and Thorpe, S.R., Role of oxidative stress in diabetic complications: A new perspective on an old paradigm, *Diabetes* 48, 1–9, 1995.
575. Reubsaet, J.L.E. et al., Analytical techniques used to study the degradation of proteins and peptides: Chemical instability, *J. Pharm. Biomed. Anal.* 17, 955–978, 1998.
576. Thornalley, P.J., Glutathione-dependent detoxification of alpha-oxoaldehyde by the glyoxalase system: Involvement in disease mechanisms and antiproliferative activity of glyoxalase I inhibitors, *Chem. Biol. Interact.* 111–112, 137–151, 1998.
577. Deyl, Z. and Miksik, I., Post-translational non-enzymatic modification of proteins. I. Chromatography of marker adducts with special emphasis on glycation reactions, *J. Chromatogr. B* 699, 287–209, 1997.
578. Henle, T. et al., Detection and identification of a protein-bound imidalone resulting from the reaction of arginine and methylglyoxal, *Z. Lebensm. Uners. Forsch.* 199, 55–58, 1994.
579. Westwood, M.E. et al., Methylglyoxal-modified arginine residues—A signal for receptor-mediated endocytosis and degradation of proteins by monocytic THP-1 cells, *Biochim. Biophys. Acta* 1316, 84–94, 1997.
580. Degenhardt, T.P., Thorpe, S.R., and Baynes, J.W., Chemical modification of proteins by methylglyoxal, *Cell. Mol. Biol.* 44, 1139–1145, 1998.
581. Pischetsrieder, M., Reaction of L-ascorbic acid with L-arginine derivatives, *J. Agric. Food Chem.* 44, 2081–2085, 1996.
582. Kim, S., Park, G.H., and Paik, W.K., Recent advances in protein methylation: Enzymatic methylation of nucleic acid binding proteins, *Amino Acids* 15, 291–306, 1998.
583. Urnov, F.D., Methylation and the genome: The power of a small amendment, *J. Nutr.* 132 (Suppl 8), 2450S–2456S, 2002.
584. Takahashi, K., Specific modification of arginine residues in proteins with ninhydrin, *J. Biochem.* 80, 1173–1176, 1976.
585. Jacob, R.T., Bhat, P.G., and Pattabiraman, T.N., Isolation and characterization of a specific enterokinase inhibitor from kidney bean (*Phaseolus vulgaris*), *Biochem. J.* 209, 91–97, 1983.
586. Rhodes, G.R. and Boppana, V.K., High-performance liquid chromatographic analysis of arginine-containing peptides in biological fluids by means of a selective post-column reaction with fluorescence detection, *J. Chromatogr.* 444, 123–131, 1988.
587. Boppana, V.K. and Rhodes, G.R., High-performance liquid chromatographic determination of an arginine-containing octapeptide antagonist of vasopressin in human plasma by means of a selective post-column reaction with fluorescence detection, *J. Chromatogr.* 507, 79–84, 1990.
588. Ruhemann, S., Triketohydrindene hydrate, *J. Chem. Soc. Trans.* 97, 2025–2031, 1910.
589. Retinger, J.M., The mechanism of the ninhydrin reaction. A contribution to the theory of color of salts of alloxantine-like compounds, *J. Am. Chem. Soc.* 39, 1059–1066, 1917.
590. Bronfenbrenner, J., The mechanism of the Abderhalden reaction: Studies on immunity I, *J. Exp. Med.* 21, 221–238, 1915.

591. Abderhalden, E. and Schmidt, H., Über die Verwendung von Triketohydrindenthydrat[1] zum Nachweis von Eiweszstoffen under deren Abbaustufen, *Hoppe-Seyler's ZeitSchrift Physiolog. Chem.* 72, 37–43, 1913.
592. Jobling, J.W., Eggstein, A.A., and Petersen, W., Serum proteases and the mechanism of the Abderhalden reaction: Studies on ferment action. XX, *J. Exp. Med.* 21, 239–249, 1915.
593. Underwood, G.E., Kethoxal for treatment of cutaneous herpes simplex, *Proc. Soc. Exp. Biol. Med.* 129, 235–239, 1968.
594. Litt, M. and Hancock, V., Kethoxal—A potentially useful reagent for the determination of nucleotide sequences in single-stranded regions of transfer ribonucleic acid, *Biochemistry* 6, 1848–1854, 1968.
595. Greenspan, C.M. and Litt, M., Characterization of the native and denatured forms of tRNA Trp by reaction kethoxal, *FEBS Lett.* 41, 297–302, 1974.
596. Kellersberger, K.A., Yu, E., Kruppa, G.M. et al., Top-down characterization of nucleic acids modified by structural probes using high-resolution tandem mass spectrometry and automated data interpretation, *Anal. Chem.* 76, 2438–2845, 2004.
597. Iijima, H., Patrzyc, H., and Bello, J., Modification of amino acids and pancreatic ribonuclease A by kethoxal, *Biochim. Biophys. Acta* 491, 305–316, 1977.
598. Brewer, L.A. and Noller, H.F., Ribonucleic acid-protein crosslinking within the intact *Escherichia coli* ribosome utilizing ethylene glycol bis[3-(2-ketobutyraldehyde)ether], a reversible, bifunctional reagent: Identification of the 30S proteins, *Biochemistry* 22, 4310–4315, 1983.

3 Modification of Hydroxyl and Carboxyl Functional Groups in Proteins

This chapter includes the modification of hydroxyl and carboxyl functions in proteins. The commonality between these functional groups is that they represent different oxidation states of carbon.[1] Serine and threonine contain hydroxyl groups, while tyrosine contains a phenolic group (defined as a hydroxyl group bound to an aromatic ring). The phenolic hydroxyl of tyrosine has a lower pK_a ($pK_a \sim 10$) than serine or threonine ($pK_a \sim 13$) and is considered more reactive; the formation of *O*-acetyl tyrosine is more likely than the formation of *O*-acetyl serine with both reversed by mild base or neutral hydroxylamine.[2] The pK_a for the tyrosine hydroxyl group is lowered ($pK_a \sim 7$) on nitration with either tetranitromethane (TNM) or peroxynitrite.[3] The formation of dinitrotyrosine would result in a further decrease in the pK_a ($pK_a \sim 4$).[4] While not directly relevant to the issue of tyrosine nitration, the pK_a for 2,4,6-trinitrophenol is 0.71.[5] Serine, threonine, and tyrosine are not considered to be ionizable groups of proteins, while the carboxyl groups of aspartic acid ($pK_a \sim 3.5$) and glutamic acid ($pK_a \sim 4.2$) are ionizable groups.[6] Carboxyl groups are unique in that there are two carbon–oxygen bonds with different bond lengths.[7]

SERINE AND THREONINE

With the exception of serine groups at enzyme active sites, the modification of serine or threonine hydroxyl groups is accomplished only with difficulty; the modification of hydroxyl groups in nucleic acids and carbohydrates is discussed elsewhere. The hydroxyl functions of serine and threonine are sites of *O*-glycosylation in proteins.[8–11] Both serine and threonine are poor nucleophiles ($pK_a > 12$).[12] The phosphorylation of serine by protein kinases is a critical part of cell regulation.[13–15] Thus, while unreactive under conditions used for modification of biological polymers, serine and threonine are sites for posttranslational modification of proteins. The nucleophilicity of serine at enzyme active sites is greatly increased by interaction with spatially adjacent residues such as histidine.[16] The function of serine or threonine as a nucleophile is dependent on the formation of the alkoxide ion. Some studies use hydroxide ion as a model for nucleophilic reactions of alkaloxides; this is largely useful but one must consider that the hydroxide ion differs from the alkoxide ion in that the hydroxide ion has a second proton available for reaction after the initial nucleophilic reaction.[17]

Modification of serine with *N*-hydroxysuccinimide(NHS)-based cross-linking agents has been reported[18,19] with such modification less favorable than reaction with lysine ε-amino groups and dependent on adjacent amino acid residues.[18] Modification of serine residues in proteins by NHS-sulfo biotin reagents has also been reported.[20,21] Miller and Kurosky have previously reported on the enhanced reactivity of serine in peptides with the sequence Ser-Xaa-His or His-Xaa-Ser.[22] Other reports on the modification of serine residues in proteins are scarce. Marzotto and Giormani[23] reported the oxidation of serine and threonine in RNase with carbodiimide in the presence of DMSO and phosphoric acid; the reaction also resulted in aggregation with formation of higher molecular weight species. Cardinaud and Baker[24] modified a serine (Ser223) outside the enzyme active site of α-chymotrypsin with sulfonyl fluoride "exosite" affinity reagent; a similar approach was used by Bing and coworkers[25] to modify exosites in thrombin. Phenylmethylsulfonyl fluoride (PMSF) is used as an inhibitor of serine proteases, frequently with little evidence. PMSF was introduced as an inhibitor for serine proteases by Fahrney and Gold.[26] Methanesulfonyl fluoride is not as potent in the inhibition of acetylcholinesterase as diisopropylphosphorofluoridate (DFP) but is much less potent than DFP in the inhibition of α-chymotrypsin, while PMSF is much more potent that methylsulfonyl fluoride. The authors argue that binding and reagent orientation are critical factors in the reaction of PMSF with α-chymotrypsin. This is consistent with observations of Cardinaud and Baker[24] discussed earlier. Gold and Fahrney[27] subsequently showed lack of reaction of chymotrypsinogen (based on incorporation of labeled reagent) with PMSF. The conversion of chymotrypsinogen to chymotrypsin results in the formation of an enzyme active site possessing a serine residue with enhanced nucleophilicity. The importance of serine nucleophilicity and/or spatial proximity for reactivity is further emphasized by the observation of that benzenesulfonyl fluoride is essentially unreactive with hydroxide ion in acetone-water.[28] Similar results were obtained for the sulfonyl fluoride by Halmann[29] who also observed the relative lack of activity of phosphorofluoridate derivatives. The reader is also directed to an excellent discussion by Cohen[30] of selectivity versus reactivity in the reaction of carbonyl, sulfonyl, and phosphoryl halides with functional groups on proteins.

The modification of N-terminal serine or threonine with periodic acid (Figure 3.1) provides an exception of sorts to the lack of reactivity of serine and/or threonine outside of the enzyme active site. Geschwind and Li[31] demonstrated that the action of periodic acid (Malaprade reaction[32]) on vicinal amino alcohols such as serine as reported by Nicolet and Shinn[33] could be used for the determination of amino-terminal serine residues in peptides and proteins. Geoghegan and Stroh[34] showed the conversion of N-terminal serine to a glyoxalic acid derivative could be used for coupling nonpeptide groups to proteins. This approach has been used by other investigators[35–38] for various ligation reactions including coupling to matrices.[35,36] Mikolajczyk and coworkers[39] used periodate oxidation of N-terminal threonine for protein ligation to a sulfhydryl group using maleimide–aminoxy chemistry (Figure 3.2).

TYROSINE

It is possible to modify tyrosyl residues in proteins under relatively mild conditions with reasonably high specificity with a variety of reagents obtaining, in turn, a variety of derivatives as described in the following. For example, *N*-acetylimidazole

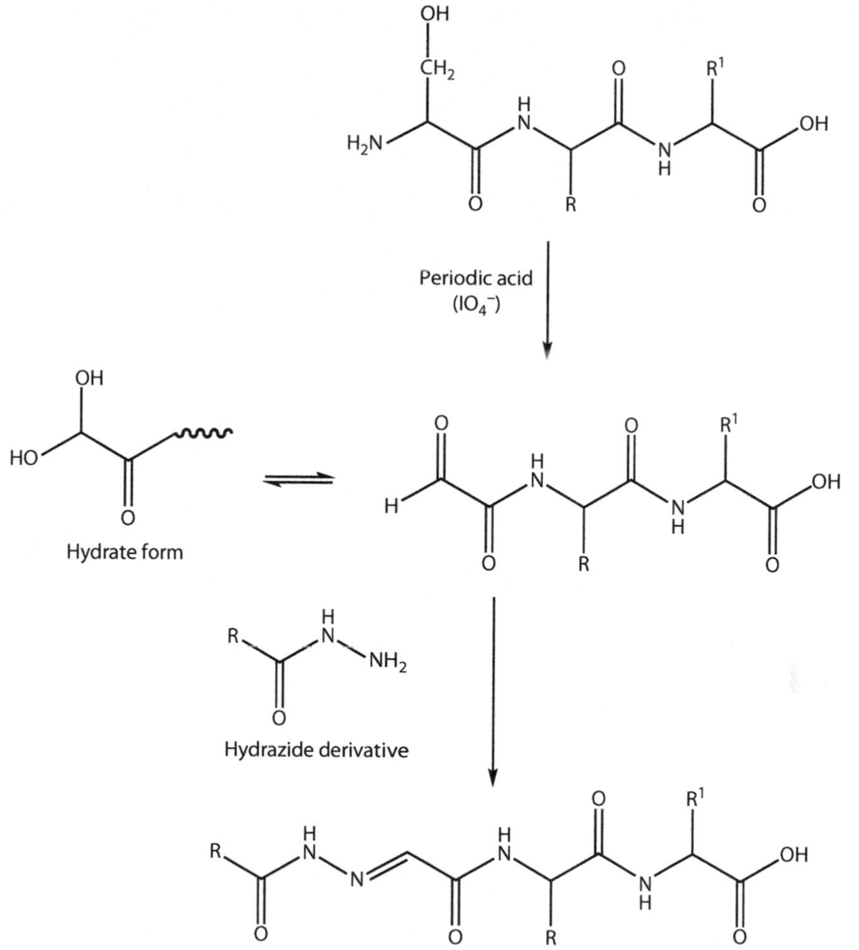

FIGURE 3.1 Modification of serine with periodic acid. Shown is the oxidation of an amino terminal serine residue with periodic acid to yield the corresponding glycolic acid derivative. This aldehyde can be coupled, as shown, with a hydrazide or similar aldehyde-reactive functional group with a probe function such as biotin hydrazide. (See Filtz, T.M., Chumpradit, S., Kung, H.F., and Molinoff, P.B., Synthesis and applications of an aldehyde-containing analogue of SCH-23390, *Bioconjug. Chem.* 1, 394–399, 1990; Chelius, D. and Shaler, T.A., Capture of peptides with N-terminal serine and threonine: A sequence-specific chemical method for peptide mixture simplification, *Bioconjug. Chem.* 14, 205–211, 2003.)

(NAI) acetylated the phenolic hydroxyl group in a reversible reaction. TNM nitrates tyrosyl residues to yield the 3-nitro derivative, which markedly lowers the pK_a of the phenolic hydroxyl group (see above). The 3-nitro function can be reduced under mild conditions to give the 3-amino derivatives, which can be subsequently modified with a variety of useful compounds such as dansyl chloride [5-(dimethylamino)-1-napthalenesulfonyl chloride] or biotinamidonexanoic acid sulfo-*N*-hydroxysuccinimide ester, sodium salt (sulfosuccinimidyl-6-biotinamidohexanoate). Reaction with

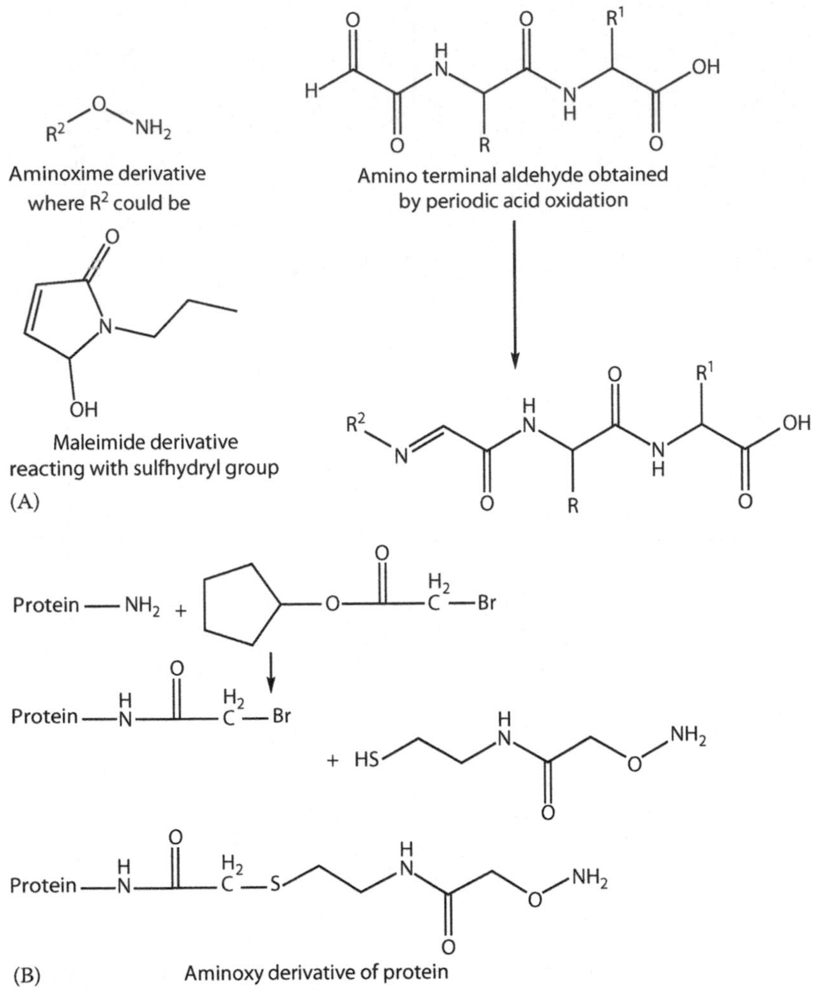

Aminoxime derivative where R^2 could be

Maleimide derivative reacting with sulfhydryl group

(A)

Amino terminal aldehyde obtained by periodic acid oxidation

(B) Aminoxy derivative of protein

FIGURE 3.2 Use of aminoxy chemistry to modify aldehyde functions in proteins. The aldehyde function generated from amino-terminal serine or threonine residues in proteins or peptides can be coupled with aminoxy derivatives for attachment of probes or introducing protein cross-links. Panel (A) shows the reaction of an amino-terminal glyoxalic derivative obtained by the oxidation of amino-terminal serine (Figure 3.1) with an aminoxime derivative. See Mikolajczyk, S.D., Meyer, D.L., Starling, J.J. et al., High-yield, site-specific coupling of N-terminally modified β-lactamase to a proteolytically derived single sulfhydryl murine Fab, *Bioconjugate Chem.* 4, 635–646, 1994; Hecker, J.G., Berger, G.O., Scarfo, K.A. et al., A flexible method for the conjugation of aminooxy ligands to preformed complexes of nucleic acids and lipids, *ChemMedChem* 3, 1356–1361, 2008. Panel (B) shows another approach to the aminooxy derivative, which involves bromoacetylation of the amino-terminal amino acid followed by reaction with a sulfhydryl derivative of an aminooxime which could then be coupled to a reducing sugar function in a polysaccharide (Lees, A., Sen, G., and LopezAcosta, A., Versatile and efficient synthesis of protein-polysaccharide conjugate vaccines using aminooxy reagents and oxime chemistry, *Vaccine* 24, 716–729, 2006).

TNM can also result in zero-length cross-linkage in proteins via the formation of dityrosine. Tyrosine residues are also modified by peroxynitrite which, as with TNM, can also modify cysteine residues and tryptophan residues in proteins. The production of peroxynitrite is complex and of considerable physiological importance but of less significance to the *in vitro*.

Factors that influence the reactivity of tyrosyl residues in proteins are, at best, poorly understood. It would seem that the reactivity is influenced most by the ionization state of tyrosine, which is the function of the microenvironment around the residue. It is extremely useful for investigators to review early literature on the factors influencing tyrosine ionization in proteins.[40–45] It is also useful to appreciate that iodination and nitration are examples of electrophilic substitution of aromatic rings.[46]

Iodination is used for the modification of tyrosyl residues in proteins.[47–49] While it could be argued that the chemistry is somewhat dated, the reaction is still of considerable value since the process of the radiolabeling of proteins with either of the iodine radioisotopes (^{125}I, ^{131}I) primarily involves the modification of tyrosine residues in proteins. It is, of course, of critical importance to appreciate the strength of the elemental halides as oxidizing agents modifying residues other than tyrosine during iodination procedures. The reaction of iodine with tyrosine is an example of electrophilic aromatic substitution with iodonium ion (I^+) as the attacking species. The processes that follow describe methods for the generation of I^+ from I_2.

Iodination has been utilized to study the reactivity of tyrosyl residues in cytochrome b_5.[50] Iodination is accomplished with a 10-fold molar excess of I_2 (15 mM I_2 in 30 mM KI) in 0.025 M sodium borate, pH 9.8. Iodination with limiting amounts of iodine is accomplished with a two- to sixfold molar excess of iodine in 0.020 M potassium phosphate, pH 7.5 at 0°C. Monoiodination and diiodination of tyrosyl residues is observed. Iodination with a 10-fold molar excess of I_2 results in the formation of 3 mol of diiodotyrosine per mole of cytochrome *c*.[51] The fourth tyrosyl residue is modified only in the presence of 4.0 M urea. Iodination of tyrosine results in a decrease in the pK_a of the phenolic hydroxyl groups similar to that observed with nitration. Iodination with a limiting amount of iodine as described earlier results first in the formation of 2 mol of monoiodotyrosine, then 1 mol of diiodotyrosine, and 1 mol of monoiodotyrosine. Tyrosyl residues that can be iodinated are also available for *O*-acetylation with acetic anhydride.

The modification of tyrosyl residues in phosphoglucomutase by iodination has been reported.[52] Modification is achieved by reaction in 0.1 M borate, pH 9.5 with 1 mM I_2 (obtained by an appropriate dilution of a stock iodine/iodide solution, 0.05 M I_2 in 0.24 M KI) at 0°C for 10 min. Complete loss of enzymatic activity was observed with these reaction conditions, but the stoichiometry of modification was not established. Nitration of 7 of 20 tyrosyl residues also resulted in an 83% loss of catalytic activity. These investigators also studied the reaction of phosphoglucomutase with diazotized sulfanilic acid and NAI.

The aforementioned modifications utilize the reaction of tyrosyl residues in proteins with iodine/iodide solutions at alkaline pH. Iodination of tyrosyl residues can also be accomplished with iodine monochloride (ICl) at mildly alkaline pH. One such study explores the modification of galactosyltransferase.[53] The modification is accomplished by reaction in 0.2 M sodium borate, at pH 8.0. The reaction is initiated by a desired amount of a stock solution of ICl.[54] A stock solution of 0.02M ICl is prepared by adding 21mL 11.8M HCl (stock concentrated HCl) to approximately 150mL of H_2O containing 0.555g KCl, 0.3567g KIO_3, and 29.23g NaCl. The solvent is taken to a final volume of 250mL with H_2O. Free iodine is then extracted with CCl_4, if necessary, and the solution is aerated to remove trace amounts of CCl_4. The resulting solution of ICl is stable for an indefinite period of time under ambient conditions. Reaction proceeds for 1 min at ambient temperature and is terminated by the addition of a 1:6 volume of 0.5M $Na_2S_2O_3$ (50μL for a 0.300mL reaction mixture). Radiolabeled sodium iodide ($Na^{125}I$) is included to provide a mechanism for establishing the stoichiometry of the reaction. The reaction mixture, after the addition of $Na_2S_2O_3$, is subjected to gel filtration on Bio-Gel P-10 (BioRad Laboratories, Richmond, California) in 0.1M Tris, pH 7.4. In experiments designed to assess the relationships between reagent (ICl) concentration and the extent of modification, a maximum of 10g-atom of iodine were incorporated into galactosyltransferase at a 40-fold molar excess of reagent. Incorporation of iodine is linear up to this excess of reagent and slowly declines at higher concentrations of ICl. Modification of tryptophanyl residues was excluded by direct analysis, and the only iodinated amino acids obtained from the modified protein were monoiodotyrosine and diiodotyrosine. Modification of other residues such as histidine and methionine by oxidation without incorporation of iodine was not excluded.

Iodination can also be accomplished by peroxidase per H_2O_2 per NaI. A recent procedure was described for the modification of tyrosyl residues in insulin.[55] In these studies, 20mg of porcine insulin in 20mL 0.4M sodium phosphate, in 6.0M urea, pH 7.8 was combined with 10mL $Na^{125}I$ (1mCi) and 3.6mg urea, and H_2O_2 (5μL 0.3mM solution) and peroxidase (Sigma Chemical Company, 0.2mg mL^{-1}; 5μL) was added. The preparative reaction was terminated by dilution with an equal volume of 40% (w/v) sucrose. It is of interest to note that the enzyme-catalyzed iodination proceeds with efficiency in 6.0M urea. Iodination of tyrosyl residues in peptides and proteins can also be accomplished with chloramine T.[56,57] The solution structure of insulin-like growth factor was investigated by iodination of tyrosyl residues mediated by either chloramine T or lactoperoxidase.[58] Chloramine T was more effective than lactoperoxidase. There continue to be improvements in the chemistry of chloramine T iodination.[59–61]

Radioisotope labeling of proteins with isotopes of iodine has been used extensively to study protein turnover (catabolism) and *in vivo* distribution.[62–65] Caution should be used in the interpretation of such results as most iodination techniques result in heterogeneous products, which are trace labeled with iodine isotopes.[66–68] Site-specific modification of tyrosine with iodine has been used to obtain protein derivatives useful in crystallographic analysis.[69,70]

The phenolic hydroxyl of tyrosine can be modified by acid anhydrides and acyl chlorides.[71–73] In general, these reactions are not as specific as the other procedures such as NAI or TNM. The preparation of *O*-acyl derivatives via the action of carboxylic acid anhydrides (i.e., acetic anhydride) has been used for some time, but it is very difficult to obtain selective modification of tyrosine as these reagents readily react with primary amines to form stable *N*-acyl derivatives.[72,74] It is, however, possible to obtain the selective modification of tyrosine in an α-amylase with acetic anhydride (Figure 3.3) by reaction at mildly acidic pH (1.0M acetate, pH 5.8 at 25°C),

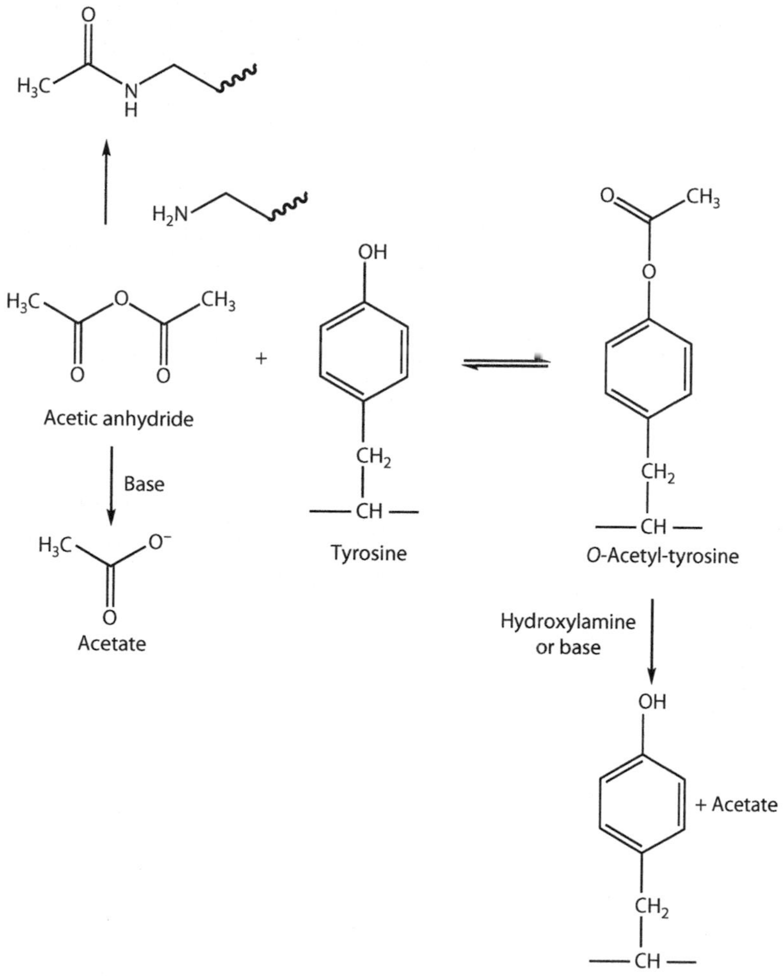

FIGURE 3.3 The modification of tyrosine with acetic anhydride. The modification of tyrosine by acetic anhydride is a readily reversible reaction. It is usually performed at pH 7.5–8.0 where the reaction rate and product stability are optimal. It is rarely used for the modification of tyrosine and more frequently for the modification of amino groups; the modification of amino groups is not reversible. (See Riordan, J.F. and Vallee, B.L., *O*-acetylation, *Methods Enzymol.* 25, 570–576, 1972.)

approximately 20,000-fold molar excess of acetic anhydride (5.1×10^{-2} M acetic anhydride, 2.9×10^{-6} M enzyme).[75] Bernad and colleagues[76] have reported on an extensive study comparing the modification of lysyl and tyrosyl residues in lysozyme with dicarboxylic acid anhydrides. In 50 mM HEPES, 1.25 M NaCl, pH 8.2, amino groups (primarily lysine residues) were far more reactive than hydroxyl groups (including tyrosine, serine, and threonine). A consideration of the recent literature suggests that the use of organic acid anhydrides for the modification of hydroxyl functions in biological polymers is not common but modification of tyrosine residues must be considered with acetic anhydride, which is used for mapping.[77] As indicated

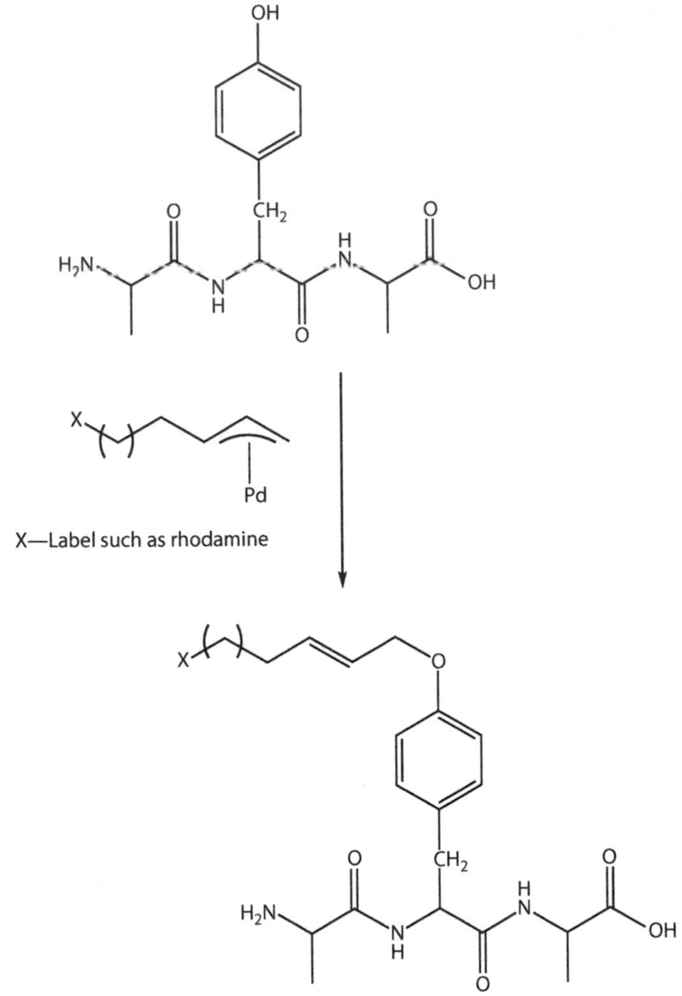

FIGURE 3.4 The modification of tyrosine by π-allylpalladium complexes. This is adapted from Tilley, S.D. and Francis, M.B., Tyrosine-selective protein alkylation using π-allylpalladium complexes, *J. Am. Chem. Soc.* 128, 1080–1081, 2006. (See also Lin, Y.Y.A., Chalker, J.M., and Davis, B.G., Olefin metathesis for site-selective protein modification, *Chem. Biol. Chem.* 10, 959–969, 2009; Basel, E., Joubert, N., and Pucheault, M., Protein chemical modification on endogenous amino acids, *Chem. Biol.* 17, 213–227, 2010.)

by the results obtained with α-amylase,[75] reaction under acid conditions would be expected to improve specificity as acid-catalyzed esterification would be favored over acetylation of amino groups at lower pH, although modification of lysine residues at mildly acid pH has been reported[78]; the acetylation of amino groups with acetic anhydride is enhanced by deprotonation.[79] Tyrosine-specific alkylation has recently been reported using π-allylpalladium complexes (Figure 3.4).[80]

The development of NAI as a reagent for the selective modification of tyrosyl residues (Figure 3.5) can, in part, be traced to the early observation[81–85] that NAI is,

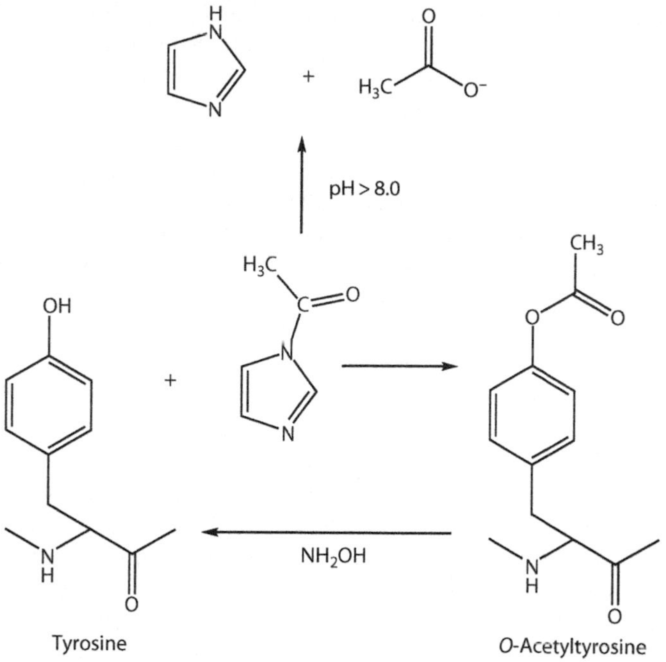

pH > 8.0

CH3

H3C

OH

NH2OH

Tyrosine

O-Acetyltyrosine

FIGURE 3.5 The *O*-acetylation of tyrosine with NAI and reversal with hydroxylamine. The modification of tyrosine by NAI proceeds optimally at pH 7.0–8.0. At value of pH greater than 8.0, base-catalyzed hydrolysis of the reagent becomes a significant problem; free imidazole also catalyzes the hydrolysis of the reagent. (See Riordan, J.F. and Vallee, B.L., *O*-acetylation, *Methods Enzymol.* 25, 570–576, 1972; Fife, T.H., Natarajan, R., and Werner, M.H., Effect of the leaving group in the hydrolysis of *N*-acylimidazoles. The hydroxide ion, water, and general-base catalyzed hydrolysis of *N*-acyl-4(5)-nitroimidazoles, *J. Org. Chem.* 52, 740–746, 1987.)

in fact, an energy-rich compound. NAI was first used as a reagent for the modification of tyrosyl residues in bovine pancreatic carboxypeptidase A.[86] This same group of investigators subsequently reported on the use of NAI for the determination of "free" tyrosyl residues in proteins[87] as opposed to "buried" residues; this has not necessarily proved to be the case.[88] *N*-acyl derivatives other than NAI have been prepared.[85,89] *N*-butyrylimidazole was demonstrated to be a more potent inhibitor of thrombin than the acetyl derivative[90] with modification occurring presumably at a histidine residue in addition to a single tyrosine residue; four residues were modified in thrombin with NAI[91] demonstrating that the specificity can be increased in the reagent. El Kabbaj and Latruffe[92] examined the membrane penetration of NAI. NAI was similar to TNM in its ability to preferentially inactivate D-3-hydroxybutyrate dehydrogenase in "inside-out" membranes when compared to intact mitochondria. Evaluation of the partition of NAI in water–organic solvent systems showed an 81%/19% distribution in an H_2O/hexane system and 31/69 in H_2O/1-octanol. Zhang and coworkers[93] have used reaction with NAI to differentiate between classes of tyrosine residues in green plant manganese stabilizing protein; five tyrosine residues were easily modified, while one to two residues were modified only at higher concentrations of NAI and one to two residues were only modified in urea-denatured protein.

Graves and coworkers have demonstrated that NAI will not acetylate 3-nitrotyrosine or 3,5-dinitrotyrosine.[94] This study also examined the rates of acetylation of tyrosine and 3-fluorotyrosine as a function of pH between 7.5 and 9.5. The formation of the *O*-acetylated derivative was evaluated by HPLC analysis. As NAI is unstable at increasing pH, it was not possible to obtain reliable second-order rate constants; thus, relative rates of reaction were reported. Since most of the reagents used for the modification of biological polymers are subject to nucleophilic attack (see Chapter 1) and water is a modest nucleophile[95–97] so a competing reaction in an aqueous system is to be expected. 3-Fluorotyrosine was acetylated more rapidly than tyrosine at pH 7.5, while the opposite was true at pH 9.5; the increased reactivity of 3-fluorotyrosine at pH 7.5 reflects the lower pK_a of 3-fluorotyrosine; at pH 9.5, both tyrosine and 3-fluorotyrosine are deprotonated and the intrinsic rate of tyrosine modification by NAI is 17 times greater than that of 3-fluorotyrosine.

Ji and Bennett[98] showed that peroxynitrite could activate (four- to fivefold) reduced rat liver microsomal glutathione S-transferase; activity was also increased on reaction with TNM and NAI. This activation does not appear to be due to sulfhydryl oxidation as treatment of the modified enzyme with dithiothreitol resulted in only 15% recovery of activity. Subsequent work from this laboratory[99] demonstrated that increase in activity was due to modification of Tyr92. This work was interesting because (1) similar results were obtained with both TNM and NAI and (2) modification increased enzyme activity. Modification of a cysteine residue (Cys69) with *N*-ethylmaleimide[100] results in increased activity and it is thought to be an indicator of oxidative stress; modification with *N*-ethylmaleimide does result in altered conformation as measured with hydrogen–deuterium exchange. While the chemical modification of proteins is frequently used to increase thermal stability[101] or, in the case of poly(ethylene) glycol, to improve therapeutic half-life,[102,103] it is rare that chemical modification increases the catalytic activity. It is noted that the reaction of peroxynitrite with cathepsin D is in association with an increase in activity.[104] Basu and Kirley[105] showed that modification of Try252 in ecto-nucleoside triphosphate diphosphohydrolase 3 with NAI resulted in an increase in enzyme activity. It is noted that the previously cited paper on increase in thermal stability of a cold-adapted lipase with an oxidized polysaccharide[101] also resulted in increased activity but, as noted by the authors, this observation defies the general principle about modification increasing stability at the expense of activity. Peroxynitrite (Figure 3.6) has come to dominate the literature on the nitration of tyrosine residues in proteins. There were only limited studies[106,107] on peroxynitrite prior to the 1992 publication by Ischiropoulos and coworkers[108] on the nitration of tyrosine catalyzed by superoxide dismutase. While it is not clear whether this was the seminal paper, it certainly is one of the earliest on the chemistry of peroxynitrite. The reaction of peroxynitrite is of great interest in the physiology of nitric oxide and biological oxidations[109–111]; peroxynitrite is of less interest to a protein chemist as it is not as convenient a reagent as TNM. There are, however, facile methods for the synthesis and storage of the reagent.[112–114] A search using SciFinder® yielded somewhat more than 12,000 citations for peroxynitrite and approximately 3,000 for TNM.

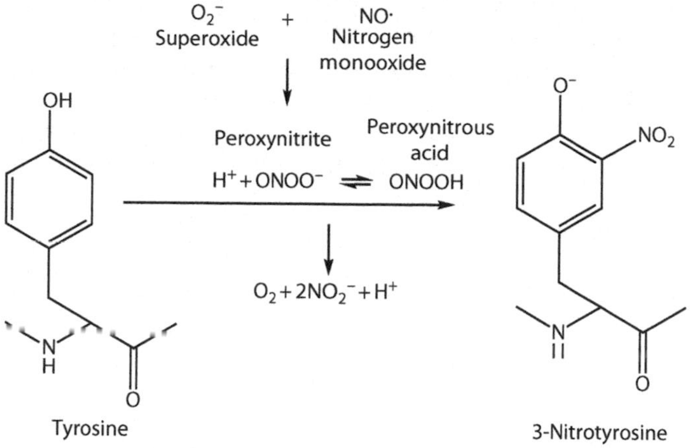

FIGURE 3.6 The reaction of peroxynitrite with tyrosine residues in proteins. Peroxynitrite is formed from superoxide and nitrogen monoxide. Decomposition is rapid forming dioxygen and nitrite. Peroxynitrate is suggested to form as an intermediate. (See Gupta, D., Harish, B., Kissner, R., and Koppenol, W.H., Peroxynitrate is formed during decomposition of peroxynitrite at neutral pH, *Dalton Trans.* 29, 5730–5736, 2009.)

The modification of proteins with peroxynitrite, while not as commonly used as TNM for structure–function studies, does deserve consideration. There is one article that deserves particular consideration as it is related to environmental aspects of tyrosine nitration.[115] This study used 23-residue transmembrane peptides with single tyrosine residues at position 4, position 8, and position 12. The peptides were inserted into a multilamellar liposome composed of 1,2-dilauroyl-sn-glycero-3-phosphatidyl choline. When peroxynitrite is generated *in situ*, nitration of the peptide in the liposome is greater than that of peptide in aqueous solution. Furthermore, nitration of the tyrosine residues increased with depth of penetration of the peptide into the liposome in that the nitration of the Tyr12 derivatives is greater than that of the Tyr8 derivative, which in turn is greater than that of the Tyr4 peptide. It is suggested that peroxynitrous acid diffuses into the membrane where it undergoes decomposition to form a nitric oxide radical, which then reacts with tyrosine to form nitrotyrosine. Hydrophobic substrates such as *N-t*-butyloxycarbonyl-tyrosine *t*-butyl ester have been developed for study of nitration by peroxynitrate in membranes.[116–118] Nitration of tyrosine in membranes is of considerable interest.[118–121]

Interest in the physiology of peroxynitrite has resulted in an increase in the number of approaches to the analysis of 3-nitrotyrosine in proteins.[109,119–122] Mass spectrometry is proving to be of increasing value in the analysis of nitrotyrosine in proteins.[123–133] Antibodies to 3-nitrotyrosine in proteins have been developed and are useful not only for the analysis of purified proteins[134–136] but also for the enrichment of nitrotyrosine-containing proteins,[136] *in situ* localization in cells,[137] and for identification on 2D gel electrophoretograms.[138] Conversion of nitrotyrosine to aminotyrosine via reduction with sodium dithionite[139] improves the specificity for the detection of nitrotyrosine-containing proteins on Western blots; nitrotyrosine-positive bands are eliminated on reduction leaving the "false-positive spots."[137,140–142] Nitrotyrosine is being developed as a biomarker for oxidative stress.[143–148] There are issues with the use of immunoassays for nitrotyrosine in plasma.[149,150]

The possible use of TNM for the modification of tyrosyl residues in proteins was advanced over 50 years ago.[151] However, it was not until some two decades later that the studies of Vallee, Riordan, Sokolovsky, and Harell established the specificity and characteristics of the reaction of TNM (Figure 3.7) with proteins.[152,153]

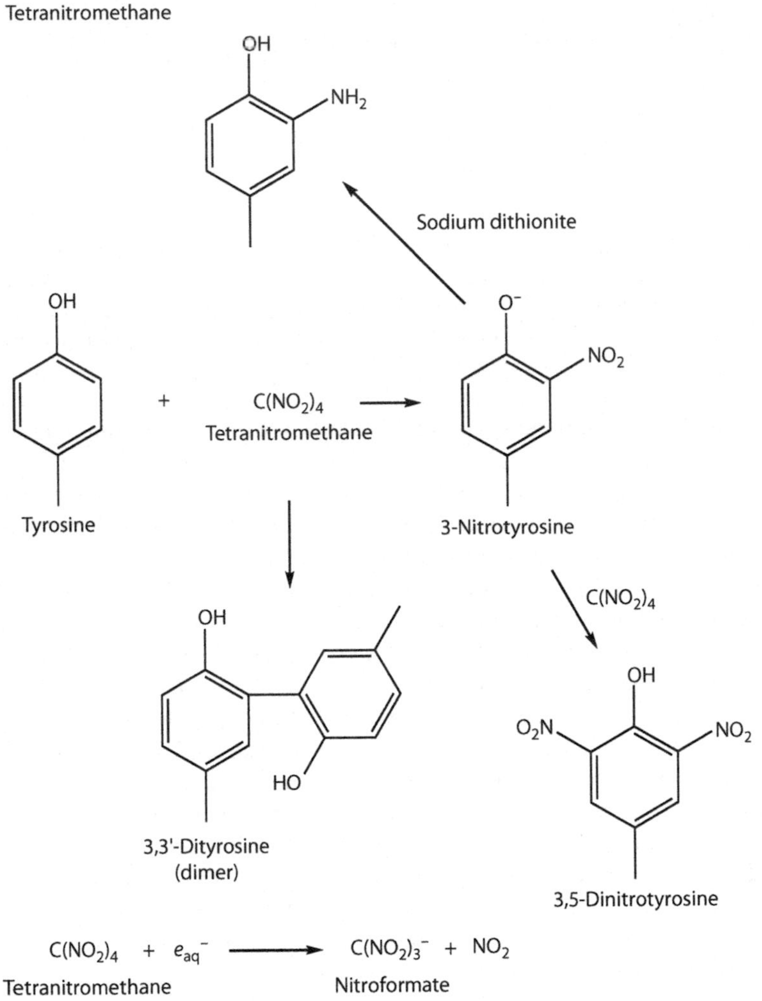

Tetranitromethane

OH

NH_2

Sodium dithionite

OH

O^-

NO_2

+ $C(NO_2)_4$ Tetranitromethane

Tyrosine

3-Nitrotyrosine

$C(NO_2)_4$

OH

OH

O_2N

NO_2

HO

3,3'-Dityrosine (dimer)

3,5-Dinitrotyrosine

$$C(NO_2)_4 + e_{aq}^- \longrightarrow C(NO_2)_3^- + NO_2$$

Tetranitromethane Nitroformate

FIGURE 3.7 The reaction of TNM with tyrosine. The major product of the reaction is 3-nitrotyrosine; there is lesser formation of 3,5-dinitrotyrosine and cross-linkage to form 3,3′-dityrosine dimer. The nitrotyrosine can be reduced to 3-aminotyrosine with dithionite. The nitration reaction is an example of an electrophilic aromatic substitution reaction with the formation of nitroform (nitroformate). (See Riordan, J.F. and Vallee, B.L., Nitration with tetranitromethane, *Methods Enzymol.* 25, 515–521, 1972; Capellos, C., Iyer, S., Liang, Y., and Gamms, L.A., Transient species and product formation from electronically excited tetranitromethane, *J. Chem. Soc. Faraday Trans.* 2 82, 2195–2206, 1986.)

The modification proceeds optimally at mildly alkaline pH. The rate of modification of *N*-acetyltyrosine is twice as rapid at pH 8.0 as at pH 7.0; it is approximately 10 times as rapid at pH 9.5 as at pH 7.0. The reaction of TNM with tyrosine produces 3-nitrotyrosine, nitroformate (trinitromethane anion), and two protons. The spectral properties of nitroformate (ε at 350 nm = 14,000)[154] suggested that monitoring the formation of this species would be a sensitive method for monitoring the time course of the reaction of TNM with tyrosyl residues.[152–154] Although determining the rate of nitroformate production appears to be effective in studying the reaction of TNM with model compounds such as *N*-acetyltyrosine, it has not proved useful with proteins.[153,155] Although the reaction of TNM with proteins is reasonably specific for tyrosine, oxidation of sulfhydryl groups has been reported[153,154] as has reaction with histidine,[153] methionine,[154] and tryptophan.[153,156]

Cysteine residues in proteins can be oxidized by TNM.[157] Palamalai and Miyagi[158] were able to modify tyrosine residues in glyceraldehyd-3-phosphate dehydrogenase and establish their role in enzyme function but were successful only after reversibly blocking the active site sulfhydryl group. Peroxynitrite also oxidizes sulfhydryl groups in proteins.[157,159–162]

The modification of tryptophan with TNM results in a variety of products.[153,156,163–171] Sokolovsky and colleagues[165] have studied the reaction of TNM with tryptophan in some detail suggesting that both nitration and oxidation of the indole ring can occur. These investigators also suggest that the tendency of tryptophan to be buried in proteins may either inhibit or facilitate modification (see above for reagent partitioning argument). Nitration of tryptophan is associated with a modest increase in absorbance in the range of 340–360.[166] There has been more recent work on the reaction of peroxynitrite with tryptophan[172–175] with the 5- and 6-nitro derivatives having been identified by Padmaja and coworkers (Figure 3.8).[176] Padmaja and coworkers[176] also reported a second-order rate constant of 130 M^{-1} s^{-1} at 25°C for the reaction. Scholz and Kwok[177] reported a second-order rate constant of 53.3 M^{-1} s^{-1} for the reaction between the tyrosine residues and TNM.

Reaction with TNM can also result in the covalent cross-linkage of tyrosyl residues resulting in inter- and intramolecular association of peptide chains.[178,179] The cross-linking of tyrosyl residues in proteins via reaction with TNM is an example of zero-length or contact-site cross-linking.[180,181] In studies of the mechanism of TNM nitration of phenols, Bruice and coworkers[182] observed more "cross-linkage" (formation of Plummer's ketone) than nitration with the reaction of TNM with *p*-cresol at neutral pH. The magnitude of this problem is dependent on variables such as protein, protein concentration, and solvent conditions (i.e., pH, presence of organic solvents).[183] For example, reaction of pancreatic deoxyribonuclease with TNM results in extensive formation of dimer.[184] The dimerization of insulin on reaction with TNM[179,183,185] has also been of interest. Both DNase and insulin are known to form dimers in solutions.[186,187]

A related reaction is the free radical-induced cross-linking between tyrosyl residues and thymine providing a basis for the formation of nucleic acid–protein conjugates occurring as a result of ionizing radiation.[188] Treatment of apo-ovotransferrin with periodate (50 mM HEPES, pH 7.4 with 5 mM sodium periodate) resulted in protein cross-linking via 3,3′-dityrosine.[189] The reader is also directed to the earlier

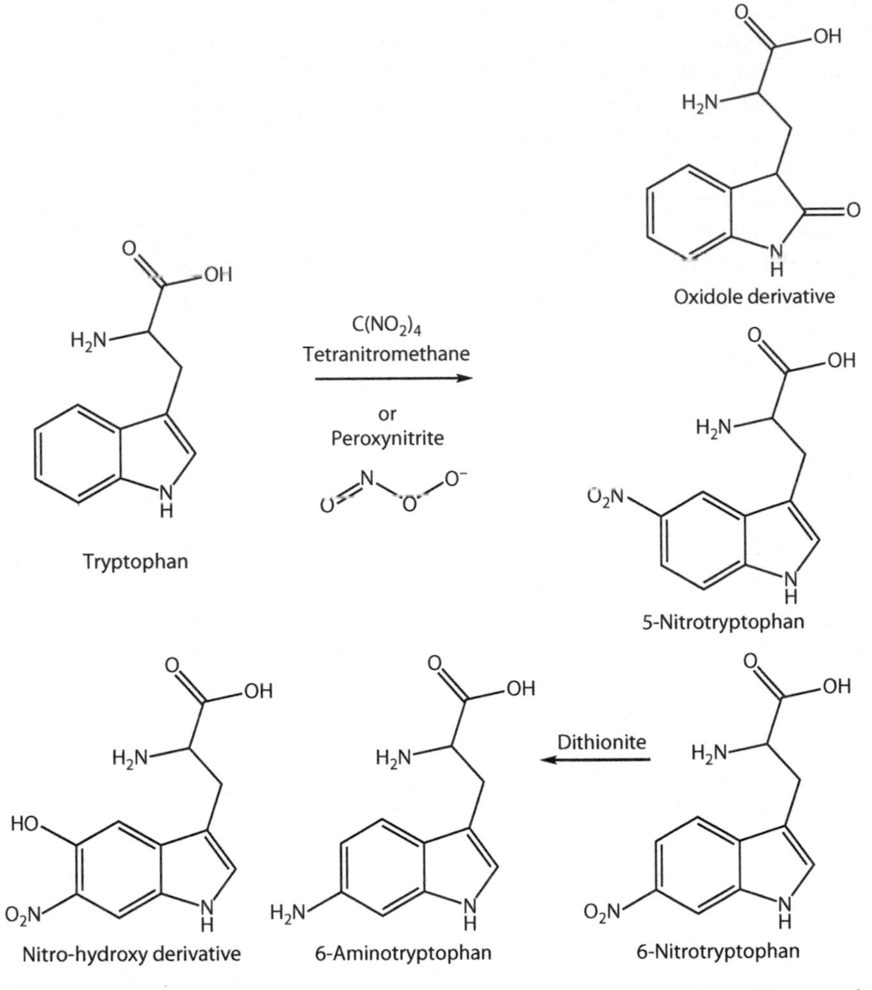

Tryptophan

$C(NO_2)_4$
Tetranitromethane
or
Peroxynitrite

Oxidole derivative

5-Nitrotryptophan

Dithionite

Nitro-hydroxy derivative

6-Aminotryptophan

6-Nitrotryptophan

FIGURE 3.8 The reaction of TNM or peroxynitrite with tryptophan. The reaction of TNM with tryptophan can provide a variety of substitution and oxidation products. (See Yamakura, F., Ikeda, K., Matsumoto, T. et al., Formation of 6-nitrotryptophan in purified proteins by a reactive nitrogen species: A possible new biomarker, *Int. Congr.* 1304, 22–32, 2007; Bregere, C., Rebrin, I., and Sohal, R.S., Detection and characterization of *in vivo* nitration and oxidation of tryptophan residues in proteins, *Methods Enzymol.* 441, 339–349, 2008.) The reader is also directed to Chapter 4 for a more thorough discussion of the oxidation of tryptophan.

cited work by Bartesaghi and coworkers[122] on the dimerization of tyrosine mediated by lipid peroxyl radicals in membranes. It is recognized that peroxynitrite does modify DNA.[190]

The use of immunochemistry either in solid-phase assays, Western blotting, or immunohistocytochemistry has been discussed earlier and is related more to the study of biological polymer modification in biological fluids and tissues. These analytical techniques are quite useful for the physiologists and of less value to those interested in solution chemistry of biological polymers where analysis is a critical part of the process of characterization. The extent of modification of tyrosyl residues by TNM in proteins can be assessed by either spectrophotometric means or by amino acid analysis.[155,191] At alkaline pH (pH ≥ 8), 3-nitrotyrosine has an absorption maximum at 428 nm with $\varepsilon = 4100\,M^{-1}\,cm^{-1}$; the absorption maximum of tyrosine at 275 nm increases from $\varepsilon = 1360$ to $4000\,M^{-1}\,cm^{-1}$. At acid pH (pH ≤ 6), the absorption maximum is shifted from 428 to 360 nm, with an isosbestic point at 381 nm ($\varepsilon = 2200\,M^{-1}\,cm^{-1}$). We have found it convenient to determine the A_{428} in 0.1 M NaOH. Amino acid analysis after acid hydrolysis has also proved to be a convenient method of assessing the extent of 3-nitrotyrosine formation. 3-Nitrotyrosine is stable to acid hydrolysis (6 N HCl, 105°C, 24 h). This approach has the added advantage that other modifications of tyrosine such as free-radical-mediated cross-linkage can be either excluded or quantitatively determined. If nitration to form 3-nitrotyrosine is the only modification of tyrosyl residues in a protein occurring on reaction with TNM, the sum of 3-nitrotyrosine and tyrosine should be equivalent to the amount of tyrosine in the unmodified protein.

There are several consequences of the nitration of a tyrosyl residue. The most obvious is the placing of a somewhat bulky substituent (the nitro group) *ortho* to the phenolic hydroxyl function. The properties of the substituent nitro group "push" electrons into the benzene ring (inductive effect), lowering the pK_a of the phenolic hydroxyl from approximately 10.3 to 7.3. This of course means that the phenolic hydroxyl of the nitrated tyrosyl residue will be in a partially ionized state at physiological pH. The nitro function can be reduced to the corresponding amine under relatively mild conditions ($Na_2S_2O_4$, 0.05 M Tris, pH 8.0).[139] The conversion of 3-nitrotyrosine to 3-aminotyrosine is associated with the loss of the absorption maximum at 428 nm and the change in the pK_a of the phenolic hydroxyl group from approximately 7.0 to 10.0. On occasion, reduction of the nitro function in this manner reverses the modification of function observed on nitration. The resultant amine function can be subsequently modified.[192] Fox and colleagues[193] modified the single tyrosine residue in *Escherichia coli* acyl carrier protein with TNM, subsequently reduced the 3-nitrotyrosyl residue to 3-aminotyrosine with sodium dithionite, and modified the 3-aminotyrosine with dansyl chloride at pH 5.0 (50 mM sodium acetate) to obtain dansyl acyl carrier protein. The dansyl acyl carrier protein was subsequently used for fluorescence anisotropy studies of enzyme–substrate complex formation in stearoyl-ACP desaturase.[194] The 3-aminotyrosine derivative has been used as a "handle" for the purification of peptides containing nitrotyrosine. Nikov and coworkers[136] coupled biotin to aminotyrosine-containing peptides using a cleavable derivative [sulfosuccinimidyl-2-(biotinamido)ethyl-1,3-dithioproprionate]; peptides were purified by affinity chromatography using streptavidin columns with elution by reductive cleavage of the label (Figure 3.9). This approach has been used by other investigators for the identification of nitrotyrosine in proteins.[132,195–197] It should be noted that the low pK_a for 3-aminotyrosine (ca. 4.7) provides potential for specific modification in a protein.[156]

Modification of protein-bound tyrosine with TNM also introduces a spectral probe (3-nitrotyrosine) that can be used to detect conformational change in the protein.

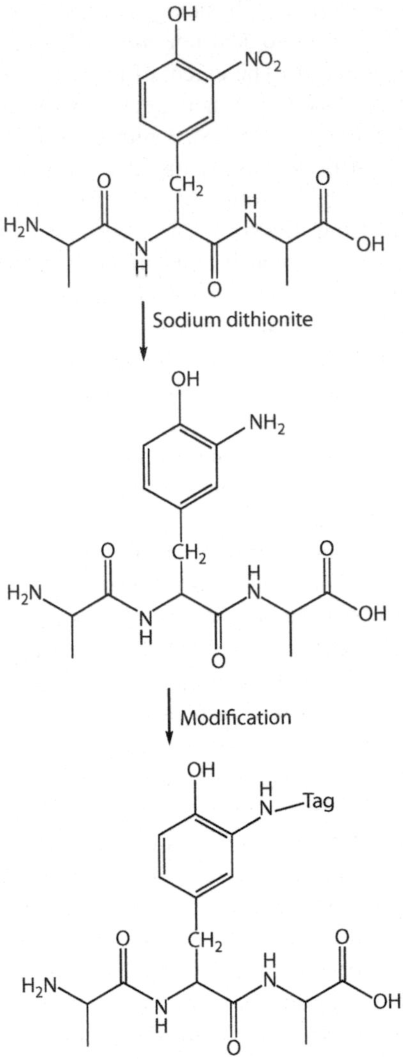

FIGURE 3.9 Coupling of biotin (or another tag) to aminotyrosine for purification of peptides. The nitrated tyrosine is reduced to the corresponding amine. The amino group has a low pK_a, which permits specific modification; however, some investigators do treat the peptide/protein with acetic anhydride prior to the reduction step to block any reaction with the tag (e.g., succinimidyl-biotin). (See Nikov, G., Bhat, V., Wishnok, J.S. and Tannenbaum, S.R., Analysis of nitrated proteins by nitrotyrosine-specific affinity probes and mass spectrometry, *Anal. Biochem.* 320, 214–222, 2003; Zhang, O., Qian, W.J., Knyushko, T.V. et al., A method for selective enrichment and analysis of nitrotyrosine-containing peptides in complex proteome samples, *J. Proteome Res.* 6, 2257–2268, 2007; Lee, J.R., Lee, S.J., Kim, T.W. et al., Chemical approaches for specific enrichment of and mass analysis of nitrated peptides, *Anal. Chem.* 81, 6620–6626, 2009; Abello, N., Barroso, B., Kerstjens, H.A.M. et al., Chemical labeling and enrichment of nitrotyrosine-containing peptides, *Talanta* 80, 1503–1512, 2010.)

3-Nitrotyrosine has an absorption maximum at 428 nm at alkaline pH. This spectral property was first used by Riordan and coworkers with studies on nitrated carboxypeptidase A[197] to study changes in the microenvironment around the modified residue. Addition of β-phenylpropionate, a competitive inhibitor of carboxypeptidase and nitrated carboxypeptidase, decreased the absorbance of mononitrocarboxypeptidase at 428 nm. This change is consistent with an increase in the hydrophobic quality of the microenvironment surrounding the modified tyrosyl residue. The spectral characteristics of nitrotyrosine are thus useful as a reporter group. Reporter groups are probes that have absorbance or fluorescence properties and can be used to monitor conformational change in a protein.[198–202] There has only been limited use of the UV spectral characteristics of 3-nitrotyrosine as a reporter group[203,204]; there has been use of the fluorescence quenching properties of nitrotyrosine as a conformational probe.[205–208] There are several unique applications that deserve specific mention. Herz and coworkers[209] coupled nitrotyrosine methyl ester to a carboxyl group in bacteriorhodopsin for use as a reporter group.

The reactivity of tyrosyl residues in proteins, as with other residues, has been suggested as a measure of exposure to solvent. In this model, "free" tyrosyl residues are considered to be in direct contact with solvent and have pK_a values between 9.5 and 10.5 while "buried" tyrosyl residues are relatively inaccessible to solvent and have pK_a values above 10.5. TNM is not considered to be water soluble[210] but is soluble in polar solvents such as ethanol and apolar solvents such as ethyl ether. This latter fact is clearly stated in a review of TNM by Riordan and Vallee.[155] As noted by these authors, TNM can be washed with water to remove impurities and high concentrations of reagent should be avoided to obviate the possibility of phase separation. Rabani and coworkers also describe an aqueous extraction process to remove impurities from TNM.[154] In his description of the synthesis of TNM from acetic anhydride and nitric acid in 1955, Liang[211] reported that TNM separates from the aqueous phase during the process. However, studies on the chemistry of TNM at concentrations of 50 μM in aqueous solutions have been performed without reported difficulty.[212] It should be noted that most experiments using TNM involve the introduction of the reagent as a solution in ethanol or less frequently with methanol. The final concentration of ethanol is frequently 5%–10% of the final reaction volume, which likely enhances the solubility of TNM in the reaction mixture. The author is not aware of any studies that specifically address the effect of ethanol (or other organic water-miscible solvents) on the solubility of TNM in an aqueous solution.

The reaction mechanism as suggested by Walters and Bruice[213] postulates the formation of an intermediate phenoxide–TNM charge transfer complex involving electron transfer from the aromatic ring. This would imply that the reaction of TNM with a tyrosine residue occurs with the ionized species. These studies were performed with TNM and *p*-methylphenol (4-methylphenol; *p*-cresol) and the reaction was followed by the disappearance of TNM as measured by the formation of trinitromethane anion (nitroformate) at 350 nm. The products of the reaction were trinitromethane anion, nitrite, and 4-methyl-3-nitrophenol. The rate of reaction (k_2 = min^{-1} 5.1 × 10^4 M^{-1}) was slower with phenol (k_2 = 2.0 × 10^3 M^{-1} min^{-1}) and much slower with either 4-chlorophenol (5.7 × 10^2 M^{-1} min^{-1}) or 4-cyanophenol (0.26 M^{-1} min^{-1}). There is no obvious correlation between these rates and the dissociation constant for the phenolic hydroxyl group; there is a correlation with reaction rate and the Hammett constant σ.[214] Analysis of the products of the reaction of TNM and 4-methylphenol demonstrated the presence of 4-methyl-3-nitrophenol in 23% yield and a 30% yield of 1,2,10,11-tetrahydro-6,11-dimethyl-2-oxodibenzfuran (Pummerer's ketone), a product derived from the free radical mediated cross-linking of the parent 4-methylphenol. Given the average pK_a value of 10.13 for tyrosine (see Table 1.1), this would imply that the average tyrosine residues would be mostly unreactive within the pH range of 6.0–8.0, which is suggested as the optimal range for the modification of tyrosine in proteins. This is supported by Bruice and coworkers[182] who demonstrated that the rate of reaction of TNM with the phenolic form must be at least 4 orders of magnitude smaller than those observed with the phenolate form. At pH values above 8, the rate of reaction of TNM with tyrosine would appear to increase consistent with the importance of the ionization of the phenolic function; however, the side reactions with methionine, tryptophan, and histidine increase. The studies on the reaction with 4-methylphenol and TNM yielded an approximate 20% yield of 3-nitro-4-methylphenol, which could be only modestly increased by excess TNM, while reaction in proteins could yield 100% of modification of a given tyrosyl residue.[215] Early studies[152,153] suggested that the formation of trinitromethane anion (nitroformate) could be used to stoichiometrically follow the reaction of TNM with tyrosyl residues in proteins (see Figure 3.7). This has not proved to be the situation as, in general, the release of trinitromethane anion greatly exceeds the extent of modification of tyrosine and continues beyond the nitration reaction in proteins. While it is not clear as to why this occurs, Fendler and Liechti[212] have shown that the certain detergent micelles catalyze the reaction of TNM with hydroxide ion. Micellar hexadecyltrimethylammonium bromide and polyoxyethylene (15) nonylphenol (Igepal® CO-730) enhance the rate constant for the reaction, while micellar sodium dodecyl sulfate has no effect. Nitroformate is derived from nitrotyrosine by reduction (addition of an electron)[154,216] and this reaction is used for the measurement of superoxide radical.[217,218] The reaction of hydroxyl radical with TNM also yields nitroformate.[219]

Photolysis is used to generate the nitrogen dioxide radical and the nitroformate anion from TNM.[220] This process can result in the nitration of suitable targets and the formation of trinitromethyl derivatives.[221] There is at least one study[222] on the photochemical reactions of aromatic compounds with TNM, which may be applicable to the interpretation of the reaction of TNM with proteins.

This author is also aware of at least one study[223] that examined the effect of light on the reaction of TNM with proteins, however the effect of light on proteins, not TNM. That study reported the effect of light on the reaction of TNM with bacteriorhodopsin with light where light influenced the intrinsic properties of the protein. Bacteriorhodopsin is a bacterial membrane protein, which is activated by light to generate an electrochemical gradient. Absorption of light by the protein initiates a cycle of reactions, which results in proton translocation across the membrane.[224] Tyr26 was preferentially nitrated in the dark, while Tyr64 was preferentially modified in the presence of light.[223] In a subsequent study,[225] selective ionization of Tyr64 ($pK_a \sim 9.0$) was observed to occur with illumination. Photochemically induced dynamic nuclear polarization NMR spectroscopy of bacteriorhodopsin suggested that only one tyrosine residue is exposed to solvent.[226] Light also influenced the modification of bacteriorhodopsin with a carbodiimide.[227]

Other studies on the chemistry of TNM that might be instructive include a study by Capellos and others,[216] which showed that, upon excitation with an excimer laser, TNM formed the NO_2 radical and the trinitromethane anion in either aerated or deaerated polar solvents (e.g., methanol), which the trinitromethyl radical was suggested to form in a nonpolar solvent (e.g., hexane). TNM also reacts with unsaturated compounds[228] and has been used as a colorimetric method[229] for the determination of unsaturation. The study on the difference of reactions in the polar and nonpolar solvents might be useful in interpreting environmental effects on the reaction of TNM with individual residues in proteins. Given the complexity of the interpretation of the aforementioned studies and extension to the reaction of tyrosine in proteins, it is clear that the reaction of TNM with biological polymers can be complex and care should be taken in the interpretation of experimental results.[230]

The next section will discuss several studies on the reaction of tyrosine with tyrosine in proteins where there is information about the microenvironment surrounding the modified residues. Strosberg[231] and coworkers reported on the reaction of TNM with hen egg white lysozyme. With a 12-fold molar excess of TNM in 50 mM Tris-HCl, 1 M NaCl, pH 8.0 at 20°C for 1 h, a derivative is obtained containing 1 mol of nitrotyrosine per mole protein as determined by amino acid analysis. A derivative containing 2 mol of nitrotyrosine per mole protein is obtained with a 47-fold molar excess of TNM, while a derivative containing 3 mol of nitrotyrosine per mole of protein is obtained with a 100-fold molar excess of TNM in 50 mM Tris-HCl, 1 M NaCl, and 8.0 M urea, pH 8.0, at 50°C for 1 h. Analysis of the various products showed that with a 12-fold molar excess of reagent, two mononitrated derivatives are obtained, the major one (69%) is modified at Tyr23 and the minor product (21%) at Tyr20. With a 47-fold molar excess of TNM, two dinitrated products are obtained. The major product (71%) is modified at both Tyr23 and Tyr20, while there is a minor product (15%) modified at Tyr23 and Tyr53. The derivative modified with 100-fold molar excess of reagent in the presence of 8.0 M urea was homogenous and was modified at Tyr23, Tyr20, and Tyr53. A later study[232] examined the characteristics of hen egg white lysozyme modified at Tyr23 or at both Tyr23 and Tyr20. Both derivatives were obtained by reaction in 50 mM Tris-HCl, pH 8.0 at 37°C for 1 h. The Tyr23/Tyr20 derivative was obtained at a 15-fold molar excess of TNM, while the Tyr23 was selectively obtained at twofold molar excess of reagent. The Tyr23 derivative could also be obtained by electrochemical nitration.[233] Skawinski and coworkers[234] have used [^{15}N]TNM to modify tyrosine residues in lysozyme. Two or three ^{15}N resonances were obtained dependent on the extent of modification; the pH dependence of the detected resonances depicted the apparent microscopic pK_a values. These investigators also studied the modification of ribonuclease with [^{15}N]TNM. Šantrůček and coworkers[235] showed only small differences in the quality and quantity of modification of lysozyme with TNM and iodine using MALDI-TOF mass spectrometry. The order of reactivity of tyrosine residues with TNM and iodine (iodine + potassium iodide at pH 7.0) is similar in hen egg white lysozyme. There was no clear correlation between surface accessibility of residues and modification with TNM or iodine.

Crambin is a small water-insoluble protein, which contains two tyrosine residues. Reaction with TNM was performed in 50% EtOH/H_2O at pH 9.0 (0.0005 M Tris) and was monitored by NMR spectroscopy.[236] Tyr29 is modified with a 50-fold molar excess of reagent in 10 min at 298 K (25°C). Addition of a second 50-fold molar excess of TNM and continuation of the reaction resulted in the additional modification of Tyr44. A reverse susceptibility is observed with iodination (I_2, pH 8.5 with 0.0005 M Tris-HCl in 50% EtOH/H_2O). The reaction of tyrosine residues with TNM is consistent with solvent-exposed residues being more susceptible to reaction, while the iodination pattern is not. It was concluded that the susceptibility to modification by either TNM or I_2 is controlled by the microenvironment of the residue independent of solvent exposure; it was concluded that the preferential iodination of Tyr44 was facilitated by an electrostatic effect from Asp43. The authors also suggest that iodination or nitration should not be taken as a single criterion for the exposure of tyrosyl residues. In a study that was discussed earlier with respect to the use of NAI to label membrane proteins, El Kabbaj and Latruffe[92] studied the ability of TNM to penetrate the inner mitochondrial membrane by evaluating the inactivation of D-3-hydroxybutyrate dehydrogenase by TNM in its normal location and in "inside-out" membranes. TNM readily inactivated the enzyme on "inside-out" membranes but was much less effective with intact mitochondrial membrane. These investigators also reported on the distribution of TNM between aqueous and organic solvents. The percentage partition was 17/83 in water/hexane and 11/89 in water/*n*-octanol. The authors suggest that the behavior of TNM is consistent with both partial amphiphilic and hydrophobic properties. The limited discussion mentioned earlier suggests that it is not possible to accurately predict the environment of a tyrosyl residue modified by TNM in a protein. Care should be taken in the interpretation of results when such observations are used to describe the topography of a protein.[237–241]

The chemical modification of biological polymers can be said to have two general objectives. One is to determine topography and spatial relationships of functional groups. A second broad purpose is to identify functional groups that participate in ligand binding or catalysis. The following studies are presented as examples of where interpretation becomes a challenge. The modification of tyrosine residues in thermolysin with TNM was performed in 40 mM Tris, pH 8.0 containing 10 mM $CaCl_2$ at 25°C for 1 h.[242] Thermolysin contains 28 tyrosine residues and approximately 9 residues were modified under these conditions as determined by absorbance at 381 nm. When the reaction was allowed to proceed for 15 h under the same conditions, approximately 16 residues were modified. Analysis of the reaction rate allowed the identification of three classes of reactive tyrosyl residues, which reacted at different rates with TNM. The apparent second-order rate constants for the three classes are 3.32, 0.52, and 0.18 M^{-1} min^{-1} compared to a second-order rate constant for the reaction of TNM with *N*-acetyltyrosine ethyl ester of 1.99 M^{-1} min^{-1} under the same reaction conditions. In the same study, spectrophotometric (295 nm) titration of thermolysin by pH jump demonstrated that 16 tyrosine residues were readily ionized, while 12 additional tyrosine residues required an apparent conformational change in the protein for ionization after adjustment to pH ≥ 12. These investigators divided the readily ionized tyrosine residues into three classes with pK_a values of 10.2, 11.4, and 11.8. These investigators concluded that the microenvironment of the more slowly reacting tyrosine residues is either negatively charged or hydrophobic. The effect of nitration on the catalytic activity of thermolysin was examined in a subsequent communication from the same research group.[243] Activity was reduced to 10% of that of the native enzyme with nitration of 16 tyrosine residues (the effect was on k_{cat}) with partial recovery on reduction of the nitro groups to amino groups; activation by NaCl was also decreased on nitration. In another study, Muta and coworkers[244] showed that alkaline pK_a for human matrilysin-7 shifted from 9.8 to 10.3–10.6 with nitration; the pK_a returned to 9.8 on reduction of the nitro groups to amino groups.

Protection of an enzyme from loss of activity and lack of modification of an amino acid residue(s) is taken to infer the importance of the said residue(s) in the catalytic activity of the enzyme. However Christen and Riordan[245] observed that aspartate aminotransferase was readily inactivated by TNM only in the presence of both substrates glutamate and α-keto-glutarate ("syncatalytic" modification). This inactivation was associated with the modification of an additional tyrosyl residue. The syncatalytic modification of other enzymes has been subsequently reported including the inactivation of chicken liver xanthine oxidase by hydrogen peroxide generated during oxidation of the substrate by oxygen,[246] the syncatalytic modification of adrenocortical cytochrome P-450$_{scc}$ with TNM,[247] and the formation of a disulfide in aldolase.[248] Another anomaly is when a modifying agent such as TNM is a substrate for enzyme such as with isocitrate lyase from germinating castor seed endosperm where the enzyme catalyzes the reaction of TNM with isocitrate to form nitroform.[249] Medda and coworkers[250] demonstrated the formation of nitroform in the oxidation of putrescine by lentil seedling amine oxidase. In an earlier study,[251] Mekhanik and Torchinskii in studies on glutamate decarboxylase showed that TNM reacted with a pyridoxal phosphate substrate carbanion during enzymatic decarboxylation with α-methylglutamate or aspartate as substrate.

Antibodies to nitrotyrosine have been of value in the study of this modification as a biomarker.[252] This has been discussed earlier and a more thorough discussion of biomarkers is available.[253] There is also interest in the of nitration to study antigen–antibody interactions as well as related interactions such as those between avidin and biotin. Tawfik and coworkers[254] showed that nitrated antibodies demonstrate pH-dependent binding near physiological pH. Modification of monoclonal antibodies with TNM resulted in the loss of antigen binding at pH 8.0, which is regained at pH 6.0. This change is related to the decrease in the pK_a of the phenolic hydroxyl group (which would be ionized at pH 8.0 but protonated at pH 6.0). It is suggested that the decrease in the activity of hen egg white lysozyme seen on modification with TNM[255] is due to the ionization of the nitrotyrosyl residue(s), which would repel the negatively charged substrate; removal of the cell wall anionic polymer, teichuronic acid, from the *Micrococcus luteus* cell wall fragment eliminated this effect. The tight binding of biotin to avidin is reduced by modification of a tyrosine residue in avidin by TNM.[256] The modified avidin can then be used for the affinity chromatography of biotinylated proteins under relatively mild conditions.[257,258]

Tyrosyl residues in proteins can also be modified by reaction with cyanuric fluoride.[259–261] The reaction proceeds at alkaline pH (9.1) via modification of the phenolic hydroxyl group with a change in the spectral properties of tyrosine. The phenolic hydroxyl groups must be ionized (phenoxide ion) for reaction with cyanuric fluoride. The modification of tyrosyl residues in elastase[262] and yeast hexokinase[263] with cyanuric fluoride has been reported. Coffe and Pudles[263] compared the effect of cyanuric fluoride, TNM, and NAI in the modification of tyrosine residues in yeast hexokinase; two tyrosyl residues are modified with either cyanuric fluoride or TNM, while only one is modified with NAI. While it is suggested that cyanuric fluoride is relatively specific for the modification of tyrosine in proteins,[259] there is no direct data available with a direct comparison to either cyanuric chloride or cyanuric bromide. Cyanuric chloride has been demonstrated to react with both tyrosine and amino groups in proteins and has been used for the coupling of dextran to proteins[264]; in this latter study, the inactivation associated with lysine residues was obviated by prior modification of the protein with citraconic anhydride. The author was not able to find information on the reaction of cyanuric bromide with any biological polymer. It is noted that the various cyanuric halides are precursors in the synthesis of various dyes. Cyanuric bromide can be an impurity in preparations of cyanogen bromide.[265,266] Cyanogen bromide is discussed in Chapter 5.

Modification of tyrosyl residues can occur as a side reaction with other residue-specific reagents such as 7-chloro-4-nitrobenzo-2-oxa-1,3-diazole (7-chloro-4-nitrobenzofurazan; NBD-Cl; Nbf-Cl).[267,268] NBD-Cl reacts primarily with amino groups and sulfhydryl groups in proteins.[269–271] The reaction product obtained with tyrosine, unlike that obtained with either amino groups or sulfhydryl groups, is not fluorescent and has an absorption maximum at 385 nm compared to 475 nm for amino derivatives and 425 nm for sulfhydryl derivatives.[267] The reaction of NBD-Cl with sulfenic acid has been reported[270] and is used for the detection of sulfenic acid in oxidized proteins,[272,273] although dimedone appears to be more specific.[272,274]

Diazonium salts readily couple with tyrosine, lysine, and histidine residues in proteins at alkaline pH (Figure 3.10) to form colored derivatives with interesting spectral properties.[275–280] The reaction of chymotrypsinogen A with diazotized arsanilic acid has been investigated.[281] The extent of the formation of monoazotyrosyl and monoazohistidyl derivatives is determined by spectral analysis.[279,280] The arsaniloazo functional group provides a spectral probe that can be used to study conformational change in proteins. In this particular study, there was a substantial change in the circular dichroism spectrum (extrinsic Cotton effect) during the activation of the modified chymotrypsinogen preparation by trypsin.[281] The reaction of α-chymotrypsin with three diazonium salt derivatives of *N*-acetyl-D-phenylalanine methyl ester[282] (Figure 3.11) has also been studied. As noted in Figure 3.11, these diazonium salts are analogues of allosteric activators of chymotrypsin. Subsequent analysis showed that Tyr146 is modified by each of the three reagents. It was observed that the peptide with the modified tyrosine residue (possessing a yellow color) absorbs to the gel filtration matrix (G-10 equilibrated with 0.001 M HCl) and was eluted with 50% acetic acid. This phenomenon is somewhat similar to that observed with tryptophan-containing peptides, which have been modified with 2-hydroxy-5-nitrobenzyl bromide.[283] A single tyrosine residue in bovine pancreatic ribonuclease has been modified by a diazonium salt derivative of uridine 2′(3′)5′-diphosphate [5′-(4-diazophenyl phosphoryl)-uridine-2′(3′)-phosphate].[284] These investigators also

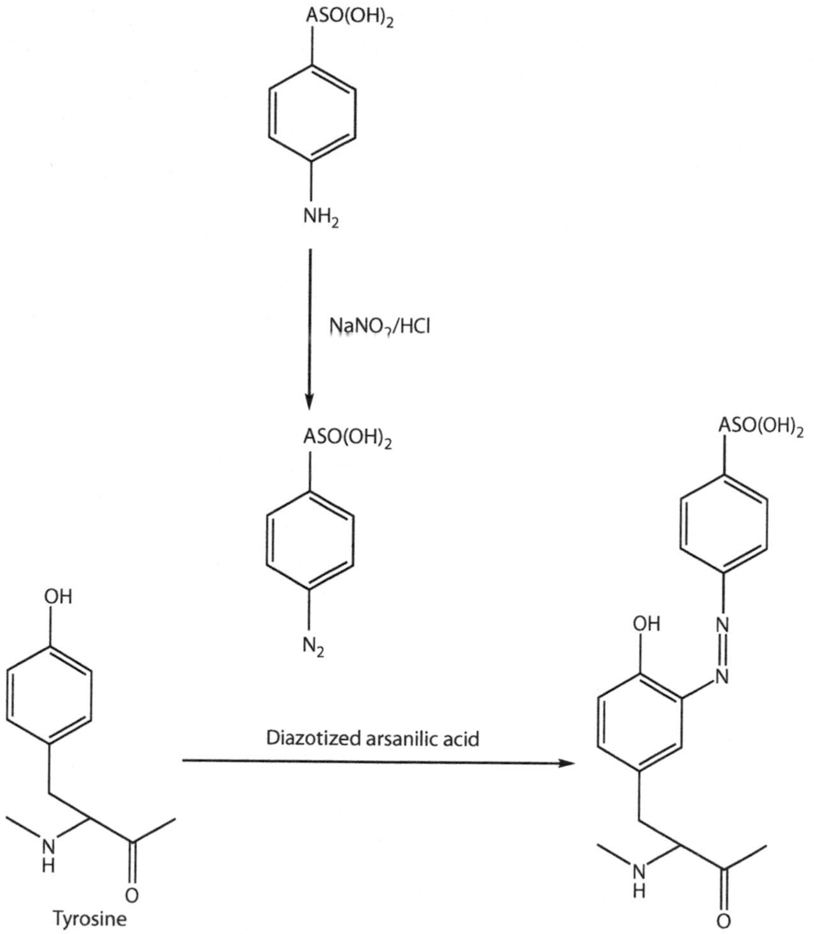

FIGURE 3.10 The reaction of tyrosine with diazotized arsanilic acid. (See Riordan, J.F. and Vallee, B.L., Diazonium salts as specific reagents and probes of protein conformation, *Methods Enzymol.* 25, 521–531, 1972. See also Pielak, G.J., Urdea, M.S., and Legg, J.I., Preparation and characterization of sulfanilazo and arsaniloazo proteins, *Biochemistry* 23, 596–603, 1984; Pielak, G.J., Gurusiddaiah, S., and Legg, J.I., Quantification of azo-coupled lysine in azo proteins by amino-acid-analysis, *Anal. Biochem.* 156, 403–405, 1986.)

examined the reaction of ribonuclease with *p*-diazophenylphosphate under the same conditions of solvent and temperature. Reaction with this reagent was far less specific, with losses of lysine, histidine, and tyrosine (3 mol mol^{-1} ribonuclease). Vallee and coworkers[285,286] used diazotized *p*-arsanilic acid to obtain specific modification of Tyr248. The reaction of bovine carboxypeptidase A with diazotized 5-amino-1H-tetrazole has also been reported.[287] Diazotized 5-amino-1H-tetrazole also specifically reacts with Tyr248 in bovine carboxypeptidase A; modification of this tyrosine residue permits the subsequent modification of Tyr198 by TNM.

p-Nitrobenzenesulfonyl fluoride (NBSF) is another reagent that can be used for the modification of tyrosyl residues in proteins (Figure 3.12). This reagent was

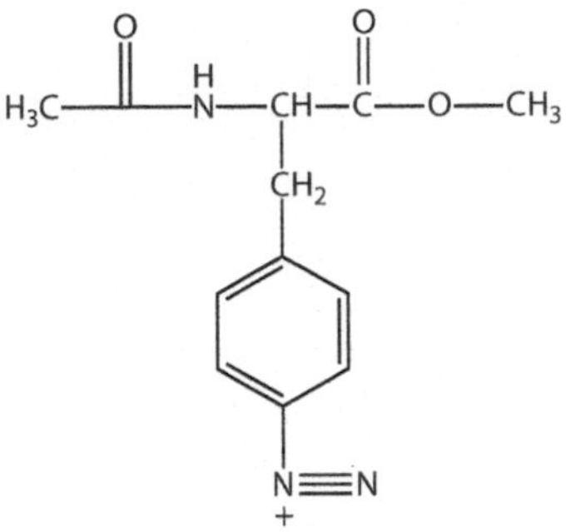

$H_3C-C(=O)-NH-CH(CH_2-C_6H_4-N^+\equiv N)-C(=O)-O-CH_3$

p-Diazo-N^α-D-phenylalanine methyl ester

$N\equiv N^+-C_6H_4-C(=O)-NH-CH(CH_2-C_6H_4-NO_2)-C(=O)-O-CH_3$

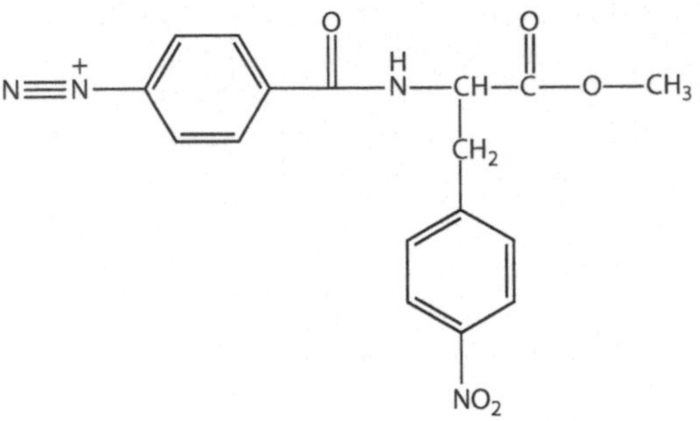

p-Nitro-N^α-(4-diazobenzoyl)-D-phenylalaninemethyl ester

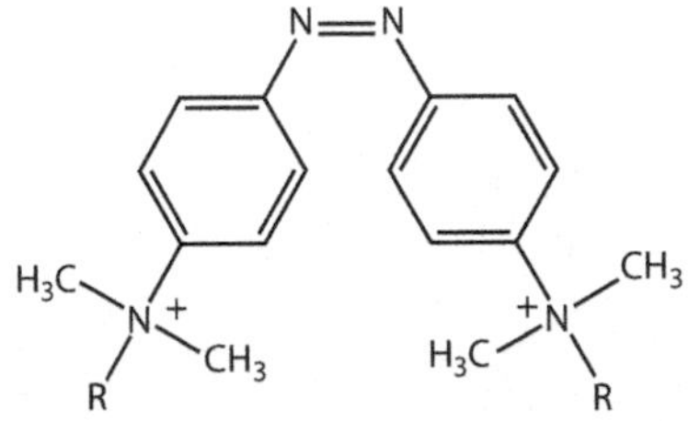

Allosteric activator as described by Erlanger et al.

FIGURE 3.11 Reaction of substrate analogue diazonium salts with chymotrypsin. (See Gorecki, M., Wilchek, M., and Blumberg, S., Modulation of the catalytic properties of α-chymotrypsin by chemical modification at Tyr 146, *Biochim. Biophys. Acta* 535, 90–99, 1978.) The reader is referred to studies on the allosteric activation of chymotrypsin with some azobenzene compounds (Erlanger, B.F., Wassermann, N.H., Cooper, A.G., and Monk, R.J., Allosteric activation of the hydrolysis of specific substrates by chymotrypsin, *Eur. J. Biochem.* 61, 287–295). In a related study, benzyloxycarbonyl derivatives of aromatic amino acids facilitated binding of chymotrypsin to an affinity column using 4-phenylbutylamine linked to succinylated agarose (Dunn, B.M. and Gilbert, W.A., Quantitative affinity chromatography of α-chymotrypsin, *Arch. Biochem. Biophys.* 198, 533–540, 1979).

developed by Liao and coworkers for the selective modification of tyrosyl residues in pancreatic DNase.[288] The modification reaction with NBSF can be performed in solvents (i.e., 0.1 M Tris-Cl, pH 8.0; 0.1 M *N*-ethylmorpholine acetate, pH 8.0) typically used for the modification of tyrosine residues by other reagents such as TNM or NAI. The rate of reagent hydrolysis is substantial and increases with increasing pH. Calcium ions are important for DNase structure[289] and influences reaction with NBSF; in the presence of calcium ions, a single tyrosyl residue is modified, while in the absence of calcium ions, approximately 3 tyrosyl residues are modified. Liao and coworkers[288] also evaluated related compounds for the inactivation of DNase; *p*-nitrobenzenesulfonyl chloride (NBSC) also inactivates DNase but both tyrosine and lysine residues are modified. NBSF reacts only with tyrosine in amino acid mixtures, while NBSC reacts with both tyrosine and lysine. NBSF did oxidize glutathione. No appreciable inhibition was observed with phenylmethylsulfonyl fluoride, which reacts with the active serine of chymotrypsin and related enzymes. Dastoli and coworkers[290]

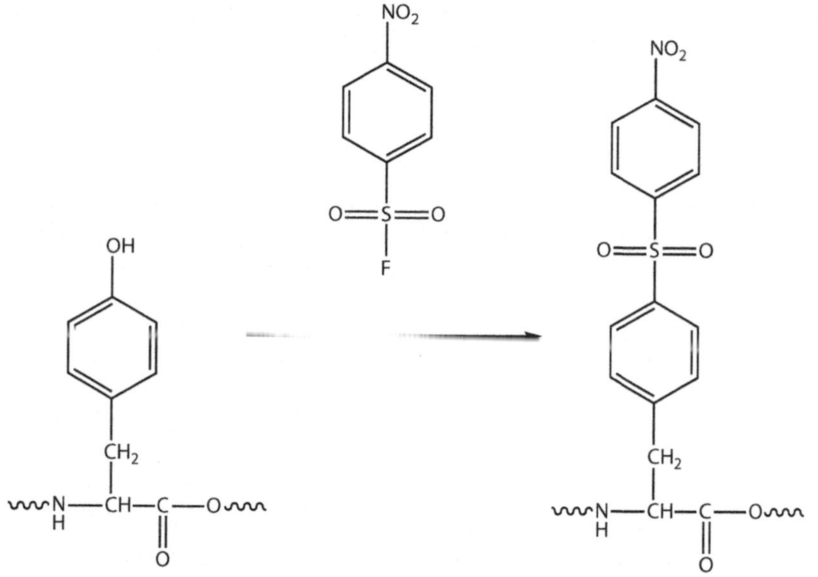

FIGURE 3.12 Reaction of nitrobenzenesulfonyl fluoride with tyrosine. (See Liao, T.H., Ting, R.S., and Yeung, J.E., Reactivity of tyrosine in bovine pancreatic deoxyribonuclease with *p*-nitrobenzenesulfonyl fluoride, *J. Biol. Chem.* 257, 5637–5644, 1982.)

observed the inactivation of chymotrypsin with *m*-nitrobenzenesulfonyl fluoride and ascribed the inactivation to be due to reaction with a serine residue as demonstrated by Gold[291] for the reaction of dansyl chloride with chymotrypsin. Dastoli and coworkers[290] performed their study with crystals of chymotrypsin suspended in methylene chloride. While not every study provides a rigorous identification of the site of modification, it is reasonable to assume that NBSF reacts specifically with tyrosyl residues in proteins.

There has been steady use of NBSF for the modification of tyrosyl residues in proteins since the work of Liao and coworkers.[288] NBSF has been used to characterize tyrosyl residues in NAD(P)H:quinone reductase.[292] Analysis of the product of the reaction showed that NBSF-modified tyrosyl residues were located in hydrophobic regions of the protein. In another study,[293] the reaction of NBSF with tyrosyl residues in human placental taurine transporter was compared with modification observed with TNM, NAI, and NBD-Cl. TNM and NBD-Cl were the most effective reagents. NBSF was an order of magnitude less effective, while NAI was 500 times less potent. Gitlin and coworkers[294] used NBSF to modify the tyrosyl residue in the biotin-binding site of avidin and streptavidin, while Suzuki and coworkers used NBSF to modify the tyrosyl residue at the active site of a catalytic antibody.[295] NBSF has also been used to modify the tyrosyl residue important for activity of snake venom phosphatases.[296–298] Modification of the phenolic hydroxyl group with 2,4-dinitrofluorobenzene has also been reported.[299] A novel reaction of PMSF with a tyrosyl residue in an archaeon superoxide dismutase has been reported[300] and it is noted that Means and Wu reported the modification of a tyrosine residue in human serum albumin with DFP.[301] The modification of albumin by related organophosphorous compounds[302,303] is suggested to be useful as biomarker of exposure to organophosphorous pesticides.[304]

CARBOXYL GROUPS

The site-specific modification of carboxyl groups in proteins is somewhat difficult to achieve, as is differentiation between aspartyl residues and glutamyl residues. Most of the published work utilizes water-soluble carbodiimides as described in the following. The reader is recommended to several recent articles addressing the modification of carboxyl groups in proteins.[305–310] Carbodiimides are also used for coupling of protein to matrices and for zero-length cross-linking.[311–314] Carbodiimide chemistry is also used for the modification of carboxyl groups in polysaccharides.[315,316]

Parsons and coworkers used triethyloxonium fluoroborate to modify the β-carboxyl group of an aspartic residue essential for the enzymatic activity of lysozyme.[317,318] Paterson and Knowles[319] used trimethyloxonium fluoroborate to determine the number of carboxyl groups in pepsin that are essential for catalytic activity. This article discusses in some depth the rigorous precautions necessary for the preparation of this reagent. This reagent is highly reactive and considerable care is required for its introduction into the reaction mixture containing protein. The reaction is performed at pH 5.0 (0.020 M sodium citrate, pH maintained at 5.0 with 2.5 M NaOH). These investigators also report the preparation of the ^{14}C-labeled reagent from sodium methoxide and [^{14}C]-methyl iodide. Llewellyn and Moczydlowski[320] have modified saxiphilin with trimethyloxonium tetrafluoroborate. As trimethyloxonium tetrafluoroborate is unstable in water,[321] the reaction was performed under stringent conditions. Trimethyloxonium tetrafluoroborate was placed in a tube, which was flushed with nitrogen and then sealed. A buffered solution (100 mM Tris-HCl, 100 mM NaCl, pH 8.6) of saxiphilin and bovine serum albumin was introduced and the reaction was allowed to proceed for 10 min. Cherbavez[322] observed that modification of batrachotoxin-activated Na channels with trimethyloxonium tetrafluoroborate altered function. Estrada and coworkers[323] used triethyloxonium tetrafluoroborate to modify carboxyl groups at the active site of a β-xylosidase from *Trichoderma reese* QM9414. The reagent in dichloromethane was added to the protein in 20 mM MES, at pH 5.1.

Woodward and coworkers[324] developed *N*-ethyl-5-phenylisoxazolium-3′-sulfonate (Woodward's Reagent K) (Figure 3.13) and various other *N*-alkyl-5-phenylisoxazolium fluoroborates as reagents for the "activation" of carboxyl groups for synthetic purposes. Anfinsen and coworkers[325] have studied the kinetics of the aqueous hydrolysis of this reagent and reaction with staphylococcal nuclease. This study demonstrated that Woodward's Reagent K is very unstable in aqueous solution above pH 3.0. Studies on the rate of enzyme inactivation by this reagent should be corrected for reagent hydrolysis to obtain accurate second-order rate constants. Bodlaender and coworkers[326] used *N*-ethyl-5-phenylisoxazolium-3-sulfonate, the *N*-methyl and *N*-ethyl derivatives of 5-phenylisoxazolium fluoroborate or *N*-methylbenzisoxazolium fluoroborate to activate carboxyl groups on trypsin for subsequent modification with methylamine or ethylamine. The extent of modification obtained ranged from approximately three

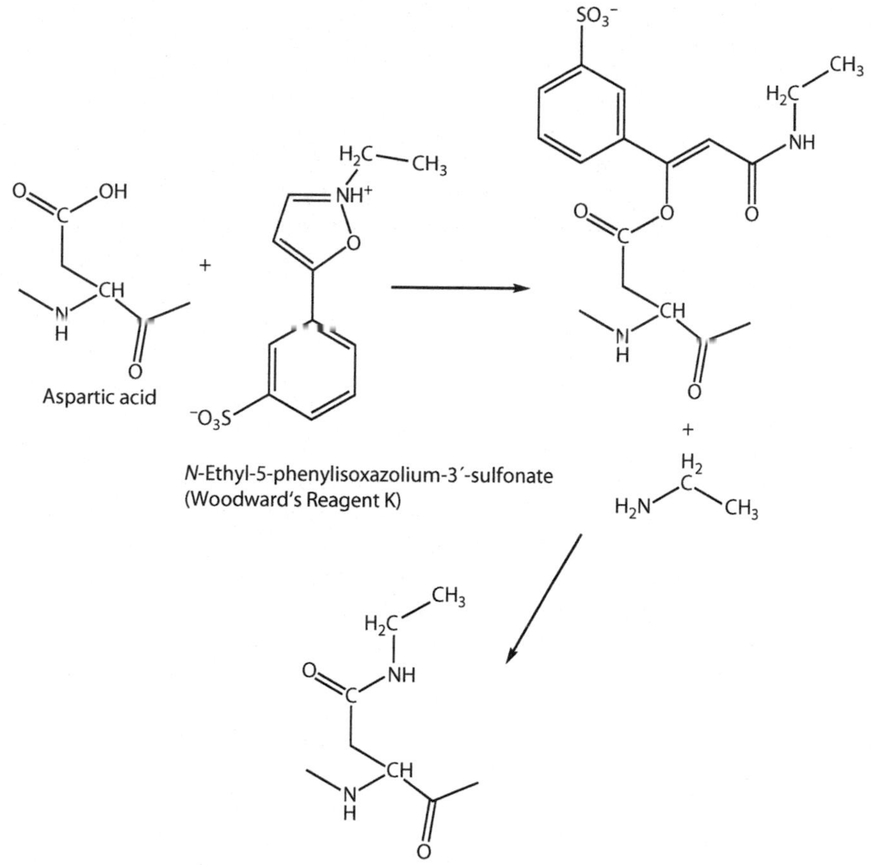

FIGURE 3.13 The modification of aspartic acid with Woodward's Reagent K (*N*-ethyl-5-phenylisoxazolium-3′-sulfonate). The formation of the ketoketimine intermediate is shown with the subsequent reaction with a nucleophile (ethyl amine) to form a stable modified derivative.

residues modified with *N*-methyl-5-phenylisoxazolium fluoroborate or *N*-ethyl-5-phenylisoxazolium fluoroborate (pH 3.80, 20°C, 80 min) to approximately 11 residues modified with *N*-methyl-5-phenylisoxazolium fluoroborate (pH 6.0, 20°C, 10 min). Reagent decomposition occurs quite rapidly, even at ice-bath temperature (2°C). The modification appears fairly selective for carboxyl groups, although some modification of lysine was observed under conditions where extensive modification was obtained (250-fold molar excess of *N*-methyl-5-phenylisoxazolium fluoroborate, pH 4.75, 72 min, 20°C, methylamine as the attacking nucleophile).

Saini and Van Etten[327] reported on the reaction of Woodward's Reagent K with human prostatic acid phosphatase. The modification was performed with a 4,000–10,000-fold molar excess of reagent in 0.020 M pyridinesulfonic acid, pH 3.6 at 25°C. Ethylamine was utilized as the attacking nucleophile to determine the extent of modification. A substantial number of carboxyl groups in the protein were modified under these experimental conditions. Arana and Vallejos[328] have compared the reaction of chloroplast coupling factor with Woodward's Reagent K and dicyclohexylcarbodiimide. Reaction with Woodward's Reagent K was accomplished at 25°C in 0.040 M Tricine, pH 7.9, while reaction with dicyclohexylcarbodiimide was accomplished at 30°C in 0.040 M 3-(*N*-morpholino)propanesulfonic acid (MOPS), pH 7.4. ATP and derivatives such as ADP and inorganic phosphate protect against the loss of activity occurring upon reaction with Woodward's Reagent K, but they do not have any effect on inactivation by dicyclohexylcarbodiimide. The reverse was seen with divalent cations such as Ca^{2+}. The modification of an essential carboxyl group in pancreatic phospholipase A_2 by Woodward's Reagent K has been reported.[329] The reaction was performed in 0.01 M sodium phosphate, pH 4.75 (pH-Stat) at 25°C. A second-order rate constant of $k_2 = 25.5\,M^{-1}\,min^{-1}$ was obtained for the loss of catalytic activity. This rate of inactivation increased more than twofold in the presence of 30 mM $CaCl_2$ (69.3 $M^{-1}\,min^{-1}$). Quantitative information on the extent of modification is obtained with [^{14}C]-glycine ethyl ester. It is of interest that treatment with a water-soluble carbodiimide, 1-(3′-dimethylaminopropyl)-3-ethylcarbodiimide, results in the loss of catalytic activity in a reaction with characteristics different from those seen with Woodward's Reagent K. Kooistra and Sluyterman[330] modified guanidated mercuripapain with *N*-ethylbenziosoxazolium tetrafluoroborate at pH 4.2 at 0°C to yield a protein modified with *N*-ethylsalicylamide esters. When the active ester groups on the protein were allowed to undergo aminolysis (2.0 M ammonium acetate, pH 9.2) and the esters were converted to the corresponding amides (an isosteric modification), the conversion to the amides resulted in the loss of a bulky substituent group with an increase in activity. Johnson and Dekker[331] have reported the modification of histidine and cysteine residues in L-threonine dehydrogenase. The lack of specificity of Woodward's Reagent K can be used to possible advantage. Puri and Colman[332] have taken advantage of the fact that Woodward's Reagent K reacts with carboxyl groups to yield enol esters, which cannot be reduced with sodium borohydride, while reaction with other functional groups such as amino groups and histidine forms unsaturated ketones, which can be reduced. The use of tritiated sodium borohydride (sodium borotritide) enables the incorporation of radiolabel into a modified protein. Woodward's Reagent K has been used by several groups[333,334] to modify Glu681 in human erythrocyte band 3 protein.

The use of carbodiimide-mediated modification of carboxyl functional groups in proteins[335,336] (Figure 3.14) is likely the most extensively used method for the study of such functional groups. Carbodiimides most likely react with protonated carboxyl groups yielding an "activated intermediate," most likely an acylisourea, which then reacts with a nucleophile such as an amine.[337]

Carbodiimides are also used for zero-length cross-linking of proteins between proximate lysine residues and carboxyl groups.[180,338] While water-insoluble carbodiimides such as *N,N′*-dicyclohexylcarbodiimide (DCC) continue to be useful for the site-specific modification of carboxyl groups,[306,339,340] most current work uses water-soluble carbodiimides such as 1-ethyl-3-(3-dimethylaminopropyl)carbodiimide (*N*-ethyl-*N′*-(dimethylaminopropyl) carbodiimide (EDC)). Several studies comparing the application of water-insoluble (hydrophobic) carbodiimides and water-soluble carbodiimides (hydrophilic) are discussed in the following. Water-soluble

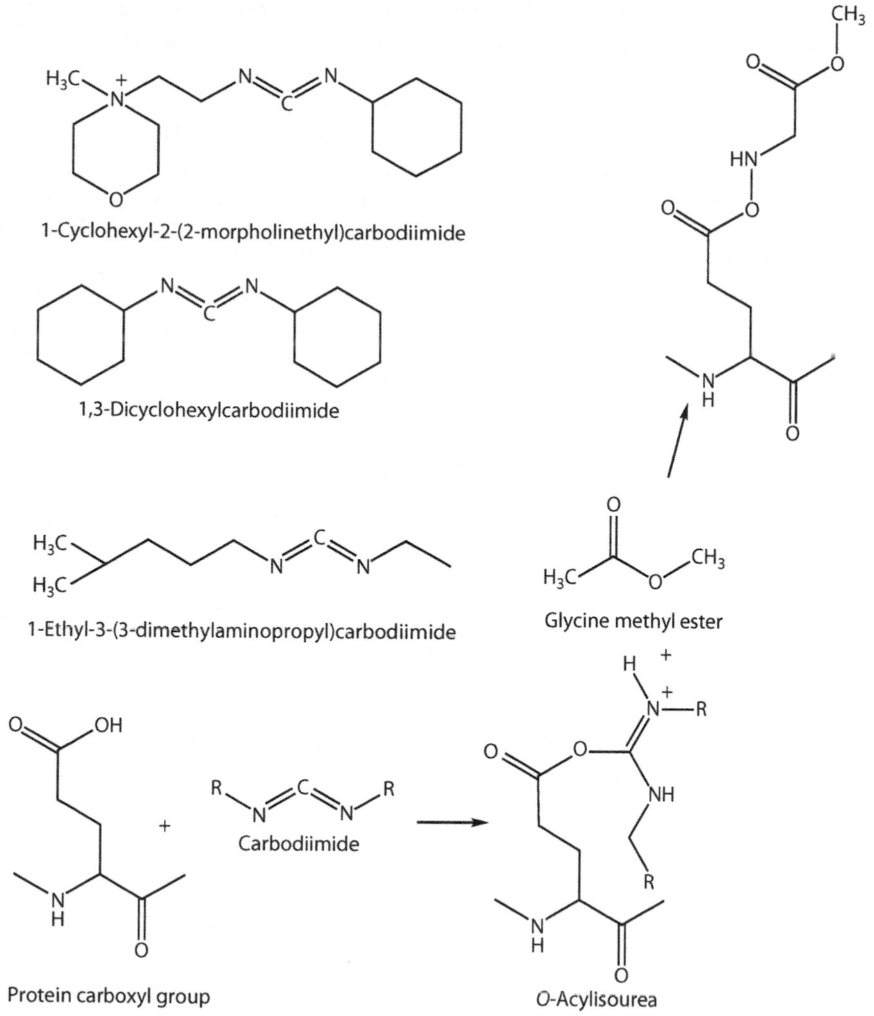

1-Cyclohexyl-2-(2-morpholinethyl)carbodiimide

1,3-Dicyclohexylcarbodiimide

1-Ethyl-3-(3-dimethylaminopropyl)carbodiimide

Glycine methyl ester

Carbodiimide

Protein carboxyl group

O-Acylisourea

FIGURE 3.14 Structures of some commonly used carbodiimides and a scheme for the reaction of carbodiimides with carboxyl groups in proteins.

carbodiimides were developed by Sheehan and Hlavka.[341,342] It is of interest that the first application of proteins was the zero-length cross-linking of collagen.[342] This study used 1-ethyl-3-(2-morpholinyl-(4)-ethyl)carbodiimide metho-*p*-toluenesulfonate in unbuffered aqueous solution. Riehm and Scheraga[343] advanced the use of water-soluble carbodiimides for proteins in a study on the modification of ribonuclease with 1-cyclohexyl-3-(2-morpholinylethyl)carbodiimide at pH 4.5 with a pH-Stat. A number of different products were obtained, which could be separated by ion-exchange chromatography (BioRex® 70). These investigators also suggest that the mechanism between a carbodiimide and a carboxyl group leads to the formation of an unstable acylisourea, which can either decompose to an acylurea derivative or react with a nucleophile. Work by Koshland and associates provided a major advance in this technology. Their approach used a water-soluble carbodiimide in the presence of an excess of amino acid nucleophile. The initial study by Hoare and Koshland[344] used *N*-benzyl-*N′*-3-dimethylaminopropylcarbodiimide. Subsequent studies used EDC[344,345] and resulted in the development of a quantitative method for the measurement of carboxyl groups in proteins.[338] These initial studies also introduced the concept of using a unique nucleophile such as norleucine methyl ester, aminomethane-sulfonic acid, and norvaline. These initial studies used 0.1 M carbodiimide, pH 4.75 with 1.0 M glycine methyl ester in a pH-Stat at 25°C. The possibility of a side reaction was discussed with reference to the possible modification of the phenolic hydroxyl of tyrosine to form the *O*-acylisourea. This general approach continues to be used today[346] with the note that most reactions are performed at pH 6–7 for reasons discussed in the following. Furthermore, few, if any, investigators perform the reaction in a pH-Stat, although, again as noted in the following, there are issues with respect to the effect of buffer components. Most investigators use the "Good" buffers developed by Good and colleagues.[347,348]

The issue of reagent stability has been mentioned previously. Border and colleagues[349] have evaluated the stability of EDC in aqueous solution. EDC has a $t½$ of 37 h (pH 7.0), 20 h (pH 6.0) and 3.9 h (pH 5.0) in 50 mM 2-(*N*-morpholino)ethanesulfonic acid at 25°C; in the presence of 100 mM glycine, the $t½$ values were 15.8 h (pH 7.0), 6.7 h (pH 6.0), and 0.73 h (pH 5.0). The authors suggest that this supports the use of EDC at pH 6.0 or 7.0 but at pH 5.0, stability would be an issue. This study also reported a major decrease in the stability of EDC under the previously mentioned solvent conditions in the presence of 10 mM phosphate or 10 mM ATP. Lei and coworkers[350] have reported kinetic studies on the hydrolysis of EDC in aqueous solution under acidic conditions. The conversion of EDC to the corresponding acylurea was measured by mass spectrometry and capillary electrophoresis. Consistent with Border's results,[349] the rate of decomposition increased with increasing acidity (decreased pH). Wrobel and colleagues[351] used the decrease in absorbance at 214 nm ($\varepsilon = 6.3 \times 10^{-3}\,M^{-1}\,cm^{-1}$) to measure the stability of EDC. These investigators also report decreased stability with decreasing pH. The presence of citrate, acetate, or phosphate increased the rate of EDC decomposition. Sehgal and Vijay[352] have optimized the conditions for EDC mediated coupling of a carboxyl-containing compound to an amine matrix (Affi-Gel® 102). These investigators noted that the presence of *N*-hydroxysuccinimide greatly improved the coupling of butyric acid to the matrix.

The modification of the active-site cysteinyl group in papain has been reported[353] as occurring under conditions (pH 4.75, 25°C) where 6:14 carboxyl groups are modified together with 9/19 tyrosyl residues. Modification of the tyrosyl residues is reversed by 0.5 M hydroxylamine, pH 7.0 (5 h at 25°C) as first demonstrated by Carraway and Koshland.[354] Despite the problems with the side reaction, modification of the carboxyl group in proteins with a water-soluble carbodiimide and an appropriate nucleophile (e.g., [^{14}C]-glycine ethyl ester, norleucine methyl ester—easily detected by amino acid analysis, aminomethylsulfonic acid) has proved extremely useful.[355] It should be noted that ammonium ions can be used as the attacking nucleophile to generate asparaginyl and glutaminyl residues from "exposed" carboxyl groups.

The modification was accomplished in 5.5 M NH_4Cl, pH 4.75 for 3 h at 25°C. Under these conditions, approximately 11 of the 15 free carboxyl groups in chymotrypsinogen were converted to the corresponding amide.[356] 1,2-Diaminoethane or diaminomethane can be coupled to aspartic acid residues to produce a trypsin-sensitive bond.[357] Another example of using this chemistry to introduce a new functional group into a protein is provided by the work of Lin and coworkers.[358] The first step involves the water-soluble carbodiimide-mediated coupling of cystamine to protein carboxyl groups. Reduction of the coupled cystamine with dithiothreitol results in 2-aminothiol functional groups bound to protein carboxyl groups.

There are several studies that have used nitrotyrosine ethyl ester as the modifying nucleophile with EDC.[359–363] Pho and coworkers[359] reported the inactivation of yeast hexokinase with 1-cyclohexyl-3-(2-morpholinoethyl)carbodiimide metho-*p*-sulfonate (CMC). The reaction was performed in 0.1 M phosphate, at pH 6.0, and was accelerated by the addition of nitrotyrosine ethyl ester. The nitrotyrosine nucleophile provides a chromophoric label for the modified carboxyl group ($\varepsilon = 4.6 \times 10^3\ M^{-1}\ cm^{-1}$, 0.1 M NaOH). The nitrotyrosine also provides for the isolation of modified peptides by immunoaffinity chromatography.[360]

A number of studies have used several different reagents for the site-specific modification of carboxyl groups in the same research study. There are several that compare a water-insoluble carbodiimide, DCC, and EDC. Sigrist-Nelson and Azzi[364] studied the effect of carboxyl-modifying reagents on proton translocation in a liposomal chloroplast ATPase complex. Proton transport was inhibited by DCC but not by EDC. Sussman and Slayman[365] reported similar differences with *Neurospora crassa* plasma membrane ATPase. This difference most likely reflects the hydrophobic nature of DCC permitting it to pass through membranes (membrane-permeant reagent). There are subsequent studies that demonstrate that EDC can inhibit the purified F_1 ATPase[366] but the reaction is complex and multiple products, including cross-linked derivatives are obtained. Shoshan-Barmatz et al.[367] showed that voltage-dependent anion channel activity in skeletal muscle sarcoplasmic reticulum was inhibited by DCC (20 mM MES, 0.2 M NaCl, pH 6.4) but not by EDC or Woodward's Reagent K. Pan and colleagues[368] modified purified vacuolar H^+-pyrophosphatase (*Vigna radiate* L.) with several carboxyl reagents [50 mM MOPS, 20% (V/V) glycerol, 1 mM EGTA, 1 mM DTT, pH 7.2, 33°C]. Rapid inactivation was observed with DCC and much slower inactivation by either EDC or Woodward's Reagent K. Bjerrum and colleagues[369] used a impermeant, water-soluble carbodiimide, 1-ethyl-3-(1-azonia-4,4-dimethylpentyl) carbodiimide to inhibit anion exchange in human red blood cell membranes. The rate and extent of inactivation was increased by the addition of tyrosine ethyl ester.

It is not necessary to have a nucleophile present in the study of the modification of a carboxyl group with a carbodiimide. It does make the identification of the modified residue less complicated but, at the same time, can introduce kinetic ambiguity. Faijes and colleagues[308] studied the modification of carboxyl groups in *Bacillus licheniformis* 1,3–1,4-β-glucanse. Modification of the native enzyme with EDC (20 mM MES, 20 mM HEPES, 40 mM 4-hydroxy-1-methylpiperidine, pH 5.5, 25°C) resulted in the rapid loss of activity to a residual level of 23%; the addition of glycine ethyl ester reduced the activity to 8%. It was not possible to use the analytical approach of

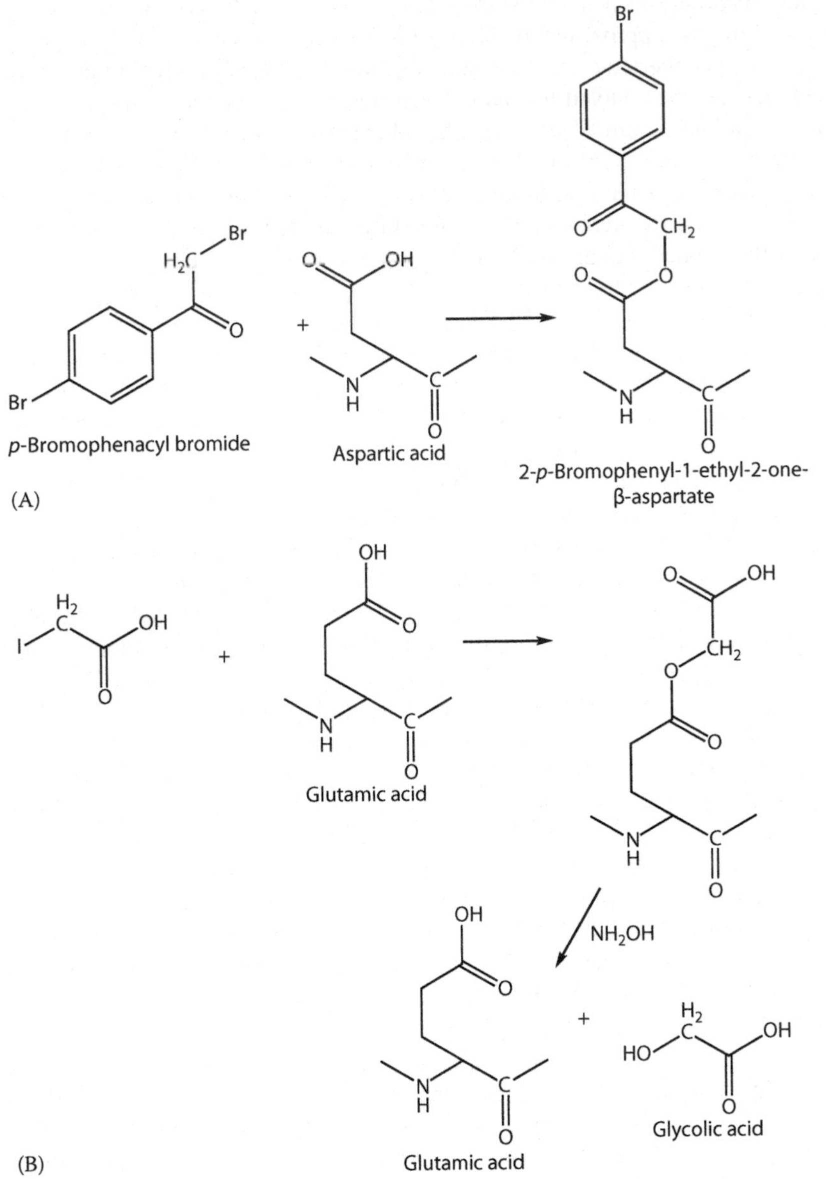

FIGURE 3.15 Reaction of *p*-bromophenacyl bromide or iodoacetate with carboxyl groups in proteins. Reaction (A) shows the reaction of *p-bromophenacyl bromide* with a β-carboxyl group of pepsin. Although this reaction forms an ester, the reaction is not reversed by hydroxylamine but is reversed by 2-mercaptoethanol (see Gross, E. and Morell, J.L., Evidence for an active carboxyl group in pepsin, *J. Biol. Chem.* 241, 3638–3639, 1966). Reaction (B) shows the reaction of iodoacetic acid with glutamic acid in ribonuclease T_1. This reaction is reversed by hydroxylamine (see Takahashi, K., Stein, W.H., and Moore, S., The identification of a glutamic acid residue as part of the active site of ribonuclease T_1, *J. Biol. Chem.* 242, 4682–4690, 1967).

Levy et al.[370] to obtain useful kinetic data and it was concluded that the loss of activity reflected the modification of two carboxyl groups at the active site. A pH dependence study showed a single titration curve with the most rapid inactivation at acid pH. It was possible to estimate a pK_a of 5.3 ± 0.2 for the modified group(s).

There are examples of carboxyl group modification with reagents expected to react far more effectively with other nucleophiles (Figure 3.15). An example of this is the reaction of iodoacetamide with ribonuclease T_1 to form the glycolic acid derivative of the glutamic acid residue as elegantly shown by Takahashi and coworkers.[371] While this modification resulted in the loss of enzymatic activity, the ability to bind substrate was retained. Subsequent work by Kojima and coworkers[372] using ^{1}H NMR demonstrated that this modification was associated with increased thermal stabilization of the modified enzyme compared to the native protein. Another example is the modification of a specific carboxyl group in pepsin by *p*-bromophenacyl bromide[373] (the use of *p*-bromophenacyl bromide in the specific modification of proteins is not uncommon, but is generally associated with the modification of cysteine, histidine, or methionine). In the study of pepsin, optimal inactivation (approximately 12-fold molar excess of reagent, 3 h, 25°C) was obtained in the pH range of 1.5–4.0 with a rapid decrease in the extent of inactivation at pH 4.5 and above (the effect of pH greater than 5.5–6.0 on the modification of pepsin cannot be studied because of irreversible denaturation of pepsin at pH 6.0 and above). In studies with a 10% molar excess of *p*-bromophenacyl bromide, at pH 2.8 at 37°C for 3 h, complete inactivation of the enzyme was obtained concomitant with the incorporation of 0.93 mol of reagent per mole of pepsin (assessed by bromide analysis). Attempts to reactivate the modified enzyme with a potent nucleophile such as hydroxylamine were unsuccessful, but reactivation could be obtained with sulfhydryl-containing reagents (i.e., β-mercaptoethanol, 2,3-dimercaptopropanol, thiophenol). It has been subsequently established that reaction occurs at the β-carboxy group of an aspartic acid residue (formation of 2-*p*-bromophenyl-1-ethyl-2-one -β-aspartate).[374] These investigators noted that the reduction of enzyme under somewhat harsh conditions ($LiBH_4$ in tetrahydrofuran) resulted in the formation of homoserine. Cury and colleagues[375] have modified ASP49 in the phospholipase A from *Bothrops asper* snake venom. This modification eliminates enzymatic activity and also abolishes the hyperalgesia effect.

REFERENCES

1. Hanson, R.W., Oxidation states of carbon as aids to understanding oxidative pathways in metabolism, *Biochem. Educ.* 18, 194–196, 1990.
2. Grossberg, A.L. and Pressman, D., Effect of acetylation on the active site of several antihapten antibodies: Further evidence for the presence of tyrosine in each site, *Biochemistry* 2, 90–96, 1963.
3. Abello, N., Kerstjens, H.A.M., Postma, D.S., and Bishoff, R., Protein tyrosine nitration: Selectivity, physicochemical and biological consequences, denitration, and proteomics methods for the identification of tyrosine-nitrated proteins, *J. Proteome Res.* 8, 3222–3238, 2009.
4. Hill, R.A., Wallace, L.J., Miller, D.D. et al., Structure–activity studies for α-amino-3-hydroxy-5-methyl-4-isoxazolepropanoic acid receptors: Acidic hydroxyphenylalanines, *J. Med. Chem.* 40, 3182–3191, 1997.

5. Vermerris, W. and Nicholson, R., in *Phenolic Compound Biochemistry*, W. Vermerris (ed.), Springer, Dordrecht, the Netherlands, Chapter 2, pp. 35–62, 2009.
6. Grimsely, G.R., Scholtz, J.M., and Pace, C.N., A summary of the measured p*K* values of the ionizable groups of folded proteins, *Protein Sci.* 19, 247–251, 2009.
7. Simonetta, M. and Carrå, S., General and theoretical aspects of the COOH and COOR groups, in *The Chemistry of Carboxyl Acids and Esters*, S. Patai (ed.), Wiley Interscience, London, U.K., Chapter 1, pp. 1–52, 1969.
8. Fülöp, N., Marchase, R.B., and Chatham, J.C., Role of protein *O*-linked *N*-acetyl-glucosamine in mediated cell function and survival in the cardiovascular system, *Cardiovasc. Res.* 73, 288–297, 2007.
9. Tarp, M.A. and Clausen, H., Mucin-type O-glycosylation and its potential use in drug and vaccine development, *Biochim. Biophyhs. Acta* 1780, 546–563, 2008.
10. Zhou, M. and Wu, H., Glycosylation and biogenesis of a family of serine-rich bacterial adhesins, *Microbiology* 155, 317–327, 2009.
11. Lommel, M. and Strahl, S., Protein O-mannosylation: Conserved from bacterial to humans, *Glycobiology* 19, 816–828, 2009.
12. Epstein, J., Michel, H.O., and Mosher, W.A., On the nucleophilicity of serine in enzymes, *J. Theor. Biol.* 19, 320–326, 1968.
13. Czech, M.P., Klarlund, J.K., Yagaloff, K.A. et al., Insulin receptor signaling. Activation of multiple serine kinases, *J. Biol. Chem.* 263, 11017–11020, 1988.
14. Sale, G.J. and Smith, D.M., Serine phosphorylations triggered by the insulin receptor, *Cell Signal.* 1, 205–218, 1989.
15. Rosen, O.M., Protein tyrosine kinases, protein serine kinases, and the mechanism of action of insulin, *Harvey Lect.* 82, 105–122, 1986–1987.
16. Spink, E., Cosgrove, S., Rogers, L. et al., ^{13}C and ^{1}H NMR studies of ionizations and hydrogen bonding in chymotrypsin-glyoxal inhibitor complexes, *J. Biol. Chem.* 282, 7852–7861, 2007.
17. Fyfe, C.A., Nucleophilic attack by hydroxide and alkoxide ion, in *The Chemistry of the Hydroxyl Group. Part 1*, S. Patai (ed.), Wiley Interscience, London, U.K., Chapter 2, pp. 51–131, 1971.
18. Mädler, S., Bich, C., Touboul, D., and Zenobi, R., Chemical cross-linking with NHS esters: A systematic study on amino acid reactivities, *J. Mass Spectrom.* 44, 694–706, 2009.
19. Kalkhof, S. and Sinz, A., Chances and pitfalls of chemical cross-linking with amine-reactive *N*-hydroxysuccinimide esters, *Anal. Bioanal. Chem.* 392, 305–312, 2008.
20. Gabant, G., Augier, J., and Armengaud, J., Assessment of solvent residues accessibility using three Sulfo-NHS-biotin reagents in parallel: Application to footprint changes of a methyltransferase upon binding its substrate, *J. Mass Spectrom.* 43, 360–370, 2008.
21. Miller, B.T., Collins, T.J., Rogers, M.E., and Kurosky, A., Peptide biotinylation with amine-reactive esters: Differential side chain reactivity, *Peptides* 18, 1585–1595, 1997.
22. Miller, B.T. and Kurosky, A., Elevated intrinsic reactivity of seryl hydroxyl groups within the linear peptide triads His-Xaa-Ser or Ser-Xaa-His, *Biochem. Biophys. Res. Commun.* 195, 461–467, 1993.
23. Marzotto, A. and Giormani, V., Attempts at chemical modification of threonine and serine residues in RNase A, *Experentia* 26, 833–834, 1970.
24. Cardinaud, R. and Baker, B.R., Irreversible enzyme inhibitors. CLXXII. Proteolytic enzymes. XVI. Covalent bonding of the sulfonyl fluoride groups to serine outside the active site of α-chymotrypsin by *exo*-site type active-site directed irreversible inhibitors, *J. Med. Chem.* 13, 467–470, 1970.
25. Bing, D.H., Cory, M., and Fenton, J.W., Exo-site affinity labeling of human thrombins—Similar labeling on A-chain and B-chain fragments of clotting alpha- and non-clotting gamma-beta-thrombins, *J. Biol. Chem.* 252, 8027–8034, 1976.

26. Fahrney, D.E. and Gold, A.M., Sulfonyl fluorides as inhibitors of esterases. I. Rates of reaction with acetylcholinesterase, α-chymotrypsin, and trypsin, *J. Am. Chem. Soc.* 85, 997–1000, 1963.
27. Gold, A.M. and Fahrney, D., Sulfonyl fluorides as inhibitors of esterases. II. Formation and reactions of phenylmethylsulfonyl fluoride α-chymotrypsin, *Biochemistry* 3, 783–791, 1964.
28. Swain, C.G. and Scott, G.E., Rates of solvolysis of some alkyl fluorides and chlorides, *J. Am. Chem. Soc.* 75, 246–248, 1953.
29. Halmann, M., Hydrolysis of dialkylphosphinic fluorides and dialkyl phosphorofluoridates. Kinetic and tracer studies, *J. Chem. Soc.* 305–310, 1959.
30. Cohen, S., Biological reactions of carbonyl halides, in *The Chemistry of Acyl Halides*, S. Patai (ed.), Wiley Interscience, London, U.K., Chapter 10, pp. 313–340, 1972.
31. Geschwind, I.I. and Li, C.H., Corticotropin (ACTH) V. The application of oxidation with periodate to the determination of N-terminal serine in α-corticotropin, *Biochim. Biophys. Acta* 15, 442–444, 1954.
32. Malaprade, L., Action of polyalcohols on periodic acid and alkaline periodate, *Bull. Soc. Chim.* 1, 833–852, 1934.
33. Nicolet, B.H. and Shinn, L.A., The action of periodic acid on α-amino alcohols, *J. Am. Chem. Soc.* 61, 1615, 1939.
34. Geoghegan, K.F. and Stroh, J.G., Site-directed conjugation of nonpeptide groups to peptides and proteins via periodate oxidation of a 2-amino alcohol. Application to modification at N-terminal serine, *Bioconjug. Chem.* 3, 138–146, 1992.
35. Sharon, J.L. and Puleo, D.A., The use of N-terminal immobilization of PTH (1-34) on PLGA to enhance bioactivity, *Biomaterials* 29, 3137–3142, 2008.
36. Pereira, H.J.V., Salgado, M.C.O., and Oliveira, E.B., Immobilized analogues of sunflower trypsin inhibitor-1 constitute a versatile group of affinity sorbents for selective isolation of serine proteases, *J. Chromatogr. B. Analyt. Technol. Biomed. Life Sci.* 877, 2039–2044, 2009.
37. Souplet, V., Desmet, R., and Melnyk, O., *In situ* ligation between peptides and silica nanoparticles for making peptide microarrays on polycarbonate, *Bioconjug. Chem.* 20, 550–557, 2009.
38. Tiefenbrunn, K.K. and Dawson, P.E., Chemoselective ligation technique: Modern application of time-honored chemistry, *Biopolymers* 94, 95–106, 2010.
39. Mikolajczyk, S.D., Meyer, D.L., Starling, J.J. et al., High yield, site-specific coupling of N-terminally modified β-lactamase to a proteolytically derived single-sulfhydryl murine Fab', *Bioconjug. Chem.* 5, 636–646, 1994.
40. Donovan, J.W., Changes in ultraviolet absorption produced by alteration of protein conformation, *J. Biol. Chem.* 244, 1961–1967, 1969.
41. Markland, F.S., Phenolic hydroxyl ionization in two subtilisins, *J. Biol. Chem.* 244, 694–700, 1969.
42. Laws, W.R. and Shore, J.D., Spectral evidence for tyrosine ionization linked to a conformational change in liver alcohol dehydrogenase ternary complex, *J. Biol. Chem.* 254, 2582–2584, 1979.
43. Kuramitsu, S. et al., Ionization of the catalytic groups and tyrosyl residues in human lysozyme, *J. Biochem.* 87, 771–778, 1980.
44. Kobayashi, J., Hagashijima, T., and Miyazawa, T., Nuclear magnetic resonance analyses of side chain conformations of histidine and aromatic acid derivatives, *Int. J. Pept. Protein Res.* 24, 40–47, 1984.
45. Poklar, N., Vesnaver, G., and Laponje, S., Studies by UV spectroscopy of thermal denaturation of beta-lactoglobulin in urea and alkylurea solutions, *Biophys. Chem.* 47, 143–151, 1993.

46. Taylor, R., *Electrophilic Aromatic Substitution*, John Wiley & Sons, Ltd., Chichester, U.K., 1990.
47. Roholt, O.A. and Pressman, D., Iodination-isolation of peptides from the active site, *Methods Enzymol.* 25, 438–449, 1972.
48. Tsomides, T.J. and Eisen, H.N., Stoichiometric labeling of peptides by iodination on tyrosyl or histidyl residues, *Anal. Biochem.* 210, 129–135, 1993.
49. Rosenfeld, R. et al., Sites of iodination in recombinant human brain-derived neurotrophic factor and its effect on neurotrophic activity, *Protein Sci.* 2, 1664–1674, 1993.
50. Huntley, T.E. and Strittmatter, P., The reactivity of the tyrosyl residues of cytochrome b_5, *J. Biol. Chem.* 247, 4648–4653, 1972.
51. McGowan, E.B. and Stellwagen, E., Reactivity of individual tyrosyl residues of horse heart ferricytochrome c toward iodination, *Biochemistry* 9, 3047–3053, 1970.
52. Layne, P.P. and Najjar, V.A., Evidence for a tyrosine residue at the active site of phosphoglucomutase and its interaction with vanadate, *Proc. Natl. Acad. Sci. USA* 76, 5010–5013, 1979.
53. Silva, J.S. and Ebner, K.E., Protection by substrates and α-lactalbumin against inactivation of galactosyltransferase by iodine monochloride, *J. Biol. Chem.* 255, 11262–11267, 1980.
54. Izzo, J.L., Bale, W.F., Izzo, M.J., and Roncone, A., High specific activity labeling of insulin with ^{131}I, *J. Biol. Chem.* 239, 3743–3748, 1964.
55. Linde, S. et al., Monoiodoinsulin labelled in tyrosine residue 16 or 26 of the insulin B-chain. Preparation and characterization of some binding properties, *Hoppe-Seyler's Z. Physiol. Chem.* 362, 573–579, 1981.
56. Hunter, W.M. and Greenwood, F.C., Preparation of iodine-131 labelled human growth hormone of high specific activity, *Nature* 194, 495–496, 1962.
57. Heber, D. et al., Improved iodination of peptides for radioimmunoassay and membrane radioreceptor assay, *Clin. Chem.* 24, 796–799, 1978.
58. Maly, P. and Lüthi, C., The binding sites of insulin-like growth factor I (IGF I) to type I IGF receptor and to a monoclonal antibody. Mapping by chemical modification of tyrosine residues, *J. Biol. Chem.* 263, 7068–7072, 1988.
59. Hussain, A.A., Jona, J.A., Yamada, A., and Dittert, L.W., Chloramine T in radiolabeling techniques. II. A nondestructive method for radiolabeling biomolecules by halogenation, *Anal. Biochem.* 224, 221–226, 1995.
60. Tashtoush, B.M. et al., Chloramine T in radiolabeling techniques. IV. Penta-*O*-acetyl-*N*-chloro-*N*-methylglucamine as an oxidizing agent, *Anal. Biochem.* 288, 16–21, 2001.
61. Nikula, T.K. et al., Impact of the high tyrosine fraction in complementarity determining regions: Measured and predicted effects of radioiodination on IgG immunoreactivity, *Mol. Immunol.* 32, 865–872, 1995.
62. Bauer, R.J. et al., Alteration of the pharmacokinetics of small proteins by iodination, *Biopharm. Drug. Dispos.* 17, 761–774, 1996.
63. Mathew, S. et al., Characterization of the interaction between α_2-macroglobulin and fibroblast growth factor-2: The role of hydrophobic interactions, *Biochem. J.* 374, 123–129, 2003.
64. Smith, C.L. and Peterson, C.L., Couple tandem affinity purification and quantitative tyrosine iodination to determine subunit stoichiometry of protein complexes, *Methods* 31, 104–109, 2003.
65. Linde, S., Hansen, B., and Lemmark, A., Preparation of stable radioiodiniated polypeptide hormones and proteins using polyacrylamide gel electrophoresis, *Methods Enzymol.* 92, 309–335, 1983.
66. Kamatso, Y. and Hayashi, H., Revaluating the effects of tyrosine iodination of recombinant hirudin on its thrombin inhibitor kinetics, *Thromb. Res.* 87, 343–352, 1997.

67. Sohoel, A., Plum, A., Frokjaer, S., and Thygesen, P., ^{125}I used for labelling of proteins in an absorption model changes the absorption rates of insulin aspart, *Int. J. Pharm.* 330, 114–120, 2007.
68. Braschi, S. et al., Role of the kidney in regulating the metabolism of HDL in rabbits: Evidence that iodination alters the catabolism of apolipoprotein A-1 by the kidney, *Biochemistry* 39, 5441–5449, 2000.
69. Ghosh, D. et al., Determination of a protein structure by iodination: The structure of iodinated acetylxylan esterase, *Acta Crystallogr. D Biol. Crystallogr.* 55, 779–784, 1999.
70. Leinala, E.K., Davies, P.L., and Jia, Z., Elevated temperature and tyrosine iodination aid in the crystallization and structure determination of an antifreeze protein, *Acta Crystallogr. D Biol. Crystallogr.* 58, 1081–1083, 2002.
71. Fraenkel-Conrat, H. and Colloms, M.D., Reactivity of tobacco mosaic virus and its protein toward acetic anhydride, *Biochemistry* 6, 2740–2745, 1967.
72. Riordan, J.F. and Vallee, B.L., Acetylation, *Methods Enzymol.* 11, 565–576, 1967.
73. Riordan, J.F. and Vallee, B.L., *O*-acetylation, *Methods Enzymol.* 25, 570–576, 1972.
74. Karibian, D., Jones, C., Gertler, A., Dorrington, K.J., and Hofmann, T., On the reaction of acetic and maleic anhydrides with elastase. Evidence for a role of the NH_2-terminal valine, *Biochemistry* 13, 2891–2897, 1974.
75. Ohnishi, M., Suganuma, T., and Hiromi, K., The role of a tyrosine residue of bacterial liquefying α-amylase in the enzymatic hydrolysis of linear substrates as studied by chemical modification with acetic anhydride, *J. Biochem.* 76, 7–13, 1974.
76. Bernad, A., Nieto, M.A., Vioque, A., and Palaciáan, E., Modification of the amino groups and hydroxyl groups of lysozyme with carboxylic acid anhydrides: A comparative study, *Biochim. Biophys. Acta* 873, 350–355, 1986.
77. Sanchez, A., Ramon, Y., Solano, Y. et al., Double acylation for identification of amino-terminal peptides of proteins isolated by polyacrylamide gel electrophoresis, *Rapid Commun. Mass Spectrom.* 21, 2237–2244, 2007.
78. Olsen, D.B., Hepburn, T.W., Lee, S.L. et al., Investigation of the substrate binding and catalytic groups of the P-C bond cleaving enzyme, phosphoacetaldehyde hydrolase, *Arch. Biochem. Biophys.* 296, 144–151, 1992.
79. Kaplan, H., Stevenson, K.J., and Hartley, B.S., Competitive labelling, a method for determining the reactivity of individual groups in proteins. The amino groups of porcine elastase, *Biochem. J.* 124, 289–299, 1971.
80. Tilley, S.D. and Francis, M.B., Tyrosine-selective protein alkylation using π-allylpalladium complexes, *J. Am. Chem. Soc.* 128, 1080–1081, 2006.
81. Wieland, T. and Schneider, G., *N*-Acylimidazoles as acyl derivatives of high energy, *Ann. Chem. Justus Liebigs* 580, 159, 1953.
82. Stadtman, E.R. and White, F.H., Jr., The enzymic synthesis of *N*-acetylimidazole, *J. Am. Chem. Soc.* 75, 2022, 1953.
83. Stadtman, E.R., On the energy-rich nature of acetyl imidazole, an enzymatically active compound, in *A Symposium on the Mechanism of Enzyme Action*, W.D. McElroy and B. Glass (eds.), Johns Hopkins Press, Baltimore, MD, 1954.
84. Fife, T.H., Steric effects in the hydrolysis of N-acylimidazoles and ester of *p*-nitrophenol, *J. Am. Chem. Soc.* 87, 4597–4600, 1965.
85. Lee, J.P., Bembi, R., and Fife, T.H., Steric effects in the hydrolysis reactions of N-acylimidazoles. Effect of aryl substitution in the leaving group, *J. Org. Chem.* 62, 2872–2876, 1997.
86. Simpson, R.T., Riordan, J.F., and Vallee, B.L., Functional tyrosyl residues in the active center of bovine pancreatic carboxypeptidase A, *Biochemistry* 2, 616–622, 1963.
87. Riordan, J.F., Wacker, W.E.C., and Vallee, B.L., N-Acetylimidazole: A reagent for determination of "free" tyrosyl residues of proteins, *Biochemistry* 4, 1758–1765, 1965.

88. Myers, B., II. and Glazer, A.N., Spectroscopic studies of the exposure of tyrosine residues in proteins with special reference to the subtilisins, *J. Biol. Chem.* 246, 412–419, 1971.
89. Cronan, J.E., Jr. and Klages, A.L., Chemical synthesis of acyl thioesters of acyl carrier protein with native structure, *Proc. Nat. Acad. Sci. USA* 78, 5440–5444, 1981.
90. Lundblad, R.L., The reaction of bovine thrombin with *N*-butyrylimidazole. Two different reactions resulting in the inhibition of catalytic activity, *Biochemistry* 14, 1033–1037, 1975.
91. Lundblad, R.L., Harrison, J.H., and Mann, K.G., On the reaction of purified bovine thrombin with *N*-acetylimidazole, *Biochemistry* 12, 409–413, 1973.
92. El Kabbaj, M.S. and Latruffe, N., Chemical reagents of polypeptide side chains. Relationship between solubility properties and ability to cross the inner mitochondrial membranes, *Cell. Mol. Biol.* 40, 41, 1994.
93. Zhang, F., Gao, J., Weng, J. et al., Structural and functional differentiation of three groups of tyrosine residues by acetylation of N-acetylimidazole in manganese stabilizing protein, *Biochemistry* 44, 719–725, 2005.
94. Martin, B.L., Wu, D., Jakes, S., and Graves, D.J., Chemical influences on the specificity of tyrosine phosphorylation, *J. Biol. Chem.* 265, 7108–7111, 1990.
95. Wehtje, E. and Adlercreutz, P., Water activity and substrate concentration effects on lipase activity, *Biotechnol. Bioeng.* 55, 796–806, 1997.
96. Kurzawa, J. and Suszka, A., Kinetics and mechanism of the nucleophilic cleavage of disulfide bond in 2,2′-dithio-diimidazoles with hydroxide ions, *Pol. J. Chem.* 8, 1487–1494, 2007.
97. Lawlor, D.A., More O'Farrall, R.A., and Rao, S.N., Stabilities and partitioning of arenonium ions in aqueous media, *J. Am. Chem. Soc.* 130, 17997–18007, 2008.
98. Ji, Y. and Bennett, B.M., Activation of microsomal glutathione S-transferase by peroxynitrite, *Mol. Pharmacol.* 63, 136–146, 2003.
99. Ji, Y., Neverova, I., Van Eck, J.E., and Bennett, B.M., Nitration of tyrosine 92 mediates the activation of rat microsomal glutathione *S*-transferase by peroxynitrite, *J. Biol. Chem.* 281, 1986–1991, 2006.
100. Busenlehner, L.S., Codreaunu, S.G., Holm, P.J. et al., Stress sensor triggers conformational response of the integral membrane protein microsomal glutathione transferase 1, *Biochemistry* 43, 11145–11152, 2004.
101. Sidddiqui, K.S. and Cavicchioli, R., Improved thermal stability and activity in the cold-adapted lipase B from *Candida antartica* following chemical modification with oxidized polysaccharides, *Extremophiles* 9, 471–476, 2005.
102. Ryan, S.M., Mantovani, G., Wang, X. et al., Advances in PEGylation of important biotech molecules: Delivery aspects, *Expert Opin. Drug Deliv.* 5, 371–383, 2008.
103. Jevsevar, S., Kunstelj, M., and Porekar, V.G., PEGylation of therapeutic proteins, *Biotechnol. J.* 5, 113–128, 2010.
104. Zaragoza, R., Torres, L., Garcia, C. et al., Nitration of cathepsin D enhances its proteolytic activity during mammary gland remodeling after lactation, *Biochem. J.* 419, 279–288, 2009.
105. Basu, S. and Kirley, T.L., Identification of a tyrosine residue responsible for *N*-acetylimidazole-induced increase of activity of ecto-nucleoside triphosphate diphosphohydrolase 3, *Purinergic Signal.* 1, 271–280, 2005.
106. Yagil, G. and Anbar, M., Formation of peroxynitrite by oxidation of chloramine, hydroxylamine, and nitrohydroxamate, *J. Inorg. Nucl. Chem.* 26, 453–460, 1964.
107. Petriconi, G.L. and Papee, H.M., Aqueous solutions of sodium "pernitrite," *Can. J. Chem.* 44, 977–980, 1966.
108. Ischiropoulos, H., Zhu, L., Chen, J. et al., Peroxynitrite-mediated tyrosine nitration catalyzed by superoxide dismutase, *Arch. Biochem. Biophys.* 298, 431–437, 1992.

109. Greenacre, S.A.B. and Ischeriopoulos, H., Tyrosine nitration: Localisation, quantification, consequences for protein function and signal transduction, *Free Radic. Res.* 34, 541–581, 2001.
110. Ischiropoulos, H., Biological tyrosine nitration: A pathophysiological function of nitric oxide and reactive oxygen species, *Arch. Biochem. Biophys.* 356, 1–11, 1998.
111. Ischiropoulos, H., Biological selectivity and functional aspects of protein tyrosine nitration, *Biochem. Biophys. Res. Commun.* 305, 776–783, 2003.
112. Beckman, J.S., Chen, J., Ischiropoulos, H., and Crow, J.P., Oxidative chemistry of peroxynitrite, *Methods Enzymol.* 233, 229–240, 1994.
113. Pryor, W.A., Cueto, R., Jin, X. et al., A practical method for preparing peroxynitrite solutions of low ionic strength and free of hydrogen peroxide, *Free Radic. Biol. Med.* 18, 75–83, 1995.
114. Uppu, R.M., Squadrito, G.L., Cueto, R., and Pryor, W.A., Selecting the most appropriate synthesis of peroxynitrite, *Methods Enzymol.* 269, 285–295, 1996.
115. Zhang, H., Bhargava, K., Kessler, A. et al., Transmigration nitration of hydrophobic tyrosyl peptides. Localization, characterization, mechanisms of nitration, and biological implications, *J. Biol. Chem.* 278, 8969–8978, 2003.
116. Bartesaghi, S., Valez, V., Trujillo, M. et al., Mechanistic studies of peroxynitrite-mediated tyrosine nitration in membranes using the hydrophobic probe *N*-t-BOC-L-tyrosine tert-butyl ester, *Biochemistry* 45, 6813–6825, 2006.
117. Romero, N., Peluffo, G., Bartesaghi, S. et al., Incorporation of the hydrophobic probe *N*-t-BOC-L-tyrosine tert-butyl ester to red cell membranes to study peroxynitrite-dependent reactions, *Chem. Res. Toxicol.* 20, 1638–1648, 2007.
118. Bartesaghi, S., Peluffo, G., Zhang, H. et al., Tyrosine nitration, dimerization, and hydroxylation by peroxynitrite in membrane studied by the hydrophobic probe *N*-t-BOC-l-tyrosine tert-butyl ester, *Methods Enzymol.* 441, 217–236, 2008.
119. Rubbo, H., Trostchansky, A., and O'Donnell, V.B., Peroxynitrite-mediated lipid oxidation and nitration: Mechanisms and consequences, *Arch. Biochem. Biophys.* 484, 167–172, 2009.
120. Ferrer-Sueta, G. and Radi, R., Chemical biology of peroxynitrite: Kinetics, diffusion, and radicals, *ACS Chem. Biol.* 4, 161–177, 2009.
121. Kennett, E.C., Rees, M.D., Malle, E. et al., Peroxynitrite modifies the structure and function of the extracellular matrix proteoglycan perlecan by reaction with both the protein core and the heparin sulfate chains, *Free Radic. Biol. Med.* 49, 282–293, 2010.
122. Bartesaghi, S., Wenzel, J., Trujillo, M. et al., Lipid peroxyl radicals mediate tyrosine dimerization and nitration in membranes, *Chem. Res. Toxicol.* 23, 821–835, 2010.
123. Schmidt, P. et al., Specific nitration at tyrosine 430 revealed by high resolution mass spectrometry as basis for redox regulation of bovine prostacyclin synthase, *J. Biol. Chem.* 278, 12813–12819, 2003.
124. Petersson, A.S., Steen, H., Kalume, D.E. et al., Investigation of tyrosine nitration in proteins by mass spectrometry, *J. Mass Spectrom.* 36, 616–625, 2001.
125. Willard, B.B., Ruse, C.I., Keightley, J.A. et al., Site-specific quantitation of protein nitration using liquid chromatography/tandem mass spectrometry, *Anal. Chem.* 75, 2370–2376, 2003.
126. Liu, B., Tewari, A.K., Zhang, L. et al., Proteomic analysis of protein tyrosine nitration after ischemia reperfusion injury: Mitochondria as a major target, *Biochim. Biophys. Acta* 1794, 476–485, 2009.
127. Chiappetta, G., Corbo, C., Palmese, A. et al., Quantitative identification of protein nitration sites, *Proteomics* 9, 1524–1537, 2009.
128. Sultana, R., Reed, T., and Butterfield, D.A., Detection of 4-hydroxy-2-nonenal- and 3-nitrotyrosine-modified proteins using a proteomics approach, *Methods Mol. Biol.* 519, 351–361, 2009.

129. Lee, J.R., Lee, S.J., Kim, T.W. et al., Chemical approach for specific enrichment and mass analysis of nitrated peptides, *Anal. Chem.* 81, 6620–6626, 2009.
130. Jones, A.W., Mikhailov, V.A., Iniesta, J., and Cooper, H.J., Electron capture dissociation mass spectrometry of tyrosine nitrated peptides, *J. Am. Soc. Mass Spectrom.* 21, 268–277, 2010.
131. Zhan, X. and Desiderio, D.M., Mass spectrometric identification of *in vivo* nitrotyrosine sites in the human pituitary tumor proteome, *Methods Mol. Biol.* 566, 137–163, 2009.
132. Abello, N., Barroso, B., Kerstjens, H.A. et al., Chemical labeling and enrichment of nitrotyrosine-containing peptides, *Talanta* 80, 1503–1512, 2010.
133. Tsumoto, H., Taguchi, R., and Kohda, K., Efficient identification and quantification of peptides containing nitrotyrosine by matrix-assisted laser desorption/ionization time-of-flight mass spectrometry after derivatization, *Chem. Pharm. Bull. (Tokyo)* 58, 488–494, 2010.
134. Daiber, A., Bachschmid, M., Kawaklik, C. et al., A new pitfall in detecting biological end products of nitric oxide—Nitration, nitros(yl)ation and nitrite/nitrate artifacts during freezing, *Nitric Oxide* 9, 44–52, 2003.
135. Irie, Y., Saekii, M., Kamisaki, Y. et al., Histone H1.2 is a substrate for dinitrase, an activity that reduces nitrotyrosine immunoreactivity in proteins, *Proc. Natl. Acad. Sci. USA* 100, 5634–5639, 2003.
136. Nikov, G., Bhat, V., Wishnok, J.S., and Tannenbaum, S.R., Analysis of nitrated proteins by nitrotyrosine-specific affinity probes and mass spectrometry, *Anal. Biochem.* 320, 214–222, 2003.
137. Ogino, K., Nakajima, M., Kadama, N. et al., Immunohistochemical artifact for nitrotyrosine in eosinophils or eosinophil containing tissue, *Free Radic. Res.* 36, 1163–1170, 2002.
138. Miyagi, M., Sakajushi, H., Darrow, R.M. et al., Evidence that light modulates protein nitration in rat retina, *Mol. Cell. Proteomics* 1, 293–303, 2003.
139. Sokolovsky, M., Riordan, J.F., and Vallee, B.L., Conversion of 3-nitrotyrosine to 3-aminotyrosine in peptides and proteins, *Biochem. Biophys. Res. Commun.* 27, 20–25, 1967.
140. Viera, L., Ye, Y.Z., Estévez, A.G., and Beckman, J.S., Immunohistochemical methods to detect nitrotyrosine, *Methods Enzymol.* 301, 373–381, 1999.
141. Aulak, K.S., Koeck, T., Crabb, J.W., and Stuehr, D.J., Proteomic method for identification of tyrosine-nitrated proteins, *Methods Mol. Biol.* 279, 151–165, 2004.
142. Koeck, T., Willard, B., Crabb, J.W. et al., Glucose-mediated tyrosine nitration in adipocytes: Targets and consequences, *Free Radic. Biol. Med.* 46, 884–892, 2009.
143. Halliwell, B., What nitrates tyrosine? Is nitrotyrosine specific as a biomarker of peroxynitrite formation *in vivo*?, *FEBS Lett.* 411, 157–160, 1997.
144. van derVliet, A., Eiserich, J.P., Kaur, H. et al., Nitrotyrosine as biomarker for reactive nitrogen species, *Methods Enzymol.* 269, 175–184, 1996.
145. Drel, V.R., Lupachyk, S., Shevalye, H. et al., New therapeutic and biomarker discovery for peripheral diabetic neuropathy: PARP inhibitor, nitrotyrosine, and tumor necrosis factor-α, *Endocrinology* 151, 2547–2555, 2010.
146. Reddy, S. and Bradley, J., Immunohistochemical demonstration of nitrotyrosine, a biomarker of oxidative stress, in islets cells of the NOD mouse, *Ann. N.Y. Acad. Sci.* 1037, 199–202, 2004.
147. Souza, J.M., Peluffo, G., and Radi, R., Protein tyrosine nitration—Functional alteration or just a biomarker?, *Free Radic. Biol. Med.* 45, 357–366, 2008.
148. Radabaugh, M.R., Nemirovskiy, O.V., Misko, T.P. et al., Immunoaffinity liquid chromatography—Tandem mass spectrometry detection of nitrotyrosine in biological fluids: Development of a clinical translatable biomarker, *Anal. Biochem.* 380, 68–76, 2008.
149. Safinowski, M., Wilhelm, B., Reimer, T. et al., Determination of nitrotyrosine concentrations in plasma samples of diabetes mellitus patients by four different immunoassays leads to contradictive results and disqualifies the majority of the tests, *Clin. Chem. Lab. Med.* 47, 483–488, 2009.

150. Tsikas, D., Measurement of nitrotyrosine in plasma by immunoassays is fraught with danger: Commercial availability is no guarantee of analytical reliability, *Clin. Chem. Lab. Med.* 48, 141–143, 2010.
151. Herriott, R.M., Reactions of native proteins with chemical reagents, *Adv. Protein Chem.* 3, 169–225, 1947.
152. Riordan, J.F., Sokolovsky, M., and Vallee, B.L., Tetranitromethane. A reagent for the nitration of tyrosine and tyrosyl residues in proteins, *J. Am. Chem. Soc.* 88, 4104–4105, 1966.
153. Sokolovsky, M., Harell, D., and Riordan, J.F., Reaction of tetranitromethane with sulfhydryl groups in proteins, *Biochemistry* 8, 4740, 1969.
154. Rabani, J., Mulae, W.A., and Matheson, M.S., The pulse radiolysis of aqueous tetranitromethane. I. Rate constants and the extinction coefficient of e_{aq}^-. II. Oxygenated solutions, *J. Phys. Chem.* 69, 53–70, 1965.
155. Riordan, J.F. and Vallee, B.L., Nitration with tetranitromethane, *Methods Enzymol.* 25, 515–521, 1972.
156. Cuatrecasas, P., Fuchs, S., and Anfinsen, C.B., The tyrosyl residues at the active site of staphylococcal nuclease. Modifications by tetranitromethane, *J. Biol. Chem.* 243, 4787–4798, 1968.
157. Kuhn, D.M. and Geddes, T.J., Peroxynitrite inactivates tryptophan hydroxylase via sulfhydryl oxidation. Coincident nitration of enzyme tyrosyl residues has minimal impact on catalytic activity, *J. Biol. Chem.* 274, 29726–29732, 1999.
158. Palamalai, V. and Miyagi, M., Mechanism of glyceraldehyde-3-phosphate dehydrogenase inactivation by tyrosine nitration, *Protein Sci.* 19, 255–262, 2010.
159. Nelson, K.J., Parsonage, D., Hall, A. et al., Cysteine pK(a) values for the bacterial peroxiredoxin AhpC, *Biochemistry* 47, 12860–12868, 2008.
160. Ying, J., Sharov, V., Xu, S. et al., Cysteine-674 oxidation and degradation of sarcoplasmic reticulum Ca^{2+} ATPase in diabetic pig aorta, *Free Radic. Biol. Med.* 45, 756–762, 2008.
161. Figtree, G.A., Liu, C.C., Bibert, S. et al., Reversible oxidative modification: A key mechanism of Na^+-K^+ pump regulation, *Circ. Res.* 105, 185–193, 2009.
162. Zhang, L., Chen, C.L., Kang, P.T. et al., Peroxynitrite-mediated oxidative modifications of complex II: Relevance in myocardial infarction, *Biochemistry* 49, 2529–2539, 2010.
163. Spande, T.F., Fontana, A., and Witkop, B., An unusual reaction of skatole with tetranitromethane, *J. Am. Chem. Soc.* 91, 6169–6170, 1969.
164. Riggle, W.L., Long, J.A., and Borders, C.L., Jr., Reaction of turkey egg-white lysozyme with tetranitromethane. Modification of tyrosine and tryptophan, *Can. J. Biochem.* 51, 1433–1439, 1973.
165. Sokolovsky, M., Fuchs, M., and Riordan, J.F., Reaction of tetranitromethane with tryptophan and related compounds, *FEBS Lett.* 7, 167–170, 1970.
166. Teuwissen, G., Masson, P.L., Osinski, P., and Heremans, J.F., Metal-combining properties of human lactoferrin. The effect of nitration of lactoferrin with tetranitromethane, *Eur. J. Biochem.* 35, 366–371, 1973.
167. Katsura, T., Lam, E., Packer, L., and Seltzer, S., Light dependent modification of bacteriorhodopsin by tetranitromethane. Interaction of a tyrosine and a tryptophan residue with bound retinal, *Biochem. Int.* 5, 445–456, 1982.
168. Lam, E., Seltzer, S., Katsura, T., and Packer, L., Light-dependent nitration of bacteriorhodopsin, *Arch. Biochem. Biophys.* 227, 321–328, 1983.
169. Atassi, M.Z. and Habeeb, A.F.S.A., Enzymic and immunochemical properties of lysozyme. I. Derivatives modified at tyrosine. Influence of nature of modification on activity, *Biochemistry* 8, 1385–1393, 1969.
170. Haddad, I.Y., Zhu, S., Ischiropoulos, H. et al., Nitration of a surfactant protein A results in decreased ability to aggregate lipids, *Am. J. Physiol.* 270, L281–L288, 1996.

171. Zhu, S., Haddad, I.Y., and Matalon, S., Nitration of surfactant protein A (SP-A) tyrosine residues results in decreased mannose binding ability, *Arch. Biochem. Biophys.* 333, 283–290, 1996.
172. Alverez, B., Rubbio, H., Kirk, M. et al., Peroxynitrite-dependent tryptophan nitration, *Chem. Res. Toxicol.* 9, 390–396, 1996.
173. Pollet, E., Martinez, A., Metha, B. et al., Role of tryptophan oxidation in peroxynitrite-dependent protein chemiluminescence, *Arch. Biochem. Biophys.* 349, 74–80, 1996.
174. Lehnig, M. and Kirsch, M., ^{15}N-CIDNP investigations during tryptophan, *N*-acetyl-L-tryptophan, and melatonin nitration with reactive nitrogen species, *Free Radic. Res.* 41, 523–535, 2007.
175. Rebrin, I., Bregere, C., Gallaher, T.K., and Sohal, R.S., Detection and characterization of peroxynitrite-induced modification of tyrosine, tryptophan, and methionine residues by tandem mass spectrometry, *Methods Enzymol.* 441, 284–294, 2008.
176. Padmaja, S., Ramazerian, M.S., Bounds, P.L., and Koppenal, W.H., Reaction of peroxynitrite with L-tryptophan, *Redox Rep.* 2, 173–177, 1996.
177. Scholz, G. and Kwok, F., Brain pyridoxal kinase: Photoaffinity labeling of the substrate-binding site, *J. Biol. Chem.* 264, 4318–4321, 1989.
178. Doyle, R.J., Bello, J., and Roholt, O.A., Probable protein crosslinking with tetranitromethane, *Biochim. Biophys. Acta* 160, 274–276, 1970.
179. Boesel, R.W. and Carpenter, F.H., Crosslinking during the nitration of bovine insulin with tetranitromethane, *Biochem. Biophys. Res. Commun.* 38, 678–682, 1970.
180. Kunkel, G.R., Mehrabian, M., and Martinson, H.G., Contact-site cross-linking agents, *Mol. Cell. Biochem.* 34, 3–13, 1981.
181. Nadeau, O.W., Traxler, K.W., and Carlson, G.M., Zero-length crosslinking of the beta subunit of phosphorylase kinase to the N-terminal half of its regulatory alpha subunit, *Biochem. Biophys. Res. Commun.* 251, 637–641, 1998.
182. Bruice, T.C., Gregory, M.J., and Walters, S.L., Reactions of tetranitromethane. I. Kinetics and mechanism of nitration of phenols by tetranitromethane, *J. Am. Chem. Soc.* 90, 1612, 1968.
183. Hass, G.M. and Gentry, L., Nitration of polypeptides using ethanol in reaction buffers minimizes crosslinking, *J. Biochem. Biophys. Methods* 1, 257–261, 1979.
184. Hugli, T.E. and Stein, W.H., Involvement of a tyrosine residue in the activity of bovine pancreatic deoxyribonuclease A, *J. Biol. Chem.* 246, 7191–7200, 1971.
185. Cutfield, S.M., Dodson, G.G., Ronco, N., and Cutfield, J.F., Preparation and activity of nitrated insulin dimer, *Int. J. Pept. Protein Res.* 27, 335–343, 1986.
186. Sorrentino, S., Yakovlev, G.I., and Libonati, M., Dimerization of deoxyribonuclease I, lysozyme, and papain. Effects of ionic strength on enzymic activity, *Eur. J. Biochem.* 124, 183–189, 1982.
187. Strazza, S., Hunter, R., Walker, E., and Darnall, D.W., The thermodynamics of bovine and porcine insulin and proinsulin association determined by concentration difference spectroscopy, *Arch. Biochem. Biophys.* 238, 30–42, 1985.
188. Margolis, S.A., Coxon, B., Gajewski, E., and Dizdaroglu, M., Structure of a hydroxyl radical induced cross-link of thymine and tyrosine, *Biochemistry* 27, 6353–6359, 1988.
189. Hsuan, J.J., The cross-linking of tyrosine residues in apo-ovotransferrin by treatment with periodate anions, *Biochem. J.* 247, 467, 1987.
190. Yermilov, V., Rubio, J., and Ohshima, H., Formation of 8-nitroguanine in DNA treated with peroxynitrite *in vitro* and its rapid removal from DNA by depurination, *FEBS Lett.* 376, 207–210, 1995.
191. Crow, J.P. and Ishiropoulos, H., Detection and quantitation of nitrotyrosine residues in proteins: *In vivo* marker of peroxynitrite, *Methods Enzymol.* 269, 185–194, 1996.
192. Riordan, J.F., Sokolovsky, M., and Vallee, B.L., Environmentally sensitive tyrosyl residues. Nitration with tetranitromethane, *Biochemistry* 6, 3582–3589, 1967.

193. Haas, J.A., Frederick, M.A., and Fox, B.G., Chemical and post-translational modifications of *Escherichia coli* acyl carrier protein for preparation of dansyl carrier protein, *Prot. Exp. Purif.* 20, 274–284, 2000.
194. Haas, J.A. and Fox, B.G., Fluorescence anisotropy studies of enzyme-substrate complex formation in stearoyl-ACP-desaturase, *Biochemistry* 41, 14472–14481, 2002.
195. Zhang, Q., Qian, W.J., Knyushko, T.V. et al., A method for selective enrichment and analysis of nitrotyrosine-containing peptides in complex proteome samples, *J. Proteome Res.* 6, 2257–2268, 2007.
196. Sharov, V.S., Dremina, E.S., Pennington, J. et al., Selective fluorogenic derivatization of 3-nitrotyrosine and 3,4-dihydroxyphenylalanine in peptides: A method designed for quantitative proteomic analysis, *Methods Enzymol.* 441, 19–32, 2008.
197. Riordan, J.F., Sokolovsky, M., and Vallee, B.L., The functional tyrosyl residues of carboxypeptidase A. Nitration with tetranitromethane, *Biochemistry* 6, 3609–3617, 1967.
198. Burr, M. and Koshland, D.E., Jr., Use of a reporter groups in structure-function studies of proteins, *Proc. Natl. Acad. Sci. USA* 52, 1017–1024, 1964.
199. Muller, W.E. and Wollert, U., Spectroscopic studies on the complex formation of suramin with bovine and human serum albumin, *Biochim. Biophys. Acta* 427, 465–480, 1976.
200. Ajtai, K., Peyser, Y.M., Park, S. et al., Trinitrophenylated reactive lysine residues in myosin detects lever arm movement during the consecutive steps of ATP hydrolysis, *Biochemistry* 38, 6428–6440, 1999.
201. Pezzementi, L., Shi, J., Johnson, D.A. et al., Ligand-induced conformational changes in residues flanking the active site gorge of acetylcholinesterase, *Chem. Biol. Interact.* 157–158, 413–414, 2005.
202. Dempski, R.E., Friedrich, T., and Bamberg, E., Voltage clamp fluorometry: Combining fluorescence and electrophysiological methods to examine the structure-function of the Na^+K^+-ATPase, *Biochim. Biophys. Acta* 1787, 714–720, 2009.
203. Garel, J.R., Evidence for involvement of proline *cis-trans* isomerization in the slow unfolding of RNase A, *Proc. Natl. Acad. Sci. USA* 77, 795–798, 1980.
204. Parker, D.M., Jeckel, D., and Holbrook, J.J., Slow structural changes shown by the 3-nitrotyrosine-237 residue in pig heart [Tyr(3-NO_2)237] lactate dehydrogenase, *Biochem. J.* 201, 465–471, 1982.
205. Rischel, C. and Poulsen, F.M., Modification of a specific tyrosine enables tracing of the end-to-end distance during apomyoglobin, *FEBS Lett.* 374, 105–109, 1995.
206. De Fillippis, V., Draghi, A., Frasson, R. et al., *o*-Nitrotyrosine and *p*-iodophenylalanine as spectroscopic probes for structural characteristics of SH3 complexes, *Protein Sci.* 16, 1257–1265, 2007.
207. Hartings, M.R., Gray, H.B., and Winkler, J.R., Probing melittin helix-coil equilibria in solutions and vesicles, *J. Phys. Chem. B* 112, 3202–3207, 2008.
208. Digambaranath, J.L., Dang, L., Dembinska, M. et al., Conformations within soluble oligomers and insoluble aggregates revealed by resonance energy transfer, *Biopolymers* 93, 299–317, 2010.
209. Herz, J.M., Hrabeta, E., and Packer, L., Evidence for a carboxyl group in the vicinity of the retinal chromophore of bacteriorhodopsin, *Biochem. Biophys. Res. Commun.* 114, 872–881, 1983.
210. Lide, D.R. (ed.), *Handbook of Chemistry and Physics*, 82nd edn., CRC Press, Boca Raton, FL, pp. 3–207, 2001 (CAS Registry Number 509-14-8).
211. Liang, P., Tetranitromethane, *Org. Synth. Coll.* 3, 803, 1955.
212. Fendler, J.H. and Liechti, R.R., Micellar catalysis of the reaction of hydroxide ion with tetranitromethane, *J. Chem. Soc. Perkin Trans.* 2, 9, 1041–1043, 1972.
213. Walters, S.L. and Bruice, T.C., Reactions of tetranitromethane. II. Kinetics and products for the reactions of tetranitromethane with inorganic ions and alcohols, *J. Am. Chem. Soc.* 93, 2269, 1971.

214. Selassie, C.D., Mekapati, S.B., and Verma, R.P., QSAR: Then and now, *Curr. Top. Med. Chem.* 2, 1357–1379, 2002.
215. Eberson, L. and Hartshorn, M.P., The formation and reactions of adducts from the photochemical reactions of aromatic compounds with tetranitromethane and other X-NO_2 reagents, *Aust. J. Chem.* 51, 1061–1081, 1998.
216. Capellos, C., Iyer, S., Liang, Y., and Gamms, L.A., Transient species and product formation from electronically excited tetranitromethane, *J. Chem. Soc. Faraday Trans.* 2, 82, 2195–2206, 1986.
217. Hodges, G.R., Young, M.J., Paul, T., and Ingold, K.U., How should xanthine oxidase-generated superoxide yields be measured?, *Free Radic. Biol. Med.* 29, 434–441, 2000.
218. Liochev, S.I. and Fridovich, I., Reversal of the superoxide dismutase reaction revisited, *Free Radic. Biol. Med.* 34, 908–910, 2003.
219. Sutton, H.C., Reactions of the hydroperoxyl radical (HO_2) with nitrogen dioxide and tetranitromethane in aqueous solutions, *J. Chem. Soc. Faraday Trans.* 1, 71, 2142–2147, 1975.
220. Naqvi, K.R. and Melø, T.B., Reduction of tetranitromethane by electronically excited aromatics in acetonitrile: Spectra and molar absorption coefficients of radical cations of anthracene, phenanthrene and pyrene, *Chem. Phys. Lett.* 428, 83–87, 2006.
221. Butts, C.P., Eberson, L., Hartshorn, M.P. et al., Photochemical nitration by tetranitromethane. Part XXXIX. The photolysis of tetranitromethane with 2,8-dimethyl and 1,3,7,9-tetramethyl-dibenzofuran, *Acta Chem. Scand.* 51, 476–482, 1997.
222. Rasmusson, M. et al., Ultrafast formation of trinitromethanide ($C(NO_2)_3^-$) by photoinduced dissociative electron transfer and subsequent ion pair coupling reaction in acetonitrile and dichloromethane, *J. Phys. Chem. B.* 105, 2027–2035, 2001.
223. Scherrer, P. and Stoeckenius, W., Selective nitration of tyrosine-26 and -64 in bacteriorhodopsin with tetranitromethane, *Biochemistry* 23, 6195–6202, 1984.
224. Brown, L.S., Reconciling crystallography and mutagenesis: A synthetic approach to the creation of a comprehensive model for proton pumping by bacteriorhodopsin, *Biochim. Biophys. Acta* 1460, 49–59, 2000.
225. Scherrer, P. and Stoeckenius, W., Effects of tyrosine-26 and tyrosine-64 nitration on the photoreactions of bacteriorhodopsin, *Biochemistry* 24, 7733–7740, 1985.
226. Mayo, K.H., Schussheim, M., Visserr, G.W. et al., Mobility and solvent exposure of aromatic residues in bacteriorhodopsin investigated by ^{1}H-NMR and photo-CIDNP-NMR spectroscopy, *FEBS Lett.* 235, 163–168, 1988.
227. Renthal, R., Cothran, M., Dawson, N., and Harris, G.J., Light activates the reaction of bacteriorhodopsin aspartic acid-115 with dicyclohexylcarbodiimide, *Biochemistry* 24, 4275–4279, 1985.
228. Eberson, L., Hartshorn, M.P., and Persson, O., The reactions of some dienes with tetranitromethane, *Acta Chem. Scand.* 52, 450–452, 1998.
229. Fieser, M. and Fieser, L., *Reagents for Organic Synthesis*, Wiley Interscience, New York, Vol. 1, p. 1147, 1967.
230. Jewett, S.W. and Bruice, T.C., Reactions of tetranitromethane. Mechanisms of the reaction of tetranitromethane with pseudo acids, *Biochemistry* 11, 3338–3350, 1972.
231. Strosberg, A.D., Van Hoeck, B., and Kanarek, L., Immunochemical studies on hen's egg white lysozyme. Effect of selective nitration of the three tyrosine residues, *Eur. J. Biochem.* 19, 36–41, 1971.
232. Richards, P.G., Walton, D.J., and Heptinstall, J., The effects of nitration on the structure and function of hen egg-white lysozyme, *Biochem. J.* 315, 473–479, 1996.
233. Richards, P.G., Coles, B., Heptinstall, J., and Walton, D.J., Electrochemical modification of lysozyme: Anodic reaction of tyrosine residues, *Enzyme Microb. Technol.* 16, 795–801, 1994.
234. Skawinski, W.J., Adebodun, F., Cheng, J.T. et al., Labeling of tyrosines in proteins with [^{15}N]tetranitromethane, a new NMR reporter for nitrotyrosines, *Biochim. Biophys. Acta* 1162, 297–308, 1993.

235. Šantrůček, J., Strohalm, M., Kadlčík, V. et al., Tyrosine residues modification studied by MALDI-TOF mass spectrometry, *Biochem. Biophys. Res. Commun.* 323, 1151–1156, 2004.
236. Lecomte, J.T.J. and Llinãs, M., Characterization of the aromatic proton magnetic resonance spectrum of crambin, *Biochemistry* 23, 4799–4807, 1984.
237. Ploug, M., Rahbek-Nielsen, H., Ellis, V. et al., Chemical modification of the Urokinase-type plasminogen activator and its receptor using tetranitromethane. Evidence for the involvement of specific tyrosine residues in both molecules during receptor-ligand interaction, *Biochemistry* 34, 12524–12534, 1995.
238. Zappacosta, F., Ingallinella, P., Scaloni, P. et al., Surface topology of minibody by selective chemical modifications and mass spectrometry, *Protein Sci.* 6, 1901–1909, 1997.
239. Leite, J.F. and Cascio, M., Probing the topology of the glycine receptor by chemical modification couple to mass spectrometry, *Biochemistry* 41, 6140–6148, 2002.
240. Cascio, M., Glycine receptors: Lessons on topology and structural effects of the lipid bilayer, *Biopolymers* 66, 359–368, 2002.
241. D'Ambrosio, C., Talamo, F., Vitale, R.N. et al., Probing the dimeric structure of porcine aminoacylase 1 by mass spectrometric and modeling procedures, *Biochemistry* 42, 4430–4443, 2003.
242. Lee, S.B., Inouye, K., and Tomura, B., The states of tyrosyl residues in thermolysin as examined by nitration and pH-dependent ionization, *J. Biochem.* 121, 231–237, 1997.
243. Inouye, K., Lee, S.B., and Tonomura, B., Effects of nitration and amination of tyrosyl residues in thermolysin on its hydrolytic activity and its remarkable activation by salts, *J. Biochem.* 124, 72–78, 1998.
244. Muta, Y., Oneda, H., and Inouye, K., Anomalous pH-dependence of the activity of human matrilysin (matrix metalloproteinase-7) as revealed by nitration and amination of its tyrosine residues, *J. Biochem.* 386, 263–270, 2005.
245. Christen, P. and Riordan, J.F., Syncatalytic modification of a functional tyrosyl residue in aspartate aminotransferase, *Biochemistry* 9, 3025–3034, 1970.
246. Betcher-Lange, S.L., Couglan, M.P., and Rajagopalan, K.V., Syncatalytic modification of chicken liver xanthine dehydrogenase by hydrogen peroxide. The nature of the reaction, *J. Biol. Chem.* 254, 8825–8829, 1979.
247. Usanov, S.A., Pikuleva, I.A., Chashchin, V.L., and Akhrem, A.A., Chemical modification of adrenocortical cytochrome P-450_{scc} with tetranitromethane, *Biochim. Biophys. Acta* 790, 259–267, 1984.
248. Heyduk, T., Michalczyk, R., and Kochman, M., Long-range effects and conformational flexibility of aldolase, *J. Biol. Chem.* 266, 15650–15655, 1991.
249. Malhotra, O.P. and Dwivedi, U.N., Formation of enzyme-bound carbanion intermediate in the isocitrate lyase-catalyzed reaction: Enzyme reaction of tetranitromethane with substrates and its dependence on effector, pH, and metal ions, *Arch. Biochem. Biophys.* 250, 236–248, 1986.
250. Medda, R., Padiglia, A., Pedersen, J.Z., and Floria, G., Evidence for α-proton abstraction and carbanion formation involving a functional histidine residue in lentil seedling amine oxidase, *Biochem. Biophys. Res. Commun.* 196, 1349–1355, 1993.
251. Mekhanik, M.L. and Torchinskii, Y.M., Carbanion detection during the decarboxylation of quasisubstrates by glutamate dehydrogenase, *Biokhimiya* 37, 1308–1311, 1972.
252. Moreira, P.I., Sayre, L.M., Zhu, X. et al., Detection and localization of markers of oxidative stress by *in situ* methods: Application in the study of Alzheimer disease, *Methods Mol. Biol.* 610, 419–434, 2010.
253. Lundblad, R.L., *Development and Application of Biomarkers*, CRC Press, Boca Raton, FL, 2011.
254. Tawfik, D.S. et al., pH on-off switching of antibody—Hapten binding by site-specific chemical modification of proteins, *Protein Eng.* 7, 431–434, 1994.

255. Richards, P.G., Walton, D.J., and Heptinstall, J., The effects of tyrosine nitration on the structure and function of hen egg-white lysozyme, *Biochem. J.* 315, 473–478, 1996.
256. Morag, E., Bayer, E.A., and Wilchek, M., Reversibility of biotin-binding by selective modification of tyrosine in avidin, *Biochem. J.* 316, 193–199, 1996.
257. Morag, E., Bayer, E.A., and Wilchek, M., Immobilized nitro-avidin and nitro-streptavidin as reusable affinity matrices for application in avidin-biotin technology, *Anal. Biochem.* 243, 257–263, 1996.
258. Bolivar, J.G., Soper, S.A., and Carley, R.L., Nitroavidin as a ligand for the surface capture and release of biotinylated proteins, *Anal. Biochem.* 80, 9336–9342, 2008.
259. Kurihara, K., Horinishi, H., and Shibata, K., Reaction of cyanuric halides with proteins. I. Bound tyrosine residues of insulin and lysozyme as identified with cyanuric fluoride, *Biochim. Biophys. Acta* 74, 678–687, 1963.
260. Gorbunoff, M.J., Cyanuration, *Methods Enzymol.* 25, 506–514, 1972.
261. Gorbunoff, M.J., The pH dependence of tyrosine cyanuration in proteins, *Biopolymers* 11, 2233–2240, 1972.
262. Gorbunoff, M.J. and Timasheff, S.N., The role of tyrosines in elastase, *Arch. Biochem. Biophys.* 152, 413–422, 1972.
263. Coffe, G. and Pudles, J., Chemical reactivity of the tyrosyl residues in yeast hexokinase. Properties of the nitroenzyme, *Biochim. Biophys. Acta* 484, 322–335, 1977.
264. Larionova, N.L., Kazanskaya, N.F., and Sakharov, I.Y., Soluble-high-molecular-weight derivatives of the pancreatic inhibitor of trypsin. Isolation and properties of the dextran-bound pancreatic inhibitor, *Biokhimiya* 42, 1237–1243, 1977.
265. Moller, M., Halogen cyanides. Molecular weight, stability and basic hydrolysis of cyanogen bromide, *Kgl. Danske Videnskab. Math.-fys. Medd.* 12, 17, 1934.
266. Perret, A. and Perrot, R., Polymerization of cyanogen bromide. Preparation of pure cyanogen bromide, *Bull. Soc. Chim.* 7, 743–750, 1940.
267. Ferguson, S.J., Lloyd, W.J., Lyons, M.H., and Radda, G.K., The mitochondrial ATPase. Evidence for a single essential tyrosine residue, *Eur. J. Biochem.* 54, 117–126, 1975.
268. Ferguson, S.J., Lloyd, W.J., and Radda, G.K., The mitochondrial ATPase. Selective modification of a nitrogen residue in the β-subunit, *Eur. J Biochem.* 54, 127–133, 1975.
269. Sesaki, H., Wong, E.F., and Siu, C.H., The cell adhesion molecule DdCAD-1 in *Dictyostelium* is targeted to the cell surface by a nonclassical transport pathway involving contractile vacuoles, *J. Cell. Biol.* 138, 939–951, 1997.
270. Denu, J.M. and Tanner, K.G., Specific and reversible inactivation of protein tyrosine phosphatases by hydrogen peroxide: Evidence for a sulfenic acid intermediate and implications for redox regulation, *Biochemistry* 37, 5633–5642, 1998.
271. Nieslanik, B.S. and Atkins, W.M., The catalytic Tyr-9 of glutathione S-transferase A1–1 controls the dynamics of the C terminus, *J. Biol. Chem.* 275, 17447–17451, 2000.
272. Carballal, S., Radi, R., Kirk, M.C. et al., Sulfenic acid formation in human serum albumin by hydrogen peroxide and peroxynitrite, *Biochemistry* 42, 9906–9914, 2003.
273. Shetty, V., Spellman, D.S., and Neubert, T.A., Characterization by tandem mass spectrometry of stable cysteine sulfenic acid in a cysteine switch peptide of matrix metalloproteinases, *J. Am. Soc. Mass Spectrom.* 18, 1544–1551, 2007.
274. Lavergne, S.N., Wang, H., Callan, H.E. et al., "Danger" conditions increase sulfamethoxazole-protein adduct formation in human antigen-presenting cells, *J. Pharmacol. Exp. Ther.* 331, 372–381, 2009.
275. de Almeida Olivera, M.G., Rogana, E., Roas, J.C. et al., Tyrosine 151 is part of the substrate activation binding site of bovine trypsin. Identification by covalent labeling with *p*-diazonium-benzamidine and kinetic characterization of TYR-151-(*p*-benzamidino)-azo-β-trypsin, *J. Biol. Chem.* 268, 26893, 1993.

276. Landsteiner, K., *The Specificity of Serological Reactions*, Harvard University Press, Cambridge, MA, 1945.
277. Fraenkel-Conrat, H., Bean, R.S., and Lineweaver, H., Essential groups for the interaction of ovomucoid (egg white trypsin inhibitor) and trypsin, and for tryptic activity, *J. Biol. Chem.* 177, 385–403, 1949.
278. Riordan, J.F. and Vallee, B.L., Diazonium salts as specific reagents and probes of protein conformation, *Methods Enzymol.* 25, 521–531, 1972.
279. Tabachnick, M. and Sobotka, H., Azoproteins. I. Spectrophotometric studies of amino acid azo derivatives, *J. Biol. Chem.* 234, 1726–1730, 1959.
280. Tabachnick, M. and Sobotka, H., Azoproteins. II. A spectrophotometric study of the coupling of diazotized arsanilic acid with proteins, *J. Biol. Chem.* 235, 1051–1054, 1960.
281. Vallee, B.L. and Fairclough, G.F., Jr., Arsanilazochymotrypsinogen. The extrinsic Cotton effects of an arsanilazotyrosyl chromophore as a conformation probe of zymogen activation, *Biochemistry* 10, 2470–2477, 1971.
282. Gorecki, M., Wilchek, M., and Blumberg, S., Modulation of the catalytic properties of α-chymotrypsin by chemical modification at Tyr 146, *Biochim. Biophys. Acta* 535, 90–99, 1978.
283. Robinson, G.W., Reaction of a specific tryptophan residue in streptococcal proteinase with 2-hydroxy-5-nitrobenzyl bromide, *J. Biol. Chem.* 245, 4832–4841, 1970.
284. Gorecki, M. and Wilchek, M., Modification of a specific tyrosine residue of ribonuclease A with a diazonium inhibitor analog, *Biochim. Biophys. Acta* 532, 81–91, 1978.
285. Johansen, J.T., Livingston, D.M., and Vallee, B.L., Chemical modification of carboxypeptidase A crystals. Azo coupling with tyrosine-248, *Biochemistry* 11, 2584–2588, 1972.
286. Harrison, L.W. and Vallee, B.L., Kinetics of substrate and product interactions with arsanilazotyrosine-248 carboxypeptidase A, *Biochemistry* 17, 4359–4363, 1978.
287. Cueni, L. and Riordan, J.F., Functional tyrosyl residues of carboxypeptidase A. The effect of protein structure on the reactivity of tyrosine-198, *Biochemistry* 17, 1834–1842, 1978.
288. Liao, T.H., Ting, R.S., and Yeung, J.E., Reactivity of tyrosine in bovine pancreatic deoxyribonuclease with *p*-nitrobenzenesulfonyl fluoride, *J. Biol. Chem.* 257, 5637–5644, 1982.
289. Chen, B., Costantino, H.R., Liu, J. et al., Influence of calcium ions on the structure and stability of recombinant human deoxyribonuclease I in the aqueous and lyophilized states, *J. Pharm. Sci.* 88, 477–482, 1999.
290. Dastoli, F.R., Musto, N.A., and Price, S., Reactivity of active sites of chymotrypsin suspended in an organic medium, *Arch. Biochem. Biophys.* 115, 44–47, 1966.
291. Gold, A.M., Sulfonyl fluorides as inhibitors of esterases. III. Identification of serine as the site of sulfonation in phenylmethylsulfonyl α-chymotrypsin, *Biochemistry* 4, 897–901, 1965.
292. Haniu, M., Yuan, H., Chen, S. et al., Structure–function relationship of NAD(P)H: quinone reductase: Characterization of NH_2-terminal blocking group and essential tyrosine and lysine residues, *Biochemistry* 27, 6877–6883, 1988.
293. Kulanthaivel, P., Leibach, F.H., Mahesh, V.B., and Ganapathy, V., Tyrosine residues are essential for the activity of the human placental taurine transporter, *Biochim. Biophys. Acta* 985, 139–146, 1989.
294. Gitlin, G., Bayer, E.A., and Wilchek, M., Studies on the biotin-binding sites of avidin and streptavidin—Tyrosine residues are involved in the binding-site, *Biochem. J.* 269, 527–530, 1990.
295. Suzuki, H., Higashi, Y., Naitoh, N. et al., Chemical modification of a catalytic antibody that accelerates the hydrolysis of carbonate esters, *J. Protein Chem.* 19, 419–424, 2000.

296. Chang, L.S. and Lin, S.R., Modification of tyrosine-3(63) and lysine 6 of Taiwan cobra phospholipase A2 affects its ability to enhance 8-anilinonapthaline-1-sulfonate fluorescence, *Biochem. Mol. Biol. Int.* 40, 235–241, 1996.
297. Chang, L.S., Lin, S.R., and Chang, C.C., The essentiality of calcium ion in the enzymic activity of Taiwan cobra phospholipase A2, *J. Prot. Chem.* 15, 701–707, 1996.
298. Soares, A.M. and Giglio, J.R., Chemical modifications of phospholipase A(2) from snake venom: Effects on catalytic and pharmacologic properties, *Toxicon* 42, 855–868, 2003.
299. Andrews, W.W. and Allison, W.S., 1-Fluoro-2,4-dinitrobenzene modifies a tyrosine residue when it inactivates the bovine mitochondrial F_1-ATPase, *Biochem. Biophys. Res. Commun.* 99, 813–819, 1981.
300. De Vendittis, E. et al., Phenylmethylsulfonyl fluoride inactivates an archaeal superoxide dismutase by chemical modification of a specific tyrosine residue. Cloning, sequencing and expression of the gene coding for *Sulfolobus solfataricus* superoxide dismutase, *Eur. J. Biochem.* 268, 1794–1801, 2001.
301. Means, G.E. and Wu, H.L., The reactive tyrosine residue of human serum albumin: Characterization of its reaction with diisopropylfluorophosphate, *Arch. Biochem. Biophys.* 194, 526–530, 1979.
302. Williams, N.H., Harrison, J.M., Read, R.W., and Black, R.M., Phosphorylated tyrosine in albumin as a biomarker of exposure to organophosphorous nerve agents, *Arch. Toxicol.* 81, 627–639, 2007.
303. Ding, S.J., Carr, J., Carlson, J.E. et al., Five tyrosine and two serines in human albumin are labeled by the organophosphorous agent FP-biotin, *Chem. Res. Toxicol.* 21, 1787–1794, 2008.
304. Tarhoni, M.H., Lister, T., Ray, D.E., and Carter, W.G., Albumin binding as a potential biomarker of exposure to moderately low levels of organophosphorous pesticides, *Biomarkers* 13, 343–363, 2008.
305. Kiss, T., Erdei, A., and Kiss, L., Investigation of the active site of the extracellular β-D-xylosidase from *Aspergillus caronarius*, *Arch. Biochem. Biophys.* 399, 188–194, 2002.
306. Meier, T. et al., Evidence for structural integrity in the undecameric c-rings isolated from sodium ATP synthases, *J. Mol. Biol.* 325, 389–397, 2003.
307. Scholz, S.R. et al., Experimental evidence for a ββα- Me-finger nuclease motif to represent the active site of the caspase-activated DNase, *Biochemistry* 42, 9288–9294, 2003.
308. Faijes, M. et al., Glycosynthase activity of *Bacillus licheniformis* 1,3–1,4-β-glucanase mutants: Specificity, kinetics, and mechanism, *Biochemistry* 42, 13304–13318, 2003.
309. Montes, T., Grazu, V., Lopez-Gallego, F. et al., Chemical modification of protein surfaces to improve their reversible enzyme immobilization on ionic exchangers, *Biomacromolecules* 7, 3052–3058, 2006.
310. Bhatti, H.N., Rashid, M.H., Asgher, M. et al., Chemical modification results in hyperactivation and thermostabilization of *Fusarium solani* glucoamylase, *Can. J. Microbiol.* 53, 177–185, 2007.
311. Maruyama, H., Yoshimura, S.H., Akita, S. et al., Covalent attachment of protein to the tip of a multiwalled carbon nanotube without sidewall decoration, *Appl. Phys. Lett.* 90, 144107/1–144107/3, 2007.
312. Gao, Y. and Kyratzis, I., Covalent immobilization of proteins on carbon nanotubes using the cross-linker 1-ethyl-3-(3-dimethylaminopropyl) carbodiimide—A critical assessment, *Bioconj. Chem.* 19, 1945–1950, 2008.
313. Takayama, M. Nogami, K., Saenko, E.L. et al., Identification of a protein S-interactive site with the A2 domain of the factor VIII heavy chain, *Thromb. Haemostas.* 102, 645–655, 2009.
314. Muskolai, A. and Kokoi, V., CE detection of *N*-(3-dimethylaminopropyl)-*N*-hydroxysuccinimide-coupled proteins after homo- and hetero-crosslinking reactions, *Electrophoresis* 31, 1097–1100, 2010.

315. Mechref, Y. and El Rassi, Z., Capillary zone electrophoresis of derivatized acidic monosaccharides, *Electrophoresis* 15, 627–634, 1994.
316. Klein, J., Kraus, M., Ticha, M. et al., Water-soluble poly(acryalamide-allylamine) derivatives of saccharides for protein-saccharide binding studies, *Glycoconj. J.* 12, 51–54, 1995.
317. Parsons, S.M., Jao, L., Dahlquist, F.W. et al., The nature of amino acid side chains which are critical for the activity of lysozyme, *Biochemistry* 8, 700–712, 1969.
318. Parsons, S.M. and Raftery, M.A., The identification of aspartic acid residue 52 as being critical to lysozyme activity, *Biochemistry* 8, 4199–4205, 1969.
319. Paterson, A.K. and Knowles, J.R., The number of catalytically essential carboxyl groups in pepsin. Modification of the enzyme by trimethyloxonium fluoroborate, *Eur. J. Biochem.* 31, 510–517, 1972.
320. Llewellyn, L.E. and Moczydlowski, E.G., Characterization of saxitoxin binding to saxiphilin, a relative of the transferrin family that displays pH-dependent ligand binding, *Biochemistry* 33, 12312–12322, 1994.
321. Mackinannon, R. and Miller, C., Functional modification of a Ca^{2+}-activated K^+ channel by trimethyloxonium, *Biochemistry* 28, 8087–8092, 1989.
322. Cherbavez, D.B., Trimethyloxonium modification of batrachotoxin-activated Na channels alters functionally important pore residues, *Biophys. J.* 68, 1337–1346, 1995.
323. Gómez, M., Isarna, P., Rojo, M., and Estrada, P., Chemical modification of β-xylosidase from *Trichoderma reesei* QM 9414: pH-dependence of kinetic parameters, *Biochimie* 83, 961–967, 2001.
324. Woodward, R.B., Olofson, R.A., and Mayer, H., A new synthesis of peptides, *J. Am. Chem. Soc.* 83, 1010–1012, 1961.
325. Dunn, B.M., Anfinsen, C.B., and Shrager, R.I., Kinetics of Woodward's Reagent K hydrolysis and reaction with staphylococcal nuclease, *J. Biol. Chem.* 249, 3717–3723, 1974.
326. Bodlaender, P., Feinstein, G., and Shaw, E., The use of isoxazolium salts for carboxyl group modification in proteins. Trypsin, *Biochemistry* 8, 4941–4949, 1969.
327. Saini, M.S. and Van Etten, R.L., An essential carboxylic acid group in human prostate acid phosphatase, *Biochim. Biophys. Acta* 568, 370–376, 1979.
328. Arana, J.L. and Vallejos, R.H., Two different types of essential carboxyl groups in chloroplast coupling factor, *FEBS Lett.* 123, 103–106, 1981.
329. Dinur, D., Kantrowitz, E.R., and Hajdu, J., Reaction of Woodward's Reagent K with pancreatic porcine phospholipase A2: Modification of an essential carboxylate residue, *Biochem. Biophys. Res. Commun.* 100, 785–792, 1981.
330. Kooistra, C. and Sluyterman, L.A.A., Isosteric acid and non-isosteric modification of carboxyl groups of papain, *Biochim. Biophys. Acta* 997, 115–120, 1989.
331. Johnson, A.R. and Dekker, E.E., Woodward's Reagent K inactivation of *Escherichia coli* L-threonine dehydrogenase: Increased absorbance at 340–350 nm is due to modification of cysteine and histidine residues, not asparate or glutamate carboxyl groups, *Protein Sci.* 5, 382–390, 1996.
332. Puri, R.N. and Colman, R.W., A novel method for the chemical modification of functional groups other than a carboxyl group in proteins with *N*-ethyl-5-phenylisooxazolim-3-sulfonate (Woodward's Reagent K): Inhibition of ADP-induced platelet responses involves covalent modification of aggregin, an ADP receptor, *Anal. Biochem.* 240, 251–261, 1996.
333. Bahar, S., Gunter, C.T., Wu, C. et al., Persistence of external chloride and DIDS binding after chemical modification of Glu681 in human band 3, *Am. J. Physiol.* 277, C791–C799, 1999.
334. Salhany, J.M., Sloan, R.L., and Cordes, K.S., The carboxyl side chain of glutamate 681 interacts with a chloride binding modifier site that allosterically modulates the dimeric conformational state of band 3 (AEI). Implications of the mechanism of anion/proton cotransport, *Biochemistry* 42, 1589–1602, 2003.

335. Hoare, D.G. and Koshland, D.E., Jr., A method for the quantitative modification and estimation of carboxyl groups in proteins, *J. Biol. Chem.* 242, 2447–2453, 1967.
336. George, A.L., Jr. and Border, C.L., Jr., Essential carboxyl groups in yeast enolase, *Biochem. Biophys. Res. Commun.* 87, 59–65, 1979.
337. Khorana, H.G., The chemistry of carbodiimides, *Chem. Rev.* 53, 145–166, 1953.
338. Iwamoto, H., Oiwa, K., Suzuki, T., and Fujisawa, T., States of thin filament regulatory proteins as revealed by combining cross-linking/X-ray diffraction techniques, *J. Mol. Biol.* 317, 707–720, 2002.
339. Cook, G.M., Keis, S., Morgan, H.W. et al., Purification and biochemical characterization of the F_1F_0 ATP synthase from thermophilic *Bacillus* sp. strain TA2.A1, *J. Bacteriol.* 185, 4442–4449, 2003.
340. Das, A. and Ljungdahl, L.G., *Clostridium pasteurianum* F_1F_0 ATP synthase: Operon, composition and some properties, *J. Bacteriol.* 185, 5527–5535, 2003.
341. Sheehan, J.C. and Hlavka, J.J., The use of water-soluble and basic carbodiimides in peptide synthesis, *J. Org. Chem.* 21, 439–440, 1956.
342. Sheehan, J.C. and Hlavka, J.J., The cross-linking of gelatin using a water-soluble carbodiimide, *J. Am. Chem. Soc.* 79, 4528–4529, 1957.
343. Riehm, J.P. and Scheraga, H.A., Structural studies on ribonuclease. XXI. The reaction between ribonuclease and a water-soluble carbodiimide, *Biochemistry* 5, 99–115, 1966.
344. Hoare, D.G. and Koshland, D.E., Jr., A procedure for the selective modification of carboxyl groups in proteins, *J. Am. Chem. Soc.* 88, 2057, 1966.
345. Lin, T.Y. and Koshland, D.E., Jr., Carboxyl group modification and the activity of lysozyme, *J. Biol. Chem.* 244, 505–508, 1969.
346. Tohri, A., Dohmae, H., Suzuki, T. et al., Identification of the domains on the extrinsic 23 kD protein possibly involved in electrostatic interaction with the extrinsic 33 kD protein in spinach photosystem II, *Eur. J. Biochem.* 271, 962–971, 2004.
347. Good, N.E., Winget, G.D., Winter, W. et al., Hydrogen ion buffers for biological research, *Biochemistry* 5, 467–477, 1966.
348. Good, N.E. and Izawa, S., Hydrogen ion buffers, *Methods Enzymol.* 24, 53–68, 1972.
349. Gilles, M.A., Hudson, A.Q., and Borders, C.L., Jr., Stability of water-soluble carbodiimides in aqueous solution, *Anal. Biochem.* 184, 244–248, 1990.
350. Lei, P.Q., Lamb, D.H., Heller, R.K. et al., Kinetic studies on the rate of hydrolysis of *N*-ethyl-*N*′-(dimethylaminopropyl)carbodiimide I aqueous solution using mass spectrometry and capillary electrophoresis, *Anal. Biochem.* 310, 122–124, 2002.
351. Wrobel, N., Schinkinger, M., and Mirsky, V.M., A novel ultraviolet assay for testing side reactions of carbodiimide, *Anal. Biochem.* 305, 135–138, 2003.
352. Sehgal, D. and Vijay, I.K., A method for the high efficiency of water-soluble carbodiimide-mediated amidation, *Anal. Biochem.* 218, 87–91, 1994.
353. Perfetti, R.B., Anderson, C.D., and Hall, P.L., The chemical modification of papain with 1-ethyl-3(3-dimethylaminopropyl) carbodiimide, *Biochemistry* 15, 1735–1743, 1976.
354. Carraway, K.L. and Koshland, D.E., Jr., Reaction of tyrosine residues in proteins with carbodiimide reagents, *Biochim. Biophys. Acta* 160, 272–274, 1968.
355. Carraway, K.L. and Koshland, D.E., Jr., Carbodiimide modification of proteins, *Methods Enzymol.* 25, 616–623, 1972.
356. Lewis, S.D. and Shafer, J.A., Conversion of exposed aspartyl and glutamyl residues in proteins to asparaginyl and glutaminyl residues, *Biochim. Biophys. Acta* 303, 284–291, 1973.
357. Wang, T.T. and Young, N.M., Modification of aspartic acid residues to induce trypsin cleavage, *Anal. Biochem.* 91, 696–699, 1978.
358. Lin, C.M., Mihal, K.A., and Krueger, R.J., Introduction of sulfhydryl groups into proteins at carboxyl sites, *Biochim. Biophys. Acta* 1038, 382–385, 1990.
359. Pho, D.B. et al., Evidence for an essential glutamyl residue in yeast hexokinase, *Biochemistry* 16, 4533–4537, 1977.

360. Desvages, G., Rousten, C., Fattoum, A., and Pradel, L.A., Structural studies on yeast 3-phosphoglycerate kinase. Identification by immuno-affinity chromatography of one glutamyl residue essential for yeast 3-phosphoglycerate kinase activity. Its location in the primary structure, *Eur. J. Biochem.* 105, 259–266, 1980.
361. Lacombe, G., Van Thiem, N., and Swynghedauw, B., Modification of myosin subfragment 1 by carbodiimide in the presence of a nucleophile. Effect on adenosinetriphosphatase activity, *Biochemistry* 20, 3648–3653, 1981.
362. Korner, M. et al., Location of an essential carboxyl group along the heavy chain of cardiac and skeletal myosin subfragments 1, *Biochemistry* 22, 5843–5847, 1983.
363. Hegde, S.S. and Blanchard, J.S., Kinetic and mechanistic characterization of recombinant *Lactobacillus viridescens* FemX (UDP-*N*-acetylmuramoyl pentapeptide-lysine N^6-alanine transferase, *J. Biol. Chem.* 278, 22861–22867, 2003.
364. Sigrist Nelson, K. and Azzi, A., The proteolipid subunit of the chloroplast adenosine triphosphatase complex. Reconstitution and demonstration of proton-conductive properties, *J. Biol. Chem.* 255, 10638–10643, 1980.
365. Sussman, M.R. and Slayman, C.W., Modification of the *Neurospora crassa* plasma membrane (H+)-ATPase with *N,N′*-dicyclohexylcarbodiimide, *J. Biol. Chem.* 258, 1839–1843, 1983.
366. Lotscher, H.R., deJay, C., and Capaldi, R.A., Inhibition of the adenosinetriphosphatase activity of *Escherichia coli* F_1 by the water-soluble carbodiimide, 1-ethyl-3-(3-dimethylaminopropyl)carbodiimide is due to modification of several carboxyls in the beta subunit, *Biochemistry* 23, 4134–4140, 1984.
367. Shafir, I., Feng, W., and Shoshan-Barmatz, V., Dicyclohexylcarbodiimide interaction with the voltage-dependent anion channel from sarcoplasmic reticulum, *Eur. J. Biochem.* 253, 627–636, 1998.
368. Yang, S.J., Jiang, S.S., Kuo, S.Y. et al., Localization of a carboxylic residue possibly involved in the inhibition of vacuolar H+-pyrophosphatase by *N,N′*-dicyclohexylcarbodiimide, *Biochem. J.* 342, 641–646, 1999.
369. Bjerrum, P.J., Anderson, O.S., Borders, C.L., Jr., and Wieth, J.O., Functional carboxyl groups in the red blood cell anion exchange protein, *J. Gen. Physiol.* 93, 813–839, 1989.
370. Levy, H.M., Leber, P.D., and Ryan, E.A., Inactivation of myosin by 2,4-dinitrophenol and protection by adenosine triphosphate and other phosphate compounds, *J. Biol. Chem.* 238, 3654–3659, 1963.
371. Takahashi, K., Stein, W.H., and Moore, S., The identification of a glutamic acid residue as part of the active site of ribonuclease T_1, *J. Biol. Chem.* 242, 4682–4690, 1967.
372. Kojima, M., Mizukoshi, T., Miyano, H. et al., Thermal stabilization of ribonuclease T1 by carboxymethylation at Glu-58 as revealed by ^{1}H nuclear magnetic resonance spectrometry, *FEBS Lett.* 351, 389–392, 1994.
373. Erlanger, B.F., Vratsanos, S.M., Wassermann, M., and Cooper, A.G., Specific and reversible inactivation of pepsin, *J. Biol. Chem.* 240, PC3447, 1965.
374. Gross, E. and Morell, J.L., Evidence for an active carboxyl group in pepsin, *J. Biol. Chem.* 241, 3638–3639, 1966.
375. Chacur, M., Longo, I., Picolo, G. et al., Hyperalgesia induced by Asp49 and Lys49 phospholipases A2 from *Bothrops asper* snake venom: Pharmacological mediation and molecular determinants, *Toxicon* 41, 667–678, 2003.

4 Modification of Heterocyclic Amino Acids: Histidine and Tryptophan

Histidine and tryptophan (Figure 4.1) are both aromatic heterocyclic amino acids that are considered to be electron-rich amino acids susceptible to oxidation.[1,2] Histidine contains the imidazole ring and tryptophan contains the indole ring. However, histidine is a polar amino acid, while tryptophan is a nonpolar amino acid. The majority of the reactions of histidine involve substitution of one or both of the ring nitrogens (while the nitrogens are equivalent in imidazole, they differ somewhat in reactivity in histidyl residues in proteins both from structure and electrostatic environment), while the majority of tryptophan modifications occur at the second carbon on the indole ring. This difference likely reflects the difference in the nucleophilicity of the imidazole nitrogens and the indole nitrogen; the pK_a for the nitrogen on the imidazole ring of histidine is approximately 6.8, while that for the pyrrole nitrogen in tryptophan is 17, which, as noted by Jack Kyte,[3] is greater than that for water (15.8).

HISTIDINE

A substantial number of studies on the modification of histidine have been directed at the study of the catalytic mechanism of enzymes and substantially less directed at protein–protein interactions or substrate/cofactor binding. Thus, many of the modifying reagents have been used as affinity labels[4] such as tosyl-phenylalanine chloromethyl ketone[5] (TPCK) or 2′(3′)-*O*-bromoacetyluridine.[6] TPCK reacts with active-site histidine (His57) in chymotrypsin, while 2′(3′)-*O*-bromoacetyluridine reacts with His12 in bovine pancreatic ribonuclease (RNase). The reaction rate of 2′(3′)-*O*-bromoacetyluridine with RNase is approximately 3000 times faster than that with free histidine and the rate of inactivation of RNase is 4.5 times faster than that observed with bromoacetate. While there is great interest in the modification of histidine residues in enzymes, a relatively small number of reagents have been studied for the group-specific modification of this amino acid in proteins. Diethyl pyrocarbonate (DEPC) is the most commonly used reagent but *p*-bromophenacyl bromide, methyl-*p*-nitrobenzenesulfonate, haloalkyl compounds such as iodoacetate as well as other alkylating agents, and cyanylation have seen limited application. Histidine is also susceptible to modification with 4-hydroxy-2-nonenal and can be modified in the Maillard reaction. Photooxidation is a relatively nonspecific modification[7] but

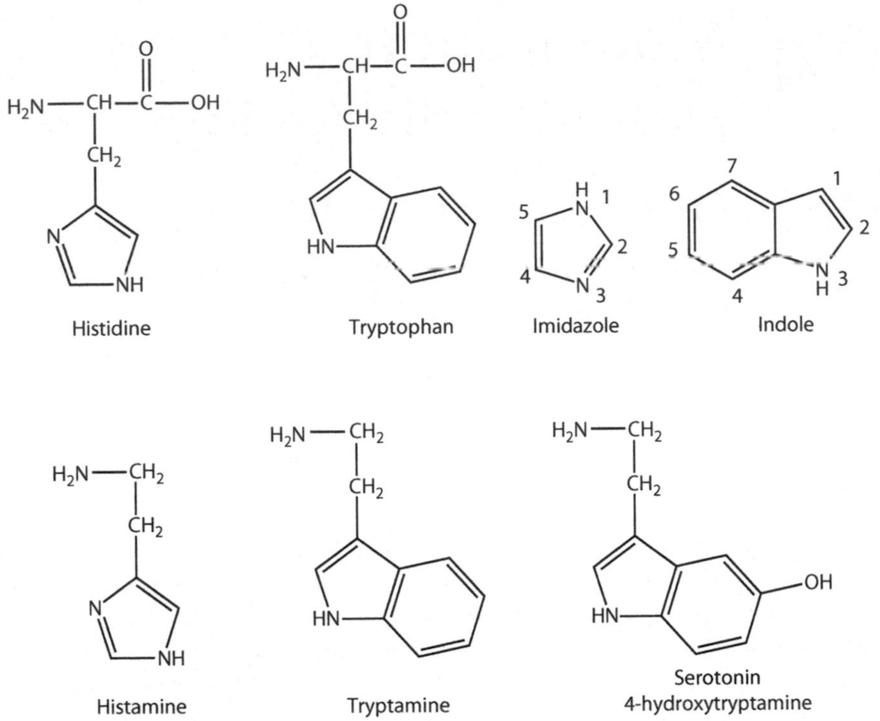

FIGURE 4.1 The structure of histidine and tryptophan and related compounds. Shown is the structure of histidine, imidazole, and histamine. Histamine is of pharmacological significance (Jutel, M., Akdis, M., and Akdis, C.A., Histamine, histamine receptors and their role in immune pathology, *Clin. Exp. Allergy* 39, 1786–1800, 2009). Also shown is the structure of tryptophan, the parent indole ring, tryptamine, and serotonin. Serotonin and, to a lesser extent tryptamine, also have pharmacological significance (Gillman, P.K., Triptans, serotonin agonists, and serotonin syndrome (serotonin toxicity): A review, *Headache* 50, 264–272, 2010; Loder, E., Triptan therapy in migraines, *N. Engl. J. Med.* 363, 63–70, 2010).

has been extensively used for the modification of electron-rich amino acids such as tryptophan and histidine (Figure 4.1) in proteins.[8,9] Rigorous analysis of the reaction can provide extremely useful information. Photooxidation is literally oxidation by light; more specifically, photooxidation is the loss of one or more electrons from a chemical species as a result of photoexcitation of that species. If a process involves oxygen and oxygen remains in the product of the reaction, the process is described as photooxygenation. Photosensitization is a process by which a photochemical or other alteration occurs in a chemical species as a result of the initial absorption of radiation by another chemical species described as a photosensitizer. In most of the studies cited below, dyes such as methylene blue serve as photosensitizers for the photooxidation of histidine and tryptophan. The reader is directed to Braslavsky and Houk[10] for further information on photochemistry terminology. In fact, most biological photooxidations involve oxygen[11,12]; type I reactions involve a radical resulting from the formation of reactive oxygen species such as superoxide, while type II reactions involve singlet oxygen that is derived from triplet oxygen. Photooxidation is also an issue with protein biopharmaceuticals.[13–15] The oxidation of proteins as an *in vivo* process is of increasing interest with respect to biological control and aging.[16–20] Histidine residues are also oxidized in the process of radiolytic protein footprinting.[21,22] Radiolytic modification of proteins is related to type I photooxidation in that radical chemistry is involved in both processes. A description of radiolysis is shown in Figure 4.2 as well as the products derived from histidine and tryptophan. Takamoto and Chance have reviewed radiolytic footprinting as used for the study of protein–protein interaction.[23] Radiolytic modification has not been extensively used as it is technically challenging and, where radiolysis of water is used for the generation of hydroxyl radical, requires significant instrument resources.

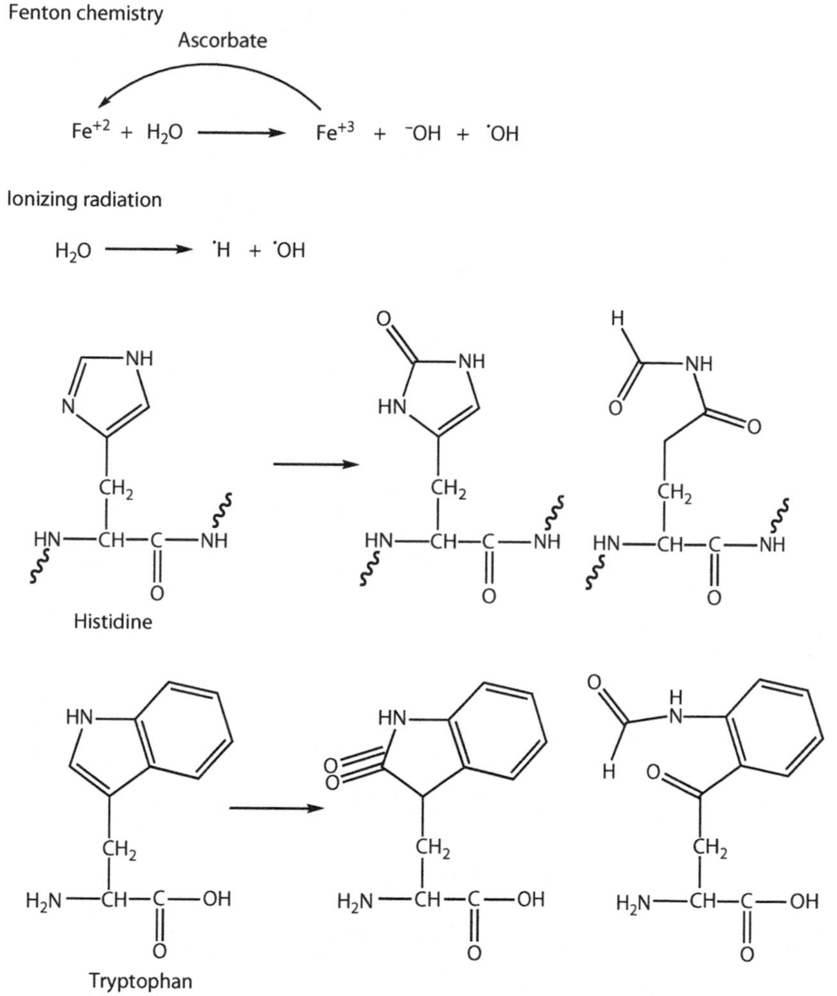

Fenton chemistry

Ascorbate

$$Fe^{+2} + H_2O \longrightarrow Fe^{+3} + {}^{-}OH + {}^{\bullet}OH$$

Ionizing radiation

$$H_2O \longrightarrow {}^{\bullet}H + {}^{\bullet}OH$$

Histidine

Tryptophan

FIGURE 4.2 The radiolytic modification of histidine and tryptophan. (See Takamoto, K. and Chance, M.R., Radiolytic protein footprinting with mass spectrometry to probe the structure of macromolecular complexes, *Ann. Rev. Biophys. Biomol. Struct.* 35, 251–276, 2006.)

Histidine residues can be modified by α-halo carboxylic acids and amides (i.e., bromoacetate and bromoacetamide) (Figure 4.3) but there has been little use of these reagents as other amino acids, most notably cysteine, are modified more rapidly than histidine. As an example, Rahimi and coworkers[24] modified cysteine in a far-red fluorescent protein with iodoacetamide and the histidine residue with DEPC (see below).

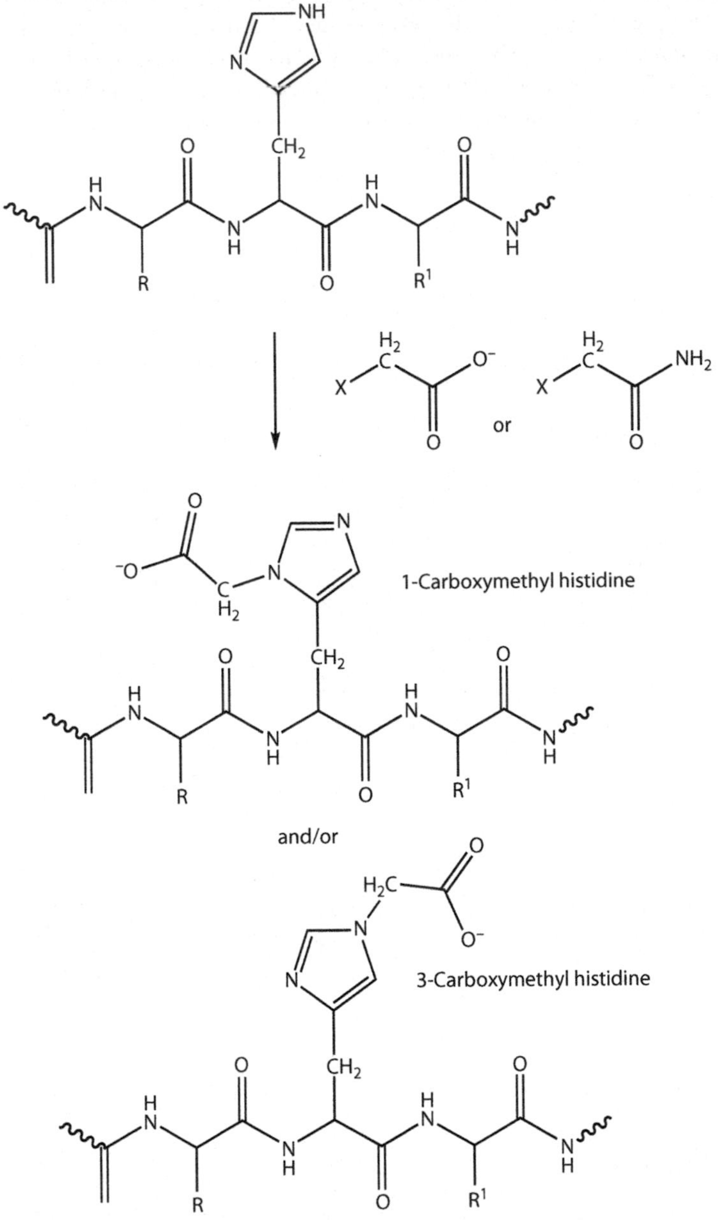

FIGURE 4.3 The reaction of histidine with 2-haloacetic acid/haloacetamide to form 1-carboxymethyl histidine and 3-carboxymethyl histidine.

There are several studies showing that histidyl residues can be modified with compounds such as iodoacetamide.[25–28] Gregory[25] showed that bovine heart mitochondrial dehydrogenase was inactivated with iodoacetamide but not iodoacetic acid; a sulfhydryl group is modified with some difficulty by *N*-ethylmaleimide. Thus, here is a situation in contradiction to the earlier statement on the preference for alkylation with cysteine residues. It should be noted that the cysteine residue in question does not appear to be available for modification at neutral pH where the histidine residue is modified with iodoacetamide; modification was accomplished by lowering the pH or including urea in the reaction mixture. Schelté and coworkers[29] examined the reaction of a polyoxylethyl bromoethyl derivative, 2-bromo-*N*-(2-(2-(2-hydroxyethoxy)ethoxy)ethyl)-acetamide (Figure 4.4), with *N*-benzoyl-glycyl-histidyl-leucine. Under conditions where there is complete modification of the terminal cysteine residue in the octapeptide (pH 9.0, 25°C), CGIRGERA, there was no modification of the histidyl residue in the tripeptide derivative. Another example of the increased reactivity of cysteine compared to histidine is provided from studies on histidine modification in sulfhydryl proteases. Liu[30] found it necessary to modify the cysteine residue at the active site of streptococcal proteinase with sodium tetrathionate in order to alkylate the active-site histidine residue. Alkylation of the active-site histidine was accomplished by using a positively charged reagent, α-*N*-bromoacetylarginine methyl ester, as histidine was resistant to modification with iodoacetic acid, iodoacetamide, *N*-chloroacetyltryptophan, or *N*-chloroacetylglycyleucine. Gleisner and Liener used a similar approach to modify the histidine residue at the active site of ficin.[31] Modification of active-site cysteine with tetrathionate and oxidation of the methionine residue with sodium periodate permitted the modification of histidine at the enzyme active site with bromoacetone in the presence of 2 M urea.

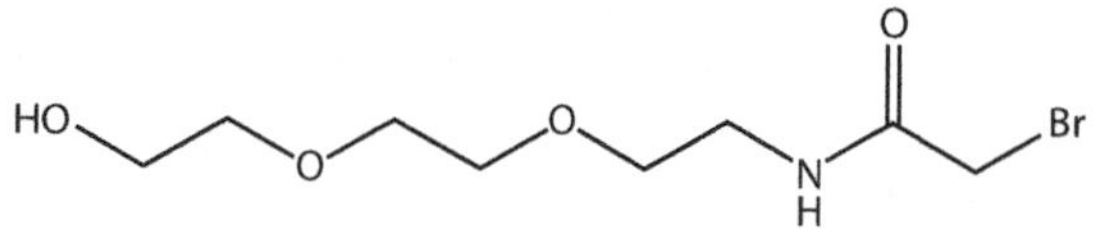

2-Bromo-*N*-(2-(2-(2-hydroxyethoxy)ethoxy)ethyl)-acetamide

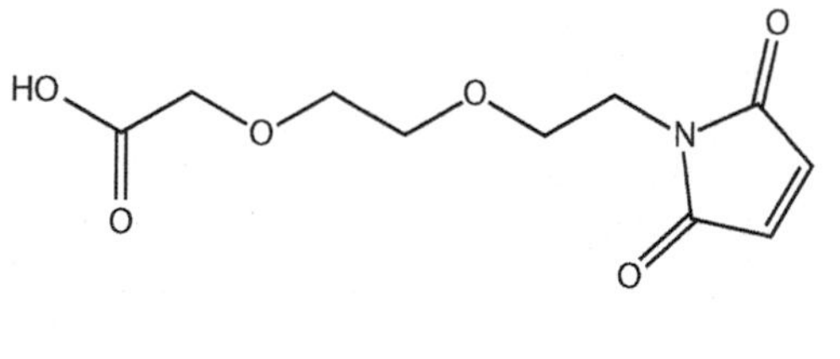

[2-[2-(2,5-dioxo-2,5-dihydropyrrol-1-yl)-ethoxy]ethoxy]acetic acid

FIGURE 4.4 The structures of 2-bromo-*N*-(2-(2-(2-hydroxyethoxy)ethoxy)ethyl)-acetamide and [2-[2-(2,5-dioxo-2,5-dihydropyrrol-1-yl)-ethoxy]ethoxy]acetic acid. These represent halo and maleimido derivatives, which differentially react with sulfhydryl and histidine. At pH 6.5, the maleimido derivative reacts with a terminal sulfhydryl function much faster (2–3 orders of magnitude) than does the bromo derivative; no insignificant reaction is observed with histidine.

It is useful to consider the history of the modification of histidine residues in bovine pancreatic RNase in more detail as it provides insight into the development of haloalkyl derivatives for protein modification. Specifically, the contributions of Stanford Moore and William H. Stein and their coworkers at the Rockefeller Institute for Medical Research at the corner of 66th Street and York Avenue in New York City (now the Rockefeller University) will be briefly reviewed.

Bromoacetic acid and iodoacetic acid were first used as metabolic inhibitors and were important in understanding glycolysis. Concomitant with this early research, it was observed that iodoform and related compounds reacted with thiol groups in biological fluids. Studies on the inhibition of glyoxalase by iodoacetate in the early 1930s demonstrated the importance of thiol groups in enzymes.[32] This early work resulted in the identification of iodoacetate and bromoacetate (and related compounds) as specific reagents for the modification of sulfhydryl groups in proteins and the potential for reaction at other functional groups was not fully appreciated.[33] It is thus not surprising that Zittle[34] ascribed the inhibition of RNase that he observed with iodoacetate to modification of a cysteinyl residue. Considering later observations, it is of interest that iodoacetamide was much less effective than iodoacetic acid. Determination of the primary structure of RNase by Werner Hirs working with Stanford Moore and William H. Stein at the Rockefeller Institute revealed the absence of free cysteine residues.[35,36] The Rockefeller group proceeded to examine the reaction of iodoacetate with RNase in the light of this information. Gundlach and coworkers[37] observed that the rate of inactivation of RNase by iodoacetate was most rapid at pH 5.5–6.0 and at pH 2.2; the rate of inactivation was less rapid both at intermediate pH values and at more alkaline pH. Amino acid analysis demonstrated modification of histidine at pH 5.5–6.0, methionine at pH 2.2, and the ε-amino group of lysine at alkaline pH. Chromatography of the modified protein on IRC-50 yielded three fractions, two of which were inactive. These investigators also reported that the reaction of iodoacetate with *N*-acetylhistidine yielded two products. George Stark and coworkers[38] then proceeded to show that the modification of histidine residues was dependent on the native conformation of the protein. No modification of histidine occurred at pH 5.5 in the presence of 8 M urea, 4 M guanidine, or after reduction with sodium borohydride. These investigators did report the modification of methionine with iodoacetic acid under these reaction conditions. The lack of reaction of iodoacetamide was also confirmed.

Another research group with a different W. Stein at King's College in London studied the reaction of bromoacetate with RNase[39] reporting that the inactivation was associated with the modification of a single histidine residue. The reaction proceeded optimally at pH 5.5–7.0 and the enzyme was protected by cytidylic acid. The rate of reaction of bromoacetate with RNase was estimated to be some 30 times more rapid than the rate of reaction of bromoacetate with free histidine[40] or bovine serum albumin.[41] Korman and Clarke[40] also demonstrated the rate of reaction of bromoacetate with histidine was more rapid than the rate of reaction with chloroacetate; iodoacetate was similar to bromoacetate. The reaction of iodoacetate is more rapid with amino groups or phenolic hydroxyl groups than the reaction of these functional groups with bromoacetate. Subsequent structural analysis by the King's College group[42] showed that modification of RNase with bromoacetate occurred at His119.

The early work of Korman and Clarke[41] on the carboxymethylation of albumin and trypsin deserves further comment. The reaction of albumin (500 mg) with bromoacetate (6 mM; an approximate 1000-fold excess of reagent) with MgO in 20 mL H_2O at 35°C resulted in the progressive denaturation of the protein; reaction with trypsin resulted in the loss of enzyme activity. Serum albumin contains a free sulfhydryl group that is readily modified by iodoacetic acid[43] and other investigators have noted structural changed on modification with iodoacetamide[44] or iodoacetic acid.[45] In summary then, at this point in time, it has been demonstrated that the inactivation of RNase by iodoacetate at neutral pH resulted in the modification of a histidine residue(s); the reaction yielded two inactive products, and the reaction of histidine in RNase with iodoacetate, but not methionine, was dependent on the native conformation of the protein. The crystal structure of the protein was not yet available and RNase, together with lysozyme and, to a lesser extent, albumin, had become model proteins to develop basic information on protein chemistry. Major laboratories such as that of Chris Anfinsen at the National Institutes of Health in Bethesda and that of Harold Scheraga were also interested in this protein.

The author joined the laboratory of Stanford Moore and William H. Stein in 1966. I was settling in after returning from the Federation Meeting in Atlantic City when the late Rachael Fruchter introduced herself and asked me to sign a chromatographic chart (see Figure 4.1 in reference 46). This chart showed the separation of the reaction products of iodoacetate and RNase on IRC-50 and was taken from a paper by Art Crestfield and coworkers.[46] The work showed that the reaction of iodoacetate with RNase yielded a major product (1-carboxymethyl-His119-RNase) and a minor product (3-carboxymethyl-His12-RNase). The 119-derivative was inactive with either cyclic cytidylic acid or RNA as substrate, while the 12-derivative had 7% activity with cyclic cytidylic acid and less than 5% activity with RNA. Polymorphism was excluded when the same results were obtained by modification of RNase derived from a single cow. As an aside, the author was informed by Stanford Moore that all of the early work on bovine pancreatic ribonuclease was performed on protein derived from Kosher tissue. As a further aside, the issue of polymorphic expression was an issue with early work on carboxypeptidase A which also required the isolation of protein from a single pancreas for resolution of the heterogeniety issue.[47] A subsequent study from this group[48] showed the modification of one of the two histidine residues precluded modification at the other residue. Alkylation of histidine also affected the ability of RNase to bind phosphate; chromatography on IRC-50 in the presence of 0.2 M NaCl instead of 0.2 M sodium phosphate (both at pH 6.47) changes the elution position of the products of the reaction; in phosphate, the native protein is eluted first followed by the 119-derivative and then the 12-derivative, while in the presence of 0.2 M NaCl, the 199-derivative is eluted first followed by the 12-derivative and then the native protein. This work, as well as various studies from other laboratories cited in reference 46, showed that both histidine-12 and histidine-119 were required for the enzymatic activity of RNase. A heterodimer formed from the 12-carboxymethyl derivative of RNase and the 119-carboxymethyl derivative of RNase could have activity as the heterodimer would contain an unmodified residue of 119 from the 12-carboxymethyl derivative and an unmodified residue 12 from the 119-carboxymethyl derivative. Crestfield and colleagues[48] prepared a dimer from equimolar quantities of the 12-derivative and 119-derivative by lyophilization from 50% acetic acid. Assay showed that the hybrid dimer has 45% of the specific activity of the native enzyme; the majority of this activity was lost on heating under conditions where there was dissociation of the dimer. These studies and other studies permitted the suggestion that the two active-site residues are approximately 5 Å apart in the native protein. Fruchter and Crestfield[49] subsequently demonstrated that RNase dimer formed by lyophilization from 50% acetic acid is inactivated by iodoacetate at the same rate as RNase monomer with predominant modification at histidine-119. The modified protein contained 1-carboxymethyl-histidine and 3-carboxymethyl-histidine as well as *S*-carboxyhomocysteine. The presence of *S*-carboxymethylhomocysteine indicates that methionine was modified under these reaction conditions. This work and later studies[50,51] suggested lyophilization from 50% acetic acid results in mutual displacement of the amino terminal region of RNase monomer in the formation of dimer bringing histidine-12 of one monomer in opposition to histidine-119 of carboxyl terminal region of other RNase monomer in the dimer.

Crestfield and colleagues[46,48] had suggested that the mutually exclusive modification of the histidine residues with preferential reaction at His119 reflected orientation of the negatively charged reagent by a positive charge in the protein and that a protonated His12 could supply that charge. These investigators suggested that the positive charge could be provided by a lysine residue, but Michael Lin and coworkers[52] eliminated the participation of either lysine-7 or lysine-41. This work provided additional support for an additional role for one of the two histidine residues in providing orientation for the iodoacetate. Implicit in this consideration would be the necessity of a higher pK_a for His12 compared to His119. This is supported by NMR[53] but not necessarily by other physical techniques.[54] The relative ineffectiveness of iodoacetamide in the inactivation of RNase was mentioned earlier; differences in the relative reactivity of iodoacetate and iodoacetamide with proteins is not uncommon and usually reflects the difference between a charged reagent and a neutral compound. Fruchter and Crestfield[55] studied the reaction of iodoacetamide with RNase in more detail and observed that the rate of inactivation was quite slow compared to iodoacetic acid but still some 10–100 times faster than reaction with free histidine. Two products were obtained from the reaction of iodoacetamide with RNase, 3-carboxyamidomethyl-His12 and fully active *S*-carboxamidomethyl-Met. It is possible that the lower pK_a value for His12 is responsible for the increased reactivity of this residue. The issue of the roles of His12 in the modification of His119 in RNase was further investigated by Bryce Plapp.[56] The study showed that a careful examination of the effect of concentration on reaction rate showed a Michaelis–Menten relationship. This work also studied the effect of various leaving groups (chloride, bromide, iodine, and tosylate) on reaction rate. The data thus obtained support the importance of the binding of iodoacetate to RNase prior to the alkylation of His119. This study also showed that the rate of reaction of tosylglycolate with RNase was similar to that observed with the various halo compounds with increased specificity for reaction at His119. The reader is referred to the Nobel comment by Stanford Moore and William H. Stein[57] and the review[58] by Garland Marshall and others for further discussion of these studies. The work by Marshall emphasizes the work of the Merrifield group at the Rockefeller in the chemical synthesis of bovine pancreatic ribonuclease.

Hirs and colleagues noted in their publication on the primary structure of RNase[35] that much had been learned in that and previous studies, which would have made the work much easier. The same can be said for the earlier story that started with the Zittle observations,[34] which were followed by the studies of Gerd Gundlach and coworkers[37] followed in turn by a series of publications, which resulted in a considerable increase in our understanding of basic protein chemistry. As with the Hirs observation mentioned earlier, the answers to the RNase problem would have come more rapidly with today's instrumentation but our understanding of solution chemistry would not be as rich. While today's protein therapeutics are efficiently prepared by recombinant DNA technology rather than being obtained from tissue sources and purified by "black box" technologies, there is knowledge that is critical to the understanding of the protein going from an active pharmaceutical ingredient to a final drug product.

The chemistry of histidine alkylation with α-halo carboxylic acids and amides provided the basis for the development of peptide chloromethylketones (Figure 4.5) for the affinity labeling of proteolytic enzymes.[59–62] These reagents were mentioned earlier and the reader is directed to the review by Plapp[4] for discussion of affinity labeling technology. Specificity can be enhanced by increasing the complexity of the reagents as demonstrated by comparing the inactivation of urokinase by Pro-Gly-Arg-chloromethyl ketone to N^{α}-tosyl-L-lysine chloromethyl ketone.[63] Inactivation with the tripeptide chloromethylketone occurred at micromolar concentrations, while millimolar concentrations of tosyl-lysylchloromethyl ketone (TLCK) provided only slow inactivation. Care must be taken with the use of peptide chloromethyl ketones in complex systems as these materials react with sulfhydryl groups via an alkylation mechanism.[64]

A compound related to α-halo carboxylic acids is *p*-bromophenacyl bromide [2-bromo-1-(4-bromophenyl) ethanone] (Figure 4.6), which has been used to modify histidyl and carboxyl residues in proteins. This reagent modifies a histidine residue in the active site of phospholipase A_2 and this modification is used as discriminating factor in identifying the physiological impact of phospholipase A_2 in complex biological systems.[65–75] A study reported by Shaw and coworkers[75] illustrates the value of *p*-bromophenacyl bromide for the inactivation of phospholipase A_2 in studies with cobra venom factor, an anticomplement protein obtained from *Naja Naja* cobra venom. These investigators note that the hydrophobic properties of phospholipase A_2 complicated separation from cobra venom factor during purification but it was possible to inhibit the phospholipase A_2 with *p*-bromophenacyl bromide permitting clear interpretation of data obtained from whole-animal studies.

p-Bromophenacyl bromide was developed for the preparation of esters of organic acids[76] as the bromophenacyl derivative absorbs light at 254 nm.[77,78] Bernard Erlanger and colleagues at Columbia[79] observed the *p*-bromophenacyl bromide inactivated pepsin via the modification of a carboxyl group at the enzyme active site; this observation was confirmed by Gross and Morrell.[80] It should be noted that iodoacetate has also been shown to react with a glutamic acid residue in ribonuclease T_1.[81] *p*-Bromophenacyl bromide would be expected to react with sulfhydryl groups in proteins but there is a limited literature on this possibility.[82] The reader might well question (as the author did), how did *p*-bromophenacyl bromide become a "signature" reagent for phospholipase A_2. There are precious few reagents that get to that

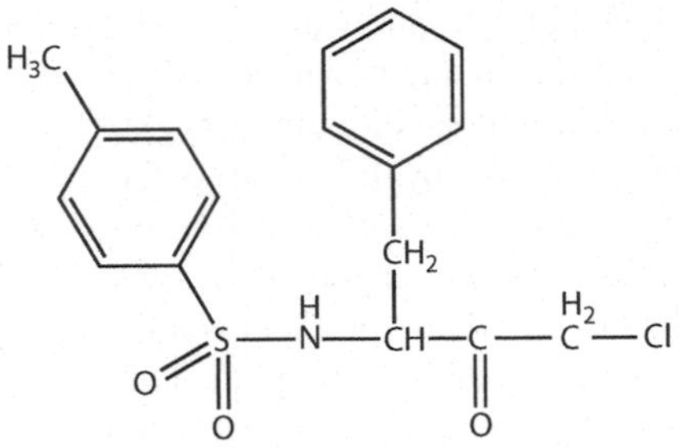

N^{α}-Tosyl-L-phenylalanine chloromethyl ketone (TPCK)

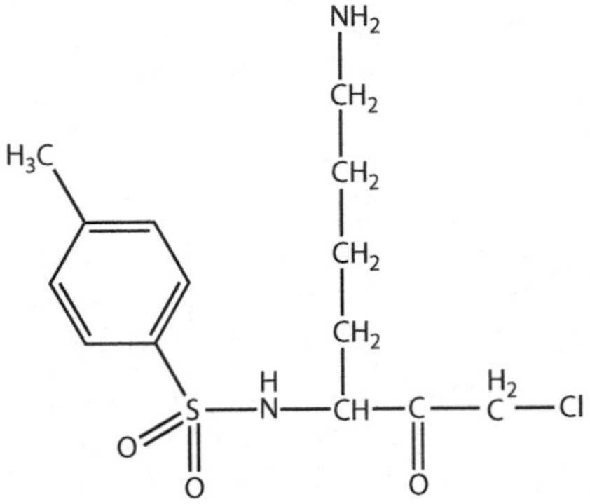

N^{α}-Tosyl-L-lysine chloromethyl ketone (TLCK)

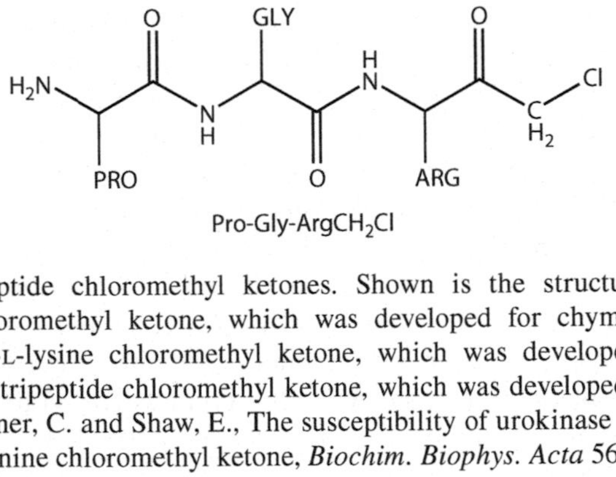

Pro-Gly-ArgCH_2Cl

FIGURE 4.5 Peptide chloromethyl ketones. Shown is the structure of N^{α}-tosyl-L-phenylalanine chloromethyl ketone, which was developed for chymotrypsin-like proteinases, N^{α}-tosyl-L-lysine chloromethyl ketone, which was developed for tryptic-like proteinases, and a tripeptide chloromethyl ketone, which was developed for the inhibition of urokinase (Kettner, C. and Shaw, E., The susceptibility of urokinase to affinity labeling by peptides of arginine chloromethyl ketone, *Biochim. Biophys. Acta* 569, 31–40, 1979).

status. A colleague of mine recalls a quote from Efraim Racker, which goes something like "a reagent is never as specific as when it is first discovered." The first citation that the author could readily locate on the reaction of *p*-bromophenacyl bromide with phospholipase A_2 appeared in 1974.[83] This is a rather nice study that showed the modification of a single histidine at the active site; a minor amount of modification occurred at another histidine residue. The enzyme was protected from inactivation by substrate and cofactors. Phospholipase A_2 occurs as a zymogen[84] and Volwerk and colleagues[83] showed that modification of this residue also occurred in the zymogen form of the enzyme. Volwerk and colleagues do refer an earlier report[85] from their

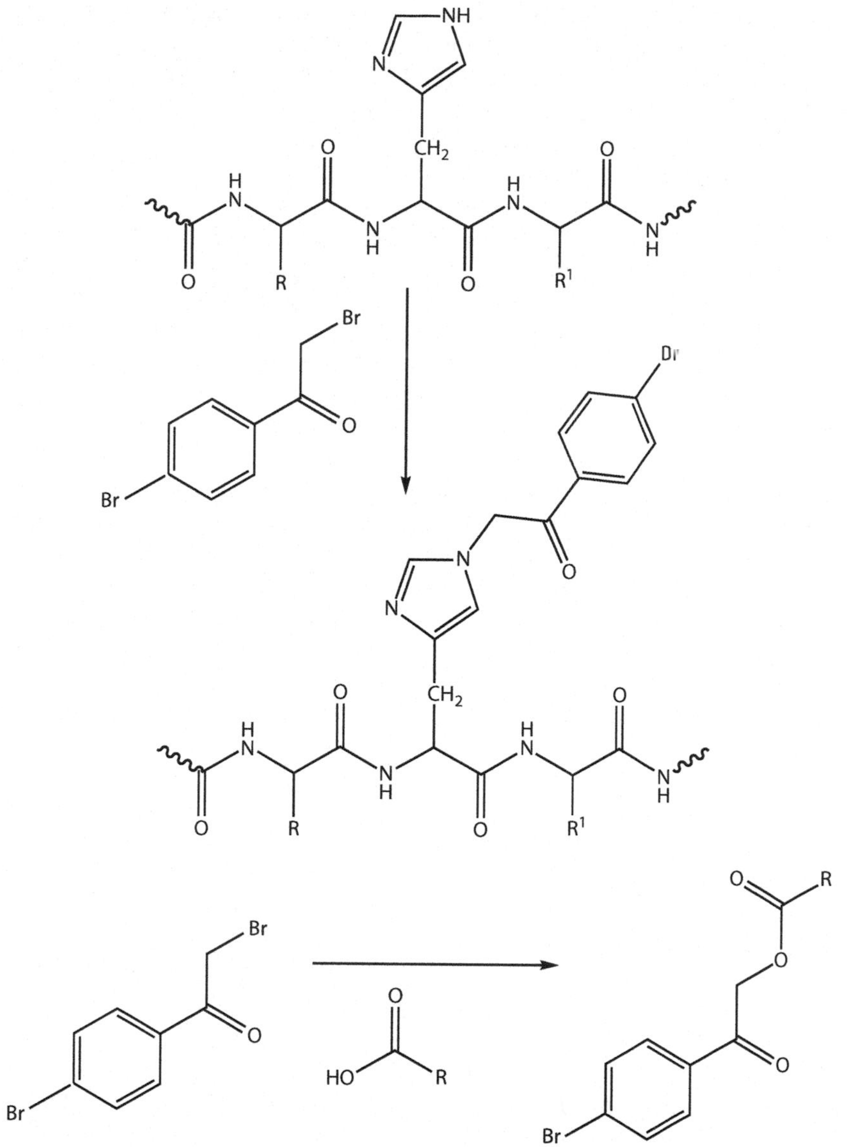

FIGURE 4.6 *p*-Bromophenacyl bromide [2-bromo-1-(4-bromophenyl)ethanone]. Shown is the reaction of *p*-bromophenacyl bromide with histidine. This reaction appears to be unique for the modification of a histidine residue at the active site of phospholipase A.

laboratory, which describes the initial experiments with a reference to Erlanger and colleagues,[79] so that it is presumed that *p*-bromophenacyl bromide was one of several reagents used to characterize phospholipase A_2 and Bonsen and coworkers[85] were astute enough to recognize that inactivation reflected the modification of histidine, not carboxyl groups. A decade, later, a review by Ed Dennis,[86] one of the more distinguished investigators in lipid biochemistry, describes *p*-bromophenacyl bromide as a specific inhibitor of phospholipase A_2. Snake venoms exhibit a spectrum of physiological responses[87] and *p*-bromophenacyl bromide continues to be used, as mentioned earlier, as an inhibitor of phospholipase A_2 in the study of these complex biological mixtures.[74] The crystal structure of dimeric Lys49-phospholipase A_2 complexed with *p*-bromophenacyl bromide has been reported[88] showing that the *p*-bromophenacyl group is bound to the N^1 in the imidazole ring of His48 making hydrophobic interactions with a cystine residue [there are seven disulfide bonds in this relatively small (14.5 kDa) protein] and Lys49 together with other nonpolar interactions. It seems reasonable the hydrophobic nature of *p*-bromophenacyl bromide contributes to the usual specificity of this reagent for phospholipase A_2. Prior studies[89,90] on the reaction of *p*-bromophenacyl bromide with phospholipase A_2 from several sources suggested that complex formation occurs between inhibitor and enzyme prior to the modification of the histidine residue; however, the data does not appear to support this suggestion as there is a lack of demonstration of saturation kinetics.

The following studies have been selected as being, at least to the author, some examples of the more useful studies on the *p*-bromophenacyl bromide modification of proteins. Taipoxin is an extremely lethal neurotoxin isolated from the venom of *Oxyranus scutellatus scutellatus* (Australian taipan).[91] While taipoxin is considered a neurotoxic phospholipase A_2, the ability to block synaptic transmission is more complex than the enzymatic activity and involves the affinity of binding to the presynaptic membrane.[92] *p*-Bromophenacyl bromide modifies a single histidine residue in taipoxin with a 350-fold decrease in neurotoxicity.[93] The extent of modification was assessed by both amino acid analysis (loss of histidine) and spectral analysis ($\Delta\varepsilon$ 271 = 17,000 M^{-1} cm^{-1}).[94] Two of seven histidine residues are modified (1 mol mol^{-1} in α-subunit; 1 mol mol^{-1} in β-subunit) under these reaction conditions. This study shows the use of the spectral properties of *p*-bromophenacyl bromide for determination of the extent of modification. As cited earlier, *p*-bromophenacyl bromide is used for the derivatization of organic acids prior to analysis. Verheij and coworkers[95] reported on the rate of inactivation of porcine pancreatic phospholipase A_2 by *p*-bromophenacyl bromide and other reagents that modified His48 at the enzyme active site. The second-order rate constants (0.1 M cacodyate-0.1 M NaCl, pH 6.0, 30°C) for the inactivation of porcine pancreatic phospholipase A_2 by *p*-bromophenacyl bromide is 125 M^{-1} min^{-1} as compared to 79 M^{-1} min^{-1} for phenacyl bromide and 75 M^{-1} min^{-1} for 1-bromooctan-2-one; inactivation of porcine pancreatic phospholipase A_2 with iodoacetamide was not detected under these reaction conditions.

Other than the pepsin papers cited earlier, there are only several studies of the reaction of *p*-bromophenacyl bromide with proteins other than phospholipase A_2. Battaglia and Radominska-Pandya[96] studied the functional role of histidine in the UDP-glucuronic acid carrier by measuring the effect of chemical modification on the uptake of radiolabeled UDP-glucuronic acid in rat liver endoplasmic reticulum. Inhibition of uptake was more pronounced with either *p*-bromophenacyl bromide or DEPC (both hydrophobic reagents) than with *p*-nitrobenzenesulfonic acid methyl ester, a hydrophilic reagent. de Vet and van den Bosch[97] examined the role of histidine in recombinant guinea pig alkyl-dihydroxyacetonephosphate synthase by modification with *p*-bromophenacyl bromide and oligonucleotide-directed mutagenesis. Modification with *p*-bromophenacyl bromide was performed in 10 mM

Tris-HCl, containing 0.15% triton X-100, pH 7.4, at room temperature. Inactivation was observed showing pseudo-first-order kinetics and the enzyme was protected by substrate. A remarkable increase in the rate of inactivation was observed at pH ≥ 8.0 (50 mM Tris-HCl). Replacement of His617 with alanine also eliminated catalytic activity. Modrow and coworkers[98] have shown that the phospholipase A_2-like activity of the VP1 region of parvovirus B19 is inactivated by *p*-bromophenacyl bromide. Parvovirus is the causative agent of erytheme infection and perhaps other diseases. Primary structure analysis showed the presence of a phospholipase Amotif. A new analytical method demonstrated the presence of enzyme activity in the VP1 region, which was Ca^{2+}-dependent and inhibited by *p*-bromophenacyl bromide.

Methyl *p*-nitrobenzenesulfonate (Figure 4.7) is a reagent that has been used infrequently to modify histidine residues in proteins. Nakagawa and Bender[99,100] used methyl *p*-nitrobenzenesulfonate as a substrate analogue to modify the active-site histidine in α-chymotrypsin. Modification of trypsin or subtilisin is not observed under these reaction conditions; reaction was not observed with free imidazole, *N*-acetyl-L-serinamide, or *N*-acetyl-L-methioninamide. Subsequently, Glick used methyl *p*-nitrobenzenesulfonate to methylate histidine residue(s) in ribosomal peptidyl transferase.[101] The author suggests only histidine residues are modified, but definitive evidence on this point is absent. Marcus and Dekker[102] examined the effect of methyl-*p*-nitrobenzenesulfonate on the activity of *Escherichia coli* L-threonine dehydrogenase. The reaction was performed in 100 mM potassium phosphate, pH 7.0 at 25°C; early reactions in this study were performed at pH 8.4 but it was observed that the reagent was more specific for the modification of histidine at the lower pH. Examination of the effect of reagent concentration on reaction rate demonstrated saturation kinetics with a limit value for the rate of inactivation of 0.01 min^{-1} suggesting the binding of inhibitor prior to the inactivation reaction. Analysis of the modified protein showed that His90 had been methylated at the N^3 position. Subsequent studies[103] with oligonucleotide directed mutagenesis confirmed the importance of this residue in enzyme function. Verheij and coworkers[95] reported the inactivation of equine, porcine, and bovine pancreatic phospholipase A_2 with methyl-*p*-nitrobenzenesulfonate. Reagent stability precluded determination of rate constants; it was possible to show that the equine enzyme was inactivated more rapidly than either the porcine or bovine enzyme; the rate of inactivation was an order of magnitude slower with methyl-*p*-toluenesulfonate. While there were minor sites of modification on the protein, the inactivation of phospholipase A_2 with methyl-*p*-nitrobenzenesulfonate is associated with methylation at the N^1 position on the imidazole ring of His48. The majority of the recent work with methyl-*p*-nitrobenzenesulfonate has focused on the use of methyl *p*-nitrobenzenesulfonate as model compound in nucleophilic substitution reactions.[104–107] Histidine residues in proteins may also be modified with methyl iodide but the reaction is not specific.[108–110] While most of the studies are based on the increased reactivity of histidyl residues at enzyme active sites, the study by Taralp and Kaplan[109] used methyl iodide to study the reactivity of lyophilized proteins in a nonaqueous (*n*-octane) environment.

The introductory statement to this chapter noted the peculiar character of the imidazole ring of histidine. It would appear that the nucleophilic character is maximum in the pH range of 5.5–7.0 reflecting the ionization of the protonated nitrogen.

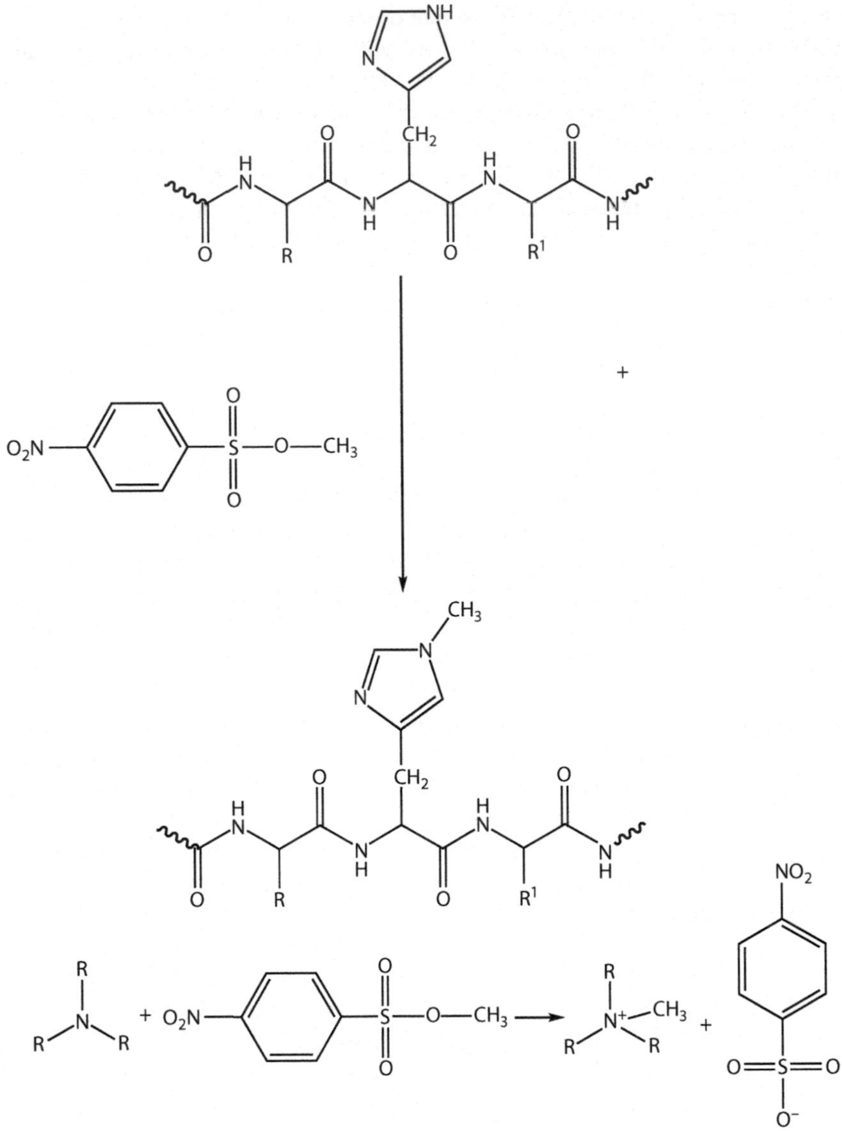

FIGURE 4.7 The *N*-methylation of histidine with methyl-*p*-nitrobenzene sulfonate. Shown is the formation of *N*-3 methyl histidine (see Nakagawa, Y. and Bender, M.L., Methylation of histidine-57 in α-chymotrypsin by methyl *p*-nitrobenzenesulfonate. A new approach to enzyme modification, *Biochemistry* 9, 259–267, 1970). At the bottom, the reaction of methyl-*p*-nitrobenzenesulfonate with a tertiary amine is shown as model for nucleophilic substitution (Ford, J.W., Janakat, M.E., Lu, J. et al., Local polarity in CO_2-expanded acetonitrile: A nucleophilic substitution reaction and solvatochromic probes, *J. Org. Chem.* 73, 3364–3368, 2008).

While the reactivity of histidine in the active sites of serine proteases is fairly well understood, the reactivity of histidine residues outside the active site is somewhat more difficult to understand. Likewise, the apparently random reactivity of histidine in proteins is equally difficult to understand. The author recalls a conversation with Stanford Moore at the Rockefeller Institute many years ago, which went something like this: "Bill and I thought that when we had the final sequence of RNase, we would know how it worked—no such luck, then when Fred Richard at Yale brought the model derived from crystallographic analysis, we were again optimistic but again disappointed." The author hopes that his nephew Nathan (physics faculty at Bates College) and his colleagues will solve the problem of how "stuff sticks together." Given our lack of understanding of molecular interactions, it is not unreasonable that there are histidine residues that one would expect to be modified but are not as well as unexpected modifications of histidine residues in proteins.

Most modifications of histidine occur most efficiently between pH 5 and 7. Lower pH decreases the reactivity of the nitrogen on the imidazole ring, while a higher pH increases the rate of modification at other sites such as the ε-amino group of lysine and decreases the stability of a number of modifications of histidine such as acetylation and ethoxyformylation, the related acyl derivatives of poly (ethylene glycol) (PEG), and the dansyl derivatives as described in the following.

The modification of histidine in interferon provides an example of an atypical modification. Succinimidyl carbonate-PEG (Figure 4.8) was developed to modify lysine residues in protein to yield a more effective biopharmaceutical product.[111] This reagent uses *N*-hydroxysuccinimide chemistry as described in Chapter 2. The succinimidyl carbonate derivative was developed to replace succinimidyl succinate-PEG. It was thought that the reagent would be specific for the modification of amino groups, although care was taken to examine the possibility of tyrosine modification. The succinimidyl carbonate-PEG form a carbamate derivative[112] with protein amino groups (Figure 4.8).

Wang and coworkers[113] reported that the reaction of interferon α-2b with succinimidyl carbonate PEG at pH 6.5 (phosphate) yielded a mixture of PEGylated products; 47% of the product was a carbamatehistidine (His34) PEG derivative of interferon α-2b (Figure 4.8). Subsequently Wylie and coworkers[114] studied the PEGylation of interferon α-2b with succinimidyl carbonate PEG and observed the pH-dependent formation of a PEG-histidine derivative. At mildly acidic pH (pH 5.4–6.5), there were several products; the major product was modified at His34. At more alkaline pH, there was more extensive modification of lysine. Both groups noted that the PEGylation of histidine was reversed under mild basic conditions consistent with the stability of *N*-acetyl histidine or ethoxyformylated histidine (see the following for more extensive discussion of these derivatives). Both studies argue that solvent exposure of His34 is important for PEGylation; however, to the best of the author's knowledge, this reaction has not been reported with the use of succinimidyl carbonate in other proteins[115,116] and was not observed with the solid-phase modification of interferon α-2a with succinimidyl carbonate-PEG.[117] In this study, PEGylation occurred at the amino-terminal amino acid. The study of Wang and coworkers[113] is instructive in this regard. These investigators compared the stability of the His34-PEGylated interferon α-2b with products prepared from the reaction of a peptide,

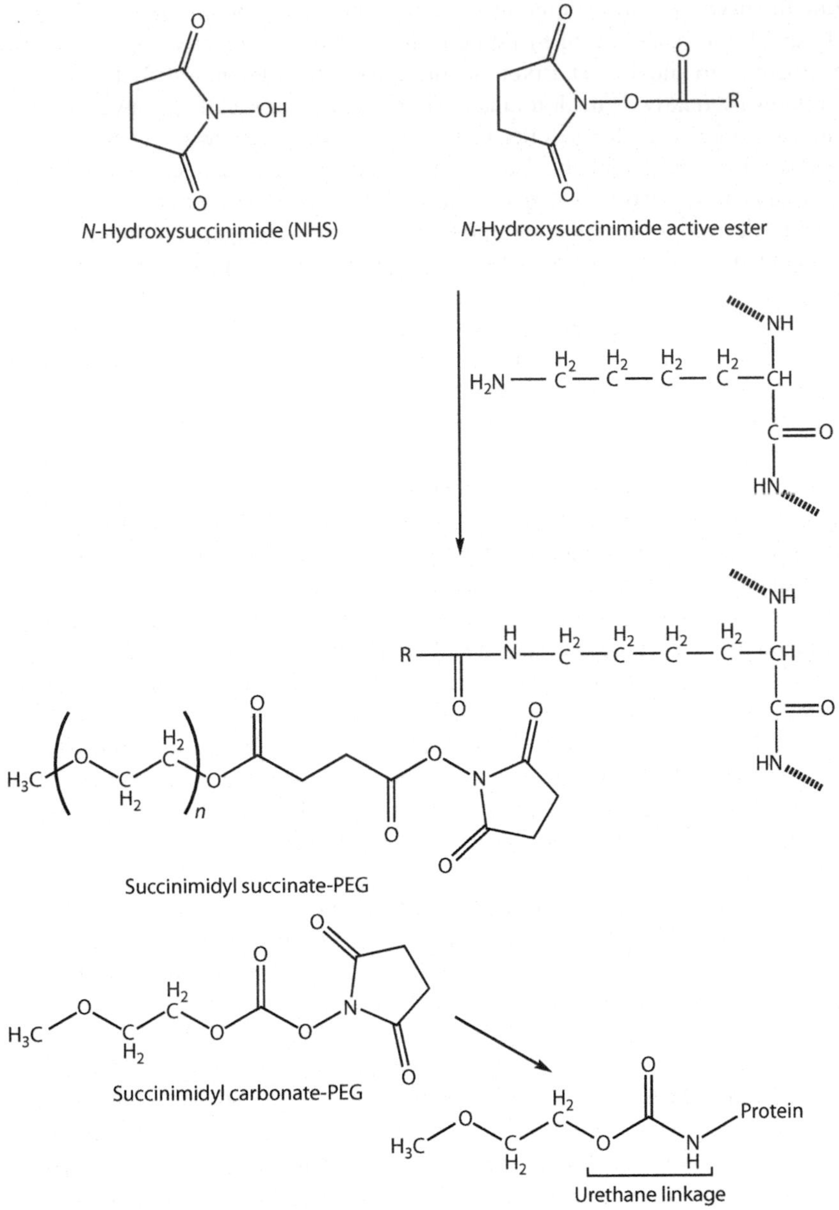

FIGURE 4.8 Acylation reactions of histidine. Shown is the reaction of a succinimidyl carbonate derivative with histidine (Wang, M., Basu, A., Palm, T. et al., Engineering an arginine catabolizing bioconjugate: Biochemical and pharmacological characterization of PEGylated derivatives of arginine deiminase from *Mycoplasma arthritidis*, *Bioconjug. Chem.* 17, 1447–1459, 2006). This derivative is labile at alkaline pH or in the presence of hydroxylamine. Shown also is an analogous modification of histidine with *N*-acetylimidazole in trypsin (Houston, L.L. and Walsh, K.A., The transient inactivation of trypsin by mild acetylation with *N*-acetylimidazole, *Biochemistry* 9, 156–166, 1970).

Ac-βDRH(PEG)DFGFPQ with succinimidyl carbonate-PEG or from the PEGHis peptide isolated from the subtilisin digestion of the modified interferon. The His34-PEGylated derivative was much more stable than either peptide derivatives. In work not cited in any of the earlier studies, Borukhov and Strongin[118] did not see any effect of DEPC or TLCK with either interferon αA-interferon or β-interferon; inhibition of β-interferon was observed with TPCK. This study lacks data on the extent of modification in the several proteins. TPCK was developed as an active-site-directed reagent, which reacts with an active-site histidine as an affinity reagent[4] but reaction at other sites such as sulfhydryl groups has been reported.[119]

Another example of the modification of histidine by reagents that, in general, react more avidly with residues other than histidine is the reaction with dansyl chloride (Figure 4.9). Hartley and Massey[120] described the inactivation of chymotrypsin by dansyl chloride. The inactivated protein was stable in the pH range of 4–6;

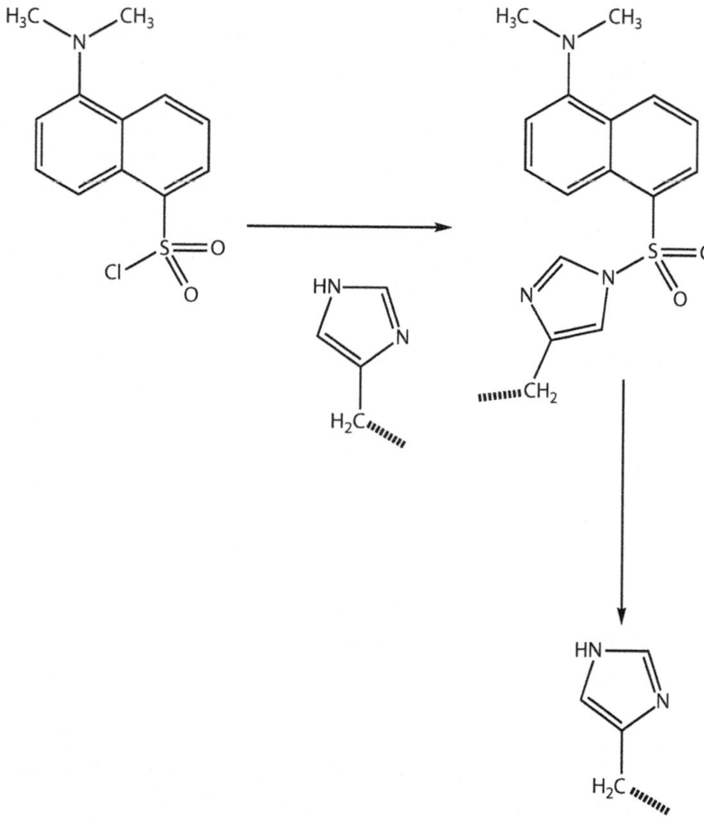

FIGURE 4.9 The reaction of dansyl chloride with histidine residues in proteins. The product of the reaction is not stable in acid or base or in the presence of hydroxylamine. (See Hartley, B.S. and Massey, V., Active centre of chymotrypsin I. Labeling with a fluorescent dye, *Biochim. Biophys. Acta* 21, 58–70, 1956; Nishino, T., Massey, V., and Williams, C.H., Jr., Chemical modification of D-amino acid oxidase. Evidence for active-site histidine tyrosine and arginine residues, *J. Biol. Chem.* 255, 3610–3615, 1980.)

enzyme activity was recovered on incubation in either more acidic or basic conditions. A comparison of the spectral property of the modified chymotrypsin and dansylated histidine suggested that the modification of chymotrypsin occurs by a modification of histidine at the enzyme active site. Subsequently Tamura and others[121] reported on the use of dansyl chloride for peptide identification and noted the instability of dansyl derivatives of the imidazole ring of histidine. Nishino and coworkers[122] reported on the inactivation of D-amino acid oxidase with dansyl chloride. Substantially complete reactivation occurred with 0.5 M hydroxylamine (NH_2OH) at pH 6.6. This reactivation excluded reaction with primary amino functional groups such as lysine, and amino acid analysis suggested the reaction had not occurred with an oxygen nucleophile such as tyrosine. Treatment of the enzyme with DEPC also resulted in the loss of catalytic activity and reduced the amount of dansyl groups incorporated in a subsequent reaction, suggesting that dansyl chloride reacts with the same functional group that reacted with DEPC. Subsequent structural analysis from this group[123] showed the modification of His217 in the active center of D-amino acid oxidase. Gadda and coworkers[124] observed the inactivation of D-amino acid oxidase from *Rhodotorula gracilis* by dansyl chloride. The reaction was performed in 50 mM phosphate, pH 6.6 containing 10% glycerol at 18°C in the dark with a 300-fold molar excess of dansyl chloride. The enzyme was protected from inactivation by benzoate. Nonlinearity was observed in the time course of inactivation reflecting the hydrolysis of dansyl chloride during the modification reaction. The modified enzyme retained activity with altered substrate specificity.

The cyanation of histidine residues in myoglobin using an equimolar ratio of cyanogen bromide and protein at pH 7.0 has been reported.[125–127] This derivative is somewhat unstable, but it has proved useful in spectral studies (NMR, IR, UV-Vis) of this protein. It was not possible to find application of cyanogen bromide for the modification of histidine in any protein except myoglobin. A search for cyanation (cyanylation) of histidine with other cyanylation reagents such as 2-nitro-5-thiocyanobenzoic acid was unsuccessful.

Competitive labeling of proteins is a method for determining residue accessibility as a measure of protein conformation and protein–protein interaction.[128] Competitive labeling of the amino-terminal histidine residue in secretin with 1-fluoro-2,4-dinitrobenzene (FDNB) has been used to study the reactivity of this residue versus other nucleophiles.[129] The amino-terminal functional group has a pK_a of 8.83 and fivefold greater reactivity than the model compound (histidylglycine), while the imidazolium ring has a pK_a value of 8.24 and a 26-fold greater reactivity than the model compound. These results were interpreted as reflecting a conformational state where the histidine is interacting with a carboxylate function. Kaplan and Oda[130] reported on the use of FDNB for the selective isolation of free and blocked amino-terminal peptides. The dinitrophenyl group on histidine imidazole ring (Figure 4.10) can be removed by thiolysis.[131] Thiolysis will also remove the dinitrophenyl group from tyrosine or cysteine.

DEPC (pyrocarbonic acid, diethyl ester, dicarbonic acid, diethyl ester, oxydiformic acid diethyl ester, ethoxyformic anhydride) was used as a pesticide and antifungal agent but its use was banned in food and food products in 1972.[132] There has been recent interest in the use of a related compound, dimethyldicarbonate (DMDC)

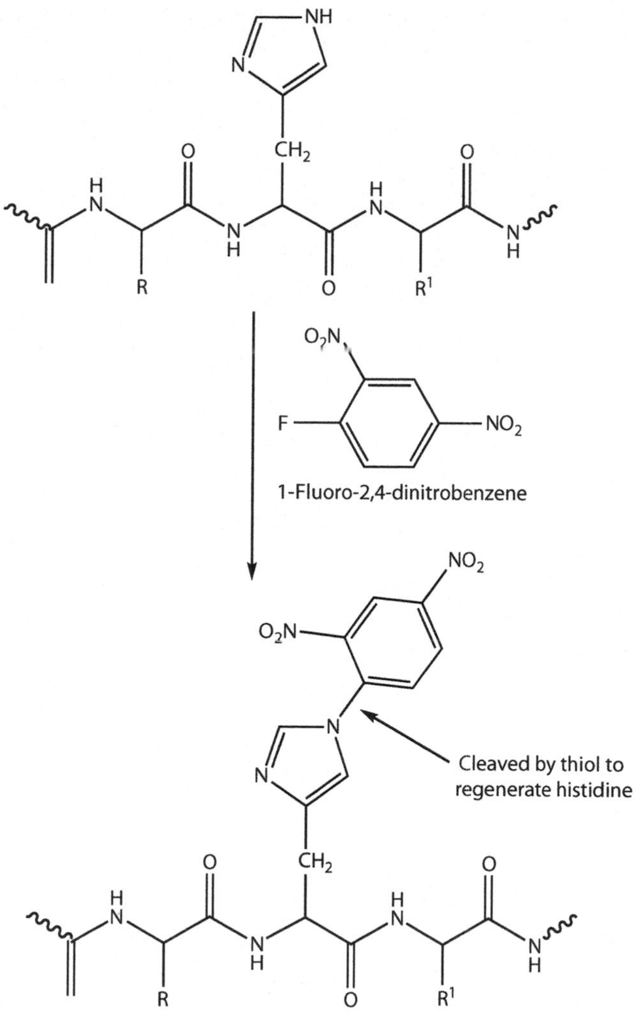

FIGURE 4.10 The reaction of histidyl residues with 1–4–2,4-dinitrofluorobenzene to yield the *N*im-dinitrophenyl derivative. Histidine can be regenerated by thiolysis to yield free histidine and the corresponding 1-thio-2,4-dintrobenzyl derivative. (See Shatiel, S., Thiolysis of some dinitrophenyl derivatives of amino acids, *Biochem. Biophys. Res. Commun.* 29, 178–183, 1967; Kaplan, H. and Oda, G., Selective isolation of free and blocked amino-terminal peptides from enzymatic digestion of proteins, *Anal. Biochem.* 132, 384–388, 1983.)

for decontamination of grape musts.[133] It has received attention as nuclease inhibitor for the isolation of undegraded nucleic acid from tissues.[134] The use of a nuclease inhibitor for isolation of nucleic acids results in the finding that DEPC modified adenine residues in nucleic acids.[135] The utility of DEPC for the preparation of undegraded nucleic acids has resulted in the use of DEPC water for processing of nucleic acids, although degradation is still possible. While useful for the initial extraction, the value of long-term use of DEPC water is questionable.[136]

DEPC is the most extensively used reagent for the specific modification of histidine in proteins (Figure 4.11). In the pH range of 5.5–7.5, DEPC is reasonably specific for reaction with histidyl residues. The reaction can be complex and modification of other functional groups such as tyrosine, the ε-amino group of lysine, and cysteine occur. With the exception of the modification of the ε-amino group of lysine and secondary products derived from histidine, the modifications are reversible in base and, in some cases, quite transient.

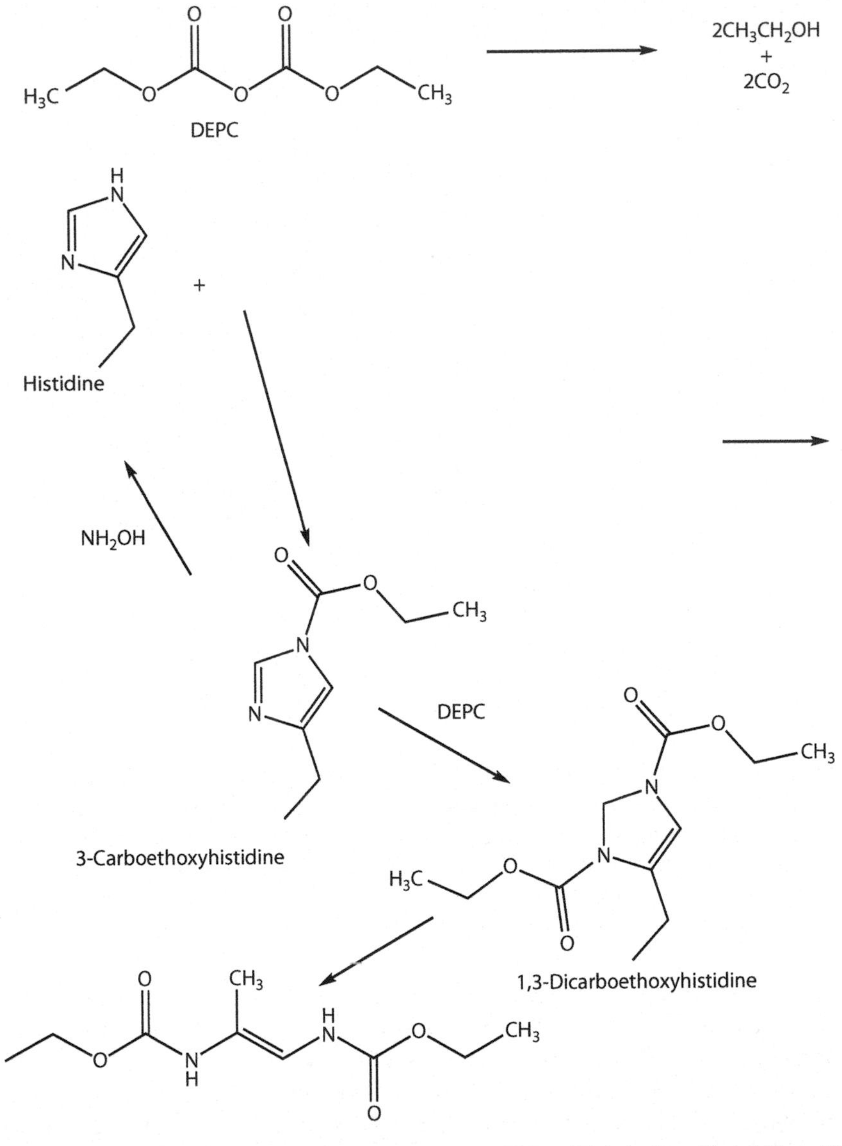

FIGURE 4.11 The chemistry of the reaction of DEPC with proteins. (See Miles, E.W., Modification of histidyl residues in proteins by DEPC, *Methods Enzymol.* 47, 431–442, 1977.)

The use of DEPC for the modification of histidine in proteins dates back to work by Fedorcsák and colleagues at the Royal University of Stockholm.[137–139] This work was performed to obtain understanding of the mechanism of action of cold sterilization.[139] The work suggested modification of tryptophan and lysine resulting in irreversible structural changes in protein upon reaction with DEPC. Additional studies emphasized the complexity of the reaction of DEPC with proteins.[140–143] Further support was provided for reaction at tryptophan[141] and a cross-linking mechanism was proposed involving isopeptide bond formation between lysine and either aspartic or glutamic acid.[143] One of the more interesting observations was protection of RNase from inactivation by DEPC by albumin.[142] As other studies were performed,[144,145] it appeared that it was possible to selectively modify histidine residues but consideration of modification at other sites was clearly necessary as frequent modification does occur at lysine.[146] Modification also can occur at tyrosine and less frequently at serine at threonine.[147] This latter study by Mendoza and Vachet[147] noted that lysine modification occurs slowly at neutral pH. Modification of arginine and cysteine by DEPC has also been reported[148] but, as with tryptophan, there is no recent documentation on modification of arginine residues in proteins with DEPC. The modification of cysteine by DEPC would be expected to yield an unstable thioester derivative. Garrison and Himes[149] did report an unusual reaction between DEPC and cysteine in carboxylate (succinate or acetate) buffers where a thioester is formed between buffer and cysteine. The spectral properties of this derivative could pose problems in the spectrometric analysis of the reaction of DEPC with histidine. It is noted that Rua and coworkers[150] reported the modification of cysteinyl residues in *Escherichia coli* isocitrate lysase with DEPC as determined by analysis with Ellman's reagent. The early work on DEPC has been reviewed by Miles.[151]

Reaction of DEPC with histidine residues at a *moderate excess* of DEPC at neutral pH results in substitution at one of the nitrogen positions on the imidazole ring. This reaction is associated with an increase in absorbance at 240 nm ($\Delta\varepsilon = 3200\,M^{-1}\,cm^{-1}$). Monosubstitution is readily reversed at alkaline pH and, in particular, in the presence of nucleophiles such as hydroxylamine. Tris and other nucleophilic buffers can also reverse the modification and their use should be avoided with DEPC. Generally, treatment with neutral hydroxylamine (0.1–1.0 M, pH 7.0) is used to regenerate histidine. As with the deacylation of *O*-acetyl tyrosine by neutral hydroxylamine, the higher the concentration of hydroxylamine, the more rapid the process of decarboxyethylation. Carboxyethylation at both N_1 and N_3 (disubstitution) results in a derivative with altered spectral properties compared to the monosubstituted derivative; this derivative does not regenerate histidine and treatment with neutral hydroxylamine or base likely results in scission of the imidazole ring similar to that suggested for diacetylation of histidine with acetic anhydride. Loss of histidine is detected by amino acid analysis after acid hydrolysis; sequence analysis using Edman degradation chemistry also shows the absence of histidine with the presence of a disubstituted derivative.[152] In these studies, a phenylthiohydantoin (PTH) derivative eluting near PTH-glycine was observed and the structure verified by mass spectrometry. The monosubstituted derivative is unstable under conditions of acid hydrolysis and yields free histidine. As with the modification of other amino acid residues in proteins, mass spectrometry is the method of choice in the analysis of the chemical modification of histidine in proteins including carboxyethylated histidine.[153–159] Glocker and coworkers[153] were able to demonstrate modification of lysine and tyrosine in addition to histidine modification; these investigators were also able to distinguish between the monosubstitution and disubstitution of histidine with DEPC. Dage and coworkers[155] used mass spectrometry to determine the extent of ethoxyformylation of two histidine residues and a lysine residue using mass spectrometry of peptides isolated from α_1-acid glycoprotein. The modification of the histidine residues showed modest pH dependence, while the modification at lysine increased markedly with increased pH. Tyrosine modification was also demonstrated in this study. Krell and coworkers[156] used mass spectrometry to measure protein modification with DEPC. These investigators reported the use of mass spectrometry to study protein modification with phenylglyoxal, tetranitromethane, and trinitrobenzenesulfonic acid. Analysis was only limited by the stability of the derivatives. These investigators did find it difficult to locate specific histidine residues modified by DEPC because of multiple sites of modification. It was possible to obtain some information on the presence of histidine binding sites from protection experiments. Protection studies were also used by Qin and coworkers[157] in the identification of histidine residues present in copper binding sites on prion proteins using MALDI-TOF mass spectrometric footprinting. Willard and Kinter used modification of histidine with DEPC to improve product ion spectra with improved sequence information.[159] More recent studies have used mass spectrometric analysis of DEPC modification to characterize the heme-coordinating histidine residues in cytochrome b_5.[160] Nakanishi and coworkers[161] used reaction with DEPC combined with EPR spectroscopy to characterize the heme-coordinating residues of a myoglobin mutant. UV-spectroscopy and MALDI-TOF mass spectrometry were used to characterize the histidine residues involved in heme binding. Konkle and coworkers[162] used modification with DEPC to study the histidine residues in Rieske protein. An elegant study[163] has examined the reaction of DEPC with histidyl residues in cytochrome b_5. Using (NMR) spectroscopy with this well-characterized protein, it has been possible to identify factors influencing histidine modification with this reagent; three major factors include (1) the pK_a of the individual histidine residue, (2) solvent exposure of the residue, and (3) hydrogen bonding of the imidazolium ring. NMR spectroscopy enabled these investigators to assign the relative reactivities of the four histidine residues modified in cytochrome b_5 by DEPC.

The use of DEPC for the modification of histidine residues at enzyme active sites has been previously reviewed by this author.[164] This is still a useful technique provided that sufficient care is taken to assure specificity of modification. It is, however, the opinion of the author that the value of DEPC for protein modification is in the use of measurement of surface-exposed residues.[146,147] Given the ability to determine extent and site of modification with mass spectrometry and, with the appropriate protein, NMR, reaction with DEPC is a useful method for studying protein–protein interaction, ligand binding, and conformational change.

4-Hydroxy-2-nonenal and 4-oxo-2-nonenal are aldehydes derived from the oxidation of lipids,[165] which can react with proteins to form a variety of products.[166–170] Mass spectrometry can be used to identify the various products including the Michael addition product with histidine (Figure 4.12).[171,172] If either sodium cyanoborohydride or sodium borohydride were included with 4-hydroxy-2-nonenal and

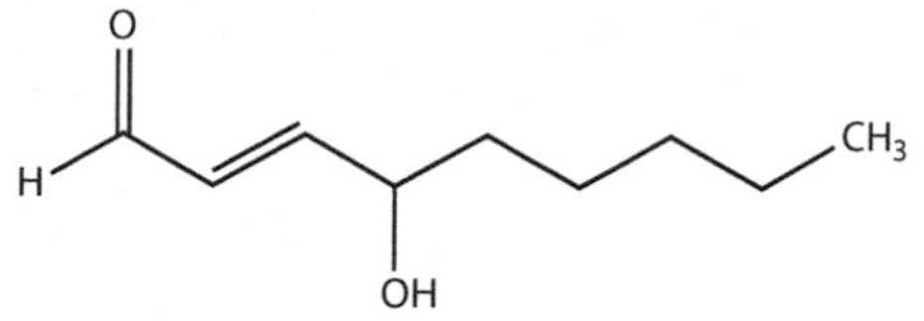

4-Hydroxy-2-nonenal (HNE)

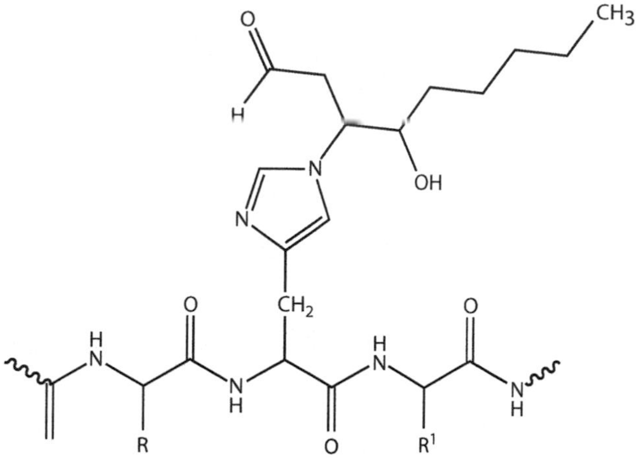

FIGURE 4.12 The modification of histidine by 4-hydroxy-2-nonenal. (See LoPachin, R.M., Gavin, T., Petersen, D.R., and Barber, D.S., Molecular mechanisms of 4-hydroxy-2-nonenal and acrolein toxicity: Nucleophilic targets and adduct formation, *Chem. Res. Toxicol.* 22, 1499–1508, 2009; Rauniyar, N., Prokai-Tatrai, K., and Prokai, L., Identification of carbonylation sites in apomyoglobin after exposure to 4-hydroxy-2-nonenal by solid-phase enrichment and liquid chromatography-electrospray ionization tandem mass spectrometry, *J. Mass Spectrom.* 45, 398–410, 2010.)

the oxidized B-chain of insulin, reaction occurs with the ε-amino group of lysine via a Schiff mechanism with the aldehyde.[173] In the absence of a reducing agent, 4-hydroxy-2-nonenal modification of histidine in the oxidized B-chain of insulin proceeded via a Michael addition as the predominant reaction. Monoclonal antibodies to the 4-hydroxy-2-nonenal adduct with histidine have been developed.[174–177]

The formation of *N*-acetylhistidine in proteins (Figure 4.13) is not unusual but the establishment of the presence of such a derivative is difficult because of product stability.[178,179] This product stability issue is also observed with other protein modifications such as the ethoxyformylation of histidine, the formation of cysteine sulfenic acid, and acylation of serine or threonine. There are, however, several examples where a stable modification of a protein by acetic anhydride has been reported. MacDonald and coworkers[179] reported that NMR spectroscopy showed the transient acetylation of histidine followed by the formation of a stable *N*-acetyllysine derivative. These investigators also observed the participation of a histidine residue in the acetylation of lysine by acetyl salicylate. Moore[180] reported the acetylation by acetic anhydride of the imidazole ring of histidine in angiotensin II. It was observed that the imidazole side chains were acetylated and deacetylated at a more rapid rate than

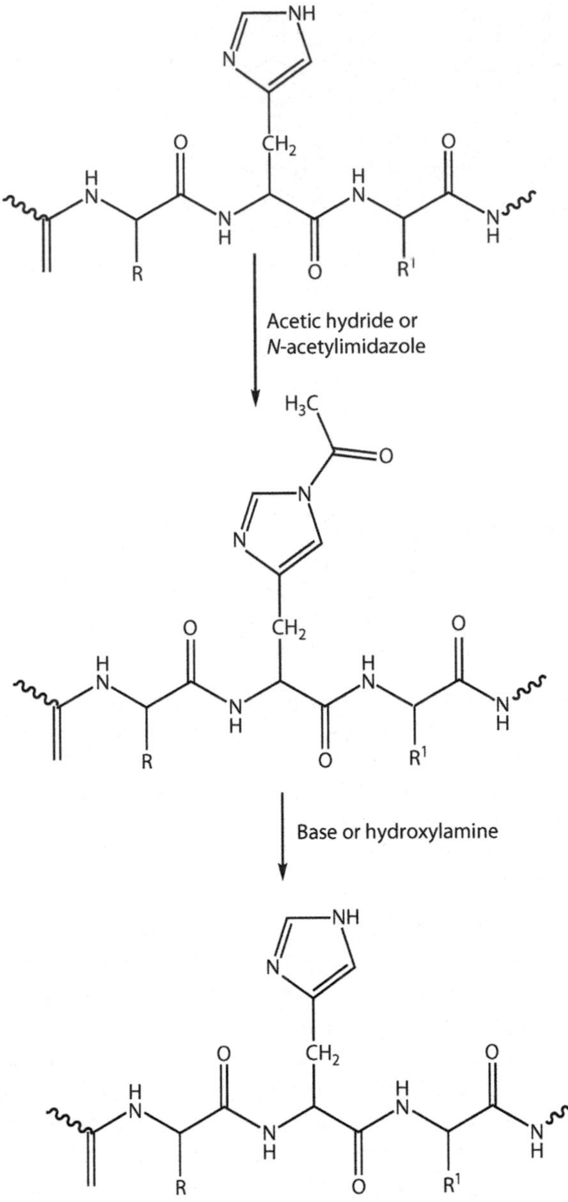

FIGURE 4.13 The acetylation of histidyl residues in proteins. Histidyl residues can be modified with various acetylating agents such as acetic anhydride and *N*-acetylimidazole. (See Macdonald, J.M., Haas, A.L., and London, R.E., Novel mechanisms of surface catalysis of protein adduct formation. NMR studies of the acetylation of ubiquitin, *J. Biol. Chem.* 275, 31908–31913, 2000; Houston, L.L. and Walsh, K.A., The transient inactivation of trypsin by mild acetylation with *N*-acetylimidazole, *Biochemistry* 9, 156–166, 1970; Lundblad, R.L., The reaction of bovine thrombin with *N*-butyrylimidazole. Two different reactions resulting in the inhibition of catalytic activity, *Biochemistry* 14, 1033–1037, 1975.)

the free amino acid. Welsch and Nelsestuen[152] observed the diacetylation of a histidine residue with the reaction of prothrombin fragment 1 with acetic anhydride. The diacetylation of histidine was associated with the opening of the imidazole ring and the loss of the C2 carbon; this modification was not reversed by hydroxylamine. This reaction is similar to the disubstitution of histidine with DEPC. Kinnunen and colleagues[181] reported the inactivation of acyl-CoA:cholesterol *O*-acyltransferase by DEPC or acetic anhydride. The inactivation by either reagent is reversed by hydroxylamine. There was a marked difference between the liver enzyme and the enzyme derived from aorta.

Several groups have reported the modification of histidine with Woodward's Reagent K (*N*-ethyl-5-phenylisoxaxolium-3-sulfonate).[182–184] This reagent is usually considered specific for carboxyl groups (see Chapter 3). One group[184] observed saturation kinetics in the inactivation of an acylphosphatase with Woodward's Reagent K suggesting the formation of a reversible complex prior to the inactivation reaction.

TRYPTOPHAN

The specific chemical modification of tryptophan in protein is one of the more challenging problems in protein chemistry as the solvent conditions for providing specificity of modification are, in general, somewhat harsh and there is considerable possibility of either the concomitant or separate modification of other amino acid residues. The extent of modification can be determined by spectral analysis as discussed in the following with the use of amino acid analysis.[185–189] Mass spectrometry has become a more important method in determining tryptophan modification in proteins.[2,190–194]

Tryptophan is an electron-rich heterocyclic aromatic amino acid (Figure 4.1). The pK_a of the indole nitrogen is above 15, making it essentially unreactive. Modification is accomplished by oxidation or alkylation of the C2 carbon, although the reaction with 2-hydroxy-5-nitrobenzyl bromide can yield a complex mixture of products.

Treatment of tryptophan with various oxidizing agents results in the modification of the indole ring resulting in a multiplicity of products (Figure 4.14).[195–200] Reubsaet and colleagues[199] have reviewed the methods for the qualitative and quantitative analysis of tryptophan oxidation in peptides and proteins including UV spectroscopy, fluorescence, and HPLC analysis. HPLC analysis of tryptophan oxidation products has been described.[200] Details are provided for the separation of kynurenine, 5-hydroxytryptophan, tryptophan, and dioxindolealanine on a C_{18} column. The reader is directed to an excellent study by Mach and coworkers[201] for the extinction coefficients for tryptophan, tyrosine, and cystine in proteins. Fluorescence has also been used to measure tryptophan modification in proteins.[202] Reshetnyak and colleagues have reviewed the fluorescence properties of tryptophan in proteins.[203] While the aforementioned studies have used hydrogen peroxide or other oxidizing agents, these reagents tend to be somewhat nonspecific. However, the power of mass spectrometry for analysis makes hydrogen peroxide quite useful also care needs to be taken with termination of reaction as, for example, modification continues in the frozen state.[204] Oxidation of proteins is a problem in the pharmaceutical industry and various approaches are being developed to protect tryptophan, methionine, and

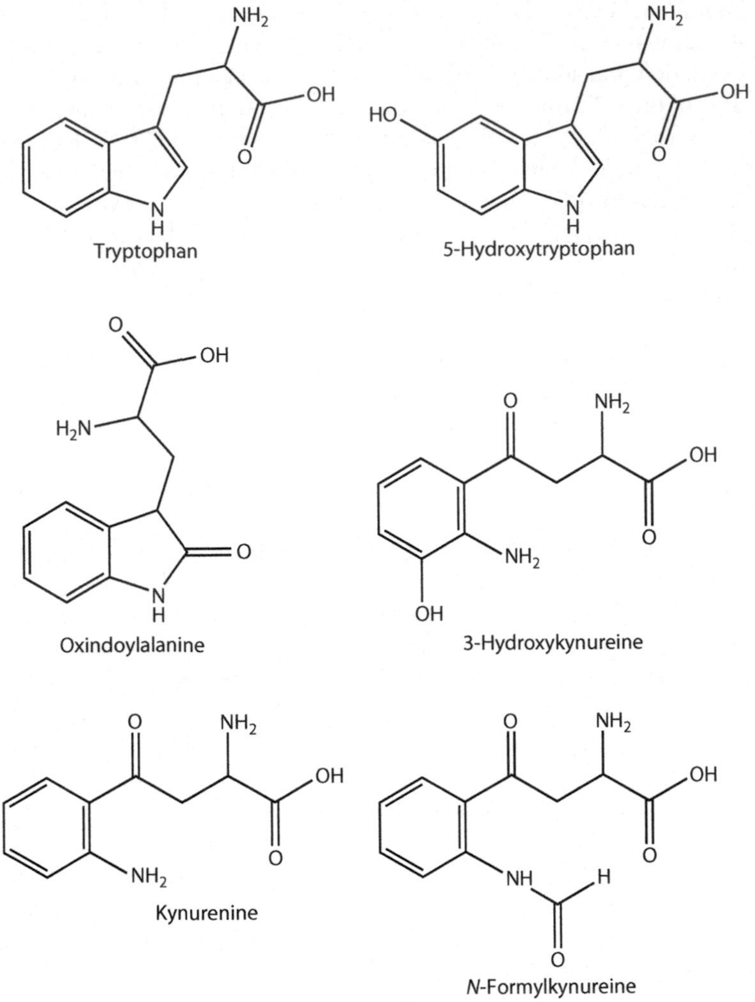

FIGURE 4.14 The structure of tryptophan and various tryptophan oxidation products. (See Simat, T.J. and Steinhart, H., Oxidation of free tryptophan and tryptophan residues in peptides and proteins, *J. Agric. Food Chem.* 46, 490–498, 1998; Finley, E.L., Dillon, J., Crouch, R.K., and Schey, K.L., Identification of tryptophan oxidation products in bovine α-crystallin, *Protein Sci.* 7, 2391–2397, 1998.)

histidine from oxidation.[205] Ji and coworkers[205] found that while free tryptophan protected the tryptophan in parathyroid hormone from oxidation with 2,2′-azobis (2-amidinopropane) dihydrochloride (a surrogate for polysorbate for generation of alkylperoxides), only a mixture of methionine and tryptophan protected the tryptophan from modification with hydrogen peroxide or hydrogen peroxide plus ferric ions (Fenton oxidation) in addition to the alkyl peroxides.

The difference in reactivity between amino acids, peptides, and intact proteins has been mentioned in the discussion of other reagents. Hypothiocyanous acid is formed by the action of myeloperoxidase on thiocyanate and hydrogen peroxide. Hypothiocyanous acid reacts with tryptophan residues to form dioxindolylalanine and *N*-formylkynurenine (Figure 4.15).[206]

Tryptophan is subject to photodegradation as discussed earlier. The products derived from the photolytic degradation of tryptophan are similar to those derived from oxidation and those derived by the same process described earlier for histidine. Photodegradation (photooxidation) uses the same basic mechanism as other oxidation processes as described earlier. Photooxidation mechanisms are also discussed earlier for histidine. It is also worth mentioning that photodegradation is an issue for protein biopharmaceuticals[13,207] and must be considered during stability testing.

L-Tryptophan as the free amino acid is stable in solution in the dark and in the absence of oxygen.[208] In the presence of oxygen and/or light, L-tryptophan is degraded to form a variety of products. Furthermore, a variety of studies suggest that the photodegradation of tryptophan in proteins depends on the location of

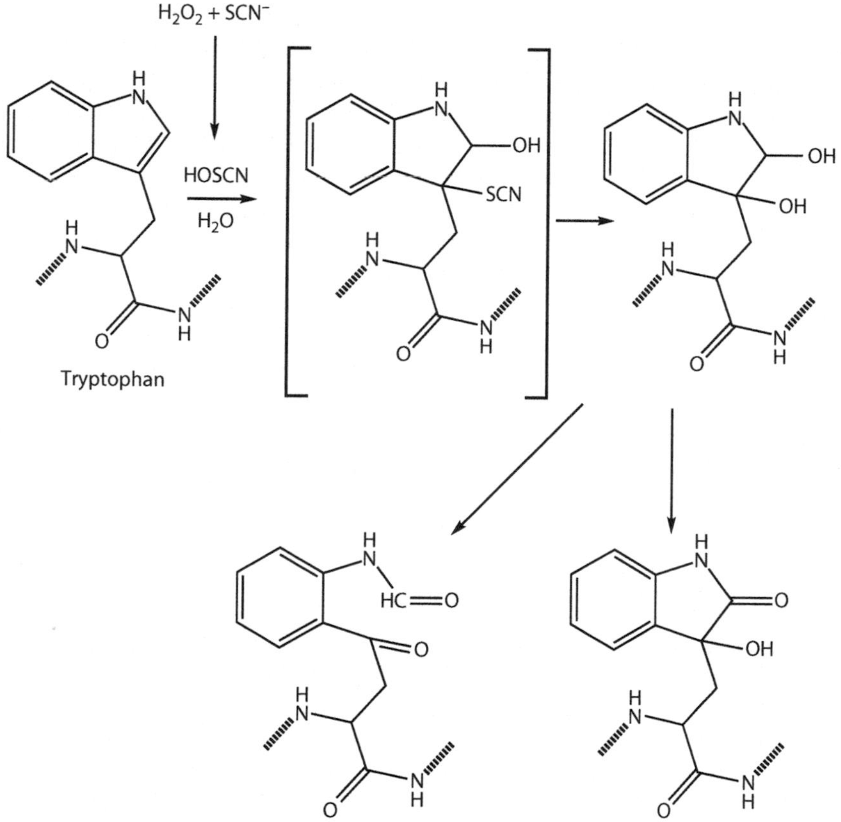

FIGURE 4.15 The reaction of hypochlorous acid and tryptophan to form dioxindoylalanine and *N*-formylkynurenine. (See Hawkins, C.L., Pattison, D.J., Stanley, N.R., and Davies, M.J., Tryptophan residues are targets in hypothiocyanous acid-mediated protein oxidation, *Biochem. J.* 416, 441–452, 2008.)

the residue in three-dimensional structure of the proteins (solvent exposure).[209–211] Pigault and Gerard[209] studied the photolysis of four proteins, each with a single tryptophan with different degree of solvent exposure. Tryptophan loss was measured by amino acid analysis, UV-spectroscopy, and fluorescence. The degree of tryptophan loss was related to exposure to solvent and surrounding amino acid; the type of photochemical product was also dependent on residue location. Rao and coworkers[210] studied photolysis in native and random coil forms of melittin and β-lactoglobulin. The results support the concept that protein conformation is a major factor in photodegradation of tryptophan in proteins. Tallmadge and Borkman[211] measured the rate of photodegradation of the four tryptophanyl residues in γ-II crystallin. The rates of degradation of the individual tryptophanyl residues do reflect the local microenvironment.

The reaction of *N*-bromosuccinimide (NBS) with tryptophan (Figure 4.16) is the best known method for the site-specific chemical modification of this residue in proteins. This reagent was used earlier for quantitative analysis of tryptophan in proteins but the development of hydrolysis methods obviated this necessity. NBS has been used for the modification of tryptophan residues in proteins. Work in this area prior to 2005 has been previously reviewed[164] and will only be discussed with respect to technique as relevant to current applications. NBS of tryptophan yields the oxindole or dioxindole derivative, while other oxidizing agents yield kynurenine and other oxidation products.[212,213] Tryptophan oxidation to oxindolylalanine is also accomplished with hydrochloric acid in dimethylsulfoxide.[214–216] Huang and coworkers[215] demonstrated that it was possible to obtain cleavage of oxidized peptides with cyanogen bromide; success does require technical expertise.

Ohnishi and coworkers[217] provided a rigorous evaluation of the reaction of NBS with model tryptophanyl and tyrosyl compounds. At ratios of NBS to *N*-acetyltryptophan ethyl ester of greater than two in acetate buffer at pH 4.5, there is an apparent reversal of the decrease in absorbance at 280 nm. The maximal decrease in absorbance occurs at a ratio of NBS to tryptophan of two. If the data are obtained by stopped-flow spectroscopy, the molar excess of NBS does not have

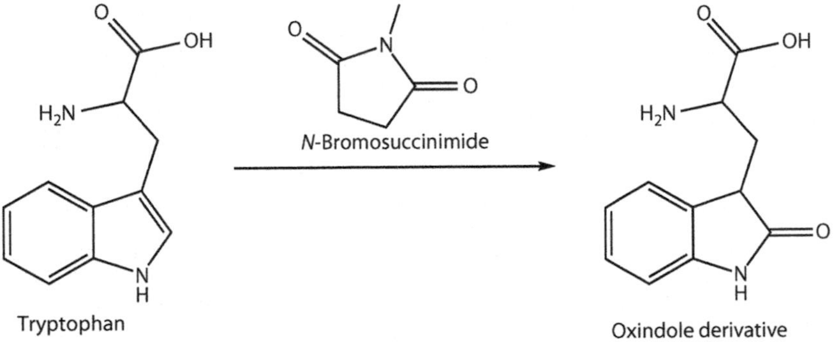

Tryptophan

N-Bromosuccinimide

Oxindole derivative

FIGURE 4.16 The reaction of NBS with tryptophan. (See Hu, H.Y., Wu, M.C., Fang, H.J. et al., The role of tryptophan in staphylococcal nuclease stability, *Biophys. Chem.* 151, 170–177, 2010.)

an effect on the maximum decrease observed, but when the spectrum is obtained 5 min after the initiation of the reaction, there is a decrease in the observed magnitude of change in absorbance at 280 nm. The evaluation of spectral changes in a protein can be further complicated by the reaction of NBS with tyrosine; however, the rate of tryptophan modification by NBS is approximately 10^3 faster than the modification of tyrosine. Daniel and Trowbridge[218] found that (at pH 4.0) the reaction of NBS with acetyl-l-tryptophan ethyl ester required 1.5 mol of NBS per mole of the acetyl-l-tryptophan ethyl ester, while trypsinogen required 2.0–2.3 mol NBS per mole of tryptophan oxidized, and trypsin required 1.5–2.0 mol NBS per mole of tryptophan oxidized. Sartin and coworkers[219] observed that the amount of NBS consumed in the modification of tryptophan in pancreatic DNAse is similar for the first two residues modified at either pH 4.0 or pH 5.5; above pH 5.5, more NBS is required for the modification of the third residue. At pH 5.5, the third residue modified is the most critical for enzyme activity, while at pH 4.0, the second residue modified is the most critical. These investigators also observed a discrepancy in the estimation of tryptophan modified between spectral analysis and amino acid analysis. However, other investigators[220] reported good agreement between spectral analysis and amino acid analysis (hydrolysis in 3.0 M toluenesulfonic acid) on the extent of modification of NBS modification of galactose oxidase. Freisheim and Huennekens[221] observed that only tryptophan in dihydrofolate reductase reacts with NBS at pH 4.0, while at pH 6.0, a sulfhydryl group apparently is preferentially oxidized by the reagent prior to the reaction of tryptophan. Oxidation of the sulfhydryl groups results in an increase in activity which modification of the tryptophan residues leads to inactivation. Ohnishi and coworkers[222] followed up on their earlier study and used stopped-flow technology to study the modification of Trp82 in lysozyme with NBS. Modification occurred below pH 6 but not above pH 7. A similar study on the effect of pH on the modification of *N*-acetyl-L-tryptophan ethyl ester did not show dependence. These results suggest that, as with other functional groups in proteins, modification of tryptophan residues with NBS does depend on local electrostatic environment. It can be concluded that usually a twofold molar excess of reagent is required for modification and specificity of modification is increased at mild acid pH (4.0).[223,224]

The conversion of tryptophanyl residues to 1-formyltryptophanyl residues (Figure 4.17) has been reported. The reaction conditions are somewhat harsh but the procedure is reversible and it may be more useful for small peptides, although it has been applied to several proteins. There is, however, no recent application to proteins. Coletti-Previero and coworkers[225] have successfully applied this procedure to bovine pancreatic trypsin. Trypsin was dissolved in formic acid saturated with HCl at a concentration of 2.5 mg mL^{-1} at 20°C. The formylation reaction is associated with an increase in absorbance at 298 nm.[226] Therefore, it is possible to follow the reaction spectrophotometrically. The reaction is judged complete when there is no further increase in absorbance at 298 nm. The earlier reaction with trypsin was complete after an incubation period of 1 h. The solvent was partially removed *in vacuo* over KOH pellets followed by lyophilization. The formyltryptophan derivative is unstable at alkaline pH. At pH 9.5 (pH-stat), conversion back to tryptophan is complete after 200 min incubation at 20°C. Holmgren has successfully

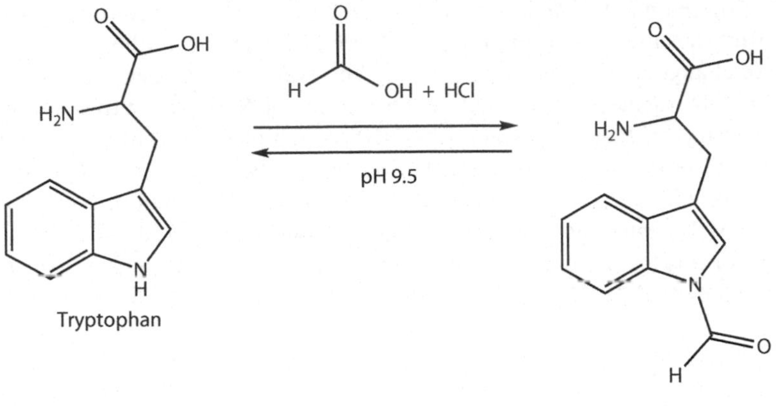

FIGURE 4.17 The formylation of tryptophan with formic acid. (See Magous, R., Bali, J.P., Moroder, L., and Previero, A., Effect and of N^{in}-formylation of the tryptophan residues on gastrin (HG-13) binding and on gastric acid secretion, *Eur. J. Pharmacol.* 77, 11–16, 1982.)

applied this procedure to thioredoxin.[227] Cooper and coworkers[228] reported that the *N*-formylation of tryptophanyl residues in proteins in cytochrome *c*. The single tryptophanyl residue was formylated with formic acid saturated with HCl. The modified protein had markedly reduced affinity for a monoclonal antibody resulting from local conformational change.

2-Hydroxy-5-nitrobenzyl bromide (Figure 4.18), frequently referred to as Koshland's reagent, was introduced by Koshland and coworkers[229,230] for the analysis and modification of tryptophanyl residues in proteins. Under appropriate reaction conditions (pH 4.0 or below), the reagent appears to be highly specific for reaction with tryptophanyl residues in proteins; this reagent also has the advantage of being a "reporter" group in the sense that the spectrum of the hydroxynitrobenzyl derivative is sensitive to changes in the microenvironment. This decrease observed in absorbance at 410 nm and associated with an increase in absorbance at 320 nm upon the addition of dioxane is similar to that seen with acidification and reflects the increase in the pK_a of the phenolic hydroxyl group. Titration curves of oxidized and reduced laccase[231] that had been modified with 2-hydroxy-5-nitrobenzyl bromide suggested that the residues modified with 2-hydroxy-5-nitrobenzyl bromide are in an essentially aqueous microenvironment. This study provides titration curves for free hydroxynitrobenzyl (HNB) and HNB bound to laccase. Titration curves were based on A_{410} and provided a pK_a of 6.83 for free HNB and 7.2 in laccase with 0.38 mol HNB suggesting the microenvironment was slightly more hydrophobic.

Figure 4.18 would suggest that there are a limited number products obtained from the reaction of HNB with tryptophan in proteins but this is likely incorrect. Disubstitution on the indole ring occurs with 2-hydroxy-5-nitrobenzyl bromide and is seen as a sudden "break" in the line when the extent of modification vs. reagent excess.[232] In the study by Barman and Koshland,[232] the "break" occurred with a 20-fold molar excess of reagent. These investigators then used this concentration of reagent (20-fold molar excess) for the determination of tryptophan to avoid complications resulting from disubstitution. The reaction of HNB with protein occurs

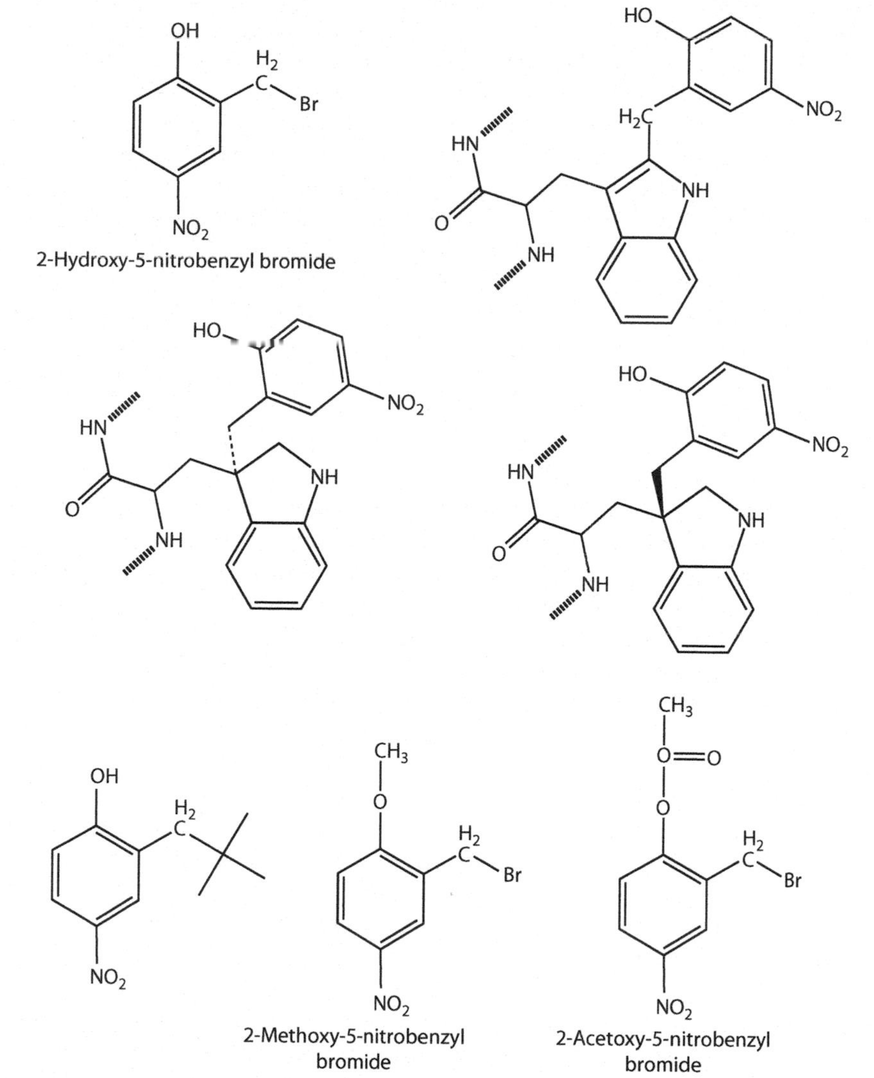

2-Hydroxy-5-nitrobenzyl bromide

2-Methoxy-5-nitrobenzyl bromide

2-Acetoxy-5-nitrobenzyl bromide

FIGURE 4.18 The reaction of tryptophan with 2-hydroxy-5-nitrobenzyl bromide. A variety of products can be obtained including substitution at the 2-carbon (A) and diastereoisomers obtained from substitution at the 1-carbon (see Loudon, G.M. and Koshland, D.E., Jr., The chemistry of a reporter group 2-hydroxy-5-nitrobenzyl bromide, *J. Biol. Chem.* 245, 2247–2254, 1970; Strohalm, M., Kodíček, M., and Pechar, M., Tryptophan modification by 2-hydroxy-5-nitrobenzyl bromide studied by MALDI-TOF mass spectrometry, *Biochem. Biophys. Res. Commun.* 312, 811–816, 2003). Derivative forms include dimethyl(2-hydroxy-5-nitrobenzyl)sulfonium chloride (Horton, H.R. and Tucker, W.P., Dimethyl (2-hydroxy-5-nitrobenzyl)sulfonium salts, *J. Biol. Chem.* 245, 3397–3401, 1970), 2-methoxy-5-nitrobenzyl bromide (Horton, H.R., Kelly, H., and Koshland, D.E., Jr., Environmentally sensitive protein reagents. 2-methoxy-5-nitrobenzyl bromide, *J. Biol. Chem.* 240, 722–724, 1965), and 2-acetoxy-5-nitrobenzyl chloride.

rapidly (less than a minute). Horton and Koshland[230] did observe the modification of cysteine with HNB with a rate considerably slower than tryptophan. There is mixed data as to whether HNB reacts with cysteine residues in proteins. Novak and colleagues[233] reported the inactivation of *Escherichia coli* core RNA polymerase with 2-hydroxy-5-nitrobenzyl bromide with data suggesting the modification of both tryptophan and cysteine. Baracca and coworkers[234] did not observe the inactivation of mitochondrial F1-ATPase with cysteine residues when inactivation was accomplished with HNB. Modification at serine or threonine was not observed by Barman and Koshland[232] nor was there reaction with carbohydrate. Modification at histidine by HNB was reported by Barman.[235] Lundblad and Noyes[236] analyzed the product of the reaction of HNB with a synthetic peptide (EAE peptide, FSWGAEGQR) that contained a single tryptophanyl residue. Chromatography on a C_8 demonstrated the presence of several products, some of which resulted from multiple substitutions on the indole rings and others that were suggested to isomers and diastereoisomers. Diastereoisomers (Figure 4.18) resulting from the reaction of HNB with tryptophan compounds have been described by Loudon and Koshland.[237]

Strohalm and coworkers[238] used MALDI-TOF mass spectrometry to characterize the products derived from the modification of tryptophan in a model peptide (GEGKGWGEGK) with 2-hydroxy-5-nitrobenzyl bromide. A total of five products were obtained, which reflected qualitative and quantitative difference in the substitution at the single tryptophan residue. The effect of 2-hydroxy-5-nitrobenzyl bromide was evaluated over a 1–200 M excess range. The degree of modification increased as the concentration of reagent increased. It observed that a disubstituted product could be obtained at an equimolar excess of reagent, while at a 200-fold molar excess, the most abundant product was a trisubstituted derivative; five different products were detected. The extent of modification increased with increasing pH with no major change in product distribution. This group[191] reported the use of HNB as reagent for determining the surface accessibility of tryptophan in proteins. Studies with lysozyme, cytochrome *c*, and myoglobin demonstrated the HNB modified "surface-accessible" residues and not "buried" residues. Denaturation of the proteins permitted modification of all tryptophanyl residues in the three proteins.

The issue of solution stability was mentioned earlier; this difficulty is avoided and the characteristics of the reaction are preserved by the use of the dimethyl sulfonium salts. Dimethyl (2-hydroxy-5-nitrobenzyl) sulfonium chloride was obtained from the reaction of 2-hydroxy-5-nitrobenzyl chloride with dimethyl sulfide.[239,240] This water-soluble sulfonium salt derivative was used to modify tryptophan in rabbit skeletal myosin subfragment-1.[241] Purification of peptides containing modified tryptophanyl residues was achieved by immunoaffinity chromatography using rabbit antibody to bovine serum albumin previously modified with dimethyl-(2-hydroxy-5-nitrobenzyl) sulfonium chloride. This allowed the identification of the most rapidly reacting residue.

2-Methoxy-5-nitrobenzyl bromide was synthesized by Horton and coworkers.[242] As would be expected, the spectra of this derivative was not sensitive to pH but was sensitive to solvent polarity. 2-methoxy-5-nitrobenzyl bromide modified both methionine and tryptophan at similar rates, both much slower than the reaction of tryptophan with HNB. While HNB reacts with tryptophan in seconds, the reaction with tryptophan was 74% complete after 80 h of reaction (pH 5, 25°C, 10-fold molar excess of reagent). Some modification of cysteine was detected but analysis was complicated by the competing oxidation of cysteine to cystine; modification of other amino acids was not observed by these investigators. The lack of reactivity of the 2-methoxy-5-nitrobenzyl bromide has been used to develop a derivative that generates the active HNB at the enzyme active site. Horton and Young[243] prepared 2-acetoxy-5-nitrobenzyl bromide (Figure 4.19). This derivative, like the methoxy derivative, is essentially unreactive. There is considerable structural identity between 2-acetoxy-5-nitrobenzyl bromide and *p*-nitrophenyl acetate, which is a nonspecific

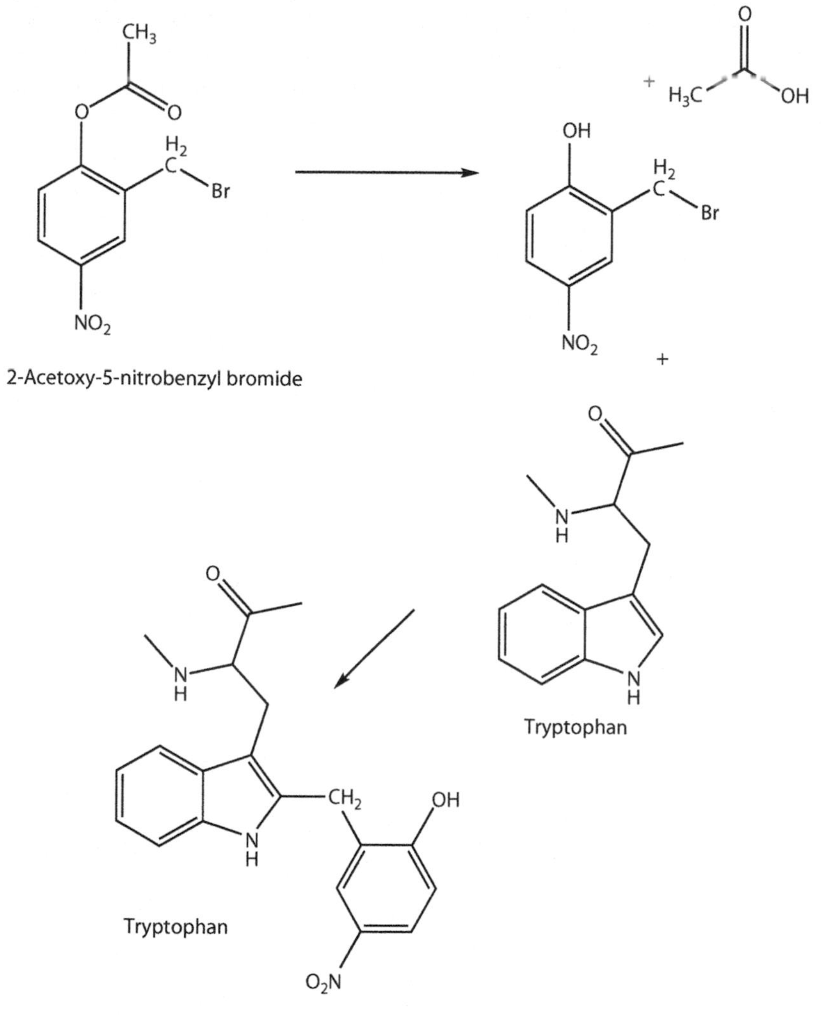

FIGURE 4.19 The hydrolysis of 2-acetoxy-5-nitrobenzyl bromide to release 2-hydroxy-5-nitrobenzyl bromide for subsequent reaction with tryptophan. (See Horton, H.R. and Young, G., 2-Acetoxy-5-nitrobenzyl chloride. A reagent designed to introduce a reporter group near the active site of chymotrypsin, *Biochim. Biophys. Acta* 194, 272–278, 1969 and Uhteg, L.C. and Lundblad, R.L., The modification of tryptophan in bovine thrombin, *Biochim. Biophys. Acta* 491, 551–557, 1977.)

substrate for chymotrypsin. α-Chymotrypsin removes the acetyl group from 2-acetoxy-5-nitrobenzyl bromide, thus generating 2-hydroxy-5-nitrobenzyl bromide at the active site, which then either rapidly reacts with a neighboring nucleophile or undergoes hydrolysis. Uhteg and Lundblad[244] have used both the acetoxy and butyroxy derivatives in the study of thrombin.

Reagents with reaction characteristics similar to 2-hydroxy-5-nitrobenzyl bromide are the nitrophenylsulfenyl derivatives[245] (Figure 4.20). The reaction product resulting from the sulfonylation of lysozyme[246] with 2-nitrophenylsulfenyl chloride (40-fold molar excess) at pH 3.5 (0.1 M sodium acetate) has spectral characteristics that can be used to determine the extent of reagent incorporation (at 365 nm $\varepsilon = 4 \times 10^3\ M^{-1}\ cm^{-1}$). These reagents show considerable specificity for the modification of tryptophan at pH ≤ 4.0. Possible side reactions with other nucleophiles such as amino groups need to be considered. In the case of human chorionic somatomammotropin and human pituitary growth hormone,[247] reaction with *o*-nitrophenyl-sulfenyl chloride (2-nitrophenylsulfenyl chloride) was achieved in 50% acetic acid, but not in 0.1 sodium acetate, at pH 4.0. Wilchek and Miron[248] have reported on the reaction of 2,4-dinitrophenylsulfenyl chloride with tryptophan in peptides and protein,

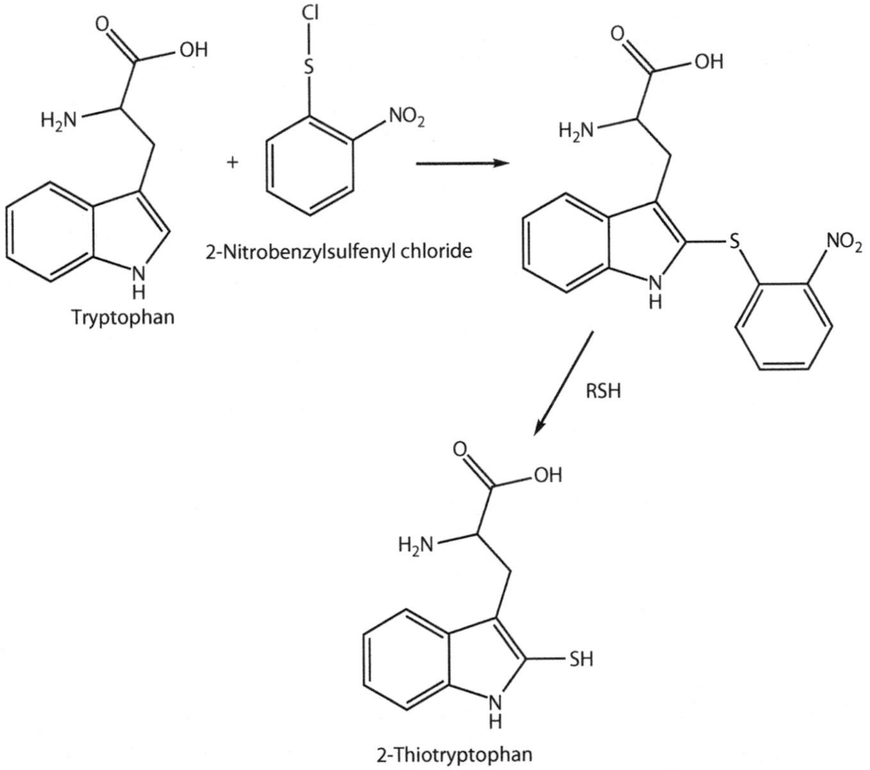

FIGURE 4.20 The reaction of tryptophan with 2-nitrophenylsulfenyl chloride. (See Matsunaga, H. and Haginaka, J., Investigation of chiral recognition mechanism on chicken α_1-acid glycoprotein using separation system, *J. Chromatogr. A* 1106, 124–130, 2006.)

and subsequent conversion of the modified tryptophan to 2-thiotryptophan by reaction with 2-mercaptoethanol at pH 8.0. The thiolysis of the modified tryptophan is responsible for changes in the spectral properties of the derivative. The characteristics of the modified tryptophan have resulted in the development of a facile purification scheme for peptides containing the modified tryptophan residues.[249,250] Mollier and coworkers[251] examined the reaction of *o*-nitrophenylsulfenyl chloride (2-nitrophenylsulfenyl chloride) with notexin (a phospholipase obtained from Notechis scutatus scutatus venom, which contains two tryptophanyl residues). Reactions with 2-nitrophenylsulfenyl chloride (twofold molar excess) in 50% (v/v) acetic acid resulted in two derivative proteins on HPLC analysis. One derivative contained two modified tryptophanyl residues (20 and 110), while the other derivative was modified only at position 20. There are several applications of this modification that are of importance for investigators working in proteomics. Nishimura and coworkers[252] have developed a heavy (^{13}C) form of 2-nitrobenzenesulfonyl chloride for the differential labeling of tryptophan residues in protein mixtures. The application of the isotope-coded affinity tag strategy[253] to tryptophanyl residues has significant advantage in that tryptophan is one of the least abundant residues in proteins.

Modification of histidine or tryptophan presents major challenges. Specificity of modification is difficult to obtain with any of the reagents described earlier. Both histidine and tryptophan are attractive targets for photooxidation, which makes them of importance in surface accessibility mapping for study of protein conformation. The ease of tryptophan modification makes light sensitivity a consideration in the preparation and distribution of the final biopharmaceutical drug product.[207] Histidine can be modified in the Maillard reaction and forms novel products,[254] which have interesting biological properties.[255,256] The Maillard reaction product from histidine has been shown to inhibit polyphenol oxidase, which is responsible for enzymatic browning.[256–259]

REFERENCES

1. Huvaere, K. and Skibsted, L., Light-induced oxidation of tryptophan and histidine reactivity of aromatic *N*-heterocycles toward triplet-excited flavins, *J. Am. Chem. Soc.* 131, 8049–8060, 2009.
2. Hara, I., Ueno, T., Ozaki, S.I. et al., Oxidative modification of tryptophan 43 in the hem vicinity of F43W/H64L myoglobin mutant, *J. Biol. Chem.* 276, 36067–36070, 2001.
3. Kyte, J., *Structure in Protein Chemistry*, 2nd edn., Garland Science, New York, pp. 0174–0176, 2007.
4. Plapp, B.V., Application of affinity labeling for studying structure and function of enzymes, *Methods Enzymol.* 87, 469–499, 1982.
5. Schoellmann, G. and Shaw, E., Direct evidence for the presence of histidine in the active center of chymotrypsin, *Biochemistry* 2, 252–255, 1963.
6. Pincus, M., Thi, L.L., and Carty, R.P., The kinetics and specificity of the reaction of 2′(3′)-*O*-bromoacetyluridine with bovine pancreatic ribonuclease A, *Biochemistry* 14, 3653–3661, 1975.
7. Ray, W.J., Jr., Photochemical oxidation, *Methods Enzymol.* 11, 490–497, 1967.
8. Fernandes, A.F., Zhou, J., Zhang, X. et al., Oxidative inactivation of the proteasome in retinal pigment epithelial cells. A potential link between oxidative stress in retinal pigment epithelial cells, *J. Biol. Chem.* 283, 20745–20753, 2008.

9. Kim, H.J., Yoon, J.H., and Yoon, S., Photooxidative coupling of thiophenol derivatives to disulfides, *J. Phys. Chem. A* 114, 12010–12015, 2010.
10. Braslavsky, S.E. and Houk, K.N., Glossary of terms used in photochemistry, in *CRC Handbook of Organic Photochemistry*, J.C. Scaliana (ed.), CRC Press, Boca Raton, FL, Vol. 2, Chapter 23, pp. 425–466, 1989.
11. Stratton, S.P. and Liebleer, D.C., Determination of singlet oxygen-specific versus radical-mediated lipid peroxidation in photosensitized oxidation of lipid bilayers: Effect of β-carotene and α-tocopherol, *Biochemistry* 36, 12911–12920, 1997.
12. Bartlett, J.A. and Indig, G.L., Effect of self-association and protein binding on the photochemical reactivity of triarylmethanes. Implications of noncovalent interactions on the competition between photosensitization mechanisms Type I and Type II, *Photochem. Photobiol.* 70, 490–498, 1999.
13. Kerwin, B.A. and Remmele, R.L., Jr., Protect from light: Photodegradation and protein biologics, *J. Pharm. Sci.* 96, 1468–1479, 2007.
14. Ehrenshaft, M., Silva, S.O., Perdivara, I. et al., Immunological detection of *N*-formylkynurenine in oxidized proteins, *Free Radic. Biol. Med.* 46, 1260–1266, 2009.
15. Kleinman, M.H., Smith, M.D., Kurali, E. et al., An evaluation of chemical photoreactivity and the relationship to phototoxicity, *Regul. Toxicol. Pharmacol.* 58, 224–232, 2010.
16. Levine, R.L. et al., Carbonyl assay for determination of oxidatively modified proteins, *Methods Enzymol.* 233, 346–357, 1994.
17. Reznick, A.Z. and Packer, L., Oxidative damage to proteins: Spectrophotometric method for carbonyl assay, *Methods Enzymol.* 233, 357–363, 1994.
18. Requena, J.R., Groth, D., Legname, G. et al., Copper-catalyzed oxidation of the recombinant SHa (29–231) prion protein, *Proc. Natl. Acad. Sci. USA* 98, 7170–7175, 2001.
19. Schöneich, C. and Williams, T.D., Cu(II)-catalyzed oxidation of β-amyloid peptide targets His13 and His14 over His6: Detection of 2-oxo-histidine by HPLC-MS/MS, *Chem. Res. Toxicol.* 15, 717–722, 2002.
20. Ghezzi, P. and Bonetto, V., Redox proteomics: Identification of oxidatively modified proteins, *Proteomics* 3, 1145–1153, 2003.
21. Rashidzadeh, H., Khrapunov, S., Chance, M.R., and Brenowitz, M., Solution structure and interdomain interactions of the *Saccharomyces cerevisiae* "TATA binding protein" (TBP) proved by radiolytic protein footprinting, *Biochemistry* 42, 3655–3665, 2003.
22. Xu, G., Takamoto, K., and Chance, M.R., Radiolytic modification of basic amino acid residues in peptides: Probes for examining protein-protein interactions, *Anal. Chem.* 75, 6995–7007, 2003.
23. Takamoto, K. and Chance, M.R., Radiolytic protein footprinting with mass spectrometry to probe the structure of macromolecular complexes, *Annu. Rev. Biophys. Biomol. Struct.* 35, 251–276, 2006.
24. Rahimi, Y., Shrestha, S., Banerjee, T., and Deo, S.K., Copper sensing based on the far-red fluorescent protein, HcRed, from *Heteractis crispa*, *Anal. Biochem.* 370, 60–67, 2007.
25. Gregory, E.M., Chemical modification of bovine heart mitochondrial malate dehydrogenase. Selective modification of cysteine and histidine, *J. Biol. Chem.* 250, 5470–5474, 1975.
26. Anderton, B.H. and Rabin, B.R., Alkylation studies in a reactive histidine in pig heart malate dehydrogenase, *Eur. J. Biochem.* 15, 568–573, 1970.
27. Hashimoto, J. and Takahashi, K., Chemical modification of ribonuclease T_1, *J. Biochem.* 81, 1175–1180, 1977.
28. Sams, C.F. and Matthews, K.S., Chemical modification of dopamine β-hydrogenase, *Biochim. Biophys. Acta* 787, 61–70, 1984.
29. Schelté, P., Boeckler, C., Frisch, B., and Schuber, F., Differential reactivity of maleimide and bromoacetyl functions with thiols: Application to the preparation of liposomal diepitope constructs, *Bioconjug. Chem.* 11, 118–123, 2000.

30. Liu, T.Y., Demonstration of a the presence of a histidine residue at the active site of streptococcal proteinase, *J. Biol. Chem.* 242, 4029–4032, 1967.
31. Gleisner, J.M. and Liener, I.E., Chemical modification of the histidine residue located at the active site of ficin, *Biochim. Biophys. Acta* 317, 482–491, 1973.
32. Dudley, H.W., Intermediary carbohydrate metabolism. The effect of sodium iodoacetate on glyoxylase, *Biochem. J.* 25, 439–445, 1931.
33. Webb, J.L., Iodoacetate and iodoacetamide, in *Enzyme and Metabolic Inhibitors*, Academic Press, New York, Vol. 3, Chapter 1, pp. 1–283, 1966.
34. Zittle, C.A., Ribonucleinase III. The behavior of copper and calcium in the purification of nucleic acid and the effect of these and other reagents on the activity of ribonucleinase, *J. Biol. Chem.* 163, 111–117, 1946.
35. Hirs, C.H., Moore, S., and Stein, W.H., The sequence of the amino acid residues in performic acid-oxidized ribonuclease, *J. Biol. Chem.* 235, 633–647, 1960.
36. Spackman, D.H., Stein, W.H., and Moore, S., The disulfide bonds of ribonuclease, *J. Biol. Chem.* 235, 648–659, 1960.
37. Gundlach, H.G., Stein, W.H., and Moore, S., The nature of the amino acid residues modified on the inactivation of ribonuclease by iodoacetate, *J. Biol. Chem.* 234, 1754–1760, 1959.
38. Stark, G.R., Stein, W.H., and Moore, S., Relationship between the conformation of ribonuclease and its reactivity with iodoacetate, *J. Biol. Chem.* 236, 436–442, 1961.
39. Barnard, E.A. and Stein, W.D., The histidine residue in the active centre of ribonuclease. I. A specific reaction with bromoacetic acid, *J. Mol. Biol.* 1, 333–349, 1959.
40. Korman, S. and Clarke, H.T., Carboxymethylamino acids and peptides, *J. Biol. Chem.* 221, 113–131, 1956.
41. Korman, S. and Clarke, H.T., Carboxymethyl proteins, *J. Biol. Chem.* 221, 133–141, 1956.
42. Stein, W.D. and Barnard, E.A., The histidine residue in the active centre of ribonuclease. II. The position of this residue in the primary protein chain, *J. Mol. Biol.* 1, 350–358, 1959.
43. Cha, M.K. and Kim, I.H., Glutathione-linked thiol peroxidase activity of human serum albumin: A possible antioxidant role of serum albumin in blood plasma, *Biochem. Biophys. Res. Commun.* 222, 619–625, 1996.
44. Huggins, C. and Jenson, E.V., Thermal coagulation of serum proteins I. The effects of iodoacetate, iodoacetamide, and thiol compounds on coagulation, *J. Biol. Chem.* 179, 845–854, 1949.
45. Batra, P.P., Sasa, K., Ueki, T., and Takeda, K., Circular dichroic study of the conformational stability of sulfhydryl-blocked bovine serum albumin, *Int. J. Biochem.* 21, 857–862, 1989.
46. Crestfield, A.M., Stein, W.H., and Moore, S., Alkylation and identification of the histidine residues at the active site of ribonuclease, *J. Biol. Chem.* 238, 2413–2420, 1963.
47. Kumar, K.S., Walsh, K.A., and Neurath, H., Chemical characterization of bovine carboxypeptidase A isolated from a single pancreas, *Biochemistry* 3, 1726–1727, 1964.
48. Crestfield, A.M., Stein, W.H., and Moore, S., Properties and conformation of the histidine residues at the active site of ribonuclease, *J. Biol. Chem.* 238, 2421–2428, 1963.
49. Fruchter, R.G. and Crestfield, A.M., Preparation and properties of two active forms of ribonuclease dimer, *J. Biol. Chem.* 240, 3868–3874, 1965.
50. Fruchter, R.G. and Crestfield, A.M., On the structure of ribonuclease dimer. Isolation and identification of monomers derived from inactive carboxymethylated dimers, *J. Biol. Chem.* 240, 3875–3882, 1965.
51. Crestfield, A.M. and Fruchter, R.G., The homologous and hybrid dimers of ribonuclease A and the carboxymethylhistidine derivatives, *J. Biol. Chem.* 242, 3279–3284, 1967.

52. Lin, M.C., Stein, W.H., and Moore, S., Further studies on the alkylation of the histidine residues of pancreatic ribonuclease, *J. Biol. Chem.* 243, 6167–6170, 1968.
53. Meadows, D.H., Jardetsky, O., Epand, R.M. et al., Assignment of histidine peaks in the nuclear magnetic resonance spectroscopy spectrum of ribonuclease, *Proc. Natl. Acad. Sci. USA* 60, 766–772, 1968.
54. Miyagi, M. and Nakazawa, T., Determination of pK_a values of individual histidine residues in proteins using mass spectrometry, *Anal. Chem.* 80, 6481–6487, 2008.
55. Fruchter, R.G. and Crestfield, A.M., The specific alkylation by iodoacetamide of histidine 12 in the active site of ribonuclease, *J. Biol. Chem.* 242, 5807–5812, 1967.
56. Plapp, B.V., Mechanisms of carboxymethylation of bovine pancreatic nucleases by haloacetates and tosylglycolate, *J. Biol. Chem.* 248, 4896–4900, 1973.
57. Moore, S. and Stein, W.H., Chemical structures of pancreatic ribonuclease and deoxyribonuclease, *Science* 180, 458–464, 1973.
58. Marshall, G.R., Fend, J.A., and Kuster, D.J., Back to the future: Ribonuclease A, *Biopolymers* 90, 259–277, 2008.
59. Kettner, C. and Shaw, E., Inactivation of trypsin-like enzymes with peptides of arginine chloromethyl ketone, *Methods Enzymol.* 80, 826–842, 1981.
60. Bock, P.E., Active site selective labeling of serine proteases with spectroscopic probes using thioester peptide chloromethyl ketones: Demonstration of thrombin labeling using N^{α} [T-[(acetylthio)acetyl]-D-Phe-Pro-Arg-CH_2Cl, *Biochemistry* 27, 6633–6639, 1988.
61. Bock, P.E., Active-site-selective labeling of blood coagulation proteinases with fluorescence probes by the use of thioester peptide chloromethyl ketones. II. Properties of thrombin derivatives as reporters of prothrombin fragment 2 binding and specificity of the labeling approach for other proteinases, *J. Biol. Chem.* 267, 14974–14981, 1992.
62. Williams, E.B., Krishnaswamy, S., and Mann, K.G., Zymogen/enzyme discrimination using peptide chloromethyl ketones, *J. Biol. Chem.* 264, 7536–7545, 1989.
63. Kettner, C. and Shaw, E., The susceptibility of urokinase to affinity labeling by peptides of arginine chloromethyl ketone, *Biochim. Biophys. Acta* 569, 31–40, 1979.
64. Perez-G, M., Cortes, J.R., and Rivas, M.D., Treatment of cells with *N*-α-tosyl-L-phenylalanine-chloromethyl ketone induces the proteolytic loss of STAT6 transcription factor, *Mol. Immunol.* 45, 3896–3901, 2008.
65. Jiménez, M., Cabanes, J., Gandía-Herrero, F. et al., A continuous spectrophotometric assay for phospholipase A_2 activity, *Anal. Biochem.* 319, 131–137, 2003.
66. Ram, A., Das, M., Gangal, S.V., and Ghosh, B., *p*-Bromophenacyl bromide alleviates airway hyperresponsiveness and modulates cytokines, IgE and eosinophil levels in ovalbumin-sensitized and-challenged mice, *Int. Immunopharmacol.* 4, 1697–1707, 2004.
67. Tariq, M., Elfaki, I., Khan, H.A. et al., Bromophenacyl bromide, a phospholipase A_2 inhibitor attenuates chemically induced gastroduodenal ulcers in rats, *World J. Gastroenterol.* 12, 5798–5804, 2006.
68. Merchant, M., Heard, R., and Monroe, C., Characterization of phospholipase A_2 activity in the serum of the American alligator (*Alligator mississippiensis*), *J. Exp. Zool. A. Ecol. Genet. Physiol.* 311, 662–666, 2009.
69. Zychar, B.C., Dale, C.S., Demarchi, D.S., and Goncalves, L.R., Contribution of metalloproteases, serin proteases and phospholipases A_2 to the inflammatory reaction induced by *Bothrops jararaca* crude venom in mice, *Toxicon* 55, 227–234, 2010.
70. Romero-Vargas, F.F., Ponce-Soto, L.A., Martins-de-Souza, D., and Marangoni, S., Biological and biochemical characterization of two new PLA2 isoforms Cdc-9 and Cdc-10 from *Crotalus durissus cumanensis* snake venom, *Comp. Biochem. Physiol. C. Toxicol. Pharmacol.* 151, 66–74, 2010.
71. Ravindran, S., Lodoen, M.B., Verhelst, S.H. et al., 4-Bromophenacyl bromide specifically inhibits rhoptry secretion during Toxoplasma invasion, *PLoS One* 4, 12, e8143, 2009.

72. Fonseca, F.V., Baldissera, L., Jr., Carmargo, E.A. et al., Effect of the synthetic coumarin, ethyl 2-oxo-2H-chromene-3-carboxylate, on activity *Crotalus durissus ruruima* sPLA2 as well as on edema and platelet aggregation induced by this factor, *Toxicon* 55, 1527–1530, 2010.
73. Berger, M., Reck, J., Jr., Terra, R.M. et al., *Lonomia obliqua* venomous secretion induces human platelet adhesion and aggregation, *J. Thromb. Thrombolysis* 30, 300–310, 2010.
74. Blacklow, B., Escoubas, P., and Nicholson, G.M., Characterization of the heterotrimeric presynaptic phospholipase A_2 neurotoxin complex from the venom of the common death adder (*Acanthophis antarcticus*), *Biochem. Pharmacol.* 80, 277–287, 2010.
75. Shaw, J.O., Roberts, M.F., Ulevitch, R.J. et al., Phospholipase A_2 contamination of cobra venom factor preparations. Biologic role in complement-dependent *in vivo* reactions and inactivation with *p*-bromophenacyl bromide, *Am. J. Pathol.* 91, 517–530, 1978.
76. Berger, J., Identification of organic compounds. I. Preparation of *p*-bromophenacyl esters of carboxylic acids, *Acta Chem. Scand.* 10, 638–642, 1956.
77. Durst, H.D., Milano, M., Kikta, E.J. et al., Phenacyl esters of fatty acids via crown ether catalysts for enhanced ultraviolet detection in liquid chromatography, *Anal. Chem.* 47, 1797–1801, 1975.
78. Zamir, I., Derivatization of saturated long-chain fatty acids with phenacyl bromide in ionic micelles, *J. Chromatogr.* 586, 347–350, 1991.
79. Erlanger, B.F., Vratsanos, S.M., Wasserman, N., and Cooper, A.G., Chemical investigation of the active center of pepsin, *Biochem. Biophys. Res. Commun.* 23, 243–245, 1966.
80. Gross, E. and Morrell, J.L., Evidence for an active carboxyl group in pepsin, *J. Biol. Chem.* 241, 3638–3639, 1966.
81. Takahashi, K., Stein, W.H., and Moore, S., The identification of a glutamic acid residue as part of the active site of ribonuclease T_1, *J. Biol. Chem.* 242, 4682–4690, 1967.
82. Kyger, E.M. and Franson, R.C., Nonspecific inhibition of enzymes by *p*-bromophenacyl bromide, *Biochim. Biophys. Acta* 794, 96–103, 1984.
83. Volwerk, J.J., Pieterson, W.A., and de Haas, G.H., Histidine at the active site of phospholipase A_2, *Biochemistry* 13, 1446–1454, 1974.
84. Abita, J.P., Lazdunski, M., Bonsen, P.P. et al., Zymogen-enzyme transformations. On the mechanism of activation of prophospholipase A, *Eur. J. Biochem.* 30, 37–47, 1972.
85. Bonsen, P.P.M., Pieterson, W.A., Volwerk, J.J., and de Haas, G.H., Phospholipase A and its zymogen from porcine pancreas, in *Current Trends in the Biochemistry of Lipids* (Biochemical Society Symposium 35), P. Ganguly and R.M.S. Smellie (eds.), Academic Press, London, U.K., pp. 189–200, 1971.
86. Dennis, E.A., Phospholipases, in *The Enzymes*, 3rd edn., P.D. Boyer (ed.), Academic Press, New York, Vol. 16, Chapter 9, pp. 307–353, 1983.
87. Mackessey, S.P., Evolutionary trends in venom composition in the western rattlesnakes (*Crotalus viridis sensu lato*): Toxicity vs. tenderness, *Toxicon* 55, 1463–1474, 2010.
88. Marchi-Salvador, D.P., Fernandes, C.A.H., Silveira, L.B. et al., Crystal structure of a phospholipase A_2 homolog complexed with *p*-bromophenacyl bromide reveals important structural changes associated with the inhibition of myotoxic activity, *Biochim. Biophys. Acta* 1794, 1583–1590, 2009.
89. Miyake, T., Inoue, S., Ikeda, K., pH Dependence of the reaction rate of His 48 with *p*-bromophenacyl bromide and the binding constant to Ca^{2+} of the monomeric forms of intact and α-NH_2 modified phospholipase A_2 from *Trimeresurus flavoviridis*, *J. Biochem.* 105, 565–572, 1989.
90. Fujii, S., Meida, M., Tani, T. et al., pH Dependence of the reaction rate of *p*-bromophenacyl bromide and of the binding constants of Ca^{2+} and an amide-type substrate analog to bovine pancreatic phospholipase A_2, *Arch. Biochem. Biophys.* 354, 73–82, 1998.

91. Fohlman, J., Eaker, D., Karlsoon, E., and Theslff, S., Taipoxin, an extremely potent presynaptic neurotoxin from the venom of the Australian snake taipan (*Oxyranus scutellatus*). Isolation, characterization, quaternary structure and pharmacological properties, *Eur. J. Biochem.* 68, 457–469, 1976.
92. Tzeng, M.C., Yen, C.H., Hseu, M.J. et al., Binding proteins on synaptic membranes for crotoxin and taipoxin, two phospholipases A_2 with neurotoxicity, *Toxicon* 33, 451–457, 1995.
93. Fohlman, J., Eaker, D., Dowdall, M.J., Lüllmann-Rauch, R., Sjödin, T., and Leander, S., Chemical modification of taipoxin and the consequences for phospholipase activity, pathophysiology, and inhibition of high-affinity choline uptake, *Eur. J. Biochem.* 94, 531–540, 1979.
94. Halpert, J., Eaker, D., and Karlsson, E., The role of phospholipase activity in the action of a presynaptic neurotoxin of *Notechis scutatus scutatus* (Australian tiger snake), *FEBS Lett.* 61, 72–76, 1976.
95. Verheij, H.M., Volwerk, J.J., Jansen, E.H. et al., Methylation of histidine-48 in pancreatic phospholipase A_2. Role of histidine and calcium ion in the catalytic mechanism, *Biochemistry* 19, 743–750, 1980.
96. Battaglia, E. and Radominska-Pandya, A., A functional role for histidyl residues of the UDP-glucuronic acid carrier in rat liver endoplasmic reticulum membranes, *Biochemistry* 37, 258–263, 1998.
97. de Vet, E.C. and van den Bosch, H., Characterization of recombinant guinea pig alkyldihydroxyacetonephosphate synthase expressed in *Escherichia coli*: Kinetics, chemical modification, and mutagenesis, *Biochim. Biophys. Acta* 1436, 299–306, 1999.
98. Dorsch, S. et al., The VPI unique region of parvovirus B19 and its constituent phospholipase A_2-like activity, *J. Virol.* 76, 2014–2018, 2002.
99. Nakagawa, Y. and Bender, M.L., Modification of α-chymotrypsin by methyl *p*-nitrobenzenesulfonate, *J. Am. Chem. Soc.* 91, 1566–1567, 1967.
100. Nakagawa, Y. and Bender, M.L., Methylation of histidine-57 in α-chymotrypsin by methyl *p*-nitrobenzenesulfonate. A new approach to enzyme modification, *Biochemistry* 9, 259–267, 1970.
101. Glick, B.R., The chemical modification of *Escherichia coli* ribosomes with methyl *p*-nitrobenzenesulfonate. Evidence for the involvement of a histidine residue in the functioning of the ribosomal peptidyl transferase, *Can. J. Biochem.* 58, 1345–1347, 1980.
102. Marcus, J.P. and Dekker, E.E., Identification of a second active site residue in *Escherichia coli* L-threonine dehydrogenase: Methylation of histidine-90 with methyl-*p*-nitrobenzenesulfonate, *Arch. Biochem. Biophys.* 316, 413–420, 1995.
103. Johnson, A.R. and Dekker, E.E., Site-directed mutagenesis of histidine-90 in *Escherichia coli* L-threonine dehydrogenase alters its substrate specificity, *Arch. Biochem. Biophys.* 351, 8–16, 1998.
104. Lancaster, N.L. and Welton, T., Nucleophilicity in ionic liquids. 3. Anion effects on halide nucleophilicity in a series of 1-butyl-3-methylimidazolium ionic liquids, *J. Org. Chem.* 80, 5986–5992, 2004.
105. Crowhurst, L., Lancaster, N.L., Perez Arlandis, J.M., and Welton, T., Manipulating solute nucleophilicity with room temperature ionic liquids, *J. Am. Chem. Soc.* 126, 11549–11555, 2004.
106. Ford, J.W., Janakat, M.E., Lu, J. et al., Local polarity in CO_2-expanded acetonitrile: A nucleophilic substitution reaction and solvatochromic probes, *J. Org. Chem.* 73, 3364–3368, 2008.
107. Hayaki, S., Kido, K., Sato, H., and Sakaki, S., *Ab initio* study on SN_2 reaction of methyl *p*-nitrobenzenesulfonate and chloride anion in [mmim][PF6], *Phys. Chem. Chem. Phys.* 12, 1822–1826, 2010.

108. Edmondson, D.E., Kenney, W.C., and Singer, T.P., Structural elucidation and properties of 8α-(N^1-histidyl)riboflavin: The flavin component of thiamine dehydrogenase and β-cyclopiazonate oxidocyclase, *Biochemistry* 15, 2937–2945, 1976.
109. Taralp, A. and Kaplan, H., Chemical modification of lyophilized proteins in nonaqueous environments, *J. Protein Chem.* 16, 183–193, 1997.
110. Kamińska, J., Wiśiewska, A., and Kościelak, J., Chemical modifications of α 1,6-fucosyltransferase define amino acid residues of catalytic importance, *Biochimie* 85, 303–310, 2003.
111. Zalipsky, S., Seltzer, R., and Menon-Rudolph, S., Evaluation of a new reagent for covalent attachment of polyethylene glycol to proteins, *Biotechnol. Appl. Biochem.* 15, 100–114, 1992.
112. Sumiyoshi, H., Shimizu, T., Katoh, M. et al., Solution-phase parallel synthesis of carbamates using polymer bound *N* hydroxysuccinimide, *Org. Lett.* 4, 3923–3926, 2002.
113. Wang, Y.S., Youngster, S., Bausch, J. et al., Identification of the major positional isomers of pegylated interferon Alpha-2b, *Biochemistry* 39, 10634–10640, 2000.
114. Wylie, D.C., Voloch, M., Lee, S. et al., Carboxyalkylated histidine is a pH dependent product of pegylation with SC-PEG, *Pharm. Res.* 18, 1354–1360, 2001.
115. Miron, T. and Wilchek, M., A simplified method for the preparation of succinimidyl carbonate polyethylene glycol for coupling to proteins, *Bioconjug. Chem.* 4, 580–589, 1993.
116. Wang, M., Basu, A., Palm, T. et al., Engineering an arginine catabolizing bioconjugate: Biochemical and pharmaceutical characterization of PEGylated derivatives of arginine deiminase from *Mycoplasma arthritidis*, *Bioconjug. Chem.* 17, 1447–1459, 2006.
117. Lee, B.K., Kwon, J.S., Kim, H.J. et al., Solid-phase PEGylation of recombinant interferon α-2b for site-specific modification: Process performance, characterization, and *in vitro* bioactivity, *Bioconjug. Chem.* 18, 1728–1734, 2007.
118. Borukhov, S.I. and Strongin, A.Ya., Chemical modification of the recombinant human αA- and β-interferons, *Biochem. Biophys. Res. Commun.* 167, 74–80, 1990.
119. Tsan, M.F., Inhibition of neutrophil sulfhydryl groups by chloromethyl ketones. A mechanism for their inhibition of superoxide production, *Biochem. Biophys. Res. Commun.* 112, 671–677, 1983.
120. Hartley, B.S. and Massey, V., Active centre of chymotrypsin I. Labeling with a fluorescent dye, *Biochim. Biophys. Acta* 21, 58–70, 1956.
121. Tamura, Z., Nakajima, T., Nakayama, T. et al., Identification of peptides with 5-dimethylaminonaphthalenesulfonyl chloride, *Anal. Biochem.* 52, 595–606, 1973.
122. Nishino, T., Massey, V., and Williams, C.H., Jr., Chemical modifications of D-amino acid oxidase. Evidence for active site histidine, tyrosine, and arginine residues, *J. Biol. Chem.* 255, 3610–3615, 1980.
123. Swenson, R.P., Williams, C.R., Jr., and Massey, V., Identification of the histidine residue in D-amino acid oxidase that is covalently modified during inactivation with 5-dimethylaminonapthalene-1-sulfonyl chloride, *J. Biol. Chem.* 258, 497–502, 1983.
124. Gadda, G., Beretta, G.L., and Pilone, M.S., Reactivity of histidyl residues in D-amino acid oxidase from *Rhodotorula gracilis*, *FEBS Lett.* 363, 307–310, 1995.
125. Morishima, I., Shiro, Y., Adachi, S., Yano, Y., and Orii, Y., Effect of the distal histidine modification (cyanation) of myoglobin on the ligand binding kinetics and the heme environmental structures, *Biochemistry* 28, 7582–7586, 1989.
126. Shiro, Y. and Morishima, I., Modification of the heme distal side chain in myoglobin by cyanogen bromide. Heme environmental structures and ligand binding properties of the modified myoglobin, *Biochemistry* 23, 4879–4884, 1984.
127. Tanguchi, I., Sonoda, K., and Mie, Y., Electroanalytical chemistry of myoglobin with modification of distal histidines by cyanated imidazole, *J. Electroanal. Chem.* 468, 9–16, 1999.

128. Oomwn, R.P. and Kaplan, H., Competitive labeling as an approach to defining the binding surfaces of proteins: Binding of monomeric insulin to lipid bilayers, *Biochemistry* 26, 303–308, 1987.
129. Hefford, M.A. and Kaplan, H., Chemical properties of the histidine residue of secretin: Evidence for a specific intramolecular interaction, *Biochim. Biophys. Acta* 998, 262–270, 1989.
130. Kaplan, H. and Oda, G., Selective isolation of free and blocked amino-terminal peptides from enzymatic digestion of proteins, *Anal. Biochem.* 132, 384–388, 1983.
131. Shatiel, S., Thiolysis of some dinitrophenyl derivatives of amino acids, *Biochem. Biophys. Res. Commun.* 29, 178–183, 1967.
132. 21 CFR 189.150, Diethyl pyrocarbonate (DEPC), August 2, 1972.
133. Delfini, C., Gaia, P., Schellino, R. et al., Fermentability of grape must after inhibition with dimethyldicarbonate, *J. Agric. Food Chem.* 50, 5601–5611, 2002.
134. Solymosy, F., Fedorcsák, I., Gulyás, A. et al., A new method based on the use of diethyl pyrocarbonate as a nuclease inhibitor for the extraction of undegraded nucleic acid, *Eur. J. Biochem.* 5, 520–527, 1968.
135. Leonard, N.J., McDonald, J.J., and Reichmann, M.E., Reaction of diethyl pyrocarbonate with nucleic acid components. I. Adenine, *Proc. Natl. Acad. Sci. USA* 67, 93–98, 1970.
136. Huang, Y.H., Leblanc, P., Apostolou, V. et al., Comparison of Milli-Q PF plus water with DEPC-treated water in the preparation and analysis of RNA, *Nucleic Acids Symp. Ser.* 33, 129–133, 1995.
137. Hullán, L., Szontagh, T., Turtóczky, I., and Fedorcsák, I., The inactivation of trypsin by diethyl pyrocarbonate, *Acta Chem. Scand.* 19, 2440–2441, 1965.
138. Fedorcsák, I. and Ehrenberg, L., Effects of diethyl pyrocarbonate and methyl methanesulfonate on nucleic acids and nucleases, *Acta Chem. Scand.* 20, 107–112, 1966.
139. Rosén, C.G. and Fedorcsák, I., Studies on the action of diethyl pyrocarbonate on proteins, *Biochim. Biophys. Acta* 130, 401–405, 1966.
140. Ovádi, J. and Keleti, T., Effect of diethyl pyrocarbonate on the conformation and enzymatic activity of d-glyceraldehyde-3-phosphate dehydrogenase, *Acta Biochim. Biophys. Sci. Hung.* 4, 365–378, 1969.
141. Rosén, C.G., Gejvall, T., and Andersson, L.O., Reaction of diethyl pyrocarbonate with indole derivatives with special reference to the reaction with tryptophan residues in a protein, *Biochim. Biophys. Acta* 221, 207–213, 1970.
142. Wiener, S.L., Wiener, R., Urivetzky, M., and Meilman, E., Inactivation of ribonuclease by diethyl pyrocarbonate and other methods, *Biochim. Biophys. Acta* 259, 378–385, 1972.
143. Wolf, B., Lesnaw, J.A., and Reichmann, M.E., A mechanism of irreversible inactivation of bovine pancreatic ribonuclease by diethylpyrocarbonate. A general reaction of diethyl pyrocarbonate with proteins, *Eur. J. Biochem.* 13, 519–525, 1970.
144. Morris, D.L. and McKinley-McKee, J.S., The histidines in liver alcohol dehydrogenase. Chemical modification with diethylpyrocarbonate, *Eur. J. Biochem.* 29, 515–520, 1972.
145. Holbrook, J.J., Lodola, A., and Illesley, N.P., Histidine residues and the enzyme activity of pig heart supernatant malate dehydrogenase, *Biochem. J.* 139, 797–800, 1974.
146. Hnízda, A., Šantrůcek, J., Šanda, M. et al., Reactivity of histidine and lysine side-chains with diethyl pyrocarbonate—A method to identify surface exposed residues in proteins, *J. Biochem. Biophys. Methods* 70, 1091–1097, 2008.
147. Mendoza, V.L. and Vachet, R.W., Protein surface mapping using diethyl pyrocarbonate with mass spectrometric detection, *Anal. Chem.* 81, 2895–2901, 2008.
148. Muhirad, A., Hegyi, G., and Toth, G., Effect of diethyl pyrocarbonate on proteins. I. Reaction of diethyl pyrocarbonate with amino acids, *Acta Biochim. Biophys. Acad. Sci. Hung.* 2, 19–29, 1967.

149. Garrison, C.K. and Himes, R.H., The reaction between diethyl pyrocarbonate and sulfhydryl groups in carboxylate buffers, *Biochem. Biophys. Res. Commun.* 67, 1251–1255, 1975.
150. Rua, J., Robertson, A.G.S., and Nimmo, H.G., Identification of the histidine residues in *Escherichia coli* isocitrate lyase that reacts with diethylpyrocarbonate, *Biochim. Biophys. Acta* 1122, 212–218, 1992.
151. Miles, E.W., Modification of histidyl residues in proteins by diethylpyrocarbonate, *Methods Enzymol.* 47, 431–442, 1977.
152. Welsch, D.J. and Nelsestuen, G.L., Irreversible degradation of histidine-96 of prothrombin fragment 1 during protein acetylation: Another unusually reactive site in the kringle, *Biochemistry* 27, 7513–7518, 1988.
153. Glocker, M.O., Kalkum, M., Yamamoto, R., and Schreurs, J., Selective biochemical modification of function residues in recombinant human macrophage colony-stimulating factor beta (rhM-CSFbeta): Identification by mass spectrometry, *Biochemistry* 35, 14625–14633, 1996.
154. Kalkum, M., Prxybylski, M., and Glocker, M.O., Structural characterization of functional histidine residues and carbethoxylated derivates in peptides and proteins by mass spectrometry, *Bioconjug. Chem.* 9, 226–235, 1998.
155. Dage, J.L., Sun, H., and Halsall, H.B., Determination of diethyl pyrocarbonate-modified amino acid residues in alpha-1-acid glycoprotein by high-performance liquid chromatography electrospray ionization mass spectrometry and matrix-assisted laser desorption/ionization time-of-flight mass spectrometry, *Anal. Biochem.* 257, 176–185, 1998.
156. Krell, T., Chackrawarthy, S., Pitt, A.R. et al., Chemical modification monitored by electrospray mass spectrometry: A rapid and simple method for identifying and studying functional residues in enzymes, *J. Pept. Res.* 51, 201–209, 1998.
157. Qin, K., Yang, Y., Mastrangelo, P., and Westaway, D., Mapping Cu(II) binding sites in prion protein by diethyl pyrocarbonate modification of matrix-assisted laser desorption time-of-flight (MALDI-TOF) mass spectrometric footprinting, *J. Biol. Chem.* 277, 1981–1990, 2002.
158. Ginotra, Y.P. and Kulkarni, P.P., Solution structure of physiological $Cu(His)_2$: Novel considerations into imidazole coordination, *Inorg. Chem.* 48, 7000–7002, 2009.
159. Willard, B.B. and Kinter, M., Effects of internal histidine residues on the collision-induced fragmentation of triply protonated tryptic peptides, *J. Am. Soc. Mass Spectrom.* 12, 1262–1271, 2001.
160. Nakanishi, N., Takeuchi, F., Okamoto, H. et al., Characterization of heme-coordinating histidyl residues of cytochrome b_5 based on the reactivity with diethylpyrocarbonate: A mechanism for the opening of axial imidazole rings, *J. Biochem.* 140, 561–571, 2006.
161. Nakanishi, N., Takeuchi, F., Park, S.Y. et al., Characterization of heme-coordinating histidyl residues of an engineered six-coordinated myoglobin mutant based on the reactivity with diethylpyrocarbonate, mass spectrometry and electron paramagnetic resonance spectroscopy, *J. Biosci. Bioeng.* 105, 604–613, 2008.
162. Konkle, M.E., Eisenheimer, K.N., Hakala, K. et al., Chemical modification of the Rieske protein from *Thermus thermophilus* using diethyl pyrocarbonate modifies ligating histidine 154 and reduces the [2Fe-2S] cluster, *Biochemistry* 49, 7272–7281, 2010.
163. Altman, J., Lipka, J.J., Kuntz, I., and Waskell, L., Identification by proton nuclear magnetic resonance of the histidine in cytochrome b_5 modified by diethyl pyrocarbonate, *Biochemistry* 28, 7516–7523, 1989.
164. Lundblad, R.L., *Chemical Reagents for Protein Modification*, 3rd edn., CRC Press, Boca Raton, FL, 2004.
165. Uchida, K., 4-Hydroxy-2-nonenal: A product and mediator of oxidative stress, *Prog. Lipid Res.* 42, 318–343, 2003.

166. Hidalgo, F.J., Alaiz, M., and Zamero, R., A spectrophotometric method for the determination of proteins damaged by oxidized lipids, *Anal. Biochem.* 262, 129–136, 1998.
167. Refsgaard, H.H.F., Tsai, L., and Stadtman, E.R., Modification of proteins by polyunsaturated fatty acid peroxidation products, *Proc. Natl. Acad. Sci. USA* 97, 611–616, 2000.
168. Oe, T. et al., A novel lipid peroxide-derived cyclic covalent modification to histone H4, *J. Biol. Chem.* 278, 42098–42105, 2003.
169. Zhu, X., Anderson, V.E., and Sayre, L.M., Charge-derivatized amino acids facilitate model studies on protein side-chain modifications by matrix-assisted laser desorption/ionization time-of-flight mass spectrometry, *Rapid Commun. Mass Spectrom.* 23, 2113–2124, 2009.
170. Grimsrud, P.A., Xie, H., Griffin, T.J., and Bernlohn, D.A., Oxidative stress and covalent modification of protein with bioactive aldehydes, *J. Biol. Chem.* 283, 21837–21841, 2008.
171. Fenaiile, F., Tabet, J.C., and Guy, P.A., Identification of 4-hydroxy-2-nonenal-modified peptides within unfractionated digests using matrix-assisted laser desorption/ionization time-of-flight mass spectrometry, *Anal. Chem.* 76, 867–873, 2004.
172. Rauniyar, N., Prokai-Tatrai, K., and Prokai, L., Identification of carbonylation sites in apomyoglobin after exposure to 4-hydroy-2-nonenal by solid-phase enrichment and liquid chromatography-electrospray ionization tandem mass spectrometry, *J. Mass Spectrom.* 45, 398–410, 2010.
173. Fenaille, F., Guy, P.A., and Tabet, J.C., Study of protein modification by 4-hydroxy-2-nonenal and other short chain aldehydes analyzed by electrospray ionization tandem mass spectrometry, *J. Am. Soc. Mass Spectrom.* 14, 215–226, 2003.
174. Toyokuni, S., Miyake, N., Hiai, H. et al, The monoclonal antibody specific for the 4-hydroxy-2-nonenal histidine adduct, *FEBS Lett.* 359, 189–191, 1995.
175. Uchida, K., Itakura, K., Kawakishi, S. et al., Characterization of epitopes recognized by 4-hydroxy-2-nonenal specific antibodies, *Arch. Biochem. Biophys.* 324, 241–248, 1995.
176. Ozeki, M., Miyagawa-Hayshino, A., Akatsuka, S. et al., Susceptibility of actin to modification by 4-hydroxy-2-nonenal, *J. Chromatogr. B.* 827, 119–126, 2005.
177. Waeg, G., Dimsity, G., and Esterhaven, H., Monoclonal antibodies for detection of 4-hydoxy-nonenal modified proteins, *Free Radic. Res.* 25, 149–159, 1996.
178. Jerfy, A. and Roy, A.B., Sulfatase of ox liver. XII. Effects of tyrosine and histidine reagents on the activity of sulfatase A, *Biochim. Biophys. Acta* 175, 355–364, 1969.
179. MacDonald, J.M., Hass, A.L., and London, R.E., Novel mechanism of surface catalysis of protein adduct formation. NMR studies of the acetylation of ubiquitin, *J. Biol. Chem.* 275, 31908–31913, 2000.
180. Moore, G.J., Kinetics of acetylation-deacetylation of angiotensin OII. Intramolecular interactions of the tyrosine and histidine side-chains, *Int. J. Pept. Protein Res.* 26, 469–481, 1985.
181. Kinnunen, P.M., DeMichele, A., and Lange, L.G., Chemical modification of acyl-CoA:cholesterol *O*-acyltransferase. 1. Identification of acyl-CoA:cholesterol *O*-acyltransferase subtypes by differential diethyl pyrocarbonate sensitivity, *Biochemistry* 27, 7344–7350, 1988.
182. Johnson, A.R. and Dekker, E.E., Woodward's Reagent K inactivation of *Escherichia coli* L-threonine dehydrogenase: Increased absorbance at 340–350 nm is due to modification of cysteine and histidine residues, not asparate or glutamate carboxyl groups, *Protein Sci.* 5, 382–390, 1996.
183. Bustos, P. et al., Woodward's Reagent K reacts with histidine and cysteine residues in *Escherichia coli* and *Saccharomyces cerevisiae* phosphoenolpyruvate carboxykinase, *J. Protein Chem.* 15, 467–472, 1996.
184. Paoli, P. et al., Mechanism of acylphosphatase inactivation by Woodward's Reagent K, *Biochem. J.* 328, 855–861, 1997.

185. Liu, T.Y. and Chang, Y.H., Hydrolysis of proteins with *p*-toluenesulfonic acid. Determination of tryptophan, *J. Biol. Chem.* 246, 2842–2848, 1971.
186. Simpson, R.J., Neuberger, M.R., and Liu, T.Y., Complete amino acid analysis of proteins from a single hydrolysate, *J. Biol. Chem.* 251, 1936–1940, 1976.
187. Molnár, I., Tryptophan analysis of peptides and proteins, mainly by liquid chromatography, *J. Chromatogr.* 763, 1–10, 1997.
188. Aiken, A. and Learmonth, M., Quantitation of tryptophan in proteins, in *Protein Protocols Handbook*, 2nd edn., J.M. Walker (ed.), Humana Press, Totowa, NJ, pp. 41–44, 2002.
189. Friedman, M., Applications of the ninhydrin reaction for analysis of amino acids, peptides, and proteins in agricultural and biomedical sciences, *J. Agric. Food Chem.* 52, 385–406, 2004.
190. Steen, H. and Mann, M., Analysis of bromotryptophan and hydroxyproline modifications by high-resolution, high accuracy precursor ion scanning utilizing fragment ions with mass-deficient mass tags, *Anal. Chem.* 74, 6230–6236, 2002.
191. Strohalm, M., Šantrůček, J., Hynek, R., and Kodíček, M., Analysis of tryptophan surface accessibility in proteins by MALDI-TOF mass spectrometry, *Biochem. Biophys. Res. Commun.* 323, 1134–1138, 2004.
192. Dyer, J.M., Bringans, S.D., and Bryson, W.G., Determination of photo-oxidation products within photoyellowed bleached wool proteins, *Photochem. Photobiol.* 82, 551–557, 2006.
193. Mouls, L., Silajdzic, E., Haroune, N. et al., Development of novel mass spectrometric methods for identifying HOCl-induced modifications of proteins, *Proteomics* 9, 1617–1631, 2009.
194. Madian, A.G. and Regnier, F.E., Profiling carbonylated proteins in human plasma, *J. Proteome Res.* 9, 1330–1343, 2010.
195. Hachimori, Y., Horinishi, H., Kurihara, K., and Shibata, K., States of amino residues in proteins. V. Different reactivities with H_2O_2 of tryptophan residues in lysozyme, proteinases and zymogens, *Biochim. Biophys. Acta* 93, 346–360, 1964.
196. Finley, E.L., Dillon, J., Crouch, R.K., and Schey, K.L., Identification of tryptophan oxidation products in bovine α-crystallin, *Protein Sci.* 7, 2391–2397, 1998.
197. Simat, T.J. and Steinhart, H., Oxidation of free tryptophan residues in peptides and proteins, *J. Agric. Food Chem.* 46, 490–498, 1998.
198. Matsuhima, A., Takiuchi, H., Salto, Y., and Inada, Y., Significance of tryptophan residues in the D-domain of the fibrin molecule in fibrin polymer formation, *Biochim. Biophys. Acta* 625, 230–236, 1980.
199. Reubsaet, J.L. et al., Analytical techniques used to study the degradation of proteins and peptides: Chemical instability, *J. Pharm. Biomed. Anal.* 17, 955–978, 1998.
200. Simat, T., Meyer, K., and Steinhart, H., Syntheses and analysis of oxidation and carbonyl condensation compounds of tryptophan, *J. Chromatogr. A* 661, 93–99, 1994.
201. Mach, H., Middaugh, C.R., and Lewis, R.V., Statistical determination of the average values of the extinction coefficients of tryptophan and tyrosine in native proteins, *Anal. Biochem.* 200, 74–80, 1992.
202. Takita, T., Nakagoshi, M., Inouye, K., and Tonomura, B., Lysyl-tRNA synthetase from *Bacillus stearothermophilus*: The tryp312 residue is shielded in a non-polar environment and is responsible for the fluorescence changes observed in the amino acid activation reaction, *J. Mol. Biol.* 325, 677–695, 2003.
203. Reshetnyak, Y.K., Koshevnik, Y., and Burstein, E.A., Decomposition of protein tryptophan fluorescence spectra into log-normal components. III. Correlation between fluorescence and microenvironmental parameters of individual tryptophan residues, *Biophys. J.* 81, 1735–1758, 2001.
204. Hambly, D.M. and Gross, M.L., Cold chemical oxidation of proteins, *Anal. Chem.* 81, 7235–7242, 2008.

205. Ji, J.A., Zhang, B., Cheng, W., and Wang, J., Methionine, tryptophan, and histidine oxidation in a model protein, PTH: Mechanisms and stabilization, *J. Pharm. Sci.* 98, 4485–4500, 2009.
206. Hawkins, C.L., Pattison, D.L., Stanley, N.R., and Davies, M.J., Tryptophan residues are targets in hypothiocyanous acid-mediated protein oxidation, *Biochem. J.* 416, 441–452, 2008.
207. Qi, P., Volkin, D.B., Zhao, H. et al., Characterization of the photodegradation of a human IgG1 monoclonal antibody formulated as a high-concentration liquid dosage form, *J. Pharm. Sci.* 98, 3117–3130, 2009.
208. Lee, M.G. and Rogers, C.M., Degradation of tryptophan in aqueous solution, *J. Parenter. Sci. Technol.* 42, 20–22, 1988.
209. Pigault, C. and Gerard, D., Influence of the location of tryptophanyl residues in proteins on their photosensitivity, *Photochem. Photobiol.* 40, 291–296, 1984.
210. Rao, S.C., Ran, C.M., and Balasubramanian, D., The conformational status of a protein influences the aerobic photolysis of its tryptophan residues: Melittin, β-lactoglobulin, and the crystallins, *Photochem. Photobiol.* 41, 357–362, 1990.
211. Tallmadge, D.H. and Borkman, R.E., The rates of photolysis of the four individual tryptophan residues in UV exposed calf γII crystallin, *Photochem. Photobiol.* 51, 363–368, 1990.
212. Nagasaka, T. and Ohki, S., Indoles. II. Oxidation of *N*-phthaloyl-1-acetyltryptophan with chromium trioxide, *Chem. Pharm. Bull.* 19, 603–611, 1971.
213. Boccu, E., Veronese, F.M., Fontana, A., and Benassi, A., Studies on the function of tryptophan-108 on lysozyme, *Acta Vitamin Enzymol.* 29, 266–269, 1975.
214. Savige, W.E. and Fontana, A., Oxidation of tryptophan to oxindoylalanine by dimethyl sulfoxide-hydrochloric acid. Selective modification of tryptophan containing peptides, *Int. J. Pept. Prot. Res.* 15, 285–297, 1980.
215. Huang, H.V., Bond, N.W., Hunkapiller, M.W., and Hood, L.W., Cleavage at tryptophanyl residues with dimethyl sulfoxide-hydrochloride acid and cyanogen bromide, *Methods Enzymol.* 91, 318–324, 1983.
216. Wagner, R.M. and Fraser, B.A., Analysis of peptides containing oxidizing methionine and/or tryptophan by fast atom bombardment mass spectrometry, *Biomed. Environ. Mass Spectrom.* 14, 69–72, 1987.
217. Ohnishi, M., Kawagishi, T., Abe, T., and Hiromi, K., Stopped-flow studies on the chemical modification with *N*-bromosuccinimide of model compounds of tryptophan residues, *J. Biochem.* 87, 273–279, 1980.
218. Daniel, V.W., III. and Trowbridge, C.G., The effect of *N*-bromosuccinimide upon trypsinogen activation and trypsin catalysis, *Arch. Biochem. Biophys.* 134, 506–514, 1969.
219. Sartin, J.L., Hugli, T.E., and Liao, T.H., Reactivity of the tryptophan residues in bovine pancreatic deoxyribonuclease with *N*-bromosuccinimide, *J. Biol. Chem.* 255, 8633–8637, 1980.
220. Kosman, D.J., Ettinger, M.J., Bereman, R.D., and Giordano, R.S., Role of tryptophan in the spectral and catalytic properties of the copper enzyme, galactose oxidase, *Biochemistry* 16, 1597–1601, 1977.
221. Freisheim, J.H. and Huennekens, F.M., Effect of *N*-bromosuccinimide on dihydrofolate reductase, *Biochemistry* 8, 2271–2276, 1969.
222. Ohnishi, M., Kawagishi, T., and Hiromi, K., Stopped-flow chemical modification with *N*-bromosuccinimide: A good probe for changes in the microenvironment of the Trp 62 residue of chicken egg white lysozyme, *Arch. Biochem. Biophys.* 272, 46–51, 1989.
223. Inokuchi, N., Takahashi, T., Yoshimoto, A., and Irie, M., *N*-Bromosuccinimide oxidation of a glucoamylase from *Aspergillus saitoi*, *J. Biochem.* 91, 1661–1668, 1982.
224. O'Gorman, R.B. and Matthews, K.S., *N*-Bromosuccinimide modification of lac repressor protein, *J. Biol. Chem.* 252, 3565–3571, 1977.

225. Coletti-Previero, M.A., Previero, A., and Zuckerkandl, E., Separation of the proteolytic and esteratic activities of trypsin by reversible structural modification, *J. Mol. Biol.* 39, 493–501, 1969.
226. Previero, A., Coletti-Previero, M.A., and Cavadore, J.C., A reversible chemical modification of the tryptophan residue, *Biochim. Biophys. Acta* 147, 453–461, 1967.
227. Holmgren, A., Reversible chemical modification of the tryptophan residues of thioredoxin from *Escherichia coli* B, *Eur. J. Biochem.* 26, 528–534, 1972.
228. Cooper, H.M., Jemmerson, R., Hunt, D.F. et al., Site-directed chemical modification of horse cytochrome c results in changes in antigenicity due to local and long-range conformation perturbations, *J. Biol. Chem.* 262, 11591–11597, 1987.
229. Koshland, D.E., Jr., Karkhanis, Y.D., and Latham, H.G., An environmentally-sensitive reagent with selectivity for the tryptophan residue in proteins, *J. Am. Chem. Soc.* 86, 1148–1150, 1964.
230. Horton, H.R. and Koshland, D.E., Jr., A highly reactive colored reagent with selectivity for the tryptophanyl residue in proteins, 2-hydroxy-5-nitrobenzyl bromide, *J. Am. Chem. Soc.* 87, 1126–1132, 1965.
231. Clemmer, J.D., Carr, J., Knaff, D.B., and Holwerda, R.A., Modification of laccase tryptophan residues with 2-hydroxy-5-nitrobenzyl bromide, *FEBS Lett.* 91, 346–350, 1978.
232. Barman, T.E. and Koshland, D.E., Jr., A colorimetric procedure for the quantitative determination of tryptophan residues in proteins, *J. Biol. Chem.* 242, 5771–5776, 1967.
233. Novak, R.L., Banerjee, K., Dohnal, J. et al., Inhibition of RNA initiation in *E. coli* core RNA polymerase with 2-hydroxy-5-nitrobenzyl bromide, *Biochem. Biophys. Res. Commun.* 60, 833–837, 1974.
234. Baracca, A., Menegatti, D., Parenti Castelli, G. et al., Does 2-hydroxy-5-nitrobenzyl bromide react with the ε-subunit of the mitochondrial F1-ATPase?, *Biochem. Int.* 21, 1135–1142, 1990.
235. Barman, T.E., The chemistry of the reaction of 2-hydroxy-5-nitrobenzyl bromide with his-32 of α-lactalbumin, *Eur. J. Biochem.* 83, 465–471, 1978.
236. Lundblad, R.L. and Noyes, C.M., Observations of the reaction of 2-hydroxy-5-nitrobenzyl bromide with a peptide-bound tryptophanyl residue, *Anal. Biochem.* 136, 93–100, 1985.
237. Loudon, G.M. and Koshland, D.E., Jr., The chemistry of a reporter group: 2-Hydroxy-5-nitrobenzyl bromide, *J. Biol. Chem.* 245, 2247–2254, 1970.
238. Strohalm, M., Kodíĉek, M., and Pechar, M., Tryptophan modification by 2-hydroxy-5-nitrobenzyl bromide studied by MALDI-TOF mass spectrometry, *Biochem. Biophys. Res. Commun.* 312, 811–816, 2003.
239. Horton, H.R. and Tucker, W.P., Dimethyl (2-hydroxy-5-nitrobenzyl) sulfonium salts. Water-soluble environmentally sensitive protein reagents, *J. Biol. Chem.* 245, 3397–3401, 1970.
240. Tucker, W.P., Wang, J., and Horton, H.R., The reaction of dimethyl (2-hydroxy-5-nitrobenzyl) sulfonium salts with tryptophan ethyl ester, *Arch. Biochem. Biophys.* 144, 730–733, 1971.
241. Peyser, Y.M., Muhlrad, A., and Werber, M.M., Tryptophan-130 is the most reactive tryptophan residue in rabbit skeletal myosin subfragment-1, *FEBS Lett.* 259, 346–348, 1990.
242. Horton, H.R., Kelly, H., and Koshland, D.E., Jr., Environmentally sensitive protein reagents. 2-Methoxy-5-nitrobenzyl bromide, *J. Biol. Chem.* 240, 722–724, 1965.
243. Horton, H.R. and Young, G., 2-Acetoxy-5-nitrobenzyl chloride. A reagent designed to introduce a reporter group near the active site of chymotrypsin, *Biochim. Biophys. Acta* 194, 272–278, 1969.
244. Uhteg, L.C. and Lundblad, R.L., The modification of tryptophan in bovine thrombin, *Biochim. Biophys. Acta* 491, 551–557, 1977.

245. Fontana, A. and Scoffone, E., Sulfenyl halides as modifying reagents for polypeptides and proteins, *Methods Enzymol.* 25, 482–494, 1972.
246. Shechter, Y., Burstein, Y., and Patchornik, A., Sulfenylation of tryptophan-62 in hen egg-white lysozyme, *Biochemistry* 11, 653–660, 1972.
247. Bewley, T.A., Kawauchi, H., and Li, C.H., Comparative studies of the single tryptophan residue in human chorionic somatomammotropin and human pituitary growth hormone, *Biochemistry* 11, 4179–4187, 1972.
248. Wilchek, M. and Miron, T., The conversion of tryptophan to 2-thioltryptophan in peptides and proteins, *Biochem. Biophys. Res. Commun.* 47, 1015–1020, 1972.
249. Chersi, A. and Zito, R., Isolation of tryptophan-containing peptides by adsorption chromatography, *Anal. Biochem.* 73, 471–476, 1976.
250. Rubinstein, M., Schechter, Y., and Patchornik, A., Covalent chromatography—The isolation of tryptophanyl containing peptides by novel polymeric reagents, *Biochem. Biophys. Res. Commun.* 70, 1257–1263, 1976.
251. Mollier, P., Chwetzoff, S., Bouet, F., Harvey, A.L., and Ménez, A., Tryptophan 110, a residue involved in the toxic activity but not in the enzymatic activity of notexin, *Eur. J. Biochem.* 185, 263–270, 1989.
252. Kuyama, H. et al., An approach to quantitative proteome analysis by labeling tryptophan residues, *Rapid Commun. Mass Spectrom.* 17, 1642–1650, 2003.
253. Hansen, K.C. et al., Mass spectrometric analysis of protein mixtures at low levels using cleavable 13C-isotope-coded affinity tag and multidimensional chromatography, *Mol. Cell. Proteomics* 2, 299–314, 2003.
254. Dai, Z., Nemet, I., Shen, W. et al., Isolation, purification and characterization of histidino-threosidine, a novel Maillard reaction protein crosslink from threose, lysine, and histidine, *Arch. Biochem. Biophys.* 463, 78–88, 2007.
255. Kundinger, M.M., Zabala-Díaz, I.B., Chalova, V.I. et al., Effects of Maillard reaction products on hi1A expression in *Salmonella typhimurium*, *J. Food Sci.* 73, M32–M35, 2008.
256. Mogol, B.A., Yildirim, A., and Gökmen, V., Inhibition of enzymatic browning in actual food systems by the Maillard reaction products, *J. Sci. Food Agric.* 90, 2556–2562, 2010.
257. Tan, B.K. and Harris, N.D., Maillard reaction products inhibit polyphenol oxidase, *Food Chem.* 53, 267–273, 1995.
258. Lee, M.K. and Park, I., Inhibition of potato polyphenol oxidase by Maillard reaction product, *Food Chem.* 91, 57–61, 2004.
259. Altrooz, O.M., The effects of Maillard reaction products on apple and potato polyphenol oxidase and their antioxidant activity, *Int. J. Food Sci. Technol.* 43, 490–494, 2008.

5 Modification of Sulfur-Containing Amino Acids in Proteins

Methionine and cysteine/cysteine (Figure 5.1) are among the less commonly occurring amino acids in proteins (cysteine, 1.8%; methionine, 2.6%).[1] Cysteine can be present as cysteine or as a disulfide (occasionally as trisulfide);[2] the low level of methionine was useful in obtaining polypeptide fragments via cyanogen bromide cleavage.[3] The ability of methionine to function as a nucleophile at low pH allows for selective modification with reagents such as iodoacetate under acidic conditions.[4]

Cysteine is the sulfur analogue of serine where the hydroxyl group is replaced with a sulfhydryl group. The bond dissociation energy for sulfhydryl groups is substantially less than that of the corresponding alcohol function providing a basis for the increased acidity of sulfhydryl groups; the pK_a for ethanethiol is approximately 10.6 while the pK_a for ethanol is approximately 18.[5,6] Hederos and Baltzer[7] studied nucleophilic selectivity in an acyl transfer reaction (acyl transfer of *S*-glutathionyl benzoate catalyzed by an engineered human glutathione transferase A1–1); ethanethiol was far more effective than ethanol while ethylamine was ineffective in stimulating the acyl transfer reaction; it is noted that ethylamine would be fully protonated at the pH of the reaction (pH 7.0). Given this information, it is not surprising that while the reaction of cysteine with haloacetates or haloacetamides such as chloroacetate or chloroacetamide is slow reaction with serine is essentially nonexistent. However, as with serine (see Chapter 3), the presence of cysteine at the enzyme active site such as in papain greatly influences reactivity. The reaction of chloroacetic acid with a cysteine residue at the active site of papain is some 30,000 times faster (150 M^{-1} min^{-1}) than that of free cysteine (5.3×10^{-3} M^{-1} min^{-1}) at pH 6.0.[8] As a comparison, the rate of reaction of the cyanate with the cysteine residue at the active site of papain is 9.4×10^3 M^{-1} min^{-1} compared to the rate of reaction of cysteine (3.4 M^{-1} min^{-1})[9]; thus rate of reaction of cyanate with the cysteine at the enzyme active site is some 3000-fold more rapid than reaction with free cysteine. Stark and coworkers[10] found a rate of reaction of 4.0 M^{-1} min^{-1} for the reaction of cyanate with free cysteine. Unlike the reaction of chloroacetic acid with cysteine, the reaction of cyanate with the active site cysteine of papain is reversible.[8] The reader is referred to an excellent review by Janssen[11] that addresses the nucleophilicity of organic sulfur where he notes that thiolates are less basic than alkoxide anions but in general are better nucleophiles. Solvent does have an effect on the relative nucleophilicity of sulfur and oxygen compounds.[12] This is shown by changes in the reactivity of halo acids depending on solvent as discussed later. The reader is directed to an excellent

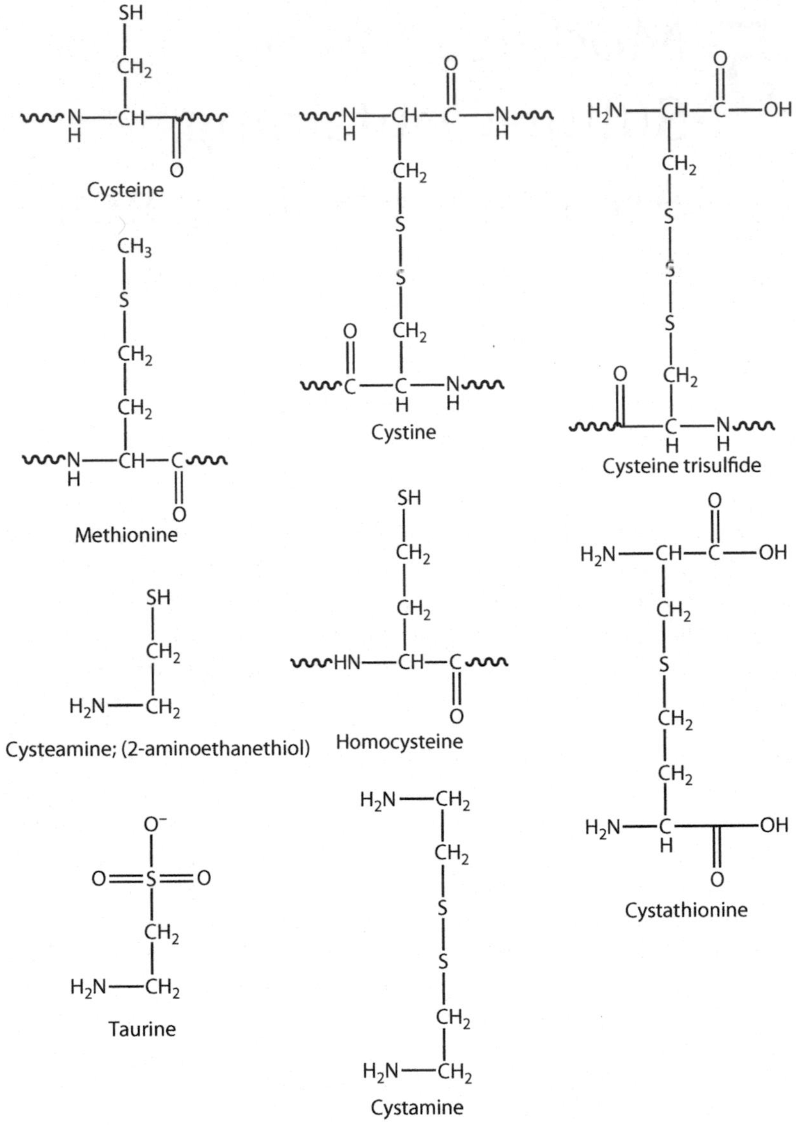

FIGURE 5.1 Structures of sulfur-containing amino acids. Cysteine, its oxidation product, cystine, and methionine are present in proteins and intracellular and extracellular fluids. Cysteine trisulfide, cystathionine, and homocysteine are found in trace quantities in proteins and extracellular fluids; cysteine trisulfide is extremely uncommon. Homocysteine is considered to be a biomarker for various cardiovascular diseases (Newton, L.A., Sandhu, K., Livingstone, C. et al., Clinical diagnostics for homocysteine: A rogue amino acid? *Expert Rev. Mol. Diagn.* 10, 489–500, 2010). Cystamine, cysteamine, and taurine are other derivative products. Taurine is formed by the oxidation of cysteine via cystine sulfenic acid as an intermediate. Taurine is considered important for a variety of physiological functions (Wójcik, O.P., Koenig, K.L., Zeleniuch-Jacquotte, A. et al., The potential protective effects of taurine on coronary heart disease, *Atherosclerosis* 208, 19–25, 2010).

discussion on the effect of solvent on organic chemical reactions;[13] there is also a discussion of solvent effects on reactions in Chapter 1 with emphasis on the effect of solvent on transition states; too often there is tendency to forget that proteins (and other biological polymers) are complex organic polymers, albeit with unique patterns of reactivity. The chemistry of cysteine has been reviewed[14–20] and the reader is directed to these articles for more information.

Thioesters (Figure 5.2) are more reactive than oxyesters and, in the case of biological thioesters such as acetyl-coenzyme A (acetyl-CoA), are high energy compounds.[21] The increased electrophilic nature of the carbonyl carbon in a thioester makes it far more susceptible to nucleophilic attack such as that seen in the thiolactone structure at the metastable binding sites in α_2-macroglobulin and the C3 and C4 components of complement.[22,23] Thioester intermediates are also involved in the process of ubiquitin-like modifications.[24] Homocysteine thiolactone can modify several different functional groups in proteins including lysine[25] resulting in pathologies.[26,27] Betaine thioester can be used to modify proteins to improve solubility.[28] This later study also emphasizes the importance of thioesters in native chemical ligation.[29] There is evidence to show that the cysteinyl residue that is involved in the trans-esterification step in the formation of the thioester intermediate in protein splicing has a low pK_a (5.8).[23,30] It is noted that a reactive histidine has a role in enhancing the nucleophilicity of the cysteine residue[31] in protein splicing.[24]

The chemical modification of cysteine residues can mostly proceed via a nucleophilic addition or displacement reaction with the thiolate anion as the nucleophile (Figure 5.3). The reaction with the α-ketohaloalkyl compounds such as iodoacetate is an example of a nucleophilic displacement reaction (S_N2 reaction) while the reaction of maleimide is a nucleophilic addition to an olefin (Michael reaction). Disulfide and disulfide-like compounds such as the methyl thiosulfonates react with cysteine to form mixed disulfides. The haloalkyl compounds, maleimides, and alkyl alkanethiosulfonates are also base structures for more complex molecules such as spectral probes and crosslinking agents, which are discussed elsewhere in the text. Metal ions such as mercuric can react with cysteine to form a presumably covalent derivative while zinc ions form a less stable linkage. Gold reacts with cysteine to form a covalent bond. Cysteine residues are also subject to oxidation (modification by reactive oxygen species) and nitrosylation (Figure 5.4). While these reactions are not generally used for the site-specific modification of cysteine, they are included as they are of significant biological interest.[32–41] The reversible modification of cysteine by oxidation or nitrosylation is considered an important process in redox regulation.[38,42] The review by Spadaro and coworkers[39] is of considerable interest as they argue that the specificity of modification by oxidation or nitrosylation is driven by the reactivity of the cysteine residues; cysteine residues with low pK_a values are targets for modification by oxidation or nitrosylation with the reversible formation of nitrosocysteine and cysteine sulfenic acid providing the basis for specific regulation. Sulfinic acid may also have a regulatory role as there are enzymes for the reduction of this derivative. The following sections will discuss the general chemistry of cysteine and reaction of cysteine with various chemical reagents.

The reaction of cysteine residues with chemical reagents depends on two major factors: physical accessibility and local electrostatic environment. It is, however, a bit

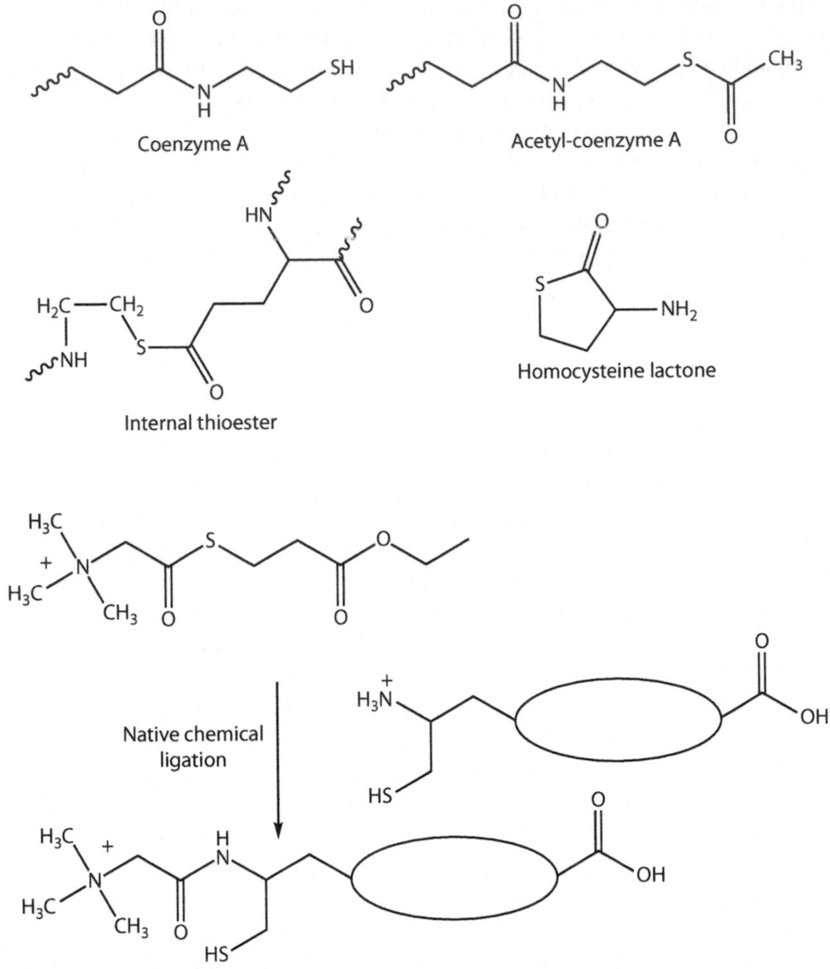

Coenzyme A

Acetyl-coenzyme A

Internal thioester

Homocysteine lactone

Native chemical ligation

FIGURE 5.2 Examples of thioesters important in protein chemistry. Coenzyme A is the basis for a number of biologically active thioesters including acetyl-coenzyme A which is involved in a large number of biological processes. Not shown is palmitoyl-coenzyme A (Linder, M.E. and Deschenes, R.J., Palmitoylation: Policing protein stability and traffic, *Nat. Rev. Cell. Mol. Biol.* 8, 74–84, 2007) which is important in membrane protein function (Wei, Y., Di Vizio, D.K., Steeen, H., and Freeman, M.R., Peptide and protein thioester synthesis by N,S acyl transfer, *Mol. Cell. Proteomics* 9, 54–70, 2010). Homocysteine lactone is an internal thioester which has the potential of modifying different functional groups in proteins (Jop, E.C.A. and Bakhtiar, R., Homocysteine thiolactone and protein homocysteinylation: Mechanistic studies with model peptides and proteins, *Rapid Commun. Mass Sp.* 16, 1049–1053, 2002). Betaine thioester is used to increase the solubility of proteins via coupling to the amino terminal via native chemical ligation (Xiao, J., Burn, A., and Tolbert, T.J., Increasing solubility of proteins and peptides by site-specific modification with betaine, *Bioconj. Chem.* 19, 1113–1118, 2008).

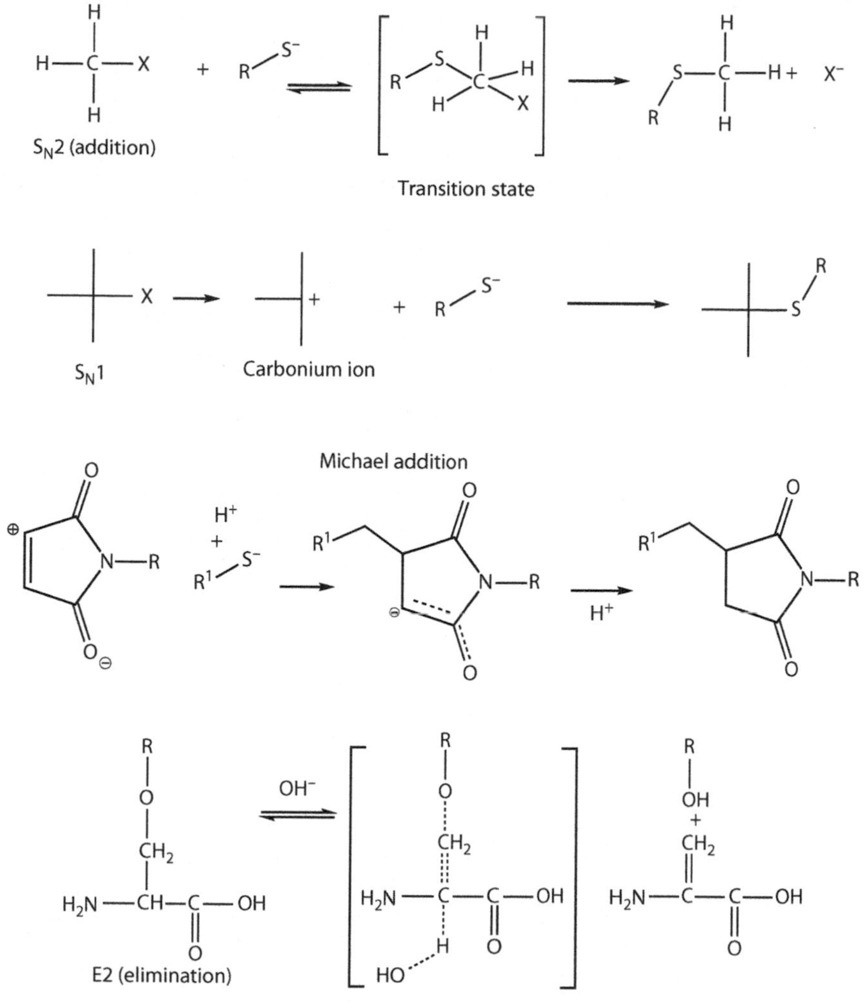

S_N2 (addition)

Transition state

S_N1

Carbonium ion

Michael addition

E2 (elimination)

FIGURE 5.3 S_N1 and S_N2 reactions in the modification of thiol groups in proteins. S_N1 reactions do not retain configuration and there is racemization of product. S_N2 reactions occur with an inversion of configuration. This is usually, but not always, not of significance with reaction of proteins nucleophiles. The alkylation reactions using iodoacetate and iodo-actamide are examples of the S_N2 reaction. The Michael addition is a conjugate addition reaction where a nucleophile adds to α, β-unsaturated carbon; modification of cysteine with *N*-ethylmaleimide is an example of a Michael addition. Also shown is the E2 elimination reaction which can result in the formation of β-alanine residues from serine or cysteine derivatives in proteins (Herbert, B., Hopwood, F., Oxley, D. et al., β-Elimination: An unexpected artifact in proteome analysis, *Proteomics* 3, 826–831, 2003). The reader is directed to an organic chemistry text such as Roberts (Roberts, J.D. and Caserio, M.C., *Basic Principles of Organic Chemistry*, p. 548, W.A. Benjamin, New York, 1964) for further information on organic reaction mechanisms.

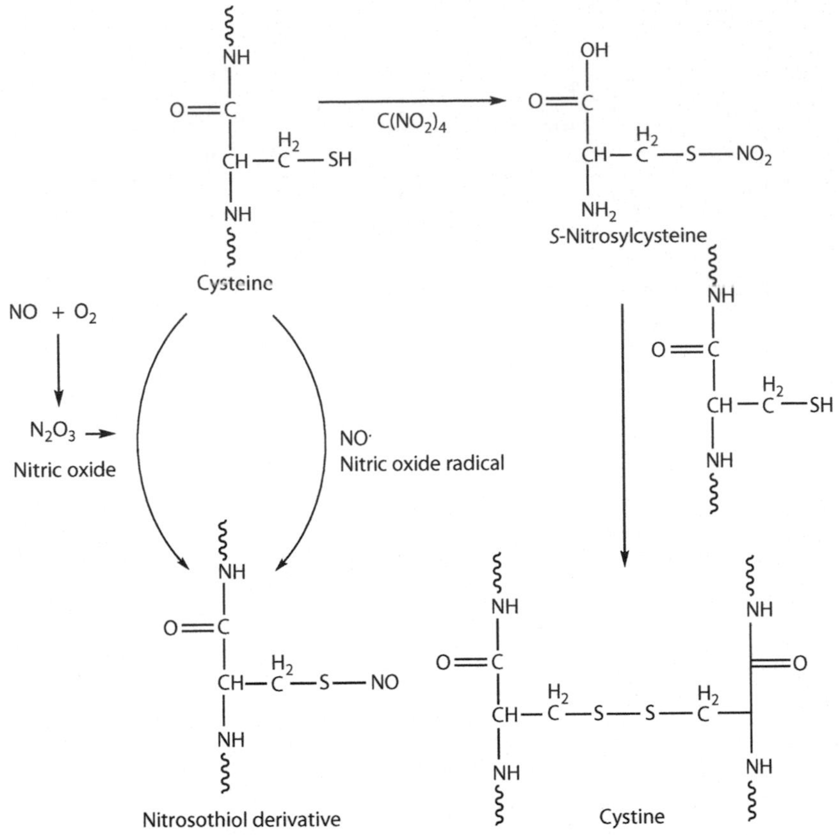

FIGURE 5.4 The nitrosylation of thiols in biological polymers. Active nitrogen compounds such as nitric oxide and tetranitromethane. Cysteinyl residues in proteins can be modified with tetranitromethane as well as by reactive nitrogen species (Hess, D.T., Matsumoto, A., Kim, S.O. et al., Protein S-nitrosylation: Purview and parameters, *Nat. Rev. Mol. Cell. Biol.* 6, 150–166, 2005; Torta, F., Usuelli, V., Malgaroli, A., and Bachi, A., Proteomic analysis of protein S-nitrosylation, *Proteomics* 8, 4484–4494, 2008; Nagahara, N., Matsumura, T., Okamoto, R., and Kajihara, Y., Protein cysteine modifications: (1) medical chemistry for proteomics, *Curr. Med. Chem.* 16, 4419–4444, 2009).

difficult to absolutely separate these two variables. The reactivity of cysteine residues of proteins can be classified as freely reacting, sluggish, and masked.[42–48] Ueland and coworkers[42] characterized the reactivity of the 24 sulfhydryl groups in an adenosine 3′,5′-monophosphate-adenosine binding protein from mouse liver. Reaction with 5,5′-dithiobis-(2-nitrobenzoic acid) (DTNB) demonstrated that 4 sulfhydryl groups that reacted with DTNB rapidly (in seconds), 4 sulfhydryl groups that reacted more slowly, and 16 sulfhydryl groups that reacted quite slowly might reflect protein denaturation. Floris and coworkers[46] studied the reactivity of sulfhydryl groups in diamine oxidase with 4,4-dithiopyridine. The two sulfhydryl groups per monomer unit are unreactive in the native enzyme but can be modified by 4,4-dithiopyridine after denaturation in 8 M urea or in the absence of oxygen in the presence of substrate.

Polyanovsky and colleagues[47] identified the exposed and buried cysteine residues (five total residues) in aspartate aminotransferase based on reaction with iodoacetate; two are exposed and the remaining three are "partially buried," buried, and buried only with great difficulty. Fuchs and coworkers[48] studied the reactivity of cysteine in bovine cardiac troponin C using DTNB. There are two cysteine residues in bovine cardiac troponin C. Modification with DTNB occurs more rapidly ($3.37\,M^{-1}\,s^{-1}$, $t_{1/2}$ 25 min) in the presence of calcium ions than in the absence of calcium ions ($1.82\,M^{-1}\,s^{-1}$) while reaction is complete within 2 min in the presence of 6 M urea. Lee and Blaber[49] report that the solution stability of fibroblast growth factor is improved by mutagenesis of buried cysteine residues. Subsequent work[50] from this laboratory suggested that while this residue is conserved in the fibroblast growth factor family, the residue is vestigial. Cysteine can be considered to be a hydrophobic amino acid[51–55] and can be "buried" without major thermodynamic consequences.[52] On the other hand, there are anomalous factors with the reactivity of cysteine in proteins.[52,55]

The relationship of chemical reactivity and whether a residue is "buried" or exposed was the subject of lively discussion some 30 or 40 years ago when the crystal structures of proteins were more a rarity than commonplace. Early protein chemists, as discussed later, equated the exposure of a chemically reactive cysteine residue with the denaturation of a protein. While crystallography or one of the more advanced solution chemical techniques such as NMR usually can assign the position of a residue in space, there can still be unanswered questions about reactivity. In the case of cysteine, while there are some crystallographic studies which demonstrate exposure of cysteine residues, the great majority of studies on cysteine exposure are based on availability for chemical modification.[56–58] There are a few studies which correlate chemical reactivity and crystallographic data.[59] Notwithstanding the lack of extensive crystallographic data supporting cysteine exposure, cysteine residues have been engineered into proteins to study conformation using chemical reactivity to measure accessiblity.[60–65]

The early work on the appearance of sulfhydryl groups on denaturation of proteins was summarized by Greenstein in 1939[66] who also noted the importance of the counterion in the use of guanidine salts. The exposure of cysteine occurring during protein denaturation was observed some 100 years ago but not associated with denaturation until some years later[67] although it is clear that understanding of protein denaturation was an evolving art.[68] What is clear is that early investigators considered native proteins devoid of sulfhydryl groups[69] and composition is usually reported as cystine.[70] These early studies used reaction with porphyridin or nitroprusside (Figure 5.5) to detect sulfhydryl groups. Egg albumin, for example, did not react with porphyridin or nitroprusside with in the native state but did react after denaturation or digestion with pepsin. Anson in 1940[71] reported that cysteine in native egg albumin, while unreactive with nitroprusside or porphyridin, could be modified with iodine or iodoacetamide. The reaction with iodine was performed as acid pH to prevent reaction with tyrosine. In a subsequent publication,[72] Anson demonstrated that *p*-mercuribenzoate reacted with denatured guanidine hydrochloride, but not native egg albumin. Bull[73] reviewed the concept of buried and exposed sulfhydryl groups in 1941.

Local electrostatic environment (see Chapter 1) has a profound effect on the reactivity of cysteine residues in proteins. Britto and colleagues[74] used chloroacetamide to

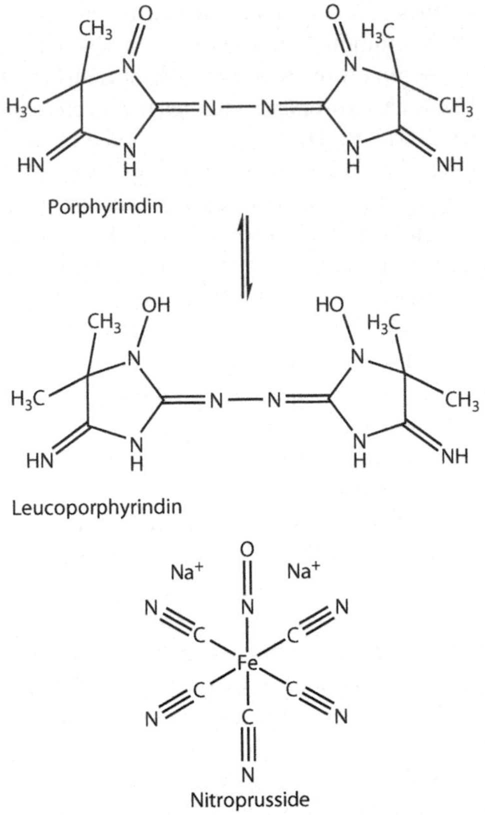

Porphyrindin

Leucoporphyrindin

Nitroprusside

FIGURE 5.5 The structures of porphyrindin and nitroprusside. These two reagents were used for the determination of sulfhydryl groups. See Greenstein, J.P. and Jenrette, W.V., Reactivity of porphyrindin in the presence of denatured proteins, *J. Biol. Chem.* 142, 175–180, 1942; Christen, P. and Gasser, A., Visual detection of fructose-1,6-diphosphate aldolase after electrophoresis by its oxidative paracatalytic reaction, *Anal. Biochem.* 109, 270–272, 1980 for more information porphyrindin and Johnson, M.D. and Wilkins, R.G., Kinetics of the primary interaction of pentacyanonitrosylferrate (2-) (nitroprusside) with aliphatic thiols, *Inorg. Chem.* 23, 231–235, 1984 for nitroprusside.

identify a single highly reactive cysteine residue in tubulin. This study also demonstrated the importance of local electrostatic potential in the modulation of the reactivity of individual cysteine residue in rat brain tubulin. Rat brain tubulin dimer contains 20 cysteine residues; 12 residues in the α-subunit and 8 in the β-subunit. The reagents evaluated included *syn*-monobromobimane, *N*-ethylmaleimide, iodoacetamide, and [5-((((2-iodoacetyl) amino) ethyl)amino) naphthalene-1-sulfonic acid] AEDANS. Reaction is slower with iodoacetamide than with *N*-ethylmaleimide; a greater number of cysteine residues are modified with *N*-ethylmaleimide than with iodoacetamide and the difference in the rates of reaction is ascribed to the differences in the chemistry of the reaction of the two compounds with the thiolate ion with the reaction with iodoacetamide being a nucleophilic displacement while the reaction with *N*-ethylmaleimide is an addition reaction. Local electrostatic potential is suggested to be responsible for the reactivity of the cysteine residues; reactive cysteinyl residues are close to arginine and/or lysine residues while unreactive cysteinyl residues are associated with negatively charged local environments. Parente and colleagues[75] suggested that the increased reactivity of cysteine residues in seminal ribonuclease at neutral pH was dependent on the proximity of a positively charged lysine residue. Heitmann[76] provided insight into the importance of local environmental factors on the reactivity of cysteine residues. The *N*-dodecanoyl derivative of cysteine was prepared and inserted into micelles and modified with either chloroacetamide, iodoacetamide, or *p*-nitrophenyl acetate. Incorporation into a cationic micelle (hexadecyltrimethylammonium bromide) increased the rate of reaction with any of the three reagents with an apparent decrease in the pK_a of the cysteine derivative suggesting the importance of an adjacent positively charged residue on enhancing cysteine reactivity. Incorporation of dodecanoyl cysteine into micelles of *N*-dodecanoyl glycine resulted in inhibition of the rate of reaction while a neutral micelle (Brij 35) has little effect. This is not due to charge repulsion as neutral reagents, chloroacetamide, iodoacetamide, and *p*-nitrophenyl acetate, were used in this study. Gitler and coworkers[77] reviewed the enhancement of the reactivity of cysteine residues by cationic detergent such as cetyltrimethylammonium bromide. Lutolf and coworkers[78] studied the rate of reaction of pentapeptide thiols with PEG-diacrylates. The pentapeptides contained arginine residues or aspartic acid residues or one aspartic and one arginine residues such that the peptides had net charges at pH 7.4 of −2, −1, 0, +1, or +2; there was variance in the placement of the charged amino acids. Net charge was the dominant factor with positively charged peptides reacting more rapidly than neutral or negatively charged peptides; the pK_a of the cysteine was lowest in the doubly charged peptide ($pK_a = 8.12$) and highest in the doubly negatively charged peptide ($pK_a = 8.93$) while the pK_a was 8.50 in the neutral peptide. Snyder and coworkers[79] studied the reaction of DTNB (Ellman's reagent, see the following) with cysteine in peptides which contained either two positively charged neighboring amino acid residues, one positively charged residue, or a neutral residue. The rate observed with the cysteine having two positively charged residues was $132{,}000\,M^{-1}\,s^{-1}$ while the rate with only neutral residues was $367\,M^{-1}\,s^{-1}$. The presence of a positively charged residue also influences the reactivity of a methionine residue as discussed later.

- Physical accessibility is a factor in the reactivity of sulfhydryl groups in proteins.
- Proximity to a positively charged residue such as lysine enhances cysteine reactivity while a negatively charged residues such as aspartic acid decreases reactivity.

Oxidation is not used as frequently for the modification of cysteine residues in proteins although increased interest in cysteine sulfenic acid has generated interest in the use of mild oxidizing agents for protein modification. It is noted that oxidation of cysteine to cysteic acid is usually required for measurement of total cysteine (cysteine + cystine) by amino acid analysis as Cys/2.[80] Oxidation of cysteine residues can occur in proteins during processing as a result of oxygen in aqueous solvents; the presence of metal ions such as copper or iron can greatly accelerate the oxidative processes. A variety of products can be obtained from the oxidation of cysteine (Figure 5.6).

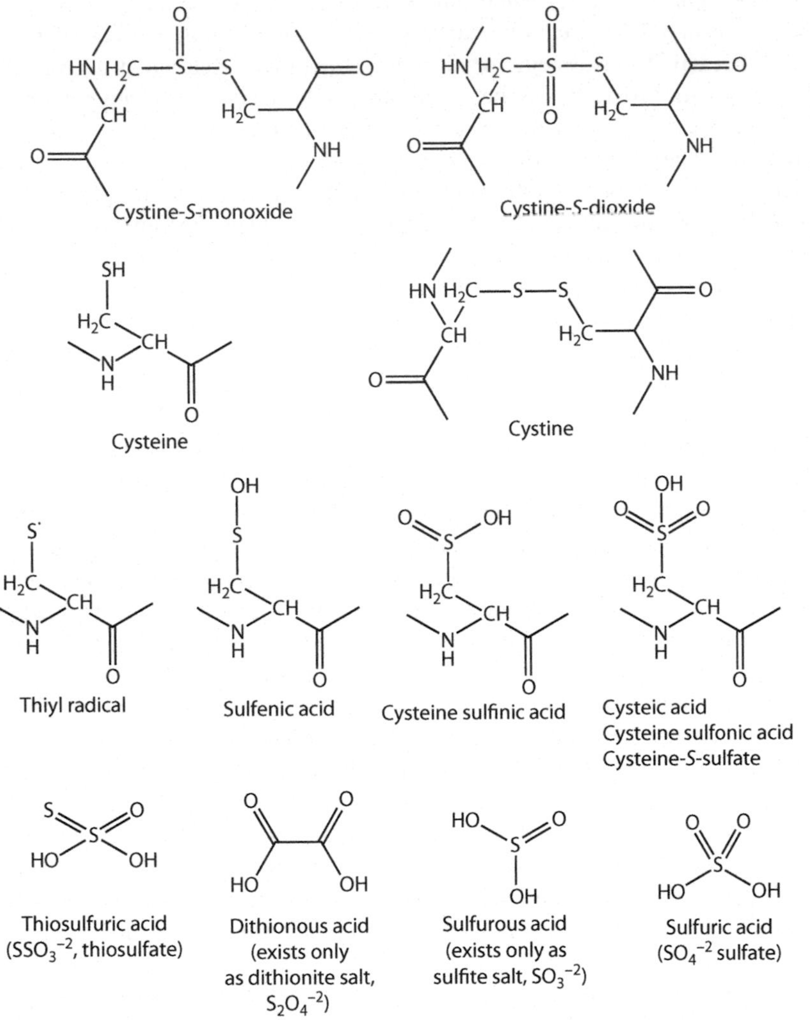

Cystine-S-monoxide

Cystine-S-dioxide

Cysteine

Cystine

Thiyl radical

Sulfenic acid

Cysteine sulfinic acid

Cysteic acid
Cysteine sulfonic acid
Cysteine-S-sulfate

Thiosulfuric acid
(SSO_3^{-2}, thiosulfate)

Dithionous acid
(exists only
as dithionite salt,
$S_2O_4^{-2}$)

Sulfurous acid
(exists only as
sulfite salt, SO_3^{-2})

Sulfuric acid
(SO_4^{-2} sulfate)

FIGURE 5.6 The oxidation products of cysteine and sulfur. Shown here are the various oxidation products of cysteine. Also shown are several relevant oxidation products of sulfur. (See Wallins, J.D., Chemistry of sulfur, in *Biological Interactions of Sulfur Compounds*, S. Mitchell (ed.), Taylor & Francis, London, U.K., Chapter 1, pp. 1–19, 1996.) Information on thiyl radicals which are not discussed in the text can be found in Wardman, P., Reactions of thiyl radicals, in *Biothiols in Health and Disease,* L. Packer and E. Cadenas (eds.), Marcel Dekker, New York, Chapter 1, pp. 1–19, 1995; Jacob, C., Knight, I., and Winyard, P.G., Aspects of the biological redox chemistry of cysteine: From simple redox responses to sophisticated signalling pathways, *Biol. Chem.* 387, 1385–1397, 2006; Mozziconacci, O., Kerwin, B.A., and Schöneich, C., Reversible hydrogen transfer between cysteine thiyl radical and glycine and alanine in model peptides: Covalent H/D exchange, radical-radical reactions, and L- to D-Ala conversion, *J. Phys. Chem. B* 114, 6752–6762, 2010.

Early understanding of the function of cysteine and cystine in proteins suggested that cysteine was found only in intracellular proteins and disulfide bonds found in extracellular proteins.[81] While there is a definite difference between intracellular proteins and extracellular proteins,[82] the difference is not absolute in that while intracellular proteins do not contain disulfide bonds, some 10% of extracellular proteins contain sulfhydryl groups which, if exposed, would be susceptible to oxidation. Bessette and coworkers[83]noted that cytoplasms (intracellular) have a reducing environment while the extracellular environment is characterized by an oxidizing environment; it is possible to make the cytoplasm of *Escherichia coli* sufficiently oxidizing to promote disulfide formation without compromising cell viability; the ability to manipulate the cytoplasmic redox environment is of great value for the use of microorganisms to produce properly folded therapeutic proteins.[84–86]

Cysteine sulfenic acid is the proximate product in the oxidation of cysteine (Figure 5.7). While organic sulfenic acids are unstable in solution,[87] cysteine sulfenic acid in proteins is somewhat stable although there is tendency to move to high oxidation states (cysteine sulfinic acid) or to form mixed disulfides.[87,88] It is noted that, of all the cysteine derivatives, only cysteine sulfenic acid has not been isolated.[89,90] The presence of cysteine sulfenic acid in proteins has been demonstrated by mass spectrometry[91] and there are a variety of assay techniques which are available.[92] The most common assays for cysteine sulfenic acid use either reaction with dimedone[93] or thionitrobenzoic acid (Figure 5.7).[94] Other novel reagents have been developed.[95] Cysteine sulfenic acid can be reduced to cysteine with hydrogen sulfide or with dithiothreitol (DTT).[96] In these latter studies,[96] the physical recovery of cysteine was not demonstrated but rather the recovery of peroxidase activity from enzyme which had been inactivated with hydrogen peroxide or peroxynitrite. A novel finding was the formation of a mixed disulfide between the oxidized enzyme and 2-mercaptoethanol. The formation of a mixed disulfide between protein-bound sulfenic acid and glutathione is an important feature of redox regulation.[97,98] Oxidation of a cysteine to, for example, cysteine sulfenic acid, would likely preclude reaction with alkylating agents such as *N*-ethylmaleimide and would likely interfere with metal ion binding. The formation of cysteine sulfenic acid by mild oxidation (and the ease of reversibility) is important in redox regulation[39] as discussed earlier. Cysteine sulfenic acid is important in the formation of mixed disulfide as discussed later.

While cysteine sulfenic acid as such is unstable (see above) this modified amino acid is found in a variety of proteins.[87,97,99] One particular striking example is glyceraldehyde-3-phosphate dehydrogenase where oxidation of the active site cysteine to cysteine sulfenic acid converts the dehydrogenase activity to acyl phosphatase activity.[100–102] Albumin is a protein which contains a single free sulfhydryl group which can be found as the sulfenic acid derivative.[92,103–105] Cysteine sulfenic acid has also been found in transerythrin.[92] It is somewhat surprising that the work of Le Gaillard and Dautrevaux[106] on transcortin (corticosteroid binding globulin) found in human blood plasma has not received greater consideration. Transcortin contains two cysteine residues which are not present in a disulfide bond.[107] Le Gaillard and Dautrevaux[106] reported that one of the two cysteine residues in transcortin appeared to be oxidized to cysteine sulfenic acid. The presence of an oxidized sulfhydryl group in transcortin that could be reduced with DTT was

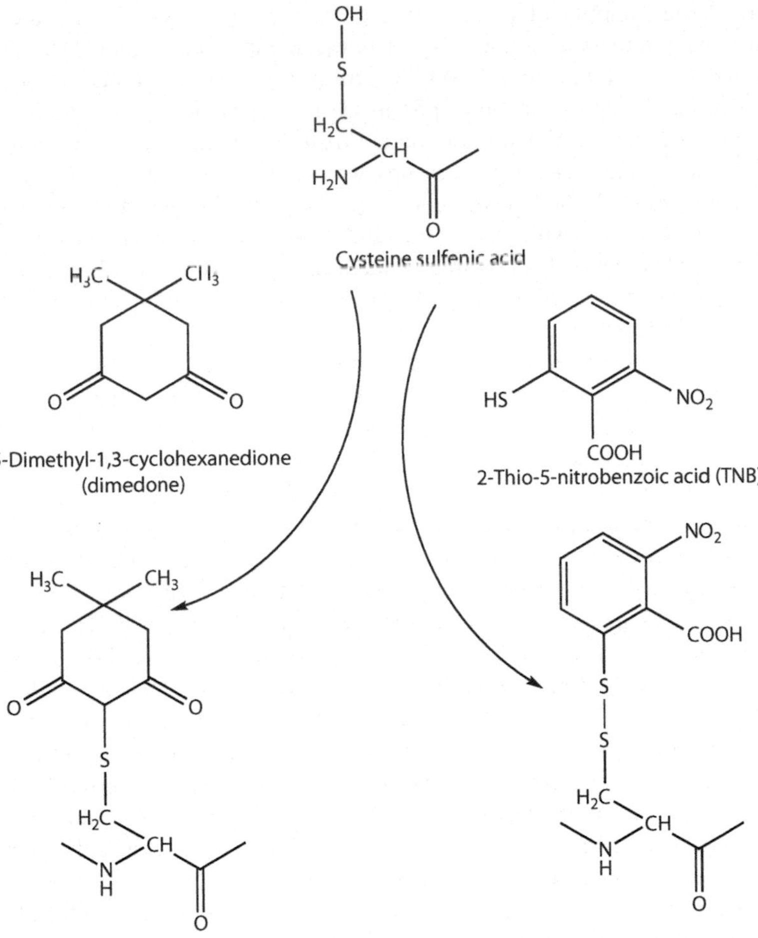

Cysteine sulfenic acid

5,5-Dimethyl-1,3-cyclohexanedione (dimedone)

2-Thio-5-nitrobenzoic acid (TNB)

FIGURE 5.7 Cysteine sulfenic acid and reaction with dimedone and 2-thio-5-nitrobenzoic acid. (See Rehder, D.S. and Borges, C.R., Possibilities and pitfalls in quantifying the extent of cysteine sulfenic acid modification of specific proteins within complex biofluids, *BMC Biochem.* 11, 25, 2010.)

confirmed subsequently by Defaye and coworkers.[108] Later work showed that one cysteine, Cys60, was important in binding steroid and may be the residue oxidized to cysteine sulfenic acid.[109] However, the role of the cysteine and the importance of oxidation is not clear as inhibition with sulfhydryl reagents appears to depend on prior treatment with reducing agent. Reaction of the isolated wild type protein with *N*-ethylmaleimide resulted in 40% inhibition of binding; pretreatment with DTT followed by *N*-ethylmaleimide resulted in 80% inhibition. Similar results were obtained for modification with tetrathionate. It was somewhat surprising (see earlier comment on reactivity of cysteine sulfenic acid) that inactivation of steroid binding with iodoacetate did not require pretreatment with DTT suggesting that cysteine sulfenic acid can react with iodoacetate.

An upsurge in recent interest in cysteine sulfenic acid has resulted from its role in redox regulation.[110–113] Finally, Nimmo and coworkers[114] used cysteine sulfenyl thiocyanate to synthesize unsymmetric disulfides with thiols. Metal ions can also influence the reactivity of cysteine residues in proteins either by influencing conformation and, therefore, the local electrostatic environment or by direct interaction with the sulfhydryl group such as that described below for zinc or mercury. Metal ions such as copper or iron promote the oxidation of cysteine as described earlier but such effect is separate from that described here. While there is evidence that metal ions can influence the chemical modification of proteins,[115,116] a careful examination of the literature failed to find many examples where the metal ion effect could be absolutely ascribed to an effect on conformation. There is considerable evidence to suggest that protein conformation can influence sulfhydryl group reactivity[117–123] but interpretation can be complicated by conformational changes which occur secondary to chemical modification.[124–127] More germane to the current discussion is the direct effect of metal ions on the chemical modification of cysteine.

Metal ions such as gold and mercury can form covalent bonds with cysteine. Mercury has been the subject of considerable investigation as it was one of the earliest described inhibitors of enzyme activity.[128] The sensitivity of enzymes to inhibition by mercuric salts proves useful as an assay for environmental mercury.[129] While as discussed later, Hg^{2+} ions bind most tightly to cysteine, binding can occur with considerable avidity to other sites in a protein.[130–132] Brouwer and coworkers[133] reported an association constant of $10^{33.9}$ for binding mercuric ion to hemocyanin; the mercuric ion bound to cysteine was not removed by EDTA. Binding also occurs at tryptophan (1:1 complex with indole ring) with an association constant of 10^{15}; binding occurs at a third class site and this binding is associated with mercury-induced self association. Aggregation or self-association is also observed in hemoglobin[132] at mercuric ion concentrations considerably higher than those associated with binding to the single sulfhydryl group. Carvalho and colleagues[134] reported that all five thiols in reduced human thioredoxin were blocked (loss of reaction with DTNB) on reaction with either $Hg(Cl)_2$ or Methylmercury (MeHg). Fully reduced thioredoxin treated with MeHg contains approximately 5.0 mol Hg mol^{-1} of thioredoxin while reduced thioredoxin treated with $Hg(Cl)_2$ contained approximately 2.5 mol Hg mol^{-1} of thioredoxin; the Hg^{2+}-treated protein also showed the presence of dimers. The use of thioredoxin from *E. coli* which contains only two active site cysteine residues did not show the presence of dimer on reaction with Hg^{2+} ions. Addition of DTT did not remove Hg^{2+} ions nor did DTT reverse dimer formation.[135] MerP is a periplasmic protein in bacteria which participates in mercury detoxification by reducing Hg(II) to Hg(0). MerP contains two cysteine residues which can be present as a disulfide or as free sulfhydryl forms.[136] One of the cysteine residues is buried (Cys17) and has a low pK_a (5.5) while the other is surface exposed (Cys14) with a normal pK_a (9.2).[137] Despite the low pK_a, Cys14 reacts much more rapidly with iodoacetamide at pH 7.5; this difference is thought to reflect the physical inaccessibility of Cys17. The binding of Hg(II) occurs with the reduced form of merP forming a mercuric ion bridge between the two cysteine residues (Figure 5.8). A similar S-Hg-S bridge was used to replace an intrachain disulfide bond in the constant fragment of immunoglobulin light chain.[138] The S-Hg-S derivative was prepared by the combination of mercuric ions and the reduced C_l fragment.

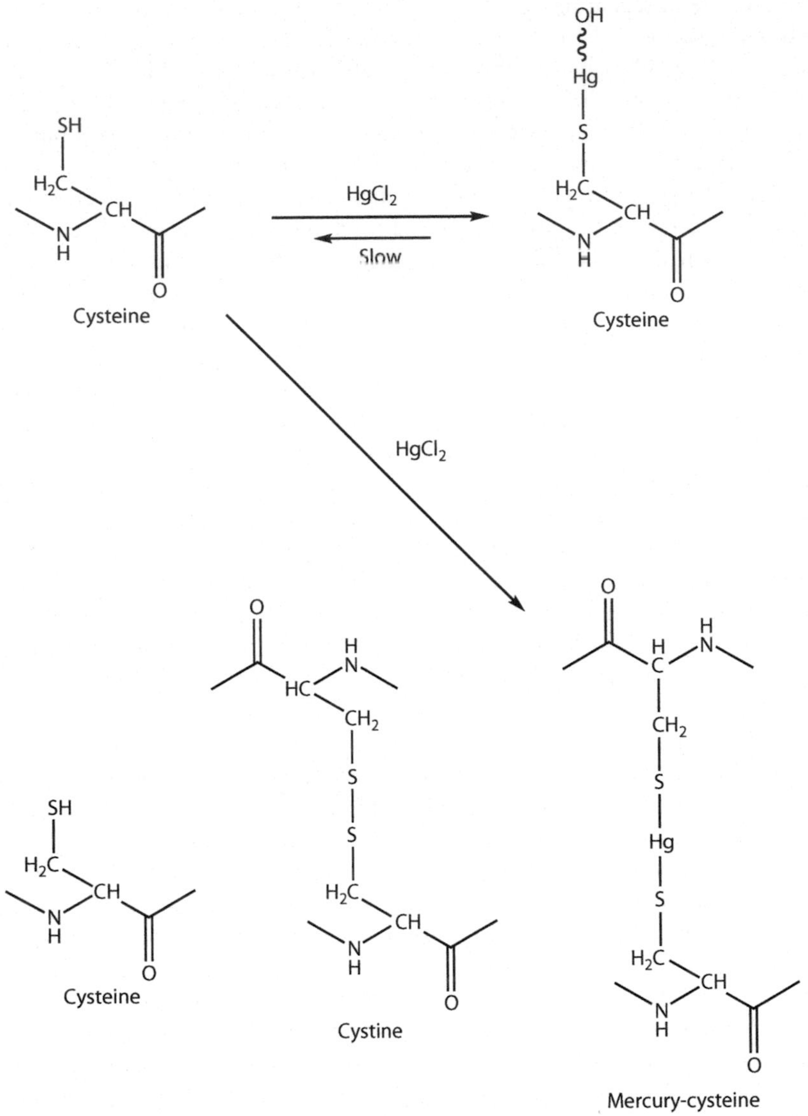

FIGURE 5.8 The binding of mercury to cysteine in proteins. A single cysteine residue reacts with cysteine to form a somewhat unstable derivative. When a second sulfhydryl group is available, a mercuric bridge is formed between two cysteine residues with a very stable linkage. This reaction is essentially irreversible and can be used for crosslinking proteins (Soskine, M., Steiner-Mordoch, S., and Schuldiner, S., Crosslinking of membrane-embedded cysteines reveals contact points in the EmrE oligomer, *Proc. Natl. Acad. Sci.* 99, 12043–12048, 2002). Also see Powlowski, J. and Sahlman, L., Reactivity of the two essential cysteine residues of the periplasmic mercuric-binding protein, merP, *J. Biol. Chem.* 274, 33320–33326, 1999; Goto, Y. and Hamaguchi, K., Conformation and stability of the constant fragment of the immunoglobulin light chain containing an intramolecular mercury bridge, *Biochemistry* 25, 2821–2828, 1986.

Marston and Wright[139] described a method for the insertion of Hg^{2+} into the disulfide bonds of proteins. Proteins are reduced in the presence of chaotropic agent and then renatured by passage over a column containing bound Hg^{2+}. The formation of dimers with thioredoxin also likely involves mercuric ion-bridging of cysteine residues. The presence of reducing agents such as DTT influences the binding of Hg^{2+} to NF-κB to DNA while no effect is seen with tris-(2-carboxyethyl)phosphine (TCEP).[130] It is of interest that TCEP can bind metal ions such copper and lead.[140] Sahlman and Skärfstad[141] showed that when one of the two sulfhydryl groups in merP was removed (replaced by serine or alanine), the mutant merP could no longer bind Hg^{2+} in the presence of exogenous thiol (cysteine). Subsequent work from this laboratory[137] suggested that the binding of Hg^{2+} to merP was associated with a conformational change; these investigators also showed the merP bound Hg^{2+} at low pH (pH 4.0) with similar avidity (apparent Kd of 4.7± 1.9 μM) to binding at pH 7.3 (3.7 ± 1.3 μM). Frasco and coworkers[142] have studied the binding of mercuric ion to cholinesterases from various species. When there is a single free sulfhydryl group (*Torpedo californica*) inactivation occurs within an hour at micromolar concentrations of mercuric ions with the reaction being irreversible.[142] The reaction (measured by loss of enzyme activity) required millimolar concentrations of mercuric ions with human butyrylcholinesterase which is also suggested to have an essential sulfhydryl group; the reaction was reversible. It should be noted that variable results have been obtained for the reactivity of cysteine sulfhydryl group in human butyrylcholinesterase where iodoacetate had no effect but inactivation was observed with *S*-mercuric-*N*-dansyl cysteine.[143] Prior reaction of cysteine with mercuric ions usually precludes reaction with another reagent such as *O*-methylisourea.[144] As a further note, it is not unusual for a protein cysteinyl residue in a protein to be modified by mercuric ions and not by other sulfhydryl reagents such as *N*-ethylmaleimide or iodoacetamide.[145,146]

The interaction of mercuric ion with sulfur[147] as well as with free cysteine is complex resulting in a variety of products[147,148] suggesting a variety of interactions including chelate formation involving the unprotonated amino group at alkaline pH.[135] Stable $Hg(Cys)_2$ complexes are formed with an excess of cysteine.[149] Glutathione forms primarily mono [Hg(Glutathione)] and di $[Hg(Glutathione)_2]$[149] but higher order complexes are reported. Hopman and coworkers developed a sulfhydryl-hapten probe to bind to nucleic acids modified by reaction with mercuric ion.[150,151] The reporter group interacting with the mercurated nucleic acid is a glutathione derivative; change of the charge on the probe from negative to positive increased the stability of the derivative. As previously discussed, the presence of positively charged residues near a cysteine residue increases the rate of reaction with alkylating agents and the rate of formation of mixed disulfide suggesting the presence of such positive charges increases the nucleophilicity of the cysteine thiol group.[78,79] Chekmeneva and coworkers[149] reported mercuric ions could form complexes (at stoichiometric or lower ratios) with cysteine or glutathione which has the composition of $Hg(Cys)_2$ or $Hg(GS)_2$ with association constants on the order of 10^{46} or higher (differential pulse voltammetry on rotating Au-disk electrode, electrospray ionization mass-spectrometry, or isothermal calorimetry); at higher ligand to Hg^{2+} ratios, the association constant decreases. This would be consistent with the formation of monosubstituted products. Simpson[152] reported an association constant for Hg^{2+} with cysteine of $10^{20.1}$ while that for methylmercury is $10^{15.9}$.

The various studies cited earlier support the following general statements regarding mercuric ions:

- Mercuric ions, usually as $Hg(Cl)_2$, bind different sites in a proteins (and nucleic acids)
- Mercuric ions bind tightly to cysteine sulfhydryl groups with tightest binding occurring when it is possible to form a disulfide
- Mercuric ions bound in a disulfide cannot be reversed by a reducing agent such as DTT or 2-mercaptoethanol
- Mercuric ions bind cysteine more tightly than do organic mercurials such as methylmercury chloride

Organic mercurials (Figure 5.9) can also be used to modify cysteine sulfhydryl groups. Methylmercury is produced in the environment from mercury.[153] While a major factor in the toxicity of mercury,[154] methylmercury is used only occasionally for the modification of cysteine in proteins. The use of methylmercury was described earlier in the study of the cysteine sulfhydryl groups in thioredoxin[134] as was the reaction of methylmercury with cysteine.[152] Keller and coworkers[155] compared the effect of methylmercury chloride, *p*-chloromercuribenzoate (see below for more comment on this reagent), phenylmercury acetate, and methyl methanethiosulfonate (MMTS) on acetylcholine and vesamicol binding sites in the vesicular acetylcholine transporter of *T. californica*. Modification with either *p*-chloromercuribenzoate or phenylmercury acetate reduced the binding affinity of vesamicol by 10^3 while only a modest sevenfold reduction was observed with either methylmercury chloride or MMTS. Modification with methylmercury chloride but not MMTS eliminated the binding of phenylmercury acetate. These investigators concluded that there are two classes of cysteine sulfhydryl groups within the vesicular acetylcholine receptor; one which reacts only with organic mercurials and the second class which reacts with both the mercurials and thiosulfonates. Subsequent work from this group[156] showed that modification with *p*-chloromercuriphenylsulfonate (Figure 5.9) was slower ($0.11\,min^{-1}$) than with phenylmercury chloride ($1.36\,min^{-1}$). *p*-Chloromercuriphenylsulfonate was used by this group earlier as a membrane-impermeant reagent.[157] *p*-Chloromercuriphenyl-sulfonate was synthesized for Vellick[158] as a more soluble organic mercurial.

Takatera and Watanabe[159] studied the relative ability of several organic mercurials to react with the sulfhydryl groups of chicken ovalbumin. Chicken ovalbumin has four cysteine residues and one disulfide bond. The four sulfhydryl groups vary in reactivity with reagents such as iodoacetate and DTNB.[160,161] Three of the four cysteinyl residues are available for modification with various organic mercurials.[162] Taketara and Watanabe[159] observed that three of the four sulfhydryl groups in chicken ovalbumin could be modified by the various organic mercurials. The most rapid rate of modification (0.3 M sodium phosphate, pH 7.0) was observed with ethylmercuric chloride ($t_{1/2}$, 1 min) with a lower rate with methylmercuric chloride ($t_{1/2}$, 3 min), *p*-chloromercuribenzoate ($t_{1/2}$, 8 min), and *p*-chloromercuribenzenesulfonate ($t_{1/2}$, 50 min).

Methylmercury is very toxic (and relatively insoluble, 0.1 g L^{-1} at 21°C) and seldom used in protein chemistry studies with the aforementioned work being more the

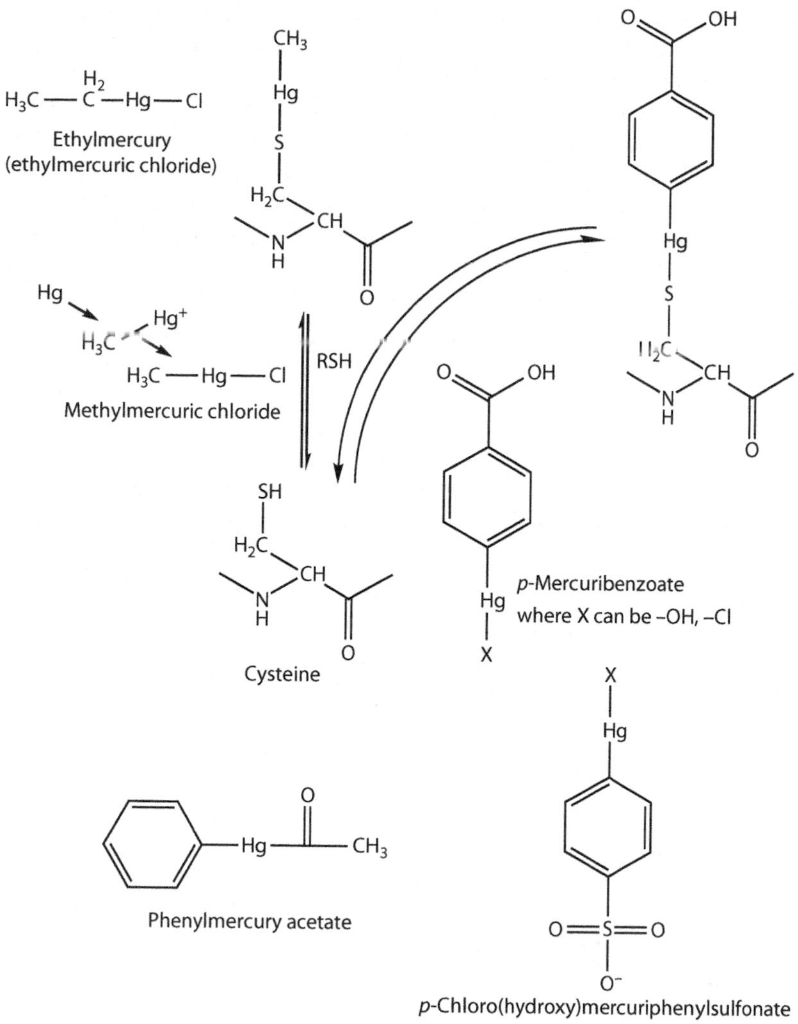

Ethylmercury (ethylmercuric chloride)

Methylmercuric chloride

RSH

Cysteine

p-Mercuribenzoate where X can be –OH, –Cl

Phenylmercury acetate

p-Chloro(hydroxy)mercuriphenylsulfonate

FIGURE 5.9 Organomercurial compounds. The structures and reactions of some organomercurial compounds. Shown are the structures and some reactions of organomercurial compounds. Shown also is the genesis of methylmercury. The reactions of most organic mercurial compounds with cysteine can be reversed by the addition of thiols such as 2-mercaptoethanol or DTT.

exception than the rule. The aromatic mercurials are also toxic and must be handled with considerable caution (*p*-hydroxylmercuribenzoate, CAS 1126-48-3, danger from skin exposure). Methylmercury is stable (3 days) on storage in the dark; exposure to UV light results in rapid decompostion.[163] These investigators also observed the importance of container on storage stability; storage in pyrex or PTFE was useful while storage in polyethylene provided less stability.

Aromatic mercurial derivatives were synthesized and characterized by Whitmore in 1921.[164] Boyer in 1954[165] described the use of organic mercurials in the study of sulfhydryl groups in proteins with emphasis on the use of *p*-chloromercuribenzoate (Figure 5.9). Boyer notes that the designation of *p*-chloromercuribenzoate is generally used in the description of the use of this reagent; however, dependent on the solution, the mercuribenzoate may be combined with hydroxyl or other monovalent anion. It is probably best to refer to the compound as *p*-mercuribenzoate.[166] Hasinoff and coworkers[167] have studied the reaction rate of *p*-mercuribenzoate with glutathione, 2-mercaptoethanol, and phosphorylase b. The rates of reactions with the model compounds with *p*-mercuribenzoate are shown in Table 5.1. The rates are quite rapid comparable to those with *N*-ethylmaleimide and much faster than those with haloalkyl derivatives. These investigators also studied the reaction of the sulfhydryl groups in phosphorylase b; one group of cysteine residues reacted as rapidly as the model compounds (3×10^6 M^{-1} s^{-1}) while the other cysteines could be divided into two rather slowly reacting groups (4×10^2 M^{-1} s^{-1} and 3.4×10^2 M^{-1} s^{-1}). *p*-Mercuribenzoate does have the advantage of providing a spectral change which can be useful to quantitate modification (and measure sulfhydryl groups). The absorbance change at 255 nm upon modification is 6200 M^{-1} cm^{-1} at pH 4.6 and 7600 M^{-1} cm^{-1} at pH 7.0. Bai and Hayashi[168] have examined the reaction of organic mercurials with yeast carboxypeptidase (carboxypeptidase Y); treatment of the modified enzyme with millimolar cysteine resulted in virtually complete recovery of catalytic activity. The inactivation of chalcone isomerase by *p*-chloromercuribenzoate and mercuric chloride has been studied by Bednar and coworkers[169] and, as with carboxypeptidase Y, the modified protein could be readily reactivated by treatment with either thiols or KCN. The reactivation by KCN is based on the formation of a tight complex between cyanide and either organic mercurials. Ojcius and Solomon[170] have examined the inhibition of erythrocyte urea and water transport by *p*-chloromercuribenzoate. Other studies with this reagent have included dissociation of erythrocyte membrane proteins[171] and NADH peroxidase.[172]

p-Mercuribenzoate continues to be of use for the modification of sulfhydryl groups in proteins.[173–189] Several of these recent studies are worthy of additional comment. Kutscher and coworkers[177,178] used *p*-hydroxymercuibenzoate (pHMB), to modify sulfhydryl groups in ovalbumin in developing a method for the quantitation of proteins by mass spectrometry. Lu and coworkers[179] have proposed the use of pHMB for the determination of exposed cysteinyl residues in proteins. This approach is based on the premise that pHMB will not react with "buried" sulfhydryl groups; while this is reasonable, the relatively hydrophobic nature of the pHMB reagent presents the possibility that partitioning of the reagent occurs (see Chapter 1). This appears to be an issue with the modification of lactose repressor protein by 2-chloromercuri-4-nitrophenol as described later.[180] Fillmonova and coworkers[181] observed that the effect of *p*-mercuribenzoate on the endonuclease activity of the endonuclease from *Serratia marcescens* was directed at substrate, not enzyme. Angeli and colleagues[182] reported the use of pHMB for the measurement of oxidized glutathione in blood. The reaction is performed in strong base under conditions known to cleave disulfide bonds.[183] *p*-Aminophenylmercuric acetate (Figure 5.10) is described best as an activator of various metalloproteinases.[184–188]

2-Chloromercuri-4-nitrophenol (Figure 5.10) is a compound related to the organic mercurial, *p*-mercuribenzoate, described above. This reagent was described by

TABLE 5.1
Mercury-Containing Reagents for the Modification of Cysteine Sulfhydryl in Proteins[a]

Reagent	Comment	References
Mercuric chloride [$Hg(Cl)_2$]	Usually very tight binding based on formation of a mercury disulfide formation. Some studies describe the use of mercuric chloride as crosslinking agent.	[1,2]
Methylmercury chloride	Rarely used as reagent; more frequent use as biomarker for mercury exposure.	[3,4]
Ethylmercury chloride	Rarely used for chemical modification of proteins; more frequent use as biomarker and a test substance for organic mercurials.	[5–7]
Phenylmercury chloride	Infrequent use of either the chloride or acetate derivative in the modification of sulfhydryl groups in proteins.	[8,9]
Phenylmercury acetate	The two studies cited for phenylmercury acetate describe "activation" of matrix metalloproteinases with phenylmercury acetate (this phenomenon has also been reported from phenylmercury chloride).	[10,11]
p-Aminophenylmercury acetate	Extensive use for the activation of matrix metalloproteinases. Also used for the preparation of a mercuric-based affinity matrix for thiol and thiol containing proteins.	[12–15]
p-Chloromercuriphenyl-sulfonate[a]	A water-soluble derivative which has seen considerable use in the study of membrane enzymes. As this reagent is a strong anion, it will not readily pass a membrane as would *P*-chloromercuribenzoate.	[16–18]
p-Chloromercuribenzoate[b]	One of the original reagents developed by Paul Boyer in 1954.	[19,20]
2-Chloromercuri-4-nitrophenol	A chromogenic agent which places a reporter group on sulfhydryl groups in proteins.	[21–24]
2-(2′-Pyridylmercapto) mercuri-4-nitrophenol	Despite the attractive features (1) providing a chromogenic leaving group and (2) introducing a chromogenic reporter group, there has not been extensive use of this reagent.	[25]

[a] The rate of mercuric ions and organic mercurials with cysteine thiol groups is rapid (see Tables 5.2 and 5.3). The organic mercurials form more stable derivatives than does mercuric ion unless the mercuric ion forms a bridge between two cysteine residues.

[b] Per comment in the text, it is unlikely that the chloro form exists in solution; it is most likely replaced by hydroxyl or other counterion in most solution use. Hence, most current work uses the term *p*-mercuribenzoate to describe the reagent.

References for Table 5.1

1. Lobo, I.A., Harris, R.A., and Trudell, J.R., Cross-linking of sites involved with alcohol action between transmembrane segments 1 and 3 of the glycine receptor following activation, *J. Neurochem.* 104, 1649–1662, 2008.
2. Jaiswal, R. and Panda, D., Cysteine 155 plays an important role in the assembly of *Mycobacterium tuberculosis* FtsZ, *Protein Sci.* 17, 846–854, 2008.

(continued)

TABLE 5.1 (continued)
Mercury-Containing Reagents for the Modification of Cysteine Sulfhydryl in Proteins[a]

3. Oppedisano, F., Galluccio, M., and Indiveri, C., Inactivation by Hg^{2+} and methylmercury of the glutamine/amino acid transporter (ASCT2) reconstituted in liposomes: Prediction of the involved of a CXXC motif by homology modelling, *Biochem. Pharmacol.* 80, 1266–1273, 2010.
4. George, G.N., Singh, S.P., Myers, G.J. et al., The chemical forms of mercury in human hair: A study using X-ray absorption spectroscopy, *J. Biol. Inorg. Chem.* 15, 709–715, 2010.
5. Takatera, K. and Watanabe, T., Determination of sulfhydryl groups in ovalbumin by high-performance liquid chromatography with inductively coupled plasma mass spectrometric detection, *Anal. Chem.* 65, 3644–3646, 1993.
6. Hempel, M., Chau, Y.K., Dutka, B.J. et al., Toxicity of organomercury compounds: Bioassay results as a basis for risk assessment, *Analyst* 120, 721–724, 1995.
7. Santucci, B., Cannistraci, C., Cristaudo, R. et al., Thimersal positivities: The role of SH groups and divalent cations, *Contact Dermatitis* 39, 123–126, 1998.
8. Konttinen, Y.T., Lindy, O., Suomalainen, K. et al., Substrate specificity and activation mechanisms of collagenases from human rheumatoid synovium, *Matrix* 11, 395–403, 1991.
9. Sørensen, S.O. and Winther, J.R., Active-site residues of procarboxypeptidase Y are accessible to chemical modification, *Biochim. Biophys. Acta* 1205, 289–293, 1994.
10. Nagase, H., Enghild, J.J., Suzuki, K., and Salvesen, G., Stepwise activation mechanisms of the precursor of matrix metalloproteinase 3 (stromelysin) by proteinases and (2-aminophenyl) mercuric acetate, *Biochemistry* 29, 5783–5789, 1990.
11. Das, S., Mandal, M., Chakraborti, T. et al., Isolation of MMP-2 from MMP-2/TIMP-2 complex: Characterization of the complex and the free enzyme in pulmonary vascular smooth muscle plasma membrane, *Biochim. Biophys. Acta* 1674, 158–174, 2004.
12. Crabbe, T., Kelly, S.M., and Price, N.C., An analysis of the conformational changes that accompany the activation and inhibition of gelatinase A, *FEBS Lett.* 380, 53–57, 1996.
13. Molina, M.A., Codony-Servat, J., Albenell, J. et al., Trastuzumab (herceptin), a humanized anti-HER2 receptor monoclonal antibody, inhibits basal and activated HER2 ectodomain cleavage in breast cancer cells, *Cancer Res.* 61, 4744–4749, 2001.
14. Emod, I. and Keil, B., Five Sepharose-bound ligands for the purification of *Clostridium* collagenase and clostripain, *FEBS Lett.* 77, 51–56, 1977.
15. Raftery, M.J., Enrichment by organomercurial agarose and identification of cys-containing peptides from yeast cell lysates, *Anal. Chem.* 80, 3334–3341, 2008.
16. Mueckler, M., Roach, W., and Makepeace, C., Transmembrane segment 3 of the Glut1 glucose transporter is an outer shell helix, *J. Biol. Chem.* 279, 46876–46881, 2004.
17. Slugoski, M.D., Ng, A.M., Yao, S.Y. et al., A proton-mediated conformational shift identifies a mobile pore-lining cysteine residue (Cys-561) in human concentrative nucleosidetransporter 3, *J. Biol. Chem.* 283, 8496–8507, 2008.
18. Chiang, W.C. and Knowles, A.F., Inhibition of human NTPDase 2 by modification of an intramembrane cysteine by *p*-chloromercuriphenylsulfonate and oxidative cross-linking of the transmembrane domains, *Biochemistry* 47, 8775–8785, 2008.
19. Brahma, A., Banerjee, N., and Bhattacharyya, D., UDP-galactose 4-epimerase from *Kluyveromyces fragilis*—Catalytic sites of the homodimeric enzyme are functional and regulated, *FEBS J.* 276, 6725–6740, 2009.
20. Akita, K., Hieda, N., Baba, N. et al., Purification and some properties of wild-type and N-terminal-truncated ethanolamine ammonia-lyase of *Escherichia coli*, *J. Biochem.* 147, 83–93, 2010.

TABLE 5.1 (continued)
Mercury-Containing Reagents for the Modification of Cysteine Sulfhydryl in Proteins[a]

21. Liu, C. and Tsou, C.L., Kinetic differentiation between enzyme inactivation involving complex-formation with the inactivator and that involving a conformation-change step, *Biochem. J.* 282, 501–504, 1992.
22. Wang, Z.X., Wang, H.R., and Zhou, H.M., Kinetics of inactivation of aminoacylase by 2-chloromercuri-4-nitrophenol: A new type of complexing inhibitor, *Biochemistry* 34, 6863–6868, 1995.
23. Colotti, G., Verzili, D., Boffi, A., and Chiancone, E., Identification of the site of ferrocyanide binding involved in the intramolecular electron transfer process to oxidized heme in *Scapharca* dimeric hemoglobin, *Arch. Biochem. Biophys.* 311, 103–106, 1994.
24. Wang, H.R., Bai, J.H., Zheng, S.Y. et al., Ascertaining the number of essential thiol groups for the folding of creatine kinase, *Biochem. Biophys. Res. Commun.* 221, 174–180, 1996.
25. Jocelyn, P.C., Spectrophotometric assay of thiols, *Methods Enzymol.* 143, 44–67, 1987.

McMurray and Trentham[190] in 1969 together with other mercurinitrophenol derivatives and their reaction with glyceraldehyde 3-phosphate dehydrogenase. It has proved useful as a reporter group where changes in A_{410} can be used to measure polarity of the solvent surrounding the nitrophenol group. McMurray and Trentham used the mercuri-4-nitrophenol group on glyceraldehyde 3-phosphate dehydrogenase to measure conformation change on binding of phosphate and divalent cations. Sanyal and Khalifah[191] reported on the effect of labile (leaving group in the reaction) ligand (hydroxyl, chloride, cyanide, etc.) on the reaction of mercuri-4-nitrophenol with glutathione. Bromide, chloride, and cyanide stimulated the reaction while phosphate, ADP, and EDTA inhibited the reaction. Tris was observed to provide a twofold stimulation at pH 7.0. These investigators reported a rate of reaction of $3 \times 10^7\ M^{-1}\ s^{-1}$ at pH 7.0 (25°C). Marshall and Cohen[192] used mercuri-4-nitrophenol as a reporter group in ornithine transcarbamylase. They also reported useful pK_a values and spectral data for various derivatives of 2-chloromercuri-4-nitrophenol including 2-mercaptoethanol and glutathione. The pK_a of 2-chloromercuri-4-nitrophenol is 6.59, 7.17 for the glutathione derivative, and as high as 8.81 when bound to protein. The alkaline spectra (phenolate ion) is maximum at 403.5 nm (ε_M = 17,900) for 2-chloromercuri-4-nitrophenol and maximum at 408.5 nm (ε_M = 18,500) for the glutathione derivative. Protein-bound mercuri-4-nitrophenol is usually redshifted. Scawen and coworkers[193] modified the sulfhydryl groups in plastocyanin from *Cucurbita pepo* with 2-chloromercuri-4-nitrophenol. The reaction with 2-chloromercuri-4-nitrophenol was more rapid in the denatured protein than with the native protein; modification of the native protein with 2-chloromercuri-4-nitrophenol is also associated with the loss of copper. Simpson[194] modified the single sulfhydryl group in yeast phosophoglycerate kinase with 2-chloromercuri-4-nitrophenol. The mercuri-4-nitrophenol group could be used to follow the conformational changes on urea denaturation (the pK_a of the nitrophenol group changed from 8.20 in the native enzyme to 7.36 in the presence of 5.0 M urea). Mercuri-4-nitrophenol bound to cysteine residues was used to monitor conformational change in creatine kinase on binding the competitive inhibitor,

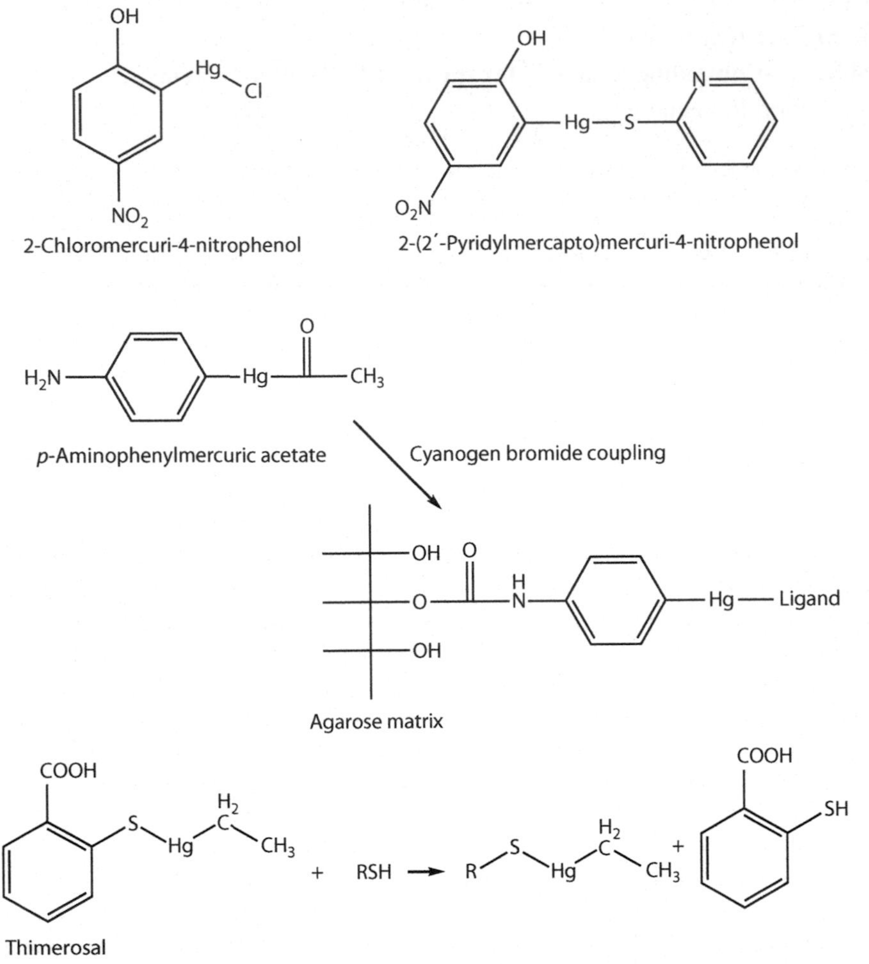

2-Chloromercuri-4-nitrophenol

2-(2′-Pyridylmercapto)mercuri-4-nitrophenol

p-Aminophenylmercuric acetate

Cyanogen bromide coupling

Agarose matrix

Thimerosal

FIGURE 5.10 Various mercury-containing reagents for the modification of cysteine sulfhydryl in protein. Shown are several chromogenic derivatives as well as the aminophenylmercurial derivative which can be coupled to an activated agarose matrix for use in affinity chromatography. Thimerosal is a compound used for the stabilization of vaccines (Wu, X., Liang, H., O'Hara, K.A. et al., Thiol-modulated mechanisms of the cytotoxicity of thimerosal and inhibition of DNA topoisomerase IIα, *Chem. Res. Toxicol.* 21, 483–493, 2008).

2′,4′,5′,66-tetraiodofluorescein.[195] A hyperchromic response was observed together with a redshift in the spectrum. It should be noted that creatine kinase modified with 2-chloromercuri-4-nitrophenol retains full enzyme activity. Previous studies from this laboratory[196] had reported that the modification of creatine kinase with 2-chloromercuri-4-nitrophenol in the absence of substrates (creatine, MgATP nitrate) was rapid ($k = 4.5 \times 10^6\ M^{-1}\ s^{-1}$) at pH 8.0 (20°C); the addition of various components of the substrate complex resulted in a modest decrease in the rate. Enzyme activity is not lost on the modification of the enzyme. Yang and coworkers[197] modified lactose repressor protein with 2-chloromercuri-4-nitrophenol; the modification could be reversed with 2-mercaptoethanol or DTT. These investigators also observed a different pattern of sulfhydryl modification with 2-bromoacetamido-4-nitrophenol. Earlier work from this laboratory[180] showed that the reaction of lactose repressor protein with 2-chloromercuri-4-nitrophenol was quite slow when compared, for example, with the rate of modification of creatine kinase.[196] Friedman and coworkers[180] also observed that the reaction of 2-chloromercuri-4-nitrophenol with the lactose repressor was strictly first-order (which runs contrary to conventional wisdom which would suggest a second-order reaction). These investigators estimated a $t_{1/2}$ of 5 s for the reaction of 2-chloromercuri-4-nitrophenol with lactose repressor compared to a $t_{1/2}$ of 5 ms for the reaction with of this reagent with creatine kinase.[196] Friedman and colleagues[180] also noted that the cysteine residues in lactose repressor are refractory to reaction with either iodoacetate or iodoacetamide.

Baines and Brocklehurst[198] have reported the synthesis and characterization of 2-(2′-pyridylmercapto)-mercuri-4-nitrophenol (Figure 5.10), a reagent which does have certain advantages. The pyridine-2-thione as the leaving group provides a measurement for the reaction based on absorbance at 343 nm (ε_{343} = 7300 M^{-1} cm^{-1}). The rate of reaction of 2-(2′-pyridylmercapto)-mercuri-4-nitrophenol with sulfhydryl groups (3.5×10^3 M^{-1} s^{-1}) is somewhat slower than that of the 2-chloromercuri-4-nitrophenol. Scott-Ennis and Noltmann[199] observed that a varying proportion of the three sulfhydryl groups in pig muscle phosphoglucose isomerase could be modified by different sulfhydryl reagents including 2-(2′-pyridylmercapto)-mercuri-4-nitrophenol. This reagent modified one of the three sulfhydryl groups under non-denaturing conditions; DTNB, iodoacetamide, and 2,2′-dithiopyridine also modified one sulfhydryl group although reaction at pH 8.5 is required for modification with DTNB. Iodoacetate did not modify sulfhydryl groups in the native enzyme at pH 7.5 or pH 8.5.

The studies mentioned earlier suggest the following general comments:

- Organic mercurials can be very useful for the specific modification of sulfhydryl groups in proteins.
- The reaction of organic mercurials with exposed sulfhydryl groups is rapid. While not as rapid as mercuric ion, the rate is still faster than most other reagents used to modify sulfhydryl groups in proteins.
- The reaction product of organic mercurials with sulfhydryl groups is stable but can be reversed by the addition of reducing agents such as 2-mercapto-ethanol or DTT.

Gold surfaces are used in biotechnological assays such as surface plasmon resonance[200–202] but this process is still poorly understood.[203] It would appear that the interaction of alkyl thiols (Figure 5.11) with gold surfaces to form self assembled monolayers involves a physical adsorption step followed by covalent bond (Au-S) formation[303] with the release of hydrogen.[204] The gold–sulfur bond is very stable and the author could not find any evidence for breakage of this bond with common biochemical reducing agents such as DTT, 2-mercaptothanol (these reagent would also react with the gold surface), or tris-(2-carboxyethyl)phosphine; one study was found which used stannous chloride to reduce a gold–sulfur bond in the preparation of

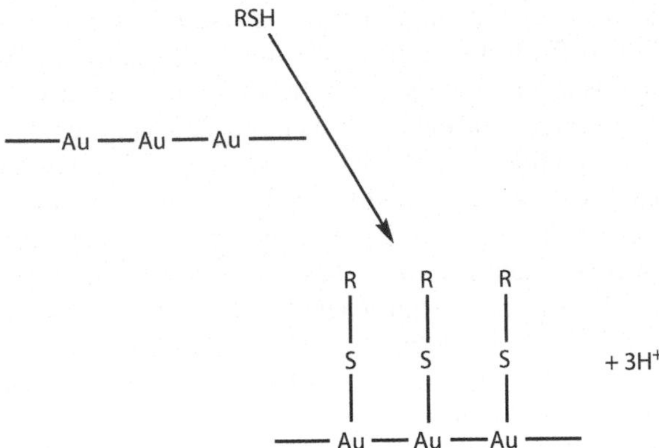

FIGURE 5.11 The reaction of thiol with a gold surface. The most reasonable mechanism is shown which involves the formation of a covalent bond between a thiol and gold with the release of protons. This is a process used for the formation of self-assembled monolayers.

a technetium-labeled therapeutic.[205] Stannous chloride has also been used to reduce a disulfide bond in the preparation of a rhenium-labeled biopharmaceutical.[206] Stannous chloride is important for the reduction-mediated labeling of monoclonal antibodies with Technetium.[207] In summary, there appears to be no facile method for the reversible interaction of protein sulfhydryl with Au(111) surfaces. It is apparent that there is reversible and irreversible binding of protein to gold surfaces[208,209] and gold-specific binding sequences which do not contain cysteine[210] can be identified by combinatorial biotechnology.[211] Gold salts also interact with cysteinyl residues in proteins[212,213] and are used in the treatment of arthritis.[214]

Metallothioneins are redox-active metalloproteins[215–219] with a high content of cysteinyl residues which can be characterized on the basis of their reactivity with alkylating agents as either reduced, oxidized, or metal-bound.[220] Metallothioneins are synthesized as the metal-free protein which binds metal ions in a posttranslational event.[221] The extent of disulfide bond formation (oxidation of cysteine) occurs during oxidative stress with the concomitant release of zinc.[222] The disappearance of free sulfhydryl groups was determined by reaction of *N*-ethylmaleimide. Kägi and Vallee[223] showed that while sulfhydryl groups in metallothionein (equine renal cortex) reacted rapidly with Ag^+ or *p*-chloromercuribenzoate, reaction is slow with *N*-ethylmaleimide; removal of metal ions produces thionein with equal reactivity toward the three reagents. If cadmium (Cd^{+2}) is added back to thionein, reactivity with Ag^{+1} is not changed while reactivity with *N*-ethylmaleimide is markedly reduced. Bernhard[224] has reviewed the use of iodoacetamide to determine the state of cysteine residues (oxidized, reduced, metal bound) in metallothionein. Dahl and McKinely-McKee[225] reported that the reaction of 2-mercaptoethanol with iodoacetate was some 200 times slower in the presence of zinc. Hunt and coworkers reported that modification of aspartate transcarbamylase with *p*-hydroxymercuriphenysulfonate resulted in the release of zinc.[226] Gardonio and Siemann reported the modification of a sulfhydryl group in IMP-1 metallo-β-lactamase with DTNB on the removal of zinc.[227] The sulfhydryl group was resistant to modification with DTNB, MMTS, or 2-nitro-5-thiocyanobenzoate; addition of dipiconlinic acid, a chelator of zinc, with DTNB resulted in the modification of the sulfhydryl group. The modification was reversed with TCEP.

A final example of the effect of metal ions is provided by Roehm and Berg[228] who studied the effect of cobalt on the reactivity of cysteine residues in variant 26-mer zinc finger peptides. Methylation of cysteine residues was accomplished with dimethyl sulfate.[229] The presence of cobalt decreased the rate of methylation while methylation in the absence of cobalt resulted in the loss of cobalt binding by the modified product.

Reaction with arsonous acid derivatives such as melarson oxide (Figure 5.12) have also proved useful for the identification of vicinal dithiols in proteins.[230,231] Phenylarsine oxide (Figure 5.12) has been used for the affinity chromatography of vicinol dithiol proteins.[232] Trivalent arsenic (arsenic oxide) is thought to react with sulfhydryl groups as part of its toxic mechanism.[233–235]

Haloalkyl compounds such as methyl iodide, iodoacetate, iodoacetamide, and more complex derivatives have been used to modify cysteine residues in proteins.

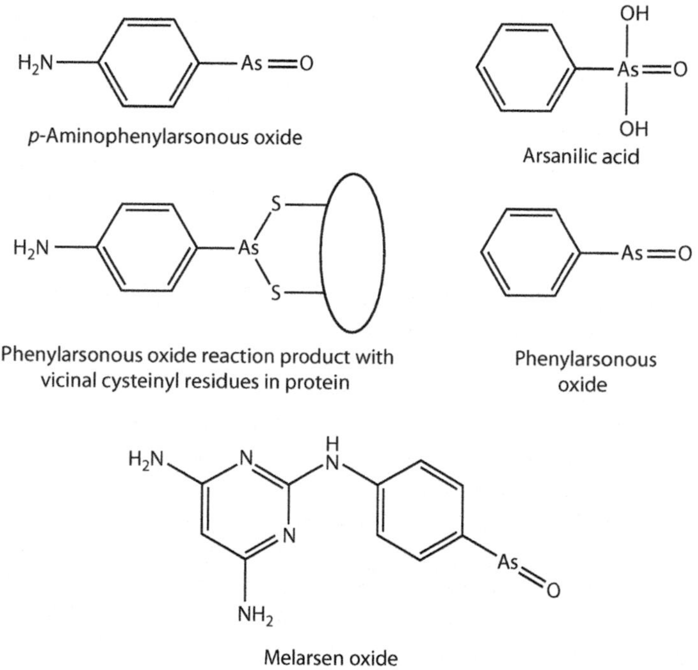

FIGURE 5.12 Trivalent arsenic forms stable bonds with vicinal diols. Shown are structures of *p*-aminophenyarsonous acid, the interaction of phenylarsonous acid, the structure of arsanilic acid, and the structure of melarsen oxide [*p*-(4,6 diamino-1,3,5-triazine-2-yl)aminophenylarsonsous acid]. (See Kalef, E. and Gitler, C., Purification of vicinal dithiol-containing proteins by arsenical-based affinity chromatography, *Methods Enzymol.* 233, 395–403, 1994; Happersberger, H.P., Przybylski, M., and Glocker, M.O., Selective bridging of bis-cysteinyl residues by arsonous acid derivatives as an approach to the characterization of protein tertiary structure and folding pathways by mass spectrometry, *Anal. Biochem.* 264, 237–250, 1998.)

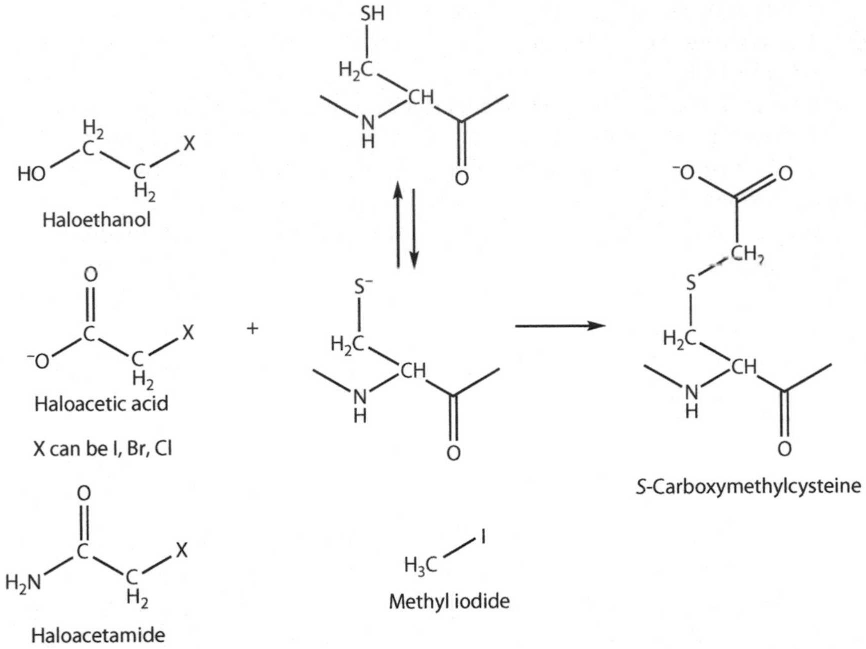

FIGURE 5.13 Modification of cysteine with haloacetic and haloacetamide. Also shown is the structure of methyl iodide which modifies cysteine to yield *S*-methylcysteine (Refsvik, T., The mechanism of biliary excretion of methyl mercury: Studies with methylthiols, *Acta Pharmacol. Toxicol.* (Copenh), 53, 153–158, 1983).

These various reagents react with cysteine via a S_N2 reaction mechanism to give the corresponding carboxymethyl or carboxamidomethyl derivatives (Figure 5.13). Specificity is an issue as α-ketoalkyl halides can also modify histidine, lysine, tyrosine, and glutamic acid. In general, the rate of reaction at a sulfhydryl group is much more rapid than that of other functional groups. Dahl and McKinley-McKee[236] have made a rather detailed study of the reaction of alkyl halides with thiols. It is emphasized that reactivity of alkyl halides not only depends on the halogen, but also on the nature of the alkyl groups. These investigators emphasized that the reactivity of an alkyl halide such as iodoacetate depends not only on the leaving potential of the halide substituent (I > Br >>> CI; 130:90:1), but also on the nature of the alkyl group (Table 5.2). The rate of reaction of 2-bromoethanol with the sulfhydryl group of L-cysteine (pH 9.0) is approximately 1000 times less than that observed with bromoacetic acid. The reactions are extremely pH dependent, emphasizing the importance of the thiolate anion in the reaction. As a general rule, when a rapid reaction is desired, the iodine-containing compounds are generally used but there can be issues with respect to specificity of modification.

Iodoacetic acid and iodoacetamide are used more often that bromo- or chloro-derivatives. Fluoroacetic acid or fluoroacetamide derivatives are rarely used. Sodium fluoroacetamide is used as a rodentocide on the basis of metabolic formation of fluorocitrate, an inhibitor of aconitase in the tricarboxylic acid cycle.[237,238] Chloroacetate

TABLE 5.2
Comparison of the Reactivity of Cysteine with Haloacetates, Haloacetamides, and Haloethanols[a]

Halo Compound	Thiol	pH-Independent Rate Constant ($M^{-1}\ s^{-1} \times 10^2$)
Chloroacetate	Cysteine	97
Chloroacetamide	Cysteine	330
Bromoacetate	Cysteine	3,200
Iodoacetate	Cysteine	9,650
Iodoacetamide	Cysteine	26,000–30,000
2-Chloroethanol	Cysteine	0.35
2-Bromoethanol	Cysteine	4.4

[a] Data from Dahl, K.H. and McKinley-McKee, J.S., The reactivity of affinity labels: A kinetic study of the reaction of alkyl halides with thiolate anion: A model reaction for protein alkylation, *Bioorganic Chem.* 10, 329–341, 1981.

is also toxic both as a strong acid and also by reaction with sulfhydryl groups.[239,240] It is of historical interest that the first use of α-haloacids was with bromoacetate in 1874[241] with iodoacetate introduced in 1930.[242] Both of these studies used whole animals and the reader is directed to Webb for more detail.[243] Some differences between acid and amide derivatives have been previously discussed. It is noted that there are some situations where iodoacetate is an affinity label.[244,245] Iodoacetate and to a somewhat lesser extent, iodoacetamide, are used to probe sulfhydryl group reactivity in proteins. Chloro- and bromo-derivatives are less reactive and are more useful in affinity labeling. The reader is recommended to an excellent review by Plapp[246] on affinity labeling. With nitrogen as the nucleophile, the order of reaction of α-haloacetic acids with 4-(*p*-nitrobenzyl) pyridine was I > Br >> Cl consistent with the order of reaction with thiols as described by Dahl and McKinley-McKee; the order of reactivity with the histidine residues in RNase was Br > I > Cl and Br > I > Cl with DNase.

More than 30 years ago, Brenda Gerwin[250] demonstrated dramatic differences in the kinetics of the reaction of chloroacetic acid and chloroacetamide with the active site cysteine of streptococcal proteinase. The curve for the pH dependence for reaction with chloroacetamide was sigmoidal, supporting the ionization of a cysteine residue as a critical factor while it was bell-shaped with chloroacetic acid with a broad optima between pH 5.0 and 8.5. The pH dependence for the reaction with glutathione is the same with either chloroacetic acid or chloroacetamide. This difference in reactivity is suggested to reflect the influence of the histidine residue at the active site which would influence the reactivity of the negatively charged chloroacetic acid without effect on the neutral chloroacetamide. It was noted that the rate of reaction of chloroacetamide at pH 9.0 was 50–100 times more rapid with streptococcal proteinase (approximately $5 \times 10^2\ M^{-1}\ min^{-1}$) than with glutathione (approximately $8\,M^{-1}\ min^{-1}$). The reaction of streptococcal proteinase with iodoacetate or iodoacetamide was too rapid to allow for accurate analysis of rates. The unique properties of this cysteine residue can be explained in part by the presence of an adjacent histidyl residue which was demonstrated by an elegant series of studies by Liu.[248] Although histidine residues will react with α-halo acids and amides, the presence of an adjacent cysteine residue precluded the direct use of this class of reagents to demonstrate the presence of a histidyl residue at the active site of streptococcal proteinase. Liu took advantage of the reversible modification of cysteinyl residues with sodium tetrathionite to modify the active-site histidine with iodoacetate.

Kallis and Holmgren[249] have examined the differences in reactivity of two sulfhydryl groups present at the active site of thioredoxin. The pH dependence of the reaction with iodoacetate suggested that one group had a pK_a value of 6.7, while the second was 9.0. Iodoacetamide showed the same pH dependence, but the rate of reaction was approximately 20-fold greater than with iodoacetate. For example, at pH 7.2, the second-order rate constant for reaction with iodoacetate was $5.2\,M^{-1}\ s^{-1}$, while it was $107.8\,M^{-1}\ s^{-1}$ for iodoacetamide. The low pK_a of one of the sulfhydryl groups was suggested to be a reflection of the presence of an adjacent lysine residue. Mikami and coworkers have examined the inactivation of soybean α-amylase with iodoacetamide and iodoacetate.[250] Inactivation with iodoacetamide occurred approximately 60 times more rapidly than with iodoacetate at pH 8.6. Hempel and Pietruszko[251] have shown that human liver alcohol dehydrogenase is inactivated by iodoacetamide, but not by iodoacetic acid. These experiments were performed in 0.030 M sodium phosphate, at pH 7.0 containing 0.001 M EDTA.

The use of iodoacetate and iodoacetamide for the modification of sulfhydryl groups appeared to peak in the 1970s (Table 5.3) and seemed to parallel the number of studies on the chemistry of enzyme active sites. There has been little use of these reagents in the past decade (since 2000); the chemistry continues to be of value in the attachment of probes. Examples include 5-iodoacetamido-fluorescein,[252] 5-[2-((iodoacetyl)amino)-ethyl]naphthalene-1-sulfonic acid (1,5-IAEDANS),[253–261] 4-(2-iodoacetamido)-TEMPO,[262,263] cyanine dyes,[264] and some newly developed solvent-sensitive fluorescent dyes.[265] The iodoacetyl function group has been used for introducing a biotin probe in proteins[266–269] in addition to its use in the development of isotope-coded affinity tags. A recent study[270] demonstrated the use of (4-iodobutyl) triphenylphosphonium for the modification of specific cysteine residues in membrane preparation. This reagent is accumulated by mitochondria permitting the selective medication of cysteine residues inside the mitochondria. This reagent was used to monitor the thiol redox state of individual mitochondrial proteins.

The inactivation of hamster arylamine *N*-acetyltransferase one (NAT1) by bromoacetamido derivates of aniline and 2-aminofluorene[271] is worth some discussion considering that the same "active-site" cysteine residue is modified by similar reagents by kinetically different processes. Arylamine *N*-transferases are cytosolic enzymes that catalyze the acetylation of arylamines in an acetyl-coenzyme A-dependent reaction. In this study (2-bromoacetylamino) fluorine (Figure 5.14) demonstrates saturation kinetics consistency while bromoacetanilide is an alkylating agent reacting by an S_N2 reaction similar to iodoacetic acid which has been

TABLE 5.3
Chronological Record of Iodoactate Use[a]

Time Period	Number of Citations	
	PubMed[b]	SciFinder©[c]
1920–1930	(4)[d]	
1930–1940	(224)[d]	
1940–1950	(145)[d]	
1950–1960	161 (764)[d]	782
1960–1970	1337	1403
1970–1980	2292	1664
1980–1990	1023	1625
1990–2000	771	1433
2000–2010	328	954

[a] The citations have not been selected for application. Thus, for example, this number would include use of an enzyme inhibitor, use in the carboxymethylation of proteins for structural analysis, or whole cell metabolism studies to list only a few examples.

[b] PubMed is a service through The National Center for Biotechnology Information in the National Library of Medicine within the National Institute of Health (Bethesda, MD, USA). PubMed contains material from the biomedical literature and, more recently on-line books.

[c] SciFinder© is a literature service provided by the American Chemical Society (Washington, DC) which has evolved from Chemical Abstracts Service (American Chemical Society, Columbus, Ohio). This database contains material from the chemistry journals (including biochemical journals), selected books, abstracts from American Chemical Society Meetings, and patents from various geographies. It is usual to obtain a larger number of citations from SciFinder© than from PubMed.

[d] Numbers in parenthesis are obtained from J.L Webb, *Enzyme and Metabolic Inhibitors*, Vol. 1, Preface (page x), Academic Press, New York, 1963.

previously demonstrated to inhibit these enzymes.[272,273] One of the vexing problems in the site-specific chemical modification of proteins is differentiating between the unique reactivity of a functional group and "affinity labeling." It is noted that a previous study suggested that bromoacetanilide demonstrated saturation kinetics with the rabbit enzyme.[274]

Reduction under denaturing conditions followed by alkylation with haloalkyl reagents or other alkylating agents is a common procedure in preparation of the protein for further analysis.[275–284] The success of analytical procedures such as two-dimensional (2-D) electrophoresis[278,279] in proteomics depends on the effectiveness

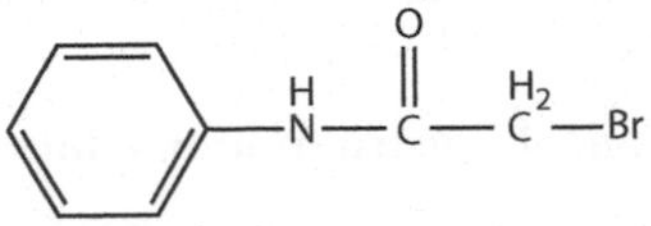

Bromoacetanilide

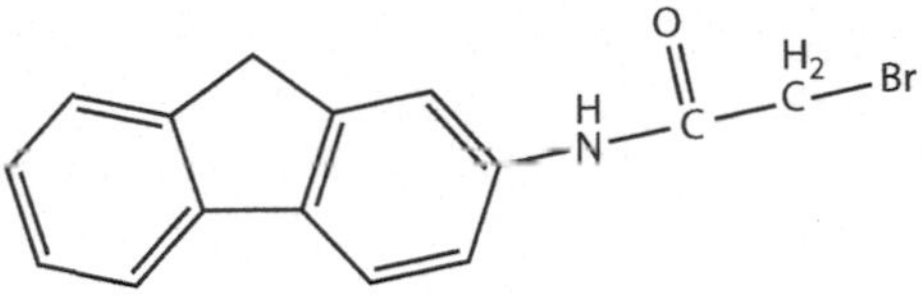

2-(Bromoacetylamino)fluorene

FIGURE 5.14 Structure of bromoacetanilide and 2-(bromoacetylamino) fluorene. Bromoacetanilide inhibits arylamine transferase with uncomplicated second-order kinetics while 2-(bromoacetylamino) fluorene shows saturation kinetics indicating binding to the enzyme before reaction consistent with an affinity labeling process.

of this process. Historically, reduction and alkylation has had an important role in the determination of the protein primary structure.[280] It is a useful step today in the determination of protein modification by mass spectrometry.[281] In 2-D electrophoresis, reduction under denaturing conditions is critical in the preparation of the sample for analysis. Alkylation prevents the formation of "spurious" spots during the subsequent electrophoretic analysis.[282,283] The reduction and alkylation step could be used in the preparation of the sample prior to the isoelectric focusing step[284,285]or between the isoelectric focusing step and the SDS/PAGE (polyacrylamide gel electrophoresis in the presence of sodium dodecyl sulfate) step. In the latter, this would involve the *in situ* treatment of the immobilized pH gradient (IPG) strip prior to the SDS/PAGE step.[286] It is probably best to perform the alkylation step prior to the isoelectric focusing step to avoid the issue of disulfide bond reformation ("disulfide scramble") during separation. Another consequence which can be avoided by alkylation is the beta-elimination of cysteine in the alkaline pH range forming dehydroalanine and consequent unwanted peptide bond cleavage.[287] Nielsen and coworkers[288] reported that modification with iodoacetamide presents an ubiquitin-like artifact on mass spectrometry analysis. Lapko and coworkers[289] have also reported an artifact in the mass spectrometric analysis of protein derivatized with iodoacetamide. It would appear that the inclusion of the alkylation step during the preparation of sample for isoelectric focusing is the preferred approach. The reduction and alkylation step must be performed under denaturing conditions. An uncharged chaotropic agent such as urea has been the denaturant of choice for this process. However, urea has the potential of creating problems through a dismutation reaction which forms cyanate which in turn can react with nucleophiles including cysteine and lysine in subject proteins increasing heterogeneity.[290,291] Recently it was recommend that thiourea be included in the sample "cocktail" (9.0 urea is the common concentration; the inclusion of 2 M thiourea is recommended) preparation for the purpose of improving protein solubilization[292] prior to the isoelectric focusing step. However, the use of thiourea complicates the alkylation step by reacting with iodoacetic acid or iodoacetamide;[292,293] such inhibition is not observed with maleimides.[294] In studies with wool proteins, carboxymethylation with iodoacetate was essentially complete after 10 min of reaction.[295] Side reactions with other amino acids occurred for longer times of reaction. Other investigators report that the alkylation process required a long period of time and was incomplete after 6 h of reaction.[296] As shown in the section on haloalkyl derivatives, it is clear that local environmental factors do influence reaction rate of cysteine residues in proteins with iodoacetate or iodoacetamide. Schrooyen and workers[297] prepared partially carboxymethylated feather keratin in the development of products for environmentally sustainable application for protein waste streams. Feather keratin was extracted in urea/2-mercaptoethanol; upon dialysis, aggregates were formed. Partial modification with iodoacetate, iodoacetamide, or bromosuccinic acid yielded a soluble product. The ability to prepare partially modified proteins was based on the use of rate constants obtained from the reaction of iodoacetate or iodoacetamide with model thiols such as glutathione or 2-mercaptoethanol.[236] The partially carboxymethylated keratin is then used for the preparation of a protein plastic fiber.[298]

The development of isotope-coded affinity tags[299,300] as a method for studying differential protein expression[301,302] has fostered the development of novel reagents based on the iodoacetyl function. Examples include *N*-(13-iodoacetamido-4,7,10-trioxatridecanyl) biotinamide,[302] *N*-ethyliodoacetamide,[30.304] *N*-*t*-butyliodoacetamide, and iodoacetanilide.[305] These reagents are prepared as the "light" form (hydrogen) and the "heavy" form (deuterium).[306] The biotin derivatives are referred to as ICAT (isotope-coded affinity tagging) reagents (Figure 5.15) and available with deuterium or carbon-13.[306,307] The biotin "tag" enables the rapid purification of peptides/proteins from mixtures after modification of cysteine residues with this reagent. Another approach to cysteine modification for proteomics is the use of ethyleneimine (Figure 5.16) for an in-gel derivatization of proteins after electrophoretic separation[308] introducing a new site for cleavage by trypsin. Vinyl pyridine (Figure 5.17) is an alternative for the alkylation of cysteine residues in proteins[309] and deuterated derivatives have been prepared for use in the study of differential protein expression.

Recent application of ICAT technology include the study of protein expression in thrombosed hemodialysis grafts[310] and the mitochondrial proteome from subjects with cognitive disorders.[311] Guaragna and coworkers[312] reported the synthesis of a more soluble ICAT reagent (β-alanine-arm-ICAT) where the polyether linkage is replaced by a polyamide linkage (Figure 5.15). Häggland and coworkers[313] used an acid-cleavable ICAT tag (Figure 5.15) to identify the thioredoxin-reducible disulfide bonds in germinated barley seeds. Wu and coworkers used ICAT to study low molecular-weight serum proteins.[314] Absorption onto a C^{18} matrix is accomplished in the presence of DTT in 9 M urea with elution with 60% acetonitrile. Pritchard and coworkers[315] have used ICAT reagents to evaluate somatropin standards for clinical chemistry. Leroy and coworkers[316] used primary amine labeling in proteins with *N*-nicotinoyloxysuccinimide in a technique referred to as isotope-coded protein labeling[317] to study microgravity effects in *Cupriavidus metallidurans* CH34. Maleimide-based ICAT reagents are discussed in the following section.

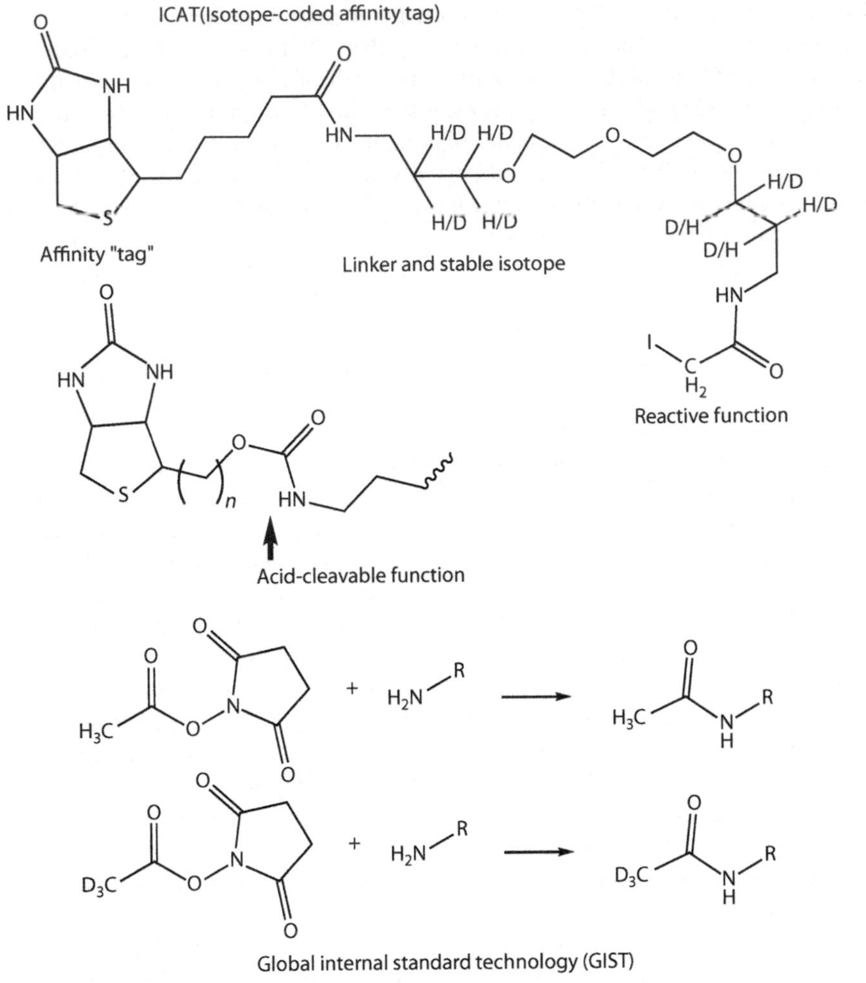

FIGURE 5.15 Reagents used for ICAT. The reagents are designed with or without isotope. Shown here is a reagent which can be prepared with or without deuterium. Other than the isotope labeling, the structures are identical and have identical reactivity with sulfhydryl groups (as shown). (Tao, W.A. and Aebersold, R., Advances in quantitative proteomics via stable isotope tagging and mass spectrometry, *Curr. Opin. Biotechnol.* 14, 110–118, 2003.) An acid cleavable derivative is shown in the lower part (Fauq, A.H., Karche, R., Khan, M.A., and Vega, I.E., Synthesis of acid-cleavable light isotope-coded affinity tags (ICAT-*L*) for potential use in proteomic expression profiling analysis, *Bioconj. Chem.* 17, 248–254, 2006). Also shown is a reagent developed for global internal standard technology (GIST) (Chakraborty, A. and Regnier, F.E., Global internal standard technology for comparative proteomics, *J. Chromatog. A* 949, 173–184, 2002). The biotin "tag" provides an avenue for purification of labeled peptides or proteins.

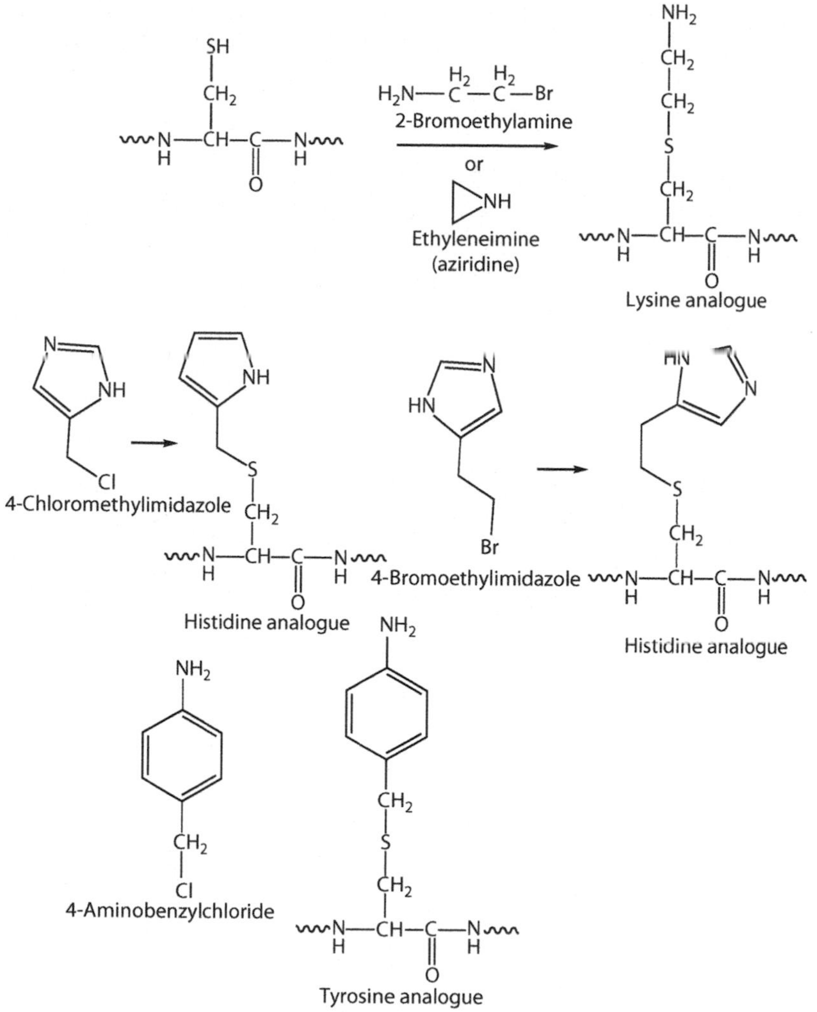

FIGURE 5.16 The formation of *S*-aminoethylcysteine and other amino acid analogues from cysteine. The reaction of bromoethylamine with cysteine provides a site for trypsin cleavage (Hopkins, C.E., Herandez, G., Lee, J.P. et al., Aminoethylation in model peptides reveals conditions for maximizing thiol specificity, *Arch. Biochem. Biophys.* 443, 1–10, 2005; Rehulková, H., Marchetti-Deschmann, M., Pittenauer, E. et al., Improved identification of hordeins by cysteine alkylation with 2-bromoethylamine, SDS-PAGE and subsequent in-gel tryptic digestion, *J. Mass Spectrom.* 44, 1613–1621, 2009.) Aminoethylation is also used for chemical rescue in site-specific mutagenesis studies (Hopkins, C.E., O'Connor, P.B., Allen, C.N. et al., Chemical-modification rescue assessed by mass spectrometry demonstrates that γ-thia-lysine yields the same activity as lysine in aldolase, *Protein Sci.* 11, 1591–1599, 2002). Other amino acid analogues can be prepared as shown (Schindler, J.F. and Viola, R.E., Conversion of cysteinyl residues to unnatural amino acid analogs. Examination in a model system, *J. Protein Chem.* 15, 737–742, 1996). 4-Methylimidazole has been shown to provide a useful derivative (Elder, I., Han, S., Tu, C. et al., Activation of carbonic anhydrase II by active-site incorporation of histidine analogs, *Arch. Biochem. Biophys.* 421, 283–289, 2004).

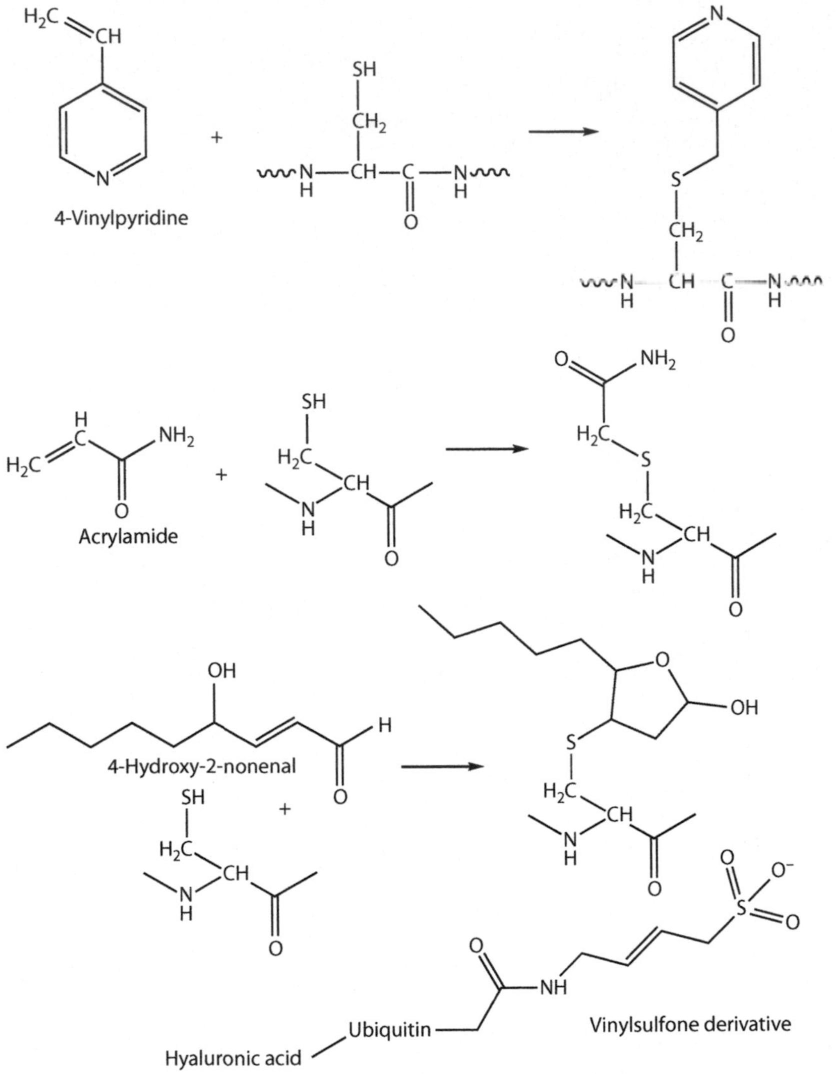

4-Vinylpyridine

Acrylamide

4-Hydroxy-2-nonenal

Ubiquitin

Hyaluronic acid

Vinylsulfone derivative

FIGURE 5.17 A variety of reagents which add to sulfhydryl groups via Michael addition. Shown is the reaction of vinyl pyridine with cysteine (Friedman, M., Application of the S-pyridylethylation reaction to the elucidation of the structures and functions of proteins, *J. Protein Chem.* 20, 431–453, 2001; Kleinova, M., Belgacem, O., Pock, K. et al., Characterization of cysteinylation of pharmaceutical-grade human serum albumin by electrospray ionization mass spectrometry and low-energy collision-induced dissociation tandem mass spectrometry, *Rapid Commun. Mass Sp.* 19, 2965–2973, 2005), the reaction of acrylamide with thiols (LoPachin, R.M., Gavin, T., Geohagen, B.C., and Das, S., Neurotoxic mechanisms of electrophilic type-2 alkenes: Soft interactions described by quantum mechanical parameters, *Toxicol. Sci.* 98, 561–570, 2007), and 4-HNE (Wakita, C., Maeshima, T., Yamazaki, A. et al., Stereochemical configuration of 4-hydroxy-2-nonenal -cysteine adducts and their stereoselective formation in a redox-regulated protein, *J. Biol. Chem.* 284, 28810–28827, 2009) with cysteine.

The Michael addition (Figure 5.17) was originally used to describe the addition of an activated methylene to an activated unsaturated system such as an α,β-ketone but has a broad definition today in describing the addition of a nucleophile, such as thiolate, across a double-bond conjugated with an electron-withdrawing group such as ketone.[318] A clever application of the use of Michael addition is the selective modification of a disulfide bond for the insertion of a poly (ethylene glycol) chain (Figure 5.18).[319–321] There are several reagents (Figure 5.17) for the modification of sulfhydryl groups in proteins including acrylamide and *N*-ethylmaleimide; 4-hydroxy-2-nonenal (4-HNE) also can modify sulfhydryl groups via a Michael addition.

Acrylamide modifies sulfhydryl groups in proteins (Figure 5.17). Originally considered to be an unwanted complication of the electrophoresis of reduced proteins in acrylamide gel systems,[322] modification with acrylamide is now considered to be a useful site-specific modification of cysteine in proteins.[323–329] Some of these studies[326] focus on the molecular mechanism for acrylamide toxicity.

4-HNE is a product produced from the oxidation of lipids which modifies sulfhydryl groups (Figure 5.17) and other residues in proteins[330–334] such as histidine in proteins (Figure 5.17). 4-HNE is somewhat nonspecific as it can modify nucleophiles such as lysine by Schiff base formation and sulfhydryl groups by Michael addition; histidine is also subject to modification via Michael addition (Chapter 4).

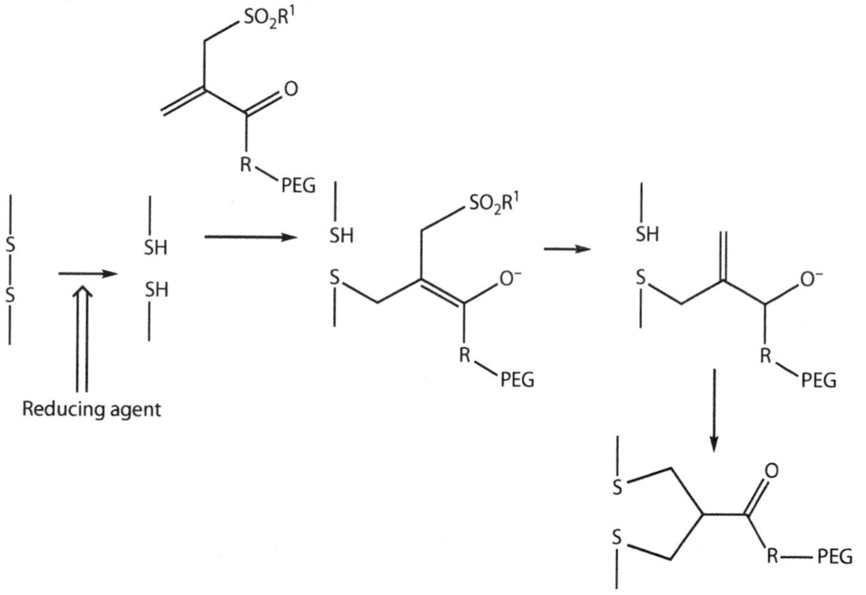

FIGURE 5.18 The insertion of a poly (ethylene glycol) moiety into a protein via a disulfide bond using successive Michael additions. The disulfide bond is reduced with a suitable reducing agent [DTT or tris-(carboxyethyl)-phosphine]. The addition of an α,β-unsaturated-β′-sulfone functionalized poly(ethylene glycol) results in the covalent linkage of the reagent via successive Michael additions. (See Brocchini, S., Godwin, A., and Balan, S., Disulfide bridge based PEGylation of proteins, *Adv. Drug Deliv. Rev.* 60, 3–12, 2008.)

N-Ethylmaleimide reacts with sulfhydryl groups in proteins (Figure 5.17) with considerable specificity.[335–337] The studies by Britto and coworkers on the modification of tubulin by *N*-ethylmaleimide[75] have been discussed earlier. These investigators noted that the reaction of *N*-ethylmaleimide was more rapid than iodoacetamide with the cysteine residues in tubulin; a similar difference in reactivity was observed with model sulfhydryl groups (see Table 5.4). Bednar has examined the chemistry of the reaction of *N*-ethylmaleimide with cysteine and other thiols in some detail.[338] The second-order rate constant for the reaction of *N*-ethylmaleimide (NEM) with the thiolate anion of 2-mercaptoethanol is in the range of 10^7 M^{-1} min^{-1}which is only slight less than the rate of aromatic mercuric compounds; it is suggested that this value is at least 5×10^{10} greater than the reaction with thiol (2×10^{-4} M^{-1} min^{-1}). This study also reports data for the decomposition of NEM in several buffers and should be considered for the determination of truly accurate kinetic data. This study[338] was focused on the reaction of *N*-ethylmaleimide with chalcone isomerase. The pH dependence of the results suggests that the reaction is dependent on protein conformation; at high pH (pH ≥ 8), the cysteinyl residue is much less reactive than a model thiol at the same pH (pH ≥ 8) (see Table 5.5) while below pH 8 the rate of reaction of NEM with the cysteinyl residue in chalcone isomerase approaches that of a model thiol. As reported earlier for papain, modification of chalcone isomerase with mercuric chloride or *p*-mercuribenzoate blocks modification with NEM.

The reaction of NEM with a cysteinyl residue can be followed by the decrease in absorbance at 300 nm, the absorbance maximum of N-ethylmaleimide. The extinction coefficient of N-ethylmaleimide is 620 M^{-1} cm^{-1} at 302 nm.[335] The spectrophotometric assay is not sensitive and the modification is usually monitored by the incorporation of radiolabeled reagent. The alkylation product (S-succinyl cysteine) is stable and can be determined by amino acid analysis following acid hydrolysis. Although the reagent is reasonably specific for cysteine, reaction with other nucleophiles such as histidine must be considered.[339–342] A "diagonal" procedure for the isolation of cysteine-containing peptides modified with N-ethylmaleimide has been reported.[343] This procedure is based on the base hydrolysis of the reaction product obtained from reaction with *N*-ethylmaleimide to cysteine-S-N-ethyl succinamic acid generating a new negative charge on the peptide/protein. Various derivatives of maleimide provide the basis for the design of cross-linking reagents. A consideration of Table 5.4 shows that the reaction of model thiols with *N*-ethylmaleimide is much faster that with haloalkyl compounds. A direct comparison of the reactivities of *N*-ethylmaleimide and iodoacetic acid/iodoacetamide with proteins in tissues was performed by Rogers and coworkers.[344] These workers observed that the reaction of NEM with protein sulfhydryl groups in tissue (murine myofibril tissue extracts) was (1) much faster than either iodoacetate or iodoacetamide, (2) could proceed at acidic pH, and (3) required less reagent.

Brown and Matthews[345,346] have studied the reaction of lactose repressor protein with N-ethylmaleimide, two spin-label derivatives of *N*-ethylmaleimides, and a fluorophore derivative. The spin-labeled compounds showed the same pattern of reaction with the three cysteinyl residues as seen with *N*-ethylmaleimide. The fluorophore-derivative (*N*-(3-pyrene) maleimide) shows a slightly different reaction

TABLE 5.4
Comparison of Rates of Reaction of Sulfhydryl Reagents with Cysteine Residues in Model Compounds[a]

Model Compound	Reagent/Experimental Conditions	Rate	Reference
2-Mercaptoethanol	2-Pyridylmercurial	3.5×10^3 M^{-1} s^{-1}	[1]
2-Mercaptoethanol	*p*-Chloromercuribenzoate in 0.02 M glycerophosphate-HCl, pH 6.8, 30°C	6.5×10^6 M^{-1} s^{-1}	[2]
2-Mercaptoethanol	*p*-Chloromercuribenzoate in potassium phosphate-potassium chloride (ionic strength of 0.2), pH 7.63, 25°C	1.0×10^7 M^{-1}·s^{-1}	[2]
N-acetyl-L-cysteine	2-Mercuri-4-nitrophenol in 5 mM bis-tris (ionic strength 0.2 obtained with potassium sulfate), pH 6.9 (8°C)	1.0×10^7 M^{-1} s^{-1}	[2]
Glutathione	*p*-Chloromercuribenzoate in 0.02 M glycerophosphate-HCl, pH 6.8, 30°C	6.2×10^6 M^{-1} s^{-1}	[2]
2-Mercaptoethanol	Iodoacetate in 0.5 M triethanolamine, pH 8.55, 23.5°C	0.134 M^{-1} s^{-1}	[3]
2-Mercaptoethanol	Iodoacetate in 0.5 M triethanolamine, pH	2.5 M^{-1} s^{-1}[b]	[3]
2-Mercaptoethanol	Iodoacetate in 0.5 M triethanolamine, pII 7.40 with 1:1 zinc[c]	11.8×10^{-3} M^{-1} s^{-1}	[3]
2-Mercaptoethanol	Iodoacetate in 0.5 M triethanolamine, pH 7.55	0.026 M^{-1} s^{-1}	[3]
2-Mercaptoethanol	Chlorodinitrobenzene (30°C, pH 8.0)	4.7 M^{-1} min^{-1}	[4]
Cysteine	Chlorodinitrobenzene (30°C, pH 8.0)	12.3 M^{-1} min^{-1}	[4]
2-Mercaptoethanol	DTNB, pH 6.0 (0.05 M phosphate, pH 6.0, 15°C)	30 M^{-1} s^{-1}	[5]
2-Mercaptoethanol	5,5-Dithiobis-(2-nitro-*N*-trimethylbenzylammonium iodide, pH 6.0 (0.05 M phosphate, 15°C)	7×10^2 M^{-1} s^{-1}	[5]
2-Mercaptoethanol	5,5′-Dithiobis-(2-nitro-*N*-2′-hydroxyethyl benzamide), pH 6.0 (0.05 M phosphate, 15°C)	3×10^2 M^{-1} s^{-1}	[5]
2-mercaptoethanol-amine	*N*-ethylmaleimide (1.0 M KCl, 0.4 mM EDTA, 25°C)	4×10^6 M^{-1} min^{-1}[b]	[6]
2-mercaptoethanol	*N*-ethylmaleimide (1.0 M KCl, 0.4 mM EDTA, 25°C)	11×10^6 M^{-1} min^{-1}[b]	[6]
L-cysteine	*N*-ethylmaleimide (1.0 M KCl, 0.4 mM EDTA, 25°C)	1.1×10^6 M^{-1} min^{-1}[b]	[6]
2-Mercaptoethanol	5,5′-Dithiobis-(2-nitrobenzoate), pH 6.65[e]	2.20×10^2 M^{-1} s^{-1}	[7]
Cysteine	5,5′-Dithiobis-(2-nitrobenzoate), pH 6.65[e]	9.40×10^2 M^{d1} s^{-1}	[7]
Glutathione	5,5′-Dithiobis-(2-nitrobenzoate), pH 6.65[e]	2.70×10^2 M^{-1} s^{-1}	[7]
2-Mercaptoethanol	MMTS	1.8×10^8 M^{-1} min^{-1}[f]	[8]

(continued)

TABLE 5.4 (continued)
Comparison of Rates of Reaction of Sulfhydryl Reagents with Cysteine Residues in Model Compounds[a]

Model Compound	Reagent/Experimental Conditions	Rate	Reference
2-Mercaptoethyl-amine	Bromoacetate in H_2O	$1.89\,M^{-1}\,s^{-1}$[g]	[9]
2-Mercaptoethyl-amine	Bromoacetate in D_2O	$1.81\,M^{-1}\,s^{-1}$[g]	[9]
L-Cysteine	Bromoacetate in H_2O	$1.09\,M^{-1}\,s^{-1}$[g]	[9]
L-Cysteine	Bromoacetate in D_2O	$1.41\,M^{-1}\,s^{-1}$[g]	[9]
2-Mercaptoethyl-amine	5,5′-Dithiobis-(2-nitrobenzoate), pH 6.65	$4.6 \times 10^3\,M^{-1}\,s^{-1}$	[10]
L-Cysteine	Chloroacetate, pH 6.0, 0.3M KCl, 25°C	$5.3 \times 10^{-3}\,M^{-1}\,min^{-1}$	[11]
L-Cysteine	Acrylonitrile, pH 8.1, 30°C	$2.15 \times 10^{-1}\,M^{-1}\,s^{-1}$	[12]
Homocysteine	Acrylonitrile, pH 8.1, 30°C	$1.2 \times 10^{-1}\,M^{-1}\,s^{-1}$	[12]
2-Mercaptothanol	5-(2-aminoethyl)dithio-2-nitrobenzoate, pH 9.0	$1.7 \times 10^4\,M^{-1}\,s^{-1}$	[13]
2-Mercaptoethanol	DTNB, pH 9.0	$2.1 \times 10^4\,M^{-1}\,s^{-1}$[f]	[13]
L-Cysteine	N-Ethylmaleimide	$8.17 \times 10^7\,M^{-1}\,s^{-1}$[f]	[13]
L-Cysteine	4-HNE[h]	$5.74 \times 10\,M^{-1}\,s^{-1}$	[13]
L-Cysteine	Acrylamide	$5.84\,M^{-1}\,s^{-1}$	[13]
Glutathione	2-mercuri-4-nitrophenol in 5 mM bis-tris (ionic strength 0.2 obtained with potassium sulfate), pH 7.0	$1.6 \times 10^7\,M^{-1}\,s^{-1}$ @ 8°C[i] $3.0 \times 10^7\,M^{-1}\,s^{-1}$ @ 25°C[i]	[14]
Glutathione	2-Vinylpyridine, 25°C	$6.6 \times 10^2\,M^{-1}\,s^{-1}$[j]	[15]
Glutathione	Chloroacetamide, pH 9.0	$8\,M^{-1}\,min^{-1}$	[16]
Glutathione	Chloroacetic acid, pH 9.0	$0.54\,M^{-1}\,min^{-1}$	[16]

[a] Also see Dahl, K.H. and McKinley-McKee, J.S., The reactivity of affinity labels: A kinetic study of the reactions of alkyl halides with thiolate anions - A model for protein alkylation, *Bioorg. Chem.* 10, 329–341, 1981; Schöneich, C., Kinetics of thiol reactions, *Methods Enzymol.* 251, 45–55, 1995.

[b] pH-Independent rate constant (see Jones, J.G., Otieno, S., Barnard, E.A., and Bhargava, A.K., Essential and nonessential thiols of yeast hexokinase reactions with iodoacetate and iodoacetamide, *Biochemistry* 14, 2376–2403, 1975).

[c] The rate of reaction of iodoacetate with the thiolate form is some 200-fold greater than reaction with the zinc-thiol complex (Dahl, K.H. and McKinley-McKee, J.S., Enzymatic catalysis in the affinity labeling of liver alcohol dehydrogenase with haloacids, *Eur. J. Biochem.* 118,507–513, 1981).

[d] Buffer information not provided.

[e] Data obtained at pH 4.0, 5.0, and 6.65 and corrected to reflect amount of thiolate anion present. It has been observed that the rate of reaction of alkyl methanethiosulfonates (alkyl thiol sulfonates) with sulfhydryl groups is rapid (Wynn, R. and Richards, F.M., Chemical modification of protein thiols: Formation of mixed disulfides, *Methods Enzymol.* 251, 351–356, 1995). Calculation of kinetic data for alkyl methanethiosulfonates must consider the rate of hydrolysis of the reagent (Karlin, A. and Akabas, M.H., Substituted-cysteine accessibility method, *Meth. Enzymol.* 293, 123–145, 1998).

[f] Rate constant with thiolate anion (pH independent rate constant).

[g] Estimated from graphical data of rate versus pH or pD (as applicable).

[h] Rate of reaction of HNE with cysteine was approximately 1000 faster than with either lysine or histidine.

TABLE 5.4 (continued)
Comparison of Rates of Reaction of Sulfhydryl Reagents with Cysteine Residues in Model Compounds[a]

[i] At pH 9.0, the rate approaches $10^9\ M^{-1}s^{-1}$ which is considered diffusion-limited (Pinto, A.F., Rodriguez, J.V., and Teixeiro, M., Reductive elimination of superoxide: Structure and mechanism of superoxide reductases, *Biochim. Biophys. Acta* 1804, 285–297, 2010). Reaction was performed at 8°C in order to obtain useful data.

[j] pH-independent rate constant assuming a pK_a of 9.2 for the glutathione thiol function.

References for Table 5.4

1. Baines, B.S. and Brocklehurst, K., A thiol-labelling reagent and reactivity probe containing electrophilic mercury and a chromophoric leaving group, *Biochem. J.* 179, 701–704, 1979.
2. Hasinoff, B.B., Madsen, N.B., and Avramovic-Zikic, O., Kinetics of the reaction of *p*-chloromercuribenzoate with the sulfhydryl groups of glutathione, 2-mercaptoethanol, and phosphorylase b, *Can. J. Biochem.* 49, 742–751, 1971.
3. Dahl, K.H. and McKinley-McKee, J.S., Enzymatic catalysis in the affinity labelling of liver alcohol dehydrogenase with haloacetate, *Eur. J. Biochem.* 118, 507–513, 1981.
4. Gold, A.M., Sulfhydryl groups of rabbit muscle glycogen phosphorylase b. Reaction with dinitrophenylating agents, *Biochemistry* 7, 2106–2115, 1968.
5. Legler, G., 4,4′-Dinitrophenyldisulfides of different charge types as probes for the electrostatic environment of sulfhydryl groups, *Biochim. Biophys. Acta* 405, 136–145, 1976.
6. Bednar, R.A., Reactivity and pH dependence of thiol conjugation to N-ethylmaleimide - detection of a conformational changes in chalcone isomerase, *Biochemistry* 29, 3684–3690, 1990.
7. Zhang, H., Le, M., and Means, G.E., A kinetic approach to characterize the electrostatic environments of thiol groups in proteins, *Bioorg. Chem.* 26, 356–364, 1998.
8. Lewis, S.D., Misra, D.C., and Shafer, J.A., Determination of interactive thiol ionizations in bovine serum albumin, glutathione, and other thiols by potentiometric difference titration, *Biochemistry* 19, 6129–6137, 1980.
9. Wandinger, A. and Creighton, D.J., Solvent isotope effects on the rates of alkylation of thiolamine models of papain, *FEBS Lett.* 116, 116–121, 1980.
10. Zhu, J., Dhimitruka, I., and Pei, D., 5-(2-aminoethyl) dithio-2-nitrobenzoate as a more base-stable alternative to Ellman's reagent, *Org. Lett.* 6, 3809–3812, 2004.
11. Sluyterman, L.A., The rate-limiting reaction in papain action as derived from the reaction of the enzyme with chloroacetic acid, *Biochim. Biophys. Acta* 151, 178–187, 1968.
12. Friedman, M., Cavins, J.F., and Wall, J.S., Relative nucleophilic reactivities of amino groups and mercaptide ions in addition reactions with α.β-unsaturated compounds, *J. Am. Chem. Soc.* 87, 3672–3682, 1965.
13. LoPachin, R.M., Gavin, T., Petersen, D.R., and Barber, D.S., Molecular mechanisms of 4-hydroxy-2-nonenal and acrolein toxicity: Nucleophilic targets and adduct formation, *Chem. Res. Toxicol.* 22, 1499–1508, 2009.
14. Sanyal, G. and Khalifah, R.G., Kinetics of organomercurial reactions with model thiols: Sensitivity to exchange of the mercurial labile ligand, *Arch. Biochem. Biophys.* 196, 157–164, 1979.
15. Lindorff-Larsen, K. and Winther, J.R., Thiol alkylation below neutral pH, *Anal. Biochem.* 286, 308–310, 2000.
16. Gerwin, B.L., Properties of a single sulfhydryl group of streptococcal proteinase. A comparison of the rate of alkylation by chloroacetic acid and chloroacetamide, *J. Biol. Chem.* 242, 451–456, 1967.

TABLE 5.5
Some Rate Constants for the Modification of Cysteine Thiol Groups in Proteins

Protein	Reagent/Conditions	Rate Constant	Reference
Phosphorylase b	Chlorodinitrobenzene (30°C, pH 8.0)	$60\,M^{-1}\,min^{-1}$	[1]
Phosphorylase b sulfhydryl groups (type 1)[a]	*p*-Chloromercuribenzoate in 0.02 M glycerophosphate-HCl, pH 6.8, 30°C	$2.0 \times 10^{6}\,M^{-1}\,s^{-1}$	[2]
Chalcone isomerase	*N*-Ethylmaleimide, pH 6.8, 25°C	$8.5\,M^{-1}\,s^{-1}$	[3]
Chalcone isomerase	*N*-Ethylmaleimide, pH 8.5, 25°C	$5.7 \times 10\,M^{-1}\,s^{-1}$	[3]
Chalcone isomerase	*N*-Ethylmaleimide, pH 5.8 (50 mM MES) with 6.0 M urea, 25°C	$2.8 \times 10^{3}\,M^{-1}\,s^{-1}$[b]	[3]
Chalcone isomerase	Iodoacetamide, pH 5.2, 25°C	$0.06\,M^{-1}\,s^{-1}$	[3]
Chalcone isomerase	*P*-mercuribenzoate, pH 5.2, 25°C	$2.5 \times 10^{6}\,M^{-1}\,s^{-1}$	[3]
Papain	Chloroacetamide at 25°C	$2.2 \times 10\,M^{-1}\,s^{-1}$[c]	[4]
Hexokinase B	Iodoacetate at 35°C	$3.3\,M^{-1}\,s^{-1}$[d]	[5]
Hexokinase B	Iodoacetamide at 35°C	$14\,M^{-1}\,s^{-1}$[d]	[5]
Papain	Iodoacetamide, pH 6.5, 26°C	$1.44 \times 10^{-1}\,M^{-1}\,s^{-1}$	[6]
Papain	Iodoacetic acid, pH 6.5, 26°C	$2.75\,M^{-1}\,s^{-1}$	[6]
Papain	*N*-ethylmaleimide, pH 6.5, 26°C	$2.55\,M^{-1}\,s^{-1}$	[6]
Papain	*N*-Ethylmaleimide, pH 6.0, 25°C	$6.61 \times 10^{-1}\,M^{-1}\,s^{-1}$	[7]
Yeast alcohol dehydrogenase	*N*-Ethylmaleimide, pH 7.0, 20°C	$1.34 \times 10\,M^{-1}\,min^{-1}$	[8]
pI258 CadC^{-} Cys7[e]	*N*-Ethylmaleimide	$1.93 \times 10^{3}\,M^{-1}\,s^{-1}$	[9]
pI258 CadC^{-} Cys11[e]	*N*-Ethylmaleimide	$\geq 5 \times 10^{3}\,M^{-1}\,s^{-1}$	[9]
C97A, V376C[f]	DTNB/100 mM Tris, pH 7.5, 20°C	$9.4 \times 10^{2}\,M^{-1}\,s^{-1}$	[10]
C97A, A377C[f]	DTNB/100 mM Tris, pH 7.5, 20°C	$3.4 \times 10^{2}\,M^{-1}s^{-1}$	[10]
Ficin	2,2′-Dipyridyl disulfide (as unprotonated disulfide) reaction with thiolate anion	$8.61 \times 10^{4}\,M^{-1}\,s^{-1}$	[11]
Ficin	4,4′-Dipyridyl disulfide (as unprotonated disulfide) reaction with thiolate anion	$8.15 \times 10^{4}\,M^{-1}\,s^{-1}$	[11]
Phosphorylase b[a]	DTNB, pH 6.8	$3.2 \times 10^{3}\,M^{-1}\,s^{-1}$ $3.3 \times 10^{2}\,M^{-1}\,s^{-1}$	[12]
Sperm whale ferric aquomyoglobin[g]	DTNB, 3.0 M guanidine, 50 mM glycine, 2 mM EDTA, pH 9.0	$2.0 \times 10^{4}\,M^{-1}\,s^{-1}$	[13]
Sperm whale ferric aquomyoglobin[g]	MMTS, 3.0 M guanidine, 50 mM glycine, 2 mM EDTA, pH 9.0	$2.2 \times 10^{5}\,M^{-1}\,s^{-1}$	[13]
Bovine Serum Albumin	2,2′-dipyridyl disulfide, pH 2.2[H]	$1.1 \times 10^{2}\,M^{-1}\,s^{-1}$	[14]
Papain	MMTS	$2.1 \times 10^{5}\,M^{-1}\,s^{-1}$	[15,16]
Leukotriene A_4	MMTS, 2°C, pH 7.0 (Mops)	$9 \times 10^{-1}\,M^{-1}\,s^{-1}$	[17]
Hydrolase	MMTS, 2°C, pH 7.0 (Mops with 0.2 M NaCl)	$4.3\,M^{-1}\,s^{-1}$	
Human serum albumin	Hypochlorous acid, pH 7.2	$3 \times 10^{7}\,M^{-1}\,s^{-1}$[i]	[18]

TABLE 5.5 (continued)
Some Rate Constants for the Modification of Cysteine Thiol Groups in Proteins

[a] There are two classes of sulfhydryl groups in phosphorylase b which differ by a factor 10 in reactivity (Damjanovich, S. and Kleppe, K., The reactivity of SH groups in phosphorylase b, *Biochim. Biophys. Acta* 122, 145–147, 1966; Kleppe, K. and Damjanovich, S., Studies on the SH groups of phosphorylase b. Reaction with 5,5′-dithiobis-(2-nitrobenzoic acid), *Biochim. Biophys. Acta* 185, 88–102, 1969).

[b] Some uncertainty in this value; at pH > 6.8, the value is > 6 × 10^3. It is necessary to correct for decomposition of *N*-ethylmaleimide at alkaline pH; protein assumed to unfolded under denaturing conditions.

[c] pH independent second order rate constant for reaction with active-site of papain.

[d] Absolute (pH-independent) rate for alkylation of the apparently essential thiol residue in hexokinase.

[e] The apo form of *Staphylococcus aureus* pI258 CadC (a metalloregulatory transcription repressor).

[f] Yeast phosphoglycerate kinase, which contains one cysteine residue at 97 which was mutated to an alanine in the cited studies. Mutations were introduced at V375 or A377 to correspond the rapidly reacting cysteine residues in horse muscle phosphoglycerate kinase. Addition of substrates markedly altered reactivity; the addition of MgATP to C97A, A377C reduced the rate of modification to 27 $M^{-1}s^{-1}$. The rate of DTNB with N-acetylcysteine was 1.2 × 103 $M^{-1}s^{-1}$ under the same conditions.

[g] Sperm whale ferric myoglobin with H_2O bound to Fe^{3+}.

[h] pH dependence showed two optima with a peak at ~pH 2–3 with a minimum at pH 4–6 with rate to rapid to accurately measure above pH 7.

[i] Rate to rapid measure with accuracy, value obtained by modeling.

References for Table 5.5

1. Ratner, V., Kahana, E., Eichler, M., and Haas, E., A general strategy of the site-specific double labeling of globular proteins for kinetic FRET studies, *Bioconj. Chem.* 12, 1163–1170, 2002.
2. Malthouse, J.P.G. and Brocklehurst, K., A kinetic method for the study of solvent environments of thiol groups in proteins involving the use of a pair of isomeric reactivity probes and a differential solvent effect. Investigation of the active centre of ficin using 2,2′ and 4,4′dipyridyl disulphides as reactivity probes, *Biochem. J.* 185, 217–222, 1980.
3. Feng, Z., Butler, M.C., Alam, S.L., and Loh, S.N., On the nature of conformational openings: Native and unfolded-state hydrogen and thiol-disulfide exchange studies in ferric aquomyoglobin, *J. Mol. Biol.* 314, 153–166, 2001.
4. Chaiken, I.M. and Smith, E.L., Reaction of chloroacetamide with the sulfhydryl group of papain, *J. Biol. Chem.* 244, 5087–5094, 1969.
5. Jones, J.G., Otieno, S., Barnard, E.A., and Bhargava, A.K., Essential and nonessential thiols of yeast hexokinase, *Biochemistry* 14, 2396–2403, 1975.
6. Brubacher, L.J. and Glick, B.R., Inhibition of papain by N-ethylmaleimide, *Biochemistry* 13, 915–920, 1974.
7. Bednar, R.A., Reactivity and pH dependence of thiol conjugation to N-ethylmaleimide: Detection of a conformational change in chalcone isomerase, *Biochemistry* 29, 3684–2690, 1990.
8. Heitz, J.R., Anderson, C.D., and Anderson, B.M., Inactivation of yeast alcohol dehydrogenase by N-alkylmaleimides, *Arch. Biochem. Biophys.* 127, 627–636, 1968.
9. Apuy, J.L., Busenlehner, L.S., Russell, D.H., and Giedroc, D.P., Radiometric pulsed alkylation mass spectrometry as a probe of thiolate reactivity in different metalloderivatives of *Staphylococcus aureus* pI258 CadC, *Biochemistry* 43, 3824–3834, 2004.
10. Minard, P., Dedsmadril, M., Ballery, N. et al., Study of the fast-reacting cysteines in phosphoglycerate kinase using chemical modification and site-directed mutagenesis, *Eur. J. Biochem.* 185, 419–423, 1989.

(continued)

TABLE 5.5 (continued)
Some Rate Constants for the Modification of Cysteine Thiol Groups in Proteins

11. Gold, A.M., Sulfhydryl groups of rabbit muscle glycogen phosphorylase *b*. Reaction with dinitrophenylating agents, *Biochemistry* 7, 2106–2115, 1968.
12. Hasinoff, B.B., Madsen, N.B., and Avramovic-Zikic, O., Kinetics of the reaction of *p*-chloromercuribenzoate with the sulfhydryl groups of glutathione, 2-mercaptoethanol, and phosphorylase *b*, *Can. J. Biochem.* 49, 742–751, 1971.
13. Sanner, T. and Tron, L., Properties of the highly reactive SH groups of phosphorylase *b*, *Biochemistry* 14, 230–235, 1975.
14. Svenson, A. and Carlsson, J., Thiol group of bovine serum albumin. High reactivity at acidic pH as measured by reaction with 2,2′-dipyridyl disulfide, *Biochim. Biophys. Acta* 400, 433–438, 1975.
15. Roberts, D.D., Lewis, S.D., Ballou, D.P. et al., Reactivity of small thiolate anions and cysteine-25 in papain toward methane methylthiosulfonate, *Biochemistry* 25, 5595–5601, 1986.
16. Evans, B.L.B., Knopp, J.A., and Horton, H.R., Effect of hydroxynitrobenzylation of tryptophan-177 on reactivity of active site cysteine-25 in papain, *Arch. Biochem. Biophys.* 206, 362–371, 1961.
17. Orning, L. and Fitzpatrick, F.A., Modification of leukotriene A_4 hydrolase/aminopeptidase by sulfhydryl-blocking reagents: Differential effects on dual enzyme activities by methyl-methane thiosulfonate, *Arch. Biochem. Biophys.* 368, 131–138, 1999.
18. Pattison, D.I. and Davies, M.I., Absolute rate constants for the reaction of hypochlorous acid with protein side chains and peptide bonds, *Chem. Res. Toxicol.* 14, 1453–1464, 2001.

pattern. Other probes (Figure 5.19) include 4-maleimido-2,2,6,6-tetramethylpiperidine-1-oxyl,[347] maleimidotetra-methylrhodamine,[347] N-(1-pyrenyl) maleimide,[348] 2-(4-maleimidoanilino) naphthalene-6-sulfonic acid,[349] 2,5-dimethoxy-4-stilbenzylmaleimide,[349] rhodamine maleimide,[349] and eosin-5-maleimide.[350]

Le-Quoc and colleagues have examined the effect of the nature of the *N*-substituent groups (Figure 5.20) on the rate of sulfhydryl group modification in membrane-bound succinate dehydrogenase.[351] The derivatives used were *N*-ethylmaleimide, *N*-butylmaleimide, and *N*-benzylmaleimide. The rate of modification (inactivation) of membrane-bound succinate dehydrogenase correlated with the octanol-water partitioning coefficient. The partition coefficient for NEM is 3 and the rate of modification was $3.67\,mM^{-1}\,min^{-1}$; the partition coefficient for *N*-benzylmaleimide was infinite (could not be measured) and the rate of modification was $15.4\,mM^{-1}\,min^{-1}$. *N*-Butylmaleimide was intermediate. Later studies by various investigators[352–357] extended information on the use of hydrophobic maleimide derivatives including N-polymethoxycarboxymaleimide,[352,353] *N,N′*-*o*-phenylenedimaleimide,[354] *N*-tetramethylrhodamine maleimide,[355] aryl-indole maleimide,[356] and *N*-arachidonylmaleimide.[357] *N*-Pyrene maleimide is also a fluorescent probe.[358]

Hydrophobic derivatives of *N*-ethylmaleimide (Figure 5.20) can be used as probes of the environment surrounding a sulfhydryl group in membrane anion channels.[359] Reaction with *N*-ethylmaleimide, *N*-benzylmaleimide, and *N,N′*-1,2-phenylenedimaleimide was evaluated and the reaction rate increased with increasing hydrophobicity. *N*-Phenylmaleimide has been used for the modification of a

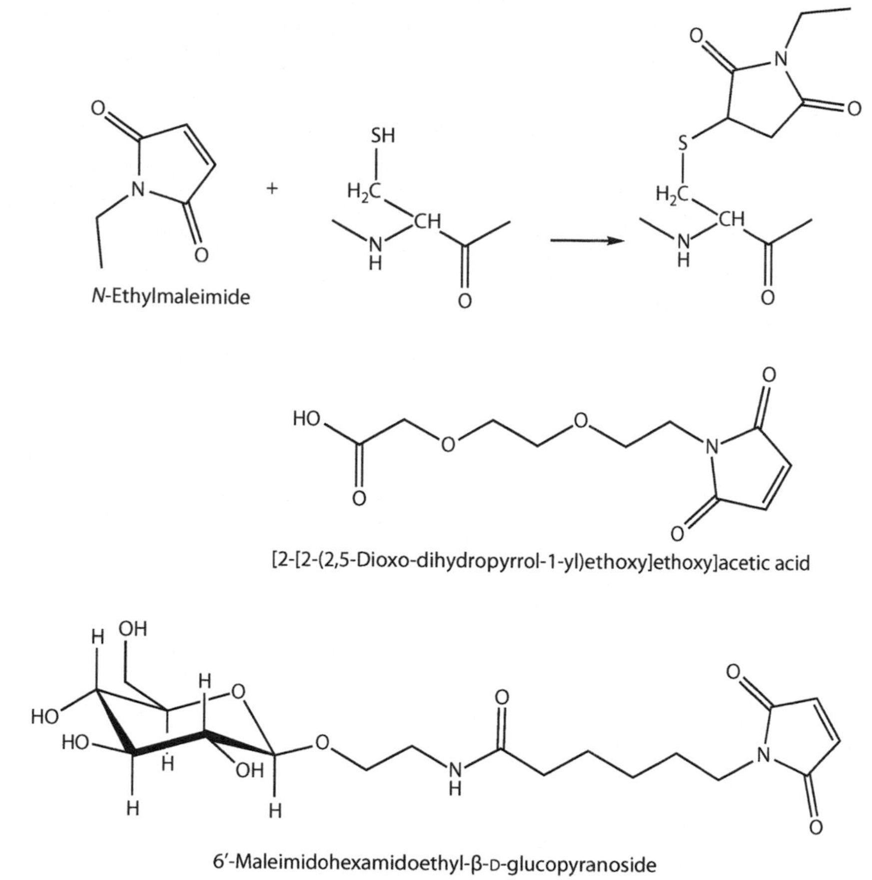

N-Ethylmaleimide

[2-[2-(2,5-Dioxo-dihydropyrrol-1-yl)ethoxy]ethoxy]acetic acid

6′-Maleimidohexamidoethyl-β-D-glucopyranoside

FIGURE 5.19 The use of *N*-alkylmaleimide chemistry for modification of cysteine. *N*-Ethylmaleimide is the best example and the radiolabeled form is quite useful (Guan, L. and Kaback, H.R., Site-directed alkylation of cysteine to test solvent accessibility of membrane proteins, *Nat. Protoc.* 2, 2012–2017, 2007). Also shown is [2[2-(2,5-dioxo-dihydropyrrol-1-yl)ethoxy]ethoxy]acetic acid which was used in a study of differential reactivity of haloalkyl and maleimido derivatives (Schelté, P., Boeckler, C., Frisch, B., and Schuber, F., Differential reactivity of maleimide and bromoacetyl functions with thiols: Application to the preparation of liposomal diepitope constructs, *Bioconj. Chem.* 11, 118–123, 2000) and 6′-maleimidohexamidoethyl-β-D-glucopyranoside (Ni, J., Singh, S., and Wang, L.-X., Synthesis of maleimide-activated carbohydrates as chemoselective tags for site-specific glycosylation of peptides and proteins, *Bioconj. Chem.* 14, 232–238, 2003).

sulfhydryl group in the acetylcholine receptor.[360,361] Detergent was required for the modification reaction (10 mM MOPS to 100 mM NaCl to 0.1 mM EDTA with 0.02% sodium azide and 1% cholate). These studies identified cysteine residues potentially important in membrane function. Subsequent studies using site-specific mutagenesis have supported the importance of these cysteinyl residues.[361] Site-directed modification of cysteine has been demonstrated to be useful in the study of the relationship between structure and function in proteins. An equal number of

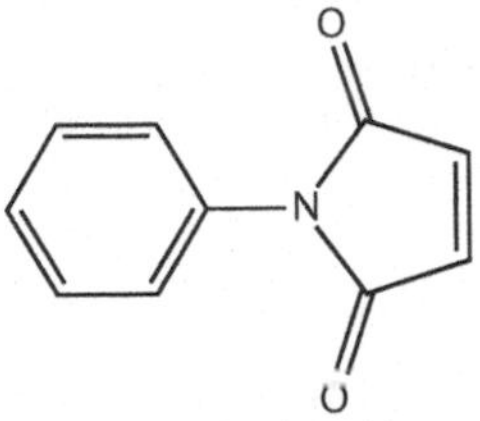

N-Phenylmaleimide

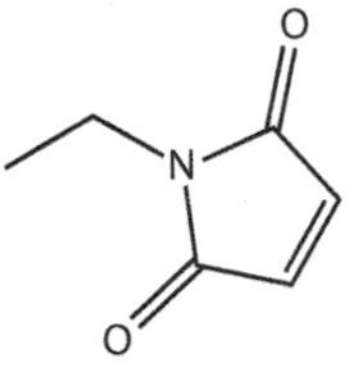

N-Ethylmaleimide

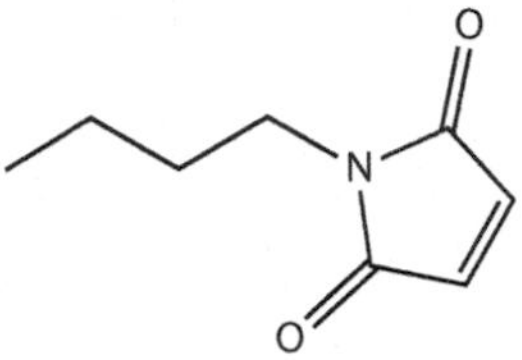

N-Butylmaleimide

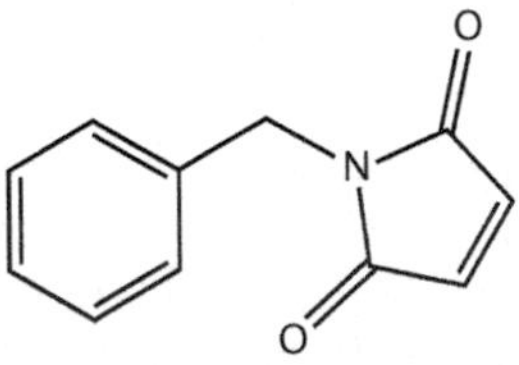

N-Benzylmaleimide

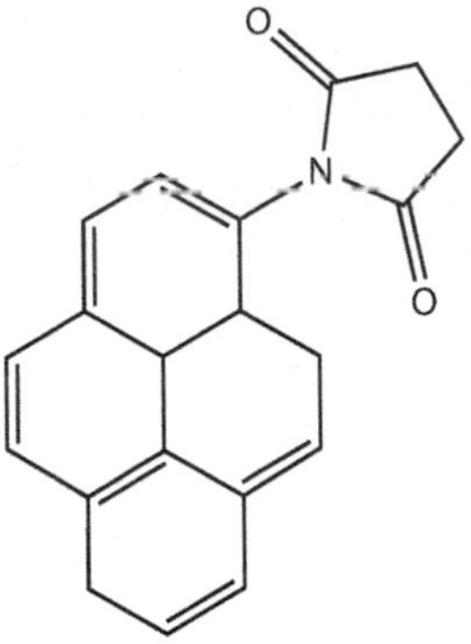

N-(3-Pyrene)-maleimide

FIGURE 5.20 Some hydrophobic maleimide chemistry for membrane protein modification. Shown is *N*-phenylmaleimide which has been used to modify cysteine residues in a hydrophobic environment (Indiveri, C., Giangregorio, N., Iacobazzi, V., and Palmeri, F., Site-directed mutagenesis and chemical modification of the six native cysteine residues of the rat mitochondrial carnitine carrier: Implications for the role of Cysteine-136, *Biochemistry* 41, 8649–8656, 2002) and three maleimide derivatives which were used to differentiate sulfhydryl reactivity with a membrane-bound succinate dehydrogenase (LéQuôc, K., LéQuôc, D., and Gauderner, Y., Evidence for the existence of two classes of sulfhydryl groups essential for membrane-bound succinate dehydrogenase activity, *Biochemistry* 20, 1705–1710, 1981). *N*-Pyrene maleimide is a hydrophobic fluorescent probe for membrane structure (Brown, R.D. and Matthews, K.S., Chemical modification of lactose repressor protein using N-substituted maleimides, *J. Biol. Chem.* 254, 5128–5134, 1979; Mancek-Keber, M., Gradisar, N., Iñigo Pestaña, M. et al., *J. Biol. Chem.* 284, 19493–19500, 2009).

proteins where endogenous cysteine residues have been modified are included in this tabulation for comparison of reaction conditions.

Localization of sulfhydryl groups within membranes has been achieved through the comparison of the reaction with membrane-permeant and membrane-impermeant maleimide derivatives.[362–364] The use of these reagents can be traced back to the original observations of Abbott and Schachter in 1976.[365] The basic concept is to provide either a polar derivative or a derivative with steric considerations which preclude passage through or into the membranes. Maleimide derivatives of glucosamine have been synthesized as affinity labels for the human erythrocyte hexose transport protein.[366] A particularly novel approach to this problem has been used by Falke and Koshland to analyze aspartate receptor structure.[367] In this study, site-specific mutagenesis was used to place cysteinyl residues at six positions in the peptide chain. A novel membrane-impermeant maleimide derivative [*N*-(6-phosphonyl-*n*-hexyl)-maleimide] was used to study the reactivity of the individual sulfhydryl residues. From these studies, it was possible to "map" the domain structure of the receptor protein. Cysteinyl residues placed in the surface area could be modified by aqueous reagents, while transmembrane areas could be excluded by lack of reaction with membrane-impermeant reagents.

Modification with *N*-ethylmaleimide has found considerable use in the modification of inserted cysteine residues in the study of transmembrane proteins.[368] The synthesis of deuterated *N*-alkylmaleimides has been reported[369] and these reagents have been used for quantitative peptide analysis in the study of differential protein expression. An electroactive probe has been developed which uses maleimide coupled to sulfhydryl groups in proteins.[370]

Pulse-chase labeling using deuterated *N*-ethylmaleimide has been used to determine the relative reactivity of cysteine residues in the zinc finger regions of the MRE-binding domain of MRE-binding transcription factor-1 (MTF-1).[371] The results are used as a surrogate measurement of the zinc chelate stability of individual zinc fingers in MTF-1. As with the reaction of metallothionein with aromatic mercurials or *N*-ethylmaleimide discussed earlier, thiol (thiolate) reactivity with maleimide in MRE-binding domain is more rapid in the absence of metal ion. In these experiments, the protein is first reacted with a 10-fold molar excess (to cysteine) of d_5-*N*-ethylmaleimide and at various time points, portions are removed and added to a solution containing H_5-*N*-ethylmaleimde (at a 100-fold molar excess to cysteine) and trypsin or chymotrypsin. Mass spectrometric analysis is used to determine the ratio of deuterium to hydrogen label at each cysteine residue which is, in turn, a measure of reactivity and can be used to derive second-order rate constants for the alkylation reaction at each individual cysteine. Cysteine reactivity was also influenced by the binding of oligonucleotide. This approach was subsequently used by Meza and coworkers[372] to study the effects of oxidation on estrogen receptor. Incomplete reaction with cysteine in the initial alkylation required the use of a chaotropic agent. Apuy and coworkers[373] applied this technology to the study of thiolate reactivity in a metal-regulated transcriptional repressor. It is noted that Apuy and coworker originally used the radiometric pulsed alkylation technique for the study of protein folding.[374]

Maleimide poly(ethylene) glycol derivatives have been used to modify cysβ93 in human adult hemoglobin.[375] Subject maleimides include maleimide phenyl carbamyl (*O″*-methyl PEG 5000) and *O′*-(2-maleimidoethyl) –*O′*–methyl-PEG5000.

Modification with the various N-alkylmaleimides was accomplished in phosphate-buffer saline, at pH 7.4 at various temperatures. Other applications of PEG-maleimide derivatives include the modification of cysteine β93 in human hemoglobin,[375] recombinant engineered human GM-CSF,[376,377] an erythropoietin mutant with an engineered cysteine,[378] and recombinant human interleukin-1 receptor.[379]

A related study[380] describes the synthesis of maleimide derivatives of carbohydrates as "chemoselective" tags for site-specific glycosylation of proteins. Amino derivatives of carbohydrates were first coupled to 6-maleimidohexanoic acid *N*-hydroxysuccinimide ester to form the 6′-maleimdohexanamidoethylglycosides which can then be coupled to a cysteinyl residue in the protein. This is an interesting approach within the area of glycoengineering.[381] Maleimide was used for the preparation of carbohydrate microarrays.[382,383] A related application is the coupling of doxorubicin to lactosamined human albumin for drug delivery to liver cells via the asialoglycopotein receptor.[384]

A novel maleimide reagent has been developed for the quantitative analysis of protein mixtures.[385] Acid-labile isotope-coded extractants (ALICE) are *N*-alkyl maleimide derivatives (Figure 5.21) where there is an acid labile function between the alkylmaleimide and a polymer matrix. Cysteine-containing peptides are "captured" by reaction with the maleimide function bound via linker which can contain either hydrogen or deuterium to a matrix via an acid-labile bond. The bound polymer-peptide can be removed by filtration and the peptides released from the matrix with 5% trifluoroacetic acid in dichloromethane. Chowdhury and coworkers[386] isolated phosphopeptides by converting phosphopeptides to thiol-containing peptides by coupling with an alkyl dithiol (either ethanedithiol or propanedithiol), followed by capture of the modified peptide with a maleimide function linked to a matrix by an acid-labile bond. A similar approach was used by other investigators[387] with an iodoacetyl-matrix for the capture of thiol peptide derived from phosphopeptides. Fauq and coworkers[388] used an acid-labile biotin reagent which can contain either maleimide or haloalkyl function.

Alkyl alkanethiosulfonates (e.g., MMTS) (Figure 5.22) have been extensively used in the past decade for the modification of cysteine residues in proteins. Methyl, ethyl, and trichloromethyl derivatives of methanethiosulfonate (MTS) and propyl-propanethiosulfonate were described by Smith and colleagues in 1975.[389] These alkyl alkanethiosulfonates form mixed disulfides with sulfhydryl group of cysteine with the release of methylsulfinic acid. The mixed disulfides are going to be of variable stability, more stable than thio esters but less stable than thio ethers. It is likely that the stability of the mixed disulfide is influence by the chemistry of the alkyl function. The modification is highly specific for sulfhydryl groups and is easily reversed by mild reduced agents such as 2-mercaptoethanol or DTT. Reaction could occur at sites other than cysteine in proteins.[390] Such reactions would most likely occur at amino groups and these modifications would not be reduced by reducing agents such as DTT or 2-mercaptoethanol. Kluger and Tsui[390] observed that the inactivation of D-3-hydroxy butyrate dehydrogenase by MMTS could not be reversed with hydroxylamine, 2-mercaptoethanol, or tributylphosphine suggesting that reaction occurred at residue other than cysteine. Modification of the enzyme with 2,3-dimethylmaleic anhydride, a reversible modifier of amino groups in proteins, protected the enzyme from inactivation. NMR studies on the reaction of MMTS with gly–gly suggested

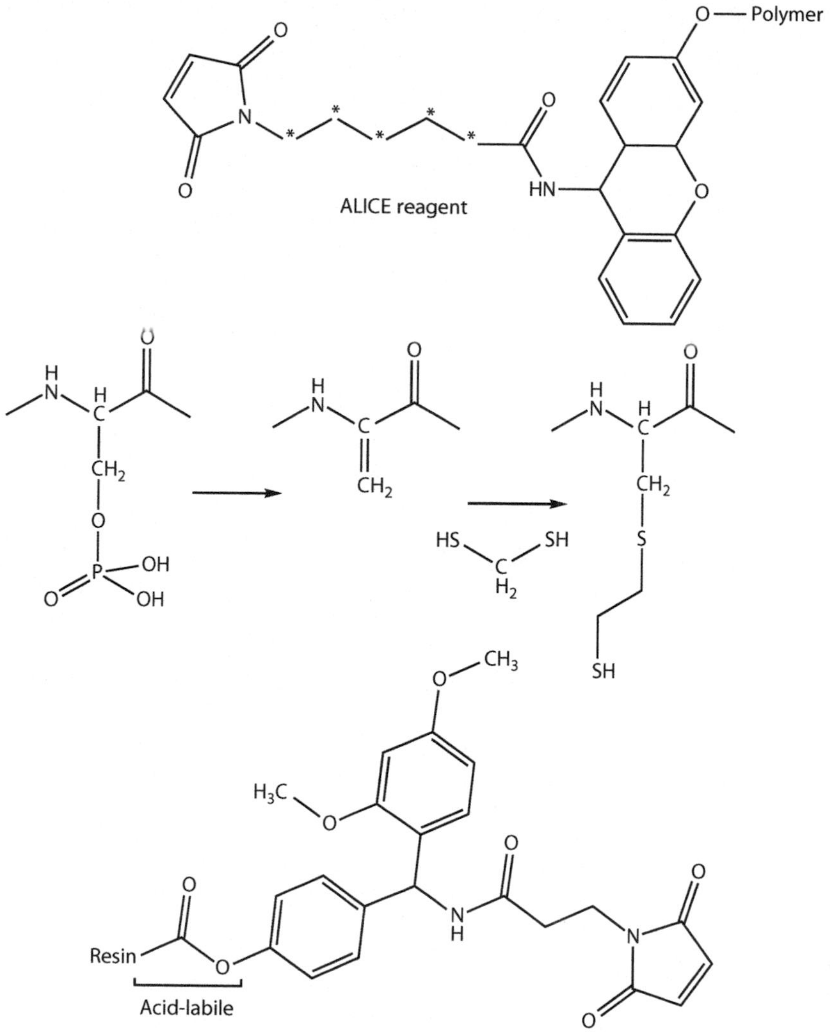

ALICE reagent

Resin

Acid-labile

FIGURE 5.21 Various acid-labile capture reagents for the selective isolation of cysteine peptides. The structure at the top of the figure is the ALICE reagent developed for the selective isolation of cysteine peptides (Qiu, Y., Sousa, E.A., Hewick, R.M., and Wang, J.H., Acid-labile isotope-coded extractants: A class of reagents for quantitative mass spectrometric analysis of complex protein mixtures, *Anal. Chem.* 74, 4969–4979, 2002). The asterisk (*) marked carbon can contain hydrogen or deuterium providing the basis for use as an isotope-coded reagent (Butler, G.S. and Overall, C.M., Proteomic validation of protease drug targets: Pharmacoproteomics of matrix metalloproteinase inhibitor drugs using isotope-coded affinity tag labelling and tandem mass spectrometry, *Curr. Pharm. Des.* 13, 263–270, 2007). The lower part of the figure shows the well-known base-catalyzed β-elimination of an *O*-substituted serine yielding a unsaturated double bond amenable to Michael addition by 1,2-dithiolethane, in turn, providing a free sulfhydryl group available for reaction with a maleimide-based reagent (Chowdhury, S.M., Munske, G.R., Siems, W.F., and Bruce, J.E., A new maleimide-bound acid-cleavable solid-support reagent for profiling phosphorylation, *Rapid Commun. Mass Sp.* 19, 899–909, 2005).

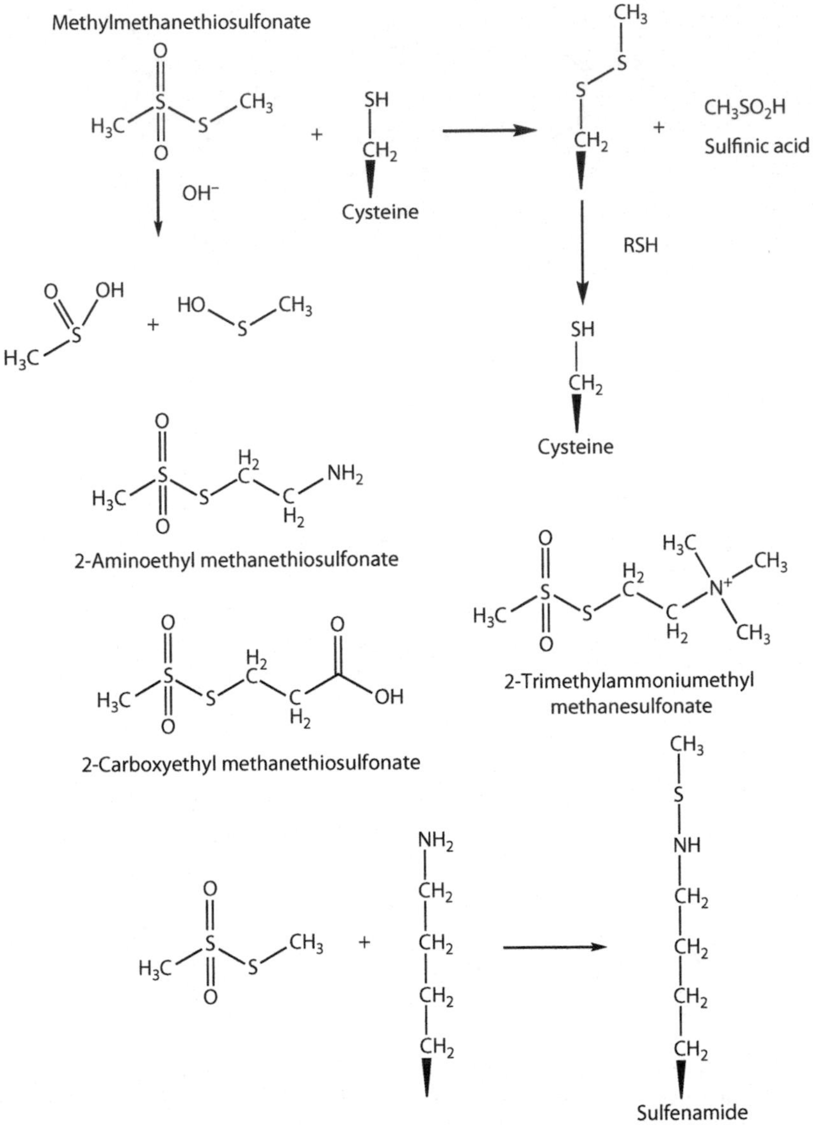

FIGURE 5.22 The reaction of alkylthiosulfonates with function groups in proteins. Shown is the reaction of MMTS with a cysteine sulfhydryl groups in proteins. Also shown are various alkyl methanethiosulfonates which are charged. The sulfonic acid and trimethylammonium derivatives are membrane-impermeant. The formation of the mixed disulfide is, in principle, reversible in the presence of added sulfhydryl reagents such as DTT, 2-mercaptoethanol, or trialkylphosphines. Also shown is the putative reaction of MMTS with an amine to form a sulfenamide derivative (Kluger, R. and Tsui, W.-C., Amino group reactions of the sulfhydryl reagent methyl methanethiosulfonate. Inactivation of D-3-hydroxybutyrate dehydrogenase and reaction with amines in water, *Can. J. Biochem.* 58, 639–632, 1980). Also shown at the top is the hydrolysis of MMTS which is a significant issue (Karlin, A. and Akabas, M.H., Substituted cysteine accessibility method, *Methods Enzymol.* 293, 123–145, 1998).

the formation of a sulfenamide derivative. While, to the best of my knowledge, there are no reports of the modification of amino groups in proteins with alkylthiosulfonates, there are modifications with alkylthiosulfonates which are only partially reversed on reduction.[391] Pathak and coworkers modified a cysteine residue in the Wbp1p subunit of *Saccharomyces cerevisiae* oligosaccharyl transferase MMTS. Catalytical activity was lost on reaction with MMTS but not by reaction with iodoacetate, iodoacetamide, or *N*-ethylmaleimide. Activity in the MMTS-modified enzyme could be partially recovered (50%) with DTT; the use of tributylphosphine resulted in the loss of activity in the modified and unmodified enzyme. Nishimura and coworkers[392] reported one of the early studies on the reactivation of MMTS-modified proteins. These investigators reported the inactivation of *E. coli* succinic thiokinase with either MMTS or methoxycarbonylmethyl disulfide and reactivation with tributylphosphine but not with DTT. Other studies do suggest that the reaction of MMTS-inactivated proteins is not a trivial matter. The failure of DTT to recover activity while recovery was observed with tributylphosphine is similar to the pattern observed with cysteine sulfenic acid (see above). Roberts[393] compared the reactivity of MMTS with several small thiols and the active site cysteine in papain. The reaction of MMTS with the thiolate is at least 5×10^9 times more rapid than the corresponding thiol and the rate of reaction depends on thiol basicity. The reader is also directed to a study by Stauffer and Karlin[394] on the effect of ionic strength on the reaction of alkylthiosulfonates with simple and protein-bound thiols. The reaction of 2-aminoethyl methanethiosulfonate, [2-(trimethylammonium) ethyl] methanesulfonate, (2-sulfonatoethyl)methane thiosulfonate, and MMTS (Figure 5.22) with the acetylcholine binding site on the acetylcholine receptor was studied as a function of ionic strength. The goal of the study was to evaluate long-range electrostatic interactions in the acetylcholine binding site. The positively charged (2-aminoethyl) methanethiosulfonate and [2-(trimethylamino) ethyl] methanethiosulfonate reacted more rapidly with simple thiols such as 2-mercaptoethanol, 2-mercaptoethylamine, and 2-mercaptoethylsulfonic acid than did the negatively charged (2-sulfonatoethyl) methane-thiosulfonate. Both the negatively and positively charged derivatives were more active than the neutral MMTS. Both positively charged MMTS derivatives reacted much more rapidly with the sulfhydryl group in the acetylcholine binding site than did either the neutral MMTS which was, in turn, far more potent than the negatively charged reagent. There was a significant effect of ionic strength on reaction of the positively charged reagents consistent with the presence of two or three negative charges at the binding site. In later work,[395,396] Kenyon and Bruice extended the understanding of the use of this class of reagents as part of a more general review of the modification of cysteine residues. Most work has used alkyl derivatives of MTS such that these reagents are frequently referred to as MTS reagents.[397]

The alkyl alkanethiosulfonates (MTS derivatives) have received much more attention as reagents for the determination of cysteine accessibility than as reagents for determination of functionally important residues in proteins. Karlin and Akabas developed the method of substituted cysteine accessibility modification[398] (SCAM[399] scanning cysteine accessibility mutations) as systematic method for inserting cysteine residues using standard methods of mutagenesis and then probing the availability of the inserted cysteine residues for modification. In the most simple terms, MTS reagents can be considered large, small, positively charged, negatively charged, and neutral. The stability of reagents must be considered as the rate of hydrolysis (Figure 5.22) does vary with the alkyl function. The rate of hydrolysis of (2-aminoethyl) methane thiolsulfonate is more rapid ($t_{1/2}$ ~ 12 min) at pH 7.0 (20°C) than at pH 6.0 (20°C) ($t_{1/2}$ ~ 92 min); at lower temperature (4°C), the $t_{1/2}$ at pH 7.0 is ~116 min. (2-Sulfonatoethyl) methane thiosulfonate is considerably more stable ($t_{1/2}$ ~ 370 min) at pH 7.0 than the positively charged (2-aminoethyl) methanethiosulfonate. As with other studies,[394] these investigators also show that the positively charged reagents react more rapidly with 2-mercaptoethanol than either negatively charged or neutral MTS reagents.

A number of studies have used MTS reagents to probe cysteine reactivity in membranes[400,401] with particular emphasis on ion channels.[397–412] Membrane permeability is dependent on charge and hydrophobicity. Karlin and Akabas do report octanol/water partitioning data. Generally a charged reagent is not membrane permeable; with the primary amino reagent, charge would be pH dependent. The use of pulse-labeling to study conformation effects on cysteine residue reactivity has been discussed earlier with the use of *N*-ethylmaleimide.[370–373] Jha and Udgaonkar[413] used MMTS to study the folding of a small protein, barstar. As they needed a "baseline" value for cysteine reaction, a value of 4.8×10^5 M^{-1} s^{-1} was obtained for a cysteine (Cys82) residue in urea-denatured proteins (pH 9.2, 25°C).

Reaction with [2-(trimethylammonium) ethyl]methanethiosulfonate, a membrane-impermeant derivative, has been used to study a neuronal glutamate transporter.[400] An externally accessible cysteine was introduced into excitatory amino transporter-1 for the site-directed introduction of a fluorescent maleimide reporter group. Net current induced by aspartate in the oocyte system is blocked with [2-(trimethylammonium) ethyl] methanethiosulfonate, (2-aminoethyl) methanethiosulfonate, and β-maleimidopropionic acid. Subsequent reaction with DTT restored current reduced by (2-aminoethyl) methanethiosulfonate but not by β-maleimidopropionic acid. This is the behavior expected for a mixed disulfide formed by reaction of cysteine with (2-aminoethyl) methanethiosulfonate and for the thioether formed by the alkylation of cysteine with the maleimide derivative. The modification reactions were performed in Tris/HEPES buffer at pH 7.5.

MTS derivatives have been used to study the mitochondrial ornithine/citrulline carrier.[401] Both the ornithine/ornithine (antiport) and ornithine/H^+ (unidirectional) transport modes were inhibited by the MTS derivatives and other sulfhydryl reagents. (2-aminoethyl) methanethiosulfonate was the most effective in the antiport transport activity (IC_{50} 0.17 μM) with [2-(trimethylammonium) ethyl] methanethiosulfonate (IC_{50} 0.42 μM) and (2-sulfonatoethyl) methanethiosulfonate (IC_{50} 200 μM). Reference to the partition coefficients obtained by Karlin and Akabas[398] will permit a correlation between the relative effectiveness of the various MTS reagents as inhibitors in this complex system and their ability to cross a membrane as determined in the water-*n*-octanol system.

The production of recombinant proteins in bacterial expression systems such as *E. coli* frequently involves reduction under denaturing conditions, renaturation, and purification.[414–416] It is useful to utilize the purification step prior to renaturation/refolding but is difficult to do with free sulfhydryl groups. Inoue and coworkers[417] blocked the sulfhydryl groups in recombinant BDNF (brain-derived neurotrophic factor) with trimethylammoniopropyl methanethiosulfonate. The derivatized protein was partially purified by chromatography on SP-Sepharose® in the presence of urea. The blocking groups are removed by reduction and allowed to refold in the presence of oxidized glutathione, DTT, and a lower urea concentration.[418] The use of trimethylammoniopropyl methanethiosulfonate is an example of protein cationization which is useful for internalizing proteins into cells.[419,420]

The reaction between the alkyl alkanethiosulfonates and cysteine residues in proteins is an example of the formation of a mixed disulfide (Figure 5.23). There are other examples of the use of this reaction for the site-specific modification of cysteine residues in proteins. Cystine or cystamine have proved effective in the inactivation of guanylate cyclase.[421] This reaction involves the formation of a mixed disulfide and is easily reversed by the addition of DTT. The formation of a mixed disulfide can be regarded as an oxidative reaction when two thiols are involved and is referred to as *S*-thiolation and is observed with a variety of proteins.[422–426] The chemistry of these reactions is poorly understood. Guanylate cyclase is activated by oxidizing agents such as nitric oxide[427,428] under conditions consistent with the formation of cysteine sulfenic acid; the tendency of cysteine sulfenic acid to form mixed disulfide bonds is discussed earlier.

The mixed disulfides discussed earlier are part of redox systems involved in biological control. Mixed disulfide formation can be used as a specific chemical modification in proteins. Cunningham and Nuenke[429,430] demonstrated that the controlled oxidation of a protein sulfhydryl group with iodine resulted in the formation of cysteine sulfenyl iodide which can be used to reversibly modify proteins via the formation of a mixed disulfide.[430] There has been limited application of this technology as (1) it is technically challenging and (2) similar derivatives can be formed via MTS technology. Cupa and Pace[431] used iodine oxidation to form a variety of mixed disulfides with the single sulfhydryl group in β-lactoglobulin B. Mixed disulfides with mercaptoethyl amine (2-mercaptoethylamine), mercaptopropyl (1-propanethiol), mercaptopropylcarboxyl (3-mercaptopropionic acid), and mercaptohydroxyethyl (2-mercaptoethanol) were obtained. All of the derivatives were less stable (urea denaturation) than the native protein with the mercaptoethyl amine being the least stable. A related approach was recently developed by Mecinović and coworkers to measure the nucleophilicity of cysteine residues in prolyl hydroxylase domain 2.[432] These investigators used controlled oxidation of cysteine with hydrogen peroxide which could then be coupled with aromatic thiols to form mixed disulfides which were measured with mass spectrometry. A single cysteine residue (Cys20) was identified as being highly susceptible to oxidation; the same residue was preferentially modified with *N*-ethylmaleimide or a spin-label MTS. This approach is based on the assumption that the nucleophilicity (pK_a) of cysteine is a major determinant of susceptibility to oxidation. Sanchez and coworkers[433] have identified cysteine pK_a as a determinant of susceptibility to oxidation; other determinants include distance to another cysteine and solvent accessibility. The reader is also directed to other studies on the relationship of cysteine oxidation and thiol pK_a.[434–436] Wynn and Richards provided a review of the various approaches in using mixed disulfide for the chemical modification of proteins in 1995.[437]

Another modification with similarity to that of alkanethiosulfonates is the reaction of sodium tetrathionate (Figure 5.24) with cysteine residues to yield the

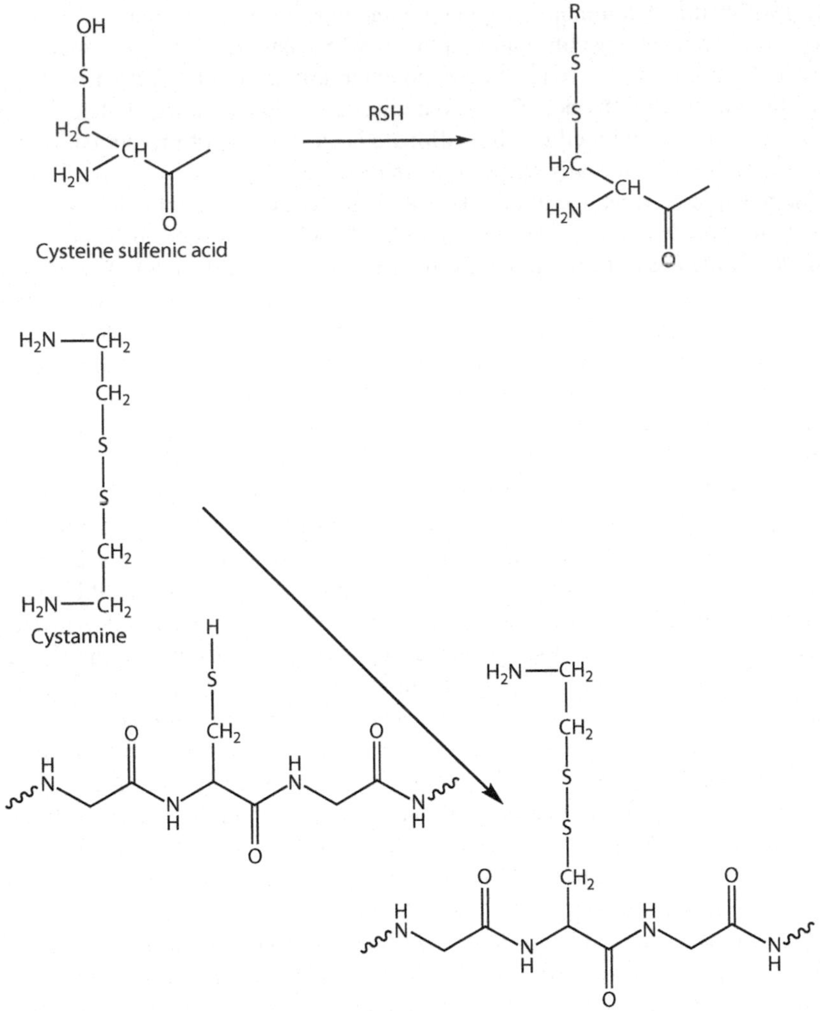

FIGURE 5.23 The formation of mixed disulfides in proteins. Shown is the formation of a mixed disulfide between cysteine sulfenic acid and a thiol such as 2-mercaptoethanol (Peshenko, I.V. and Shichi, H., Oxidation of active center cysteine of bovine 1-Cys peroxiredoxin to the cysteine sulfenic acid form by peroxide and peroxynitrite, *Free Radic. Biol. Med.* 31, 292–303, 2001). Cystamine has been demonstrated to form mixed disulfides with a variety of peptides and proteins (Tamai, K., Shen, H.X., Tsuchida, S. et al., Role of cysteine residues in the activity of rat glutathione transferase P (7–7): Elucidation by oligonucleotide site-directed mutagenesis, *Biochem. Biophys. Res. Commun.* 179, 790–797, 1991; Cappiello, M., Del Corso, A., Camici, M., and Mura, U., Thiol and disulfide determination by free zone capillary electrophoresis, *J. Biochem. Biophys. Methods* 26, 335–341, 1993; Maret, M., Metallothionein/disulfide interactions, oxidative stress, and the mobilization of cellular zinc, *Neurochem. Int.* 27, 111–117, 1995).

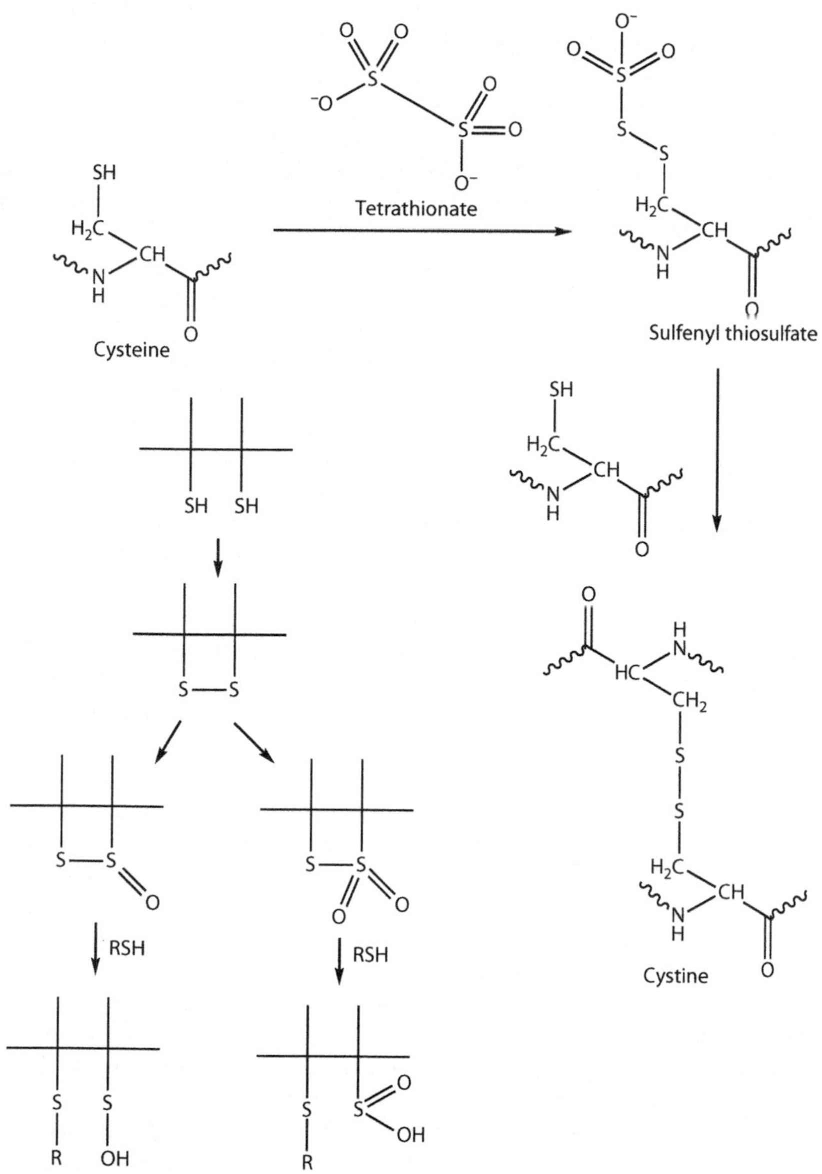

FIGURE 5.24 The formation of *S*-sulfonate, disulfide formation with tetrathionate, and *S*-cysteinyl protein derivatives. Shown is the reaction of sodium tetrathionate with proteins to form *S*-sulfonyl derivatives and subsequent formation of disulfide bonds. Also shown is the electrochemical formation of thiosulfinates and thiosulfonates by electroxidation and subsequent formation of matrix-bound mixed disulfides (Pavlovic, E., Quist, A.P., Gelius, U. et al., Generation of thiolsulfinates/thiosulfonates by electrooxidation of thiols on silicon surfaces for reversible immobilization of molecules, *Langmuir* 19, 4217–4221, 2003).

S-sulfonate derivative[438–441] which is primarily used for the "protection" of cysteinyl residues.[442,443] Kings and coworkers[444,445] have shown that the *S*-sulfonation of cysteinyl residues in transthyretin stabilizes tetramer formation while *S*-cysteinylation promotes dissociation of the native tetramer. Previous work by Zhang and Kelly[446] suggested that the formation of a mixed disulfide of Cys10 with cysteine, GlyCys, or glutathione increased the amyloidicity rate of transthyretin above pH 4.6. Transthyretin modified with sodium tetrathionate (*S*-sulfonation) was less amyloidogenic than native transthyretin. Mixed disulfide formation can be used for the reversible immobilization of thiol proteins. Solid-phase bound thiolsulfonate groups (Figure 5.24) were proposed by Batista-Viera and coworkers[447] almost 20 years ago for the reversible immobilization of thiol proteins. The solid-phase thiol/disulfides were converted to the reactive thiolsulfonate groups with hydrogen peroxide. More recent work by Pavlovic and coworkers[448] has used electrooxidation of vicinal thiols on a silicon surface to produce thiolsulfinate and thiosulfonates. Suárez and coworkers[449] introduced a disulfide into glycoproteins which was subsequently used to anchor the modified protein to a gold surface. The glycoprotein was oxidized with periodate and coupled to cystamine. The imine bond was reduced to a stable amine with sodium cyanoborohydride (which does not reduce disulfide bonds) and allowed to couple with a gold surface. Sodium tetrathionate will also crosslink proteins via sulfhydryl groups.[450–452]

A final example of the formation of mixed disulfide bonds in proteins is provided by the various reagents which form a colorimetric derivative on reaction with cysteinyl residues (Figure 5.25). George Ellman developed bis (*p*-nitrophenyl) disulfide (PNPD) for the measurement of thiols in 1958.[453] This reagent was developed in response to a need for a more quantitative method for the measurement of mercaptans in biological samples. This reagent was effective and sensitive. Reaction with a thiol-containing compound (*p*-nitrobenzylthiol, 2-mercaptoethanol, glutathione) yielded a mixed disulfide and *p*-nitrobenzylthiol. At pH values above 8, the nitrobenzylthiol anion was yellow with a molar extinction coefficient of 13,600 M^{-1} cm^{-1}. While PNPD was useful, solubility problems precluded extensive use. In 1959, Ellman developed DTNB (Ellman's reagent)[454] which today is one of the more popular reagents for the modification and determination of the sulfhydryl group (Figure 5.25). Reaction with sulfhydryl groups in proteins results in the release of 2-nitro-5-mercaptobenzoic acid which has a molar extinction coefficient of 13,600 M^{-1} cm^{-1} at 410 nm. The chemistry of DTNB has been the subject of continuing investigation after the original description with most of the concern focusing on the extinction coefficient.[455–459] General consensus[459] would suggest that a value of 14.15×10^3 M^{-1} cm^{-1} at 25°C and 13.8×10^3 M^{-1} cm^{-1} at 37°C, 412 nm, 0.1 M sodium phosphate, at pH 7.4. While it is important to have accurate values for extinction coefficients, the difference between the original value determined by Ellman in 1959 and the most recent value obtained in 2003 is on the order of 4%.

Riddles and coworkers[457] studied the chemistry of DTNB in alkaline solutions as well as the reaction of DTNB with thiols. They concluded that the rate of reaction of DTNB was dependent on the pH and the pK_a of the thiol residue. The steric and electrostatic considerations which have been discussed previously are also applicable to this reaction. Zhang and coworkers[460] have prepared a positive charged

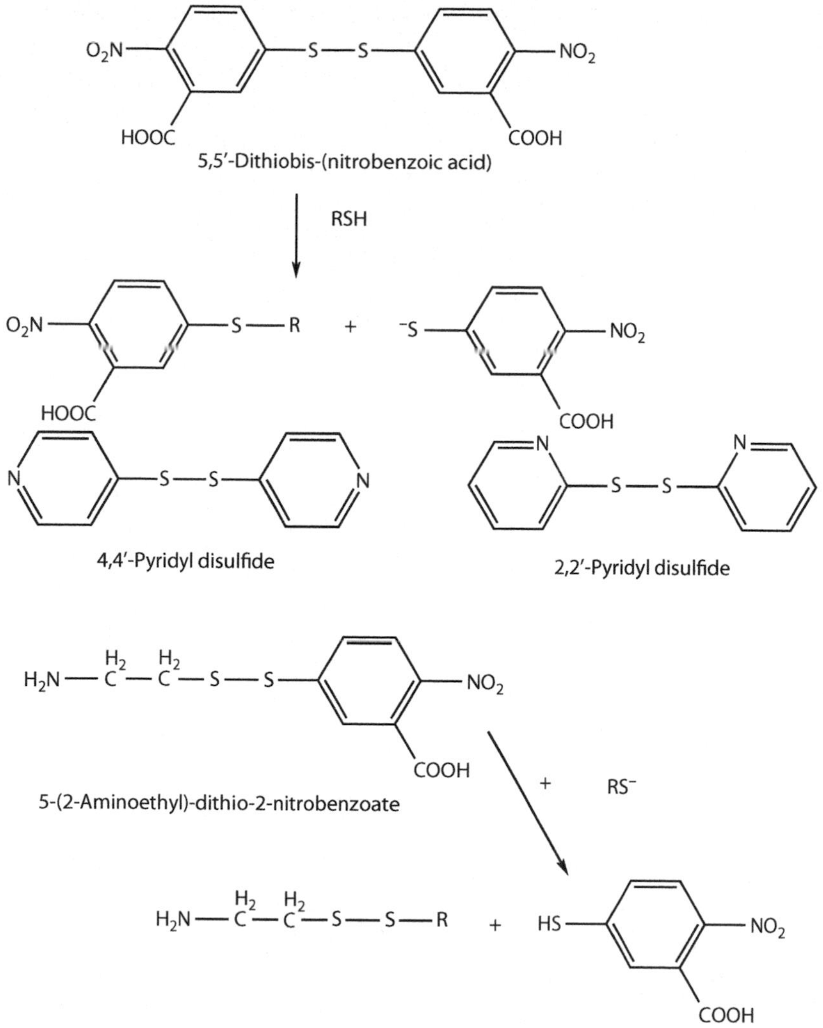

FIGURE 5.25 The structure of Ellman's reagent and some related reagents. ADNB was developed to probe the electrostatic environment around cysteine residues in proteins (Zhang, H., Le, M., and Means, G.E., A kinetic approach to characterize the electrostatic environments of thiol groups in proteins, *Bioorg. Chem.* 26, 356–364, 1998).

derivative of DTNB, 5-(2-aminoethyl)-dithio-2-nitrobenzoate (ADNB) (Figure 5.25). Depending on the model thiol, the reaction with ADNB was either faster (Glutathione, 3-mercaptopropionic acid, thioglycolic acid) or slower (2-mercaptoethylamine, 2-mercaptothanol, cysteine, homocysteine) than the corresponding reaction with DTNB at pH 6.65. Increasing ionic strength markedly enhanced the rate of reaction of DTNB with mercaptoacetic acid with a lesser effect on the rate of DTNB reaction with 2-mercaptoethanol; inhibition was observed with the reaction of DTNB with 2-mercaptoethylamine. The reaction of ADNB was far less sensitive to ionic strength and suggested to be of value in modifying sulfhydryl groups in anionic environments. This suggestion was tested by a study of the modification of the single sulfhydryl group (Cys34) in serum albumin. ADNB reacted far more rapidly than did DTNB with the sulfhydryl group in bovine serum albumin where there is glutamic acid residue (Glu82) close to Cys34; the rates of reaction with the sulfhydryl are essentially equal in equine albumin where the Glu82 is replaced with an alanine. Stewart and colleagues[461] argued for a role of Tyr84 in the reactivity of Cys34 in human albumin where there is also a glutamic residue at position 82 in

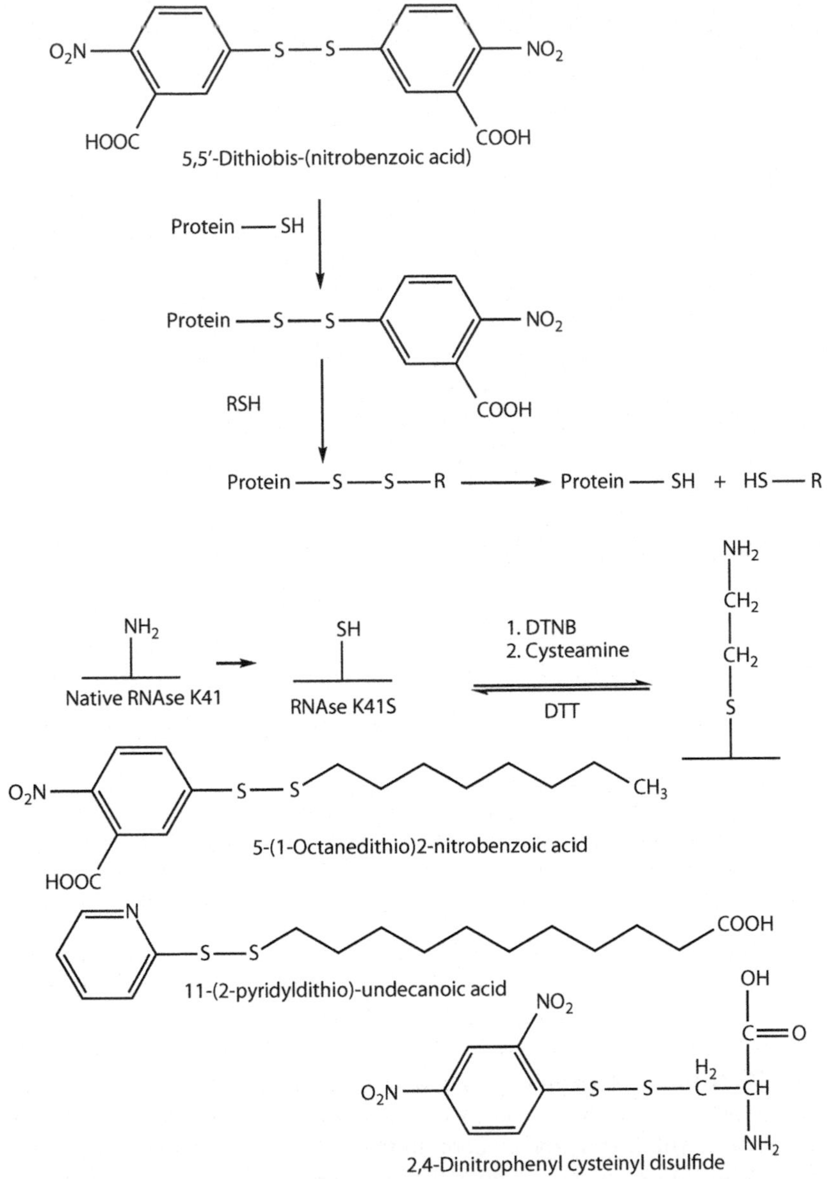

FIGURE 5.26 (See caption on facing page.)

the sequence. Replacing Tyr84 with phenylalanine resulted in a fourfold increase in the rate of reaction of Cys34 with DTNB. Zhu and coworkers observed a 70-fold difference in the second-order rate constant for the reaction of DTNB with bovine albumin ($1.1\,M^{-1}\,s^{-1}$) and equine albumin ($70.1\,M^{-1}\,s^{-1}$). Zhu and coworkers[462] observed the ADNB is more stable than DTNB under basic conditions. Legler[463] has reported on the synthesis and characterization of a positively charged (5,5′-dithiobis-(2-nitro-*N*-trimethylbenzyl ammonium iodide) and neutral (5,5′-dithiobis-(2-nitro-*N*-2′-hydroxyethyl benzamide) analog of Ellman's reagent. Legler[463] reported the necessity of performing experiments for the determination of reaction rates with model thiols at pH 6.0 [0.05 M phosphate (counterion not stated)] and 15°C; rates were too rapid at higher pH for accurate measurement. The reader is also directed to a study by Whitesides and coworkers[464] on the rate of reaction of Ellman's reagent with monothiols and dithiols. The rate of reaction correlated with pK_a of the thiol. As with the reaction of cysteine with other reagents as discussed previously, the presence of a positive change close to the thiol function results in enhanced reactivity.

DTNB and related derivatives react rapidly with model thiols and many sulfhydryl groups in proteins. The driving force for the reaction of DTNB is the release of the thionitrobenzoate anion.[465] Faulstich and Heintz[465] discuss the use of DTNB for the reversible introduction of thiols into protein for the formation of mixed disulfides (Figure 5.26). Messmore and coworkers[466] used this chemistry for the preparation of some novel derivatives of bovine pancreatic ribonuclease A (RNase A). These investigators prepared a mutant where Lys41 is replaced by cysteine reducing catalytic efficiency (K_{cat}/K_m) by 10^5. The mutant protein is modified with DTNB (there are no other free cysteinyl residues other than the cysteine introduced at Lys41). It is possible to introduce cysteamine which has the net effect of introducing an amino function at position 41 resulting in a 10^3 increase in catalytic activity; the enzyme is inactivated by treatment with DTT which removes the cysteamine. If mercaptopropylamine is introduced by the same chemistry, activity is further reduced below that found in the mutant (K41C) enzyme. Most model thiols and some protein thiols react rapidly (second-order rate constants of $10^3\,M^{-1}\,s^{-1}$; Table 5.4). However, the rate of reaction with protein are variable and some protein thiols react quite

---

FIGURE 5.26 Various approaches to the preparation of protein mixed disulfides. Shown at the top is the preparation of the thionitrobenzoate derivative of a protein which can then be used to the preparation of a protein mixed disulfide with a selected thiol (Faulstich, H. and Heintz, D., Reversible introduction of thiol compounds into proteins by use of active mixed disulfides, *Methods Enzymol.* 251, 357–366, 1995). Shown at the bottom is the application the DTNB technology for "rescue" of a mutant RNase (Messmore, J.M., Holmgren, S.K., Grilley, J.E., and Raines, R.T., Sulfur shuffle: Modulating enzymatic activity by thiol disulfide interchange, *Bioconj. Chem.* 11, 401–413, 2000). Shown at the bottom is 5-(1-octanedithio)-2-nitrobenzoic acid and 11-(2-pyridyldithio)-undecanoic acid. These are two hydrophobic reagents prepared for the study of thiol groups in biological membranes (Czerski, L. and Sanders, C.R., Thiol modification of diacylglycerol kinase: Dependence upon site membrane disposition and reagent hydrophobicity, *FEBS Lett.* 472, 225–229, 2000) and 2,4-dinitrophenyl cysteine disulfide (Drewes, G. and Faulstich, H., 2,4-dintrophenyl [^{14}C] cysteinyl disulfide allows selective radioactive labeling of protein thiols under spectrophotometric control, *Anal. Biochem.* 188, 109–113, 1990).

slowly.[462] In a related study, Drewes and Faulstich[467] prepared 2,4-dinitrophenyl-^{14}C-cysteinyl disulfide (Figure 5.26) via a facile synthetic method as a means for introducing radiolabeled cysteine into proteins via disulfide exchange with free thiols. The reaction can be monitored by following the release of 2,4-dinitrophenol at 408 nm (ε = 12,700 M^{-1}·cm^{-1}). The specificity of this reagent corresponded to that obtained with DTNB. Reaction with the sulfhydryl groups of papain was more rapid than that observed with DTNB. The resulting derivative can be easily reversed with thiols, but is stable to cyanogen bromide degradation and peptide purification. The activating ability of the thionitrobenzoyl anion as an activating group permits the development of reagents such as 5-(1-octanedithio)-2-nitrobenzoic acid which was used for the modification of cysteine residues in proteins.[468]

It would appear that most recent studies with DTNB have used this reagent to measure cysteine residues in proteins rather than for the site-specific chemical modification of proteins. The reader is directed to several recent studies where DTNB was used to measure the extent of cysteine modification by other sulfhydryl reagents.[120,469] Other studies on the use of DTNB to modify sulfhydryl groups in proteins have included the modification of fibronectin,[470] It was observed that DTNB was less potent than the more hydrophobic dithionitropyridine derivative (see below). As with other reagents, reaction of protein sulfhydryl groups with DTNB is dependent on protein conformation.[471] The study by Narasimhan and coworkers[470] demonstrates the importance of conformation on cysteinyl reactivity to DTNB in proteins. The sulfhydryl groups of fibronectin are not available for reaction in soluble fibronectin in the absence of chaotropic agents. One sulfhydryl group in fibronectin is available for reaction upon binding to polystyrene beads. Subsequent studies with monoclonal antibodies also showed a conformational change in fibronectin upon binding to a polystyrene surface.[472]

4,4′-Dithiodipyridine (Figure 5.26) is similar to 5,5′-dithiobis-(5-nitrobenzoic acid) in that a mixed disulfide is formed between a cysteinyl residue in the protein and the reagent with the concomitant release of pyridine-4-thione.[473] The reaction of 4,4′-dithiodipyridine with protein sulfhydryl groups can be followed by spectroscopy (ε_{324nm} = 19,800 M^{-1} cm^{-1}). The reaction is readily reversed by the addition of a reducing agent such as DTT. Kimura and coworkers[474] introduced methyl 3-nitro-2-pyridyl disulfide and methyl 2-pyridyl disulfide. Both of these reagents modify sulfhydryl groups forming the thiomethyl derivative. The spectrum of 3-nitro-2-pyridone is pH dependent. There is an isosbestic point at 310.4 nm which can be used to determine the extent of the reaction of methyl-3-nitro-2-pyridyl disulfide with sulfhydryl groups. The difference in spectrum obtained does not show the pH dependence of the nitropyridyl derivative. At 343 nm, the change in extinction coefficient is 7060 M^{-1} cm^{-1}. The extinction coefficient (7600 M^{-1} cm^{-1}) of the 2-thiopyridinone at 343 nm is relatively stable from pH 3 to 8.0.[475] There is a marked decrease in absorbance above pH 8.0 reflecting the loss of a proton. Reaction with the sulfhydryl group in the protein clearly proceeds more rapidly at alkaline pH. Malthouse and Brocklehust[476] reported that while the rate of reaction of 2,2′-dipyridyldisulfide with 2-mercaptoethanol decreases in the presence of either dimethylformamide or dioxan, there was no effect of the solvent on the rate of reaction of either 4,4′-dipyridyldisulfide or DTNB with 2-mercaptoethanol.

The synthesis of a selenium analog of this class of reagents, 6,6-diselenobis (3-nitrobenzoic acid), has been reported.[477] The selenium-containing reagent has similar reaction characteristics as the sulfur-containing compound in terms of specificity of reaction with cysteinyl residues in proteins. The reaction is monitored by spectroscopy following the release of 6-seleno-3-nitrobenzoate which has a maximum at 432 nm. The extinction coefficient for the 6-seleno-3-nitrobenzoate anion varies slightly from 9,532 (with excess reagent) to 10,200 M^{-1} cm^{-1} (with either excess cysteine or excess 2-mercaptoethanol). Other than commenting that the reaction of, 6,6-diselenobis (3-nitrobenzoic acid) with sulfhydryl groups was rapid, there is no mention of reaction rate. It is noted that Pleasants and coworkers[478] made a comparative study of the rates of selenol/diselenide (selenocysteamine/selenocystamine) and thiol/disulfide (cysteamine/cystamine) exchange using NMR spectroscopy and reported that the rate observed with selenocysteamine/selenocystamine was 1.2×10^7 faster than cysteamine/cystamine. The differences in rates reflect the higher nucleophilicity of selenium as well as serving as a better leaving group in the S_N2 reaction. Alkyl selenium compounds are considered to be considerably more nucleophilic than the corresponding sulfur compounds.[479,480] Xu and coworkers[481] have reported the modification of thiols with Se–N compounds (Figure 5.27) forming a Se–S bond. The modification is selective and very rapid and reversed with DTT.

A number of other modifications of sulfhydryl groups have proved useful. *O*-Methylisourea reacts with cysteinyl residues to form the *S*-methyl derivative.[482] *S*-methylation of proteins does occur *in vivo*.[483] *O*-Methylisourea is used far more frequently for the modification of lysine.[484–486]

Cyanate also can modify sulfhydryl groups (Figure 5.28).[487] While this reaction is more rapid for sulfhydryl groups than for other nucleophiles such as amino groups, side reactions are a complication. The carbamoyl derivative of cysteine is stable at acid pH, but rapidly decomposes at alkaline pH. Given the stability issues, it is not clear that the carbamylation of cysteine is an issue[488,489] in the practice of proteomic analysis.

S-Cyanocysteine is a useful derivative of cysteine. First the reaction of 2-nitro-5-thiocyanobenzoic acid with cysteine in protein results in the formation of *S*-cyanocysteine and the formation of 2-mercapto-5-nitrobenzoic acid (Figure 5.28), which can be used for the quantitative determination of sulfhydryl groups. 2-Mercapto-5-nitrobenzoic acid has an absorbance maximum at 412 nm with a molar extinction coefficient of 13,600 M^{-1} cm^{-1}.[454] Pecci and coworkers[490] have characterized the reaction of rhodanese with 2-nitro-5-thiocyanobenzoic acid. These investigators used a 1.3 molar excess of reagent in 0.050 M phosphate buffer, at pH 8.0 at 18°C. The reaction was followed spectrophotometrically by the release of 2-mercapto-5-nitrobenzoic acid and was complete after 6 h. Second, peptide bond cleavage can be accomplished on the amino terminal side of *S*-cyanocysteine. Cleavage at *S*-cyanocysteinyl residues was first studied by Vanaman and Stark.[491] These investigators modified the catalytic subunit of *E. coli* asparate transcarbamylase with DTNB and modified the thionitrobenzoyl derivative with cyanide resulted in the *S*-thiocyano derivative. Witkowska and coworkers[492] used the approach for the modification of a cysteinyl residue in thioesterase II and subsequent cleavage at the modified cysteine. More recent work has used 2-nitro-5-thiocyanobenzoic acid for the modification of cysteine for peptide bond cleavage.[493,494] Wu and Watson[495]

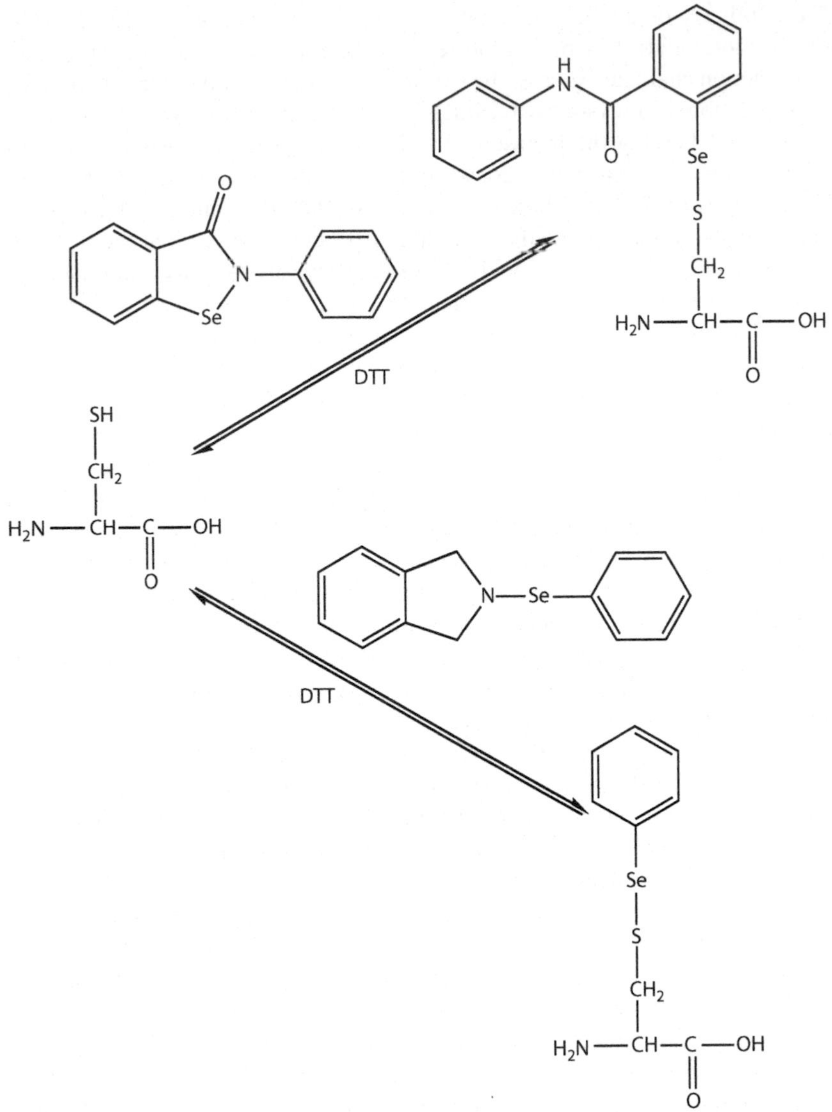

FIGURE 5.27 Reversible modification of cysteine with nitrogen-selenium compounds. Shown at the top is 2-phenyl-1,2-benzisoselenazol-3(2H)-one while *N*-(phenylseleno) phthalimide is shown at the bottom. Reversal of the modification is accomplished with DTT. (See Zu, K., Zhang, Y., Tang, B. et al., Study of highly selective and efficient thiol derivatization using selenium reagents by mass spectrometry, *Anal. Chem.* 82, 6926–6932, 2010.)

used 1-cyano-4-dimethylaminopyridinium tetrafluoroborate for modification of the cysteine residues prior to peptide bond cleavage. *S*-cyanocysteine can be used as an infrared chromophore for measuring protein conformation.[496,497]

4-Chloro-7-nitrobenzo-2-oxa-1,3-diazole (4-chloro-7-nitrobenzofurazan (Nbf-Cl)) (Figure 5.29) is a reagent developed for the modification of amino groups.[498] but has

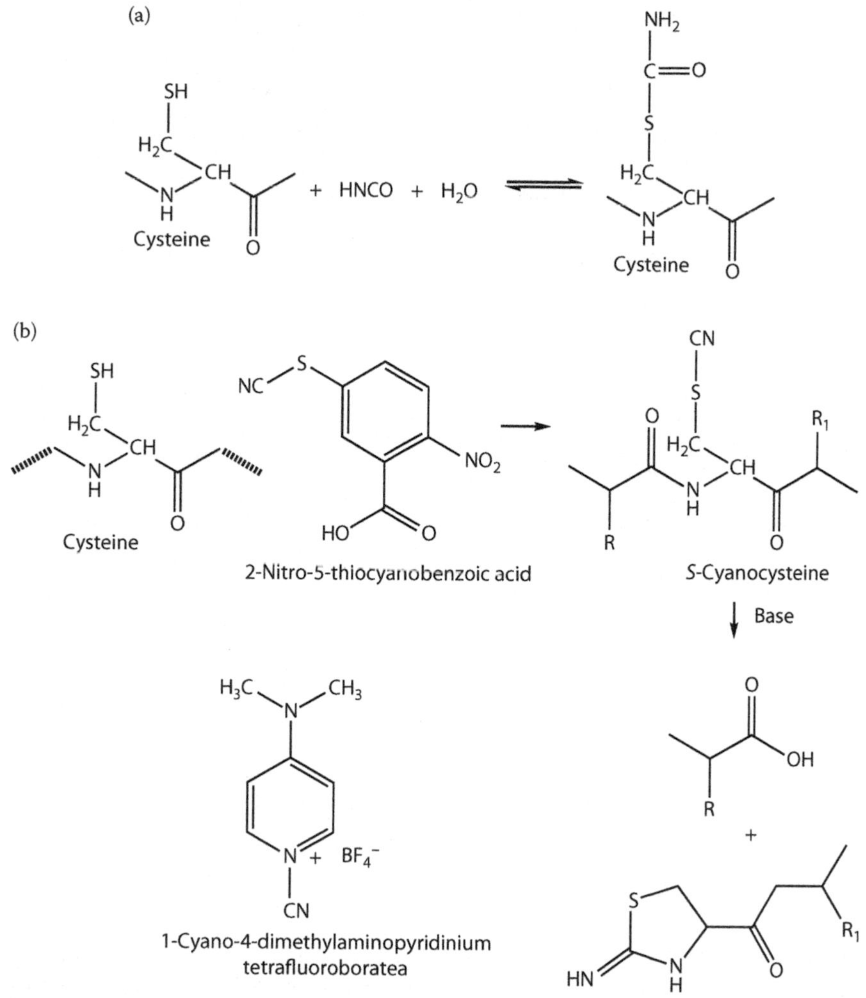

FIGURE 5.28 Various mechanisms for the reaction of cyanate and organic cyanates with cysteine. (a) The reversible reaction of cyanate with cysteine to form the carbamoyl derivative (see Stark, G.R., Reversible reaction of cyanate with sulfhydryl groups and the determination of NH_2-terminal cysteine and cystine in proteins, *J. Biol. Chem.* 239, 1411–1414, 1964). (b) The reaction of 2-nitro-5-thiocyanobenzoate with cysteine to yield *S*-cyanocysteine, a derivative that is cleaved is based on an iminothiazolidine derivative; not shown is the β-elimination reaction resulting in dehydroalanine (Iwasaki, M., Masuda, T., Tomita, M., and Ishihama, Y., Chemical cleavage-assisted tryptic digestion for membrane proteome analysis, *J. Proteome Res.* 8, 3169–3175, 2008). Also shown is the structure of 1-cyano-4-dimethylaminopyridinium tetrafluoroborate (Wu, J. and Watson, J.T., Optimization of the cleavage reaction for cyanylated cysteinyl proteins for efficient and simplified mass mapping, *Anal. Biochem.* 258, 268–276, 1998).

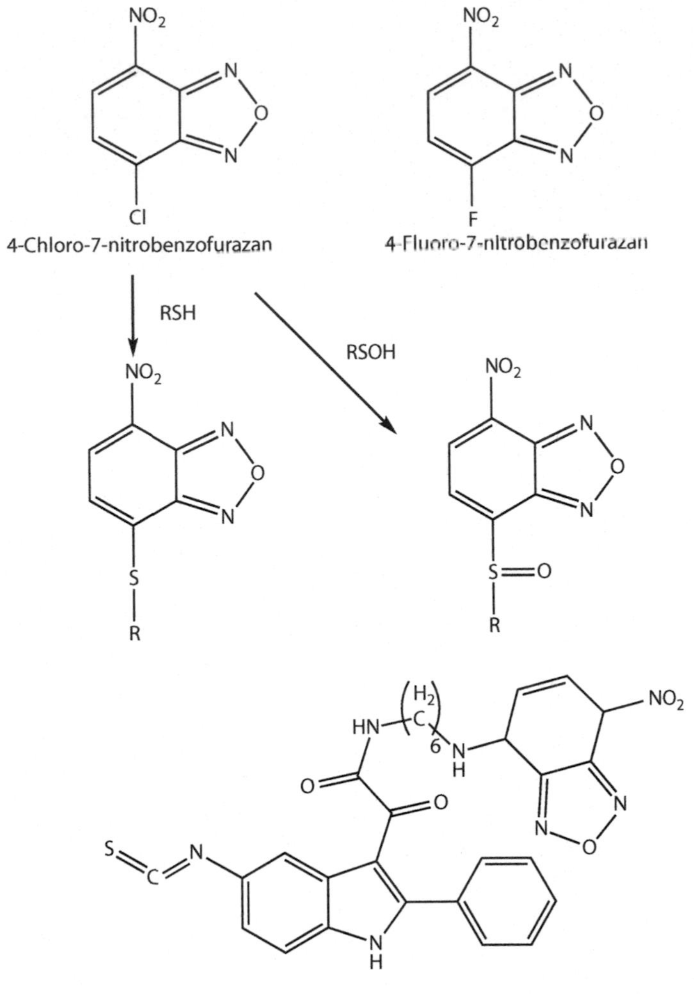

FIGURE 5.29 The structure of Nbf-Cl and 4-fluoro-7-nitrobenzofurazan and the reaction of Nbf-Cl with a thiol. Also shown is the reaction of Nbf-Cl with a sulfenic acid (Poole, L.B. and Ellis, H.R., Identification of cysteine sulfenic acid in AhpC of alkyl hydroperoxide reductase, *Methods Enzymol.* 348, 297–305, 2002). 4-Fluoro-7-nitrobenzofurazan has been used to modify lysine and tyrosine in proteins (Luo, J., Fukuda, E., Takase, H. et al., Identification of the lysine residue responsible for coenzyme A binding in the heterodimeric 2-oxoacid: Ferredoxin oxidoreductase from *Sulfolobus tokodaii*, a thermophilic archaeon, using 4-fluoro-7-nitrobenzofurazan as an affinity label, *Biochim. Biophys. Acta* 1794, 335–340, 2009; Zhang, W.Z., Lang, C., and Kaye, D.M., Determination of plasma free 3-nitrotyrosine and tyrosine by reversed-phase liquid chromatography with 4-fluoro-7-nitrobenzofurazan derivatization, *Biomed. Chromatogr.* 21, 273–278, 2007). Also shown is the structure of a fluorescent probe based on the Nbf fluorophore used for targeting a 18 kDa translocator protein (Taliani, S., Da Pozzo, E., Bellandi, M. et al., Novel irreversible fluorescent probes targeting the 18kDa translocator protein: Synthesis and biological characterization, *J. Med. Chem.* 53, 4085–4093, 2010).

found more application for the modification of sulfhydryl groups and is useful in that it introduces a fluorescent probe.[499–503] Nitta and coworkers[502] have noted that there are several possible reaction products of Nbf-Cl with protein nucleophiles including sulfhydryl groups. The reaction of Nbf-Cl with sulfhydryl groups in glutathione reductase and lipoamide dehydrogenase has also been reported.[504] Nbf-Cl is also useful for the determination of cysteine sulfenic acid[97,505,506] but does not distinguish between cysteine and cysteine sulfenic acid.

CYSTINE

Cystine is the oxidation product of cysteine and is critical for maintaining the native structure of an extracellular protein; as discussed earlier, most extracellular proteins have only disulfide bonds while intracellular proteins have free sulfhydryl groups. The extent to which these two related concepts are accurate is not clear. In support of the critical importance of disulfide bonds, it is noted that the ABA-1 allergen of the nematode *Ascaris* has a single cystine. Reduction of this cystine results in the loss of structural integrity and function.[507] Further support is obtained from studies where the insertion of an additional disulfide into a protein increases stability.[508] However, there are also proteins such as tubulin which contain a relatively large number of cysteine residues without a single disulfide bond.[74] Bull and Breese studied the denaturation of various proteins in guanidine hydrochloride and observed that protein expands on denaturation with an increase in viscosity; the presence of disulfide bonds limited the expansion of the protein molecules during denaturation. Hen egg white lysozyme is one of several proteins that has a sweet taste.[509] Both sweetness and enzymatic activity are lost on reduction and alkylation while it was possible to inhibit enzyme activity without losing sweetness.[510] The sweetness was also lost by maintaining at 95°C for 18 h. This suggests that the sweetness is a product of the tertiary structure of the protein stabilized by disulfide bonds and not a reflection of enzymatic activity.

The author would be remiss, for both personal and scientific reasons, if lima bean trypsin inhibitor was not mentioned at this juncture. Lima bean trypsin inhibitor is a group of six or more closely related proteins of low molecular weight (ca 10 kDa) and some 20% cysteine/cystine by weight. The cysteine is present as cystine so the protein is highly cross-linked (one half cystine every six residues) and very stable; the inhibitory activity is stable to boiling. Jones and coworkers[511] purified several of the variants and noted that while inhibitory activity was lost on reduction and carboxymethylation by classical methods, the protein was still resistant to trypsin. As told to the author of the current text (RLL), one of senior authors on the Jones paper (SM) was quite bothered by that such that he encouraged a subsequent fellow, William Ferdinand, to take another look at the problem. Ferdinand and coworkers[512] did indeed solve the problem. Reduction and carboxymethylation was determined to be incomplete under the conditions used in the Jones study;[511] Ferdinand and coworkers found it necessary to use 5 M guanidine hydrochloride in place of 8 M urea and the concentrations of 2-mercaptoethanol and iodoacetamide were increased 10-fold. Even under these conditions, reduction and carboxymethylation were incomplete and an increase in temperature was suggested. It is clear that the stability of lima bean trypsin inhibitor is a product of the high disulfide bond content. It would be equally remiss not to mention the need to validate reaction conditions for reduction/alkylation prior to subsequent structural analysis.

Thiol-disulfide exchange refers to a reversible reaction between a thiol and a protein disulfide process resulting in an equilibrium mixture of thiol, disulfide, and mixed disulfide consisting of the attacking thiol and one of the two cysteine partners in the protein disulfide (Figure 5.30). A mixed disulfide may also result from the reaction of an activated disulfide compound such as DTNB or MMTS. Thiol-disulfide exchange has an important role in redox regulation and protein folding with glutathione having a central role.[513–519]

The study of disulfide bond formation is of interest for the problem of protein folding[520–522] where it is clear that there are metastable intermediates. In this oxidative folding pathway, disulfide bonds are formed and then corrected through disulfide exchange until the final native conformation is acheived.[523] Selective insertion of cysteine into proteins permits targeted disulfide crosslinking which is of value in studying protein interaction.[524–527] The reader is directed to early reviews by Cecil[14] and Gilbert[528] for more information on thiol-disulfide exchange.

The thiol-disulfide reaction is an S_N2 reaction with nucleophilic addition by a thiolate function occurring along the S–S axis of the disulfide bond. Eldgam and Phil[529] measured equilibrium constants for several thiol-disulfide pairs and observed that none were far from unity implying that there will be a large amount of mixed disulfides in addition to the starting materials. This study was published in 1957 and was an early use of affinity (paper) chromatography. The progress of disulfide exchange was measured by paper chromatography on mercuric ion-impregnated paper; the free thiol would be selectively retarded during the development of the chromatogram. Keire and coworkers[530] used NMR to study the kinetics of reaction between thiols and oxidized glutathione. The equilibrium constant was dependent on the basicity of the thiolate anion; the smallest (0.022) was determined for *N*,*N*-dimethylcysteamine and the largest (760) was determined for 3-mercaptopropionic acid. There is considerable additional comment on the reduction of cystine below.

Cystine is an oxidation product of cystine and in turn is susceptible to oxidation (Figure 5.31). Cystine can be oxidized to the disulfide-*S*-monoxide and disulfide *S*-disulfide.[531] Disulfide oxides are products of the *in vivo* oxidation of biological thiols such as glutathione[532] which form under conditions of oxidative stress.[533] Disulfide oxides are reactive species[533,534] which can modify proteins.[535–537]

The conversion of cystine to cysteic acid by iodate in acid has been suggested.[538] This reaction has been applied to insulin with millimolar concentrations of iodate in 0.5 M HCl. The reaction product was not completely characterized but the consumption of iodate and the nature of the product is consistent with the oxidation of cystine to cysteic acid. The reaction of cystine with iodate is complete within 30 min; after longer periods of reaction, the iodination of tyrosine residues occurred. Nowduri and coworkers[539] have studied the oxidation of cystine with hexacyanoferrate(III) in alkaline medium suggesting a mechanism proceeding through cysteine sulfenic acid to cysteic acid as the terminal product. Oxidation of cystine can be accomplished under more vigorous conditions with reagents such as performic acid.[540] This is a procedure used somewhat infrequently for the cleavage of disulfide bonds

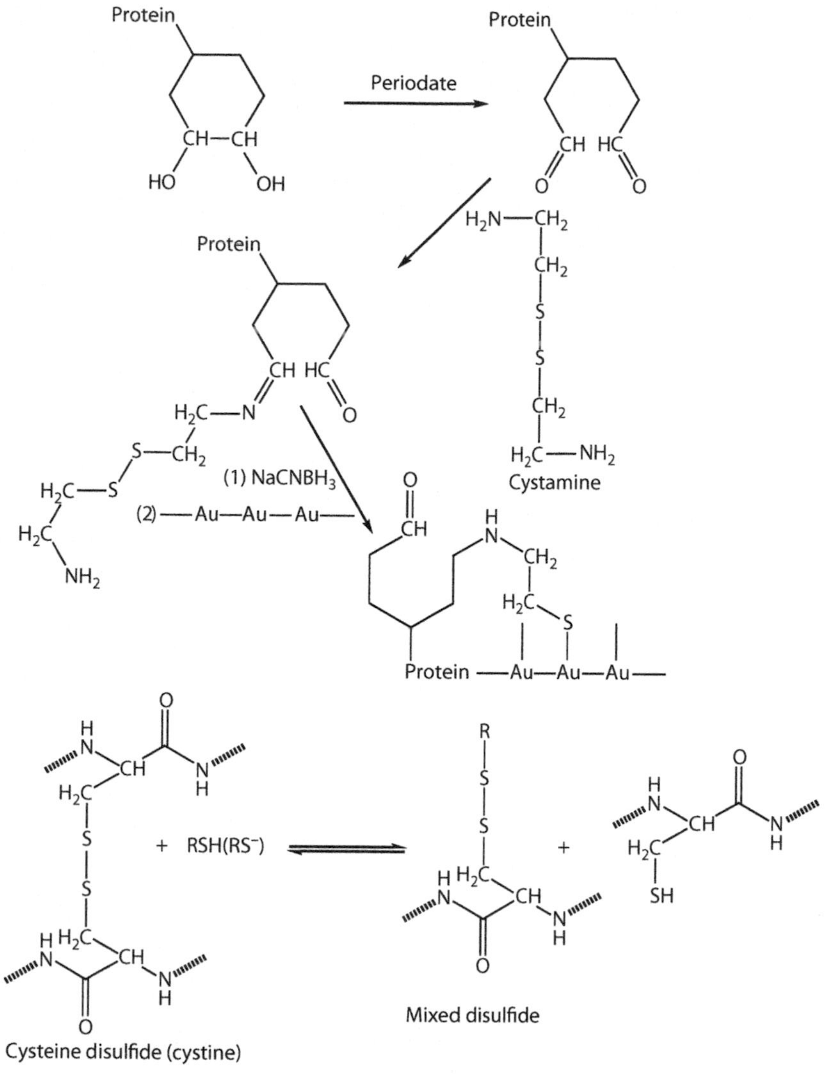

FIGURE 5.30 The formation of *S*-cysteinyl proteins and the coupling of proteins to gold surfaces via mixed disulfides. Shown is the process of formation of *S*-cysteinyl derivatives of proteins via the intermediate formation of *S*-(2-thiopyridyl)cysteine from the reaction of 2.2′-dithiopyridine and cysteine (Zhang, Q. and Kelly, J.W., Cys10 mixed disulfides make transthyretin more amyloidogenic under mildly acidic conditions, *Biochemistry* 42, 8756–8761, 2003). Figure disulfide introduction. Introduction of external disulfide bond into protein for coupling to gold. Shown is the coupling of cystamine to proteins with subsequent binding to gold surface (Suárez, G., Jackson, R.J., Spoors, J.A., and McNeil, C.J., Chemical introduction of disulfide groups on glycoproteins: A direct protein anchoring scenario, *Anal. Chem.* 72, 1961–1969, 2007). The carbohydrate is oxidized with periodate to expose vicinal dialdehydes. Cystamine is then coupled to the aldehyde function with the initial formation of an imine linkage which is stabilized with sodium cyanoborohydride to yield an amine. The disulfide is then taken to a gold surface where an Au–S bond is formed.

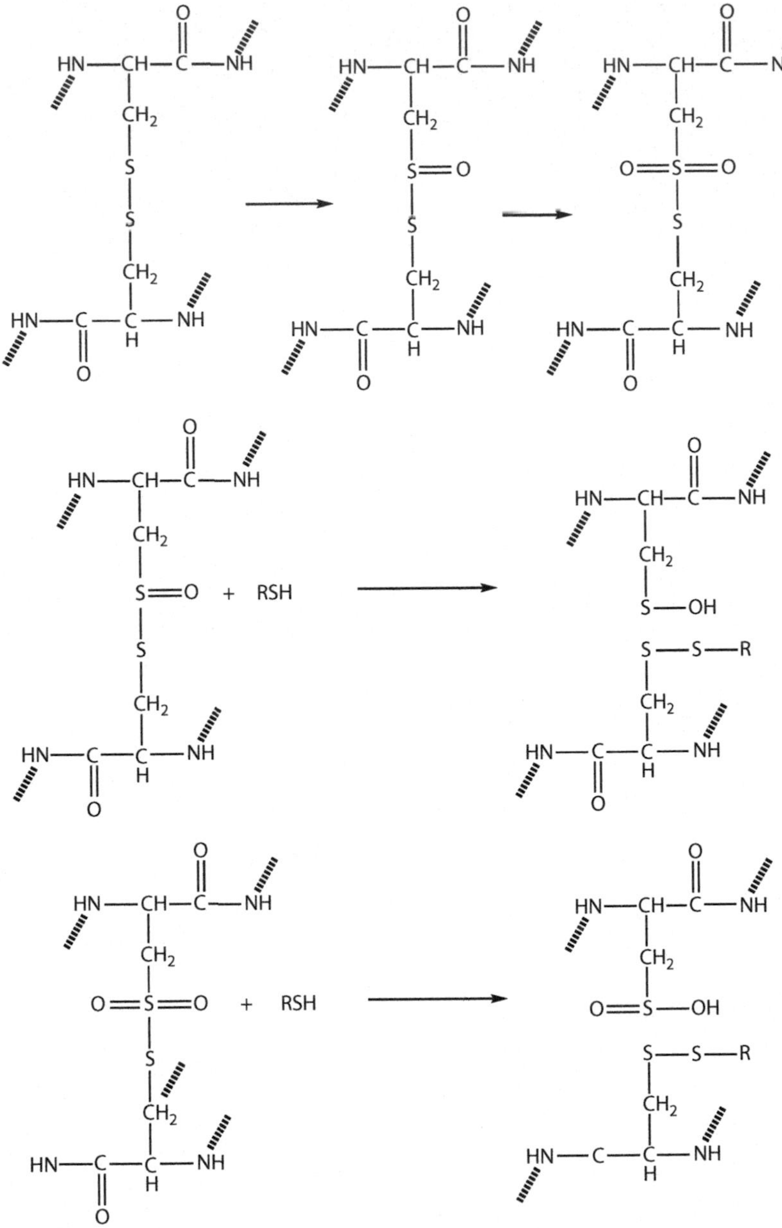

FIGURE 5.31 The oxidation of cystine to cystine *S*-monoxide and cystine *S*-dioxide and reaction with thiols. Shown is the oxidation of cystine to cystine *S*-monoxide and cystine *S*-dioxide and the reaction of the species with thiols to form mixed disulfide and cystine sulfenic acid (from cystine *S*-monoxide) and cysteine sulfinic acid (Giles, G.I. and Jacob, C., Reactive sulfur species: An emerging concept in oxidative stress, *Biol. Chem.* 363, 375–388, 2002).

in proteins for structural analysis[541,542] and somewhat more for accurate determination of cysteine/cystine in proteins using amino acid analysis.[543,544]

The cleavage of disulfide bonds by base was reported by Challenger and Rawlings in 1937[183] who suggested the products were a mercaptan (thiol) and a sulfenic/sulfinic acid. Some years later, Donovan[545] observed a time-dependent increase in absorbance of phenolic groups in ovomucoid at high pH (≥12.5). Subsequently, Donovan[546] reported that protein (ovomucoid) disulfide bonds were cleaved under those conditions resulting in an increase in absorbance at 240 and 290 nm. There was production of material which reacted with Elman's reagent at the rate similar to the rate of change in absorbance. The results were consistent with the mechanism proposed by Challenger and Rawlings although the possibility of the formation of dehydroalanine and a persulfide were mentioned. Cavallini and coworkers[547] established that persulfide was formed in insulin under alkaline conditions; these investigators also reviewed the various products from the alkaline treatment of disulfide bonds (Figure 5.32). Helmerhorst and Stokes[548] presented more data to support the heterolytic cleavage of the disulfide bonds with the formation of a protein-bound persulfide group. Some 10 years later, Florence[549] extended these observations substantiating the heterolytic nature of disulfide bond cleavage in base with the formation of persulfide and dehydroalanine; dehydroalanine can subsequently form lanthionine by reaction with cysteine and lysinoalanine by reaction with lysine. Both of these modifications are examples of a Michael addition. The persulfide could form cystine trisulfide,[550] decompose to cysteine sulfenic acid (which then combines with the persulfide to form the trisulfide), or to cysteine sulfenic acid. Florence[549] observed that the degradation of disulfide bonds in base was enhanced by the presence of a chaotropic agent suggesting that, as with reduction, alkaline cleavage of disulfide bonds in proteins is sensitive to conformation. The degradation of the disulfide bonds to form the persulfide is associated with an increase in absorbance at 335 nm.[548,549] Lu and Chang[551] recently reported that the reaction of protein disulfide with DTT in base resulted in quantitative formation of 2 mol of dehydroalanine which could then be coupled, as previously mentioned, with lysine or cysteine.

The reaction of sulfite with cystine yields the *S*-sulfo derivative of cysteine and cysteine (Figure 5.33). The reaction of cystine with sulfite to form cysteine and *S*-sulfocysteine was described by Clarke[552] and Lugg[553] in 1932. Stricks and Kolthoff[554] report that reaction between sulfite and cystine to yield the thiolate and *S*-sulfonyl derivative is reversible in alkaline solution. Carter[555] used sulfite to measure disulfide bonds in proteins in 8 M urea. Cecil and McPhee[556] studied the kinetics of the reaction and reported that the reaction of cystine and sulfite is a simple reversible reaction above pH 9.0 (below pH 9.0, the reaction is more complex reflecting the ionization of HSO_3^{-1}; disulfide bonds react with SO_3^{-2}, not with HSO_3^{-1}); the reaction can be taken to completion by recycling the thiolate anion by use of an oxidizing agent. These investigators noted that the presence of a negative charge (ionized carboxyl group) markedly inhibited the reaction. The earlier literature and chemistry of this reaction has been reviewed by Cole.[557] As previously noted, the reaction of sulfite and cystine is readily reversible but can be taken to completion by recycling the thiolate anion with an oxidizing agent such as cupric ions or *o*-iodosobenzoate (oxidative sulfitolysis).[558] As with the analogous process of disulfide bond reduction,

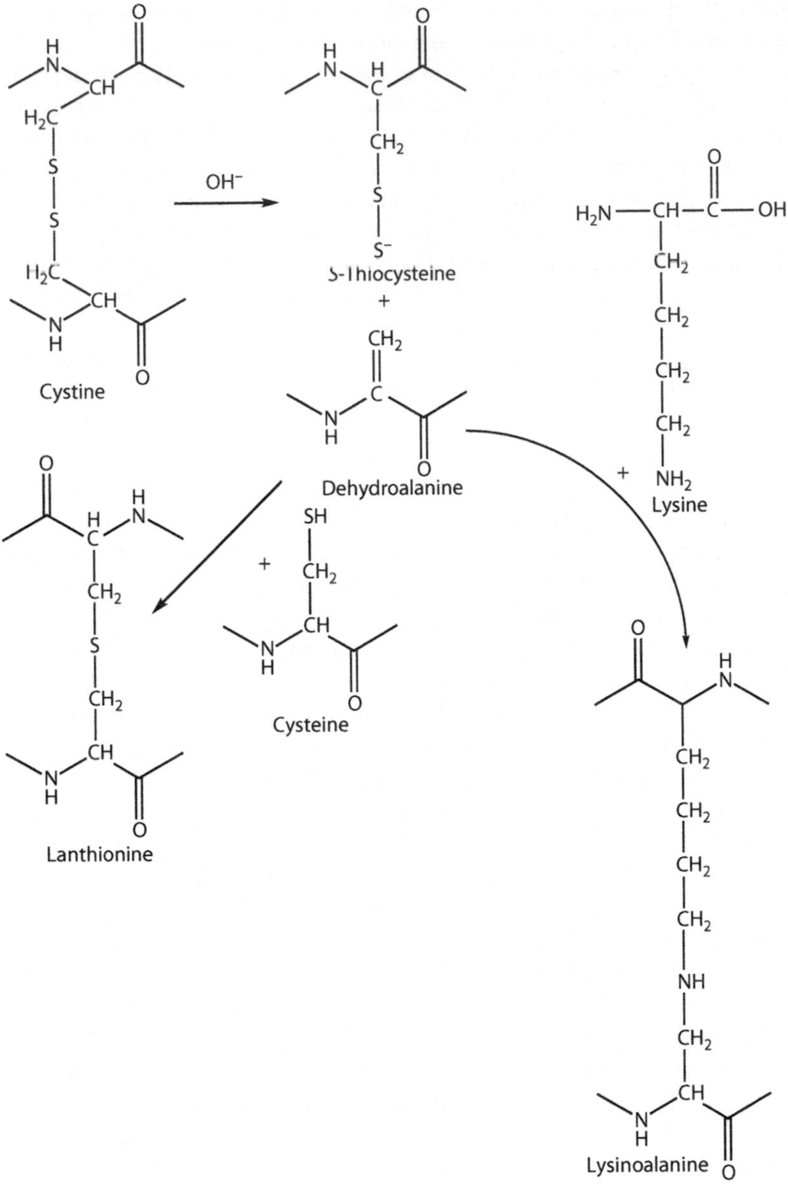

Cystine

OH^-

S-Thiocysteine

+

Dehydroalanine

+

Lysine

+

Cysteine

Lanthionine

Lysinoalanine

FIGURE 5.32 The alkaline degradation of cystine residues in proteins. Shown is the heterolytic cleavage of cystine in base yielding thiocysteine (a persulfide) and dehydroalanine; dehydroalanine could then react with cysteine to form lanthionine or with lysine to form lysinoalanine. (See Cavallini, D., Federici, G., Barboni, E., and Marcucci, M., Formation of persulphide groups in alkaline treated insulin, *FEBS Lett.* 10, 125–128, 1970; Florence, T.M., Degradation of protein disulphide bonds in dilute alkali, *Biochem. J.* 189, 507–520, 1980; Helmerhorst, E. and Stokes, G.B., Generation of an acid-stable and protein-bound persulfide-like residue in alkali- or sulfhydryl-treated insulin by a mechanism consonant with the β-elimination hypothesis of disulfide bond lysis, *Biochemistry* 22, 69–75, 1983.)

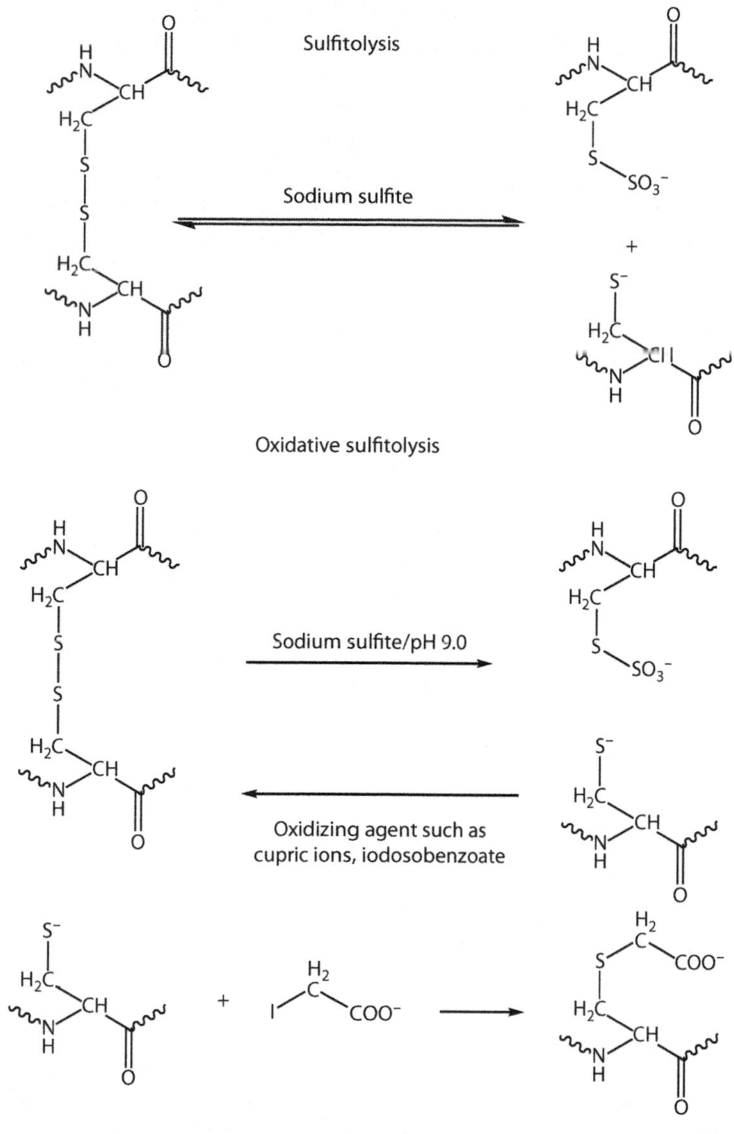

Sulfitolysis

Sodium sulfite

Oxidative sulfitolysis

Sodium sulfite/pH 9.0

Oxidizing agent such as cupric ions, iodosobenzoate

FIGURE 5.33 The reaction of sulfite with cystine to yield cysteine and *S*-sulfocysteine. The reaction of sulfite with cystine to form cysteine and *S*-sulfocysteine, an equilibrium reaction referred to as sulfitolysis; inclusion of an oxidizing agent is referred to as oxidative sulfitolysis. (See Raftery, M.J., Selective detection of thiosulfate-containing peptides using tandem mass spectrometry, *Rapid Commun. Mass Sp.* 19, 674–682, 2005; Patrick, J.S. and Lagu, A.L., Determination of recombinant proinsulin fusion protein produced in *Escherichia coli* using oxidative sulfitolysis and two-dimensional HPLC, *Anal. Chem.* 64, 507–511, 1992.) Shown at the bottom is the reaction of the free thiol formed in sulfitolysis with iodoacetate providing the basis for quantitation of disulfide bonds (Würfel, M.W., Haberlein, I., and Follmann, H., Facile sulfitolysis of the disulfide bonds in oxidized thioredoxin and gluaredoxin, *Eur. J. Biochem.* 211, 609–614, 1999).

complete reaction of disulfide bonds to sulfite usually requires protein denaturation.[555] Würfel and colleagues studied the reaction of sodium sulfite with a number of bacterial and plant thioredoxins.[559] The sulfitolysis reaction was performed in the presence or absence of guanidine hydrochloride. The process of sulfitolysis was measured by the reaction of the liberated thiol group with radiolabeled iodoacetate (Figure 5.33). The extent of sulfitolysis was, in general, more marked in the presence of guanidine and appeared to depend on the primary structure. A subsequent study from this group showed that the carboxymethylation yielded low results reflecting accessibility issues for the liberated cysteine residues.[560] There is an increase in fluorescence at 345 nm (excitation of tryptophan at 280 nm) reflecting the cleavage of disulfide bonds and the concomitant loss of the quenching effect on tryptophan fluorescence such that changes in fluorescence can be used to measure the sulfitolysis of the disulfide bond. Change of Asp26 to an alanine residues (D26A) had a small effect of sulfitolysis yield while the change of Lys36 to glutamic acid (K36E) leads to a decrease in sulfitolysis yield and requires a higher concentration of sulfite for the reaction. The enhancement of sulfitolysis observed when a positive charge is close to the cystine is consistent with the enhancement of cysteine nucleophilicity by adjacent positive charges as discussed earlier. Furthermore, the replacement of lysine by glutamic acid places a negative charge adjacent to the cysteine residue which as noted by Cecil and McPhee[556] would decrease the rate of modification of cysteine. The influence of primary structure on the susceptibility of cystine to sulfitolysis is provided by a study by Kristjánsson and coworkers comparing a subtilisin-like proteinase from a psychrotrophic *Vibrio* species, proteinase K, and aqualysin I.[561] Psychrotropic proteases are obtained from cold-adapted organisms and tend to be more flexible (less rigid) than corresponding proteases from mesophilic or thermophilic organisms.[562–564] It is suggested that there are less intramolecular contacts in psychrotrophic proteins. The disulfide bonds of the *Vibrio* protease were cleaved by sulfitolysis (0.2 M Tris-HCl, 20 mM EDTA, 0.1 M Na_2SO_3, pH 9.5) in the presence of 2-nitro-5-sultothiobenzoate where the disulfide bonds of either proteinase K or aqualysin were resistant to sulfitolysis. The *Vibrio* protease contains four disulfide bonds as determined with 2-nitro-5-thiosulfobenzoate[561] in the presence of 3.0 M guanidine hydrochloride. DTT (10 mM) caused a small loss of activity in the *Vibrio* protease which was partially reversed on oxidation. Incubation of *Vibrio* protease in the presence of sodium sulfite and absence of guanidine caused approximately 90% loss of activity with almost total modification of cystine; approximately 3.5 of the four cystine groups are available for modification within 15 min. In comparison, 2.5 M guanidine hydrochloride was required for complete sulfitolysis of the disulfide bonds of proteinase K and 5.5 M guanidine hydrochloride for aqualysin I.

An ingenious approach to the use of the sulfitolysis reaction was developed by Thannhouser and colleagues.[565] These investigators included 2-nitro-5-sulfothiobenzoate in the reaction of sodium sulfite with protein disulfide (Figure 5.34). The process of sulfitolysis yields *S*-sulfocysteine and cysteine thiolate; 2-nitro-5-thiosulfobenzoate reacts with the thiolate anions resulting in a second *S*-sulfocysteine and 2-nitro-5-thiobenzoate can be measured at 412 nm and is proportional to the cystine residues in the protein. The reaction is reversible to form cysteine upon treatment with a thiol such as 2-mercaptoethanol or DTT. Guanidine hydrochloride

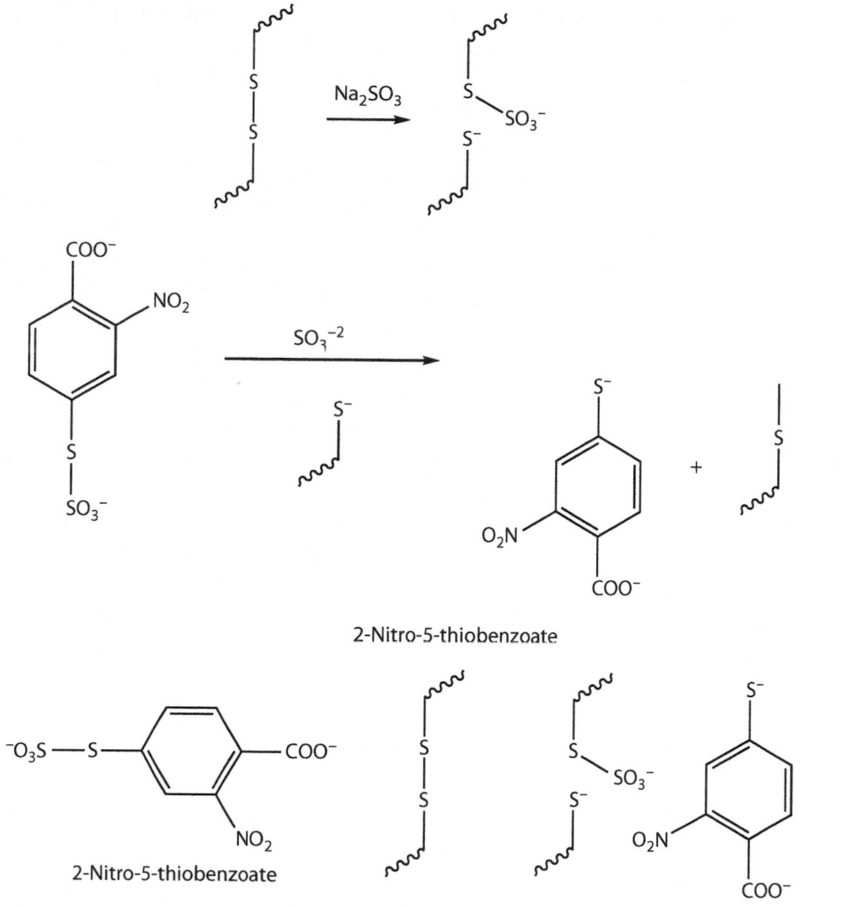

FIGURE 5.34 The use of 2-nitro-5-thiosulfobenzoate for the colorimetric determination of cystine in proteins. Shown is the sulfitolysis of a disulfide bond yielding the *S*-sulfocysteine derivative and cysteine thiolate anion. 2-Nitro-5-thiosulfobenzoate, in the presence of sulfite, reacts with the thiolate anion yielding a second *S*-sulfocysteine and 2-nitro-5-thiobenzoate which is a strong chromophore ($\varepsilon = 13{,}900\,M^{-1}\,cm^{-1}$ at 412nm). (See Thannhauser, T.W., Konishi, Y., and Scheraga, H.A., Analysis for disulfide bonds in peptides and proteins, *Methods Enzymol.* 143, 115–119, 1987.)

(3M) was required for determination of the disulfide bonds in RNase A. Urea was unsatisfactory for use in this procedure as were higher concentrations of guanidine hydrochloride. Damodaran[566] reported that the 2-nitro-5-thiobenzoate anion which is measured in the aforementioned reaction is sensitive to light in the presence of sulfite; it is important to perform this assay in the dark.

It is possible to obtain complete sulfitolysis of disulfide bonds in proteins in the absence of denaturing agent by using oxidative sulfitolysis. Kella and Kinsella[567] used sodium sulfite in the presence of cupric ionics for the progressive cleavage of disulfide bonds in bovine serum albumin and several other proteins. The time required varied from 2.66h for chymotrypsin to 6h for trypsin; approximately

3 h were required for the complete sulfitolysis of bovine serum albumin. The extent of sulfitolysis was measured with 2-nitro-5-thiosulfobenzoate as described below. The characteristics of *S*-sulfonated bovine serum albumin were subsequently reported by this group.[568] The *S*-sulfonated bovine serum albumin has decreased solubility and a lower isoelectric point as well as changes in conformation. Modification of proteins with sodium sulfite is of considerable interest for food processing[569–571] and the preparation of protein films.[572,573] 2-Nitro-5-sulfothiobenzoate with sodium sulfite (pH 9.0, 200-fold molar excess of sodium sulfite) was also used for the processing of a fusion protein containing the extracellular domain of the P_2X_2 ion channel.[574] Sulfitolysis in the presence of either cupric ions or sodium tetrathionate has proved useful for the quantitative conversion of cystine to *S*-sulfocysteine in the processing of biotherapeutics.[442,575–578] This is mostly used when there is the need to reduce and reoxidize a protein expression in a bacterial expression system. Given the history of use of sulfite in food processing (see above), there is somewhat less concern from a product safety point of view. Sato and Aimoto[579] used *S*-sulfonation for protection of thiol groups in peptide ligation using thioesters.

The reduction and alkylation of cysteine residues in proteins for structural analysis has been discussed earlier. It is possible to selectively reduce cystine disulfide bonds in proteins as a selective chemical modification. Reduction of cystine in proteins is usually accomplished with a mild reducing agent such as β-mercaptoethanol, DTT, TCEP, or cysteine and generally in the absence of chaotropic agents such as urea or guanidine hydrochloride.

Disulfide groups were originally considered to be masked in the sense of being unavailable for chemical reaction.[580–582] The reduction of a disulfide bond is a specific process and, perhaps with the reduction of cysteine sulfenic acid, there are no side reactions with the use of the various reducing agents (Figure 5.35).[582,583] On the other hand, the sulfhydryl group which is the product of disulfide bond reduction is likely the most reactive functional group in a protein. Olcutt and Fraenkel-Conrat[583] reviewed the early work on the reduction of disulfide bonds in proteins. They reported that the rate of reduction of disulfide bonds varied for the various proteins under the same reaction conditions and was greatly influenced by what was described as the degree of denaturation. The data at that time suggested that proteins which have a high content of cystine were sensitive to and adversely affected (solubility) by low concentration of reducing agent. It must be emphasized that the assays were somewhat less sophisticated than those currently available. As an example, Walker[584] added cyanide to the nitroprusside assay for cysteine. Cyanide had been described as a reagent which would "reduce" disulfide bonds.[585] Walker also reported that the cyanide/nitroprusside reaction was negative for egg albumin or serum unless the sample had been heated. This early observation is consistent with later studies showing that many proteins contain disulfide bonds which are not available unless the protein is denatured. The reaction of cyanide with disulfide bonds deserves additional comment. The formation of *S*-cyanocysteine has been previously described; *S*-cyanocysteine is a product of the reaction of cyanide with cystine and is used for peptide bond cleavage[586,587] (Figure 5.36). Catsimpoolos and Wood[587] also used 2-nitro-5-thiocyanobenzoate in their studies. Catsimpoolos and Wood[588] had previously studied the reaction of cyanide with bovine serum albumin (approximate 3000-fold molar excess,

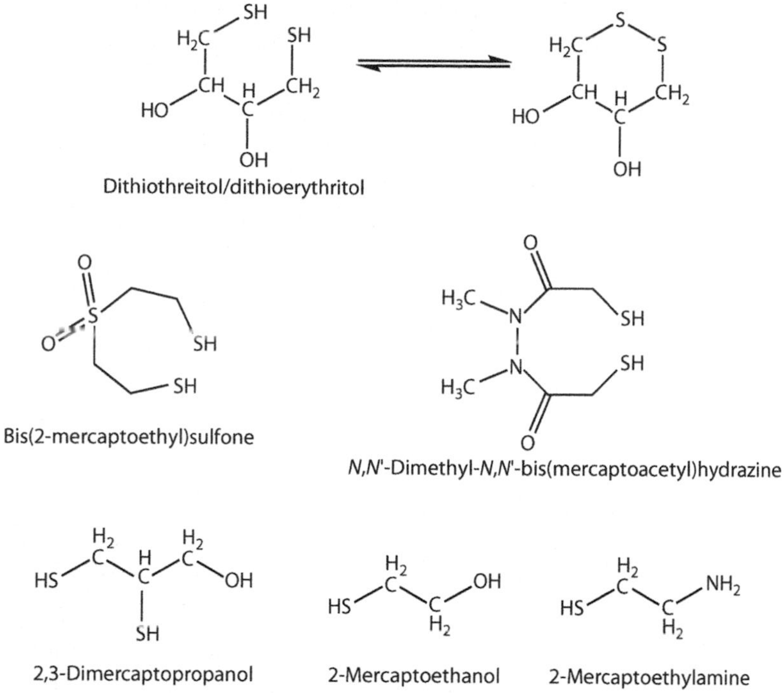

Dithiothreitol/dithioerythritol

Bis(2-mercaptoethyl)sulfone

N,N'-Dimethyl-*N,N'*-bis(mercaptoacetyl)hydrazine

2,3-Dimercaptopropanol

2-Mercaptoethanol

2-Mercaptoethylamine

FIGURE 5.35 The structure and mechanism of reducing agents used for disulfide bonds in proteins. Shown are the structures of various reagents used for the reduction of disulfide bonds in proteins. Shown is DTT (and its stereoisomer dithioerythritol) (Cleland, W.W., Dithiothreitol, a new protective reagents for SH groups, *Biochemistry* 3, 480–482, 1964), bis-(2-mercaptoethyl)sulfone and *N,N′*-dimethyl-*N,N′*-bis(mercaptoacetyl)hydrazine (Singh, R., Lamoureux, G.V., Lees, W.J., and Whitesides, G.M., Reagents for rapid reduction of disulfide bonds, *Methods Enzymol.* 251, 167–173, 1885), and 2,3-dimercaptopropanol (Simpson, S.D. and Young, L., Biochemical studies of toxic agents; experiments with radioactive 2:3-dimercaptopropanol (British anti-lewisite), *Biochem. J.* 46, 634–640, 1950).

50 mg albumin, 20 mg sodium cyanide, 18 h at 37°C) and observed the formation of thiocyanate only above pH 8.0. At pH 7.0, no thiocyanate is formed and is consistent with the formation of iminothiazolidine instead of thiocyanoalanine.

DTT was one of several polyhydric alcohols synthesized by Evans and coworkers in 1949.[589] Some years later, William Cleland, best known for his excellent work on kinetics, observed that DTT was useful for the reduction of disulfide bonds in proteins.[590] DTT and the isomeric form, dithioerythritol, are each capable of the quantitative reduction of disulfide bonds in proteins (Figure 5.35). Furthermore, the oxidized form of DTT has an absorbance maximum at 283 nm ($\Delta\varepsilon = 273$), which can be used to determine the extent of disulfide bond cleavage.[591] DTT has been extensively used for the reduction of disulfide bonds in proteins under native conditions; 2-mercaptoethanol continues to be used more in the structural analysis of proteins as previously described. The earlier discussions support the concept that the susceptibility of a disulfide bond to reduction is a function of disulfide bond location

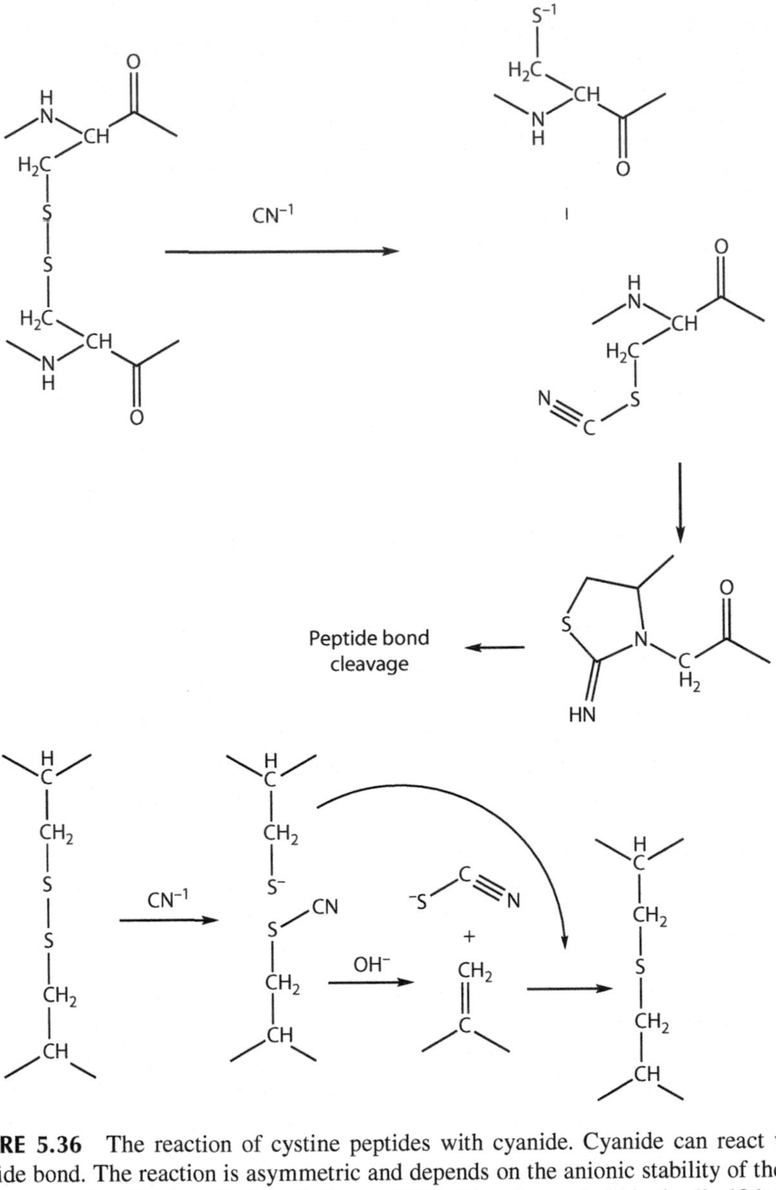

FIGURE 5.36 The reaction of cystine peptides with cyanide. Cyanide can react with a disulfide bond. The reaction is asymmetric and depends on the anionic stability of the leaving thiolate anion (Hiskey, R.G. and Harpp, D.N., Chemistry of aliphatic disulfides. VII. Cyanide cleavage in the presence of thiocyanates, *J. Am. Chem. Soc.* 86, 2014–2018, 1964). The scheme for peptide bond cleavage is adapted from Catsimpoolas, N. and Wood, J.L., The cleavage of cystine peptides by cyanide, *J. Biol. Chem.* 241, 1790–1796, 1966. The bottom scheme shows a suggestion for the reaction of cyanide with proteins at alkaline pH (Catsimpoolas, N. and Wood, J.L., The reaction of cyanide with bovine serum albumin, *J. Biol. Chem.* 239, 4132–4137, 1964).

(primary, secondary, and tertiary structure) and reaction conditions (pH, temperature, solvent). The following studies are provided as examples of work on the selective reduction of disulfide bonds over the past 30 years.

The selective or limited reduction of protein disulfide bonds can be used to study structure and function as well as providing interesting derivatives. Mise and Bahl[592] used limited reduction with DTT to assign disulfide bonds in the α subunit of human chorionic gonadotropin. Since disulfide bonds in a protein will demonstrate different susceptibility to reduction, distribution of radiolabeled iodoacetate will permit the assignment of disulfide bonds. However, the results are not as consistent as one would like as shown by the various reports on the reduction of HEWL. Gorin and coworkers[593] have examined the rate of reaction of HEWL with various thiols. At pH 10.0 (0.025 M borate), the relative rates of reaction were 1.0 for DTT, 0.4 for 3-mercapto-proptionate, and 0.2 for 2-mercaptoethanol. The rate of disulfide bond reduction with 2-mercaptoethanol greatly decreased from pH 10.0 to 9.0 while there was little change in the rate of disulfide bond reduction with DTT until the pH was decrease from 9.0 to 7.8. These experiments were performed in the absence of chaotropic agents. Bewley and coworkers[594] demonstrated the quantitative reduction of disulfide bonds in human pituitary growth hormone, bovine serum albumin, and HEWL with DTT. Warren and Gordon[595] reported that HEWL in water was resistant to reduction with 002 M DTT or 0.09 M 2-mercaptoethanol. One group[596] states that the disulfide bonds in HEWL are resistant to reduction without the addition of chaotropic agents such as dimethylforamide.[597] Radford and coworkers[598] used DTT at pH 7.8 to reduce one of the four disulfide bonds of HEWL and blocked the two cysteinyl residues (Cys6, Cys127) with carboxymethylation. It is noted that the partially reduced HEWL is in a nonnative conformation based by reaction with monoclonal antibody.[599] While the Cys6–Cys127 disulfide bond in HEWL is not considered hyperactive, the corresponding disulfide bonds in equine lysozyme and bovine α-lactalbumin are considered superactive.[600] Bewley and Li[601] used a 20-fold molar excess of DTT to reduce disulfide bonds in various proteins under native conditions (25°C, 0.1 M KCl). Quantitative reduction was observed with lysozyme, prolactin, and insulin while 14 of 18 disulfide bonds were reduced in bovine serum albumin; bovine pancreatic RNase was not reduced to any extent in the absence of chaotropic agents. White[602] used dithreitol to obtain full reduction of HEWL while 2-mercaptoethanol is used for partial reduction of HEWL. It is noted that Bradshaw and coworkers[603] observed that the mixed disulfide of cysteine with HEWL can be easily reactivated in the presence of 2-mercaptoethanol, cysteine, or 2-mercaptoethylamine.

There have been a number of efforts to selectively reduce disulfide bonds in various proteins for the purpose of obtaining new derivative proteins. The problem confronting this approach is described best by disulfide scrambling. Ryle and Sanger described disulfide interchange reactions in their early structural work on insulin.[604] Limited hydrolysis of insulin provided many more cystine peptides than could be accounted for by a unique structure of insulin. Disulfide scrambling can result in the processing and storage of therapeutic proteins.[605–611] A source of reducing agent is required for disulfide scrambling, either a free cysteine or exogenous reducing agent. Thus it was somewhat surprising when Kaneko and Kiabatake[612] observed disulfide scrambling in thaumatin, a protein that does not contain cysteine, with thermal stress.

Analysis of the products of the thermal stress supported a hypothesis where β-elimination of a cystine yielding dehydroalanine and thiocysteine/cysteine which would then have provided the initial reducing power for the disulfide scrambling. A subsequent report[613] on the thermal cleavage of cystine supported this mechanism. This heterolytic cleavage of cysteine is observed with degradation of disulfide bonds in base which is discussed earlier. Disulfide scrambling can also confound the electrophoretic analysis of proteins[614–616] as well as mass spectrometry.[616–619] Wu and Watson[620] presented an approach using cyanylation of cysteine sulfhydryl groups obtained from partial reduction of proteins for the assignment of disulfide linkages in proteins of known sequence.

Homandberg and Wai[621] demonstrated that the reduction of urokinase by DTT in the presence of arginine allows the selective reduction of a disulfide bond joining the catalytically active chain to a nonessential 13 amino acid peptide. A synthetic peptide may then be coupled to the free sulfhydryl group. This work was based on the ability to more selectively reduce interchain disulfide bond between the catalytic B-chain of urokinase and the A-chain. This was accomplished by including arginine with the dithreitol which permitted some specificity in the reduction of the interchain disulfide bonds; intrachain disulfide bonds were still reduced in the A-chain but it was possible to refold the protein. Synthetic peptide added to the refolded protein was incorporated at the free sulfhydryl group with a yield of approximately 40%. The success of this work was based on the ability of arginine to slow the rate of disulfide bond resulting in loss of activity (presumably interchain disulfide bonds in the catalytic B chain) compared to the lesser effect on the disulfide bond between the B-chain and the A-chain. Arginine was required for the correct refolding of the protein. The difficulty in obtained selective reduction of the interchain disulfide bond complicated the success of this approach. Singh and Chang[622] used disulfide scrambling to evaluate the structural stability of human α-thrombin. Reduction of native human α-thrombin with dithreitol in the absence of chaotropic agents resulted in the initial reduction of an internal B-chain disulfide bond followed by the reduction of the intrachain disulfide between the catalytic B-chain and the A-chain; reduction of the intrachain disulfide occurred with 5 mM DTT while 50 mM DTT was required for the intrachain disulfide.

The use of trivalent phosphorus nucleophiles (Figure 5.37) to reduce organic disulfides has been known for some time.[623] Tri-*n*-butylphosphine will reduce disulfide bonds in proteins[624] and will also convert *S*-sulfocysteine to cysteine.[625] The reaction is performed under alkaline conditions (pH 8.0, 0.1 M Tris or 0.5 M bicarbonate) with *n*-propanol added (50/50, v/v) to dissolve tri-*n*-butylphosphine which is insoluble in strictly aqueous solutions. The reduction of disulfide bonds with thiols such as 2-mercaptoethanol or DTT is a reversible reaction; while it could be "driven" by concomitant modification of the newly formed sulfhydryl groups, this is not feasible in the presence of large quantities of competing thiols. The cleavage of disulfide bonds with alkylphosphines with the formation of phosphines is an irreversible reaction and the presence of the alkylphosphines does not compete with alkylation of the newly formed sulfhydryl groups. Kirly,[626] and Chin and Wold[627] used the simultaneous reaction of tri-*n*-butylphosphine to reduce disulfide bonds and alkylation with 4-(aminosulfonyl)-7-fluoro-2,1,3-benzoxadiazole for the fluorescent labeling of

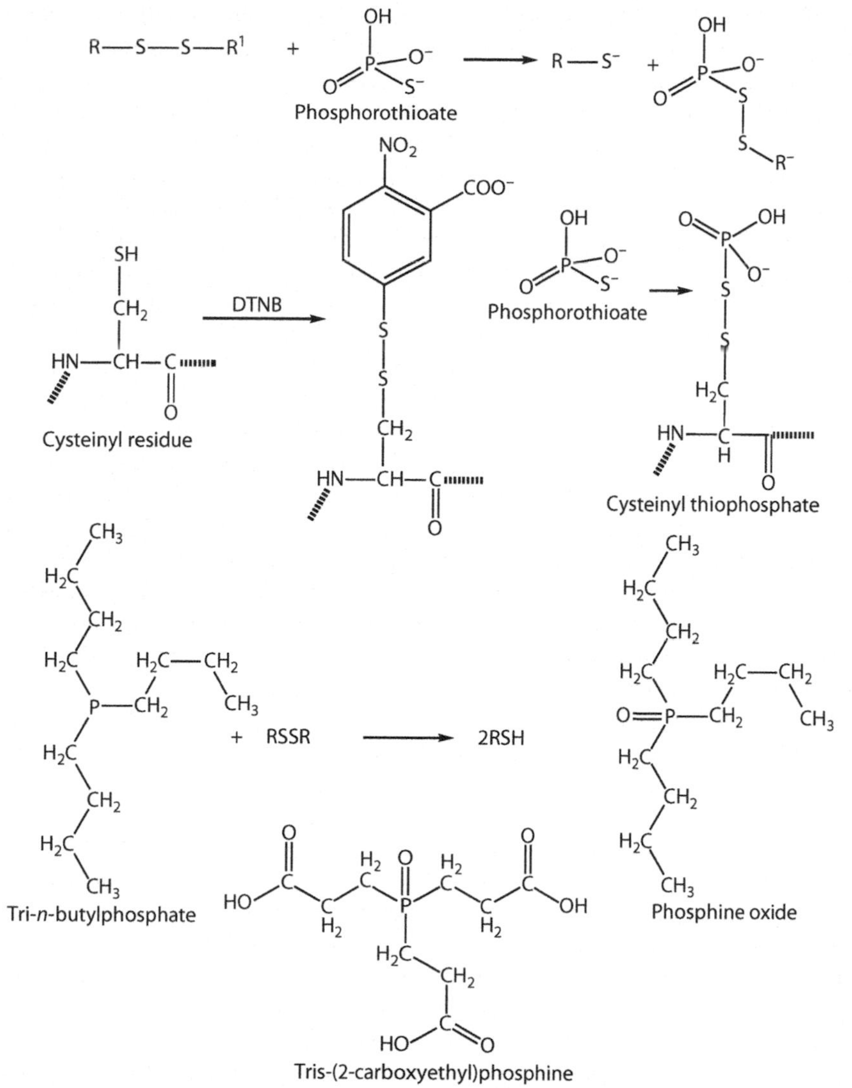

FIGURE 5.37 The use of phosphorothioate and alkyl phosphines for reduction of disulfide bonds in proteins. Shown is the reduction of a disulfide bond by phosphorothioate (Neumann, H. and Smith, R.A., Cleavage of the disulfide bonds of cystine and oxidized glutathione by phosphorothioate, *Arch. Biochem. Biophys.* 122, 354–361, 1967). Also shown is the formation of cysteine thiophosphate by reaction of phosphorothioate with DTNB-modified protein (Saxl, R.L., Anand, G.S., and Stock, A.M., Synthesis and biochemical characterization of a phosphorylated analogue of the response regulator CheB, *Biochemistry* 40, 12896–12903, 2001). Shown at the bottom is tri-*n*-butylphosphine and the reduction of a disulfide with the formation of phosphine oxide (Rüegg, U.T. and Rudinger, J., Reductive cleavage of cystine disulfide with tributylphosphine, *Methods Enzymol.* 47, 111–116, 1977); also shown is the structure of TCEP (Burns, J.A., Butler, J.C., Moran, J., and Whitesides, G.M., Selective reduction of disulfides by tris-(2-carboxyethyl) phosphine, *J. Org. Chem.* 56, 2648–2650, 1991).

the newly formed sulfhydryl groups. Lack of reaction of 4-(aminosulfonyl)-7-fluoro-2,1,3-benzoxadiazole with tri-*n*-butylphosphine permits the simultaneous use of the two reagents to modify proteins. Smejkal and coworkers[628] reported the simultaneous reduction and alkylation of proteins with tri-*n*-butylphosphine and acrylamide. The reader is recommended to a review by Overman and coworkers[623] for additional references to early work on the trialkylphosphines and triarylphosphines. The advantages of alkyl phosphines over thiols such as DTT or 2-mercaptoethanol for the preparation of samples for 2-D electrophoresis has been discussed by Herbert and colleagues.[629] The extensive application of trialkylphosphines/triarylphosphines for the modification of cystine residues in proteins was hampered by insolubility of reagents such as tri-*n*-butylphosphine and there are no current citations to the use of this reagent in protein chemistry. Current activity on phosphines in protein chemistry focuses on TCEP; there are studies comparing tri-*n*-butylphosphine and TCEP in the preparation of samples for proteomics.[630] TCEP has effectively replaced tri-*n*-butylphosphine in the preparation of samples for analysis of homocysteine in plasma.[631–633] There has been some comparison of TCEP with DTT for the reduction of proteins in tissue extracts.[344] Rogers and coworkers[344] found that DTT and TCEP were equivalent in reducing proteins in myofibrillar tissue extracts but that the resolution of fluorescently labeled proteins (monobromobimane) on electrophoresis was superior with the DTT-reduced samples. Humpreys and coworkers[634] showed that TCEP was far more potent than 2-mercaptoethanol, 2-mercaptoethylamine, reduced glutathione, or DTT in reduced disulfide bonds of an Fab fragment.

The synthesis of a water-soluble phosphine, TCEP, was a significant advance.[629] The early development of this reagent[635] described the properties of the reagent. It is quite soluble in water (310 g L^{-1}) and dilute solutions (5 mM) are reasonably stable at acid pH values; at pH values above 7, the rate of conversion of the reagent to the oxide is significant. Unlike thiols such as DTT, the reduction of disulfides with TCEP proceeds very rapidly at pH 4.5 and below. Gray[636] extended his early work in developing the use of TCEP reduction to establish the position of disulfide bonds in proteins. Since the reduction is performed at low pH (stock solution of 20 mM TCEP in 0.17 M citrate, pH 3.0, is stable for weeks at 23°C; the reduction is performed in 0.1% trifluoroacetic acid with 1–10 μM TCEP), it possible to obtain partially reduced peptides by HPLC separation. Alkylation of the free thiols in the isolated peptides with 4-vinylpyridine (Figure 5.16) permitted subsequent structural analysis of the peptide and disulfide bond assignment. In a subsequent study,[637] Gray applied the use of TCEP to the assignment of disulfide bonds in echistatin. Other studies have used trialkylphosphines for the reduction of disulfide bonds prior to electrophoretic analysis in proteomic research.[285,638,639] Komives and coworkers[640] were able to use the approach of Gray[636] to map the disulfide bonds in the fourth and fifth EGF-like domains of thrombomodulin. Watson and colleagues[620,640] have used TCEP reduction followed by modification with cyanate[579,641,642] and subsequent peptide bond cleavage in ammonia[643] to assign disulfide bonds in proteins. While TCEP is suggested to be unstable at pH 7.0 and above,[635] useful studies with this reagent have been performed at pH 7.0–8.0.[644–649] While reference can be found for the reduction of aromatic sulfenic acids to sulfhydryl groups,[650,651] citation could not be found for the reduction of cysteine sulfenic acids by alkylphosphines.

The bulk of the use of TCEP has been for the preparation of samples for proteomic analysis. There are, however, several studies which have used TCEP for the chemical modification of proteins. Tetenbaum and Miller[644] used TCEP to reduce disulfide bonds in soybean trypsin inhibitor (SBTI). The use of TCEP permitted the use of sulfur x-ray absorption spectroscopy to follow the reduction of the disulfide bonds of the protein in real time. The disulfide bonds in SBTI were reduced in a non-cooperative manner within 5 min. Circular dichroism (CD) and FTIR-spectroscopy established that the protein structure collapsed after the reduction of the disulfide bonds. Maciel and coworkers[645] used x-ray photoelectron spectroscopy to measure sulfur oxidation in self-assembled monolayers emphasizing the value of advanced spectral techniques for real-time measurements of oxidation of thiol groups. Zhang and coworkers[646] reported the use of TCEP for reducing of disulfide bonds with simultaneous proteolysis to measure H/D exchange. sRAGE proteins were taken into a deuterium solvent and exchange allowed to proceed. At various times, a portion of the hydrogen-deuterium exchange (HDX) reaction is taken into a mixture of TCEP, urea, and protease (pepsin or *Aspergillus saitoi* protease type XIII) at pH 2.3–2.5. H/D exchange is determined by mass spectrometry. This work is intended to extent the use of HDX to proteins with disulfide bonds. Side reactions with TCEP have been observed including cleavage at cysteine residues.[647] These investigators suggest that a "few" percent of the total protein population is degraded with TCEP (1–10 mM, 37°C, 16 h); the pH optimum for degradation is 8. Degradation can occur at lower temperatures over longer periods of time (40% of a protein cleaved after 2 weeks at 4°C).

DTT, TCEP, and 2-mercaptoethanol are the most frequently used reagents for reduction of disulfide bonds. There are other reagents for the reduction of disulfide bonds in proteins which are of interest. Although infrequently used, phosphorothioate (Figure 5.35) had been demonstrated to effectively cleave disulfide bonds in proteins forming the *S*-phosphorothioate derivative.[652,653] Related reducing agents including bis (2-mercaptoethyl) sulfone and *N,N′*-dimethyl-*N,N′*-bis(mercaptoacetyl)-hydrazine have been described (Figure 5.35)[654] but have not been used despite being somewhat more effective that DTT. Sodium borohydride can be used for the reduction of disulfide bonds in proteins[655] but has seen little use; it is somewhat difficult to use and while it has an advantage in terms of being able to be destroyed rather than removed before subsequent reactions, it has a disadvantage in the stability in aqueous solvent (in addition, there may be foaming). Light and coworkers have examined the susceptibility of disulfide bonds in trypsinogen to reduction.[656] At pH 9.0 (0.1 M sodium borate), a single disulfide bond (Cys179–Cys203) is cleaved in trypsinogen by 0.1 M $NaBH_4$. The resulting sulfhydryl groups are "blocked" by alkylation. The characterization of the modified protein has been performed by the same group.[657] The disulfide bond which is modified under these conditions is critical in establishing the structure of the primary specificity site in trypsin.

It is usually necessary to remove the reducing reagents from the reaction mixture before further process. Efficiency would be improved by avoiding this step by the use of a reagent which could be readily removed from the reaction mixture. Insolubilized dihydrolipoic acid (Figure 5.38) has also been proposed for use in the quantitative reduction of disulfide bonds.[658] Scouten and coworkers[659] developed

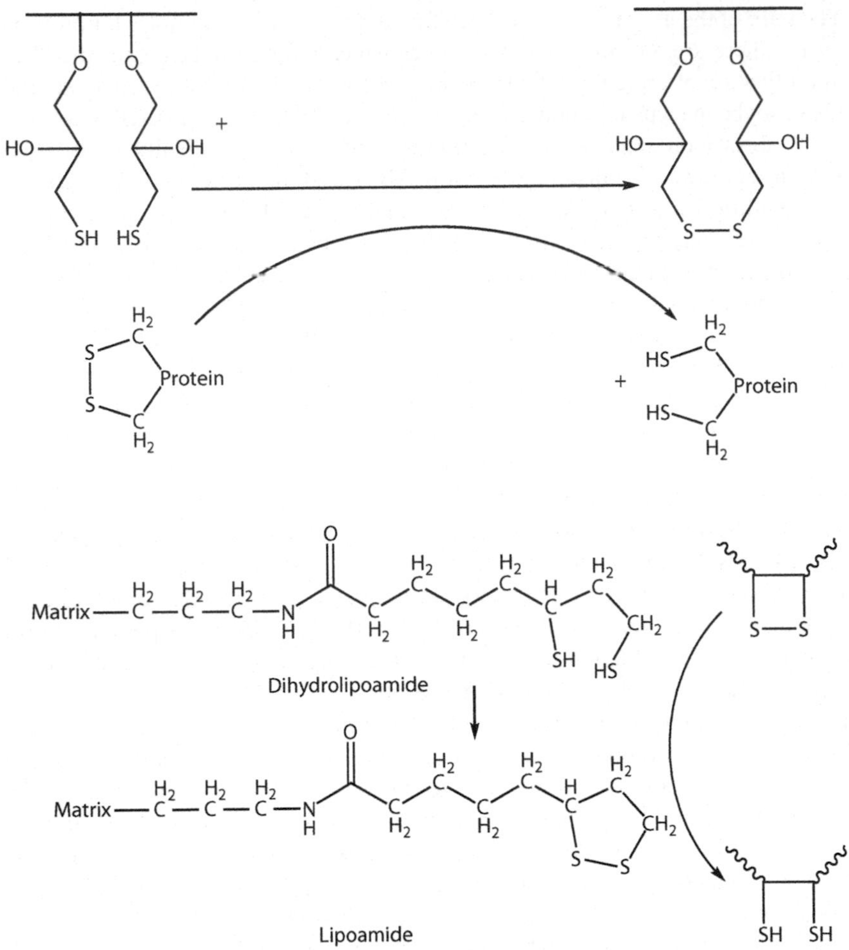

FIGURE 5.38 The structures and mechanism of insoluble reagents for protein disulfide reduction. At the top is thiopropyl agarose (Ferraz, N., Leverrier, J., Batista-Viera, F., and Mantu, C., Thiopropyl-agarose as a solid phase reducing agent for chemical modification of IgG and F(ab')$_2$, *Biotechnol. Prog.* 24, 1154–1159, 2008) while the bottom of the figure shows the mechanism for insoluble lipoic acid (Gorecki, M. and Patchornik, A., Polymer-bound dihydrolipoic acid: A new insoluble reducing agent for disulfides, *Biochim. Biophys. Acta* 303, 36–43, 1973; Scouten, W.H. and Firestone, G.L., N-Propylhydrolipoamide glass beads an immobilized reducing agent, *Biochim. Biophys. Acta* 453, 277–284, 1976).

lipoic acid coupled to a glass matrix for the affinity purification of lipoamide dehydrogenase and later suggested its use as an insoluble reducing agent.[660] Bienvenu and coworkers[661] developed a recyclable matrix (Aminomethyl-ChemMatrix®) containing dihydrolipoamide for the reduction of disulfides. Al-Dubai and coworkers[662] used insoluble dihydrolipoamide to reduce the disulfide bonds in antibody proteins. Thiopropyl-agarose (Figure 5.38) is another insoluble reagent for the reduction of disulfide bonds.[663]

The earlier studies have used the reaction of a reducing agent (soluble or insoluble) in a solution with a protein. Since reduction in its most basic form increases the electronegativity of an atom (e.g., $Cu^{2+} \rightarrow Cu^{1+}$), the process can be considered as the addition of an electron with the hydrogen coming along for the ride. Thus, any process which would permit the electron transfer would be a reduction. A unique study of the reduction of a disulfide bond is provided by the study of the Rieske [2Fe-2S] center of *Thermos thermophilus*.[664] It is possible to reduce this disulfide bond by protein-film voltammetry. Reversible oxidation and reduction of the soluble Riseke domain of *T. thermophilus* was observed with direct absorption of the protein (the soluble Rieske protein fragment obtained by heterologous expression in *E. coli*) on a pryolytic graphite-edge electrode. It has been demonstrated that disulfide bridges in α-lactalbumin are reduced by electron transfer from an excited tryptophan residue.[665,666] In studies with human α-lactalbumin,[666] UV light (270–290 nm, 1 mW cm^{-2}, 2–4 h, 50 mM HEPES, 150 mM KCl, pH 7.8, 20°C) irradiation resulted in a 10 nm redshift of its tryptophan fluorescence emission spectrum (324–334 nm). Reaction of the irradiated protein with 5,5′-dithiobis(2-nitrobenzoic acid) demonstrated the presence of free sulfhydryl groups. It is possible that a similar phenomenon is involved in the changes in a cyclic-nucleotide-gated ion channel by UV irradiation.[667] The cleavage of a disulfide bond mediated by UV light has also been reported for bovine somatropin.[668] Lyophilized recombinant bovine somatotropin was photolyzed by UV light (305–410 nm, λ_{max} = 350 nm). The protein had been lyophilized from carbonate buffer and the cake contained 6% moisture. Unlike the other examples of disulfide bond reduction mediated via tryptophan, the cysteine residues in photolyzed somatotropin appear to donate electrons back to tryptophan leaving a pair of thiyl radicals which subsequently add oxygen to form the sulfonate. Disulfide bonds can also be cleaved by X-radiation (synchrotron radiation).[669] This study reported the reduction of a redox-active disulfide in a tryporedoxin. The radiation dose was less than that required to break a structural disulfide in lysozyme.[670]

The electrolytic reduction of proteins has been explored by Leach and coworkers.[671] These investigators recognized that although small peptides containing disulfide bonds could be reduced using cathodic reduction, there would likely be problems with proteins because of size and tertiary structure considerations. Therefore, a small thiol (2-mercaptoethanol) was used as a catalyst for the reduction. Electrolysis markedly increased the extent of disulfide bond reduction in the presence of 70 mM 2-mercaptoethanol. The extent of disulfide bond reduction did increase with increasing pH; as with other reducing agents, the susceptibility of disulfide bonds to reduction is dependent on their location within the protein. More vigorous electrolytic reduction of proteins has been recently reported by Lee and Chang[672] who observe the reduction of acids to aldehydes or alcohols with an increase in the isoelectric point during 2-D electrophoresis. The extent of reduction depended on the salts present in the electrophoresis system and resulted in artifacts in 2-D electrophoresis. Li and coworkers[673] reported that the combination of a thin-layer electrochemical flow cell (electrochemistry, EC) and liquid sample desorption electrospray ionization mass spectrometry (DESI-MS) resulted in disulfide bond cleavage of insulin. Huang and coworkers[674] reported the reduction of disulfide bonds in soybean trypsin inhibitor with ultrasound; there were also conformational changes observed with CD.

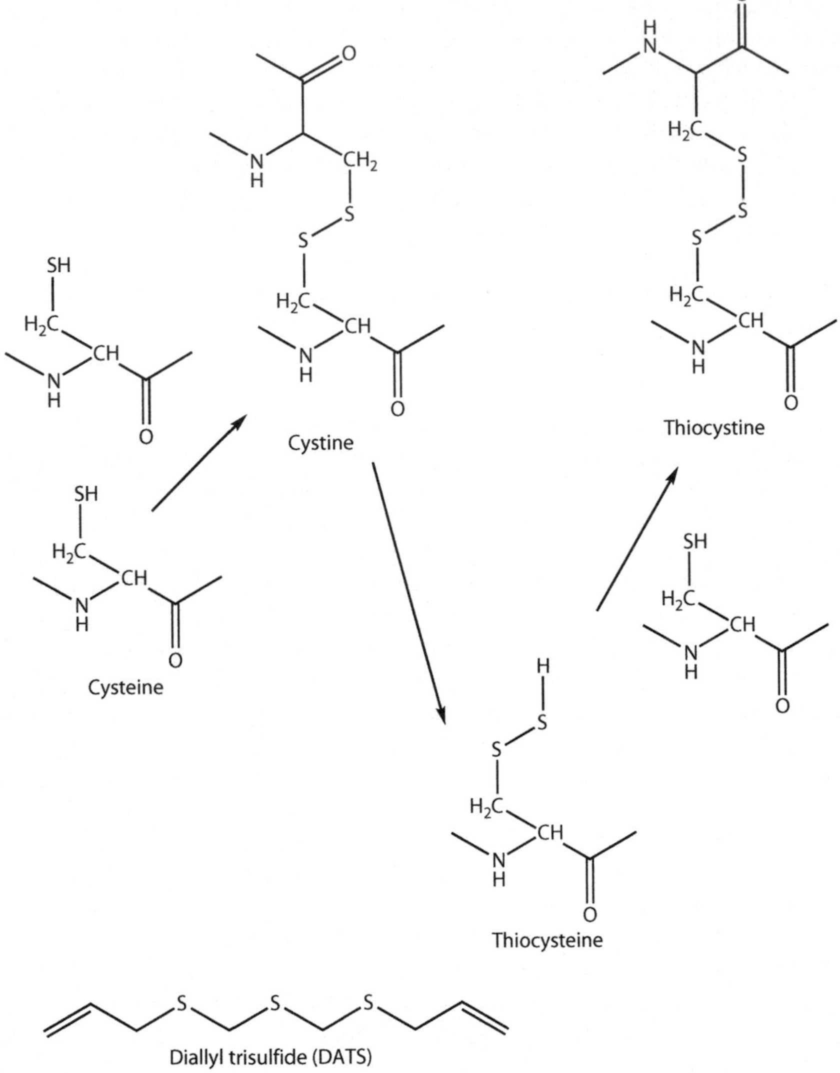

FIGURE 5.39 The structure of cysteine trisulfide and diallyl trisulfide. Shown is the structure of cysteine trisulfide with a putative mechanism for its formation. (See Breton, J., Avanzi, N., Valsasina, B. et al., Detection of traces of a trisulphide derivative in the preparation of a recombinant truncated interleukin-6 mutein, *J. Chromatog. A* 709, 135–146, 1995; Gu, S., Wen, D., Weinreb, P.H. et al., Characterization of trisulfide modification in antibodies, *Anal. Biochem.* 400, 89–98, 2010.) Also shown is the structure of DATS which is considered to be a nutraceutical derived from garlic (see Koizumi, K., Iwasaki, Y., Narukawa, M., Diallyl sulfides in garlic activate both TRPA1 and TRPV1, *Biochem. Biophys. Res. Commun.* 382, 545–548, 2009).

Ascorbic acid (vitamin C) is considered to be an antioxidant which can control free radicals.[675] It also has been used on occasion as a potential reducing agent for disulfide bonds in proteins[676–678] although it is recognized that it is extremely weak ($E^o \approx 0.06$) when compared to cysteine ($E^o \approx -0.33$).[679] Ascorbic acid has been used to selectively reduce *S*-nitroso groups in the biotin-switch assay but the ability of ascorbate to reduce disulfide bonds has created problems.[680]

Finally, a brief consideration of trisulfide bonds in proteins (Figure 5.39). Low molecular trisulfide compounds such as diallyl trisulfide (DATS) (Figure 5.39) are found in garlic and related vegetables and are of interest as nutraceutical products.[681–684] Trisulfide in proteins can be traced to studies showing the presence of such derivatives of cysteine in the acid hydrolysates of wool.[685] Breton and coworkers[686] detected traces of cysteine trisulfide in a mutein of interleukin-6. An anionic form of the mutein where the N-terminal 22 residues were deleted and Cys23 and Cys29 were replaced with serine residues was isolated by isoelectric focusing. This anionic (p*I* 6.7 compared to native p*I* of 6.56) form contained a trisulfide linkage (mass spectrometry) which could be converted to the native form by mild reduction (10 mM DTT). Pristatsky and coworkers[687] found trisulfide bonds in a recombinant variant of a human IgG2 monoclonal antibody which, as with the interleukin-6 mutein, could be eliminated by mild reduction.

METHIONINE

The modification of methionine (2-amino-4-thiomethylbutanoic acid) in a protein is generally accomplished with considerable difficulty. This is possibly a reflection of the fact that, as a relatively hydrophobic residue, methionine is frequently buried in a protein. It is some what more challenging to obtain the specific modification of methionine under mild physiological conditions but advances in the last decade have greatly improved potential in this area. It is possible to obtain highly selective oxidation with some reagents in certain proteins and the results have been useful. Since the dissociation of a proton from sulfur is unnecessary to generate the nucleophile, relatively specific derivatization by alkylating agents can be accomplished at low pH. While other residues such as cysteine and histidine are susceptible to alkylation, these residues are protonated and resist modification under acid conditions.

Interest in the oxidation of methionine (Figure 5.40) stems in part from issues associated with the manufacture of recombinant proteins[688–700] and part from the increase in interest in biological oxidation.[701–705] The reader is directed to an excellent review by Vogt[706] for a discussion of the chemical and biological oxidation of methionine. The oxidation of methionine proceeds initially to the sulfoxide which is a reversible process.[706] Reaction of proteins with hydrogen peroxide at acid pH results in the preferential modification of methionine residues to the sulfoxide although formation of sulfone is not unknown.[707] Hydrogen peroxide together with other oxidizing agents such as superoxide anion are likely responsible for the oxidation of methionine residues in proteins in activated neutrophils.[708] These investigators also reported that methionine sulfoxide bonds are not cleaved by cyanogen bromide (see following) providing an assay for methionine oxidation in proteins. Yamasaki and coworkers[709] demonstrated some specificity in the oxidation of methionine residues

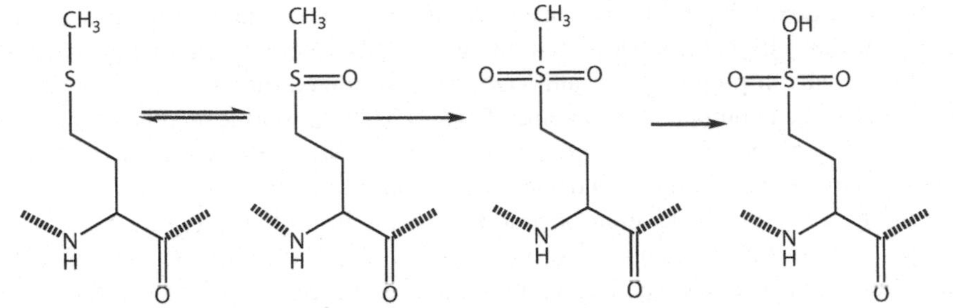

FIGURE 5.40 The structures of methionine, methionine sulfoxide, and methionine sulfone. (See Vogt, W., Oxidation of methionyl residues in proteins: Tools, targets, and reversal, *Free Radic. Biol. Med.* 18, 93–105, 1995; Ohkubo, T., Inagaki, S., Min, J.Z. et al., Rapid determination of oxidized methionine residues in recombinant human basic fibroblast growth factor by ultra-performance liquid chromatography and electrospray ionization quadrupole time-of-flight mass spectrometry with in-source collision-induced dissociation, *Rapid Commun. Mass Spectrom.* 23, 2053–2060, 2009.) Homocysteic acid (homocysteine sulfonic acid) can be formed under more rigorous conditions (Floyd, N.F., Cammaroti, M.S., and Lavine, T.E., The decomposition of DL-methionine sulfoxide in 6N hydrochloric acid, *Arch. Biochem. Biophys.* 103, 343–345, 1963).

in proteins with periodate. Careful consideration of the concentration of peroxide permits the selective formation of the sulfoxide and the sulfone. The authors note that when periodate is used for the oxidation of vicinal diols (carbohydrates) in glycoproteins, possible oxidation of methionine should be considered. Methionine sulfoxide can be reduced to methionine; the extent to which reduction occurs appears to be quite variable. A systematic study[710,711] of the reduction of methionine sulfoxide has shown that of four reducing agents tested, mercaptoacetic acid, 2-mercaptoethanol, DTT, and *N*-methylmercaptoacetamide, the last was the most effective. The reactions demonstrated little pH dependence, but did not proceed well at concentrations of acetic acid above 50% (v/v). Complete regeneration of methionine could be accomplished with 0.7–2.8 M reagent at 37°C for 21 h. There are also methionine sulfoxide reductases[712–717] which can catalyze the conversion of methionine sulfoxide to methionine. The stability of methionine sulfoxide on acid hydrolysis prior to amino acid analysis is not clearly understood.[718] Methionine sulfoxide is converted back to methionine on acid hydrolysis (6N HCl) in the presence of reducing agents such as DTT.[719] Keutmann and Potts[718] had previous observed that the presence of 2-mercaptoethanol in 6N HCl hydrolysis of proteins improved the recovery of methionine. Joergensen and Thestrup[720] reported that thorough degassing and the addition of thioglycolic acid improved the recovery of methionine (and tryptophan) after 6N HCl hydrolysis. Most current work on amino acid analysis for the measurement of methionine sulfoxide uses *p*-toluenesulfonic acid,[721] or methanesulfonic acid[722,723] for the hydrolysis of proteins under conditions where methionine sulfoxide is stable. Other approaches for maintaining methionine sulfoxide during hydrolysis of proteins for amino acid analysis used alkaline,[724] microwave,[725] or flash[726] hydrolysis. Antibodies against methionine sulfoxide have been used to detect this modification in proteins.[727,728] While not directly related to the current work, there has been interest in the formation of methionine sulfoxide in Alzheimer disease.[728,729]

Conversion of methionine sulfoxide to methionine sulfone is essentially irreversible under common solvent conditions and usually requires more vigorous reagents such as performic acid.[730] Homocysteic acid can be formed from methionine under fairly drastic conditions.[731–733] Homocysteic acid may also be formed from the oxidation of homocysteine.[734]

Chemical oxidation is one method for mapping surface residues in proteins; mass spectrometry is an analytical method which has greatly increased the value of this approach. The current approach to oxidative surface mapping used hydroxyl radical generated either from water by radiolysis or by Fenton's reagent.[735] The high reactivity of hydroxyl radicals has provided the basis for use in surface mapping.[736] Sharp and coworkers[33] used Fenton's reagent to catalyze the formation of hydroxyl radicals from hydrogen peroxide. These investigators showed that Met131 was modified with hydroxyl radicals despite a lack of surface accessibility suggesting either a different mechanism (radical transfer from an exposed residue) or that such oxidation occurred in the processing of the sample. Subsequent work[737] from this group supported the concept that methionine could be oxidized via hydroxyl radical without solvent exposure. More recently, Bern and coworkers[738] have shown the presence of homocysteic acid (Figure 5.40) in highly oxidized samples for proteomic analysis. As mentioned earlier, homocysteic acid is formed from methionine under what can be considered rigorous conditions. Methionine-containing peptides and protein samples for mass spectrometry have been oxidized by H_2O_2 directly on MALDI-MS target plates.[739] This approach permits the more facile identification of proteins from database searches; the oxidation of methionine by hydrogen peroxide increases the mass by 16 Da per methionine residue. The technique allows spectra to be obtained before and after methionine oxidation. Pan and coworkers[740] used insertional mutagenesis to place methionine residues in bacteriorhodopsin; hydroxyl radicals were then used to oxidize the inserted methionine residues in a study of folding. Methionine sulfoxide residues resulting from hydroxyl radical was identified by mass spectrometry.

Methionine oxidation is of great interest to the biopharmaceutical industry as the oxidation of therapeutic proteins may result in defective product with reduced stability and conformational change.[741] Examples of where methionine oxidation may result in conformational change include recombinant human interferon α-2b,[742] recombinant human interleukin-1 receptor antagonist,[743] immunoglobulin light chain LEN,[744] and oxidatively stressed monoclonal antibodies.[745] It should be noted that considering that methionine residues are buried, oxidation may well be following conformational change. There are examples where methionine oxidation has occurred without measured conformational change.[746,747]

Hydrogen peroxide has been frequently used to modify methionine residues in proteins as it is considered to be a mild oxidizing agent; as noted earlier hydrogen peroxide will modify cysteine. Kim and coworkers[748] have compared the effect of methionine oxidation with H_2O_2 (25 mM sodium phosphate, 100 mM NaCl, pH 7.0, 1.1% H_2O_2, room temperature, 30 min) and oligonucleotide-directed mutagenesis on the stability of staphylococcal nuclease. Oxidation of two methionine residues

(M65, M95) resulted in a considerable loss of protein stability while modification of the other two residues (M26, M32) had little effect. Substitution of a leucine residue at position 95 (M95L) resulted in a loss of stability similar to that seen with oxidation while substitution at position 65 (M65L) was less deleterious. A similar experimental approach was taken by Chien and coworkers[749] to an *N*-carbamoyl D-amino acid aminohydrolase from *Agrobacterium radiobacter.* This protein contains nine methionine residues which were individually replaced with leucine. Two of the mutants had activity similar to wild type while the other seven had reduced activity. The three mutants with solvent accessible residues retained activity on oxidation with H_2O_2 (0.1–1.0 mM H_2O_2 in 200 mM sodium phosphate, pH 7.0, 25°C, 15 min). The other mutants were more resistant to loss of activity on oxidation than wild type. Caldwell and coworkers[750] used hydrogen peroxide to oxidize the methionine residues in *E. coli* ribosomal protein L12; the extent of methionine oxidation was measured by reaction with radiolabeled iodoacetate at pH 3.0 (see below). The loss of activity on the oxidation of L12 by H_2O_2 can be reversed by 2-mercaptoethanol; oxidation of the methionine residues in L12 did convert the dimer to monomer form. Kachurin and coworkers[751] used H_2O_2 to oxidize a methionine residue in a α-galactosidase from *Trichoderma reesei.* The modified enzyme demonstrated an increase in activity with *p*-nitrophenyl-α-D-galactoside; a similar effect was observed with periodate and permanganate. One of the five methionine residues was modified; there was no modification of tryptophan, histidine, tyrosine, and cysteine. Further oxidation of the methionine residues could be accomplished with a longer reaction time or denaturation of the protein. Kornfelt and coworkers[752] used H_2O_2 to oxidize two of the four methionine residues in blood coagulation Factor VIIa. Three oxidation products were obtained from the reaction of H_2O_2 with Factor VIIa which could be separated by RP-HPLC using a C_4 column. Nomura and coworkers[753] have observed that methionine oxidation decreases the stability of the p53 tetramer; the oxidized protein is more susceptible to tryptic digestion.

Periodate is another relatively mild oxidizing agent which can selectively modify methionine and cysteine residues although tryptophan and tyrosine can also be modified. Periodate is most often used to oxidize vicinal diols in carbohydrates yielding aldehydes; periodate can also modify *N*-terminal serine residues as discussed earlier. The resulting aldehydes can be coupled to amines as used for the preparation of a conjugate vaccine[754] or converted to the hydrazide;[755] this approach is frequently used for the coupling of proteins to agarose matrices. Antibodies can be coupled to matrices via periodate oxidation of the carbohydrate in the Fc domain.[756] The periodic acid Schiff reagent is used for the staining of glycoproteins on gel electrophoresis.[757] Yamasaki and coworkers[709] established that periodate under mild conditions modified a variable number of methionine residues in various proteins under various conditions. With pancreatic ribonuclease, two out of four methionine residues were oxidized at pH 4.0 (5 mM sodium periodate at 25°C) while all 16 methionine residues in chicken ovalbumin were oxidized under the same conditions; one of the two methionine residues in chymotrypsin was modified by 5 mM periodate at pH 5.0 (5°C). Knowles[758] showed that the modification of a single methionine in chymotrypsin resulted in a partially active enzyme with an increased K_m for aromatic substrates such as *N*-acetyl-L-tryptophan ethyl ester while the K_m is unchanged for non-aromatic substrates such as *N*-acetyl-L-valine ethyl ester. De la Llosa and coworkers[759] used periodate or chloramine-T (see following) to oxidize methionine residues in bovine luteinizing hormone. Periodate oxidized seven methionine residues while chloramine-T approximately six methionine residues; periodate oxidized most of the fucose association with the peptide hormone while a small amount was lost with chloramine-T. Most of the activity is lost with the oxidation of three methionine residues (presumably to the sulfoxide); modification of methionine residues with cyanogen bromide which forms homoserine with associated peptide bond cleavage resulted in smaller loss of activity. Periodate was used by Gleisner and Liener[760] to oxidize a single methionine residue in ficin after protection of the active-site sulfhydryl group with tetrathionate permitting the modification of the active site histidine residue with bromoacetone in 2.0M urea. Methionine oxidation in prion protein has been suggested to be involved in neurodegenerative disease.[761] Wolschner and coworkers[762] used periodate to oxidize methionine residues in recombinant cellular prion protein PrP(C) as part of an effort to understand oxidative stress in neurodegenerative diseases. The extent of methionine oxidation was determined by mass spectrometry and was correlated with the tendency to aggregate.

Chloramine-T has been used for the selective oxidation of methionine in proteins. Shechter and coworkers[719] demonstrated that chloramine-T could oxidize methionine residues in proteins at neutral or slightly alkaline pH (pH 8.5); cysteine was also oxidized to cysteine under these conditions but modification of other amino acid was not observed. Modification of tryptophan with chloramine-T was observed at pH 2.2. Similar results were obtained with *N*-chlorosuccinimide while *N*-bromosuccinimide (NBS) was more promiscuous with tyrosine and histidine also modified at pH 2.2; tyrosine was also modified with NBS at pH 8.5 with histidine modification occurring only with higher excesses of reagent. Methionine sulfoxide was converted to methionine sulfone by NBS at pH 8.5 only after complete modification of histidine. This study also reported that oxidized methionine residues were not cleaved by cyanogen bromide providing for an assay for oxidized methionine residues in proteins. Trout[763] reported a spectral method for the measurement of microgram quantities of methionine by reaction with chloramine-T; the assay is based on the spectral change of chloramine-T at 244–246nm. This assay cannot be used in the presence of sulfhydryl groups, urea, or guanidine hydrochloride and, as such, has not been applied to the measurement of methionine in proteins. Chloramine-T has been for the selective oxidation of methionine in human recombinant secretory leukocyte proteinase inhibitor,[764] large conductance calcium-activated potassium channels,[765] actin,[766] and kininogens.[767] Oxidation of methionine residues occurring when chloramine-T is used for protein iodination can pose a problem for subsequent analysis.[768,769] The oxidation of methionine in recombinant human interleukin-2 by potassium peroxodisulfate has been reported.[770]

The development of *t*-butyl hydroperoxide by Keck[771] as a selective oxidizing agent for methionine in proteins represented a significant advance. Results obtained with native recombinant interferon and recombinant tissue-type plasminogen activator showed that this reagent was selective for the oxidation of exposed methionine residues in proteins. The reaction was performed in 5mM succinate, 0.1% Tween 20, at pH 5.0 (recombinant interferon) or 0.2M arginine, 0.1M sodium phosphate, 0.1% Tween 80, at pH 7.0 (recombinant tissue-type plasminogen activator). *t*-Butyl hydroperoxide

(0–73 mM) was added and the reaction allowed to proceed overnight at room temperature. Tryptic peptides were separated by HPLC and analyzed by mass spectrometry. Two methionine residues were oxidized in recombinant interferon with *t*-butyl hydroperoxide, while all five residues were oxidized to a varying extent by H_2O_2 under the same reaction conditions. Three methionine residues were oxidized in native tissue-type plasminogen activator; all five residues were oxidized to varying degrees in the presence of 8.0 M urea. *t*-Butyl hydroperoxide has been successfully used for recombinant human leptin[772] and recombinant human granulocyte colony-stimulating factor.[773] Chumsae and coworkers[774] used *t*-butyl hydroperoxide for the oxidation of methionine in a fully human monoclonal antibody as a probe for conformational change during thermal stressing. The rationale for these studies is based on the susceptibility of methionine to oxidation; exposure of methionine is considered an index of conformational change. Subsequent work by this group[775] established that this modification decreased binding affinity to protein A or protein G affinity columns. This latter observation is of significance given the importance of protein A and or protein G affinity columns in antibody purification. Further work from this group[776] compared the photooxidation and *t*-butylperoxide oxidation of the antibody. Another group[777] confirmed that the oxidation of antibody with *t*-butylperoxide decreased affinity to protein A; it was also observed that affinity for the FcRn receptor was reduced suggesting that protein A binding is a surrogate measure for FcRn binding of antibody.

Methionine can be modified with various alkylation agents such as the α-halo acetic acids and their derivatives. The reaction of iodoacetate with methionine has been examined in some detail by Gundlach and coworkers.[778] The reaction of iodoacetate with methionine was not pH dependent and proceeds much slower than the reaction with cysteine under the mildly alkaline conditions used for reduction and carboxymethylation. The resulting sulfonium salt yields homoserine and homoserine lactone when heated at 100°C at pH 6.5. On acid hydrolysis (6N HCl, 110°C, 22 h), a mixture of methionine and *S*-carboxymethyl homocysteine together with a small amount of homoserine lactone was obtained. In general, methionine residues only react with the α-halo acids after the disruption of the secondary and tertiary structure of a protein.[4] Selectivity in the modification of methionine in proteins by α-halo acids can be achieved by performing the reaction at acid pH (pH 3.0 or less) but there are examples of modification under less rigorous conditions. Cummings and coworkers[779] reported the modification of a single methionine residue in medium-chain acyl-CoA dehydrogenase; this residue is not directly involved in catalysis but the modification has a major effect on K_m and a less effect on K_{cat}. The modification of methionine by ethyleneimine has been reported in a reaction producing a sulfonium salt derivative.[780] The modification of methionine in azurin with bromoacetate has been reported.[781] In the protein, four of six methionine residues were modified at pH 4.0, while all methionine residues were reactive at pH 3.2. These modification reactions were performed in 0.1 M sodium formate at ambient temperatures for 24 h with 0.16 M bromoacetate. The modification of methionine in porcine kidney acyl-CoA dehydrogenase occurs with iodoacetate (0.030 M) in 0.1 M phosphate, at pH 6.6 at ambient temperature.[782] The identification of methionine as the residue modified by iodoacetate in this protein was supported by the comparison of the chromatogram of the acid hydrolyzate of the modified protein (reacted with ^{14}C-iodacetate) with that of the acid hydrolyzate of synthetic *S*-([1-^{14}C]carboxymethyl)-methionine.[778] This is necessary since the *S*-carboxymethyl derivative yielded several different compounds on acid hydrolysis.[778,783]

Naider and Bohak[784] have reported that the sulfonium salt derivatives of methionine (e.g., *S*-carboxymethyl methionine, the reaction product of methionine and iodoacetic acid) can be converted to methionine by reaction with a suitable nucleophile. For example, reaction of *S*-carboxamidomethyl methionine (in the peptide Gly-Met-Gly) with a sixfold molar excess of mercaptoethanol at pH 8.9 at a temperature of 30°C resulted in the complete regeneration of methionine after 24 h of reaction. The *S*-phenacyl derivative of methionine (in the peptide Gly-Met-Gly) was converted to methionine in 1 h under the same reaction conditions. These investigators also showed that chymotrypsin previously treated with phenacyl bromide under conditions which inactivate the enzyme concomitant with the alkylation of Met192[785] could be reactivated by treatment with 3-mercaptoethanol at pH 7.5 (sodium phosphate). It is of interest that the *S*-phenacyl methionine in chymotrypsin is converted to methionine at a substantially faster rate than the tripeptide derivative. The authors speculate that the increased reactivity of the chymotrypsin derivative is a reflection of interaction of the phenacyl moiety with the substrate-binding site. Alkylation of methionyl residues in pituitary thyrotropin and lutropin with iodoacetic acid has been reported.[786] Differential reactivity of various methionyl residues was reported on reaction with iodoacetate in 0.2 M formate, at pH 3.0 for 18 h at 37°C. The reversible alkylation of methionine by iodoacetate in dehydroquinase has been reported by Kleanthous and coworkers.[787] In this reaction, iodoacetate behaves kinetically as an affinity label with a K_i of 30 μM and a k_{inact} of 0.014 min^{-1}, at pH 7.0 (50 mM potassium phosphate). There is no reaction with iodoacetamide. Two methionine residues are modified during the reaction of dehydroquinase with iodoacetate. In a companion study, Kleanthous and Coggins[788] demonstrated that 2-mercaptoethanol treatment under alkaline conditions (0.5% ammonium bicarbonate, 37°C) could reverse modification at one of the two residues. If the modified protein is denatured, there is no reversal of modification at either residue. The results are interpreted in terms of the proximity of a positive charge (i.e., lysine) in close proximity to one of the two methionyl residues which (1) provides the basis for the affinity labeling and (2) provides the basis for the 2-mercaptoethanol-mediated reversal of modification.

The ability to reverse the alkylation of methionine under relatively mild conditions as described earlier has resulted in the development of a clever affinity approach to the purification of methionine peptides. Several groups[304,789,790] have reported the isolation of methionine peptide by reaction with bead containing a bromoacetyl function under acidic conditions (e.g., 25% acetic acid) and subsequent reducing agent under alkaline conditions as described previously.

REFERENCES

1. White, S.H., Global statistics of protein sequence: Implications for the origin, evolution, and prediction of structures, *Ann. Rev. Biophys. Biomol. Struct.* 23, 407–439, 1994.
2. Gu, S., Wen, D., Weinreb, P.H. et al., Characterization of trisulfide modification in antibodies, *Anal. Biochem.* 400, 89–98, 2010.

3. Moerman, P.P., Sergeant, K., Debyser, G. et al., A new chemical approach to differentiate carboxy terminal peptide fragments in cyanogen bromide digests of proteins, *J. Proteomics* 73, 1454–1460, 2010.
4. Gurd, F.R.N., Carboxymethylation, *Methods Enzymol.* 25, 424–438, 1972.
5. Ohno, A. and Oae, S., Thiols, in *Organic Chemistry of Sulfur*, S. Oae (ed.), Plenum Press, New York, Chapter 4, pp. 119–181, 1977.
6. Silva, C.O., da Silva, E.C., and Nascimento, M.A.C., Ab initio calculations of absolute p*Ka* values in aqueous solutions II. Aliphatic alcohols, thiols, and halogenated carboxylic acids, *J. Phys. Chem. A* 104, 2402–2409, 2000.
7. Hederos, S. and Baltzer, L., Nucleophile selectivity in the acyl transfer reaction of a designed enzyme, *Biopolymers* 79, 292–299, 2005.
8. Sluyterman, L.A.A., The rate-limiting reaction in papain action as derived from the reaction of the enzyme with chloroacetic acid, *Biochim. Biophys. Acta* 151, 178–187, 1968.
9. Sluyterman, L.A.A., Reversible inactivation of papain by cyanate, *Biochim. Biophys. Acta* 139, 439–449, 1967.
10. Stark, G.R., Stein, W.H., and Moore, S., Reactions of cyanate present in aqueous urea with amino acids and proteins, *J. Biol. Chem.* 235, 3177–3181, 1960.
11. Janssen, M.J., Nucleophilicity of organic sulfur compounds, in *Sulfur in Organic and Inorganic Chemistry*, A. Senning (ed.), Dekker, New York, Chapter 3, pp. 355–377, 1972.
12. Pienta, N.J. and Kessler, R.J., Pentaenyl cations from the photolysis of retinyl acetate. Solvent effects on the leaving group ability and relative nucleophilicities: An unequivocal and quantitative determination of the importance of hydrogen bonding, *J. Am. Chem. Soc.* 114, 2419–2428, 1992.
13. Reichart, C., *Solvent and Solvent Effects in Organic Chemistry*, 3rd edn., Wiley-VCH, Weinheim, Germany, 2003.
14. Cecil, R. and McPhee, J.R., The sulfur chemistry of proteins, *Adv. Prot. Chem.* 14, 255–389, 1959.
15. Liu, T.-Y., The role of sulfur in proteins, in *The Proteins*, Vol. 3, 3rd edn., H. Neurath and R.L. Hill, (eds.), Academic Press, New York, 1977.
16. Torchinsky, Y.M., *Sulfur in Proteins*, Pergamon Press, Oxford, U.K., 1981.
17. Creighton, T.E., Chemical nature of polypeptides, in *Proteins. Structure and Molecular Principles*, W.H. Freeman and Company, New York, Chapter 1, 1993.
18. Modena, G., Paradisi, C., and Scorrano, G., Solvation effects on basicity and nucleophilicity, in *Organic Sulfur Chemistry. Theoretic and Experimental Advances*, F. Bernardi, I.G. Csizmadia, and A. Mongini (eds.), Elsevier, Amsterdam, the Netherlands, Chapter 10, pp. 569–597, 1985.
19. Kooyman, E.C., Some characteristics of organosulfur compound and intermediates, in *Organosulfur Chemistry*, M.J. Janssen (ed.), Wiley Interscience, New York, Chapter 1, pp. 1–10, 1967.
20. Davies, R.E., The oxibase scale, in *Organosulfur Chemistry*, M.J. Janssen (ed.), Wiley Interscience, New York, Chapter 18, pp. 311–328, 1967.
21. Ogino, K. and Fujihara, H., Biochemical reactions involving thioesters, in *Organic Sulfur Chemistry*, S. Oae and T. Okuyama (eds.), CRC Press, Boca Raton, FL, Chapter 3, pp. 71–136, 1992.
22. Khan, S.A., Sekulski, J.M., and Erickson, B.W., Peptide models of protein metastable binding sites: Competitive kinetics of isomerization and hydrolysis, *Biochemistry* 25, 5165–5171, 1986.
23. Isenman, D.E., The role of the thioester bond in C3 and C4 in the determination of the conformational and functional states of the molecule, *Ann. N.Y. Acad. Sci.* 421, 277–290, 1983.

24. Song, J., Wang, J., Jazwiak, A.A. et al., Stability of thioester intermediates in ubiquitin-like modifications, *Protein Sci.* 18, 2482–2499, 2009.
25. Hop, C.E. and Bakhtier, R., Homocysteine thiolactone and protein homocysteinylation: Mechanistic studies with model peptides and proteins, *Rapid Commun. Mass Sp.* 16, 1049–1059, 2002.
26. Jakubowski, H., Homocysteine thiolactone: Metabolic origin and protein homocysteinylation in human, *J. Nutr.* 130, 377S–381S, 2000.
27. Paoli, P., Sbrana, E., Tiribilli, B. et al., Protein N-homocysteinylation induces formation of toxic amyloid-like protofibrils, *J. Mol. Biol.* 400, 889–897, 2010.
28. Xiao, J., Burn, A., and Tolbert, T.J., Increasing solubility of proteins and peptides by site-specific modifications with betaine, *Bioconj. Chem.* 19, 1113–1118, 2008.
29. Rohde, H. and Seitz, O., Ligation-desulfurization: A powerful combination in the synthesis of peptides and glycopeptides, *Biopolymers* 94, 551–559, 2010.
30. Shingledecker, K., Jiang, S.-Q., and Paulus, H., Reactivity of the cysteine residues in the protein splicing active center of the *Mycobacterium tuberculosis* RecA intein, *Arch. Biochem. Biophys.* 375, 138–145, 2000.
31. Du, Z., Shemella, P.T., Liu, Y. et al., Highly conserved histidine plays a dual catalytic role in protein splicing: A p*K*a shift mechanism, *J. Am. Chem. Soc.* 131, 11581–11589, 2009.
32. Furuta, S., Ortriz, F., Zhu Sun, X. et al., Copper uptake is required for pyrrolidone dithiocarbamate-mediated oxidation and protein level increase of p53 in cells, *Biochem. J.* 365, 639–648, 2002.
33. Sharp, J.S., Becker, J.M., and Hettich, R.L., Protein surface mapping by chemical oxidation: Structural analysis by mass spectrometry, *Anal. Biochem.* 313, 216–225, 2003.
34. Mannick, J.B. and Schonhoff, C.M., Nitrosylation: The next phosphorylation, *Arch. Biochem. Biophys.* 408, 1–6, 2002.
35. Tao, L. and English, A.M., Mechanism of S-nitrosylation of recombinant human brain calbindin D_{28K}, *Biochemistry* 42, 3326–3334, 2003.
36. Nogueira, L., Figueriredo-Freitas, C., Casimiro-Lopes, G. et al., Myosin is reversibly inhibited by S-nitrosylation, *Biochem. J.* 424, 221–231, 2009.
37. Ralat, L.A., Ren, M., Schilling, A.B., and Tang, W.J., Protective role of Cys-178 against the inactivation and oligomerization of human insulin-degrading enzyme by oxidation and nitrosylation, *J. Biol. Chem.* 284, 34005–34018, 2009.
38. Marino, S.M. and Gladyshev, V.N., Structural analysis of cysteine S-nitrosylation: A modified acid-based motif and the emerging role of trans-nitrosylation, *J. Mol. Biol.* 395, 844–859, 2010.
39. Spadaro, D., Yun, B.W., Spoel, S.H. et al., The redox switch: Dynamic regulation of protein function by cysteine modifications, *J. Physiol. Plant.* 138, 360–371, 2010.
40. Spickett, C.M. and Pitt, A.R., Protein oxidation: Role in signalling and detection by mass spectrometry, *Amino Acids*, in press, 2010.
41. Aesif, S.W., Janssen-Heininger, Y.N., and Reynaert, N.L., Protocols for the detection of S-glutathinoylated and S-nitrosylated proteins *in situ*, *Methods Enzymol.* 474, 289–296, 2010.
42. Ueland, P.M., Skoolard, T., Doskeland, S.O., and Flatmark, T., An adenosine 3′,5′-monophosphate-adenosine binding protein from mouse liver: Some physicochemical properties, *Biochem. Biophys. Acta* 533, 57–65, 1978.
43. Arai, K., Arai, T., Kawakita, M., and Kaziro, Y., Conformational transitions of polypeptide elongation factor Tu. I. Studies with hydrophobic probes, *J. Biochem.* 77, 1096–1108, 1975.
44. Nakamura, S., Ohta, S., Arai, K., and Arai, N., Studies on polypeptide chain-elongation factors from an extreme thermophile, *Thermus thermophilus* HB8. 3. Molecular properties, *Eur. J. Biochem.* 92, 533–543, 1978.

45. Whitfield, C.E. and Schworer, M.E., Locus of N-ethylmaleimide action on sugar transport in nucleated erythrocytes, *Am. J. Physiol.* 241, C33–C41, 1981.
46. Floris, G., Giartosio, A., and Rinaldi, A., Essential sulfhydryl groups in diamine oxidase from *Euphoria characias* latex, *Arch. Biochem. Biochem.* 220, 623–627, 1983.
47. Polyanovsky, O.L., Novikov, V.V., Deyev, S.M. et al., Location of exposed and buried cysteine residues in the polypeptide chain of aspartate aminotransferase, *FEBS Lett.* 35, 322–326, 1973.
48. Fuchs, F., Liou, Y.-M., and Grabarek, Z., The reactivity of sulfhydryl groups of bovine cardiac troponin C, *J. Biol. Chem.* 264, 20344–20349, 1989.
49. Lee, J. and Blaber, M., The interaction between thermodynamic stability and buried free cysteines in regulating the functional half-life of fibroblast growth factor-1, *J. Mol. Biol.* 393, 113–127, 2009.
50. Lee, J. and Blaber, M., Structural basis of conserved cysteine in the fibroblast growth factor family: Evidence for a vestigial half-cysteine, *J. Mol. Biol.* 393, 129–139, 2009.
51. Nagano, N., Ota, M., and Nishikawa, K., Strong hydrophobic nature of cysteine residues in proteins, *FEBS Lett.* 458, 69–71, 1999.
52. Zhou, H. and Zhou, Y., Quantifying the effect of burial of amino acid residues on protein stability, *Proteins Struct. Funct. Bioinf.* 54, 315–322, 2003.
53. Heitmann, P., A model for sulfhydryl groups in protein. Hydrophobic interactions of the side chain in micelles, *Eur. J. Biochem.* 3, 346–350, 1968.
54. Carter, J.R., Jr., Role of sulfhydryl groups in erythrocyte membrane structure, *Biochemistry* 12, 171–176, 1973.
55. Mima, J., Jung, G., and Onizuki, T. et al., Amphipathic property of free thiol group contributes to an increase in the catalytic efficiency of carboxypeptidase Y, *Eur. J. Biochem.* 269, 3220–3225, 2002.
56. Kaplan, H., Long, B.G., and Young, N.M., Chemical properties of functional groups of immunoglobulin IgA, IgG2, and IgM mouse myeloma proteins, *Biochemistry* 19, 2821–2827, 1980.
57. Shaw, C.F. 3rd., Coffer, M.T., Klingbell, J., and Mirabelli, C.K., Application of phosphorous-31 NMR chemical shift: Gold affinity correlation to hemoglobin-gold binding and the first inter-protein gold transfer reaction, *J. Am. Chem. Soc.* 110, 729–734, 1988.
58. Paul, I., Cul, J., and Maynard, E.L., Zinc binding to the HCCH motif of HIV-1 virion infectivity factor induces a conformational change that mediates protein–protein interactions, *Proc. Natl. Acad. Sci. USA* 103, 18475–18480, 2006.
59. Ingraham, R.H. and Hodges, R.S., Effects of calcium and subunit interactions on surface accessibility of cysteine residues in cardiac troponin, *Biochemistry* 27, 5891–5898, 1988.
60. Polpot, J.L. and Saraste, M., Engineering membrane proteins, *Curr. Opin. Biotechnol.* 6, 394–402, 1995.
61. Brown, L.S., Needleman, R., and Lanyl, J.K., Conformational changes of the E-F interhelical loops in the M photointermediate of bacteriorhodopsin, *J. Mol. Biol.* 317, 471–478, 2002.
62. Visudtiphole, V., Chalton, D.A., Hong, Q., and Lakey, J.H., Determining OMP topology by computation, surface plasmon resonance and cysteine labelling: The test case of OMPG, *Biochem. Biophys. Res. Commun.* 351, 113–117, 2006.
63. Nagler, C., Nagler, G., and Kuhn, A., Cysteine residues in the transmembrane regions of M13 procoat protein suggest that oligomeric coat proteins assemble onto phage progeny, *J. Bacteriol.* 189, 2897–2905, 2007.
64. Nair, M.S. and Dean, D.H., All domains of Cry1A toxins insert into insect brush border membranes, *J. Biol. Chem.* 283, 26324–26331, 2008.

65. Girard, F., Vachon, V., Labele, G. et al., Chemical modification of *Bacillus thuringiensis* Cry1Aa toxin single-cysteine mutants reveals the importance of domain I structural elements in the mechanism of pore formation, *Biochim. Biophys. Acta* 1788, 575–580, 2008.
66. Greeenstein, J.P., Sulfhydryl groups in proteins. III. Effect on egg albumin of various salts of guanidine, *J. Biol. Chem.* 130, 519–526, 1939.
67. Hopkins, F.G., Denaturation of proteins by urea and related substances, *Nature* 126, 328–330, 1930.
68. Putnam, F.W., Protein denaturation, in *The Proteins*, Vol. 1, Pt. B, H. Neurath and K. Bailey (eds.), Academic Press, New York, Chapter 9, pp. 807–892, 1953.
69. Hopkins, F.G. CXXI., Glutathione. Its influence in the oxidation of fats and proteins, *Biochem. J.* 19, 787–819, 1925.
70. Osborne, T.B., *The Proteins of the Whole Wheat Kernel*, Carnegie Institute of Washington, Washington, DC, 1902.
71. Anson, M.L., The reactions of iodine and iodoacetamide with native egg albumin, *J. Gen. Physiol.* 21, 321–331, 1940.
72. Anson, M.L., The sulfhydryl groups of egg albumin, *J. Gen. Physiol.* 24, 399–421, 1941.
73. Bull, H.B., Protein structure, *Adv. Enzymol.* 1, 1–42, 1941.
74. Britto, P.J., Knipling, L., and Wolff, J., The local electrostatic environment determines cysteine reactivity of tubulin, *J. Biol. Chem.* 277, 29018–29027, 2002.
75. Parente, A., Merrifield, B., Geraci, G., and D'Allessio, G., Molecular basis of superreactivity of cysteine residues 31 and 32 of seminal ribonuclease, *Biochemistry* 24, 1098–1104, 1985.
76. Heitmann, P., Reactivity of sulfhydryl groups in micelles. A model for proteins, *Eur. J. Biochem.* 5, 305–315, 1968.
77. Gitler, C., Zarmi, B., and Kalef, E., Use of cationic detergents to enhance reactivity of protein sulfhydryls, *Methods Enzymol.* 251, 366–375, 1995.
78. Lutolf, M.P., Tirelli, N., Cerritelli, S. et al., Systematic modulation of Michael type reactivity of thiols through the use of charged amino acids, *Bioconj. Chem.* 12, 1051–1056, 2001.
79. Snyder, G.H., Cennerazzo, M.J., Karalis, A.J., and Field, D., Electrostatic influence of local cysteine environments on disulfide exchange kinetics, *Biochemistry* 20, 6509–6519, 1981.
80. Rutherfurd, S.M., Schneuwly, A., and Moughan, P.J., Analyzing sulfur amino acids in selected feedstuffs using least-squares nonlinear regression, *J. Agric. Food Chem.* 55, 8019–8024, 2007.
81. Thornton, J.M., Disulphide bridges in globular proteins, *J. Mol. Biol.* 151, 261–287, 1981.
82. Fiser, A. and Simon, I., Predicting the redox states of cysteine in proteins, *Methods Enzymol.* 353, 10–44, 2002.
83. Bessett, P.H., Aslund, F., Beckwith, J., and Georgiou, G., Efficient folding of proteins with multiple disulfide bonds in the *Escherichia coli* cytoplasm, *Proc. Natl. Acad. Sci. USA* 96, 13703–13708, 1999.
84. Woycechowsky, K.J. and Raines, R.T., Native disulfide bond function in proteins, *Curr. Opin. Chem. Biol.* 4, 533–539, 2000.
85. Baneyx, F. and Mujacic, M., Recombinant protein folding and misfolding in *Escherichia coli*, *Nat. Biotechnol.* 22, 1399–1408, 2004.
86. Kadokura, H. and Beckwith, J., Mechanism of oxidative protein folding in the bacterial cell envelope, *Antioxid. Redox Signals* 13, 1231–1246, 2010.
87. Allison, W.S., Formation and reactions of sulfenic acids in proteins, *Acc. Chem. Res.* 9, 293–299, 1976.

88. Luo, D., Smith, S.W., and Anderson, B.D., Kinetics and mechanism of the reaction of cysteine and hydrogen peroxide in aqueous solution, *J. Pharm. Sci.* 94, 304–316, 2005.
89. Genevois, L. and Cayron, P., Combination of aldehyde and thio amine derivatives, *Bull. Soc. Chim.* 60, 1223–1230, 1939.
90. Nagy, P. and Ashby, M.T., Reactive sulfur species: Kinetics and mechanism of the hydrolysis of cysteine thiosulfinate ester, *Chem. Res. Toxicol.* 20, 1364–1372, 2007.
91. Prudent, M. and Girault, H.H., The role of copper in cysteine oxidation: Study of intra- and inter-molecular reactions in mass spectrometry, *Metallomics* 1, 157–165, 2009.
92. Rehder, D.S. and Borges, C.R., Possibilities and pitfalls in quantifying the extent of cysteine sulfenic acid modification of specific proteins within complex biofluids, *BMC Biochem.* 11, 25, 2010.
93. Heinecke, J. and Ford, P.C., Formation of cysteine sulfenic acid by oxygen atom transfer from nitrite, *J. Am. Chem. Soc.* 132, 9240–9243, 2010.
94. Poole, L.B. and Claiborne, A., The non-flavin redox center of the streptococcal NADH peroxidase. II. Evidence for a stabilized cysteine-sulfenic acid, *J. Biol. Chem.* 264, 12330–12338, 1989.
95. Poole, L.B., Klomsiri, C., Knaggs, S.A. et al., Fluorescent and affinity-based tools to detect cysteine sulfenic acid formation in proteins, *Bioconj. Chem.* 18, 2004–2017, 2007.
96. Peshenko, I.V. and Shichi, H., Oxidation of active center cysteine of bovine 1-cys peroxiredoxin to the cysteine sulfenic acid form by peroxide or peroxynitrite, *Free Radic. Biol. Med.* 31, 292–303, 2001.
97. Demasi, M., Silva, G.M., and Netto, L.E., 20S proteosome from *Saccharomyces cerevisiae* is responsible to redox modifications and its S-glutathionylated, *J. Biol. Chem.* 278, 679–685, 2003.
98. Silva, G.M., Netto, L.E., Discola, K.F. et al., Role of glutaredoxin 2 and cystosolic thioredoxins in cysteinyl-based redox modification of the 20S proteasome, *FEBS J.* 275, 2942–2955, 2008.
99. Hilgers, M.T. and Ludwig, M.L., Crystal structure of the local quorum-sensing protein LuxS reveals a catalytic metal site, *Proc. Natl. Acad. Sci. USA* 98, 11169–11174, 2001.
100. Benitez, L.V. and Allison, W.B., The inactivation of the acyl phosphatase activity catalyzed by the sulfenic acid form of glyceraldehyde 3-phosphate dehydrogenase by dimedone and olefins, *J. Biol. Chem.* 249, 6234–6243, 1974.
101. Tomova, N. and Ivanova, V., Significance of cys-153 for the phosphatase activity of glyceraldehyde-3-phosphate dehydrogenase, *Biomed. Biochim. Acta* 44, 1295–1302, 1985.
102. Schmalhausen, E.V., Pleten', A.P., and Muronetz, V.I., Ascorbate-induced oxidation of glyceraldehyde-3-phosphate dehydrogenase, *Biochem. Biophys. Res. Commun.* 308, 492–496, 2003.
103. Carballal, S., Alvarez, B., Turell, L. et al., Sulfenic acid in human serum albumin, *Amino Acids* 32, 543–551, 2007.
104. Turell, L., Botti, H., Carballal, S. et al., Reactivity of sulfenic acid in human serum albumin, *Biochemistry* 47, 358–367, 2008.
105. Turell, L., Bott, H., Carballal, S. et al., Sulfenic acid—A key intermediate in albumin thiol oxidation, *J. Chromatogr. B* 877, 3384–3392, 2009.
106. Le Gaillard, F. and Dautrevaux, M., The accessible cysteine residue of human transcortin. Evidence for oxidation of the sulfhydryl group, *FEBS Lett.* 94, 63–67, 1978.
107. Le Gaillard, F., Han, K.-K., and Dautrevaux, M., Caractérisation et propriétés, physicochimiques de la transcortine humaine, *Biochimie* 57, 559–568, 1975.
108. Defaye, G., Basset, M., Monnier, N., and Chambaz, E.M., Electron spin resonance study of human transcortin: Thiol groups and binding side topography, *Biochim. Biophys. Acta* 623, 280–294, 1980.

109. Ghose-Dastidar, J., Green, R., and Rose, J.B., Identification of the cysteine in the steroid-binding site of human corticosteroid binding globulin by site-directed mutagenesis and site-specific chemical modification, *J. Steroid Biochem. Mol. Biol.* 48, 139–144, 1994.
110. Blackinton, J., Lakshminarasimhan, M., Thomas, K.J. et al., Formation of a stabilized cysteine sulfinic acid is critical for the mitochondrial function of the parkinsonism protein DJ-1, *J. Biol. Chem.* 284, 6476–6485, 2009.
111. Paulsen, C.E. and Carroll, K.S., Chemical dissection of an essential redox switch in yeast, *Chem. Biol.* 16, 217–225, 2009.
112. Hutson, S.M., Poole, L.B., Coles, S., and Conway, M.E., Redox regulation and trapping sulfenic acid in the peroxide-sensitive human mitochondrial branched chain aminotransferase, *Methods Mol. Biol.* 476, 135–148, 2009.
113. Enami, S., Hoffman, M.R., and Colussi, A.J., Simultaneous detection of cysteine sulfon ate, sulfinate, and sulfonate during cysteine interfacial oxonolysis, *J. Phys. Chem. B* 113, 9356–9358, 2009.
114. Nimmo, S.L.A., Lemma, K., and Ashby, K.T., Reactions of cysteine sullfenyl thiocyanate with thiols to give unsymmetrical disulfides, *Heteroatom Chem.* 18, 467–471, 2007.
115. Yang, S.-H., Wu, C.-H., and Lin, W.-Y., Chemical modification of aminopeptidase isolated from Pronase, *Biochem. J.* 302, 595–600, 1994.
116. Guerra, D.G., Verkommen, D., Forthergill-Gilmore, L.A. et al., Characterization of the cofactor-independent phosphoglycerate mutase from *Leischmania mexicana mexicana.* Histidines that coordinate the two metal ions in the active center show different susceptibilities in irreversible chemical modification, *Eur. J. Biochem.* 271, 1798–1810, 2004.
117. Gallagher, T.M., Murine cornovirus membrane fusion is blocked modification of thiols buried within the spike protein, *J. Virol.* 70, 4683–4690, 1996.
118. Lee, H., Cha, M.K., and Kim, I.H., Activation of thiol-dependent antioxidant activity of human serum albumin by alkaline pH is due to the B-like conformational change, *Arch. Biochem. Biophys.* 380, 309–318, 2000.
119. Ortiz, J.O. and Bubis, J., Effects of differential sulfhydryl group-specific labeling on the rhodopsin and guanine nucleotide binding activities of transducin, *Arch. Biochem. Biophys.* 387, 233–242, 2001.
120. Tremblay, J.M., Li, H., Yarbrough, L.R., and Helmkamp, G.M., Jr., Modifications of cysteine residues in the solution and membrane-associated conformations of phosphatidylinositol transfer protein have differential effects on lipid transfer activity, *Biochemistry* 40, 9151–9158, 2001.
121. Chaudhuri, A.R., Seetharamalu, P., Schwarz, P.M. et al., The interaction of the B-ring of colchicine with α-tubulin: A novel footprinting approach, *J. Mol. Biol.* 303, 679–692, 2000.
122. Begaye, A. and Sackett, D.L., Measurement of ligand binding to tubulin by sulfhydryl reactivity, *Methods Cell Biol.* 95, 391–403, 2010.
123. Vöpel, T., Kunzelmann, S., and Herrmann, C., Nucleotide dependent cysteine reactivity of hGBP1 uncovers a domain movement during GTP hydrolysis, *FEBS Lett.* 583, 1923–1927, 2009.
124. Millard, C.B., Shnyrov, V.L., Newstead, S.I. et al., Stabilization of a metastable state of *Torpedo californica* acetylcholinesterase by chemical chaperones, *Protein Sci.* 12, 2337–2347, 2003.
125. Ogunjimi, E.O., Pokalsky, C.N., Shroyer, L.A., and Prochaska, L.J., Evidence for a conformational changes in subunit III of bovine heart mitochondrial cytochrome c oxidase, *J. Bioenerg. Biomembr.* 32, 617–626, 2000.
126. Bodo, J.M. and Foucault, G., Comparative study of the chemical modification of sulfhydryl groups from yeast and muscle glyceraldehyde-3-phosphate dehydrogenase. Chemical structure–activity relation, *Biochimie* 64, 477–486, 1982.

127. Cheng, Y., Shen, T.-J., Simplaceanu, V., and Ho, C., Ligand binding properties and structural studies of recombinant and chemically modified hemoglobins altered at β93 cysteine, *Biochemistry* 41, 11901–11913, 2002.
128. Webb, J.L., *Enzyme and Metabolic Inhibitors*, Vol. II, Mercurials, Academic Press, New York, Chapter 7, 1965.
129. Deshpande, K., Mishra, R.K., and Bhand, S., A high sensitivity micro format chemiluminescence enzyme inhibition assay for determination of Hg(II), *Sensors* 10, 6377–6494, 2010.
130. Diguez-Acuña, F.J. and Woods, J.S., Inhibition of NF-κB-DNA binding by mercuric ion: Utility of the non-thiol reductant, Tris (2-carboxyethyl)phosphine hydrochloride (TCEP), on detection of impaired NF-κB-DNA binding by thiol-directed agents, *Toxicol. In Vitro* 14, 7–16, 2000.
131. Myshkin, A.E. and Konyaeva, V.S., Mechanism of mercurial perturbation in proteins, *Appl. Biochem. Biotechnol.* 88, 185–194, 2000.
132. Myshkin, A.E. and Konyaeva, V.S., A new insight into mercurized hemoglobin aggregation mechanism, *Biochim. Biophys. Acta* 1749, 1–6, 2005.
133. Brouwer, M., Bonaventura, C., and Boneventura, J., Heavy metal ion interactions with *Callinectes sapidus* hemocyanin: Structural and functional changes induced by a variety of heavy metal ions, *Biochemistry* 21, 2529–2538, 1982.
134. Corvalho, C.M.L., Chew, E.-H., Hashemy, S.I. et al., Inhibition of the human thioredoxin system. A molecular mechanism of mercury toxicity, *J. Biol. Chem.* 283, 11913–11923, 2008.
135. Jalilehvand, F., Leung, B.O., Izadifard, M., and Damian, E., Mercury (II) cysteine complexes in alkaline aqueous solutions, *Inorg. Chem.* 45, 66–73, 2006.
136. Steele, R.A. and Opella, S.J., Structures of the reduced and mercury-bound forms of merP, the periplasmic protein from the bacterial mercury detoxification system, *Biochemistry* 36, 6885–6895, 1997.
137. Powlowski, J. and Sahlman, L., Reactivity of the two essential cysteine residues of the periplasmic mercuric-binding protein, merP, *J. Biol. Chem.* 274, 33320–33326, 1999.
138. Goto, Y. and Hamaguchi, K., Conformation and stability of the constant fragment of the immunoglobulin light chain containing an intramolecular mercury bridge, *Biochemistry* 25, 2821–2828, 1986.
139. Marston, A.W. and Wright, H.T., A method for covalent insertion of mercury into the cysteine disulfide bridges of proteins, *J. Biochem. Biophys. Methods* 9, 307–314, 1984.
140. Krężel, A., Latajka, R., Bujecz, D., and Bar, W., Coordination properties of tris(2-carboxyethyl)phosphine, a newly introduced thiol reductant, and its oxide, *Inorg. Chem.* 42, 1994–2003, 2003.
141. Sahlman, L. and Skärfstad, E.G., Mercuric ion binding abilities of merP variants containing only one cysteine, *Biochem. Biophys. Res. Commun.* 196, 583–588, 1993.
142. Frasco, M.F., Colletier, J.P., Weik, M. et al., Mechanisms of cholinesterase inhibition by inorganic mercury, *FEBS J.* 274, 1849–1861, 2007.
143. Tomlinson, G. and Kinsch, E.M., S-Mercuric-N-dansyl-cysteine labels the free sulfhydryl groups of human serum cholinesterase, *Biochem. Biophys. Res. Commun.* 158, 503–507, 1989.
144. Banks, T.E. and Shafer, J.A., Isoureas as alkylating agents. 2. Inactivation of papain by S-methylation of its cysteinyl residue with O-methylisourea, *Biochemistry* 11, 10–14, 1972.
145. Coty, W.A., Klooster, T.A., Griest, R.E., and Profita, J.A., Resolution of the effects of sulfhydryl-blocking reagents on hormone- and DNA-binding activities of the chick oviduct progesterone receptor, *Arch. Biochem. Biophys.* 225, 748–757, 1983.
146. Maerki, F. and Stanton, J.L., Inhibition of human phospholipase C by aurothiomalate and (triethylphosphine) gold complexes, *Arznei.-Forschung* 42, 328–333, 1992.

147. Manceau, A. and Nagy, K.L., Relationships between Hg (II)-S bond distances and Hg (II) coordination in thiolates, *Dalton Trans.* (11), 1421–1425, 2008.
148. Mah, V. and Jalilehvand, F., Mercury (II) complexes with glutathione in aqueous solution, *J. Biol. Inorg. Chem.* 13, 541–553, 2008.
149. Chekmeneva, E., Diaz-Critz, J.M., Arino, C., and Estaban, M., Binding of Hg^{2+} by Cys, Cys-Gly, and reduced glutathione: Study by differential probe voltammetry on rotating Au-disk electrode, electrospray ionization mass-spectrometry and isothermal calorimetry, *J. Electroanal. Chem.* 644, 20–24, 2010.
150. Hopman, A.H.N., Wagner, J., and van Duijn, F., A new hybridocytochemical method based on mercurated nucleic acid probes and sulfhydryl-hapten ligands. I. Stability of the mercury-sulfhydryl bond and influence of the ligand structure on immunochemical detection of the hapten, *Histochemistry* 84, 169–178, 1986.
151. Hopman, A.H.N., Wagner, J., and van Duijn, F., A new hybridocytochemical method based on mercurated nucleic acid probes and sulfhydryl-hapten ligands. II. Effects of variations in ligand structure on the situ detection of mercurated probes, *Histochemistry* 84, 179–185, 1986.
152. Simpson, R.B., Association constants of methylmercury with sulfhydryl and other bases, *J. Am. Chem. Soc.* 83, 4271–4277, 1961.
153. Crespo-L6pez, M.E., Macêdo, G.L., Pereira, S.I.D. et al., Mercury and human genotoxicity: Critical considerations and possible molecular mechanisms, *Pharmacol. Res.* 60, 212–220, 2009.
154. National Research Council, *Toxicological Effects of Methylmercury*, National Academies of Science, Washington, DC, 2000.
155. Keller, J.R., Bravo, D.T., and Parsons, S.M., Modification of cysteines reveals linkage to acetylcholine and vesamicol binding sites in the vesicular acetylcholine transporter of *Torpedo californica*, *J. Neurochem.* 74, 1739–1748, 2000.
156. Keller, J.E. and Parsons, S.M., Diffusion pathways to critical cysteines in the vesicular acetylcholine transporter of *Torpedo*, *Neurochem. Res.* 26, 477–482, 2003.
157. Kornreich, W.D. and Parsons, S.M., Sidedness and chemical and kinetic properties of the versamicol receptor of cholinergic synaptic vesicles, *Biochemistry* 27, 5262–5267, 1988.
158. Velick, S.F., Coenzyme binding and the thiol groups of glyceraldehyde-3-phosphate dehydrogenase, *J. Biol. Chem.* 203, 563–573, 1953.
159. Takatera, K. and Watanabe, T., Determination of sulfhydryl groups in ovalbumin by high-performance liquid chromatography with inductively coupled plasma mass spectrometric detection, *Anal. Chem.* 65, 3644–3646, 1993.
160. Webster, D.M. and Thompson, E.O., Carboxymethylation of thiol groups in ovalbumin: Implications for proteins that contain both thiol and disulfide groups, *Aust. J. Biol. Sci.* 35, 125–135, 1982.
161. Takahashi, N. and Hirose, M., Determination of sulfhydryl groups and disulfide bonds in a protein by polyacrylamide gel electrophoresis, *Anal. Biochem.* 188, 359–365, 1990.
162. Joniau, M., Bloemmen, J., and Lontie, R., The reaction of 4-(*p*-sulfophenylazo)-2-mercuriphenol with the thiol groups of proteins, *Biochim. Biophys. Acta* 214, 468–477, 1970.
163. Ahmed, R. and Stoeppler, M., Decomposition and stability studies of methylmercury in water using cold vapour atomic absorption spectrometry, *Analyst* 111, 1371–1374, 1986.
164. Whitmore, F.C., *Organic Compounds of Mercury*, American Chemical Society, The Chemical Catalogue Company, New York, 1921.
165. Boyer, P.D., Spectrophotometric study of the reaction of protein sulfhydryl groups with organic mercurials, *J. Am. Chem. Soc.* 76, 4331–4437, 1954.

166. Riordan, J.F. and Vallee, B.L., Reactions with N-ethylmaleimide and *p*-chlorobenzoate, *Methods Enzymol.* 11, 541–548, 1967.
167. Hasinoff, B.B., Madsen, N.B., and Avramovic-Zikic, O., Kinetics of the reaction of *p*-chloromercuribenzoate with the sulfhydryl groups of glutathione, 2-mercaptoethanol, and phosphorylase *b*, *Can. J. Biochem.* 49, 742–751, 1971.
168. Bai, Y. and Hayashi, R., Properties of the single sulfhydryl group of carboxypeptidase Y. Effects of alkyl and aromatic mercurials on activities toward various synthetic substrates, *J. Biol. Chem.* 254, 8473, 1979.
169. Bednar, R.A., Fried, W.B., Lock, Y.W., and Pramanik, B., Chemical modification of chalcone isomerase by mercurials and tetrathionate. Evidence for a single cysteine residue in the active site, *J. Biol. Chem.* 264, 14272, 1989.
170. Ojcius, D.M. and Solomon, A.K., Sites of p-chloromercuribenzenesulfonate inhibition of red cell urea and water transport, *Biochim. Biophys. Acta* 942, 73, 1988.
171. Clark, S.J. and Ralston, G.B., The dissociation of peripheral proteins from erythrocyte membranes brought about by p-mercuribenzenesulfonate, *Biochim. Biophys. Acta* 1021, 141, 1990.
172. Poole, L.B., and Claiborne, A., The non-flavin redox center of the Streptococcal NADH peroxidase. I. Thiol-reactivity and redox balance in the presence of urea, *J. Biol. Chem.* 264, 12322–12329, 1989.
173. Yao, S.Y.M. et al., Identification of cys^{140} in helix 4 as an exofacial cysteine residue within the substrate-translocation channel of rat equilibrative nitrobenzylthioinosine (NBMPR)-insensitive nucleoside transporter rENT2, *Biochem. J.* 353, 387–393, 2001.
174. Fann, M.C., Busch, A., and Maloney, P.C., Functional characterization of cysteine residues in G1pT, the glycerol 3-phosphate transporter of *Escherichia coli*, *J. Bacteriol.* 185, 3863–3870, 2003.
175. Ding, Z. et al., Inactivation of the human $P2Y_{12}$ receptor by thiol reagents requires interaction with both extracellular cysteine residues, Cys17 and Cys 270, *Blood* 101, 3908–3914, 2003.
176. Jagtap, S. and Rao, M., Conformation and microenvironment of the active site of a low molecular weight 1,4-β-D-glucan glucanohydrolase from an alkalothermophilic *Thermomonospora* sp.: Involvement of lysine and cysteine residues, *Biochem. Biophys. Res. Commun.* 347, 428–432, 2006.
177. Kutscher, D.J., Estele del Castillo Busto, M., Zinn, N. et al., Protein labelling with mercury tags: Fundamental studies on ovalbumin derivatised with *p*-hydroxymercuribenzoic acid (pHMB), *J. Anal. Atom. Spectrom.* 23, 1359–1364, 2008.
178. Kutshcer, D.J. and Bettmer, J., Absolute and relative protein quantification with the use of isotopically labeled *p*-hydroxymercuribenzoic acid and complementary MALDI-MS and ICPMS detection, *Anal. Chem.* 81, 9172–9177, 2009.
179. Lu, M., Li, X.-F., Le, X.C. et al., Identification and characterization of cysteinyl exposure in proteins by selective mercury labeling and nano-electrospray ionization quadrupole time-of-flight mass spectrometry, *Rapid Commun. Mass Spectrom.* 24, 1523–1532, 2010.
180. Friedman, B.E., Olson, J.S., and Matthews, K.S., Kinetics studies of inducer binding to lactose repressor protein, *J. Biol. Chem.* 251, 1171–1174, 1976.
181. Fillmonova, M.N., Gubskaya, V.P., Nuretdinov, I.A. et al., Study of the mechanism of action of *p*-chloromercuribenzoate on endonuclease from the bacterium *Serratia marcescens*, *Biochemistry (Moscow)* 66, 399–404, 2001.
182. Angeli, V., Chen, H., Mester, Z. et al., Derivatization of GSSG by pHMB in alkaline media. Determination of oxidized glutathione in blood, *Talanta* 82, 815–820, 2010.
183. Challenger, F. and Rawllings, A.A., The formation of organo-metalloidal and similar compounds by micro-organisms. Part V. Methylated alkyl sulphides. The fission of the disulphide link, *J. Chem. Soc.* 868–875, 1937.

184. Azzo, W., Woessner, J.F., Jr., Purification and characterization of an acid metalloproteinase from human articular cartilage, *J. Biol. Chem.* 261, 5434–5441, 1986.
185. Murphy, G., Ward, R., Hambry, R.M. et al., Characterization of gelatinase from pig polymorphonuclear leukocytes. A metalloproteinase resembling tumor type IV collagenase, *Biochem. J.* 258, 463–472, 1989.
186. Freimark, B.D., Fenser, W.S., and Rosenfeld, S.A., Multiple sites of the propeptide region of human stromelysin-1 are required for maintaining a latent form of the enzyme, *J. Biol. Chem.* 269, 26962–26988, 1994.
187. Sanderson, M.P., Keller, S., Alonso, A. et al., Generation of novel, secreted epidermal growth factor receptor (EGFR/ErbB1) isoforms via metalloprotease-dependent ectodomain shedding and exosome secretion, *J. Cell. Biochem.* 103, 1783–1797, 2008.
188. Pavlaki, N., Giannopoulou, E., Niariakis, A. et al., Walker256 (W256) cancer cells secrete tissue inhibitor of metalloproteinase-free metalloproteinase-9, *Mol. Cell. Biochem.* 328, 189–199, 2009.
189. Estivill, G., Guardado, P., Buser, R. et al., Identification of an essential cysteinyl residue for the structure of glutamine synthetase alpha from *Phaseolus vulgaris*, *Planta* 231, 1101–1111, 2010.
190. McMurray, C.H. and Trentham, D.R., A new class of chromophoric organomercurials and their reactions with D-glyceraldehyde 3-phosphate dehydrogenase, *Biochem. J.* 115, 913–921, 1969.
191. Sanyal, G. and Khalifah, R.G., Kinetics of organomercurial reactions with model thiols: Sensitivity to exchange of the mercurial labile ligand, *Arch. Biochem. Biophys.* 196, 157–164, 1979.
192. Marshall, M. and Cohen, P.P., The essential sulfhydryl group of ornithine transcarbamylases—pH dependence of the spectra of its 2-mercuri-4-nitrophenol derivative, *J. Biol. Chem.* 255, 7296–7300, 1980.
193. Scawen, M.D., Hewitt, E.J., and James, D.M., Preparation, crystallization and properties of *Cucurbita pepo* plastocyanin and ferredoxin, *Phytochemistry* 14, 1225–1233, 1975.
194. Simpson, R.A., Use of isoelectric focusing and a chromophoric organomercurial to monitor urea-induced conformational changes of yeast phosphoglycerate kinase, *Biochem. J.* 167, 65–70, 1977.
195. Somerville, L.L. and Quiocho, F.A., The interaction of tetraiodofluorescein with creatine kinase, *Biochim. Biophys. Acta* 481, 493–499, 1977.
196. Quiocho, F.A. and Olson, J.S., The reaction of creatine kinase with 2-chloromercuri-4-nitrophenol, *J. Biol. Chem.* 249, 5885–5888, 1974.
197. Yang, D.S., Burgum, A.A., and Matthews, K.S., Modification of the cysteine residues of the lactose receptor protein using chromophoric probes, *Biochim. Biophys. Acta* 493, 24–36, 1977.
198. Baines, B.S. and Brocklehurst, K., A thiol-labelling reagent and reactivity probe containing electrophilic mercury and a chromophoric leaving group, *Biochem. J.* 179, 701–704, 1979.
199. Scott-Ennis, R.J. and Noltmann, E.A., Differential response of cysteine residues in pig muscle phosphoglucose isomerase to seven sulfhydryl modifying reagents, *Arch. Biochem. Biophys.* 239, 1–11, 1985.
200. Yoshimoto, K., Nishio, M., Sugasawa, H., and Nagasaki, Y., Direct observation of adsorption-induced inactivation of antibody fragments surrounded by mixed-PEG layer on a gold surface, *J. Am. Chem. Soc.* 132, 7982–7989, 2010.
201. Kausaite-Minkstimiene, A., Ramanaviciene, A., Kirlyte, J., and Ramanavicius, A., Comparative study of random and oriented antibody immobilization techniques on the binding capacity of immunosensor, *Anal. Chem.* 82, 6401–6408, 2010.
202. Otsuki, S. and Ishikawa, M., Wavelength-scanning surface plasmon resonance imaging for label-free multiplexed protein microarray assay, *Biosens. Bioelectron.* 26, 202–206, 2010.

203. Vericat, C., Vela, M.E., Benitez, G. et al., Self-assembled monolayers of thiols and dithiols on gold: New challenges for a well-known system, *Chem. Soc. Rev.* 39, 1805–1834, 2010.
204. Kankate, L., Turchanin, A., and Gölzhäuser, A., On the release of hydrogen from the S-H groups in the formation of self-assembled monolayers of thiols, *Langmuir* 25, 10435–10438, 2009.
205. Pollak, A., Roe, D.G., Pollock, C.M. et al., A convenient method of preparing high specific activity Technetium complexes using thiol-containing chelators adsorbed on gold, *J. Med. Chem.* 121, 11593–11594, 1999.
206. Park, S.H., Gwon, H.J., and Park, K.B., Microwave assisted facile one-pot synthesis of ^{188}Re complex using a tetrahydroborate exchange resin. A bifunctional chelating agent for radiobiopharmaceuticals, *Chem. Lett.* 33, 1278–1279, 2004.
207. Mather, S.J. and Ellison, D., Reduction-mediated Technetium-99m labeling of monoclonal antibodies, *J. Nucl. Med.* 31, 692–697, 1990.
208. Du, Y.J., Cornelius, R.M., and Brush, J.L., Measurement of protein adsorption to gold surface by radioiodination methods: Suppression of free iodide sorption, *Colloid. Surf. B* 17, 59–67, 2000.
209. Neff, H., Beeby, T., Lima, A.M.N. et al., dc-Sheet resistance as sensitive monitoring tool of protein immobilization on thin metal films, *Biosens. Bioelectron.* 21, 1746–1752, 2008.
210. Zheng, S., Kim, D.-K., Park, T.J. et al., Label-free optical diagnosis of hepatitis B virus with genetically engineered fusion proteins, *Talanta* 82, 803–809, 2010.
211. Sarikaya, M., Tamerler, C., Jen, A.K. et al., Molecular biomimetics: Nanotechnology through biology, *Nat. Mater.* 2, 577–585, 2003.
212. Brown, D.H. and Smith, D.E., Complexes of L-cysteine and D-penicillamine with Gold (I) and Gold (III), *Proc. R. Soc. Med.* 70(suppl 3), 41–42, 1977.
213. Karver, M.R., and Barrios, A.M., Identifying and characterizing the biological targets of metallotherapeutics: Two approaches using Au (I)-protein interactions in model systems, *Anal. Biochem.* 382, 63–85, 2008.
214. Corthey, G., Giovanetti, L.J., Ramallo-Lopez, J.M. et al., Synthesis and characterization of Gold@Gold(I)-thiomalate core@shell nanoparticles, *ACS Nano* 4, 3413–3421, 2010.
215. Romero-Isart, N. and Vasak, M., Advances in the structure and chemistry of metallothioneins, *J. Inorg. Biochem.* 88, 388–396, 2002.
216. Theocharis, S.E., Margeli, A.F., and Koutselinis, A., Metallothionein: A multifunctional protein from toxicity to cancer, *Int. J. Biol. Markers* 18, 162–169, 2003.
217. Han, Q. and Maret, W., Imbalance between pro-oxidant and pro-antioxidant functions of zinc in disease, *J. Alzheimers Dis.* 8, 161–170, 2005.
218. Maret, W., Fluorescent probes for the structure and function of metallothionein, *J. Chromatogr. B* 877, 3378–3383, 2009.
219. Stillman, M.J., Shaw, III, C.F., and Suzuki, K.T. (eds.), *Metallothioneins. Synthesis, Structure, and Properties of Metallothioneins, Phytochelatins and Metal-Thiolate Complexes*, VCH Publishers, New York, 1992.
220. Krezel, A. and Moret, W., Different redox states of metallothionein/thionein in biological tissues, *Biochem. J.* 408, 551–558, 2007.
221. Rigby, K.E., Chen, J., Mackie, J., and Stilliman, M.J., Molecular dynamics studies on the folding and metallation of the individual domains of metallothionein, *Protein Struct. Funct. Bioinf.* 62, 159–172, 2005.
222. Feng, W., Benz, F.W., Cai, J. et al., Metallothionein disulfides are present in metallothionein-overexpressing transgenic mouse heart and increase under conditions of oxidative stress, *J. Biol. Chem.* 281, 681–687, 2006.
223. Kägi, J.H.H. and Vallee, B.L., Metallothionein, a cadmium and zinc-containing protein from equine renal cortex, *J. Biol. Chem.* 236, 2435–2442, 1961.
224. Bernhard, W.R., Differential modification of metallothionein with iodoacetamide, *Methods Enzymol.* 205, 426–433, 1991.

225. Dahl, K.H. and McKinley-McKee, J.S., Enzymatic catalysis in the affinity labeling of liver alcohol dehydrogenase with haloacids, *Eur. J. Biochem.* 118, 507–513, 1981.
226. Hunt, J.B., Neece, S.H., Schachman, H.K., and Ginsburg, A., Mercurial-promoted Zn^{2-} release from *Escherichia coli aspartate transcarbamylase*, *J. Biol. Chem.* 259, 14793–14803, 1984.
227. Gardonio, D. and Siemann, S., Chelator-facilitated chemical modification of IMP-1 metallo-β-lactamase and its consequences on metal binding, *Biochem. Biophys. Res. Commun.* 381, 107–111, 2009.
228. Roehm, P.O. and Berg, J.M., Selectivity of methylation of metal-bound cystinates and its consequences, *J. Am. Chem. Soc.* 120, 13083–13087, 1998.
229. Eyem, J., Sjödahl, J., Sjöquist, J., S-Methylation of cysteine residues in peptides and proteins with dimethylsulfide, *Anal. Biochem.* 74, 359–368, 1976.
230. Happersberger, H.P., Przybylski, M., and Glocker, M.O., Selective bridging of bis-cysteinyl residues by arsonous acid derivatives as an approach to the characterization of protein tertiary structure and folding pathways by mass spectrometry, *Anal. Biochem.* 24, 237–250, 1998.
231. Happersberger, H.P., Cowgill, C., and Glocker, M.O., Structural characterization of monomeric folding intermediates of recombinant human macrophage-colony stimulating factor β (rhM-CSFβ) by chemical trapping, chromatographic separation and mass spectrometric peptide mapping, *J. Chromatogr. B* 782, 393–404, 2002.
232. Kalef, E. and Gitler, C., Purification of vicinal dithiol-containing proteins by arsenical-based affinity chromatography, *Methods Enzymol.* 233, 395–403, 1994.
233. Dustin, P., Some new aspects of mitotic poisoning, *Nature* 159, 794–797, 1947.
234. Han, K.K., Delcourte, A., and Herman, B., Chemical modification of thiol group(s) in proteins: Application to the study of anti-microtubular drug binding, *Comp. Biochem. Physiol.* 88B, 1057–1065, 1987.
235. Jan, K.Y., Wang, T.C., Ramanathan, B., and Guff, J.R., Dithiol compounds at low concentrations increase arsenite toxicity, *Toxicol. Sci.* 90, 432–439, 2006.
236. Dahl, K.S. and McKinley-McKee, J.S., The reactivity of affinity labels: A kinetic study of the reaction of alkyl halides with thiolate anions—A model reaction for protein alkylation, *Bioorg. Chem.* 10, 329–341, 1981.
237. Proudfoot, A.T., Bradberry, S.M., and Vale, J.A., Sodium fluoroacetate poisoning, *Toxicol. Rev.* 25, 213–217, 2006.
238. Tsuji, H., Shimizu, H., Dote, T. et al., Effects of sodium monofluoroacetate on glucose, amino-acid, and fatty-acid metabolism and risk assessment of glucose supplementation, *Drug. Chem. Toxicol.* 32, 353–361, 2009.
239. Kulling, P., Andersson, H., Boström, K. et al., Fatal systemic poisoning after skin exposure to monochloroacetic acid, *J. Toxicol. Clin. Toxicol.* 30, 643–652, 1992.
240. Pirson, J., Toussaint, P., and Sejers, N., An unusual cause of burn injury: Skin exposure to monochloroacetic acid, *J. Burn Care Rehabit.* 24, 407–409, 2003.
241. Steinauer, E., Untersuchungen über die physiologische wirkung der bromoprärate, *Archiv. F. Pathol. Anat.* 59, 65–113, 1874.
242. Lundsgaard, E., Studies on muscle contraction without lactic acid production, *Biochem. Z.* 217, 162–177, 1930.
243. Webb, J.L., Iodoacetate and iodoacetamide, in *Enzyme and Metabolic Inhibitors*, Vol. III, Academic Press, New York, Chapter 1, pp. 1–283, 1966.
244. Kleanthous, C., Campbell, D.G., and Coggins, J.R., Active site labeling of the shikimate pathway enzyme, dehydroquinase. Evidence of a common substrate binding site within the dehydroquinone and dehydroquinate synthase, *J. Biol. Chem.* 262, 10929–10934, 1990.
245. Syvertsen, C. and McKinly-McKee, J.S., Binding of ligands to the catalytic zinc in the horse liver alcohol dehydrogenase, *Arch. Biochem. Biophys.* 299, 159–169, 1984.

246. Plapp, B.V., Application of affinity labeling for studying structure and function of enzymes, *Methods Enzymol.* 87, 469–499, 1982.
247. Gerwin, B.I., Properties of the single sulfhydryl group of streptococcal proteinase. A comparison of the rates of alkylation by chloroacetic acid and chloroacetamide, *J. Biol. Chem.* 242, 451–456, 1967.
248. Liu, T.Y., Demonstration of the presence of a histidine residue at the active site of streptococcal proteinase, *J. Biol. Chem.* 242, 4029–4032, 1967.
249. Kallis, G.-B. and Holmgren, A., Differential reactivity of the functional sulfhydryl groups of cysteine-32 and cysteine-35 present in the reduced form of thioredoxin from *Escherichia coli*, *J. Biol. Chem.* 255, 10261–10265, 1980.
250. Mikami, B., Aibara, S., and Morita, Y., Chemical modification of sulfhydryl groups in soybean α-amylase, *J. Biochem.* 88, 103–111, 1980.
251. Hempel, J.D. and Pietruszko, R., Selective chemical modification of human liver aldehyde and dehydrogenases E_1 and E_2 by iodoacetamide, *J. Biol. Chem.* 256, 10889–10996, 1981.
252. Seifried, S.E., Wang, Y., and Von Hippel, P.H., Fluorescent modification of the cysteine 202 residue of *Escherichia coli* transcription termination factor rho, *J. Biol. Chem.* 263, 13511–13514, 1988.
253. Pardo, J.P. and Slayman, C.W., Cysteine 532 and cysteine 545 are the N-ethylmaleimide-reactive residues of the *Neurospora* plasma membrane H^+-ATPase, *J. Biol. Chem.* 264, 9373–9379, 1989.
254. Miyanishi, T. and Borejdo, J., Differential behavior of two cysteine residues on the myosin head in muscle fibers, *Biochemistry* 28, 1287–1294, 1989.
255. Bishop, J.E., Squier, T.C., Bigelow, D.J., and Inesi, G., (Iodoacetamido) fluorescein labels a pair of proximal cysteines on the Ca^{2+}-ATPase of sarcoplasmic reticulum, *Biochemistry* 27, 5233–5240, 1988.
256. First, E.A. and Taylor, S.S., Selective modification of the catalytic subunit of cAMP-dependent protein kinase with sulfhydryl-specific fluorescent probes, *Biochemistry* 28, 3598–3605, 1989.
257. Ramalingam, T.S., Das, P.K., and Podder, S.K., Ricin-membrane interaction: Membrane penetration depth by fluorescence quenching and resonance energy transfer, *Biochemistry* 33, 12247–12254, 1994.
258. Hamman, B.D., Oleinikov, A.V., Jokhadze, G.G. et al., Rotational and conformational dynamics of *Escherichia coli* ribosomal protein L7/L12, *Biochemistry* 35, 16672–16679, 1996.
259. Karlstrom, A. and Nygren, P.A., Dual labeling of a binding protein allows for specific fluorescence detection of native protein, *Anal. Biochem.* 295, 22–30, 2001.
260. Teilum, K. et al., Early kinetic intermediate in the folding of acyl-CoA binding protein detected by fluorescence labeling and ultrarapid mixing, *Proc. Natl. Acad. Sci. USA* 99, 9807–9812, 2002.
261. Tripet, B., De Crescenzo, G., Grothe, S. et al., Kinetic analysis of the interactions between troponin C and the C-terminal troponin I regulatory region and validation of a new peptide delivery/capture system used for surface plasmon resonance, *J. Mol. Biol.* 323, 345–362, 2002.
262. Xie, L., Li, W.X., Barnett, V.A., and Schoenberg, M., Graphical evaluation of alkylation of myosin's SH1 and SH2: The N-phenylmaleimide reaction, *Biophys. J.* 72, 858–865, 1997.
263. Holyoak, T. and Nowak, T., Structural investigation of the binding of nucleotide to phosphoenolpyruvate carboxykinase by NMR, *Biochemistry* 40, 11037–11047, 2001.
264. Toutchkine, A., Nalbant, P., and Hahn, K.M., Facile synthesis of thiol-reactive Cy3 and Cy5 Derivatives with enhanced water solubility, *Bioconj. Chem.* 13, 387–391, 2002.
265. Toutchkine, A., Kraynov, V., and Hahn, K., Solvent-sensitive dyes to report protein conformational changes in living cells, *J. Am. Chem. Soc.* 125, 4132–4145, 2003.

266. Chu, Q. and Fukui, Y., In vivo dynamics of myosin II in *Dictyostelium* by fluorescent analogue cytochemistry, *Cell. Motil. Cytoskeleton* 35, 254–268, 1996.
267. Szalecki, W., Synthesis of norbiotinamine and its derivatives, *Bioconj. Chem.* 7, 271–273, 1996.
268. Thomas, T.H., Rutherford, P.A., Vareesangthip, K. et al., Erythrocyte membrane thiol proteins associated with changes in the kinetics of Na/Li countertransport: A possible molecular explanation of changes in disease, *Eur. J. Clin. Invest.* 28, 259–265, 1998.
269. Kim, J.R., Yoon, H.W., Kwon, K.S. et al., Identification of proteins containing cysteine residues that are sensitive to oxidation by hydrogen peroxide at neutral pH, *Anal. Biochem.* 283, 214–221, 2000.
270. Lin, T.-K., Hughes, G., Muratovska, A. et al., Specific modification of mitochondrial protein thiols in response to oxidative stress. A proteomics approach, *J. Biol. Chem.* 277, 17048–17056, 2002.
271. Guo, Z., Vath, G.M., Wagner, C.R. et al., Arylamine N-acetyltransferases: Covalent modification and inactivation of hamster NAT1 by bromoacetamido derivatives of aniline and 2-aminofluorene, *J. Prot. Chem.* 22, 631–642, 2003.
272. Chung, J.G., Kuo, H.M., Wu, L.T. et al., Evidence for arylamine N-acetyltransferase in *Hymenolepis nana, J. Microbiol. Immunol. Infect.* 30, 1–17, 1997.
273. Chung, J.G., Purification and characterization of an arylamine N-acetyltransferase in the nematode *Enterobius vermicularis, Microbios* 98, 15–25, 1999.
274. Andres, H.H., Klem, A.J., Schofer, L.M. et al., On the active site of liver acetyl-CoA arylamine N-acetyltransferase from rapid acetylator rabbits (III/J), *J. Biol. Chem.* 263, 7521–7527, 1988.
275. Crestfield, A.M., Moore, S., and Stein, W.H., The preparation and enzymatic hydrolysis of reduced and S-carboxymethylated proteins, *J. Biol. Chem.* 238, 622–627, 1963.
276. Friedman, M., Krull, L.H., and Cavins, J.F., The chromatographic determination of cystine and cysteine residues in proteins as S-β-(4-pyridyl-ethyl) cysteine, *J. Biol. Chem.* 245, 3868–3871, 1970.
277. Mak, A.S. and Jones, B.L., Application of S-pyridylethylation of cysteine to the sequence analysis of proteins, *Anal. Biochem.* 84, 432–440, 1978.
278. Pflouq, M., Stoffer, B., and Jensen, A.L., In situ alkylation of cysteine residues in a hydrophobic membrane protein immobilized on polyvinylidene difluoride membranes by electroblotting prior to microsequence and amino acid analysis, *Electrophoresis* 13, 148–153, 1992.
279. Lundell, N. and Schreitmüller, T., Sample preparation of peptide-mapping—A pharmaceutical quality-control perspective, *Anal. Biochem.* 266, 31–47, 1999.
280. Görg, A., Obermaier, C., Boguth, G. et al., The current state of two-dimensional electrophoresis with immobilized pH gradients, *Electrophoresis* 21, 1037–1053, 2000.
281. Jahn, O., Hoffman, B., Brauns, O. et al., The use of multiple ion chromatograms in online HPLC-MS for the characterization of post-translational and chemical modification of proteins, *Int. J. Mass Spectrom.* 214, 37–51, 2002.
282. Sechi, S. and Chait, B.T., Modification of cysteine residues by alkylation. A tool in peptide mapping and protein identification, *Anal. Chem.* 70, 5150–5158, 1998.
283. Herbert, B. et al., Reduction and alkylation of proteins in preparation of two-dimensional map analysis. Why, when, and how? *Electrophoresis* 22, 2046–2057, 2001.
284. Hoving, S., Gerrits, B., Voshol, H. et al., Preparative two-dimensional gel electrophoresis at alkaline pH using narrow range immobilized pH gradients, *Proteomics* 2, 127–134, 2002.
285. Shaw, M.M. and Riederer, B.M., Sample preparation for two-dimensional gel electrophoresis, *Proteomics* 3, 1408–1417, 2003.
286. Malloy, M.P., Brezenski, E.E., Hang, J. et al., Overcoming technical variation and biological variation in quantitative proteomics, *Proteomics* 3, 1912–1919, 2003.

287. Herbert, B., Hopwood, F., Oxley, D. et al., β-elimination: An unexpected artifact in proteome analysis, *Proteomics* 3, 826–831, 2003.
288. Nielsen, M.L., Vermeulan, M., Bonaldi, T. et al., Iodoacetamide-induced artifact mimics ubiquitinylation in mass spectrometry, *Nat. Methods* 5, 459–460, 2008.
289. Lapko, V.N., Smith, D.L., and Smith, J.B., Identification of an artifact in the mass spectrometry of proteins derivatized with iodoacetamide, *J. Mass Spectrom.* 35, 572–575, 2000.
290. Koshiishi, I. and Imanari, T., State analysis of endogenous cyanate in human plasma, *J. Pharmacobio. Dyn.* 13, 254–258, 1990.
291. Lu, M.F., Williams, C., Murray, M.V. et al., Ion chromatographic quantification of cyanate in urea solutions: Estimation of the efficiency of cyanate scavengers in recombinant protein manufacturing, *J. Chromatogr. B* 803, 353–362, 2004.
292. Rabilloud, T., Adessi, C., Giraudel, A. et al., Improvement of the solubilization of proteins in two-dimensional electrophoresis with immobilized pH gradients, *Electrophoresis* 18, 307–316, 1997.
293. Galvani, M., Hamdan, M., Herbert, B., and Righetti, P.G., Protein alkylation in the presence/absence of thiourea in proteome analysis: A matrix assisted laser desorption/ionization-time of flight-mass spectrometry investigation, *Electrophoresis* 22, 2066–2074, 2001.
294. Tyagarajon, K., Pretzer, E., and Wiktorowicz, J.E., Thiol-reactive dyes for fluorescence labeling of proteomic samples, *Electrophoresis* 24, 2348–2358, 2003.
295. Plowman, J.E., Flanagen, L.M., Paton, L.N. et al., The effect of oxidation or alkylation on the separation of wool keratin proteins by two-dimensional gel electrophoresis, *Proteomics* 3, 942–950, 2003.
296. Galvani, M., Hamdan, M., Herbert, B., and Righetti, P.G., Alkylation kinetics of proteins in preparation for two-dimensional maps. A matrix, *Electrophoresis* 22, 2058–2065, 2001.
297. Schrooyen, P.M.M., Dijkstra, P.J., Oberthür, R.C. et al., Partially carboxymethylated feather keratin I. Properties in aqueous systems, *J. Agric. Food Chem.* 48, 4326–4334, 2000.
298. Schrooyen, P.M.M., Dijkstra, P.J., Oberthür, R.C. et al., Partially carboxymethylated feather keratin II. Thermal and mechanical properties of fibers, *J. Agric. Food Chem.* 49, 221–230, 2001.
299. Jeffrey, D.A. and Bogyo, M., Chemical proteomics and its application to drug discovery, *Curr. Opin. Biotechnol.* 14, 87–95, 2003.
300. Mann, M., Quantitative proteomics? *Nat. Biotechnol.* 17, 954–955, 1999.
301. Smolka, M., Zhou, H., and Aebersold, R., Quantitative protein profiling using two-dimensional gel electrophoresis, isotope-coded affinity tag labeling, and mass spectrometry, *Mol. Cell. Proteomics* 1, 19–29, 2002.
302. Gygi, P., Rist, B., Gerber, S.A. et al., Quantitative analysis of complex protein mixtures using isotope-coded affinity tags, *Nat. Biotechnol.* 17, 994–999, 1999.
303. Tao, W.A. and Aebersold, R., Advances in quantitative proteomics via stable isotope tagging and mass spectrometry, *Curr. Opin. Biotechnol.* 14, 100–118, 2003.
304. Shen, M., Guo, L., Wallace, A. et al., Isolation and isotope labeling of cysteine- and methionine-containing tryptic peptides: Application to the study of cell surface proteolysis, *Mol. Cell. Proteomics* 2, 315–324, 2003.
305. Pasquarello, C., Sanchez, J.C., Hochstrasser, D.F., and Corthalls, G.L., N-*t*-butylidoacetamide and iodoacetanilide: Two new cysteine alkylating reagents for relative quantitation of proteins, *Rapid Commun. Mass Spectrom.* 18, 117–127, 2004.
306. Sebastiano, R., Citterio, A., Lapadula, M., and Righetti, P.G., A new deuterated alkylating agent for quantitative proteomics, *Rapid Commun. Mass Spectrom.* 17, 2380–2386, 2003.
307. Hansen, K.C., Schmitt-Ulms, G., Chalkley, R.J. et al., Mass spectrometric analysis of protein mixtures at low levels using cleavable ^{13}C-isotope-coded affinity tag and multidimensional chromatography, *Mol. Cell. Proteomics* 2, 299–314, 2003.

308. Thevis, M., Loo, R.R.O., and Loo, J.A., In-gel derivatization of proteins for cysteine-specific cleavages and their analysis by mass spectrometry, *J. Proteome Res.* 2, 163–172, 2003.
309. Friedman, M., Application of the S-pyridylethylation reaction to the elucidation of the structure and function of proteins, *J. Protein Chem.* 20, 431–453, 2001.
310. Misra, S., Lee, N., Fu, A.A. et al., Increased expression of a disintegrin and metalloproteinase thrombospondin 1 in thrombosed hemodialysis grafts, *J. Vasc. Interv. Radiol.* 19, 111–119, 2008.
311. Lynn, B.C., Wang, J., Markesery, W.R., and Lowell, M.A., Quantitative changes in the mitochondrial proteome from subjects with mild cognitive impairment, early stage, and late stage Alzheimer's disease, *J. Alzheimers Dis.* 19, 325–339, 2010.
312. Guaragna, A., Amoresano, A., Pinto, V. et al., Synthesis and proteomic activity evaluation of a new isotope-coded affinity tagging (ICAT) reagent, *Bioconj. Chem.* 19, 1095–1104, 2008.
313. Hägglund, P., Bunkenborg, J., Maeda, K., and Svensson, B., Identification of thioredoxin disulfide targets using a quantitative proteomics approach based on isotope-coded affinity tags, *J. Proteome Res.* 7, 5270–5276, 2008.
314. Wu, J., An, Y., Pu, H. et al., Enrichment of serum low-molecular-weight proteins using C18 absorbent under urea/dithiothreitol denatured environment, *Anal. Biochem.* 398, 34–44, 2010.
315. Pritchard, C., Quaglia, M., Mussell, C. et al., Fully traceable absolute protein quantification of somatropin that allows independent comparison of somatropin standards, *Clin. Chem.* 55, 1984–1990, 2009.
316. Leroy, B., Rosier, C., Erculisse, V. et al., Differential proteomic analysis using isotope-coded protein-labeling strategies: Comparison, improvements and application to simulated microgravity effect of *Cupriavidus metallidurans* CH34, *Proteomics* 10, 2281–2291, 2010.
317. Schmidt, A., Kellermann, J., and Lottspeich, F., A novel strategy for quantitative proteomics using isotope-coded protein labeling, *Proteomics* 5, 4–15, 2005.
318. Sharp, D.W.A. (ed.), *Penguin Dictionary of Chemistry*, 2nd edn., Penguin, London, U.K., p. 261, 1990.
319. Shaunak, S., Godwin, A., Choi, J.W. et al., Site-specific PEGylation of native disulfide bonds in therapeutic proteins, *Nat. Chem. Biol.* 2, 312–313, 2006.
320. Brocchini, S., Godwin, A., Balan, S. et al., Disulfide bridge based PEGylation of proteins, *Adv. Drug Deliv. Rev.* 60, 3–12, 2008.
321. Balan, S., Choi, J.W., Godwin, A. et al., Site-specific PEGylation of protein disulfide bonds using a three-carbon bridge, *Bioconj. Chem.* 18, 61–76, 2007.
322. Yan, J.X., Keet, W.C., Herbert, B.R. et al., Identification and quantitation of cysteine in proteins separated by gel electrophoresis, *J. Chromatogr.* 813, 187–200, 1998.
323. Bordini, E., Hamdan, M., and Righetti, P.G., Probing acrylamide alkylation sites in cysteine-free proteins by matrix-assisted laser desorption/ionization time-of-flight, *Rapid Commun. Mass Spectrom.* 14, 840–848, 2000.
324. Mineki, R. et al., *In situ* alkylation with acrylamide for identification of cysteinyl residues in proteins during one- and two-dimensional sodium dodecyl sulphate-polyacrylamide gel electrophoresis, *Proteomics* 2, 1672–1681, 2002.
325. Cahill, M.A. et al., Analysis of relative isopologue abundances for quantitative profiling of complex protein mixtures labelled with acrylamide/D-3-acrylamide alkylation tag system, *Rapid Commun. Mass Spectrom.* 17, 1283–1290, 2003.
326. LoPachin, R.M., Molecular mechanisms of the conjugated α,β-unsaturated carbonyl derivatives: Relevance to neurotoxicity and neurodegenerative diseases, *Toxicol. Sci.* 104, 235–249, 2008.
327. Lü, Z.R., Zou, H.C., Park, S.J. et al., The effects of acrylamide on brain creatine kinase: Inhibition kinetics and computational docking simulation, *Int. J. Biol. Macromol.* 44, 128–132, 2009.

328. Ghahghaei, A., Rekas, A., Carver, J.A., and Augusteyn, R.C., Structure/function studies on dogfish α-crystallin, *Mol. Vis.* 15, 2411–2420, 2009.
329. Sciandrelio, G., Mauro, M., Cardonna, F. et al., Acrylamide catalytically inhibits topoisomerase II in V79 cells, *Toxicol. In Vitro* 24, 830–834, 2010.
330. Stevens, S.M., Jr., Rauniyar, N., and Prokai, L., Rapid characterization of covalent modifications to rat brain mitochondrial proteins after ex vivo exposure to 4-hydroxy-2-nonenal by liquid chromatography-tandem mass spectrometry using data-dependent and neutral-driven MS3 acquisition, *J. Mass Spectrom.* 42, 1599–1605, 2007.
331. Zhu, X., Gallogly, M.M., Mieyal, J.J. et al., Covalent cross-linking of glutathione and carnosine to proteins by 4-oxy-2-nonenal, *Chem. Res. Toxicol.* 22, 1050–1059, 2009.
332. Wakita, C., Maeshima, T., Yamazaki, A. et al., Stereochemical configuration of 4-hydroxy-2-nonenal-cysteine adducts and their stereoselective formation in a redox-regulated protein, *J. Biol. Chem.* 284, 28810–28822, 2009.
333. Rauniyar, N. and Prokai, L., Detection and identification of 4-hydroxy-2-nonenal Schiff-base adducts along with products of Michael addition using data-dependent neutral loss-driven MS3 acquisition: Method evaluation through an in vitro study on cytochrome c oxidase modifications, *Proteomics* 9, 5188–5193, 2009.
334. Chavez, J., Chung, W.G., Miranda, C.L. et al., Site-specific protein adducts of 4-hydroxy-2(E)-nonenal in human THP-1 monocytic cells: Protein carbonylation is diminished by ascorbic acid, *Chem. Res. Toxicol.* 23, 37–47, 2010.
335. Gregory, J.D., The stability of N-ethylmaleimide and its reaction with sulfhydryl groups, *J. Am. Chem. Soc.* 77, 3922–3923, 1955.
336. Leslie, J., Spectral shifts in the reaction of N-ethylmaleimide with proteins, *Anal. Biochem.* 10, 162–167, 1965.
337. Gorin, G., Martic, P.A., and Doughty, G., Kinetics of the reaction of N-ethylmaleimide with cysteine and some congeners, *Arch. Biochem. Biophys.* 115, 593–597, 1966.
338. Bednar, R.A., Reactivity and pH dependence of thiol conjugation to *N*-ethylmaleimide; Detection of a conformational change in chalcone isomerase, *Biochemistry* 29, 3684–3690, 1990.
339. Smyth, D.G., Nagamatsu, A., and Fruton, J.S., Reactions of N-ethylmaleimide, *J. Am. Chem. Soc.* 82, 4600–4604, 1960.
340. Smyth, D.G., Blumenfeld, O.O., and Konigsberg, W., Reaction of N-ethylmaleimide with peptides and amino acids, *Biochem. J.* 91, 589–595, 1964.
341. Morodor, L., Musiel, H.J., and Scharf, R., Aziridine-2-carboxylic acid: A reactive amino acid unit for a new class of cysteine proteinase inhibitors, *FEBS Lett.* 299, 51–53, 1992.
342. Papini, A., Rudolph, S., Siglmueller, G. et al., Alkylation of histidine with maleimido-compounds, *J. Pept. Prot. Res.* 39, 345–355, 1992.
343. Gehring, H. and Christen, P., A diagonal procedure for isolating sulfhydryl peptides alkylated with N-ethylmaleimide, *Anal. Biochem.* 107, 358–341, 1980.
344. Rogers, L.K., Leinweber, B.L., and Smith, C.V., Detection of reversible protein thiol modifications in proteins, *Anal. Biochem.* 358, 171–184, 2006.
345. Brown, R.D. and Matthews, K.S., Chemical modification of lactose repressor proteins using N-substituted maleimides, *J. Biol. Chem.* 254, 5128–5134, 1979.
346. Brown, R.D. and Matthews, K.S., Spectral studies on Lac repressor modified with N-substituted maleimide probes, *J. Biol. Chem.* 254, 5135–5143, 1979.
347. Perussi, J.R., Tinto, M.H., Nascimento, O.R., and Tabak, M., Characterization of protein spin labeling by maleimide: Evidence for nitroxide reduction, *Anal. Biochem.* 173, 289–295, 1988.
348. Marquez, J., Iriarte, A., and Martinez-Carrion, M., Covalent modification of a critical sulfhydryl group in the acetylcholine receptor: Cysteine-222 of the α-subunit, *Biochemistry* 28, 7433–7439, 1989.

349. Mills, J.S., Walsh, M.P., Nemcek, K., and Johnson, J.D., Biologically active fluorescent derivatives of spinach calmodulin that report calmodulin target protein binding, *Biochemistry* 27, 991–996, 1988.
350. Jezek, P. and Drahota, Z., Sulfhydryl groups of the uncoupling protein of brown adipose tissue mitochondria—Distinction between sulfhydryl groups of the H^+-channel and the nucleotide binding site, *Eur. J. Biochem.* 183, 89–95, 1989.
351. Le-Quoc, K., Le-Quoc, D., and Gaudemer, Y., Evidence for the existence of two classes of sulfhydryl groups essential for membrane-bound succinate dehydrogenase activity, *Biochemistry* 20, 1705–1710, 1981.
352. Griffiths, D.G., Partis, M.D., Sharp, R.N., and Beechey, R.B., N-Polymethylenecarboxymaleimides—A new class of probes for membrane sulfhydryl groups, *FEBS Lett.* 134, 261–263, 1981.
353. Griffiths, D.G., Partis, M.D., Churchill, P. et al., The use of amphipathic maleimides to study membrane-associated proteins, *J. Bioeng. Biomembr.* 22, 691–707, 1990.
354. Rial, E., Arechaga, I., Sainzdelamaza, E., and Nicholls, D.G., Effect of hydrophobic sulfhydryl-reagents on the uncoupling protein and inner-membrane anion channel of brown-adipose-tissue mitochondria, *Eur. J. Biochem.* 182, 187–193, 1989.
355. Hirano, M., Takeuchi, Y., Aoki, T. et al., Rearrangements in the cytoplasmic domain underlie its gatings, *J. Biol. Chem.* 385, 3777–3783, 2010.
356. Lu, Q., Chen, Z., Perumattam, J. et al., Aryl-indole maleimides as inhibitors of CAMKIIδ Part 3: Importance of the indole orientation, *Bioorg. Med. Chem. Lett.* 18, 2399–2403, 2008.
357. Labar, G., Bauvois, C., Bore, F. et al., Crystal structure of the human monoacylglycerol lipase, a key actor in endocannabinoid signalling, *ChemBioChem.* 11, 218–227, 2010.
358. Mann, R.S., Usova, E.V., Cass, C.E., and Eriksson, S., Fluorescence energy transfer studies of human deoxycytidine kinase: Role of cysteine 185 in the conformational changes that occur upon substrate binding, *Biochemistry* 45, 3534–3541, 2006.
359. Rial, E., Aréchaga, I., Sainz-de-la-Maza, E., and Nicholls, D.G., Effect of hydrophobic sulphydryl reagents on the uncoupling protein and inner-membrane anion channel of brown-adipose-tissue mitochondria, *Eur. J. Biochem.* 182, 187–193, 1989.
360. Yee, A.S., Corley, D.E., and McNamee, M.G., Thiol-group modification of *Torpedo californica* acetylcholine receptor: Subunit localization and effects on function, *Biochemistry* 25, 2110–2119, 1986.
361. Pradier, L., Yee, A.S., and McNamee, M.G., Use of chemical modifications and site-directed mutagenesis to probe the functional role of thiol groups on the gamma subunit of *Torpedo californica* acetylcholine receptor, *Biochemistry* 28, 6562–6571, 1989.
362. Abbott, R.E. and Schachter, D., Topography and functions of sulfhydryl groups of the human erythrocyte glucose transport mechanism, *Mol. Cell. Biochem.* 82, 85–90, 1988.
363. May, J.M., Reaction of an exofacial sulfhydryl group on the erythrocyte hexose carrier with an impermeant maleimide. Relevance to the mechanism of hexose transport, *J. Biol. Chem.* 263, 13635–13640, 1988.
364. May, J.M., Interaction of a permeant maleimide derivative of cysteine with the erythrocyte glucose carrier. Differential labelling of an exofacial carrier thiol group and its role in the transport mechanism, *Biochem. J.* 263, 875–881, 1989.
365. Abbott, R.E. and Schachter, D., Impermeant maleimides. Oriented probes of erythrocyte membrane proteins, *J. Biol. Chem.* 251, 7176–7183, 1976.
366. May, J.M., Selective labeling of the erythrocyte hexose carrier with a maleimide derivative of glucosamine: Relationship of an exofacial sulfhydryl to carrier conformation and structure, *Biochemistry* 28, 1718–1725, 1989.
367. Falke, J.J., Dernburg, A.F., Sternberg, D.A. et al., Structure of a bacterial sensory receptor. A site-directed sulfhydryl study, *J. Biol. Chem.* 263, 14850–14858, 1988.

368. Niwayama, S., Kurano, S., and Matsumoto, N., Synthesis of *d*-labeled N-alkylmaleimides and application to quantitative peptide analysis by isotope differential mass spectrometry, *Bioorg. Med. Chem. Lett.* 11, 2257–2261, 2001.
369. Jones, P.C., Sivaprasadarao, A., Wray, D., and Findlay, J.B., A method for determining transmembrane protein structure, *Mol. Membr. Biol.* 13, 53–60, 1996.
370. Di Gleria, K. et al., N-(2-Ferrocene-ethyl)maleimide: A new electroactive sulfhydryl-specific reagent for cysteine-containing peptides and proteins, *FEBS Lett.* 390, 142–144, 1996.
371. Apuy, J.L., Chen, X., Russell, D.H. et al., Radiometric pulsed alkylation/mass spectrometry of the cysteine pairs in individual zinc fingers of MRE-binding transcription factor 1 (MTF-1) as a probe of zinc chelate stability, *Biochemistry* 40, 15164–15175, 2001.
372. Meza, J.E., Scott, C.K., Benz, C.C., and Baldwin, M.A., Essential cysteine-alkylation strategies to monitor structurally altered estrogen receptor as found in oxidant-stressed breast cancers, *Anal. Biochem.* 320, 21–31, 2003.
373. Apuy, J.L., Busenlehner, L.A., Russell, D.H., and Giedroc, D.P., Radiometric pulsed alkylation mass spectrometry as a probe of thiolate reactivity in different metalloderivatives of *Staphylococcus aureus* p1258, CadC, *Biochemistry* 43, 3824–3834, 2004.
374. Apuy, J.L., Park, Z.Y., Swartz, P.D. et al., Pulsed-alkylation mass spectrometry for the study of protein folding and dynamics: Development and application to the study of a folding/unfolding intermediate of bacterial luciferase, *Biochemistry* 40, 15153–15163, 2001.
375. Juszczak, L.J., Manjula, B., Bonaventura, C. et al., UV resonance Raman study of β93-modified hemoglobin A: Chemical modifier-specific effects and added influences of attached poly(ethylene glycol) chains, *Biochemistry* 41, 376–385, 2002.
376. Doherty, D.H., Rosendahl, M.S., Smith, D.J. et al., Site-specific PEGylation of engineered cysteine analogues of recombinant human granulocyte-macrophage colony-stimulated factor, *Bioconj. Chem.* 16, 1291–1298, 2005.
377. Salmaso, S., Bersani, S., Scomparin, A. et al., Tailored PEG for rh-G-CSF analogue site-specific conjugation, *Bioconj. Chem.* 29, 1179–1189, 2009.
378. Long, D.L., Doherty, D.H., Eisenberg, S.P. et al., Design of homogeneous, nonpegylated erythropoietin analogues with preserved *in vitro* bioactivity, *Exp. Hematol.* 34, 697–704, 2006.
379. Yu, P., Zheng, C., Chen, J. et al., Investigations on PEGylation strategy of recombinant human interleukin-1 receptor antagonist, *Bioorg. Med. Chem.* 15, 5396–5405, 2007.
380. Ni, J., Singh, S., and Wang, L.-X., Synthesis of maleimide-activated carbohydrates as chemoselective tags for site-specific glycosylation of peptides and proteins, *Bioconj. Chem.* 14, 232–238, 2003.
381. Gamblin, D.P., Scanlan, E.M., and Davis, B.G., Glycoprotein synthesis: An update, *Chem. Rev.* 109, 131–163, 2009.
382. Disney, M.D. and Seeberger, P.H., The use of carbohydrate microarrays to study carbohydrate-cell interactions to detect pathogens, *Chem. Biol.* 11,1701–1707, 2004.
383. Yatawara, A.K., Achani, K., Tiruchinapally, G. et al., Carbohydrate surface attachment characterized by sum frequency generation spectroscopy, *Langmuir* 25, 1901–1904, 2009.
384. Di Stefano, G., Lanza, M., Kratz, F. et al., A novel method for coupling doxorubicin to lactosaminated human albumin by an acid sensitive hydrazone bond: Synthesis, characterization and preliminary biological properties of the conjugate, *Eur. J. Pharm. Sci.* 23, 393–397, 2004.
385. Qiu, Y., Sousa, E.A., Hewick, R., and Wang, J.H., Acid-labile isotope-coded extractants: A class of reagents for quantitative mass spectrometric analysis of complex protein mixtures, *Anal. Chem.* 74, 4969–4979, 2002.
386. Chowdhury, S.M., Munske, G.R., Siems, W.F., and Bruce, J.E., A new maleimide-bound acid-cleavable solid-support reagent for profiling phosphorylation, *Rapid Commun. Mass Spectrom.* 19, 899–909, 2005.

387. Qian, W.J., Gosche, M.B., Camp, D.G. et al., Phosphoprotein isotope-coded solid-phase tag approach for enrichment and quantitative analysis of phosphoproteins from complex mixtures, *Anal. Chem.* 75, 5441–5450, 2003.
388. Fauq, A.H., Kache, R., Khan, M.A., and Vega, J.E., Synthesis of acid-cleavable light isotope-coded affinity tags (ICAT-L) for potential use in proteomic expression profiling analysis, *Bioconj. Chem.* 17, 248–254, 2006.
389. Smith, D.J., Maggio, E.T., and Kenyon, G.L., Simple alkane thiol groups for temporary blocking of sulfhydryl groups of enzymes, *Biochemistry* 14, 766, 1975.
390. Kluger, R. and Tsue, W.-C., Amkino group reactions of the sulfhydryl reagent methyl methanethiosulfonate. Inactivation of D-3-hydroxybutyrate and reaction with amines in water, *Can. J. Biochem.* 58, 629–632, 1980.
391. Pathak, R., Hendrickson, T.L., and Imperiali, B., Sulfhydryl modification of the yeast Wbp1p inhibits oligosaccharide transferase activity, *Biochemistry* 34, 4179–4181, 1995.
392. Nishimura, J.S., Kenyon, G.L., and Smith, G.L., Reversible modification of the sulfhydryl groups of *Escherichia coli* succinic thiokinase with methane thiolating reagents, 5,5′-dithio-bis(2-nitrobenzoic acid), *p*-hydroxymercuribenzoate, and ethylmercuri thiosalicylate, *Arch. Biochem. Biophys.* 170, 461–467, 1975.
393. Roberts, D.D., Lewis, S.D., Ballou, D.D. et al., Reactivity of small thiolate anions and cysteine-25 in papain toward methyl methanethiosulfonate, *Biochemistry* 25, 5595–5601, 1986.
394. Stauffers, D.A. and Karlin, A. Electrostatic potential of the acetylcholine binding site in the nicotinic receptor probed by reaction of binding-site cysteines with charged methanethiosulfonates, *Biochemistry* 33, 6840–6849, 1994.
395. Kenyon, G.L. and Bruice, T.W., Novel sulfhydryl reagents, *Methods Enzymol.* 47, 407–430, 1977.
396. Bruice, T.W. and Kenyon, G.L., Novel alkyl alkanethiolsulfonate sulfhydryl reagents. Modification of derivatives of l-cysteine, *J. Protein Chem.* 1, 47–58, 1982.
397. Toronto Research Chemicals, Inc.; http://www.trc-canada.com/white_papers.lasso: Methanethiosulfonate reagents: Application to the study of protein topology and ion channels.
398. Karlin, A. and Akabas, M.H., Substituted-cysteine accessibility method, *Methods Enzymol.* 293, 123–145, 1998.
399. Karlin, A., Scam feels the pinch, *J. Gen. Physiol.* 117, 235–238, 2001.
400. Barre, L., Kavanaugh, M.P., and Kanner, B.I., Dynamic equilibrium between coupled and uncoupled modes of neuronal glutamate transporter, *J. Biol. Chem.* 277, 13501–13507, 2002.
401. Tonazzi, A. and Indiveri, C., Chemical modification of the mitochondrial ornithine/citrulline carrier by SH reagents: Effects on the transport activity and transition from carrier to pore-like function, *Biochim. Biophys. Acta* 1611, 123–130, 2003.
402. Endeward, B., Butterwick, J.A., MacKinnon, R., and Prisner, T.F., Pulsed electron-electron double-resonance determination of spin-label distances and orientations on the tetrameric potassium ion channel KcsA, *J. Am. Chem. Soc.* 131, 15246–15250, 2009.
403. Passero, C.J., Okumura, S., and Carattino, M.D., Conformational changes associated with proton-dependent gating of ASIC1a, *J. Biol. Chem.* 284, 36473–36481, 2009.
404. Matsuda, J.J., Filali, M.S., Collins, M.M. et al., The C1C-3 Cl^-/H^+ antiporter becomes uncoupled at low extracellular pH, *J. Biol. Chem.* 285, 2569–2579, 2010.
405. McNally, B.A., Yamashita, M., Engh, A., and Prakriya, M., Structural determinants of ion permeation in CBAC channels, *Proc. Natl. Acad. Sci. USA* 106, 22516–22521, 2009.
406. Nichols, A.S. and Luetje, C.W., Transmembrane segment 3 of *Drosophilia melanogaster* oderant receptor subunit 85b contributes to ligand-receptor interactions, *J. Biol. Chem.* 285, 11854–11862, 2010.

407. Zhang, X.D., Yu, W.P., and Chen, T.Y., Accessibility of the CLC-0 pore to charged methanethiosulfonate reagents, *Biophys. J.* 98, 377–385, 2010.
408. Belvy, V., Anishkin, A., Kamaraju, K. et al., The tension-transmitting 'clutch' in the mechanosensitive channel MscS, *Nat. Struct. Mol. Biol.* 17, 451–458, 2010.
409. Bruhova, L. and Zhorov, B.S., A homology model for the pore domain of a voltage-gated calcium channel is consistent with available SCAM data, *J. Gen. Physiol.* 135, 261–274, 2010.
410. Bargeton, B. and Kellenberger, S., The contact region between the three domains of the extracellular loop of ASIC1a is critical for channel function, *J. Biol. Chem.* 285, 13816–13826, 2010.
411. Kurata, H.T., Zhu, E.A., and Nichols, C.S., Locale and chemistry of spermine binding in the archetypal inward rectifier Kir2.1, *J. Gen. Physiol.* 135, 495–508, 2010.
412. Wu, D., Delaloye, K., Zaydman, M.A. et al., State-dependent electrostatic interactions of S4 arginines with E1 in S2 during Kv7.1 activation, *J. Gen. Physiol.* 135, 595–606, 2010.
413. Jha, S.K. and Udagaonkar, J.B., Exploring the cooperativity of the fast folding reaction of a small protein using pulsed thiol labeling and mass spectrometry, *J. Biol. Chem.* 282, 37479–37491, 2007.
414. Patra, A.K., Mukhopadhyay, R., Mukhija, R. et al., Optimization of inclusion body solubilization and renaturation of recombinant human growth hormone from *Escherichia coli*, *Protein Express. Purif.* 18, 182–192, 2000.
415. Razeghifard, M.R., On-column refolding of recombinant human interleukin-4 from inclusion bodies, *Protein Express. Purif.* 37, 180–186, 2004.
416. Cindrić, M., Cepo, T., Marinc, S. et al., Determination of dithiothreitol in complex protein mixtures by HPLC-MS, *J. Sep. Sci.* 31, 3489–3496, 2008.
417. Inoue, M., Akimaru, J., Nishikawa, T. et al., A new derivatizing agent, trimethylammoniopropyl methanethiosulfonate, is efficient for preparation of recombinant brain-derived neurotrophic factor from inclusion bodies, *Biotechnol. Appl. Biochem.* 28, 207–213, 1998.
418. Gazaryan, I.G., Doseeva, V.V., Galkin, A.G., and Tishkov, V.I., Effect of single-point mutations $Phe^{41}\rightarrow His$ and $Phe^{143}\rightarrow Glu$ on folding and catalytic properties of recombinant horseradish peroxidase expressed in *E. coli*, *FEBS Lett.* 354, 248–250, 1994.
419. Murata, H., Sakaguchi, M., Futami, J. et al., Denatured and reversibly cationized p53 readily enters cells and simultaneously folds to the functional protein in the cells, *Biochemistry* 45, 6124–6132, 2006.
420. Futami, J., Kitazoe, M., Murata, H., and Yamada, H., Exploiting protein cationization techniques in future drug development, *Expert Opin.* 2, 261–269, 2007.
421. Brandwein, H.J., Lewicki, J.A., and Murad, F., Reversible inactivation of guanylate cyclase by mixed disulfide formation, *J. Biol. Chem.* 256, 2958–2962, 1981.
422. Riddles, P.W., Andrews, R.K., Blakeley, R.L., and Zerner, B., Jack Bean urease 6. Determination of thiol and disulfide content—Reversible inactivation of the sulfhydryl-group oxidation and reduction, *Biochem. Pharmacol.* 32, 811–818, 1983.
423. Ziegler, D.M., Role of reversible oxidation-reduction of enzyme thiols-disulfides in metabolic regulation, *Annu. Rev. Biochem.* 54, 305–329, 1985.
424. Thomas, J.A., Chai, Y.-C., and Jung, C.-H., Protein S-thiolation and dethiolation, *Methods Ezymol.* 233, 385–394, 1994.
425. Lim, A. et al., Identification of S-sulfonation and S-thiolation of a novel transthyretin Phe33Cys variant from a patient diagnosed with familial transthyretin amyloidosis, *Protein Sci.* 12, 1775–1785, 2003.
426. Shenton, D. and Grant, C.M., Protein S-thiolation targets glycolysis and protein synthesis in response to oxidative stress in the yeast *Saccharomyces cerevisiae*, *Biochem. J.* 374, 513–519, 2003.

427. Braughler, J.M., Soluble guanylate-cyclase activation by nitric oxide and its reversal—Involvement of sulfhydryl group oxidation and reduction, *Biochem. Pharmacol.* 32, 811–818, 1983.
428. Wu, X.B., Brune, B., Vonappen, F., and Ullrich, V., Reversible activation of soluble guanylate cyclase by oxidizing agents, *Arch. Biochem. Biophys.* 294, 75–82, 1992.
429. Cunningham, L.W. and Nuenke, B.J., Physical and chemical studies of a limited reaction of iodine with proteins, *J. Biol. Chem.* 234, 1447–1451, 1959.
430. Cunningham, L.W. and Nuenke, B.J., Analysis of modified β-lactoglobulins and ovalbumins prepared from sulfenyl iodide intermediates, *J. Biol. Chem.* 235, 1711–1715, 1960.
431. Cupo, J.F. and Pace, C.N., Conformational stability of mixed disulfide derivatives of β-lactoglobulin B, *Biochemistry* 22, 2654–2658, 1983.
432. Mecinović, J., Chowdhury, R., Flashman, E., and Schofield, C.J., Use of mass spectrometry to probe the nucleophilicity of cysteinyl residues of prolyl hydroxylase domain 2, *Anal. Biochem.* 393, 215–221, 2009.
433. Sanchez, R., Riddle, M., Woo, J., and Momand, J., Prediction of reversibly oxidized protein cysteine thiols using protein structure properties, *Protein Sci.* 17, 473–481, 2008.
434. Peskin, A.V. and Winterbourn, C.C., Taurine chloramine is more selective than hypochlorous acid at targeting critical cysteines and inactivating creatine kinase and glyceraldehyde-3-phosphate dehydrogenase, *Free Radical Bio. Med.* 40, 45–53, 2006.
435. Tew, K., Redox in redux: Emergent roles of glutathione S-transferase P in regulation of cell signalling and S-glutathioylation, *Biochem. Pharmacol.* 73, 1257–1269, 2007.
436. Ochoa, G., Gutierrez, C., Ponce, I. et al., Reactivity trends of surface-confined co-tetraphenyl porphyrins and vitamin B12 for the oxidation of 2-aminoethanethiol; Comparison with co-phthalocyanines and oxidation of other thiols, *J. Electroanal. Chem.* 639, 88–94, 2010.
437. Wynn, R. and Richards, F.M., Chemical modification of protein thiols: Formation of mixed disulfides, *Methods Enzymol.* 251, 351–356, 1995.
438. Pihl, A. and Lange, R., The interaction of oxidized glutathione, cystamine, monosulfoxide, and tetrathionate with the -SH groups of rabbit muscle, D-glyceraldehyde 3-phosphate dehydrogenase, *J. Biol. Chem.* 237, 1356–1362, 1962.
439. Parker, D.J. and Allison, W.S., The mechanism of inactivation of glyceraldehyde 3-phosphate dehydrogenase by tetrathionate, *o*-iodosobenzoate, and iodine monochloride, *J. Biol. Chem.* 244, 180–189, 1969.
440. Barbehenn, E.K. and Kaufman, B.T., Activation of chicken liver dihydrofolate reductase by tetrathionate, *Biochem. Biophys. Res. Commun.* 85, 402–407, 1978.
441. Church, J.S. and Evans, D.J., A spectroscopic investigation into the reaction of sodium tetrathionate with cysteine, *Spectrochim. Acta A Mol. Biomol. Spectrosc.* 69, 256–262, 2008.
442. Mukhopadhyay, A., Reversible protection of disulfide bonds followed by oxidative folding render recombinant hCGbeta highly immunogenic, *Vaccine* 18, 1802–1810, 2000.
443. Tikhonov, R.V. et al., Recombinant human insulin. VIII. Isolation of fusion proteins -*S*-sulfonate, biotechnological precursor of human insulin, from the biomass of transformed *Escherichia coli* cells, *Protein Express. Purif.* 21, 176–182, 2001.
444. Kingsbury, J.S., Laue, T.M., Klimtchuk, E.S. et al., The modulation of transethyretin tetramer stability by cysteine 10 adducts and the drug diflunisal. Direct analysis by fluorescence-detected analytical ultracentrifugation, *J. Biol. Chem.* 283, 11887–11895, 2008.
445. Kingsbury, J.S., Klimtchuk, E.S., Théberge, R. et al., Expression, purification, and *in vitro* cysteine-10 modification of native sequence recombinant human transethyretin, *Protein Express. Purif.* 53, 370–377, 2007.
446. Zhang, Q. and Kelly, J.W., Cys10 mixed disulfide make transethyretin more amyloidogenic under mildly acidic conditions, *Biochemistry* 42, 8756–8761, 2003.

447. Batista-Viera, F., Barbieri, M., Ovsejevi, K. et al., A new method for reversible immobilization of thiol biomolecules based on solid-phase bound thiolsulfonate groups, *Appl. Biochem. Biotechnol.* 31, 175–195, 1991.
448. Pavlovic, E., Quist, A.P., Gelius, U. et al., Generation of thiolsulfinates and thiolsulfonates by electrooxidation of thiols on silicon surfaces for reversible immobilization of molecules, *Langmuir* 19, 4217–4221, 2003.
449. Suárez, G., Jackson, R.L., Spoors, J.A., and McNeil, C.J., Chemical introduction of disulfide groups on glycoproteins: A direct protein anchoring scenario, *Anal. Chem.* 79, 1961–1969, 2007.
450. Chung, S.I. and Folk, J.E., Mechanism of the inactivation of guinea pig liver transglutaminase by tetrathionate, *J. Biol. Chem.* 245, 681–689, 1970.
451. Kaufmann, S.H. and Shaper, J.H., A subset of non-histone nuclear proteins reversibly stabilized by the sulfhydryl cross-linking reagent tetrathionate - polypeptides of the internal nuclear matrix, *Exp. Cell Res.* 155, 477–495, 1984.
452. Hahn, S.K., Park, J.K., Tomimatsu, T. et al., Synthesis and degradation test of hyaluronic acid hydrogels, *Int. J. Biol. Macromol.* 40, 374–380, 2007.
453. Ellman, G.L., A colorimetric method for determining low concentrations of mercaptans, *Arch. Biochem. Biophys.* 74, 443–450, 1958.
454. Ellman, G.L., Tissue sulfhydryl groups, *Arch. Biochem. Biophys.* 82, 70–77, 1959.
455. Habeeb, A.F.S.A., Reaction of protein sulfhydryl groups with Ellman's reagent, *Methods Enzymol.* 25, 457–464, 1972.
456. Collier, H.B., A note on the molar absorptivity of reduced Ellman's reagent, 3-carboxylato-4-nitrothiophenolate, *Anal. Biochem.* 56, 310–311, 1973.
457. Riddles, P.W., Blakeley, R.L., and Zerner, B., Ellman's reagent: 5,5′-dithiobis(2-nitrobenzoic acid)—A reexamination, *Anal. Biochem.* 94, 75–81, 1979.
458. Riddles, P.W., Blakeley, R.L., and Zerner, B., Reassessment of Ellman's reagent, *Methods Enzymol.* 91, 49–60, 1983.
459. Eyer, P. et al., Molar absorption coefficients for the reduced Ellman reagent: Reassessment, *Anal. Biochem.* 312, 224–227, 2003.
460. Zhang, H., Le, M., and Means, G.E., A kinetic approach to characterize the electrostatic environments of thiol groups in proteins, *Bioorg. Chem.* 26, 356–364, 1998.
461. Stewart, A.J., Blindauer, C.A., Berezenko, S. et al., Role of Tyr84 in controlling the reactivity of Cys34 of human albumin, *FEBS J.* 272, 353–362, 2004.
462. Zhu, J., Dhimitruka, I., and Pei, D., 5-(2-aminoethyl) dithio-2-nitrobenzoate as a more base-stable alternative to Ellman's reagent, *Org. Lett.* 6, 3809–3812, 2004.
463. Legler, G., 4,4′-dintriophenydisulfides of different charge types as probes for the electrostatic environment of sulfhydryl groups, *Biochim. Biophys. Acta* 405, 136–143, 1975.
464. Whitesides, G.M., Lilburn, J.E., and Szajewski, R.P., Rates of thiol-disulfide interchange reactions between mono- and dithiols and Ellman's reagent, *J. Org. Chem.* 42, 332–338, 1977.
465. Faulstich, H. and Heintz, D., Reversible introduction of thiol compounds into proteins by use of activated mixed disulfides, *Methods Enzymol.* 251, 357–366, 1995.
466. Messmore, J.M., Holmgren, S.K., Grilley, J.E., and Raines, R.T., Sulfur shuttle: Modulating enzymatic activity by thiol disulfide interchange, *Bioconj. Chem.* 11, 408–413, 2000.
467. Drewes, G. and Faulstich, H., 2,4-dinitrophenyl[^{14}C]cysteinyl disulfide allows selective radiolabeling of protein thiols under spectrophotometric control, *Anal. Biochem.* 188, 109–113, 1990.
468. Czerski, L. and Sanders, C.R., Thiol modification of diacylglycerol kinase: Dependence upon site membrane disposition and reagent hydrophobicity, *FEBS Lett.* 472, 225–229, 2000.
469. Liu, M. et al., Effect of cysteine residues on the activity of arginyl-tRNA synthetase from *Escherichia coli*, *Biochemistry* 38, 11006–11011, 1999.

470. Narasimhan, C., Lai, C.-S., Haas, A., and McCarthy, J., One free sulfhydryl group of plasma fibronectin becomes titratable upon binding of the protein to solid substrates, *Biochemistry* 27, 4970–4973, 1988.
471. Takeda, K., Shigemura, A., Hamada, S. et al., Dependence of reaction rate of 5,5′-dithio-(2-nitrobenzoic acid) to free sulfhydryl groups of bovine serum albumin and ovalbumin on the protein conformation, *J. Protein Chem.* 11, 187–192, 1992.
472. Underwood, P.A., Steele, J.G., and Dalton, B.A., Effects of polystyrene surface chemistry on the biological activity of solid phase fibronectin, analyzed with monoclonal antibodies, *J. Cell Sci.* 104, 793–803, 1993.
473. Talgoy, M.M., Bell, A.W., and Duckworth, H.W., The reactions of *Escherichia coli* citrate synthase with the sulfhydryl reagents 5,5′-dithiobis-(2-nitrobenzoic acid) and 4,4′-dithiodipyridine, *Can. J. Biochem.* 57, 822, 1979.
474. Kimura, T., Matsueda, R., Nakagawa, Y., and Kaiser, E.T., New reagents of the introduction of the thiomethyl group at sulfhydryl residues of proteins with concomitant spectrophotometric titration of the sulfhydryl: Methyl 3-nitro-2-pyridyl disulfide and methyl 2-pyridyl disulfide, *Anal. Biochem.* 122, 274–282, 1982.
475. Kitson, T.M. and Loomes, K.M., Synthesis of methyl 2- and 4-pyridyl disulfide from 2- and 4-thiopyridone and methyl methanethiosulfonate, *Anal. Biochem.* 146, 429–430, 1985.
476. Malthouse, J.P. and Brocklehurst, K., A kinetic method for the study of solvent environments of thiol groups in proteins involving the use of a pair of isomeric reactivity probes and a differential solvent effect. Investigation of the active centre of ficin by using 2,2′- and 4,4′-dipyridyl disulphides as reactivity probes, *Biochem. J.* 185, 217–222, 1980.
477. Luthra, M.P., Dunlap, R.B., and Odom, J.D., Characterization of a new sulfhydryl group reagent: 6,6′-diselenobis-(3-nitrobenzoic acid), a selenium analog of Ellman's reagent, *Anal. Biochem.* 117, 94–102, 1981.
478. Pleasants, J.C., Guo, W., and Rabenstein, D.L., A comparative study of the kinetics of selenol/diselenide and thiol/disulfide exchange reactions, *J. Am. Chem. Soc.* 111, 6553–6558, 1989.
479. Kang, S.I. and Spears, C.P., Structure-activity studies on organoselenium alkylating agents, *J. Pharm. Sci.* 79, 57–62, 1990.
480. Arnér, E.S., Selenoproteins—What unique properties can arise with selenocysteine in place of cysteine? *Exp. Cell Res.* 316, 1296–1303, 2010.
481. Xu, K., Zhang, Y., Tang, B. et al., Study of highly selective and efficient thiol derivatization using selenium reagents by mass spectrometry, *Anal. Chem.* 82, 6926–6932, 2010.
482. Banks, T.E. and Shafer, J.A., Inactivation of papain by S-methylation of its cysteinyl residue with O-methylisourea, *Biochemistry* 11, 110–114, 1972.
483. Lapko, V.N., Smith, D.L., and Smith, J.B., Methylation and carbamylation of human γ-crystallins, *Protein Sci.* 12, 1762–1774, 2003.
484. Beardsley, R.L., Karty, J.A., and Reilly, J.P., Enhancing the intensities of lysine-terminated tryptic peptide ions in matrix-assisted laser desorption/ionization mass spectrometry, *Rapid Commun. Mass Spectrom.* 14, 2147–2153, 2000.
485. Brancia, F.L., Montgomery, H., Tanaka, K., and Kumashiro, S., Guanidino labeling derivatization strategy for global characterization of peptide mixtures by liquid chromatography matrix-assisted laser desorption/ionization mass spectrometry, *Anal. Chem.* 76, 2748–2755, 2004.
486. Carbetta, V.J., Lit, T., Shakya, A. et al., Integrating Lys-N-proteolysis and N-terminal guanidination for improved fragmentation and relative quantification of singly-charged ions, *J. Am. Soc. Mass Spectrom.* 21, 1050–1060, 2010.
487. Stark, G., Modification of proteins with cyanate, *Methods Enzymol.* 11, 590–594, 1967.
488. Lippincot, J. and Apostol, I., Carbamylation of cysteine: A potential artifact in peptide mapping of hemoglobins in the presence of urea, *Anal. Biochem.* 267, 57–64, 1999.
489. Hebert, B., Hopwood, F., Oxley, D. et al., β-Elimination: An unexpected artefact in proteome analysis, *Proteomics* 3, 826–831, 2003.

490. Pecci, L., Cannella, C., Pensa, B., Costa, M., and Cavallini, D., Cyanylation of rhodanese by 2-nitro-5-thiocyanobenzoic acid, *Biochim. Biophys. Acta* 623, 348–353, 1980.
491. Vanaman, T.C. and Stark, G.C., A study of the sulfhydryl groups of the catalytic subunit of *Escherichia coli* aspartate transcarbamylase. The use of enzyme-5-thio-2-nitrobenzoate mixed disulfides as intermediates in modifying enzyme sulfhydryl groups, *J. Biol. Chem.* 245, 3565–3573, 1970.
492. Witkowska, H.E., Green, B.N., and Smith, S., The carboxyl-terminal region of thioesterase II participates in the interaction with fatty acid synthetase. Use of electrospray ionization mass spectrometry to identify a carboxyl-terminally truncated form of the enzyme, *J. Biol. Chem.* 265, 5662–5665, 1990.
493. Tang, H.Y. and Speicher, D.W., Identification of alternative products and optimization of 2-nitro-5-thiocyanatobenzoic acid cyanylation and cleavage at cysteine residues, *Anal. Biochem.* 334, 48–61, 2004.
494. Koehn, H., Clerens, S., Deb-Choudhury, S. et al., Higher sequence coverage and improved confidence in the identification of cysteine-rich proteins from the wool cuticle using combined chemical and enzymatic digestion, *J. Proteomics* 73, 323–330, 2009.
495. Wu, J. and Watson, J.T., Optimization of the cleavage reaction for cyanylated cysteinyl peptides for efficient and simplified mass mapping, *Anal. Biochem.* 258, 268–276, 1998.
496. Fafarman, A.T., Webb, L.J., Chuang, J.I., and Boxer, S.G., Site-specific conversion of cysteine thiols into thiocyanate creates at IR probe for electric fields in proteins, *J. Am. Chem. Soc.* 128, 13356–13357, 2006.
497. Edelstein, L., Stetz, M.A., McMahon, H.A., and Londergan, C.H., The effects of α-helical structure and cyanylated cysteine on each other, *J. Phys. Chem. B* 114, 4931–4936, 2010.
498. Ghosh, P.B. and Whitehouse, M.W., 7-Chloro-4-nitrobenzo-2-oxa-1,3-diazole: A new fluorogenic reagent for amino acids and other amines, *Biochem. J.* 108, 155–156, 1968.
499. Birkett, D.J., Price, N.D., Radda, G.K., and Salmon, A.G., The reactivity of SH groups with a fluorogenic reagent, *FEBS Lett.* 6, 346–348, 1970.
500. Birkett, D.J., Dwek, R.A., Radda, G.K. et al., Probes for the conformational transitions of phosphorylase b. Effect of ligands studied by proton relaxation enhancement, fluorescence and chemical reactivities, *Eur. J. Biochem.* 20, 494–508, 1971.
501. Lad, P.M., Wolfman, N.M., and Hammes, G.G., Properties of rabbit muscle phosphofructokinase modified with 7-chloro-4-nitrobenzo-2-oxa-1,3-diazole, *Biochemistry* 16, 4802–4806, 1977.
502. Nitta, K., Bratcher, S.C., and Kronman, M.J., Anomalous reaction of 4-chloro-7-nitrobenzofurazan with thiol compounds, *Biochem. J.* 177, 385–392, 1979.
503. Dwek, R.A., Radda, G.A., Richards, R.E., and Salmon, A.G., Probes for the conformational transitions of phosphorylase a. Effect of ligands studied by proton-relaxation enhancement, and chemical reactivities, *Eur. J. Biochem.* 29, 494–508, 1972.
504. Carlberg, I., Sahlman, L., and Mannervik, B., The effect of 2,4,6-trinitrobenzenesulfonate on mercuric reductase, glutathione reductase, and lipoamide dehydrogenase, *FEBS Lett.* 180, 102–106, 1985.
505. Silva, G.M., Netto, L.E.S., Discola, K.E. et al., Role of gluaredoxin 2 and cytosolic thioredoxins is cysteinyl-based redox modification of the 20S proteosome, *FEBS J.* 275, 2942–2955, 2008.
506. Godat, E., Herve-Grepinet, V., and Veillard, F., Regulation of cathepsin K activity by hydrogen peroxide, *Biol. Chem.* 389, 1123–1126, 2008.
507. McDermott, L. et al., Mutagenic and chemical modification of the ABA-1 allergen of the nematode *Ascaris*: Consequences for structure and lipid binding properties, *Biochemistry* 40, 9918–9926, 2001.
508. Ikegaya, K. et al., Kinetic analysis of enhanced thermal stability of an alkaline protease with engineered twin disulfide bridges and calcium-dependent stability, *Biotechnol. Bioeng.* 81, 187–192, 2003.

509. Maehashi, K. and Udaka, S., Sweetness of lysozymes, *Biosci. Biotechnol. Biochem.* 62, 605–606, 1998.
510. Masuda, T., Ueno, Y., and Kitabatake, N., Sweetness and enzymatic activity of lysozyme, *J. Agric. Food Chem.* 49, 4937–4947, 2001.
511. Jones, G., Moore, S., and Stein, W.H., Properties of chromatographically purified trypsin inhibitors from lima beans, *Biochemistry* 2, 66–71, 1963.
512. Ferdinand, W., Moore, S., and Stein, W.H., Susceptibility of reduced, alkylated trypsin inhibitors from lima beans to tryptic action, *Biochim. Biophys. Acta* 96, 524–527, 1965.
513. Ghezzi, P., Regulation of protein function by glutathionylation, *Free Radical Res.* 39, 573–580, 2005.
514. Toledano, M.B., Kumar, C., Le Moan, N. et al., The system biology of thiol redox system in *Escherichia coli* and yeast: Differential functions in oxidative stress, iron metabolism and DNA synthesis, *FEBS Lett.* 581, 3598–3607, 2007.
515. Dalle-Donne, I., Milzani, A., Gagliano, N. et al., Molecular mechanisms and potential clinical significance of S-glutathioylation, *Antiox. Redox Signal.* 10, 445–473, 2008.
516. Gallogly, M.M., Starke, D.W., and Mieyal, J.J., Mechanistic and kinetics details of catalysis of thiol-disulfide by glutaredoxins and potential mechanisms of regulation, *Antiox. Redox Signal.* 11, 1059–1081, 2009.
517. Regazzoni, L., Panusa, A., and Yeum, K.J., Hemoglobin glutathionylation can occur through cysteine sulfenic acid intermediate: Electrospray ionization LTQ-Orbitrap hybrid mass spectrometry studies, *J. Chromatogr. B* 877, 3456–3461, 2009.
518. Dalle-Donne, I., Rossi, R., Colombo, G. et al., Protein S-glutathionylation: A regulatory device from bacteria to humans, *Trends Biochem.* 34, 85–96, 2009.
519. Iversen, R., Andersen, P.A., Jensen, K.S. et al., Thiol-disulfide exchange between glutaredoxin and glutathione, *Biochemistry* 49, 810–820, 2010.
520. Creighton, T.E., Disulfide bonds as probes of protein folding pathways, *Methods Enzymol.* 131, 83–106, 1986.
521. Wedemeyer, W.J. et al., Disulfide bonds and protein folding, *Biochemistry* 39, 4207–4216, 2000.
522. Scheraga, H.A., Konishi, Y., and Ooi, T., Multiple pathways for regenerating ribonuclease A, *Adv. Biophys.* 18, 21–41, 1984.
523. Gross, G., Gallopin, M., Vandame, M. et al., Conformational exchange is critical for the productivity of an oxidatively folding intermediate with buried free cysteines, *J. Mol. Biol.* 403, 299–312, 2010.
524. Cai, S.J., Khorchid, A., Ikura, M., and Inouye, M., Probing catalytically essential domain interactions in histidine kinase EnvZ by targeted disulfide crosslinking, *J. Mol. Biol.* 328, 409–418, 2003.
525. Bunn, M.W. and Ordal, G.W., Transmembrane organization of the *Bacillus subtilis* chemoreceptor MCpB deduced by cysteine disulfide crosslinking, *J. Mol. Biol.* 331, 941–949, 2003.
526. Klco, J.M., Lassere, T.B., and Baranski, T.J., C5a receptor oligomerization- I. Disulfide trapping reveals oligomers and potential contact surfaces in a G protein-coupled receptor, *J. Biol. Chem.* 278, 35345–35353, 2003.
527. van der Sluis, E.O., Nouwen, N., and Driessen, A.J.M., SecY-SecY and SecY-SecG contacts revealed by site-specific crosslinkage, *FEBS Lett.* 527, 159–165, 2002.
528. Gilbert, H.F., Molecular and cellular aspects of thiol-disulfide exchange, *Adv. Enzymol.* 63, 69–172, 1990.
529. Eldjarn, L. and Pihl, A., The equilibrium constants and oxidation-reduction potentials of some thiol-disulfide systems, *J. Am. Chem. Soc.* 79, 4584–4593, 1957.
530. Keire, D.A., Strauss, E., Guo, W. et al., Kinetics and equilibria of thiol/disulfide interchange reactions of selected biological thiols and related molecules with oxidized glutathione, *J. Org. Chem.* 57, 123–127, 1992.

531. Jacob, C., Giles, G.I., Giles, N.M., and Sies, H., Sulfur and selenium: The role of oxidation state in protein structure and function, *Angew. Chem. Int. Ed.* 42, 4742–4758, 2003.
532. Huang, K.-P. and Huang, F.L., Glutathionylation of proteins by glutathione disulfide S-oxide, *Biochem. Pharmacol.* 64, 1049–1056, 2002.
533. Giles, G.I. and Jacob, C., Reactive sulfur species: An emerging concept in oxidative stress, *Biol. Chem.* 383, 375–383, 2002.
534. Brannan, R.G., Reactive sulfur species act as prooxidants in liposomal and skeletal model systems, *J. Agric. Food Chem.* 58, 3767–3771, 2010.
535. Huang, K.-P., Huang, F.L., Shetty, P.K., and Yergey, A.L., Modification of protein by disulfide S-monoxide and disulfide S-oxide: Distinctive effects on PKC, *Biochemistry* 46, 1961–1971, 2007.
536. Sakuma, S., Fujita, J., Nakanishi, M. et al., Disulfide S-monoxides convert xanthine dehydrogenase into oxidase in rat liver cytosol more potently than their respective disulfides, *Biol. Pharmaceut. Bull.* 31, 1013–1015, 2008.
537. Bowles, E., Alves d Sousa, R., Galardon, E. et al., Direct synthesis of a thiolate-S and thiolate-S Co^{III} complex related to the active site of nitrile hydratase: A pathway to the post-translational oxidation of the protein, *Angew. Chem. Int. Ed.* 44, 6162–6165, 2005.
538. Gorin, G. and Godwin, W.E., The reaction of iodate with cystine and with insulin, *Biochem. Biophys. Res. Commun.* 25, 227, 1966.
539. Nowduri, A., Adari, K.K., Gollapalli, N.R., and Parvataneni, V., Kinetics and mechanism of oxidation of *L*-cystine by hexacyanoferrate (III) in alkaline medium, *E. J. Chem.* 6, 93–98, 2009.
540. Moore, S., On the determination of cystine as cysteic acid, *J. Biol. Chem.* 239, 235–237, 1963.
541. Kuromizo, K., Abe, O., and Maeda, H., Location of the disulfide bonds in the antitumor neocarzionostatin, *Arch. Biochem. Biophys.* 286, 569–573, 1991.
542. Cobb, K.A. and Novolny, M.V., Peptide mapping of complex proteins at the low-picomole level with capillary electrophoresis separations, *Anal. Chem.* 64, 879–886, 1992.
543. Gehrke, C.W., Wall, L., Absheer, J.S. et al., Sample preparation for chromatography of amino acids: Acid hydrolysis of proteins, *J. Assoc. Anal. Chem.* 68, 811–821, 1991.
544. Varga-Visi, E., Terlaky-Balla, E., Pohn, G. et al., RPHPLC determination of L- and C-cystine and cysteine as cysteic acid, *Chromatographia* 51(suppl 1), S325–S327, 2000.
545. Donovan, J.W., A spectrophotometric and spectrofluorometric study of intramolecular interactions of phenolic groups in ovomucoid, *Biochemistry* 6, 3918–3927, 1967.
546. Donovan, J.W., Spectrophotometric observation of the alkaline hydrolysis of protein disulfide bonds, *Biochem. Biophys. Res. Commun.* 29, 734–740, 1967.
547. Cavallini, D., Federici, G., Barboni, E., and Marcucci, M., Formation of persulfide groups in alkaline treated insulin, *FEBS Lett.* 10, 125–128, 1970.
548. Helmerhorst, E. and Stokes, G.B., Generation of an acid-stable and protein-bound persulfide-like residue in alkali- or sulfhydryl-treated insulin by a mechanism consistent with the β-elimination hypothesis of disulfide bond lysis, *Biochemistry* 22, 69–75, 1983.
549. Florence, T.M., Degradation of protein disulphide bonds in dilute alkali, *Biochem. J.* 189, 507–520, 1980.
550. Abdolrasulnia, R. and Wood, J.L., Persulfide properties of thiocystine and related trisulfides, *Biooorgn. Chem.* 9, 253–260, 1980.
551. Lu, B.-Y. and Chang, J.-Y., Rapid and irreversible reduction of protein disulfide bonds, *Anal. Biochem.* 405, 67–72, 2010.
552. Clarke, H.T., The action of sulfite upon cystine, *J. Biol. Chem.* 97, 235–248, 1932.
553. Lugg, J.W.H., CCLIII., The application of phospho-18-tungstic acid (Folin's reagent) to the colorimetric determination of cysteine, cystine, and related substances, *Biochem. J.* 26, 2144–2159, 1932.

554. Stricks, W. and Kolthoff, I.M., Equilibrium constants of the reactions of sulfite with cystine and with dithioglycolic acid, *J. Am. Chem. Soc.* 72, 4569–4574, 1951.
555. Carter, J.R., Amperometric titration of disulfide and sulfhydryl in proteins in 8 M urea, *J. Biol. Chem.* 234, 1705–1709, 1959.
556. Cecil, R. and McPhee, J.R., A kinetic study of the reaction on some disulphides with sodium sulphite, *Biochem. J.* 60, 496–506, 1955.
557. Cole, R.D., Sulfitolysis, *Methods Enzymol.* 11, 206–208, 1967.
558. Tikhonov, R.V., Pechenov, S.E., Belacheu, I.A., Recombinant human insulin, *Protein Express. Purif.* 21, 176–182, 2001.
559. Würfel, M., Häberlein, I., and Follman, H., Facile sulfitolysis of the disulfide bonds in oxidized thioredoxin and glutaredoxin, *Eur. J. Biochem.* 211, 609–614, 1993.
560. Häberlein, I., Structure requirements for disulfide bridge sulfitolysis of oxidized *Escherichia coli* thioredoxin studied by fluorescence spectroscopy, *Eur. J. Biochem.* 223, 473–479, 1994.
561. Kristjánsson, M.M., Magnússon, O.Th., Gudmundsson, H.M. et al., Properties of a subtilisin-like proteinase from a psychrotropic *Vibrio* species. Comparison with proteinase K and aqualysin I, *Eur. J. Biochem.* 260, 752–760, 1999.
562. Tadokoro, T., You, D.J., Abe, Y. et al., Structural, thermodynamic, and mutational analyses of a psychrotrophic RNase H1, *Biochemistry* 46, 7460–7468, 2007.
563. Gráczer, E., Varga, A., Hajdu, I. et al., Rates of unfolding, rather than refolding, determine thermal stabilities of thermophilic, mesophilic, and psychrotropic 3-isopropylmalate dehydrogenases, *Biochemistry* 46, 11536–11549, 2007.
564. Kasana, R.C., Proteases from psychrotrophs: An overview, *Crit. Rev. Microbiol.* 36, 134–145, 2010.
565. Thannhauser, T.W., Konishi, Y., and Scheraga, H.A., Sensitive quantitative analysis of disulfide bonds in polypeptides and proteins, *Anal. Biochem.* 138, 181–188, 1984.
566. Damodaran, S., Estimation of disulfide bonds using 3-nitro-5-thiosulfobenzoic acid: Limitations, *Anal. Biochem.* 145, 200–205, 1985.
567. Kella, N.K.D. and Kinsella, J.E., A method for the controlled cleavage of disulfide bonds in proteins in the absence of denaturants, *J. Biochem. Biophys. Meth.* 11, 251–263, 1985.
568. Kell, N.K.D., Kang, Y.J., and Kinsella, J.E., Effect of oxidative sulfitolysis of disulfide bonds of bovine serum albumin on its structural properties: A physicochemical study, *J. Protein Chem.* 7, 535–548, 1988.
569. Kella, N.K.D., Yang, S.Y., and Kinsella, J.E., Effect of disulfide bond cleavage on structural and interfacial properties of whey proteins, *J. Agric. Food Chem.* 37, 1203–1210, 1989.
570. Petruccelli, S. and Añón, M.C., Partial reduction of soy protein isolate disulfide bonds, *J. Agric. Food Chem.* 43, 2001–2006, 1995.
571. Wang, H., Faris, R.J., Wang, T. et al., Increased in vitro and in vivo digestibility of soy proteins by chemical modification of disulfide bonds, *J. Am. Oil Chem. Soc.* 86, 1093–1099, 2009.
572. Morel, M.H., Bonicel, J., Micard, V., and Guilbert, S., Protein insolubilization and thiol oxidation in sulfite-treated wheat gluten films during aging at various temperatures and relative humidities, *J. Agric. Food Chem.* 48, 186–192, 2000.
573. Taylor, D.P., Carpenter, C.E., and Walsh, M.K., Influence of sulfonation on the properties of expanded extrudates containing 32% whey protein, *J. Food Sci.* 71, E17–E21, 2006.
574. Kim, M., Yoo, O.J., and Choe, S., Molecular assembly of the extracellular domain of $P2X_2$, an ATP-gated ion channel, *Biochem. Biophys. Res. Commun.* 240, 618–622, 1997.
575. Patrick, J.S. and Lagu, A.L., Determination of recombinant human proinsulin fusion protein produced in *Escherichia coli* using oxidative sulfitolysis and two-dimensional HPLC, *Anal. Chem.* 64, 507–511, 1992.
576. Nilsson, J. et al., Integrated production of human insulin and its C-peptide, *J. Biotechnol.* 48, 241–259, 1996.

577. Tikhonov, R.V., Pechenov, S.E., Belacheu, I.A. et al., Recombinant human insulin. VII. Isolation of fusion protein-S-sulfonate, biotechnological precursor of human insulin from the biomass of transformed *Escherichia coli* cells, *Protein Express. Purif.* 21, 176–182, 2001.
578. Tikhonov, R.V., Pechenov, S.E., Belacheu, I.A. et al., Recombinant human insulin. IX. Investigation of factors, influencing the folding of fusion protein-S-sulfonates, biotechnological precursors of human insulin, *Protein Express. Purif.* 26, 187–193, 2002.
579. Sato, T. and Aimoto, S., Use of thiosulfonate for the protection of thiol groups in peptide ligation by the thioester method, *Tetrahedron Lett.* 44, 8085–8087, 2003.
580. Harris, L.J., On the existence of an unidentified sulphur grouping in the protein molecule. Part I, On the denaturation of proteins, *Proc. Roy. Soc. Lond. B* 94, 426–441, 1923.
581. Harris, L.J., On the existence of an unidentified sulphur grouping in the protein molecule. Part II. On the estimation of cystine in certain protein, *Proc. Roy. Soc. Lond. B* 94, 441–450, 1923.
582. Putnam, F.W., Chemical modification of proteins, in *The Proteins*, Vol. 1, Pt., B., H. Neurath and K. Bailey (eds.), Academic Press, New York, 1953.
583. Olcott, H.S. and Fraenkel-Conrat, H., Specific group reagents for proteins, *Chem. Rev.* 41, 151–197, 1947.
584. Walker, E., CLIV., A colour reaction for disulphides, *Biochem. J.* 19, 1082–1084, 1925.
585. Mauthner, J., Cystin, *Z. Physiol. Chem.* 78, 28–36, 1912.
586. Lotter, H. and Timpl, R., Disulfide-linked cyanogen bromide peptides of bovine fibrinogen. I. Isolation of peptide F-CB3 and characterization of its single disulfide bond by cleavage with cyanide, *Biochim. Biophys. Acta* 427, 558–568, 1976.
587. Catsimpoolas, N. and Wood, J.L., Specific cleavage of cystine peptides by cyanide, *J. Biol. Chem.* 241, 1790–1796, 1966.
588. Catsimpoolas, N. and Wood, J.L., The reaction of cyanide with bovine serum albumin, *J. Biol. Chem.* 239, 4132–4136, 1964.
589. Evans, R.M., Fraser, J.B., and Owen, L.N., Dithiols. III. Derivatives of polyhydric alcohols, *J. Chem. Soc.* 248–255, 1949.
590. Cleland, W.W., Dithiothreitol, a new protective reagent for SH groups, *Biochemistry* 3, 480–483, 1964.
591. Iyer, K.S. and Klee, W.A., Direct spectrophotometric measurement of the rate of reduction of disulfide bonds. The reactivity of the disulfide bonds of bovine α-lactalbumin, *J. Biol. Chem.* 248, 707, 1973.
592. Mise, T. and Bahl, O.P., Assignment of disulfide bonds in the α-subunit of human chorionic gonadotropin, *J. Biol. Chem.* 255, 8516, 1980.
593. Gorin, G., Fulford, R., and Deonier, R.C., Reaction of lysozyme with dithiothreitol and with other mercaptans, *Experientia* 24, 26–27, 1968.
594. Bewley, T.A., Dixon, J.S., and Li, C.H., Human pituitary growth hormone. XVI. Reduction with dithiothreitol in the absence of urea, *Biochim. Biophys. Acta* 154, 420–421, 1968.
595. Warren, J.R. and Gordon, J.A., Denaturation of globular proteins. II. The interaction of urea with lysozyme, *J. Biol. Chem.* 245, 4097–4104, 1979.
596. Yamanaka, K., Nakajima, H., and Wada, Y., Kinetic study of denaturation and subsequent reduction of disulfide bonds of lysozyme by the rapid ultrasonic absorption measurement, *Biopolymers* 17, 2159–2169, 1978.
597. Hamaguchi, K., Structure of muramidase (lysozyme). V. Effect of N,N-dimethylformamide and the role of disulfide bonds in the stability of muramidase, *J. Biochem.* 55, 333–339, 1964.
598. Radford, S.E., Woolfson, D.N., Martin, S.R. et al., A three-disulphide derivative of hen lysozyme, *Biochem. J.* 273, 211–217, 1991.
599. Oda, M., Kitai, A., Murakami, A. et al., Evaluation of the conformational equilibrium of reduced hen egg white lysozyme by antibodies to the native form, *Arch. Biochem. Biophys.* 494, 145–150, 2010.

600. Golda, S., Shimizu, A., Ikeguchi, M., and Sugai, S., The superreactive disulfide bonds in α-lactalbumin and lysozyme, *J. Protein Chem.* 14, 731–737, 1995.
601. Bewley, T.A. and Li, C.H., The reduction of protein disulfide bonds in the absence of denaturants, *Int. J. Protein Res.* 1, 117–124, 1969.
602. White, F.H., Jr., Studies on the relationship of disulfide bonds to the formation and maintenance of secondary structure in chicken egg white lysozyme, *Biochemistry* 21, 967–977, 1982.
603. Bradshaw, R.A., Kanarek, L., and Hill, R.L., The preparation, properties, and reactivation of the mixed disulfide derivative of egg white lysozyme and L-cystine, *J. Biol. Chem.* 242, 3789–3798, 1967.
604. Ryle, A.P. and Sanger, F., Disulphide interchange reactions, *Biochem. J.* 60, 535–560, 1960.
605. Costantino, H.R., Langer, R., and Klibanov, A.M., Solid phase aggregation of proteins under pharmaceutically relevant conditions, *J. Pharm. Sci.* 83, 1662–1669, 1994.
606. Fransson, J.R., Oxidation of human insulin-like growth factor I in formulation studies. 3. Factorial experiments of the effects of ferric ions, EDTA, and visible light on methionine oxidation and covalent aggregation in aqueous solution, *J. Pharm. Sci.* 86, 1046–1050, 1997.
607. Costantino, H.R., Schwendeman, S.P., Langer, R., and Klibanov, A.M., Deterioration of lyophilized pharmaceutical proteins, *Biochemistry (Moscow)* 63, 357–363, 1998.
608. Bartkowski, R., Kitchel, R., Peckham, N., and Margulis, L., Aggregation of recombinant bovine granulocyte colony stimulating factor in solution, *J. Protein Chem.* 21, 137–143, 2002.
609. Mousavi, S.H., Bordbar, A.K., and Haertlé, T., Changes in structure and in interactions of heat-treated bovine β-lactoglobulin, *Protein Pept. Lett.* 15, 818–825, 2008.
610. Thing, M., Zhang, J., Lawrence, J., and Topp, E.M., Thiol-disulfide interchange in the tocinoic acid/glutathione system during freezing and drying, *J. Pharm. Sci.* 99, 4849–4856, 2010.
611. *Aggregation of Therapeutic Proteins*, W. Wang and C.J. Roberts (eds.), J.W. Wiley, Hoboken, NJ, 2010.
612. Kaneko, R. and Kitabatake, N., Heat-induced formation of intermolecular disulfide linkages between thaumatin molecules that do not contain cysteine residues, *J. Agric. Food Chem.* 47, 4950–4955, 1999.
613. Kim, J.-S. and Kim, H.-J., Matrix-assisted laser desorption/ionization time-of-flight mass spectrometry observation of a peptide triplet induced by thermal cleavage of cystine, *Rapid Commun. Mass Spectrom.* 15, 2296–2300, 2001.
614. Taylor, F.R., Prentice, H.L., Garber, E.A. et al., Suppression of sodium dodecyl sulfate-polyacrylamide gel electrophoresis sample preparation artifacts for analysis of IgG4 half-antibody, *Anal. Biochem.* 353, 204–208, 2006.
615. Liu, H., Gaza-Bulseco, G., Chumsae, C., and Newby-Kew, A., Characterization of lower molecular weight artifact bands of recombinant monoclonal IgG1 antibodies on non-reducing SDS-PAGE, *Biotechnol. Lett.* 29, 1611–1622, 2007.
616. Pompach, P., Man, P., Kavan, D. et al., Modified electrophoretic and digestion conditions allow a simplified mass spectrometric evaluation of disulfide bonds, *J. Mass Spectrom.* 44, 1571–1578, 2009.
617. Wu, S.L., Jiang, H., Lu, Q. et al., Mass spectrometric determination of disulfide linkages in recombinant therapeutic proteins using online LC-MS with electron-transfer dissociation, *Anal. Chem.* 91, 112–122, 2009.
618. Zhao, L., Almaraz, R.T., Ziang, F. et al., Gas–phase scrambling of disulfide bonds during matrix-assisted laser desorption/ionization mass spectrometry analysis, *J. Am. Soc. Mass Spectrom.* 20, 1603–1616, 2009.
619. Xia, Y. and Cooks, R.G., Plasma induced oxidative cleavage of disulfide bonds in polypeptides during nanospray ionization, *Anal. Chem.* 82, 2856–2864, 2010.

620. Wu, J. and Watson, J.T., A novel methodology for assignment of disulfide bond pairings in proteins, *Protein Sci.* 6, 391–398, 1997.
621. Homandberg, G.A. and Wai, T., Reduction of disulfides in urokinase and insertion of a synthetic peptide, *Biochim. Biophys. Acta* 1038, 209, 1990.
622. Rajesh Singh, R. and Chang, J.Y., Structural stability of human α-thrombin studied by disulfide reduction and scrambling, *Biochim. Biophys. Acta* 1651, 85–92, 2003.
623. Overman, L.E., Matzinger, D., O'Connor, E.M., and Overman, J.D., Nucleophilic cleavage of the sulfur–sulfur bond by phosphorus nucleophiles. Kinetic study of the reduction of aryl disulfides with triphenylphosphine and water, *J. Am. Chem. Soc.* 96, 6081–6089, 1975.
624. Rüegg, U.T. and Rudinger, J., Reductive cleavage of cystine disulfides with tributylphosphine, *Methods Enzymol.* 47, 111–116, 1977.
625. Rüegg, U.T., Reductive cleavage of S-sulfo groups with tributylphosphine, *Methods Enzymol.* 47, 123–126, 1977.
626. Kirley, T.L., Reduction and fluorescent labeling of cyst(e)ine-containing proteins for subsequent structural analysis, *Anal. Biochem.* 180, 231–236, 1989.
627. Chin, C.C.Q. and Wold, F., The use of tributylphosphine and 4-(aminosulfonyl)-7-fluoro-2,1,3-benzoxadiazole in the study of protein sulfhydryls and disulfides, *Anal. Biochem.* 214, 128–134, 1993.
628. Smejkal, G.B., Li, C., Robinson, M.H. et al., Simultaneous reduction and alkylation of protein disulfides in a centrifugal ultrafiltration device prior to two-dimensional gel electrophoresis, *J. Proteome Res.* 5, 963–967, 2006.
629. Herbert, B.D. et al., Improved protein solubility in two-dimensional electrophoresis using tributylphosphine as reducing agent, *Electrophoresis* 19, 845–851, 1998.
630. Valcu, C.-M. and Schlink, K., Reduction of proteins during sample preparation and two-dimensional gel electrophoresis of woody plant samples, *Proteomics* 6, 1599–1605, 2006.
631. Gilfix, B.M., Evans, D.W., and Rosenblatt, D.S., Novel reductants for determination of total plasma homocysteine, *Clin. Chem.* 43, 687–688, 1997.
632. Pfeiffer, C.M., Huff, D.L., and Gunter, E.W., Rapid and accurate HPLC assay for plasma total homocysteine and cysteine in a clinical laboratory setting, *Clin. Chem.* 45, 290–292, 1999.
633. Krijt, J., Vackova, M., and Kozich, V., Measurement of homocysteine and other aminothiols in plasma: Advantages of using tris-(2-carboxyethyl) phosphine as reductant compared with tri-*n*-butylphosphine, *Clin. Chem.* 47, 1821–1828, 2003.
634. Humphreys, D.P., Heywood, S.P., Henry, A. et al., Alternative antibody Fab' fragment PEGylation strategies: Combination of strong reducing agents, disruption of the interchain disulphide bond and disulphide engineering, *Protein Eng. Des. Sel.* 20, 227–234, 2007.
635. Burns, J.A., Butler, J.C., Moran, J., and Whitesides, G.M., Selective reduction of disulfides by tris-(2-carboethoxyethyl)-phosphine, *J. Org. Chem.* 56, 2648–2650, 1991.
636. Gray, W.R., Disulfide structures of highly bridged peptides: A new strategy for analysis, *Protein Sci.* 2, 1732–1748, 1993.
637. Gray, W.R., Echistatin disulfide bridges: Selective reduction and linkage assignment, *Protein Sci.* 2, 1749–1755, 1993.
638. Vuong, G.L. et al., Improved sensitivity proteomics by postharvest alkylation and radioactive labeling of proteins, *Electrophoresis* 21, 2594–2605, 2000.
639. Shaw, J., Rowlinson, R., Nickson, J. et al., Evaluation of saturation labelling two-dimensional difference gel electrophoresis fluorescent dyes, *Proteomics* 3, 1181–1195, 2003.
640. White, C.E., Hunter, M.J., Meininger, D.P. et al., The fifth epidermal growth factor-like domain of thrombomodulin does not have an epidermal growth factor-like disulfide bonding pattern, *Proc. Natl. Acad. Sci. USA* 93, 10177–10182, 1996.
641. Qi, J., Wu, J., Somkuti, G.A., and Watson, J.T., Determination of the disulfide structure of sillucin, a highly knotted cysteine-rich peptide by cyanylation/cleavage mass trapping, *Biochemistry* 40, 4531–4538, 2001.

642. Stark, G.R., Cleavage at cysteine after cyanylation, *Methods Enzymol.* 47, 129–132, 1977.
643. Daniel, R., Caminade, E., Martel, A. et al., Mass spectrometric determination of the cleavage sites in *Escherichia coli* dihydroorotase induced by a cysteine-specific agent, *J. Biol. Chem.* 272, 26934–26939, 1997.
644. Tetenbaum, J. and Miller, L.M., A new spectroscopic approach to examining the role of disulfide bonds in the structure and unfolding of soybean trypsin inhibitor, *Biochemistry* 40, 12215–12219, 2001.
645. Maciel, J., Martins, M.C.L., and Barbosa, M.A., The stability of self-assembled monolayers with time and under biological conditions, *J. Biomed. Mater. Res. A* 94A, 833–843, 2010.
646. Zhang, H.M., McLoughlin, S.M., Frausto, S.D. et al., Simultaneous reduction and digestion of proteins with disulfide bonds for hydrogen/deuterium exchange monitored by mass spectrometry, *Anal. Chem.* 82, 1450–1454, 2010.
647. Liu, P., O'Mara, B.W., Warrack, B.M. et al., A tris(2-carboxyethyl) phosphine (TCEP) related cleavage on cysteine-containing proteins, *J. Am. Soc. Mass Spectrom.* 21, 837–844, 2010.
648. Singh, R. and Maloney, E.K., Labeling of antibodies by *in situ* modification of thiol groups generated from selenol-catalyzed reduction of native disulfide bonds, *Anal. Biochem.* 304, 147–156, 2002.
649. Carl, P., Kwok, C.H., Manderson, G. et al., Forced unfolding modulated by disulfide bonds in the Ig domain of a cell adhesion molecule, *Proc. Natl. Acad. Sci. USA* 98, 1565–1570, 2001.
650. Goto, K., Shimada, K., Nagabama, M. et al., Reaction of stable sulfenic and selenic acids containing a bowl-type steric protection group with a phosphine. Elucidation of the mechanism of reduction of sulfenic and selenic acids, *Chem. Lett.* 32, 1080–1081, 2003.
651. Dansette, P.M., Thibault, S., Bertho, G., and Mansuy, D., Formation and fate of a sulfenic acid intermediate in the metabolic activation of the antithrombotic prodrug Prasugrel, *Chem. Res. Toxicol.* 12, 1268–1274, 2010.
652. Neumann, H. and Smith, R.L., Cleavage of the disulfide bonds of cystine and oxidized glutathione by phosphorothioate, *Arch. Biochem. Biophys.* 122, 354–361, 1967.
653. Borman, C.D., Wright, C., Twitchett, M.B. et al., Pulse radiolysis studies on galactose oxidase, *Inorg. Chem.* 41, 2158–2163, 2002.
654. Singh, R., Lamoureux, G.V., Lees, W.J., and Whitesides, G.M., Reagents for rapid reduction of disulfide bonds, *Methods Enzymol.* 251, 167–178, 1995.
655. Hansen, R.L., Østergaard, H., Nørgaard, P., and Winther, J.R., Quantitation of protein thiols and dithiols in the picomolar range using sodium borohydride and 4,4′-dithiodipyridine, *Anal. Biochem.* 363, 77–82, 2007.
656. Light, A., Hardwick, B.C., Hatfield, L.M., and Sondack, D.L., Modification of a single disulfide bond in trypsinogen and the activation of the carboxymethyl derivative, *J. Biol. Chem.* 244, 6289, 1969.
657. Knights, R.J. and Light, A., Disulfide bond-modified trypsinogen. Role of disulfide 179–203 on the specificity characteristics of bovine trypsin toward synthetic substrates, *J. Biol. Chem.* 251, 222–228, 1976.
658. Gorecki, M. and Patchornik, A., Polymer-bound dihydrolipoic acid: A new insoluble reducing agent for disulfides, *Biochim. Biophys. Acta* 303, 36–43, 1973.
659. Scouten, W.H., Torok, F., and Gitmer, W., Purification of lipoamide dehydrogenase by affinity chromatography on propyllipoamide glass columns, *Biochim. Biophys. Acta* 309, 521–524, 1973.
660. Scouten, W.H. and Firestone, G.L., N-Propylhydrolipoamide glass beads. An immobilized reducing agent, *Biochim. Biophys. Acta* 453, 227–283, 1976.
661. Bienvenu, C., Greiner, J., Vierling, P., and Di Giorgio, C., Convenient supported recyclable material based on dihydrolipoyl-residue for the reduction of disulfide derivatives, *Tetrahedron Lett.* 51, 3309–3311, 2010.

662. Al-Dubai, H., Oberhofer, G., Kerleta, V. et al., Cleavage of antibodies using dihydrolipoamide and anchoring of antibody fragments on a to biocompatibly coated carriers, *Monatschefte fur Chemie* 141, 485–490, 2010.
663. Ferraz, N., Leverrier, J., Batista-Viera, F., and Manta, C., Thiopropyl-agarose as a solid phase reducing agent for chemical modification of IgG and F(ab')$_2$, *Biotechnol. Prog.* 24, 1154–1159, 2008.
664. Zu, Y., Fee, J.A., and Hirst, J., Breaking and re-forming the disulfide bond at the high potential respiratory-type Rieske [2Fe-2S] center of *Thermos thermophilus*: Characterization of the sulfhydryl states by protein-film voltammetry, *Biochemistry* 41, 14054–14065, 2002.
665. Vanhooren, A., Devreese, B., Vanhee, K. et al., Photoexcitation of tryptophan groups induces reduction of two disulfide bonds in goat α-lactalbumin, *Biochemistry* 41, 11035–11043, 2002.
666. Permyakov, E.A., Permyakov, S.E., Deikus, G.Y. et al. Ultraviolet illumination-induced reduction of α-lactalbumin disulfide bridges, *Proteins* 51, 498–503, 2003.
667. Middendoft, T.R., Aldrich, R.W., and Baylor, P.A., Modification of cyclic nucleotide-gated ion channels by ultraviolet light, *J. Gen. Physiol.* 116, 227–252, 2000.
668. Miller, B.L. et al., Solid-state photodegradation of bovine somatotropin (bovine growth hormone): Evidence for tryptophan-mediated photooxidation of disulfide bonds, *J. Pharm. Sci.* 92, 1698–1709, 2003.
669. Alphey, M.S., Gabrielsen, M., Micossi, E. et al., Tryporedoxins from *Crithidia fasciculata* and *Trypanosoma brucei*: Photoreduction of the redox disulfide using synchrotron radiation and evidence for a conformational switch implicated in function, *J. Biol. Chem.* 278, 25919–25925, 2003.
670. Ravelli, R.B.G. and McSweeney, S.M., The 'fingerprint' that X-rays can leave on structures, *Structure* 8, 315–328, 2000.
671. Leach, S.J., Meschers, A., and Swanepoel, O.A., The electrolytic reduction of proteins, *Biochemistry* 4, 23–27, 1965.
672. Lee, D.-Y. and Chang, G.-D., Electrolytic reduction: Modification of proteins occurring in isoelectric focusing electrophoresis and in electrolytic reactions in the presence of high salts, *Anal. Chem.* 81, 3957–3964, 2009.
673. Li, J., Dewald, H.D., and Chen, H., Online coupling of electrochemical reactions with liquid sample desorption electrospray ionization-mass spectrometry, *Anal. Chem.* 81, 9716–9722, 2009.
674. Huang, H., Kwok, K.-C., and Liang, H.-H., Inhibitory activity and conformation changes of soybean trypsin inhibitors induced by ultrasound, *Ultrasound Sonochem.* 15, 724–730, 2008.
675. Foti, M.C. and Amorati, R., Non-phenolic radical-trapping antioxidants, *J. Pharm. Pharmacol.* 61, 1435–1448, 2009.
676. Elliott, S.P., Proteolytic enzymes produced by group A Streptococci with special reference to its effect on the type-specific M antigen, *J. Exp. Med.* 81, 573–592, 1945.
677. Caldwell, M.J. and Seegers, W.H., Inhibition of prothrombin, thrombin, and autoprothrombin C with enzyme inhibitors, *Thromb. Diath. Haemorrh.* 13, 373–386, 1965.
678. Iznaga Escobar, N., Morales, A., and Nuñez, G., A computer program for quantification of SH groups generated after reduction of monoclonal antibodies, *Nucl. Med. Biol.* 23, 635–639, 1996.
679. Grob, D., Proteolytic enzymes: I. The control of their activity, *J. Gen. Physiol.* 29, 219–247, 1946.
680. Giustarini, D., Dalle-Donne, I., Colombo, R. et al., Is ascorbate able to reduce disulfide bridges? A cautionary note, *Nitric Oxide* 19, 252–258, 2008.
681. Yeh, Y.Y. and Liu, L., Cholesterol-lowering effect of garlic extracts and organosulfur compounds: Human and animal studies, *J. Nutr.* 131(3s), 989S–993S, 2001.

682. Powolny, A.A. and Singh, S.V., Multitargeted prevention and therapy of cancer by diallyl trisulfide and related *Allium* vegetable-derived organosulfur compounds, *Cancer Lett.* 269, 305–314, 2008.
683. Seki, T., Hosono, T., Hosono-Fukao, T. et al., Anticancer effects of diallyl trisulfide derived from garlic, *Asia Pac. J. Clin. Nutr.* 17(suppl 1), 249–252, 2008.
684. Wang, H.C., Yang, J.H., Hsieh, S.C., and Sheen, L.Y., Allyl sulfides inhibit cell growth of skin cancer cells through induction of DNA damage mediated G2/M arrest and apoptosis, *J. Agric. Food Chem.* 58, 7096–7103, 2010.
685. Fletcher, J.C. and Robson, A., The occurrence of bis-(2-amino-2-carboxyethyl)trisulphide in hydrolysates of wool and other proteins, *Biochem. J.* 87, 553–557, 1963.
686. Breton, J., Avanzi, N., Valsasina, B. et al., Detection of traces of a trisulphide derivative in the preparation of a recombinant truncated interleukin-6 mutein, *J. Chromatogr. A* 709, 135–146, 1995.
687. Pristasky, P., Cohen, S.L., Krantz, D. et al., Evidence for trisulfide bonds in a recombinant variant of a human IgG2 monoclonal antibody, *Anal. Chem.* 81, 6148–6155, 2009.
688. Jensen, J.L., Kolvenbach, C., Roy, S. et al., Metal-catalyzed oxidation of brain-derived neurotrophic factor (BDNF): Analytical challenge for the identification of modified sites, *Pharm. Res.* 17, 190–196, 2000.
689. Duenas, E.T., Keck, R., De Vos, A. et al., Comparison between light induced and chemically induced oxidation of rhVEGF, *Pharm. Res.* 18, 1455–1460, 2001.
690. Shapiro, R.I., Wen, D., Levesque, M. et al., Expression of sonic hedgehog-Fc fusion protein in *Pichia pastoris*. Identification and control of post-translational, chemical, and proteolytic modifications, *Protein Express. Purif.* 29, 272–283, 2003.
691. Pan, B., Abel, J., Ricci, M.S. et al., Comparative oxidation studies of methionine residues reflect a structural effect on chemical kinetics, *Biochemistry* 45, 15430–15443, 2006.
692. Liu, H., Gaza-Bulseco, G., Xiang, T., and Chumsae, C., Structural effect of deglycosylation and methionine oxidation on a recombinant monoclonal antibody, *Mol. Immunol.* 45, 701–708, 2008.
693. Pipes, G.D., Campbell, P., Bondarenko, P.V. et al., Middle-down fragmentation for the identification and quantitation of site-specific methionine oxidation in an IgG1 molecule, *J. Pharm. Sci.* 99, 4469–4476, 2010.
694. Time, V., Gruber, P., Wasilu., M. et al., Identification and characterization of oxidation and deamidation sites in monoclonal rat/mouse hybrid antibodies, *J. Chromatogr. B* 878, 777–784, 2010.
695. Schneiderheinze, J., Walden, Z., Dufield, R., and Demarest, C., Rapid online proteolytic mapping of PEGylated rhGH for identity confirmation, quantitation of methionine oxidation and quantitation of UnPEGylated N-terminus using HPLC with UV detection, *J. Chromatogr. B* 877, 4065–4070, 2009.
696. Ohkubo, T., Inagaki, S., Min, J.Z. et al., Rapid determination of oxidized methionine residues in recombinant human basic fibroblast growth factor by ultra-performance liquid chromatography and electrospray ionization quadrupole time-of-flight mass spectrometry with in-source collision-induced dissociation, *Rapid Commun. Mass Spectrom.* 23, 2053–2060, 2009.
697. Jiang, H., Wu, S.L., Karger, B.L., and Hancock, W.S., Mass spectrometric analysis of innovator, counterfeit, and follow-on recombinant human growth hormone, *Biotechnol. Prog.* 25, 207–218, 2009.
698. Ren, D., Ratnaswarmy, G., Beierle, J., Treuheit, M.J. et al., Degradation products analysis of an Fc fusion protein using LC/MS methods, *Int. J. Biol. Macromol.* 44, 81–85, 2009.
699. Silva, M.M., Lamarre, B., Cerasoli, E. et al., Physicochemical and biological assays for quality control of biopharmaceuticals: Interferon α-2 case study, *Biologicals* 36, 383–392, 2008.

700. Zamani, L., Andersson, F.O., Edebrink, P. et al., Conformational studies of a monoclonal antibody, IgG1, by chemical oxidation: Structural analysis by ultrahigh-pressure LC-electrospray ionization time-of-flight MS and multivariate data analysis, *Anal. Biochem.* 380, 155–163, 2008.
701. Tien, M., Berlett, B.S., Levine, R.L. et al., Peroxynitrite-mediated modification of protein at physiological carbon dioxide concentration: pH dependence of carbonyl formation, tyrosine nitration, and methionine oxidation, *Proc. Natl. Acad. Sci. USA* 96, 7809–7814, 1999.
702. Hawkins, C.L. and Davies, M.J., Hypochloride-induced oxidation of proteins in plasma: Formation of chloramines and nitrogen centered radicals and their role in protein fragmentation, *Biochem. J.* 340, 539–545, 1999.
703. Davies, M.J., Singlet oxygen-mediated damage to proteins and its consequences, *Biochem. Biophys. Res. Commun.* 305, 761–770, 2003.
704. Imlay, J.A., Pathways of oxidative damage, *Annu. Rev. Microbiol.* 57, 395–418, 2003.
705. Droge, W., Oxidative stress and aging, *Adv. Exp. Med. Biol.* 543, 191–200, 2003.
706. Vogt, W., Oxidation of methionyl residues in proteins: Tools, targets, and reversal, *Free Radical Bio. Med.* 18, 93–105, 1995.
707. Neumann, N.P., Oxidation with hydrogen peroxide, *Methods Enzymol.* 25, 393–400, 1972.
708. Fliss, H., Weissbach, H., and Brot, N., Oxidation of methionine residues in proteins of activated human neutrophils, *Proc. Natl. Acad. Sci. USA* 90, 7160–7164, 1983.
709. Yamasaki, R.B., Osuga, D.T., and Feeney, R.E., Periodate oxidation of methionine in proteins, *Anal. Biochem.* 126, 183–189, 1982.
710. Houghten, R.A. and Li, C.H., Reduction of sulfoxides in peptides and proteins, *Anal. Biochem.* 98, 36–46, 1979.
711. Houghten, R.A. and Li, C.H., Reduction of sulfoxides in peptides and proteins, *Methods Enzymol.* 91, 549–559, 1983.
712. Brot, N., Weissbach, L., Werth, J., and Weissbach, H., Enzymatic reduction of protein-bound methionine sulfoxide, *Proc. Natl. Acad. Sci. USA* 78, 2155–2158, 1981.
713. Brot, N. and Weissbach, H., Peptide methionine reductase: Biochemistry and physiological role, *Biopolymers* 55, 288–296, 2000.
714. Hoshi, T. and Heinemann, S., Regulation of cell function by methionine oxidation and reduction, *J. Physiol.* 531, 1–11, 2001.
715. Weissbach, H., Etienne, F., Hoshi, T. et al., Peptide methionine sulfoxide reductase: Structure, mechanism of action, and biological function, *Arch. Biochem. Biophys.* 397, 172–178, 2002.
716. Stadtman, E.R., Moskovitz, J., Berlett, B.S., and Levine, R.L., Cyclic oxidation and reduction of protein methionine residues is an important antioxidant mechanism, *Mol. Cell. Biochem.* 234–235, 3–9, 2002.
717. Antoine, M., Boschi-Muller, S., and Branlant, G., Kinetic characterization of the chemical steps involved in the catalytic mechanism of methionine sulfoxide reductase A from *Neisseria menignitidis*, *J. Biol. Chem.* 278, 45352–45357, 2003.
718. Keutmann, H.T. and Potts, J.T., Jr., Improved recovery of methionine after acid hydrolysis using mercaptoethanol, *Anal. Biochem.* 29, 175–183, 1969.
719. Shechter, Y., Burstein, Y., and Patchornik, A., Selective oxidation of methionine residues in proteins, *Biochemistry* 14, 4497–4503, 1975.
720. Joergensen, L. and Thestrup, H.N., Determination of amino acids in biomass and protein samples by microwave hydrolysis and ion-exchange chromatography, *J. Chromatogr. A* 706, 412–428, 1995.
721. Hayashi, R. and Suzuki, F., Determination of methionine sulfoxide in protein and food by hydrolysis with *p*-toluenesulfonic acid, *Anal. Biochem.* 149, 521–528, 1985.

722. Weiss, M., Manneberg, M., Juranville, J.-F. et al., Effect of hydrolysis of method on the determination of the amino acid composition of proteins, *J. Chromatogr. A* 795, 263–275, 1998.
723. Sochaski, M.A., Jenkins, A.J., Lyons, T.J. et al., Isotope dilution gas chromatography/mass spectrometry method for the determination of methionine sulfoxide in protein, *Anal. Chem.* 73, 4662–4667, 2001.
724. Todd, J.M., Marable, N.L., and Kehrberg, N.L., Methionine sulfoxide determination after alkaline hydrolysis of amino acid mixtures, model protein systems, soy products, and infant formulas, *J. Food Sci.* 49, 1547–1551, 1984.
725. Molnar-Perl, I., HPLC of amino acids as phenylthiocarbamoyl derivatives, *J. Chromatogr. Library* 70, 137–162, 2005.
726. Fujii, K., Yahashi, Y., Nakano, T. et al., Simultaneous detection and determination of the absolute configuration of thiazole-containing amino acids in a peptide, *Tetrahedron* 58, 6873–6879, 2002.
727. Wang, X.S., Shao, B., Oda, M.N. et al., A sensitive and specific ELISA detects methionine sulfoxide-containing apolipoprotein A-1 in HDL, *J. Lipid Res.* 50, 586–594, 2009.
728. Oien, D.B., Canello, T., Gabizon, R. et al., Detection of oxidized methionine in selected proteins, cellular extracts and blood serums by novel anti-methionine sulfoxide antibodies, *Arch. Biochem. Biophys.* 485, 35–40, 2009.
729. Schöneich, C., Methionine oxidation by reactive oxygen species: Reaction mechanisms and relevance to Alzheimer's disease, *Biochim. Biophys. Acta* 1703, 111–119, 2005.
730. Hirs, C.H.W., Performic acid oxidation, *Methods Enzymol.* 11, 197–199, 1967.
731. Butz, L. and du Vigneaud, V., The formation of a homolog of cystine by decomposition of methionine with sulfuric acid, *J. Biol. Chem.* 99, 135–142, 1932.
732. Kopoldova, J., Kolvesek, J., Babicky, A., and Liebster, J., Degradation of DL-methionine by radiation, *Nature* 182, 1074–1076, 1958.
733. Floyd, N.F., Cammaroti, M.S., and Lavine, T.F., The decomposition of DL-methionine in 6N hydrochloric acid, *Arch. Biochem. Biophys.* 102, 343–345, 1963.
734. Hayward, M.A., Campbell, E.B., and Griffith, O.W., Sulfonic acids: L-homocysteinesulfonic acid, *Methods Enzymol.* 143, 279–281, 1987.
735. Xu, G. and Chance, M.R., Radiolytic modification of acidic amino acid residues in peptides: Probes for examining protein-protein interaction, *Anal. Chem.* 76, 1213–1221, 2004.
736. Chance, M.R., Unfolding of apomyglobin examined by synchrotron footprinting, *Biochem. Biophys. Res. Commun.* 287, 614–621, 2001.
737. Sharp, J.S., Becker, J.M., and Hettich, R.L., Analysis of protein surface accessible residues by photochemical oxidation and mass spectrometry, *Anal. Chem.* 76, 672–683, 2004.
738. Bern, M., Saladino, J., and Sharp, J.S., Conversion of methionine into homocysteic acid in heavily oxidized proteomics samples, *Rapid Commun. Mass Spectrom.* 24, 768–772, 2010.
739. Corless, S. and Cramer, R., On-target oxidation of methionine residues using hydrogen peroxide for composition-restricted matrix-assisted laser desorption/ionization peptide mass-mapping, *Rapid Commun. Mass Spectrom.* 17, 1212–1215, 2003.
740. Pan, Y., Brown, L., and Konermann, L., Site-directed mutagenesis combined with oxidative methionine labeling for probing structural transitions of a membrane protein by mass spectrometry, *J. Am. Soc. Mass Spectrom.* 21, 1947–1956, 2010.
741. Jenkins, N., Modifications of therapeutic proteins: Challenges and prospects, *Cytotechnology* 53, 121–125, 2007.
742. Cindrić, M., Galić, N., Vuletić, M. et al., Evaluation of recombinant human interferon α-2b structure and stability by in-gel tryptic digestion, H/D exchange and mass spectrometry, *J. Pharm. Biomed. Anal.* 40, 781–787, 2006.

743. Thirumangalathu, R., Krishnan, S., Bondarenko, P. et al., Oxidation of methionine residues in recombinant human interleukin-1 receptor antagonist: Implications of conformational stability on protein oxidation kinetics, *Biochemistry* 46, 6213–6224, 2007.
744. Hu, D., Qin, Z., Zue, B. et al., Effects of methionine oxidation on the structural properties, conformational stability, and aggregation of immunoglobulin light chain LEN, *Biochemistry* 47, 8665–8677, 2008.
745. Burkitt, W., Domann, P., and O'Connor, G., Conformational changes in oxidatively stressed monoclonal antibodies studied by hydrogen exchange mass spectrometry, *Protein Sci.* 19, 826–835, 2010.
746. Labrenz, S.R., Calmann, M.A., Heavner, G.A., and Tolman, G., The oxidation of methionine-54 of epotinum α does not affect molecular structure or stability, but does decrease activity, *PDA J. Pharm. Sci.* 62, 211–223, 2008.
747. Mulinacci, F., Bell, S.E., Capelle, M.A. et al., Oxidized recombinant human growth hormone that maintains conformational integrity, *J. Pharm. Sci.* 100, 110–122, 2010.
748. Kim, Y.H., Berry, A.H., Spencer, D.S., and Stiles, W.E., Comparing the effect on protein stability of methionine oxidation versus mutagenesis: Steps toward engineering oxidation resistance in proteins, *Protein Eng.* 14, 343–347, 2001.
749. Chien, H.-C. et al., Enhancing oxidative resistance of *Agrobacterium radiobacter* N-carbamoyl D-amino acid aminohydrolase by engineering solvent-accessible methionine residues, *Biochem. Biophys. Res. Commun.* 297, 282–287, 2002.
750. Caldwell, P., Luk, D.C., Weissbach, H., and Brot, N., Oxidation of the methionine residues of *Escherichia coli* ribosomal protein L12 decreases the protein's biological activity, *Proc. Natl. Acad. Sci. USA* 75, 5349–5352, 1978.
751. Kachurin, A.M., Golubev, A.M., Geisow, M.M. et al., Role of methionine in the active site of α-galactosidase from *Trichoderma reesei*, *Biochem. J.* 308, 955–964, 1995.
752. Kornfelt, T., Persson, E., and Palm, L., Oxidation of methionine residues in coagulation factor VIIa, *Arch. Biochem. Biophys.* 363, 43–54, 1999.
753. Nomura, T., Kamada, R., Ito, I. et al., Oxidation of methionine residue at hydrophobic core destabilizes p53 tetramer, *Biopolymers* 91, 78–84, 2009.
754. Lee, C.J., Quality control of polyvalent pneumococcal polysaccharide-protein conjugate vaccine by nephelometry, *Biologicals* 30, 97–103, 2002.
755. Junowicz, E. and Charm, S.E., The derivatization of oxidized polysaccharides for protein immobilization and affinity chromatography, *Biochim. Biophys. Acta* 428, 157–165, 1976.
756. Hage, D.S., Wolfe, C.A., and Oates, M.R., Development of a kinetic model to describe the effective rate of antibody oxidation by periodate, *Bioconj. Chem.* 8, 914–920, 1997.
757. Riebe, D. and Thorn, W., Influence of carbohydrate moieties of human serum transferrin on the determination of its molecular mass by polyacrylamide gradient gel electrophoresis and staining with periodic acid-Schiff reagent, *Electrophoresis* 12, 287–293, 1991.
758. Knowles, J.R., The role of methionine in α-chymotrypsin-catalyzed reactions, *Biochem. J.* 95, 180–190, 1965.
759. de la Llosa, P., El Abed, A., and Roy, M., Oxidation of methionine residues in lutropin, *Can. J. Biochem.* 58, 745–748, 1980.
760. Gleisner, J.M. and Liener, I.E., Chemical modification of the histidine residue located at the active site of ficin, *Biochim. Biophys. Acta* 317, 482–491, 1973.
761. Silva, C.J., Onisko, B.C., Dynin, I. et al., Assessing the role of oxidized methionine at position 213 in the formation of prions in hamsters, *Biochemistry* 9, 1854–1861, 2010.
762. Wolschner, C., Giese, A., Kretzchmar, H.A. et al., Design of anti- and pro-aggregation variants to assess the effects of methionine oxidation in human prion protein, *Proc. Natl. Acad. Sci. USA* 106, 7756–7761, 2009.
763. Trout, G.E., The estimation of microgram amounts of methionine by reaction with chloroamine-T, *Anal. Biochem.* 93, 419–422, 1979.

764. Tomova, S., Cutruzzolá, F., Barra, D. et al., Selective oxidation of methionyl residues in the recombinant human secretory leukocyte proteinase inhibitor. Effect on inhibitor binding properties, *J. Mol. Recogn.* 7, 31–37, 1994.
765. Tang, X.D., Daggett, H., Hanner, M. et al., Oxidative regulation of large conductance calcium-activated potassium channels, *J. Gen. Physiol.* 117, 253–274, 2001.
766. Daile-Donne, I., Rossi, R., Giustarini, D. et al., Methionine oxidation as a major cause of the functional impairment of oxidized actin, *Free Radical Bio. Med.* 32, 927–937, 2002.
767. Nieziolek, M., Kot, M., Pyka, K. et al., Properties of chemically oxidized kininogens, *Acta Biochim. Pol.* 50, 753–763, 2003.
768. Bauer, R.J., Leigh, S.D., Birr, C.A. et al., Alteration of the pharmacokinetics of small proteins by iodination, *Biopharm. Drug Dispos.* 17, 761–774, 1996.
769. Kumar, C.C., Nie, H., Armstrong, L. et al., Chloramine T-induced structural and biochemical changes in echistatin, *FEBS Lett.* 429, 239–248, 1998.
770. Cadée, J.A., van Steenbergen, M.J., Versluis, C. et al., Oxidation of recombinant human interleukin-2 by potassium peroxodisulfate, *Pharm. Res.* 18, 1461–1467, 2001.
771. Keck, R.G., The use of t-butyl hydroperoxide as a probe for methionine oxidation in proteins, *Anal. Biochem.* 236, 56–62, 1996.
772. Liu, J.L., Lu, K.V., Eris, T. et al., *In vitro* methionine oxidation of recombinant human leptin, *Pharm. Res.* 15, 632–640, 1998.
773. Lu, H.S., Fausset, P.R., Narhi, L.O. et al., Chemical modification and site-directed mutagenesis of methionine residues in recombinant human granulocyte colony-stimulating factor: Effect on stability and biological activity, *Arch. Biochem. Biophys.* 362, 1–11, 1999.
774. Chumsae, C., Gaza-Bulseco, G., Sun, J., and Liu, H., Comparison of methionine oxidation in thermal stability and chemically stressed samples of a fully human monoclonal antibody, *J. Chromatogr. B* 850, 285–294, 2007.
775. Gaza-Bulseco, G., Faldu, S., Hurkmans, K. et al., Effect of methionine oxidation of a recombinant monoclonal antibody on the binding affinity to protein A and protein G, *J. Chromatogr. B* 870, 55–62, 2008.
776. Liu, H., Gaza-Bulseco, G., and Zhou, L., Mass spectrometry analysis of photo-induced methionine oxidation of a recombinant human monoclonal antibody, *J. Am. Soc. Mass Spectrom.* 20, 525–528, 2009.
777. Fan, H., Chen, K., Chu, L. et al., Methionine oxidation in human IgG2 Fc decreases binding affinities to protein A and FcRn, *Protein Sci.* 18, 424–433, 2009.
778. Gundlach, H.G., Moore, S., and Stein, W.H., The reaction of iodoacetate with methionine, *J. Biol. Chem.* 234, 1761–1764, 1959.
779. Cummings, J.G., Lau, S.-M., Powell, P.J., and Thorpe, C., Reductive half-reaction in medium-chain acyl-CoA dehydrogenase: Modulation of internal equilibrium by carboxymethylation of a specific methionine residue, *Biochemistry* 31, 8523–8529, 1992.
780. Schroeder, W.A., Shelton, J.R., and Robberson, B., Modification of methionyl residues during aminoethylation, *Biochim. Biophys. Acta* 147, 590–592, 1967.
781. Marks, R.H.L. and Miller, R.D., Chemical modification of methionine residues in azurin, *Biochem. Biophys. Res. Commun.* 88, 661–667, 1979.
782. Mizzer, J.P. and Thorpe, C., An essential methionine in pig kidney general acyl-CoA dehydrogenase, *Biochemistry* 19, 5500–5504, 1980.
783. Goren, H.J., Glick, D.M., and Barnard, E.A., Analysis of carboxymethylated residues in proteins by an isotopic method and its application to the bromoacetate-ribonuclease reaction, *Arch. Biochem. Biophys.* 126, 607–623, 1968.
784. Naider, F. and Bohak, Z., Regeneration of methionyl residues from their sulfonium salts in peptides and proteins, *Biochemistry* 11, 3208–3211, 1972.
785. Schramm, H.J. and Lawson, W.B., Über das activ Zentrum von Chymotrypsin. II. Modifizierung eines Methioninrestes in Chymotrypsin durch einfache Benzolderivate, *Hoppe-Seyler's Z. Physiol. Chem.* 332, 97–100, 1963.

786. Goverman, J.M. and Pierce, J.G., Differential effects of alkylation of methionine residues on the activities of pituitary thyrotropin and lutropin, *J. Biol. Chem.* 256, 9431–9435, 1981.
787. Kleanthous, C., Campbell, D.G., and Coggins, J.R., Active site labeling of the shikimate pathway enzyme, dehydroquinase. Evidence for a common substrate binding site within dehydroquinase and dehydroquinate synthase, *J. Biol. Chem.* 265, 10929–10934, 1990.
788. Kleanthous, C. and Coggins, J.R., Reversible alkylation of an active site methionine residue in dehydroquinase, *J. Biol. Chem.* 265, 10935–10939, 1990.
789. Weinberger, S.R., Viner, R.J., and Ho, P., Tagless extraction-retentate chromatography: A new global protein digestion strategy for monitoring differential protein expression, *Electrophoresis* 23, 3182–3192, 2002.
790. Grunert, T., Pock, K., Buchacher, A., and Allmaier, G., Selective solid-phase isolation of methionine-containing peptides and subsequent matrix-assisted laser desorption mass spectrometric detection of methionine- and methionine-sulfoxide-containing tryptic peptides, *Rapid Commun. Mass Spectrom.* 17, 1815–1824, 2003.

6 Chemical Modification of Nucleic Acids

This chapter is directed toward the chemistry of nucleic acids (oligonucleotides and polynucleotides); the chemistry of component nucleobases is considered only as necessary to understand mechanism and structure. The biology of the nucleic acids is, for the most part, ignored except where as, for example, with "footprinting" or mutagenesis, an understanding of the biology is useful for both the understanding of experimental design and interpretation of results. The chemistry for the various modification reactions is based on factors similar to those considered in the modification of protein functional groups—nucleophilicity and local electrostatic effects. The reader is directed to Chapter 1 for a discussion of factors governing the reactivity of functional groups and a listing of pK_a values for nucleic acid functional groups.

Early work on nucleic acids is reviewed by Davidson and Chargaff[1] and Hotchkiss[2] a decade or so before the double helix.[3] It is clear in retrospect[4] that Avery, MacLeod, and McCarty[5] defined the role of DNA as genetic material well before the work of Watson and Crick. The term nucleic acid was used by R. Altmann in 1889[8] to describe material obtained from salmon sperm and calf thymus. Clear recognition that there are two types of nucleic acids dates to the work of P.A. Levine and others more than 80 years ago.[6] This work demonstrated that sugar content of thymonucleic acid was deoxyribose (deoxypentose) with adenine, guanine, cytosine, and thymine as the base components (Figure 6.1). Hammarsten and others[7–9] had previously defined ribonucleic acid as containing pentose (ribose) with adenine, guanine, cytosine, and uracil as base components (Figure 6.1). RNA was also described as PNA (pentose nucleic acid) in early work.[10] In similar manner, DNA is described as deoxypentose nucleic acid.[11] Early work also suggested that DNA was a product of animal tissue, while RNA was derived from plant tissue. These early impressions stem from the finding of DNA in animal tissue such as thymus and the discovery of plant tissue that contained RNA. As described by Jones,[9] animal nucleic acid was composed of adenine, guanine, cytosine, thymine, phosphoric acid, and levulinic acid, while plant nucleic acid contained adenine, guanine, cytosine, uracil, phosphoric acid, and ribose. Our understanding of the complexity of nucleic acids has increased with the sophistication of our analytical technologies[12] and elucidation of specialized function.[13–16] Nucleic modifications such as splicing and methylation,[17,18] while of great important for biological function, are not included except as when such mention relates to the chemical modification of nucleic acids. These various reactions are critical for the regulation of nucleic acid function and do increase structural complexity[19–22] and are part of the discipline known as epigenetics.[23–25] Epigenetics is discussed again in the following with respect to the use of sodium bisulfite to identify sites of cytosine methylation.

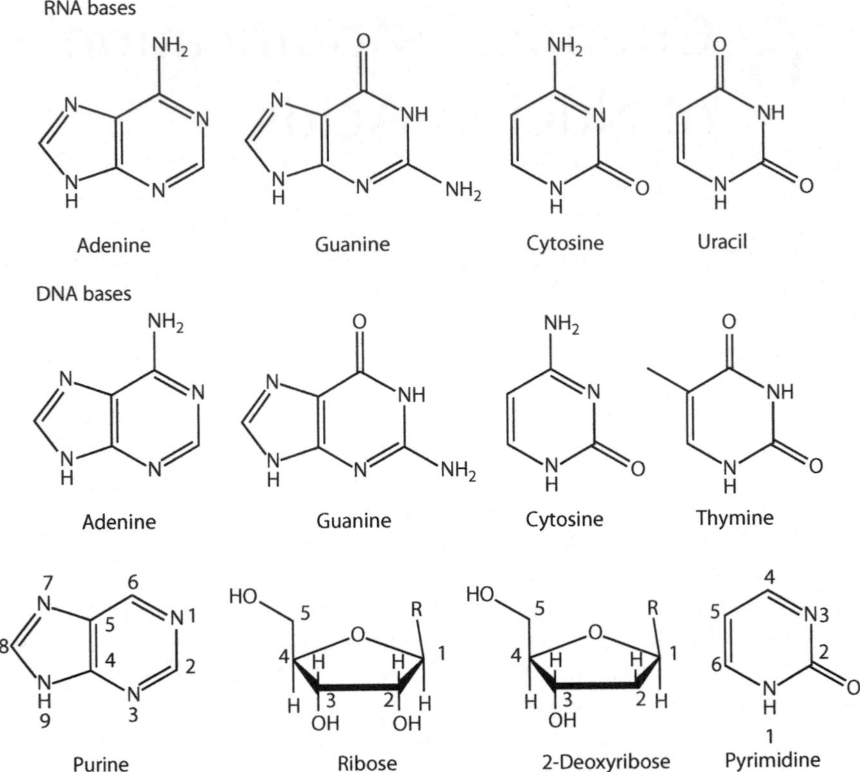

FIGURE 6.1 The structures of nucleic acid bases. Shown are the structures of adenine, guanine, cytosine, and uracil as component bases of RNA and adenine, guanine, cytosine, and thymine as component bases of DNA. Also shown are ribose (RNA) and 2-deoxyribose (DNA) as well as the ring numbering system for purine, pyrimidine, and ribose.

Some of our information on chemical modification of nucleic acids comes from early studies to characterize these biological polymers in tissues. Much of the early work on nucleic acids was based on histochemistry using color reactions and staining to identify nucleic acid components.[26,27] The Feulgen reaction continues to be used for the histochemical analysis of DNA.[28] The Feulgen reaction that is based on the conversion of the sugar backbone to aldehyde does not detect RNA. It has been suggested that the presence of the 2-hydroxyl function provides the difference in the response to this reaction. The methyl green–pyronine stain that is based on charge detects both DNA and RNA and is suggested to be the most benign for subsequent analysis by polymerase chain reaction (PCR).[29] Chemical color reactions are related to histochemistry and were of great value to early biochemists in the characterization of biological tissues and fluids including the study of nucleic acid components.[30] One such reaction is the diazotization of the exocyclic amino group of adenine and subsequent coupling to *N*-(1-naphthyl)ethylene diamine as developed by Woodhouse[31] and used to define adenine distribution in DNA in 1963.[31] In the work on the distribution of adenine, Jones and Walker[32] used the observation[33] that potassium permanganate at 37°C oxidized all of the bases except adenine in the DNA sample.

Other early work that contributed to current studies on the chemical modification of nucleic acid included spectroscopy and photochemical degradation.[34] The ability of nucleic acids and components to absorb ultraviolet (UV) light is used for the analysis of these substances.[35]

UV absorbance was the original technique used to measure thermal denaturation of nucleic acids,[36] although fluorescence and other technologies can be used.[37] The denaturation of DNA disrupts the helical structure resulting in a hyperchromic response at 260 nm.[38] Many of the chemical modifications result in a decrease in absorbance (hypochromic response) as a result of the loss of the aromatic structure of the nucleic acid base.

Photochemical degradation of nucleic acids involves photooxidation as well as hydration and cross-linking reactions (Figure 6.2).[39–42] It is useful to separate photochemical events into at least two categories, one of which involves the excitation of bases such as uracil with UV light resulting in addition of water across a double bond in the pyrimidine ring.[43] The loss of the double bond in the pyrimidine ring eliminates the resonance within the ring with the loss of the chromophore property. The photochemical reactions of pyrimidines are discussed in more detail in the following. Purines are much less sensitive to far UV absorbance than pyrimidines.[40]

Beaven and others[39] reviewed the early work on the spectral properties of nucleic acid components. It is fair to say that technical improvements in instrumentation rather than increases in conceptual understanding are responsible for today's elegant work. When the author was in graduate school, the Beckman DU was the workhouse for spectroscopy in the laboratory while the elite had access to a Cary. The determination of a spectra was a manual process as compared to today's rapid digital output. I won't even mention using the Klett–Summerson colorimeter[44]; to my surprise, I have heard that these are still in use.

Photooxidation (Figure 6.3) is different from the far UV-induced changes in pyrimidines. Guanine is the base most sensitive to photooxidation.[45,46] The photooxidation of guanine provides a diverse group of products and depends on the nature of the oxidant as well as the position of guanine in sequence.[46–48] Xu and coworkers[49] have reported that the formation of a cross-link between guanine and lysine depends on mechanism of oxidation. DNA obtained from patients receiving azothioprine or 6-mercaptopurine as part of cancer chemotherapy has increasing sensitivity to sunlight and skin cancer. Ren and coworkers[50] showed that DNA from these patients contained 6-thioguanine (Figure 6.2) that UVA converted to guanine sulfinate, which can be converted back to 6-thioguanidine or oxidized to form the sulfonate in an irreversible reaction.

The chemical modification of nucleic acids is not as complex as that of proteins since there are fewer monomer units and less diversity of nucleophilic reactive group. Reaction at the primary amine groups of adenine is referred to as an exocyclic modification, whereas reaction at the imine nitrogens or at carbon atoms in pyrimidine or purine rings is referred to as an endocyclic modification. There are also ring-opening reactions and cross-linking reactions.

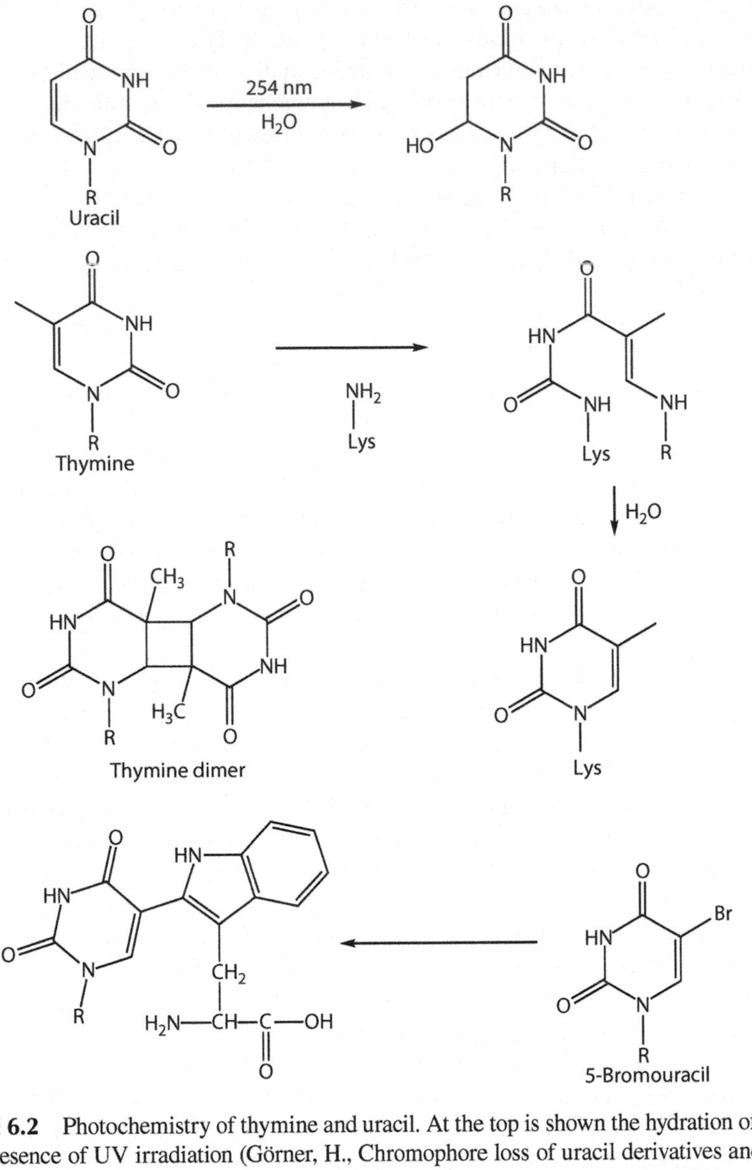

FIGURE 6.2 Photochemistry of thymine and uracil. At the top is shown the hydration of uracil in the presence of UV irradiation (Görner, H., Chromophore loss of uracil derivatives and polyuridylic acid in aqueous solution caused by 248 nm laser pulses and continuous UV irradiation: Mechanism of the photohydration of pyrimidines, *J. Photochem. Photobiol.* 10, 91–110, 1991). Also shown is the cross-link between thymidine in lysine and the photochemical reaction product of tryptophan with 5-bromouridine (Saito, I. and Matsuura, T., Chemical aspects of UV-induced cross-linking of proteins to nucleic acids. Photoreactions with lysine and tryptophan, *Acc. Chem. Res.* 18, 134–141, 1985). A mechanism for the transfer of thymine to protein is shown (Saito, I., Sugiyama, H., Ito, S. et al., A novel photoreaction of thymidine with lysine. Photoinduced migration of thymine from DNA to lysine, *J. Am. Chem. Soc.* 103, 1598–1600, 1981). Thymine also forms a dimer with irradiation at 254 nm (Beukers, R. and Berends, W., Isolation and identification of the irradiation product of thymine, *Biochim. Biophys. Acta* 41, 550–551, 1960).

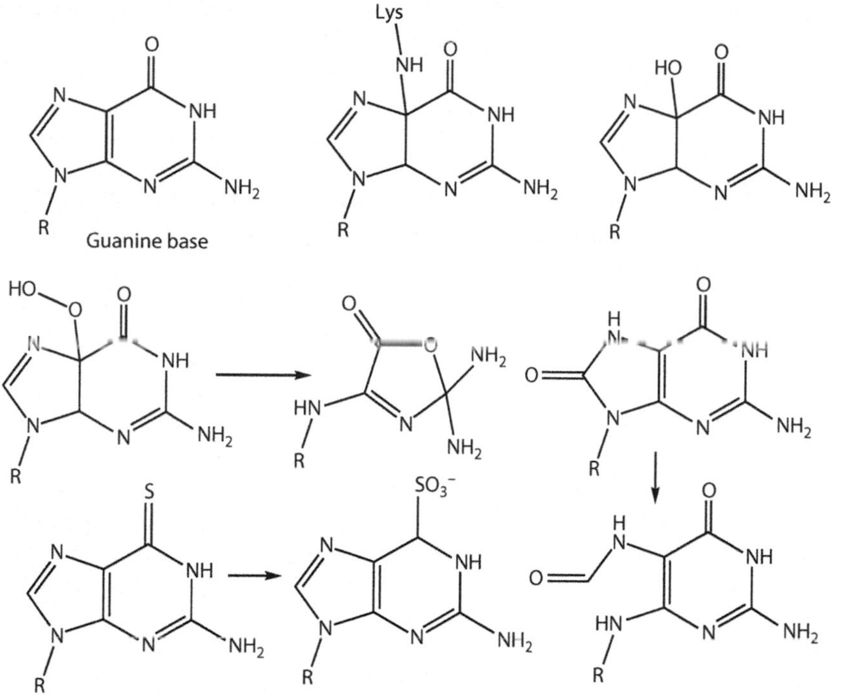

FIGURE 6.3 Photooxidation of guanine in nucleic acids. Shown is the structure of the guanine base, an addition product at C5 with lysine or water suggested for the reaction of guanine with oxygen radical (Xu, X., Muller, J.G., Ye, Y., and Burrows, C.J., DNA-protein cross-links between guanine and lysine depend on mechanism of oxidation for formation of C5 vs. C8 guanosine adducts, *J. Am. Chem. Soc.* 130, 703–709, 2008). Further degradation can occur by the addition of a peroxy group at C5 resulting in the formation of 2,5-diamino-4*H*-imidazolone derivative (Shao, J., Geacintov, N.E., and Shafirovich, V., Oxidative modification of guanine bases initiated by oxyl radicals from photolysis of azo compounds, *J. Phys. Chem.* 114, 6685–6692, 2010). Hydroxyl radicals can also yield the 8-oxo derivative; shown is the 8-oxo derivative of guanine, which can be further degraded to 2,6-diamino-5-formamido-4-hydroxypyrimidine (Margolin, Y., Shafirovich, V., Geacintov, N.E. et al., DNA sequence context as a determinant of the quality and chemistry of guanine oxidation produced by hydroxyl radicals and one-electron oxidants, *J. Biol. Chem.* 283, 35569–35578, 2008). Also shown is the structure of 6-thioguanine, which is found in the DNA of patients receiving azathioprine or 6-mercaptopurine; photochemical oxidation can yield the sulfonate derivative (Ren, X., Li, F., Jeffs, G. et al., Guanine sulphinate is a major stable product of photochemical oxidation of DNA 6-thioguanine by UVA irradiation, *Nucleic Acids Res.* 38, 1832–1840, 2010).

The term chemical modification is also used with the preparation of synthetic oligonucleotides represented by microRNAs, antisense oligonucleotides, and aptamers.[51–56] MicroRNAs are small RNA species that interfere with RNA expression (small interfering RNA, siRNA). Synthetic siRNAs have potential for therapeutic development.[57,58] Antisense oligonucleotides are 20–100 mer oligonucleotides designed to react with the sense strand of nucleic acids to block expression.[59] Antisense oligonucleotides and siRNA oligonucleotides are both "chemically modified oligonucleotides similar in concept and structure."[60,61] Fluiter and coworkers[61] discuss the importance of information gathered in the development of antisense technology to the development of therapeutic biopharmaceuticals based on siRNAs. This includes the mention of locked nucleic acid as another approach to a stable RNA therapeutic.[62] Aptamers are single-stranded DNA or RNA oligonucleotides that bind to specific proteins[63,64] and have promise as therapeutics.[65–68] In these applications, the nucleobase is modified before incorporation into the synthetic oligonucleotide; this also includes the use of chemistry other than phosphodiester linkages (Figure 6.4). Thus, when the term chemical modification is used with microRNA products, antisense oligonucleotides, or aptamers, this refers to the replacement of the phosphate diester backbone with a phosphorthioate.[69,70] It should be mentioned that peptide nucleic acids[71–73] use a peptide bond backbone (Figure 6.4).

Photochemical modification of nucleic acids is considered as chemical modification. Photooxidation of nucleic acids has been discussed previously. Photochemical reactions have proved useful in the demonstration of nucleic acid contact with proteins providing information complementary to DNA footprinting as well as information on protein–RNA interaction. Some of the proposed photochemical products of pyrimidines are shown in Figure 6.2 as is the hydration product demonstrated by Görner.[43] UV-induced cross-linking of thymine to phenylalanine has also been demonstrated by Shamoo and coworkers[41] and uracil cross-linking to protein by Banerjee and Singh.[42] Saito and Matsuura[74] reviewed the chemical cross-linking of proteins to nucleic acids; some proposed structures of products are shown in Figure 6.2. There is other evidence to support the formation of a cross-link between the thymine base and proteins via a lysine[75,76] with the subsequent release of the nucleic acid.[77] Thymine is the most photochemically sensitive base in DNA, although 5-bromouracil is about twice as sensitive (254 nm).[78] Thymine has also been shown to form a photoproduct with tyrosine[79] and forms a cyclobutane dimer on irradiation (Figure 6.3).[80–83]

Current interest in the "classical" chemical modification of nucleic acids is concerned with nucleic structure analysis including footprinting and environmental chemistry. Most of the basic chemistry underlying current work was developed some 30–50 years ago.[84–92]

The modification of guanine residue with dimethyl sulfate (Figure 6.5) and the modification of cytosine with hydrazine (Figure 6.6) are two of the several reactions developed by Allan Maxam and Walter Gilbert for determining DNA sequence.[93–95] Methylation of N7 of guanine results in a product unstable in base resulting in ring opening forming an aldehyde, which then condensing with piperidine as well as opening of the deoxyribose ring forms an additional piperidine product; this product in base decomposes with cleavage of both 5′ and 5′ phosphodiester bond (Figure 6.5). Hydrazine reacts with cytidine base in DNA, which can be taken through a cleavage cycle (Figure 6.6) analogous to that described for guanosine.

Footprinting is a method of mapping DNA–ligand interaction,[96–102] RNA structure, and RNA–protein interactions.[103–108] The technique relies on the ability of a ligand such as a protein to block a segment of nucleic acid from modification or hydrolysis by a nuclease. Protection from nuclease digestion was used in the first footprinting studies by Galas and Schmitz[96] who used DNAse to identify the region of DNA protected by protein binding; an example was provided by binding of lac

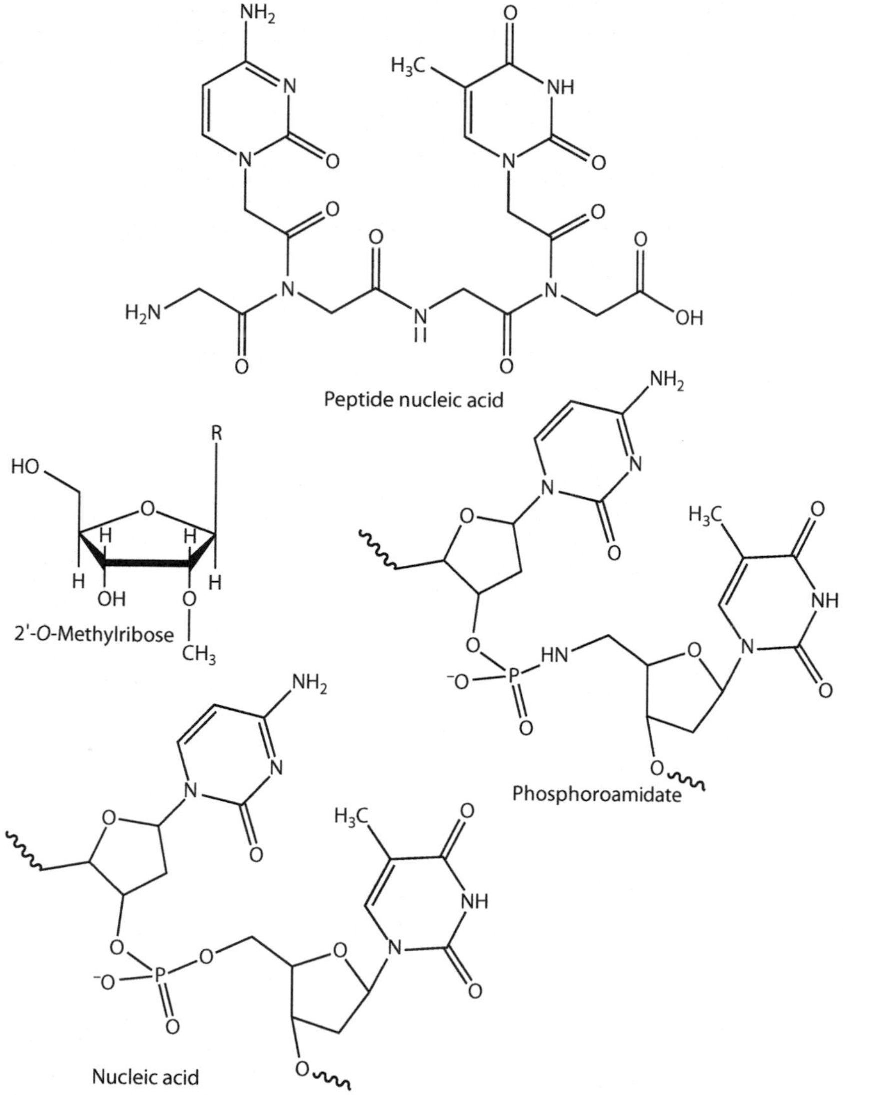

Peptide nucleic acid

2'-*O*-Methylribose

Phosphoroamidate

Nucleic acid

FIGURE 6.4 Peptide nucleic acids and phosphoramidate backbones. Shown is the structure for a peptide nucleic acid (Nielsen, P.E., Egholm, M., and Buchardt, O., Peptide nucleic acid (PNA). A DNA mimic with a peptide backbone, *Bioconjug. Chem.* 5, 3–7, 1994; Pensato, S., Saviano, M., and Romanelli, A., New peptide nucleic acid analogues: Synthesis and applications, *Expert Opin. Biol. Ther.* 7, 1219–1232, 2007; Hudson, R.H.E., Liu, Y., and Wojciechowski, F., Hydrophilic modifications in peptide nucleic acid—Synthesis and properties of PNA possessing 5-hydroxymethyluracil and 5-hydroxymethylcytosine, *Can. J. Chem.* 85, 302–312, 2007). Also shown is the structure of a phosphoramidate backbone (Letsinger, R.L. and Mungall, W.S., Phosphoramidate analogs of oligonucleotides, *J. Org. Chem.* 35, 3800–3803, 1970; Zielinska, D., Pongracz, K., and Gryaznov, S., A new approach to oligonucleotide N3′→P5′ phosphoroamidate building blocks, *Nucleos. Nucleot. Nucleic Acids* 24, 1063–1067, 2005).

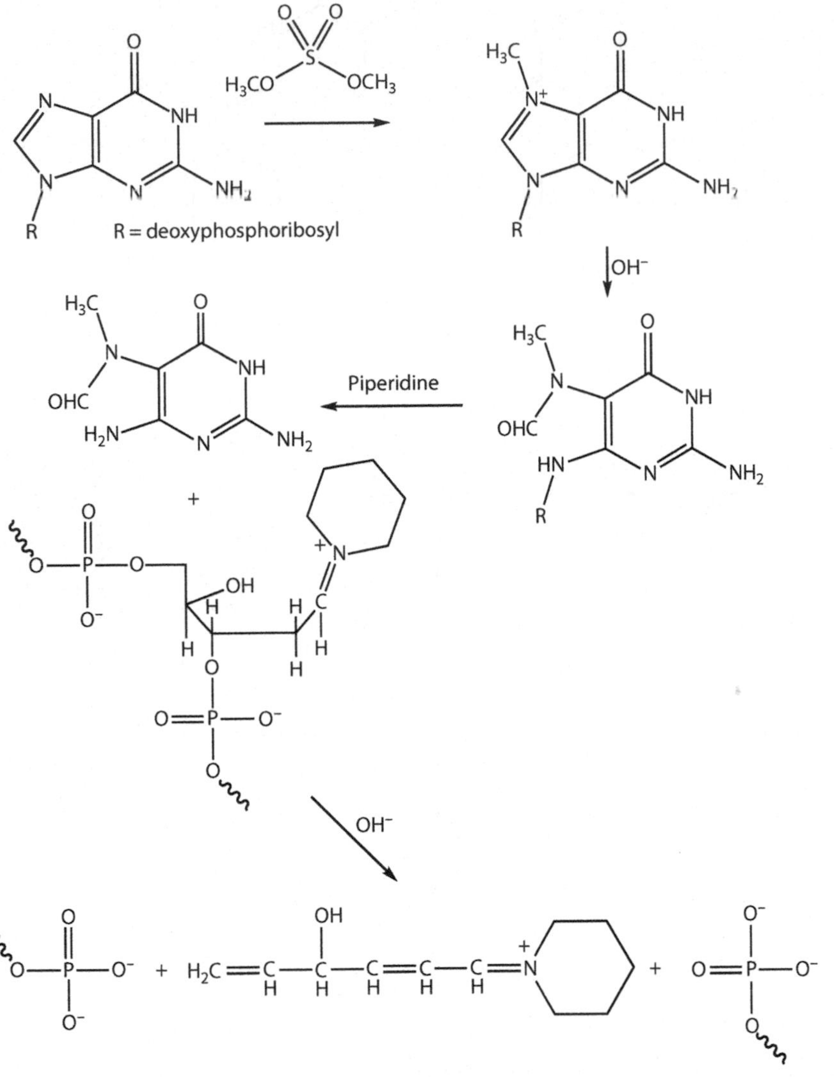

FIGURE 6.5 The reaction of dimethyl sulfate with guanine base in DNA and use in DNA sequence analysis. Shown is the reaction of dimethyl sulfate with guanine in DNA and subsequent cleavage in the presence of piperidine and base (Maxam, A.M. and Gilbert, W., A new method for sequencing DNA, *Proc. Natl. Acad. Sci. USA* 74, 560–564, 1977; Isola, N.R., Allman, S.L., Golovlov, V.V., and Chen, C.H., Chemical cleavage sequencing of DNA using matrix-assisted laser desorption/ionization time-of-flight mass spectrometry, *Anal. Chem.* 71, 2266–2269, 1999; Franca, L.T., Carrilho, E., and Kist, T.B., A review of DNA sequencing techniques, *Q. Rev. Biophys.* 35, 169–200, 2002).

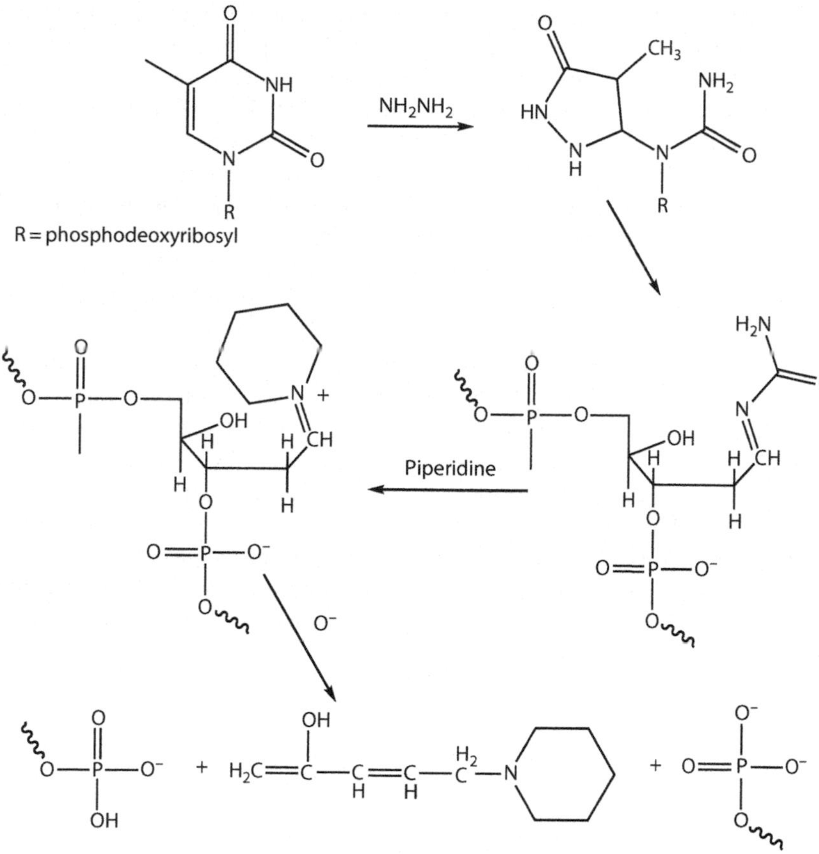

FIGURE 6.6 The reaction of hydrazine with cytidine in DNA for use in DNA sequence analysis.

repressor to the lac promoter region. Engelke and coworkers[103] used this technique to study the specific interaction between a purified transcription factor and an internal control region of 5S RNA genes. Protection from nuclease digestion can also be used for the mapping of ribosome interaction with messenger RNA.[104]

Hydroxyl radicals are useful for mapping sites in biological polymers as discussed in previous chapters. Hydroxyl radicals can be generated either by photolysis of water or in the Fenton reaction. The advantage is hydroxyl radicals react very quickly after generation; second-order rate constants of 10^9 M^{-1} s^{-1} are obtained for a number of organic chemicals in aqueous solution at 25°C.[109] Thus, hydroxyl radicals have the potential of detecting "loose" interactions of ligands with nucleic acids. The disadvantage is that hydroxyl radicals are promiscuous so methods have been developed for the local generation of hydroxyl radicals.[110–112] In these applications, Fe(II) has been "tethered" near the putative site of interaction and used to generate hydroxyl radical. Reaction of hydroxyl radicals with nucleic acids also yields a variety of products derived from base and ribose moieties as well as backbone cleavage.[113,114]

The reader is directed to the review by Tullius and Greenbaum[115] for a more through consideration of use of hydroxyl radical for mapping nucleic acid structure.

Kethoxal (3-ethoxy-1,1-dihydroxy-3-butanone; Figure 6.7) is used for the modification bases in RNA. Kethoxal was first introduced as an antiviral agent.[116–118] Other work showed that kethoxal was an effective agent for modifying guanine bases in RNA for structural analysis.[119–121] Kethoxal seems to be specific for reaction with guanine bases in RNA[122] and has been used to study the conformation of RNA in macromolecular complexes using guanine accessibility to modification.[123–127] As with the modification of proteins, the assumption in study with kethoxal is that reaction of the guanine base is an accurate measure of solvent accessibility. Mortimer and coworkers[128] point that, as with proteins, reactivity of the guanine base is a reflection of both solvent accessibility and local electrostatic environment. These investigators presented data to show that reaction with *N,N*-(dimethylamino)dimethylchlorosilane (Figure 6.7) is a more accurate measure of the accessibility of guanine base to solvent. Several approaches to RNA footprinting used several chemical modifications including kethoxal for guanine bases, hydroxyl radicals, and UV cross-linking of nucleic acid to protein.[105,129]

Formaldehyde has a long history of use in the fixation of tissue for histological study[130–135] and modifies functional groups in a variety of biological polymers. The reaction of formaldehyde with tissues was considered to be an irreversible process but its reversibility has been demonstrated by various studies on antigen recovery.[136,137]

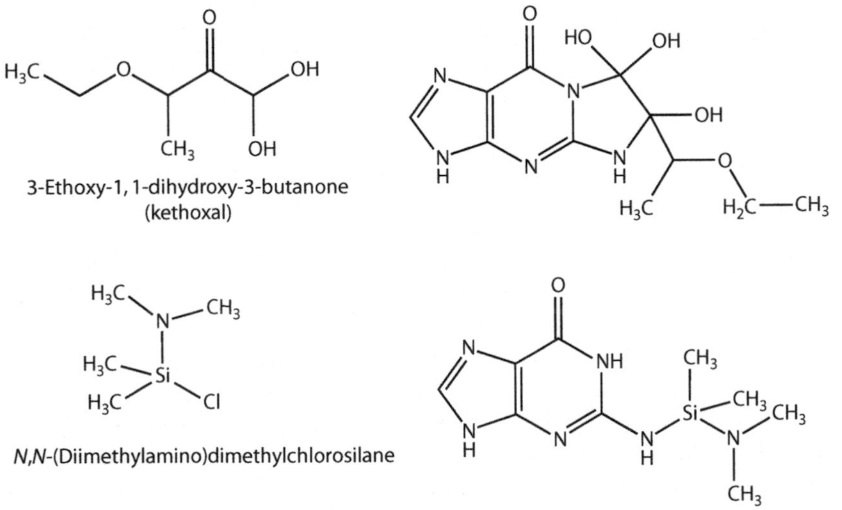

3-Ethoxy-1, 1-dihydroxy-3-butanone (kethoxal)

N,N-(Diimethylamino)dimethylchlorosilane

FIGURE 6.7 The structure and reaction of kethoxal with nucleic acid bases. Kethoxal (3-ethoxy-1, 1-dihydroxy-3-butanone; β-ethoxy-α-ketobutyraldehyde). The hydrate form is shown. Also shown is the structure of the product obtained between the guanine base and kethoxal (Shapiro, R., Cohen, B.I., Shiuey, S.J., and Maurer, H., On the reaction of guanine with glyoxal, pyruvaldehyde, and kethoxal, and the structure of the acylguanines. A new synthesis of *N*2-alkylguanines, *Biochemistry* 8, 238–245, 1969). Also shown are *N,N*-(dimethylamino)dimethylchlorosilane and the reaction product of this material with a guanine base (Mortimer, S.A., Johnson, J.S., and Weeks, K.M., Quantitative analysis of RNA solvent accessibility by *N*-silylation of guanosine, *Biochemistry* 48, 2109–2114, 2009).

The complexity of the reaction of formaldehyde with nucleophiles including cross-linking reactions is discussed in Chapter 2. Fowler and coworkers[138] showed that treatment of RNAse with 10% formaldehyde yielded a monomeric product from which antigen activity could be recovered. If, however, the protein was treated in a surrogate tissue system, a more complex group of products would be obtained. Thus, the modification of nucleic acid in tissue with formaldehyde may well yield a product qualitatively different than that obtained with purified nucleic acid or component bases, nucleosides, or nucleotides. It is also possible that the reaction of formaldehyde with a nucleic acid in tissue or in a purified nucleoprotein particle yields a product with stability characteristics different from those obtained with nucleobases or derivatives.

A variety of studies have shown that formaldehyde will react with exocyclic amines, endocyclic imines, and ring carbons in bases; bases are electron-rich heterocyclic aromatics. Alegria[139] described the formation of 5-hydroxy derivatives of cytosine and uracil (Figure 6.8) with formaldehyde; a 10% yield was obtained (5 M KOH, 75°C, 4 days); there is a suggestion that the hydroxymethyl derivative is not stable under mild basic

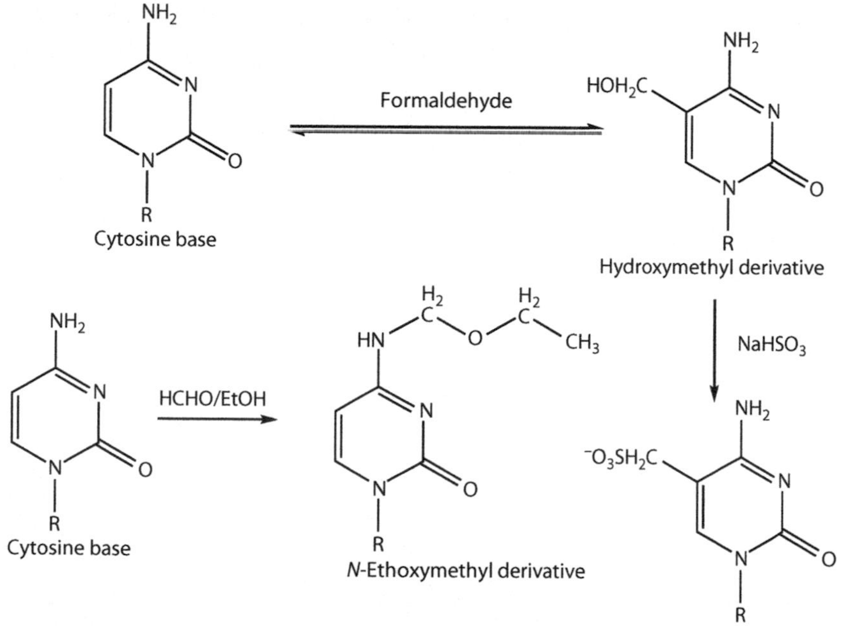

FIGURE 6.8 The reaction of formaldehyde with some pyrimidine bases. Shown is the reaction of formaldehyde with the cytidine base to yield 5-hydoxy derivative; 5-hydroxymethyluracil can also be obtained by reaction with formaldehyde (Alegria, A.H., Hydroxymethylation of pyrimidine mononucleotides with formaldehyde, *Biochim. Biophys. Acta* 149, 317–324, 1967). Reaction of 5-hydroxymethylcytidine with bisulfite does not result in deamination as it does with free cytidine or 5-methylcytidine but rather in formation of cytidine-5-methylene sulfonate (Hayatsu, H. and Shiragami, M., Reaction of bisulfite with the 5-hydroxymethyl group in pyrimidine and in phage DNAs, *Biochemistry* 18, 632–637, 1979). Also shown is the reaction of cytosine base with ethanol in the presence of formaldehyde to yield the *N*-ethoxymethyl derivative (Bridson, P.K., Jiricny, J., Kemal, O., and Reese, C.B., Reaction between ribonucleoside derivatives and formaldehyde in ethanol, *J. Chem. Soc. Chem. Commun.* 5, 208–210, 1980).

conditions. The hydroxymethyl derivative of uracil could be obtained under the same reaction conditions. It should be mentioned that 5-hydroxymethyl uridine is formed during the synthesis of the cytidine derivative reflecting deamination, which occurs under the harsh conditions of synthesis. Hudson and coworkers[140] have developed a milder and more efficient method of synthesis using triethylamine as the catalyst with paraformaldehyde. These investigators used the derivatives in peptide nucleic acids showing that 5-hydroxylmethyl group is compatible with triplex and duplex formation.

Mcghee and von Hippel reported on the reaction of formaldehyde with exocyclic amino groups of adenine, cytosine, and guanine[141] and with the endocyclic imino groups of uracil and thymine.[142] The exocyclic hydroxymethyl adduct at the N6-amino group is labile.[141,143] Yamazaki and Suzuki[144] were able to react 3-mercaptopropionic acid with adenine in the presence of formaldehyde yielding N^6-[(2-carboxyethyl)thiomethyl]-AMP (Figure 6.9), which was then condensed with 1,3-diaminopropane to yield a derivative that could be coupled to dextran to yield an affinity matrix. Bridson and coworkers[145] prepared the *N*-ethoxymethyl derivates from guanine, cytosine and adenine (Figure 6.8). Similar chemistry has been used to prepare a series of aminomethylated derivatives based on modification of the endocyclic amines of adenine, cytosine, and guanine.[146]

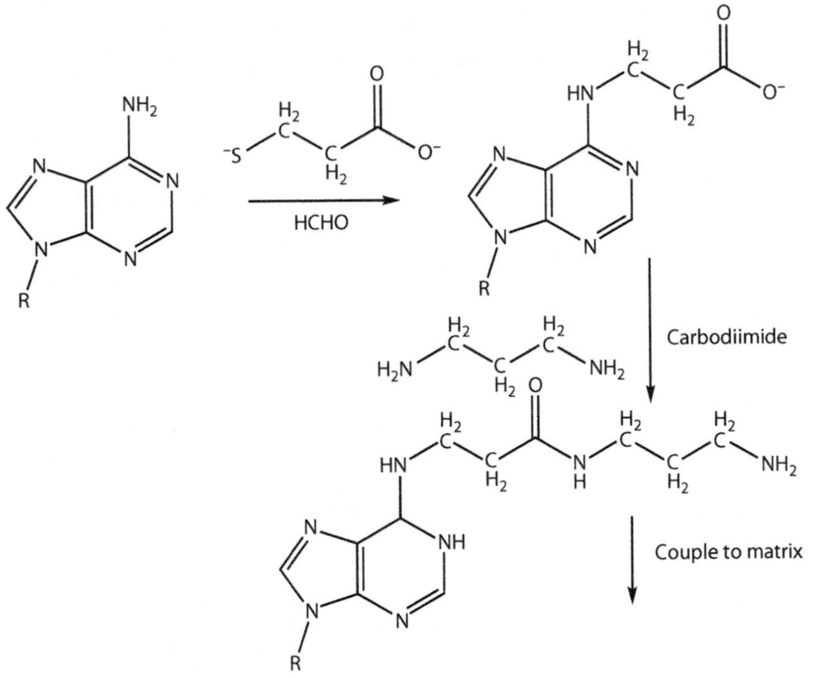

FIGURE 6.9 Modification of adenine with formaldehyde for preparation of an affinity matrix. Shown is the reaction of adenine base with 3-mercaptopropionic acid in the presence of formaldehyde to yield the propionylthiomethy derivative, which in turn is coupled to 1,3-diaminopropane to yield an amino-terminal tether for coupling to a matrix (Yamazaki, Y. and Suzuki, H., A new method of chemical modification of N6-amino group in adenine nucleotides with formaldehyde and a thiol and its application to preparing immobilized ADP and ATP, *Eur. J. Biochem.* 92, 197–207, 1978).

Notwithstanding the difficulty in the modification of nucleic acid bases, nucleosides, and nucleotides, it has been able to use formaldehyde for the modification of nucleic acids. McGhee and von Hippel[147] used their previous studies[141,142] to use formaldehyde as probe of DNA structure during denaturation and found the reaction of formaldehyde with the exocyclic amino group of adenine and the endocyclic imino group of thymine valuable in development of a mechanism. There is the related issue of cross-linking reactions that occurs with the use of formaldehyde. An example involving adenine is shown in Figure 6.10 showing the initial and reversible formation of the hydroxymethyl

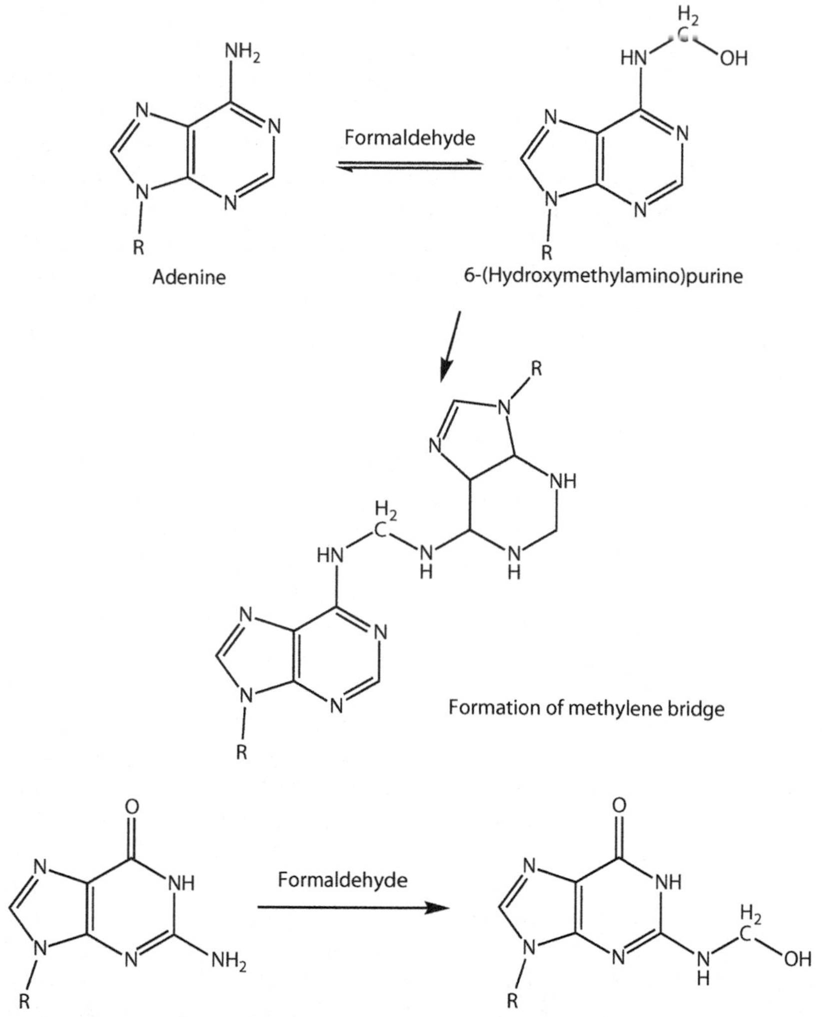

Adenine

Formaldehyde

6-(Hydroxymethylamino)purine

Formation of methylene bridge

Formaldehyde

FIGURE 6.10 The reaction of formaldehyde with pyrimidine bases. Shown is the cross-linking reaction between two adenine bases after the initial formation of a hydroxymethyl derivative (Anderson, T., Mechanism of formaldehyde-induced mutagenesis. The uniqueness of adenylic acid in the mutagenic activity of formaldehyde, *Nature* 187, 485–489, 1960).

derivative at the N6-amino group followed by formation of a methylene bridge with another adenine base as originally suggested by Anderson.[148] Grossman and coworkers[149] extended the observations of Anderson in a study on the reaction of formaldehyde with nucleotides and bacteriophage T2 (a DNA phage). These investigators observed that with the exception of guanine, there was no difference in reaction of ribonucleotides and deoxyribonucleotides. The reaction of nucleotides is pH dependent (base catalyzed). Hydroxymethylation of nucleic acid is preceded by denaturation consistent with the suggestion that hydrogen bonding prevented hydroxymethylation. Ohba and coworkers[150] observed that the reaction of formaldehyde to form the hydroxymethyl derivative in DNA in calf thymus nucleohistone was rapid; the subsequent cross-linking reaction with histone protein was slow. Craft and coworkers[151] studied the formaldehyde mutagenesis of human lymphoblasts in culture and observed the formation of DNA–protein cross-links (alkaline elution technique[152,153]) correlated with mutagenesis. Formaldehyde is considered an environmental pathogen and such toxicity is related to reaction of nucleic acids.[154] Shaham and coworkers[155] have proposed that protein–DNA cross-links are biomarkers for environmental exposure to formaldehyde. In more recent work, Grass and coworkers[156] have used DNA–protein cross-links for exposure to particulates.

Considering the effect of formaldehyde on nucleic acids in the presence of proteins, it is somewhat remarkable that gene expression information can be obtained from formalin-fixed, paraffin-embedded tissues. Gilbert and colleagues[157] published an excellent analysis of the issues associated with recovery of genetic information from fixed tissues. The reader is also directed to several recent articles on this technology.[134,158–160]

The alkylation of nucleic acids has been important for the structural analysis of nucleic acids and is currently an important issue for environmental toxicology. Epigenetics is a poorly defined area of investigation but could be described as the study of factors other than the base sequence in DNA important in the process of transcription. Broadly speaking then, epigenetics includes the various transcription factors as well as the covalent modification of nucleic acids by the process of methylation[25,161]; epigenetics is defined by the *Oxford English Dictionary* as "the causal analysis of development."[162] Epigenetics is a separate area of investigation from the current work; however, chemical modification such as reaction with sodium bisulfite[163–167] is used to study nucleic acid modification such as methylation. Cytosine methylation in a CpG dinucleotide sequence or island[25,168] may be the best known epigenetic modification and is useful for cancer diagnosis.[169] Sodium bisulfite converts cytosine to uracil[170] in a deamination reaction (Figure 6.11); the deamination reaction with 5-methylcytosine is much slower[171] so reaction with sodium sulfite combined with PCR technology can determine sites of cytosine methylation.[172,173] There are issues with approach[174,175] but it is useful for determining genomic DNA methylation.

Alkylation of nucleic acids also occurs as a deliberate modification as exemplified by dimethyl sulfate. In addition to the previously cited use for DNA sequencing, dimethyl sulfate is also used in nucleic acid mapping studies. Xu and Culver have an excellent review of the use of dimethyl sulfate to modify RNA.[176] These investigators emphasize the need to use care with dimethyl sulfate because of toxicity. Dimethyl sulfate reacts with guanine, cytidine, and adenine bases in RNA (Figure 6.12).[177–185] A word of warning—methylating agents are mutagenic agents with high toxicity.[186–190]

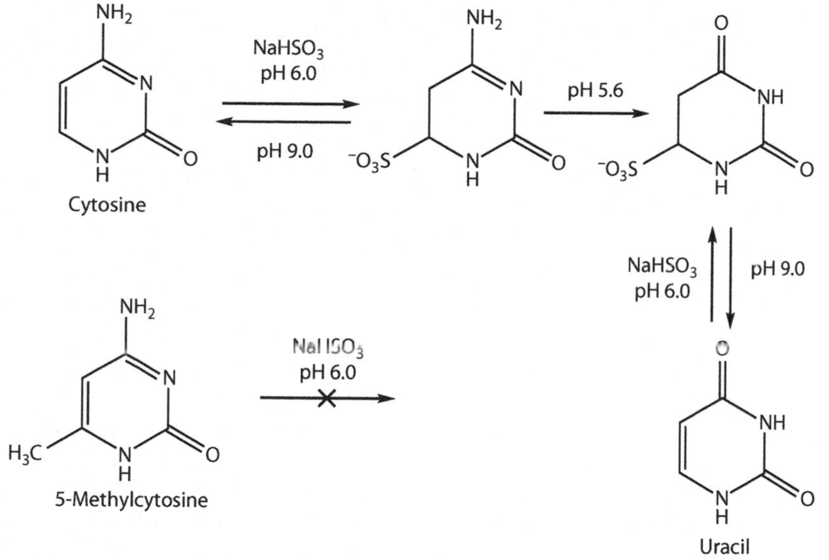

FIGURE 6.11 The sodium bisulfite-mediated conversion of cytosine to uracil (Hayatsu, H., Wataya, Y., Kai, K., and Iida, S., Reaction of sodium bisulfite with uracil, cytosine and their derivatives, *Biochemistry* 9, 2858–2864, 1970; Hayatsu, H. and Shiragami, M., Reaction of bisulfite with the 5-hydroxymethyl group in pyrimidines and in phage DNAs, *Biochemistry* 18, 632–637, 1979).

Methyl iodide can also modify nucleic acids *in vivo* by methylation.[191] Inoue and coworkers[192] have used methyl iodide to prepare 2′-*O*-methyl derivatives of ribonucleotides. 2′-*O*-methylribonucleoside phosphorothioates are more resistant to nuclease and thus provide more stability to antisense oligonucleotides.[193] Methyl iodide does not appear to be as potent as other alkylating agents.[194]

Much of the early work on the alkylation of nucleic acids was performed by Singer.[195] *N*-Methyl *N*′-nitro-*N*-nitrosoguanidine, dimethyl sulfate, and methyl methanesulfonate modify guanine, adenine, and cytosine at the same site(s) (Figure 6.12).[196] *N*′-Nitro-*N*-nitrosoguanidine reacts slowly with tobacco mosaic virus (TMV) RNA compared to the other methylating agents but is the only reagent of the three (*N*-methyl-*N*′-nitro-*N*-nitrosoguanidine, dimethyl sulfate, and methyl methanesulfonate) that results in mutagenisis.[197] There was a difference in the reaction products obtained from TMV on reaction with the three agents; reaction with dimethyl sulfate and methyl methanesulfonate yielded 7-methylguanidine as the major (97%) product, while reaction with the nitrosoguanidine derivative yields 3-methyladenine (5%) and 3-methylcytosine (35%) as significant products in addition to 7-methylguanidine (48%). These investigators also noted differences in reactivity between the intact virus and the isolated RNA. Early work by these investigators[198] had shown that reaction of the nitrosoguanidine with TMV-RNA resulted in the rapid loss of infectivity but relatively little mutagenesis, while reaction with the intact virus was slowly inactivated but there was a high level of mutagenicity. Modification of TMV-RNA with nitrosoguanidine in organic solvent (formamide) results in slower inactivation and

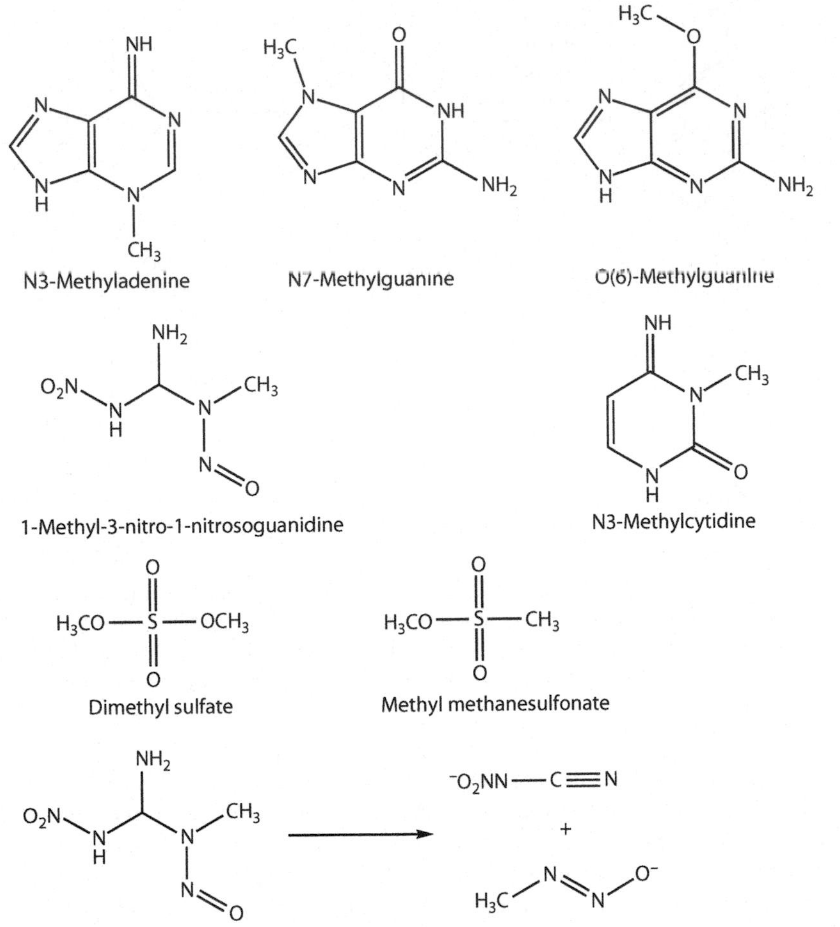

N3-Methyladenine

N7-Methylguanine

O(6)-Methylguanine

1-Methyl-3-nitro-1-nitrosoguanidine

N3-Methylcytidine

Dimethyl sulfate

Methyl methanesulfonate

$^{-}O_2NN—C≡N$ + $H_3C—N=N—O^-$

FIGURE 6.12 Methylation of bases in nucleic acids. Shown are N3-methyladenosine, N7-methylguanine, and O6-methylguanine obtained from the reaction of dimethyl sulfate with salmon testes DNA; modified bases were separated by HPLC and identified by mass spectrometry (Chadt, J., Sykora, D., Nilsson, R., and Vodicka, P., Monitoring of dimethyl sulphate-induced N3-methyladenine, N7-methylguanine, O (6)-methyl guanine adducts using reversed-phase high performance liquid chromatography and mass spectrometry, *J. Chromatog. B. Analyt. Technol. Biomed. Life Sci.* 867, 43–48, 2008). Cytidine is subject to methylation at N3 position in single-stranded DNA (Di Capua, E. and Müller, B., The accessibility of DNA to dimethyl sulfate in complexes with recA protein, *EMBO J.* 6, 2493–2498, 1987; Jensen, S.S., Ariza, X., Nielsen, P. et al., Collision-induced dissociation of cytidine and its derivatives, *J. Mass Spectrom.* 42, 49–57, 2007). Methyl methanesulfonate and 1-methyl-3-nitro-1-nitrosoguanidine (nitrosoguanidine) have also been used to methylate nucleic acid bases (Singer, B. and Fraenkel-Conrat, H., Chemical modification of viral ribonucleic acid. VIII. The chemical and biological effects of methylating agents and nitrosoguanidine on tobacco mosaic virus, *Biochemistry* 8, 3266–3269, 1969). The nitrosoguanidine derivative is suggested to form a methyl diazonium derivative as the active methylating species (Galtress, C.L., Morrow, P.R., Nag, S. et al., Mechanism for the solvolytic decomposition of the carcinogen *N*-methyl-*N′*-nitro-*N*-nitrosoguanidine in aqueous solutions, *J. Am. Chem. Soc.* 114, 1406–1411, 1992).

but a higher rate of mutagenesis. Thus, reaction of the TMV-RNA in organic solvents resembles the reaction of the intact virus. Subsequent work by Singer and Fraenkel-Conrat[199] showed that 5′-adenylic acid or 5′-guanylic acid, were essentially unreactive with nitrosoguanidine, while there was reaction with 5-cytidylic acid. These results support the influence of conformation on the reaction of nitrosoguanidine with RNA, while the reaction with other methylating agents does not appear to be influenced by conformation. It was also concluded that the methylation of cytidine is responsible for the observed mutagenicity.[197]

Methyl methanesulfonate (Figure 6.12) is another reagent used to induce aberrations in nucleic acids.[200–203] Chang and Lee[204] have reported on the reaction of methyl methanesulfonate with yeast RNA and observed the formation of six methylated products including 7-methylguanidine, 1-methyladenine, 3-methylcytidine, 1-methylguanosine, 3-methyluridine, and methyl phosphodiester. These investigators reported the hydrolysis of methyl methanesulfonate to form methanol and methanesulfonic acid, which results in a pH change. Methyl methanesulfonate also reacts with phosphate buffer to form methyl phosphate and dimethyl phosphate; the reaction of methyl methanesulfonate with phosphate is approximately 600 times faster than the hydrolysis reaction. Dimethyl sulfate also reacts with phosphate as well as with citrate.

Some workers have separated DNA mutagens into S_N1 and S_N2 reagents[205,206] based on organic reaction mechanisms (see Chapter 1). *N*-Methyl-*N*′-nitro-*N*-nitrosoguanidine was thought to react with nucleic acid via an S_N1 mechanism yielding O6-methylguanidine as major product, while methyl methanesulfonate reacts via an S_N2 mechanism with exocyclic and endocyclic amines.[207] More recent work[208–210] on the chemistry of nitrosoguanidine and nitrosoureas has suggested the formation of a methyldiazonium cation, which is the reactive species (Figure 6.12). Wyatt and Pittman[210] note that the assignment of an S_N1 reaction mechanism would require somewhat different chemistry such as the formation of a carbocation as opposed to a diazoalkyl derivative. Loechler[211] suggested that the terms high oxyphilic and low oxyphilic for classification of the different methylating agents might be more useful than describing such as S_N1 and S_N2 reactions. The point here is that reaction mechanism does not drive specificity of reaction. Gold and colleagues[212] reviewed the importance of sequence in the determination of DNA alkylation patterns. As noted above, some alkylating agents (S_N1) which react with DNA to form intermediate cations such as methanediazonium in the process of forming the adduct product. Cations such as methanediazonium show sequence specificity in reacting with DNA; such sequence specificity is not observed with dimethylsulfate (S_N2), which is a less reactive but less specific alkylating agent. Gold and coworkers[212] proposed that the formation of a double helix from single-stranded DNA establishes a sequence-specific electrostatic landscape which provides a microenvironment influencing reactivity with cationic intermediates.

Diethylpyrocarbonate (DEPC) (Figure 6.13) reacts with N7 on the purine ring resulting in rupture in a manner observed with disubstitution with histidine (Chapter 4). DEPC was introduced into nucleic acid purification protocols as a general nuclease inhibitor.[213] The finding that there was a reaction between nucleic acid and DEPC prompted an investigation of the mechanism(s) of a possible reaction.[214] The conformational sensitivity of the reaction of DEPC with nucleic acids[215] is useful

Diethylpyrocarbonate

5(4)-*N*-carbethoxyaminoimidazole-4(5)-*N'*carbethoxycarboxamidine

2-Amino-5-carbethoxyamino-4-hydroxy-6-*N*-ribofuranosylaminopyrimidine

4,5-Dicarbethoxyamino-6-*N*-ribofuranosyl-aminopyrimidine

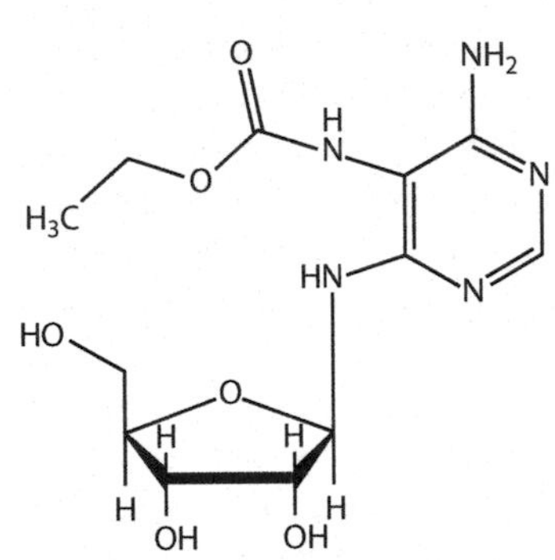

4-Amino-5-carbethoxamino-6=*N*-ribofuranosyl-aminopyrimidine

FIGURE 6.13 Reaction of DEPC with nucleic acids. Shown is 5(4)-*N*-carbethoxy-aminoimidazole-4(5)-*N'*-carbethoxycarboxamidine obtained from the reaction of adenine and DEPC (Leonard, N.J., McDonald, J.J., and Reichmann, M.E., Reaction of diethyl pyrocarbonate with nucleic acid components, I. Adenine, *Proc. Natl. Acad. Sci. USA* 67, 93–98, 1970). The reaction of adenosine yields different products arising from opening of the imidazole ring (Leonard, N.J., McDonald, J.J., Henderson, R.E.L., and Reichmann, M.E., Reaction of diethyl pyrocarbonate with nucleic acid components. Adenosine, *Biochemistry* 10, 3335–3342, 1971), which are shown at the bottom of the figure. 2-Amino-5-carbethoxyamino-4-hydroxy-6-*N*-ribofuranosylaminopyrimidine is derived from the reaction of DEPC from guanine (Vincze, A., Henderson, R.E.L., McDonald, J.J., and Leonard, N.J., Reaction of diethyl pyrocarbonate with nucleic acid components. Bases and nucleosides derived from guanine, cytosine, and uracil, *J. Am. Chem. Soc.* 95, 2677–2682, 1973).

in determining DNA distortions.[216] Adenine is the most sensitive base for DEPC modification followed by guanine; cytidine is very poorly reactive. The chemistry of the base modification by DEPC has been developed by Leonard and coworkers at the University of Illinois. Studies on the reaction of adenine with DEPC[217] suggested a disubstituted derivative (Figure 6.13) in a reaction not unlike that observed for histidine (Chapter 4). Subsequent studies on the reaction of DEPC with adenosine[218] showed different chemistry with the reaction opening the imidazole ring to yield a monosubstituted and disubstituted derivative (Figure 6.13). Further studies[219] from the Illinois group provided the structure of product obtained from the reaction of DEPC with guanine (Figure 6.13) as well as those from uracil and cytosine. Recent studies on the reaction of DEPC with nucleic acids have focused on conformational analyses.[220,221] Notwithstanding the effect on nucleic acids, DEPC is still used in inhibit nuclease activity during nucleic acid purification.[222]

Aldehydes other than formaldehyde react with nucleic acid functional groups.[223–226] Studies have focused on aldehydes such as acrolein and the closely related crotonaldehyde (Figure 6.14), which are environmental hazards or acetaldehyde and 4-hydroxy-2-nonenal (Figure 6.14). Acetaldehyde is a natural product arising, for example, from the detoxification of alcohol. 4-Hydroxy-2-nonenal arises from the oxidation of fatty acids during oxidative stress. The various aldehydes form complex derivatives with nucleobases[227,228] (Figure 6.14) including cross-linking to proteins.[225] DNA is also cross-linked by aldehydes such as acrolein and 4-hydroxy-2-nonenal.[229] Aldehydes are not used for the characterization of nucleic acids in the same manner as alkylating agents.

Studies on the use of hydrazine for the modification of nucleobases (Figure 6.15) can be traced to structural analysis of uracil by Fosse and coworkers[230] and subsequently by Levine and Bass[231] on uridine. Takamura treated RNA[232] and DNA[233,234] with anhydrous hydrazine with the goal of making apyrimidinic nucleic acids for analysis of nucleic acid structure. Apurinic nucleic acids were produced by mild acid hydrolysis (pH 1.6; 37°C) for the purpose of determination of nucleic structure.[235,236] The action of hydrazine has been developed over the years[237–241] for use in nucleic acid sequence analysis.[93,242]

Sverdlov and coworkers[243] observed that the reaction of hydrazine with cytosine in the presence of bisulfite at pH 5–7 resulted in the formation of N4-amino derivative of cytosine (Figure 6.16). Nitta and coworkers[244] used the combination of bisulfite and hydrazine in neutral solution to modify cytidine nucleobases in single-stranded polynucleotides. The emphasis here is neutral solvent as bisulfite in mild acid results in the deamination of cytosine base to form uracil. The hydrazide function on the cytosine can be condensed with the keto group on bromopyruvic acid; the alkylbromo function can then react with a sulfhydryl group on a protein to provide a cross-link (Figure 6.16). Lefever and coworkers[245] reacted the N4-amino with an *N*-hydroxysuccinimide PEG space linked to a hapten to provide a DNA probe (N4-aminodeoxycytidine-PEG-hapten).

Freese and coworkers[246,247] described hydroxylamine as both a mutagenic agent and an inactivator based on studies with bacteriophage T4. These investigators also demonstrated a specificity for cytosine in DNA with little reaction with thymidine; in RNA, there was reaction with both uracil and cytosine. The chemistry of the

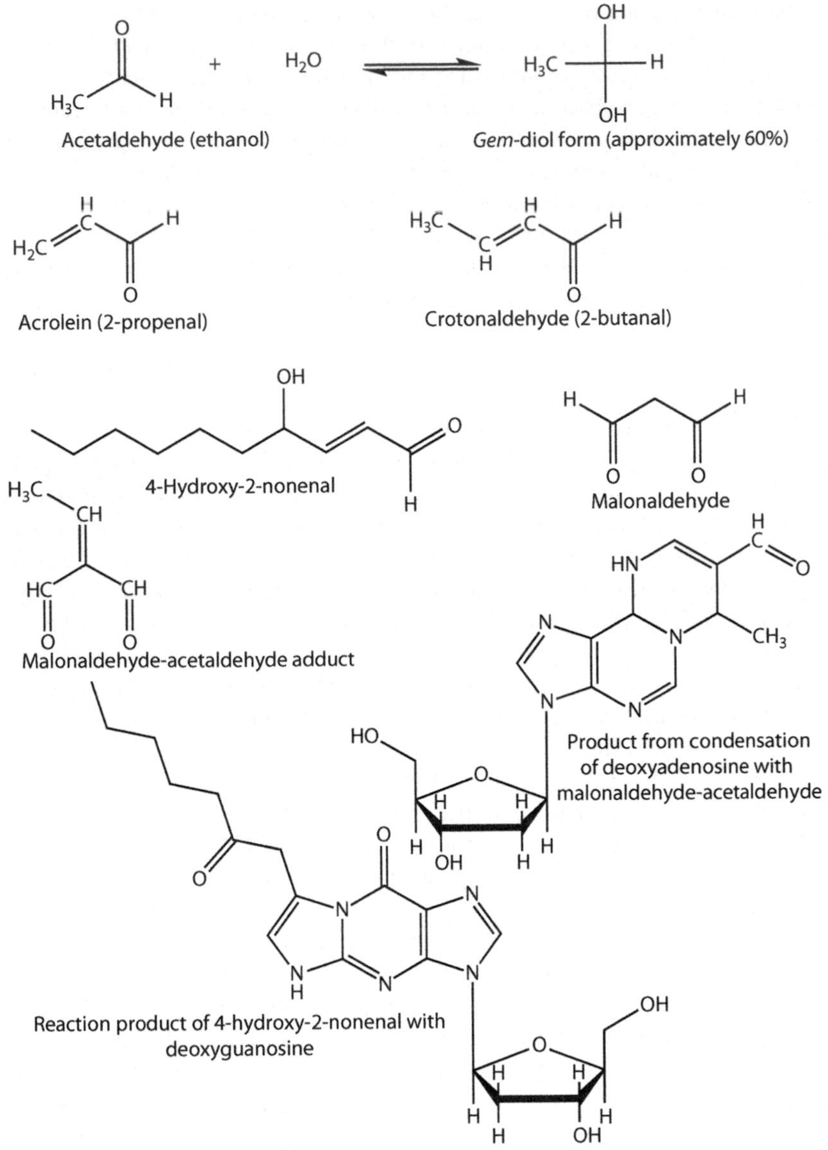

Acetaldehyde (ethanol)

Gem-diol form (approximately 60%)

Acrolein (2-propenal)

Crotonaldehyde (2-butanal)

4-Hydroxy-2-nonenal

Malonaldehyde

Malonaldehyde-acetaldehyde adduct

Product from condensation of deoxyadenosine with malonaldehyde-acetaldehyde

Reaction product of 4-hydroxy-2-nonenal with deoxyguanosine

FIGURE 6.14 Aldehydes reacting with nucleobases. Shown are the structures of acrolein, crotonaldehyde, acetaldehyde, malonaldehyde, and 4-hydroxy-2-nonenal as well as the adduct product obtained from malonaldehyde and acetaldehyde and subsequent reaction of the adduct with deoxyadenosine (Pluskota-Karwatka, D., Pawlowicz, A.J., and Kronberg, L., Formation of malonaldehyde-acetaldehyde conjugate adducts in calf thymus DNA, *Chem. Res. Toxicol.* 19, 921–926, 2006).

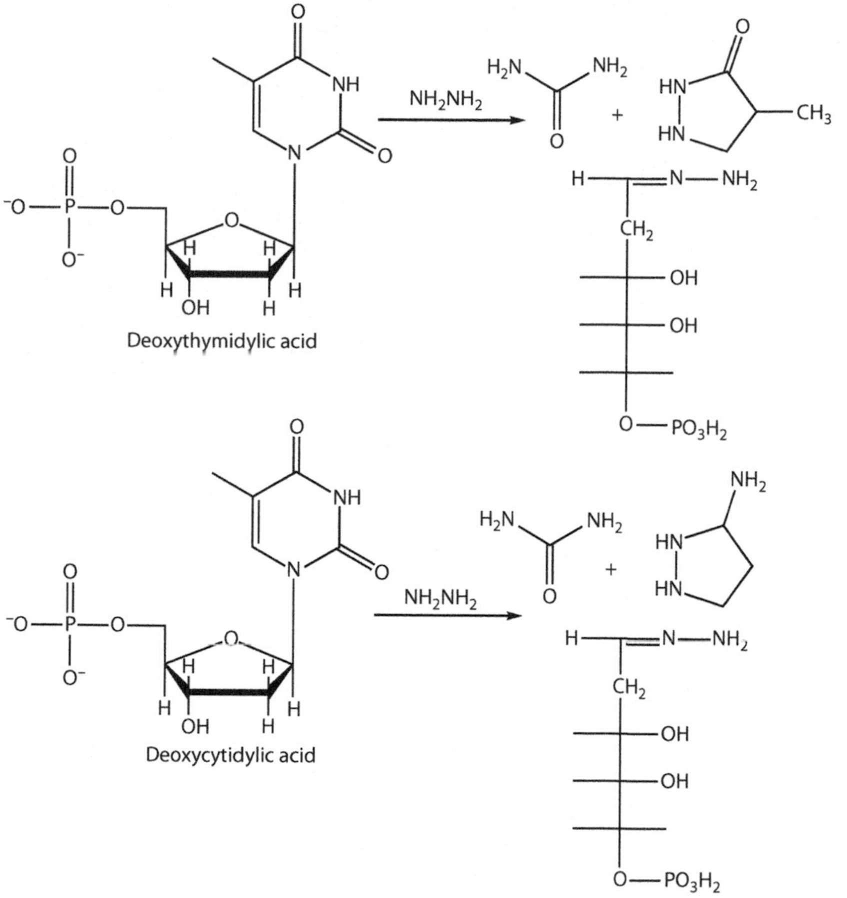

FIGURE 6.15 The reaction of hydrazine with nucleobases. Shown is the suggested mechanism of degradation of deoxythymidylic acid and deoxycytidylic acid (Temperli, A., Türler, H., Rüst, P. et al., Studies of the nucleotide arrangement in deoxyribonucleic acids. IX. Selective degradation of pyrimidine deoxyribonucleotides, *Biochim. Biophys. Acta* 91, 462–476, 1964).

reaction of hydroxylamine using 3-methylcytosine (Figure 6.17) was elucidated by Brown and Schell.[248] These investigators observed the same chemistry for cytosine, cytidine, and deoxycytidine. They did report that the reaction rate of hydroxylamine with cytosine was approximately five times greater than that observed for 3-methylcytosine. Additional work on the mechanism of reaction and structure of product has been provided by these and other investigators.[249–252] Johnston[253] has reviewed the various studies on the reaction of hydroxylamine with nucleic acids. Reaction conditions and polynucleotide influence the reaction of hydroxylamine. Johnston[253] notes that in the presence of oxygen, reaction of DNA with low concentrations (0.01–0.1 M) of hydroxylamine results in modification of all four bases and cleavage of the polynucleotide chain. This is likely a reflection of the production of hydroxyl radicals. Hydroxylamine also reacts with adenosine (Figure 6.17) but the rate is 200 times less.[254] Rubin and Schmid[255] extended the use of hydroxylamine in

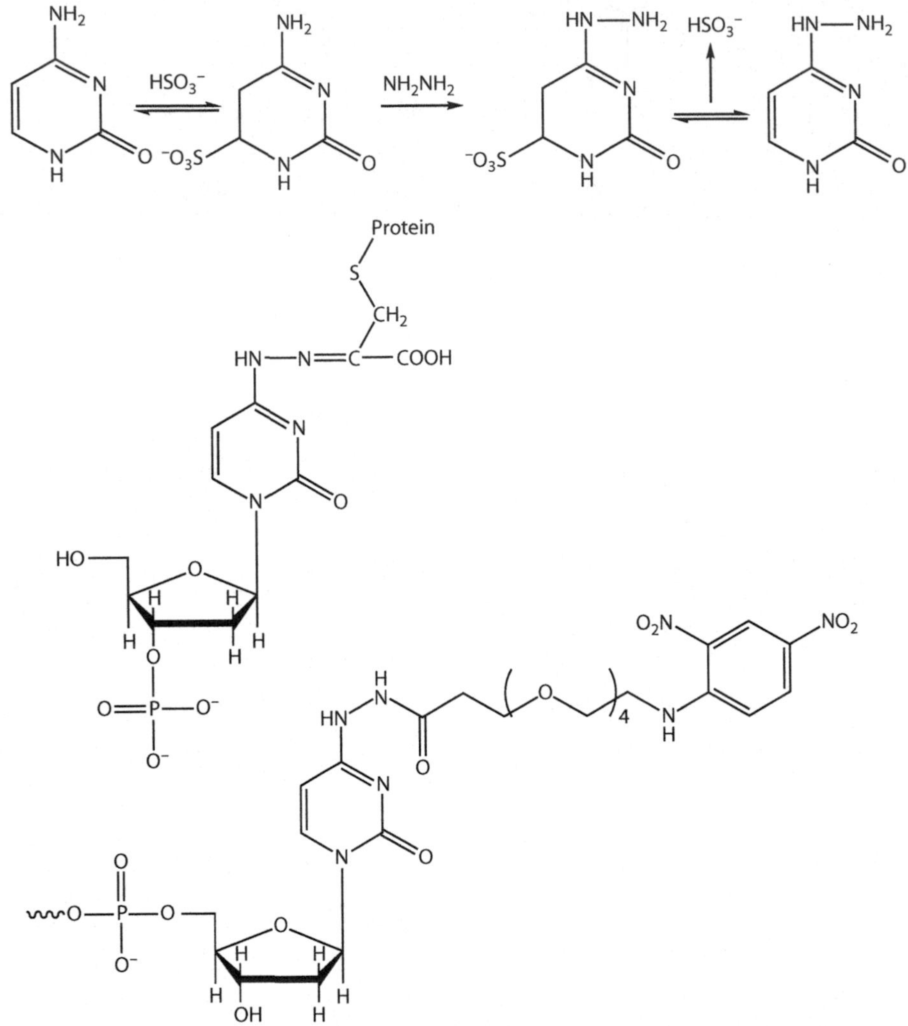

FIGURE 6.16 The reaction of hydrazine and bisulfite with cytidine base. At the top is the reaction of bisulfite and hydrazine with an N4-amino derivative (Sverdlov, E.D., Monastyrskaya, G.S., Tarabakina, N.S., and Budowsky, E.I., Reaction of the cytosine nucleus with hydrazine in the presence of bisulfite, *FEBS Lett.* 62, 212–214, 1976). Shown also are the reaction of modified cytosine base with bromopyruvate to cross-link with the sulfhydryl groups on proteins (Nitta, N., Kuge, O., Yui, S. et al., A new reaction useful for chemical cross-linking between nucleic acids and proteins, *FEBS Lett.* 166, 194–198, 1984) and formation of a labeled DNA probe with the modified cytosine (Lefever, M., Kosmeder, J.W., Farrell, M., and Bieniarz, C., Microwave-mediated synthesis of labeled nucleotides with utility in the synthesis of DNA probes, *Bioconjug. Chem.* 21, 1773–1778, 2010).

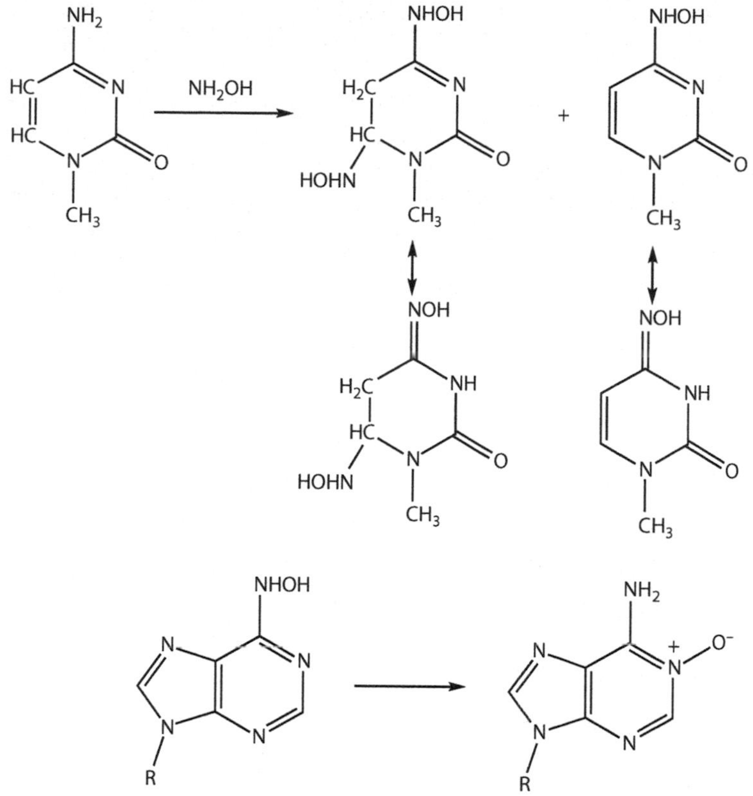

FIGURE 6.17 The reaction of hydroxylamine with cytosine. Shown is the suggested mechanism of reaction of cytosine with hydroxylamine (Brown, D.M. and Schell, P., The reaction of hydroxylamine with cytosine and related compounds, *J. Mol. Biol.* 3, 709–710, 1961). Shown also is the tautomeric form (Nagata, C. and Mårtensson, O., On the mechanism of mutagenic action of hydroxylamine, *J. Theor. Biol.* 19, 133–146, 1968). Two products are obtained from the reaction, the N6-aminohydroxyl derivative and the 5,6-dehydro-N4-hydroxy-6-hydroxyaminocytosine (Blackburn, G.M., Jarvis, S., Ryder, M.C. et al., Kinetics and mechanism of reaction of hydroxylamine with cytosine and its derivatives, *J. Chem. Soc. Perkins Trans. I* 4, 370–375, 1975). Also shown is the structure of N6-hydroxyadenosine and adenosine-1-oxide (Brown, D.M. and Osborne, M.R., The reaction of adenosine with hydroxylamine, *Biochim. Biophys. Acta* 247, 514–518, 1971).

the presence of piperidine resulting in chain cleavage as described for hydrazine by Maxam and Gilbert.[93] As hydroxylamine will not modify bases in sequence with complementary pair, hydroxylamine in combination with chain cleavage has been used to define area of mismatch in mutation detection analysis.[256–258]

There are several sites with nucleic acids that have unusual lability, which can be used to advantage in the modification of these biological polymers. The relative lability of purine glycosidic bonds permitted the preparation of apurinic derivatives of nucleic acids as described earlier. The other is the 5,6-double bond in pyrimidines,[259] which is involved in reactions with bisulfite and hydroxylamine as described earlier and with potassium permanganate.

Permanganate oxidation of DNA for structural analysis was described by Bayley and Jones.[260] These investigators used potassium permanganate in bicarbonate buffer to oxidize DNA; cytosine was the most sensitive to oxidation followed by thymine and guanine. Adenine was only slightly susceptible to oxidation under these conditions. Subsequent work[261] from this laboratory showed that the product of the oxidation was a ureido residue (Figure 6.18). Hayatsu and Ukita[262] modified the reaction conditions to improve specificity for the oxidation of thymidine residues

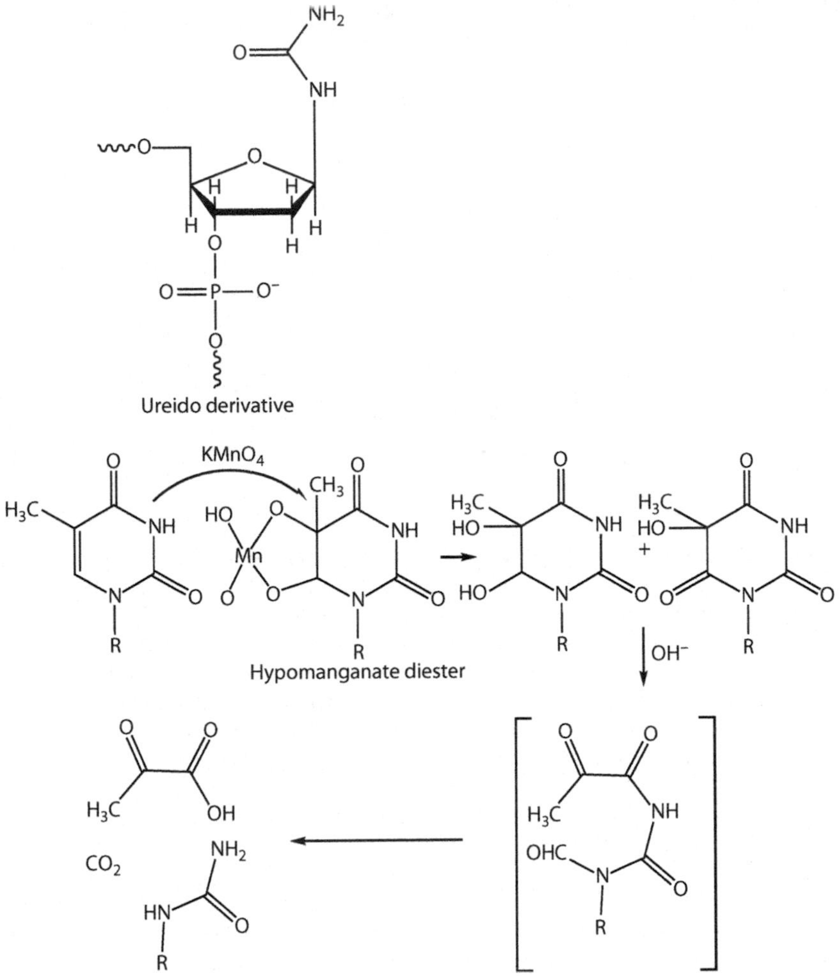

FIGURE 6.18 The oxidation of nucleobases with potassium permanganate. A degradation mechanism for the permanganate oxidation of thymine is shown (Bui, C.T., Rees, K., and Cotton, R.G.H., Permanganate oxidation reactions of DNA: Perspective in biological studies, *Nucleos. Nucleot. Nucleic Acids* 22, 1835–1855, 2003). The hypopermanganate intermediate has spectral properties (Freeman, F., Fuselier, C.O., and Karchefski, E.M., Permanganate ion oxidation of thymine spectrophotometric detection of a stable organomanganese intermediate, *Tet. Lett.* 25, 2133–2136, 1975).

in deoxyribonucleic acids. The chemistry of permanganate oxidation (Figure 6.18) has been reviewed by Bui and coworkers[263] supporting specificity for pyrimidines, although there is some reaction with guanine. The reaction was more rapid with thymine than with uracil, which was in turn more rapid than that with cytosine base. The oxidation of guanine is quite slow and Nawamura and coworkers[264] observed that 8-hydroxyguanine is not derived from DNA by the action of permanganate. Thus, while other bases are sensitive to permanganate oxidation, it is clear that the only significant reaction is that with thymine in DNA.

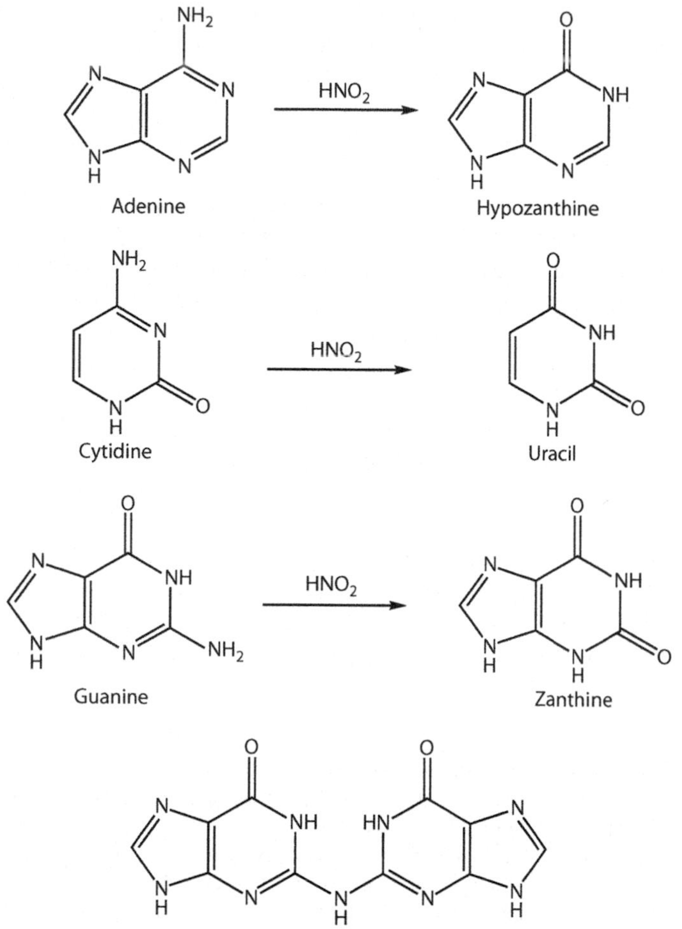

FIGURE 6.19 The reaction of nitrous acid with nucleic acid bases. Shown is the conversion of adenine to hypoxanthine, cytosine to uracil, and guanine to xanthine (Schuster, H., The reaction of nitrous acid with deoxyribonucleic acid, *Biochem. Biophys. Res. Commun.* 2, 320–323, 1960; Kotaka, T. and Baldwin, R.L., Effects of nitrous acid on the dAT copolymer as a template for DNA polymerase, *J. Mol. Biol.* 9, 323–339, 1964). Also shown is a nitrous acid-induced guanine–guanine cross-link (Edfeldt, N.B.F., Harwood, E.A., Sigurdssoon, S.T. et al., Solution structure of a nitrous acid induced DNA interstrand cross-link, *Nucleic Acids Res.* 32, 2785–2794, 2004).

Bui and coworkers[263] also reviewed the application of permanganate in the detection of mutations, footprinting, and assays for thymine dimer. The key to these various applications is the reaction of permanganate with an unpaired or mismatched thymine; in other words, the fully complementary site is resistant under mild reaction condition, while the exposed or reactive site is highly susceptible to oxidation. Figure 6.18 shows a hypopermanganate diester intermediate preceding the formation of the mixture of *cis*-5,6-dihydroxy-5,6-dihydrothymine and 5-hydroxy-5-methylbarbituric acid as demonstrated by Freeman and coworkers.[265] As these authors note, thymine base exists in the lactam form a neutral pH and the proposed reaction is then consistent with oxidation with a 2,4-cycloaddition followed by reduction of diester to manganese dioxide. The hypomanganese diester was shown to absorb light at 420 nm. This observation was extended by Bui and cowokers[266] validating the use of absorbance at 420 nm as method of measuring the reaction of permanganate with thymine. Tabone and coworkers[267] used the absorbance of the hypomanganese diester to developed a microplate assay for mutation scanning.

Stuy[268] reported on the inactivation of transforming DNA from *Haemophilus influenzae* with nitrous acid. Stuy also reviewed earlier studies on the mutagenic activity of nitrous acid on several viruses. Nitrous acid results in the deamination of three bases in DNA, adenine, cytosine, and guanine[269] resulting in the formation of hypoxanthine, uracil, and xanthine, respectively (Figure 6.19). This results in mutagenesis by changes in the base pairing during replication. Kotaka and Baldwin[270] studied the effect of nitrous acid on the dAT copolymer as a template for DNA polymerase and observed a transition from an AT to GC base pair. Cross-linkage of DNA occurs after nitrous acid deamination.[271] Edfeldt and coworkers[272] established a structure for one such cross-link that involves two guanine bases linked by a single exocyclic imino group (Figure 6.19).

The author would be remiss not to mention DNA microarrays. DNA microarrays are an assembly of DNA probes on the surface, which then can be used to "probe" a sample to determine the level of expressed genetic material.[273] It is not of direct interest for the current work and the reader is directed to other work in this area.

REFERENCES

1. Davidson, J.N. and Chargaff, E., Introduction, in *The Nucleic Acids. Chemistry and Biology*, E. Chargaff and J.N. Davidson (eds.), Academic Press, New York, Vol. 1, Chapter 1, pp. 1–8, 1955.
2. Hotchkiss, R.D., The biological role of the deoxypentose nucleic acids, in *The Nucleic Acid. Chemistry and Biology*, E. Chargaff and J.N. Davidson (eds.), Academic Press, New York, Vol. 2, Chapter 27, pp. 435–519, 1955.
3. Watson, J.M., *The Double Helix*, Simon and Shuster, New York, 1968.
4. Dubos, R.J., *The Professor, The Institute, and DNA*, The Rockefeller University Press, New York, 1978.
5. Avery, O.T., MacLeod, C.M., and McCarty, M., Studies on the chemical nature of the substance inducing transformation of pneumococcal types, *J. Exp. Med.* 79, 137–158, 1949.
6. Levine, P.A. and London, E.S., The structure of thymonucleic acid, *J. Biol. Chem.* 83, 793–802, 1929.
7. Hammarsten, O., Zur Kenntniss der nucleoproteide, *Zeit. Physiol. Chem.* 19, 19–37, 1894.

8. Eşanu, Y., Quelques moments du l'historie des acides nucléiques. I. Chimie des acides nucléiques, *Rev. Roum. Mêd. -Virol.* 39, 221–234, 1988.
9. Jones, W., *Nucleic Acids, Their Chemical Properties, and Physiological Conduct*, Longmans, Green, and Co., London, U.K., 1920.
10. Magasanik, B., Isolation and composition of the pentose nucleic acid and of corresponding nucleoproteins, in *The Nucleic Acids. Chemistry and Biology*, E. Chargaff and J.N. Davidson (eds.), Academic Press, New York, Vol. 1., Chapter 11, pp. 373–407, 1955.
11. Chargaff, E., Isolation and composition of the deoxypentose nucleic acids of the corresponding nucleoprotein, in *The Nucleic Acids. Chemistry and Biology*, E. Chargaff and J.N. Davidson (eds.), Academic Press, New York, Chapter 10, pp. 307–371, 1955.
12. Saenger, W., *Principles of Nucleic Acid Structure*, Springer-Verlag, New York, 1984.
13. Blackburn, G.M., *Nucleic Acids in Chemistry and Biology*, Royal Society of Chemistry, Cambridge, U.K., 2006.
14. Adams, R.L.P., Knowler, J.T., and Leader, D.P., *The Biochemistry of the Nucleic Acids*, 10th edn., Chapman and Hall, London, U.K., 1986.
15. Rice, P.A. and Correll, C.C., *Protein-Nucleic Acid Interactions: Structural Biology*, Royal Society of Chemistry, Cambridge, U.K., 2008.
16. Mayer, G., *The Chemical Biology of Nucleic Acids*, Wiley, Chichester, U.K., 2010.
17. Dandecker, T. and Sharma, K., *Regulatory RNA*, Springer, Berlin, Germany, 1998.
18. De Carvalho, D.D., You, J.S., and Jones, P.A., DNA methylation and cellular reprogramming, *Trends Cell Biol.* 20, 609–617, 2010.
19. Huda, A. and Jordan, I.K., Epigenetic regulation of mammalian genomes by transposable elements, *Ann. N. Y. Acad. Sci.* 1178, 276–284, 2009.
20. Brooks, W.H., Le Dantex, C., Pers, J.O. et al., Epigenetic and autoimmunity, *J. Autoimmun.* 34, 1207–1219, 2010.
21. Kirkland, J.G. and Kamakaka, R.T., tRNA insulator function: Insight into inheritance of transcriptional states?, *Epigenetics* 16, 96–99, 2010.
22. Zaidi, S.K., Young, D.W., Montecino, M. et al., Architectural epigenetics: Mitotic retention of mammalian transcriptional regulatory information, *Mol. Cell. Biol.* 30, 4758–4766, 2010.
23. Mani, S. and Heceg, Z., DNA demethylating agents and epigenetic therapy of cancer, *Adv. Genet.* 70, 327–340, 2010.
24. Ongeaert, M., Epigenetic databases and computational methodologies in the analysis of epigenetic datasets, *Adv. Genet.* 71, 259–295, 2010.
25. Bonasio, R., Tu, S., and Reinberg, D., Molecular signals of epigenetic states, *Science* 330, 612–616, 2010.
26. Stevens, A. and Bancroft, J.P., Proteins and nucleic acids, in *Theory and Practice of Histological Techniques*, J.D. Bancroft and A. Stevens (eds.), Churchill Livingstone, Edinburgh, U.K., Chapter 9, pp. 113–120, 1977.
27. Kiernan, J.A., *Histological and Histochemical Methods. Theory and Practice*, Pergamon Press, Oxford, U.K., 1981.
28. Fleskens, S.J., Takes, R.P., Otte-Höller, I. et al., Simultaneous assessment of DNA ploidy and biomarker expression in paraffin-embedded tissue sections, *Histopathology* 57, 14–26, 2010.
29. Wang, H., Owens, J.D., Shih, J.H. et al., Histological staining methods preparatory to laser capture microdissection significantly affect the integrity of the cellular RNA, *BMC Genomics* 7, 97, 2006.
30. Dische, Z., Color reactions of nucleic acid components, in *The Nucleic Acids. Chemistry and Biology*, E. Chargaff and J.N. Davidson (eds.), Academic Press, New York, Vol. 1, Chapter 9, pp. 285–305, 1955.
31. Woodhouse, D.L., The colorimetric determination of adenine, *Arch. Biochem. Biophys.* 25, 347–349, 1950.

32. Jones, A.S. and Walker, R.T., Nucleotide sequence in deoxyribonucleic acids: The determination of the distribution of adenine residues, *Nature* 202, 24–27, 1963.
33. Jones, A.S. and Bayley, C.R., *Trans. Faraday Soc.* 55, 497, 1959.
34. Shugar, D., Photochemistry of nucleic acids and their constituents, in *The Nucleic Acids. Chemistry and Biology*, E. Chargaff and J.N. Davidson (eds.), Academic Press, New York, Vol. 3, Chapter 30, pp. 39–104, 1960.
35. Loring, H.S., Hydrolysis of nucleic acids and procedures for the direct estimation of purine and pyrimidine fractions by absorption spectrophotometry, in *The Nucleic Acids. Chemistry and Biology*, E. Chargaff and J.N. Davidson (eds.), Academic Press, New York, Vol. 1, Chapter 5, pp. 191–209, 1955.
36. Mergny, J.L. and Lacroix, L., Analysis of thermal melting curves, *Oligonucleotides* 13, 515–537, 2003.
37. Wittwer, C.T., High-resolution DNA melting analysis: Advancements and limitations, *Hum. Mutat.* 30, 857–859, 2009.
38. Beers, W., Cerami, A., and Reich, E., Model for internal denaturation of DNA molecules, *Proc. Natl. Acad. Sci.* 58, 1624–1631, 1967.
39. Beaven, G.H., Holiday, E.R., and Johnson, E.A. Optical properties of nucleic acids and their components, in *The Nucleic Acids. Chemistry and Biology*, E. Chargaff and J.N. Davidson (eds.), Academic Press, New York, Vol. 1, Chapter 14, pp. 493–553, 1955.
40. Cadet, J. and Vigny, P., The photochemistry of nucleic acids, in *Bioorganic Photochemistry. Photochemistry of the Nucleic Acids*, H. Morrison (ed.), John Wiley & Sons, New York, Chapter 1, pp. 1–272, 1990.
41. Shamoo, Y., Williams, K.R., and Konigsberg, W.H., Photochemical crosslinking of bacteriophage T4 single-stranded DNA-binding protein (gp32) to oligo-p(dT)8: Identification of phenylalanine-183 as the site of crosslinking, *Proteins* 4, 1–6, 1988.
42. Banerjee, H. and Singh, R., A simple crosslinking method, CLAMP, to map the sites of RNA-contacting domains within a protein, *Methods Mol. Biol.* 488, 181–190, 2008.
43. Görner, H., Chromophore loss of uracil derivatives and polyuridylic acid in aqueous solution caused by 248 nm laser pulses and continuous UV irradiation: Mechanism of the photohydration of pyrimidines, *J. Photochem. Photobiol.* 10, 91–110, 1991.
44. Garver, L.S., Synopsin of the theory and use of the Klett-Summerson photo-electric colorimeter, *Am. J. Med. Technol.* 17, 23–32, 1951.
45. Feldberg, R.S., Brown, C., Carew, J.A., and Lucas, J.L., Probing photodynamic damage in nucleic acids with a damage-specific DNA binding protein: A comparison of the B and Z DNA conformation, *Photochem. Photobiol.* 37, 521–524, 1983.
46. Odem, D.T. and Barton, J.K., Long-range oxidative damage to DNA/RNA duplexes, *Biochemistry* 40, 8727–8737, 2001.
47. Margolin, Y., Shafirovich, V., Geacintov, N.E. et al., DNA sequence context as a determinant of the quantity and chemistry of guanine oxidation produced by hydroxyl radicals and one-electron oxidants, *J. Biol. Chem.* 283, 35569–35578, 2008.
48. Shao, J., Geacintov, N.E., and Shafirovich, V., Oxidative modification of guanine bases initiated by oxyl radicals derived from photolysis of azo compounds, *J. Phys. Chem. B* 114, 6685–6692, 2010.
49. Xu, X., Muller, J.G., Ye, Y., and Burrows, C.J., DNA protein cross-links between guanine and lysine depend on the mechanism of oxidation for formation of C5 Vs. C8 guanosine adducts, *J. Am. Chem. Soc.* 130, 703–709, 2008.
50. Ren, X., Li, F., Jeffs, G. et al., Guanine sulfinate is a major stable product of photochemical oxidation of DNA 6-thioguanine by UVA irradiation, *Nucleic Acid Res.* 38, 1832–1840, 2010.
51. Agarwal, S. (ed.), *Protocols for Oligonucleotides and Analogues*, Humana Press, Totowa, NJ, 1993.

52. Weiss, B. (ed.), *Antisense Oligonucleotides and Antisense RNA. New Pharmacological and Therapeutic Agents*, CRC Press, Boca Raton, FL, 1997.
53. Hartman, G. and Endres, S. (eds.), *Manual of Antisense Methodology*, Kluwer Academic, Boston, MA, 1999.
54. Ying, S.Y. (ed.), *MicroRNA Protocols*, Humana Press, Totowa, NJ, 2006.
55. Klussman, S., *The Aptamer Handbook: Functional Oligonucleotides and Their Application*, Wiley-VCH, Weinheim, Germany, 2006.
56. Janson, C.G. and During, M.J., *Peptide Nucleic Acids, Morpholinos, and Related Antisense Biomolecules*, Landes Bioscience, Georgetown, TX, 2008.
57. Paroo, Z. and Corey, D.R., Challenges of RNAi *in vivo*, *Trends Biotechnol.* 22, 390–394, 2004.
58. Singh, Y., Murat, P., and Defrancq, E., Recent development in oligonucleotide conjugation, *Chem. Soc. Rev.* 39, 2054–2070, 2010.
59. Zamecnik, P.C., History of antisense oligonucleotides, in *Antisense Therapeutics*, S. Agrawal (ed.), Humana Press, Totowa, NJ, Chapter 1, pp. 1–11, 1996.
60. Egli, M. and Pallan, P.S., Crystallographic studies of chemically modified nucleic acids: A backwards glance, *Chem. Biodivers.* 7, 60–89, 2010.
61. Fluiter, K., Mook, O.R., and Baas, F., The therapeutic potential of LNA-modified siRNAs: Reduction of off-target effects by chemical modification of the siRNA sequence, *Methods Mol. Biol.* 487, 189–203, 2009.
62. Veedu, R.N. and Wengel, J., Locked nucleic acid as a novel class of therapeutic agents, *RNA Biology* 6, 321–323, 2009.
63. Ellington, A.D. and Szostak, J.W., *In vitro* selection of RNA molecules that bind to proteins, *Nature* 346, 818–822, 1990.
64. Bock, L.C., Griffin, L.C., Latham, J.A. et al., Selection of single-stranded DNA molecule that bind and inhibit human thrombin, *Nature* 355, 564–566, 1992.
65. Stull, R.A. and Szoka, F.C., Jr., Antigene, ribozyme and aptamer nucleic acid drugs: Progress and prospects, *Pharm. Res.* 12, 465–483, 1995.
66. Barbas, A.S., Mi, J., Clary, B.M., and White, R.R., Aptamer applications for targeted cancer therapy, *Future Oncol.* 6, 1117–1126, 2010.
67. Eikelboom, J.W., Zelenkofske, S.L., and Rusconi, C.P., Coagulation factor IXa as a target for treatment and prophylaxis of venous thromboembolism, *Arterioscler. Thromb. Vasc. Biol.* 30, 382–387, 2010.
68. Bouchard, P.R., Hutabarat, R.M., and Thompson, K.M., Discovery and development of therapeutic aptamers, *Annu. Rev. Pharmacol. Toxicol.* 50, 237–257, 2010.
69. Stein, C.A., Iversen, P.L., Subasinghe, C. et al., Preparation of ^{35}S-labelled polyphosphorothioate oligonucleotides by use of hydrogen phosphate chemistry, *Anal. Biochem.* 188, 11–16, 1990.
70. Geary, R.S., Antisense oligonucleotide pharmacokinetics and metabolism, *Expert Opin. Drug Metab. Toxicol.* 5, 381–391, 2009.
71. Sefah, K., Phillips, J.A., Xiong, X. et al., Nucleic acid aptamers for biosensors and bioanalytical applications, *Analyst* 134, 1765–1775, 2009.
72. Chen, J., Peterson, K.R., Iancu-Rubin, C., and Bieker, J.J., Design of embedded chimeric peptide nucleic acids that efficiently enter and accurately reactivate gene expression *in vivo*, *Proc. Natl. Acad. Sci. USA* 107, 16846–16851, 2010.
73. Joshi, R., Jha, D., Su, W., and Engelmann, J., Facile synthesis of peptide nucleic acids and peptide nucleic acid-peptide conjugates on an automated peptide synthesizer, *J. Pept. Sci.* 17, 13–18, 2010.
74. Saito, I. and Matsuura, T., Chemical aspects of UV-induced cross-linking of proteins to nucleic acids. Photoreactions with lysine and tryptophan, *Acc. Chem. Res.* 18, 134–141, 1985.

75. Kurochkina, L. and Kolomijtseva, G., Photo-induced crosslinking of histones H3 and H1 to DNA in deoxyribonucleoprotein: Implications in studying histone-DNA interactions, *Biochem. Biophys. Res. Commun.* 187, 261–267, 1992.
76. Russman, C., Stollhof, J., Weiss, C. et al., Two wavelength femtosecond laser induced DNA-protein crosslinking, *Nucleic Acids Res.* 26, 3967–3970, 1998.
77. Saito, I., Sugiyama, H., and Matsuura, T., Mechanism of DNA-histone crosslinks in UV-irradiated nuclei, *Nucleic Acids Symp. Ser.* 11, 225–229, 1982.
78. Smith, K.C., The photochemistry and thymine and bromouracil *in vivo*, *Photochem. Photobiol.* 3, 1–10, 1964.
79. Shaw, A.A., Falick, A.M., and Shetlar, M.D., Photoreactions of thymine and thymidine with *N*-acetyltyrosine, *Biochemistry* 31, 10976–10983, 1992.
80. Setlow, R.B., Action spectrum for the reversal of the dimerization of thymine by ultraviolet light, *Biochim. Biophys. Acta* 49, 237–238, 1963.
81. Wulff, D.L. and Fraenkel, G., Nature of thymine photoproduct, *Biochim. Biophys. Acta* 51, 332–339, 1961.
82. Lamola, A.A., Triplet photosensitization and the photobiology of thymine dimers in DNA, *Pure Appl. Chem.* 24, 599–610, 1970.
83. Chinnapen, D.J.F. and Sen, D., A deoxyribozyme that harnesses light to repair thymine dimers in DNA, *Proc. Natl. Acad. Sci. USA* 101, 65–69, 2004.
84. Staehelm, M., Chemical modification of virus infectivity: Reactivity of tobacco virus and its nucleic acid, *Experentia* 16, 473–483, 1960.
85. Metz, D.H. and Brown, G.L., The tertiary structure of transfer ribonucleic acid investigated by a chemical modification technique, *Biochem. J.* 114, 35, 1969.
86. Metz, D.H. and Brown, G.L., Investigation of nucleic acid structure by means of chemical modification with a carbodiimide reagent. II. Reactions between *N*-cyclohexyl *N*'-β-(4-methylmorpholinium)ethylcarbodiimide and transfer ribonucleic acid, *Biochemistry* 8, 2329–2342, 1969.
87. Hayatsu, H. and Iida, S., Chemical modification of nucleic acids. Permanganate oxidation of thymine, *Tetrahedron Lett.* 13, 1031–1034, 1969.
88. Kochetkov, N.K., Chemical modification of nucleotides and its application to the investigation of nucleic acid structure, *Pure Appl. Chem.* 11, 257–273, 1969.
89. Hayatsu, H., Bisulfite modification of nucleic acids and their constituents, *Prog. Nucleic Acid Res. Mol. Biol.* 16, 75–124, 1976.
90. Phillips, T.L., Chemical modification of radiation effects, *Cancer* 39 (2 Suppl), 987–998, 1977.
91. Grunberger, D. and Weinstein, I.B., Biochemical effects of the modification of nucleic acids by certain polycyclic aromatic carcinogens, *Prog. Nucleic Acid Res. Mol. Biol.* 23, 105–149, 1979.
92. Bosshard, H.R., Mapping of contact areas in protein-nucleic acid and protein-protein complexes by differential chemical modification, *Methods Biochem. Anal.* 25, 273–301, 1979.
93. Maxam, A.M. and Gilbert, W., A new method for sequencing DNA, *Proc. Natl. Acad. Sci. USA* 74, 560–564, 1977.
94. Maxam, A.M. and Gilbert, W., Sequencing end-labeled DNA with base-specific chemical cleavages, *Methods Enzymol.* 65, 499–560, 1980.
95. Franca, L.T., Carrilho, E., and Kist, T.B., A review of DNA sequencing techniques, *Q. Rev. Biophys.* 35, 169–200, 2002.
96. Galas, D.J. and Schmitz, A., DNAse footprinting: A simple method of the detection of protein-DNA binding specificity, *Nucleic Acids Res.* 5, 3157–3170, 1978.
97. Kallenbach, N.R., *Chemistry & Physics of DNA-Ligand Interactions*, Adenine Press, Schenectady, NY, 1990.
98. Sauer, R.T., *Protein: DNA Interactions*, Academic Press, San Diego, CA, 1992.

99. Revzin, A., *Footprinting of Nucleic Acid-Protein Complexes*, Academic Press, San Diego, CA, 1993.
100. Kneale, G.G., *DNA-Protein Interactions: Principles and Protocols,* Humana Press, Totowa, NJ, 1994.
101. Moss, T., *DNA-Protein Interactions: Principles and Protocols*, Humana Press, Totowa, NJ, 2001.
102. Seitz, H., *Analysis of Protein-DNA Interactions*, Springer, Berlin, Germany, 2007.
103. Engelke, D.R., Ng, S.Y., Shastry, B.S., and Roeder, R.G., Specific interaction of a purified transcription factor with an internal control region of 5S RNA genes, *Cell* 19, 717–728, 1980.
104. Ingolia, N.T., Genome-wide translational profiling by ribosome footprinting, *Methods Enzymol.* 470, 119–143, 2010.
105. Karaduman, R., Fabrizio, P., Hartmuth, K. et al., RNA structure and RNA-protein interactions in purified yeast U6 snRNPs, *J. Mol. Biol.* 356, 1248–1262, 2006.
106. Bailey, J.M. and Tapprich, W.E., Structure of the 5′ nontranslated region of the coxsackievirus b3 genome: Chemical modification and comparative sequence analysis, *J. Virol.* 81, 650–668, 2007.
107. Stepanova, E., Want, M., Severinov, K., and Borukhov, S., Early transcriptional arrest at *Escherichia coli* rp1N and ompX promoters, *J. Biol. Chem.* 284, 35702–35713, 2009.
108. Wan, Y., Suh, H., Russell, R., and Herschlag, D., Multiple unfolding events during native folding of the *Tetrahymena* group I ribozyme, *J. Mol. Biol.* 400, 1067–1077, 2010.
109. Ervens, S., Gligorovski, S., and Herrmann, H., Temperature-dependent rate constants for hydroxyl radical reactions with organic compounds in aqueous solutions, *Phys. Chem. Chem. Phys.* 5, 1811–1824, 2003.
110. Heilek, G.M. and Noller, H.F., Site-directed hydroxyl radical probing of the RNA neighborhood of ribosomal protein S5, *Science* 272, 1659–1662, 1996.
111. Gotte, M., Maier, G., Gross, H.J., and Heumann, H., Localization of the active site of HIV-1 reverse transcriptase-associated RNase H domain on a DNA template using site-specific generated hydroxyl radicals, *J. Biol. Chem.* 263, 10139–10148, 1998.
112. Culver, G.M. and Noller, H.F., Directed hydroxyl radical probing of RNA from iron(II) tethered to proteins in ribonucleoprotein complexes, *Methods Enzymol.* 318, 461–475, 2000.
113. Gates, K.S., An overview of chemical processes that damage cellular DNA: Spontaneous hydrolysis, alklylation, and reactions with radicals, *Chem. Rev. Toxicol.* 22, 1747–1760, 2009.
114. Dizdaroglu, M., Kirkali, G., and Jaruga, P., Formamidopyrimidines in DNA: Mechanisms of formation, repair, and biological effects, *Free Radic. Biol. Med.* 45, 1610–1621, 2008.
115. Tullius, T.D. and Greenbaum, J.A., Mapping nucleic acid structure by hydroxyl radical cleavage, *Curr. Opin. Chem. Biol.* 9, 127–134, 2005.
116. Underwood, G.E., Siem, R.A., Gerpheide, S.A., and Hunter, J.H., Binding of an antiviral agent (kethoxal) by various metabolites, *Proc. Soc. Exp. Biol. Med.* 100, 312–315, 1959.
117. Underwood, G.E. and Nichol, F.R., Clinical evaluation of kethoxal against cutaneous herpes simplex, *Appl. Microbiol.* 22, 588–592, 1971.
118. Sabina, L.R., Morrow, D.G., and Szabo, R.A., Diminished virucidal activity of kethoxal against vesicular stomatitis virus pretreated with guanidinating reagent and proteases, *Acta Virol.* 20, 119–125, 1976.
119. Litt, M. and Hancock, V., Kethoxal—A potentially useful reagent for the determination of nucleotide sequences in single-stranded regions of transfer ribonucleic acid, *Biochemistry* 6, 1848–1854, 1967.
120. Shapiro, R., Cohen, B.I., Shiuey, S.J., and Maurer, H., On the reaction of guanine with glyoxal, pyruvaldehyde, and kethoxal, and the structure of the acylguanines. A new synthesis of N2-alkylguanines, *Biochemistry* 8, 238–245, 1969.

121. Poletayev, A.I., Avdonina, T.A., and Kisselev, L.L., Physical consequences of chemical modification: Circular dichroism of the kethoxalated oligoribonucleotides, *Nucleic Acids Res.* 1, 267–278, 1974.
122. Litt, M., Inactivation of yeast phenylalanine transfer ribonucleic acid by kethoxal, *Biochemistry* 10, 2223–2227, 1971.
123. Avdonina, T.A., Safronova, E.E., Gnuchev, N.V., and Kiselev, L.L., Conformational heterogeneity of tRNA developing in the reaction of guanine bases with kethoxal, *Biokhimiya* (Moscow) 41, 1548–1555, 1976.
124. Brow, D.A. and Noller, H.F., Protection of ribosomal RNA from kethoxal in polyribosomes. Implication of specific sites in ribosome function, *J. Mol. Biol.* 163, 27–46, 1983.
125. Moazed, D., Stern, S., and Noller, H.F., Rapid chemical probing of conformation in 16 S ribosomal RNA and 30 S ribosomal subunits using primer extension, *J. Mol. Biol.* 187, 399–416, 1986.
126. Banerjee, A.R. and Turner, D.R., The time dependence of chemical modification reveals slow steps in the folding of a group I ribozyme, *Biochemistry* 34, 6504–6512, 1995.
127. Kellersberger, K.A., Yu, E., Kruppa, G.H. et al., Top-down characterization of nucleic acids modified by structural probes using high-resolution tandem mass spectrometry and automated data interpretation, *Anal. Chem.* 76, 2438–2445, 2004.
128. Mortimer, S.A., Johnson, J.S., and Weeks, K.M., Quantitative analysis of RNA solvent accessibility by N-silylation of guanosine, *Biochemistry* 48, 2109–2114, 2009.
129. Ivanova, N., Pavlov, M.Y., Bouakaz, E. et al., Mapping the interaction of SmpB with ribosomes by footprinting of ribosomal RNA, *Nucleic Acids Res.* 33, 3529–3539, 2005.
130. Hopwood, D., Fixatives and fixation: A review, *Histochem. J.* 1, 323–360, 1969.
131. Miller, H.R., Fixation and tissue preservation for antibody studies, *Histochem. J.* 4, 305–320, 1972.
132. Rahn, R.O., Denaturation in ultraviolet-irradiated DNA, *Photophysiology* 8, 231–255, 1973.
133. Srinivasan, M., Sedmak, D., and Jewell, S., Effect of fixatives and tissue processing on the content and integrity of nucleic acids, *Am. J. Pathol.* 161, 1961–1971, 2002.
134. Farragher, S.M., Tanney, A., Kennedy, R.D., and Harkin, D., RNA expression analysis from formalin-fixed paraffin embedded tissues, *Histochem. Cell Biol.* 130, 435–445, 2008.
135. Matsuda, K.M., Chung, J.Y., and Hewitt, S.M., Histo-proteomic profiling of formalin-fixed, paraffin-embedded tissue, *Expert Rev. Proteomics* 7, 227–237, 2010.
136. Sompuram, S.R., Vani, K., Messana, E., and Bogen, S.A., A molecular mechanism of formalin fixation and antigen retrieval, *Am. J. Clin. Pathol.* 121, 190–199, 2008.
137. Liu, J.Y., Martinian, L., Thom, M., and Sisodiya, S.M., Immunolabeling recovery in archival, post-mortem, human brain tissue using modified antigen retrieval and the catalyzed signal amplification system, *J. Neurosci. Methods* 190, 49–56, 2010.
138. Fowler, C.B., O'Leary, T.J., and Mason, J.T., Modeling formalin fixation and histological processing with ribonuclease A: Effects of ethanol dehydration on reversal of formaldehyde cross-links, *Lab. Invest.* 88, 785–791, 2008.
139. Alegria, A.H., Hydroxymethylation of pyrimidine mononucleotides with formaldehyde, *Biochim. Biophys. Acta* 149, 317–324, 1967.
140. Hudson, R.H.E., Liu, Y., and Wojciechowski, F., Hydrophilic modifications in peptide nucleic acid—Synthesis and properties of PNA possessing 5-hydroxymethyuracil and 5-hydroxymethylcytosine, *Can. J. Chem.* 85, 302–311, 2007.
141. McGhee, J.D. and von Hippel, P.H., Formaldehyde as a probe of DNA structure I. Reaction with exocyclic amine groups of DNA bases, *Biochemistry* 14, 1281–1297, 1975.
142. McGhee, J.D. and von Hippel, P.H., Formaldehyde as a probe of DNA structure. II. Reaction with endocyclic imine groups of DNA bases, *Biochemistry* 14, 1297–1303, 1975.

143. Feldman, M.Y., Reactions of nucleic acids and nucleoproteins with formaldehyde, *Prog. Nucl. Acid Res. Mol. Biol.* 13, 1–49, 1973.
144. Yamazaki, Y. and Suzuki, H., New method of chemical modification of N6-amino group in adenine nucleotides with formaldehyde and a thiol and its application to preparing immobilized ADP and ATP, *Eur. J. Biochem.* 92, 197–207, 1978.
145. Bridson, P.K., Jiricny, J., Kemal, O., and Reese, C.B., Reactions between ribonucleoside derivatives and formaldehyde in ethanol, *J. Chem. Soc. Chem. Commun.* 5, 208–210, 1980.
146. Sloan, K.B. and Siver, K.G., The aminomethylation of adenine, cytosine, and guanine, *Tetrahedron* 40, 3997–4001, 1984.
147. McGhee, J.D. and von Hippel, P.H., Formaldehyde as a probe of DNA structure. 4. Mechanism of the initial reaction of formaldehyde with DNA, *Biochemistry* 16, 3276–3293, 1977.
148. Anderson, T., Mechanism of formaldehyde-induced mutagenesis. The uniqueness of adenylic acid in the mediation of the mutagenic activity of formaldehyde, *Nature* 187, 485–489, 1960.
149. Grossman, L., Levine, S.S., and Allison, W.S., The reaction of formaldehyde with nucleotides and T2 bacteriophage DNA, *J. Mol. Biol.* 3, 47–60, 1961.
150. Ohba, Y., Morimitsu, Y., and Watarai, A., Reaction of formaldehyde with calf thymus nucleohistone, *Eur. J. Biochem.* 100, 285–293, 1979.
151. Craft, T.R., Bermudez, E., and Skopek, T.R., Formaldehyde mutagenesis and formation of DNA protein cross-links in human lymphoblasts *in vitro. Mutat. Res.* 176, 147–155, 1987.
152. Koln, K.W., DNA as a target in cancer chemotherapy: Measurement of macromolecular damage produced in mammalian cells by anticancer agents and carcinogens, *Methods Canc. Res.* 16, 291–345, 1979.
153. Vangald, R.R., Laib, R.J., and Bolt, H.M., Evaluation of DNA damage by alkaline elution technique after inhalation exposure of rats and mice to 1,3-butadiene, *Arch. Toxicol.* 67, 34–38, 1993.
154. National Toxicology Program, Final Report on Carcinogens Background Document for Formaldehyde, *Rep. Carcinog. Backgr. Doc.* 10–5981, i-512, 2010.
155. Shaham, J., Bornstein, Y., Meltzer, A. et al., DNA-protein crosslinks, a biomarker of exposure to formaldehyde-*in vitro* and *in vivo* studies, *Carcinogenesis* 17, 121–125, 1996.
156. Grass, D.S., Ross, J.M., Family, F. et al., Airborne particulate metals in the New York City subway: A pilot study to assess the potential for health impacts, *Environ. Res.* 110, 1–11, 2010.
157. Gilbert, M.T.P., Haselkorn, T., Bunce, M. et al., The isolation of nucleic acids from fixed, paraffin-embedded tissues—Which methods are useful when?, *PLoS One* 2, e537, June 20, 2007.
158. Pena, J.T.G., Sohn-Lee, C., Rouhanifard, S.H. et al., mRNA *in situ* hybridization in formaldehyde and EDC-fixed tissues, *Nature Methods* 6, 139–141, 2009.
159. Bonin, S., Hlubek, F., Benhattar, J. et al., Multicentre validation study of nucleic acid extraction from FFPE tissues, *Virchows Arch.* 457, 309–317, 2010.
160. Waddell, N., Coccidardi, S., Johnson, J. et al., Gene expression profiling of formalin-fixed, paraffin-embedded familial breast tumours using the whole genome-DASL assay, *J. Pathol.* 221, 452–461, 2010.
161. Fazzari, M.J. and Greally, J.M., Introduction to epigenomics and epigenome-wide analysis, *Methods Mol. Biol.* 620, 243–265, 2010.
162. Waddington, C.H., (*Endeavor*, 1942) quoted in the *Oxford English Dictionary*, 2nd edn., Oxford University Press, Oxford, U.K., 1989.
163. Brena, R.M., Huang, T.H., and Plass, C., Quantitative assessment of DNA methylation: Potential applications for disease diagnosis, classification, and prognosis in clinical settings, *J. Mol. Med.* 84, 365–377, 2006.

164. Volpe, P., The language of methylation in genomics of eukaryotes, *Biochemistry* 70, 584–595, 2005.
165. Beier, V., Mund, C., and Hoheisel, J.D., Monitoring methylation changes in cancer, *Adv. Biochem. Eng. Biotechnol.* 104, 1–11, 2007.
166. Zilberman, D. and Henikoff, S., Genome-wide analysis of DNA methylation patterns, *Development* 134, 3959–3965, 2007.
167. Qureshi, S.A., Bashir, M.U., and Yaqinuddin, A., Utility of DNA methylation markers for diagnosing cancer, *Int. J. Surg.* 8, 194–198, 2010.
168. Issa, J.P., CpG-island methylation in aging and cancer, *Curr. Top. Microbiol. Immunol.* 249, 101–118, 2000.
169. Lechner, M., Boshoff, C., and Beck, S., Cancer epigenome, *Adv. Genet.* 70, 247–276, 2010.
170. Hayatsu, H., Wataya, Y., Kai, K., and Iida, S., Reaction of sodium bisulfite with uracil, cytosine, and their derivatives, *Biochemistry* 9, 2858–2865, 1970.
171. Hayatsu, H. and Shiragami, M., Reaction of bisulfite with the 5-hydroxymethyl group in pyrimidines and in phage DNAs, *Biochemistry* 18, 632–637, 1979.
172. Frommger, M., McDonald, L.E., Miller, D.S. et al., A genomic sequencing protocol that yields a positive display of 5-methylcytosine residues in individual DNA strands, *Proc. Natl. Acad. Sci. USA* 89, 1827–1831, 1992.
173. Thomassin, H., Kress, C., and Grange, T., MethylQuant: A sensitive method for quantifying methylation of specific cytosines within the genome, *Nucleic Acids Res.* 32, e168, 2004.
174. Oakeley, E.J., DNA methylation analysis: A review of current methodologies, *Pharmacol. Ther.* 84, 389–400, 1999.
175. Hayatsu, N., Negishi, K., and Wataya, Y., Progress in the bisulfite modification of nucleic acids, *Nucleic Acids Symp. Ser.* (Oxf.) 53, 217–218, 2009.
176. Xu, Z. and Culver, G.M., Chemical probing of RNA and RNA/protein complexes, *Methods Enzymol.* 468, 147–165, 2009.
177. Moore, G., Chemical modification of ribosomes with dimethyl sulfate: A probe to the structural organization of ribosomal proteins and RNA, *Can. J. Biochem.* 53, 328–337, 1975.
178. Peattie, D.A., Direct chemical method for sequencing RNA, *Proc. Natl. Acad. Sci. USA* 76, 1760–1764, 1979.
179. Fraenkel-Conrat, H. and Singer, B., Effect of introduction of small alkyl groups on mRNA function, *Proc. Natl. Acad. Sci. USA* 77, 1983–1985, 1980.
180. Peattie, D.A. and Herr, W., Chemical probing of the tRNA-ribosome complex, *Proc. Natl. Acad. Sci. USA* 78, 2273–2277, 1981.
181. Yamakawa, M., Shatkin, A.J., and Furuichi, Y., Chemical methylation of RNA and DNA viral genomes as a probe of *in situ* structure, *J. Virol.* 40, 482–490, 1981.
182. Mankin, A.S., Kopylov, A.M., and Bogdanov, A.A., Modification of 18 S rRNA in the 40 S ribosomal subunit of yeast with dimethyl sulfate, *FEBS Lett.* 134, 11–14, 1981.
183. Rairkar, A., Rubino, H.M., and Lockard, R.E., Chemical probing of adenine residues within the secondary structure of rabbit 18S ribosomal RNA, *Biochemistry* 27, 582–592, 1988.
184. Laughrea, M. and Tam, J., *In vivo* chemical footprinting of the *Escherichia coli* ribosome, *Biochemistry* 31, 12035–12041, 1992.
185. Tijerina, P., Mohrn, S., and Russell, R., DMS footprinting of structured RNAs and RNA-protein complexes, *Nat. Protoc.* 2, 2608–2623, 2007.
186. Hite, M., Rinehart, W., Braun, W., and Peck, H., Acute toxicity of methyl fluorosulfonate (Magic Methyl), *Am. Ins. Hyg. Assoc. J.* 40, 600–603, 1979.
187. Pegg, A.E., Methylation of the 06 position of guanine in DNA is the most likely initiating event in carcinogenesis by methylating agents, *Cancer Invest.* 2, 223–231, 1984.

188. Milligan, J.R., Skotnicki, S., Lu, S.J., and Archer, M.C., Specificity in the methylation of DNA by N-nitroso compounds, *IARC Sci. Publ.* 105, 329–331, 1991.
189. Farmer, P.B. and Shuker, D.E., What is the significance of increases in background levels of carcinogen-derived protein and DNA adducts? Some considerations for incremental risk assessment, *Mutat. Res.* 424, 275–286, 1998.
190. Bignami, M., O'Driscoll, M., Aquilina, G., and Karran, P., Unmasking a killer: DNA O(6)-methylguanidine and the cytotoxicity of methylating agents, *Mutat. Res.* 462, 71–82, 2000.
191. Bolt, H.M. and Gansewendt, B., Mechanisms of carcinogenicity of methyl halides, *Crit. Rev. Toxicol.* 23, 237–253, 1993.
192. Inoue, H., Hayase, Y., Asaka, M. et al., Synthesis and properties of novel nucleic acid probes, *Nucleic Acids Symp. Ser.* 16, 165–168, 1985.
193. Agrawal, S. and Zhang, R., Pharmacokinetics of phosphorothioate oligonucleotides and their novel analogs, in *Antisense Oligodeoxynucleotides and Antisense RNA. New Pharmacological and Therapeutic Agents*, B. Weiss (ed.), CRC Press, Boca Raton, FL, Chapter 3, pp. 57–78, 1997.
194. Wang, G., Palejwala, V.A., Dunman, P.M. et al., Alkylating agents induce UVM, a recA-independent inducible mutagenic phenomenon in *Escherichia coli*, *Genetics* 141, 813–823, 1995.
195. Singer, B., The chemical effects of nucleic acid alkylation and their relation to mutagenesis and carcinogenesis, *Prog. Nucl. Acid Res. Mol. Biol.* 15, 219–284, 1975.
196. Singer, B. and Fraenkel-Conrat, H., Chemical modification of viral ribonucleic acid. VIII. The action of methylating agents and nitrosoguanidine on polynucleotides including tobacco mosaic virus ribonucleic acid, *Biochemistry* 8, 3260–3266, 1969.
197. Singer, B. and Fraenkel-Conrat, H., Chemical modification of viral ribonucleic acid. VIII. The chemical and biological effects of methylating agents and nitrosoguanidine on tobacco mosaic virus, *Biochemistry* 8, 3266–3269, 1969.
198. Singer, B. and Fraenkel-Conrat, H., Chemical modification of viral RNA. VI. The action of *N*-methyl-*N'*-nitro-*N*-nitrosoguanidine, *Proc. Natl. Acad. Sci. USA* 58, 234–239, 1967.
199. Singer, B. and Fraenkel-Conrat, H., Chemical modification of viral ribonucleic acid VII. The action of methylating agents and nitrosoguanidine on polynucleotides including tobacco mosaic virus ribonucleic acid, *Biochemistry* 8, 3260–3266, 1969.
200. Strauss, B., Scudiero, D., and Henderson, E., The nature of the alklylation lesion in mammalian cells, *Basic Life Sci.* 5A, 13–24, 1975.
201. Strauss, B.S., Response of *Escherichia coli* auxotrophs to heat after treatment with mutagenic alkyl methanesulfonates, *J. Bacteriol.* 83, 241–249, 1962.
202. Ludlum, D.B., Warner, R.C., and Wahba, A.J., Alkylation of synthetic polynucleotides, *Science* 145, 397–399, 1964.
203. Buhl, S.N. and Regan, J.D., DNA replication in human cells treated with methyl methanesulfonate, *Mutat. Res.* 18, 191–197, 1973.
204. Chang, C.J. and Lee, C.G., Chemical modification of ribonucleic acid. A direct study by carbon-13 nuclear magnetic resonance spectroscopy, *Biochemistry* 20, 2657–2661, 1981.
205. Sega, G.A., Wolfe, K.W., and Owens, J.G., A comparison of the molecular action of an S_N1-type methylating agent, methyl nitrosourea and an S_N2-type methylating agent, methyl methanesulfonate, in the germ cells of male mice, *Chem. -Biol. Interactions* 33, 253–269, 1981.
206. Boffa, L.C., Bolognesi, C., and Marioni, M.R., Specific targets of alkylating agents in nuclear proteins of target hepatocytes, *Mutat. Res.* 190, 119–123, 1987.
207. Kaina, B., Mechanisms and consequences of methylating agent-induced SCEs and chromosomal aberrations: A long road traveled and still a far way to go, *Cytogenet. Genomic Res.* 104, 77–86, 2004.

208. Galtress, C.L., Morrow, P.R., Nag, S. et al., Mechanism for the solvolytic decomposition of the carcinogen *N*-methyl-*N*'-nitro-*N*-nitrosoguanidine in aqueous solution, *J. Am. Chem. Soc.* 114, 1406–1411, 1992.
209. Garcia-Río, L., Leis, J.R., Moreira, J.A. et al., Mechanism for basic hydrolysis of *N*-nitrosoguanidines in aqueous solution, *J. Org. Chem.* 68, 4330–4337, 2003.
210. Wyatt, M.D. and Pittman, D.L., Methylating agents and DNA repair responses: Methylated bases and sources of strand breaks, *Chem. Res. Toxicol.* 19, 1580–1594, 2006.
211. Loechler, E.L., A violation of the Swain-Scott principle, and *Not* S_N1 versus S_N2 reaction mechanisms, explains why carcinogenic alkylating agents can form different proportions of adducts at oxygen versus nitrogen in DNA, *Chem. Res. Toxicol.* 7, 277–280, 1994.
212. Gold, B., Marky, L.M., Stone, M.P., and Williams, L.D., A review of the sequence-dependent electrostatic landscape in DNA alkylation patterns, *Chem. Res. Toxicol.* 19, 1402–1419, 2006.
213. Solymosy, F., Fedorcsák, I., Gulyáas, A. et al., A new method based on the use of diethyl pyrocarbonate as a nuclease inhibitor for the extraction of undegraded nucleic acid from plant tissues, *Eur. J. Biochem.* 5, 520–527, 1968.
214. Solymosy, F., Hüvösm, O., Gulyás, A. et al., Diethyl pyrocarbonate, a new tool in the chemical modification of nucleic acids?, *Biochim. Biophys. Acta* 238, 406–416, 1971.
215. Guo, Q., Lu, M., and Kallanbach, N.R., Effect of hemimethylation and methylation of adenine on the structure and stability of model DNA duplexes, *Biochemistry* 34, 16359–16364, 1995.
216. Kahl, B.F. and Paule, M.R., The use of diethyl pyrocarbonate and potassium permanganate as probes for strand separation and structural distortions in DNA, *Methods in Molecular Biology, DNA-Protein Interactions*, T. Moss and B. Leblanc (eds.), Humana Press, Totowa, NJ, Vol. 543, Chapter 6, pp. 73–85, 2009.
217. Leonard, N.J., McDonald, J.J., and Reichmann, M.E., Reaction of diethyl pyrocarbonate with nucleic acid components. I. Adenine, *Proc. Natl. Acad. Sci. USA* 67, 93–98, 1970.
218. Leonard, N.J., McDonald, J.J, Henderson, R.E.L., and Reichmann, M.E., Reaction of diethyl pyrocarbonate with nucleic acid components. Adenosine, *Biochemistry* 10, 3335–3342, 1971.
219. Vincze, A., Henderson, R.E.L., McDonald, J.J., and Leonard, N.J., Reaction of diethyl pyrocarbonate with nucleic acid components. Bases and nucleosides derived from guanine, cytosine, and uracil, *J. Am. Chem. Soc.* 95, 2677–2682, 1973.
220. Messmer, M., Pütz, J., Suzuki, T. et al., Tertiary network in mammalian mitochondrial tRNAAsp revealed by solution probing and phylogeny, *Nucleic Acids Res.* 37, 6881–6895, 2009.
221. Shirokikh, N.E., Agalarov, S.Ch., and Spirin, A.S., Chemical and enzymatic probing of spatial structure of the omega leader of tobacco mosaic virus RNA, *Biochemistry* (Moscow) 75, 405–411, 2010.
222. Nadkarni, M.A., Martin, F.E., Hunter, N., and Jacques, N.A., Methods for optimizing DNA extraction before quantifying oral bacterial numbers by real-time PCR, *FEMS Microbiol. Lett.* 296, 45–51, 2009.
223. Chung, F.L., Young, R., and Hecht, S.S., Formation of cyclic 1,N^2-propanodeoxyguanosine adducts in DNA upon reaction with acrolein or crotonaldehyde, *Cancer Res.* 44, 990–995, 1984.
224. Hecht, S.S., McIntee, E.J., and Wang, M., New DNA adducts of crotonaldehyde and acetaldehyde, *Toxicology* 166, 31–36, 2001.
225. Kurtz, A.J. and Lloyd, R.S., 1, N^2-deoxyguanosine adducts of acrolein, crotonaldehyde, and *trans*-4-hydroxynonenal cross-link to peptides via Schiff base linkage, *J. Biol. Chem.* 278, 5970–5976, 2003.

226. Winter, C.K., Segall, H.J., and Haddon, W.F., Formation of cyclic adducts of deoxyguanosine with the aldehydes *trans*-4-hydroxy-2-hexenal and *trans*-4-hydroxynonenal *in vitro*, *Cancer Res.* 46, 5682–5686, 1986.
227. Kawai, Y., Uchida, K., and Osaka, T., 2-Deoxycytidine in free nucleosides and double-stranded DNA as the major target of lipid peroxidation products, *Free Rad. Biol. Med.* 36, 529–541, 2004.
228. Pluskota-Karwatka, D., Pawlowicz, A.J., and Kronberg, L., Formation of malonaldehyde-acetaldehyde conjugate adducts in calf thymus DNA, *Chem. Res. Toxic.* 91, 921–926, 2006.
229. Kozekov, I.D., Turesky, R.J., Alas, G.R. et al., Formation of deoxyguanosine cross-links from calf thymus DNA treated with acrolein and 4-hydroxy-2-nonenal, *Chem. Res. Toxicol.* 23, 1701–1713, 2010.
230. Fosse, R., Hieulle, A., and Bass, L.W., Action of hydrazine on uracil, *Compt. Rend. Acad.* 179, 811–813, 1924.
231. Levine, P.A. and Bass, L.W., The action of hydrazine hydrate on uridine, *J. Biol. Chem.* 71, 167–172, 1926.
232. Takamura, S., Hydrazinolysis of yeast ribonucleic acid. Formation of "riboapyrimidinic acid", *J. Biochem.* 44, 321–326, 1957.
233. Takamura, S., Hydrazinolysis of herring-sperm deoxyribonucleic acid, *Biochim. Biophys. Acta* 29, 447–448, 1958.
234. Takamura, S., Hydrazinolysis of nucleic acids I. The formation of deoxyriboapyrimidinic acid from herring sperm deoxyribonucleic acid, *Bull. Chem. Soc. Japan* 32, 920–926, 1959.
235. Tamm, C., Hodes, M.E., and Chargaff, E., The formation of apurinic acid from the deoxyribonucleic acid of calf thymus, *J. Biol. Chem.* 195, 49–63, 1952.
236. Tamm, C. and Chargaff, E., Physical and chemical properties of the apurinic acid of calf thymus, *J. Biol. Chem.* 203, 689–694, 1953.
237. Chargaff, E., Rüst, P., Temperli, A. et al., Investigation of the purine sequences in deoxyribonucleic acids, *Biochim. Biophys. Acta* 76, 149–151, 1963.
238. Temperli, A., Türler, H., Rüst, P., Danon, A., and Chargaff, E., Studies on the nucleotide arrangement in deoxyribonucleic acids. IX. Selective degradation of pyrimidine deoxyribonucleotides, *Biochim. Biophys. Acta* 91, 462–476, 1964.
239. Cashmore, A.E. and Petersen, G.B., The degradation of DNA by hydrazine: A critical study of the suitability of the reaction for the quantitative determination of purine nucleotide sequences, *Biochim. Biophys. Acta* 174, 591–603, 1969.
240. Türler, H. and Chargaff, E., Studies on the nucleotide arrangement in deoxyribonucleic acids, XII. Apyrimidinic acid from calf-thymus deoxyribonucleic acid: Preparation and properties, *Biochim. Biophys. Acta* 195, 446–455, 1969.
241. Cashmore, A.E. and Peterson, G.B., The degradation of DNA by hydrazine: Identification of 3-ureidopyrazole as a product of the hydrazinolysis of deoxycytidylic acid residues, *Nucleic Acids Res.* 5, 2485–2491, 1978.
242. Tolson, D.A. and Nicholson, N.H., Sequencing RNA by a combination of exonuclease digestion and uridine specific chemical cleavage using MALDI-TOF, *Nucleic Acids Res.* 26, 446–451, 1998.
243. Sverdlov, E.D., Monastryrskaya, G.S., Tarabakina, N.S., and Budowsky, E.I., Reaction of the cytosine nucleus with hydrazine in the presence of bisulfite, *FEBS Lett.* 62, 212–214, 1976.
244. Nitta, N., Kuge, O., Yui, S. et al., A new technique for chemical cross-linking between nucleic acid and proteins, *FEBS Lett.* 166, 194–198, 1984.
245. Lefever, M., Kosmeder, J.W., Farrell, M., and Bieniarz, C., Microwave-mediated synthesis of labeled nucleotides with utility in the synthesis of DNA probes, *Bioconjug. Chem.* 21, 1773–1778, 2010.

246. Freese, E., Bautz-Freese, E., and Bautz, E., Hydroxylamine as a mutagenic and inactivating agent, *J. Mol. Biol.* 3, 133–143, 1961.
247. Freese, E., Bautz, E., and Bautz-Freese, E., The chemical and mutagenic specificity of hydroxylamine, *Proc. Natl. Acad. Sci. USA* 67, 845–855, 1961.
248. Brown, D.M. and Schell, P., The reaction of hydroxylamine with cytosine and related compounds, *J. Mol. Biol.* 3, 709–710, 1961.
249. Brown, D.M. and Hewlins, M.J.E., The reaction between hydroxylamine and cytosine derivatives, *J. Chem. Soc. Organic Chem. C.* 1922–1924, 1968.
250. Nagata, C. and Mårtensson, O., On the mechanism of mutagenic action of hydroxylamine, *J. Theor. Biol.* 19, 133–146, 1968.
251. Blackburn, G.M., Jarvis, S., Ryder, M.C. et al., Kinetics and mechanism of reaction of hydroxylamine with cytosine and its derivatives, *J. Chem. Soc. Perkins Trans. I* 4, 370–375, 1975.
252. Fraenkel-Conrat, H. and Singer, B., The chemical basis for the mutagenicity of hydroxylamine and methoxyamine, *Biochim. Biophys. Acta* 262, 264–268, 1972.
253. Johnston, B.H., Hydroxylamine and methoxylamine as probes of DNA structure, *Methods Enzymol.* 212, 180–194, 1992.
254. Brown, D.M. and Osborne, M.R., The reaction of adenosine with hydroxylamine, *Biochim. Biophys. Acta* 247, 514–518, 1971.
255. Rubin, C.M. and Schmid, C.W., Pyrimidine-specific chemical reactions useful for DNA sequencing, *Nucleic Acids Res.* 8, 4613–4619, 1980.
256. Bui, C.T., Rees, K., Lambrinakos, A., Bedir, A., and Cotton, R.G., Site-selective reactions of imperfectly matched DNA with small chemical molecules: Applications in mutation detection, *Bioorg. Chem.* 30, 216–232, 2002.
257. Tabone, T., Sallman, G., Chiotis, M. et al., Chemical cleavage of mismatch (CCM) to locate base mismatches in heteroduplex DNA, *Nat. Protoc.* 1, 2297–2304, 2006.
258. Yakubovskaya, M.G., Belyakova, A.A., Gasanova, V.K. et al., Comparative reactivity of mismatched and unpaired bases in relation to their surroundings. Chemical cleavage of DNA mismatches in mutation detection analysis, *Biochimie* 92, 762–771, 2010.
259. Hayatsu, H., The 5,6-double bond of pyrimidine nucleosides, a fragile site in nucleic acids, *J. Biochem.* 119, 391–395, 1996.
260. Bayley, C.R. and Jones, A.S., Nucleotide sequence: The oxidation of DNA with potassium permanganate, *Trans. Faraday Soc.* 55, 492, 1959.
261. Jones, A.S., Ross, C.W., Takemura, S. et al., The nucleotide sequence in deoxyribonucleic acids. Part VI. The preparation and reactions of permanganate-oxidized deoxyribonucleic acid, *J. Chem. Soc.* 373–378, 1964.
262. Hayatsu, H. and Ukita, T., The selective degradation of pyrimidines in nucleic acids by permanganate oxidation, *Biochem. Biophys. Res. Commun.* 29, 556–561, 1967.
263. Bui, C.T., Rees, K., and Cotton, R.G.H., Permanganate oxidation reactions of DNA: Perspective in biological studies, *Nucleosides, Nucleotides, & Nucleic Acids* 22, 1835–1855, 2003.
264. Nawamura, T., Negishi, K., and Hayatsu, H., 8-Hydroxyguanine is not produced by permanganate oxidation of DNA, *Arch. Biochem. Biochem.* 311, 523–524, 1994.
265. Freeman, F., Fuselier, C.O., and Karchefski, E.M., Permanganate ion oxidation of thymine spectrophotometric detection of a stable organomanganese intermediate, *Tet. Lett.* 25, 2133–2136, 1975.
266. Bui, C.T., Sam, L.A., and Cotton, R.G.H., UV-visible spectral identification of the solution-phase and solid-phase permanganate oxidation reactions of thymine acetic acid, *Bioorgan. Med. Chem. Lett.* 14, 1313–1315, 2004.
267. Tabone, T., Sallmann, G., Webb, E., and Cotton, R.G.H., Detection of 100% of mutations in 124 individuals using a standard UV/Vis microplate reader: A novel concept for mutation scanning, *Nucleic Acids Res.* 34, e45, 2006.

268. Stuy, J.H., Inactivation of transforming deoxyribonucleic acid by nitrous acid, *Biochem. Biophys. Res. Commun.* 6, 328–333, 1961.
269. Schuster, H., The reaction of nitrous acid with deoxyribonucleic acid, *Biochem. Biophys. Res. Commun.* 2, 320–323, 1960.
270. Kotaka, T. and Baldwin, R.L., Effects of nitrous acid on the dAT copolymer as a template for DNA polymerase, *J. Mol. Biol.* 9, 323–339, 1964.
271. Verly, W.G. and Lacroix, M., DNA and nitrous acid, *Biochim. Biophys. Acta.* 414, 185–192, 1975.
272. Edfeldt, N.B.F., Harwood, E.A., Sigurdsson, S.T. et al., Solution structure of a nitrous acid induced DNA interstrand cross-link, *Nucleic Acids Res.* 32, 2785–2794, 2004.
273. Garaizar, J., Brena, S., Bikandi, J. et al., Use of DNA microarray technology and gene expression profiles to investigate the pathogenesis, cell biology, antifungal susceptibility and diagnosis of *Candida*, *FEMS Yeast Res.* 6, 987–998, 2006.

7 Chemical Modification of Polysaccharides

The chemical modification of oligosaccharides and polysaccharides is of interest to food science[1,2] and the broad area of biomaterials; biomaterials would include applications in regenerative medicine[3,4] as well as hydrogels in drug delivery.[5,6] Chemical degradation is used for structural analysis[7–10] and for the preparation of derivative forms from biological tissue.[11,12] Heparin, for example, is derived from porcine intestine or bovine lung by mild alkaline hydrolysis followed by other purification steps.[13–15] Low-molecular-weight heparin (LMWH) is derived from unfractionated heparin by fractionation after enzymatic or chemical degradation.[16] Methylation (permethylation)[17] and acetylation (peracetylation)[18,19] of polysaccharides have been used to prepare samples for compositional analysis by gas-liquid chromatography (GLC)[20] and are used today to prepare samples for mass spectrometry.[21,22]

Polysaccharides are composed of monomer units, which are polyols and are similar in structure. The term glycan is used to describe polysaccharides and is most frequently used to describe a polysaccharide moiety to a protein or lipid. The term protein polysaccharide refers to a glycoconjugate obtained from a plant source, while the term proteoglycan is used to refer to a glycoconjugate derived from an animal source.[23] A glycoside is a mixed acetal with a glycosyl reacting with non-acyl group (RO). A glycosamine is an *N*-glycoside, while a *C*-glycosyl has a *C*-glycosidic linkage. The reader is directed to several sources for a consideration of carbohydrate nomenclature.[24–26]

The monomer units of polysaccharides may be modified *in vivo* to contain amino groups and may be sulfated. A glycosaminoglycan is a polysaccharide that contains amino sugars. The hydroxyl groups in the monomer monosaccharide unit are quite unreactive and can only be modified under rigorous conditions.[27,28] Polysaccharides are polar molecules capable of hydrogen bonding, interaction with solvent and can exist in conformation equlibirum.[29] Morris[30] has discussed the levels of conformation of polysaccharides in solution showing that, as with proteins, there are four levels of structure: primary, secondary, tertiary, and quaternary. The behavior of polysaccharides in solution is critical for their function in foods. In particular, for fabricated foods, thickening and jellying properties[31] are characteristics that can be modulated by processing including chemical modification.[1,2,32] The rheological properties of nonfood carbohydrates can also be modulated by chemical modification.[33,34] In the case of a water-insoluble polysaccharide such as cellulose, function depends on physical strength and chemical inertness.[35]

Glycoconjugate is a term used to a complex molecule where a carbohydrate is linked via a glycosidic bond to a protein, peptide, lipid, or other non-carbohydrate molecule.[36] Glycolipids are important constituents of cell membranes[37]; endotoxin is a glycolipid derived from bacterial cell membranes.[38–40] Glycoproteins or proteoglycans are the best known glycoconjugates. There is considerable interest in chemical glycosylation where a glycan moiety can be coupled to a protein. While the significance of protein glycosylation is poorly understood, chemical glycosylation, as with modification with poly (ethylene glycol) can be seen as a method for improving product pharmacokinetics[41–43] as well as direct tissue localization of therapeutic products.[44–47] One of the more useful approaches has been the covalent attachment of galactose for the purpose of targeting the asialoglycoprotein receptor on hepatocytes.[47–49] Chemical glycosylation would prove most useful for the modification of recombinant proteins produced in bacteria, which cannot accomplish glycosylation. The ability of *Pichia pastoris* to produce glycosylated proteins provides a partial solution to this problem.[50] Another approach has been described by Pandhal and Wright[51] where the glycosylation mechanism in *Campylobacter jejuni* can be transferred to *Escherichia coli,* which permit the production of glycosylated protein. Synthetic carbohydrate-based vaccines[52–54] are also glycoconjugates. The area of chemical neoglycosylation where an oligosaccharide is coupled to a protein via chemical conjugation is reviewed by Nicotra and coworkers.[55] Neoglycosylation uses chemistry, which is discussed in Chapter 2 through 5, with the primary issue being the synthesis of the oligosaccharide moiety. Neoglycosylation is also used to prepare unique low-molecular-weight glycosides.[56]

A review of the past 20 years of *Advances in Carbohydrate Chemistry and Biochemistry* provided relatively few articles directly related to the chemical modification of polysaccharides. Several articles illustrated the difficulty in the chemical modification of polysaccharides. Fitremann[57] provided a recent discussion of the chemistry of sucrose. Sucrose is described as a complex, polyfunctional organic molecule, which is highly oxygenated and chemically sensitive. Sucrose contains eight hydroxyl groups, which can be modified and two anomeric carbons. Unlike the various functional groups in proteins and nucleic acids, the major functional group in carbohydrates, primarily the hydroxyl group, does not possess "unusual" reactivity, although the 6'-hydroxyl group is considered more reactive than ring hydroxyl groups reflecting a lack of steric hindrance. However, like the hydroxyl groups in serine and threonine, the hydroxyl groups of sucrose are only modified by vigorous conditions. Queneau and coworkers[58,59] have presented recent studies showing that the nucleophilic characteristics of the various hydroxyl groups in sucrose and isomaltulose can be modified by solvent, reagent, and catalyst. The importance of nucleophilicity in functional group reactivity and the relationship of pK_a to nucleophilicity are discussed in Chapter 1. Given that pK_a is an imperfect measure of functional group reactivity, it is still a useful approximation. It was then a little surprising to find much information other than the 2′-hydroxyl group that functions in ribozyme catalysis.[60,61] Lyne and Karplus[62] estimated a pK_a of 14.9 for the hydroxyl group in phosphorylated ribose, while Acharya and coworkers[63] found a pK_a of approximately 12.7 for the internucleotide 2′-hydroxyl group in diribonucleoside (3′→5′) monophosphate. The kinetic pK_a for the activity of a ribozyme is lower (8–8.5) depending on the metal ion.[64] It would appear that the pK_a of hydroxyl groups of alcohols is approximately 14.[65] Ray and coworkers[66] have suggested that the 6-hydroxyl group of the glucose-1-phosphate should resemble that of using ethylene glycol methyl ether suggesting a pK_a of 14.8 for that functional group. Silva and coworkers[67] determined the pK_a values of the hydroxyl group in citrate (14.4), malic acid (14.5) and lactic acid (15.1). For reference, the pK_a value of methanol is 15.1, while that for ethanol is 16–18. There are polysaccharides where specificity of modification in the sense of modifying few sites in a population is not a consideration and modification is designed to change the physical and chemical properties. Starch is an excellent example of such a polysaccharide and modification is a property of the constituent glucose residues. Tomasik and Schilling[68] have recently reviewed the various chemical modifications of starch including oxidation, ether and ester formation, acetylation, halogenation, amination, and carbamoylation.

Other reactions that can be of interest for carbohydrates are vicinal diol cleavage with periodate and condensation of terminal reducing sugar with an amine. Reaction of periodic acid with a vicinal diol cleavage is the well-known Malaprade reaction,[69] which yields a dialdehyde. This reaction was used by McManus[70] to couple with Schiff reagent to detect mucin. This reaction was subsequently shown to be of great value in histology.[71] The Schiff reagent dates back to the work of Scott and Clayton,[72] Hotchkiss,[73] and McManus and Cason[74] for use in histological staining of polysaccharides after oxidation. The periodic acid-Schiff (PAS) reagent is used for the detection of glycoproteins in gel electrophoresis.[75] Hotchkiss's article deserves additional mention. Rollin Hotchkiss was an exceptional individual who was a member of the Rockefeller Institute for Medical Research; there was a time when it was rumored that it was more difficult to become a member of the Rockefeller Institute than to gain membership in the National Academy of Sciences. Hotchkiss made contributions in a variety of areas most notably on the role of DNA with Oswald Avery. In fairness to Hotchkiss, the work cited herein on histological detection of polysaccharide was performed during World War II as defense research. As noted in the 1948 article, Hotchkiss's research was performed in parallel with that of McManus at Oxford, which was published as a short note in *Nature*. Hotchkiss's article provides considerable insight into the development of the use of the PAS reaction for the detection of polysaccharide. While the action of periodic acid on vicinal hydroxyl groups was described by Malaprade,[69] Jackson and Hudson[76] established the chemistry in 1938 providing the necessary background for Hotchkiss's work together with earlier work on the chemistry of Schiff reagent.[77] A mechanism for the PAS reaction is shown in Figure 7.1. There are examples of spurious reaction with PAS reagent. MacCallum[78] reported that the α-aminoadipic-δ-semialdehyde in elastin provided a positive reaction with Schiff reagent without periodic acid oxidation. Kimura and Stadtman[79] reported that the positive PAS reaction with glycine reductase selenoprotein A is a result of peptide bond cleavage and carbonyl group formation.

The aldehyde function at the anomeric carbon in monosaccharides can react with amines in the Maillard reaction. This reaction is best known for reducing sugars such as glucose and some disaccharides[80,81]; only limited information is available for large oligosaccharides.[82] Oligosaccharides with a terminal reducing sugar can be coupled to an aminophenyl disulfide, which can be bound to a gold surface (Figure 7.2).[83] Coupling may also be accomplished by use of a hydrazide

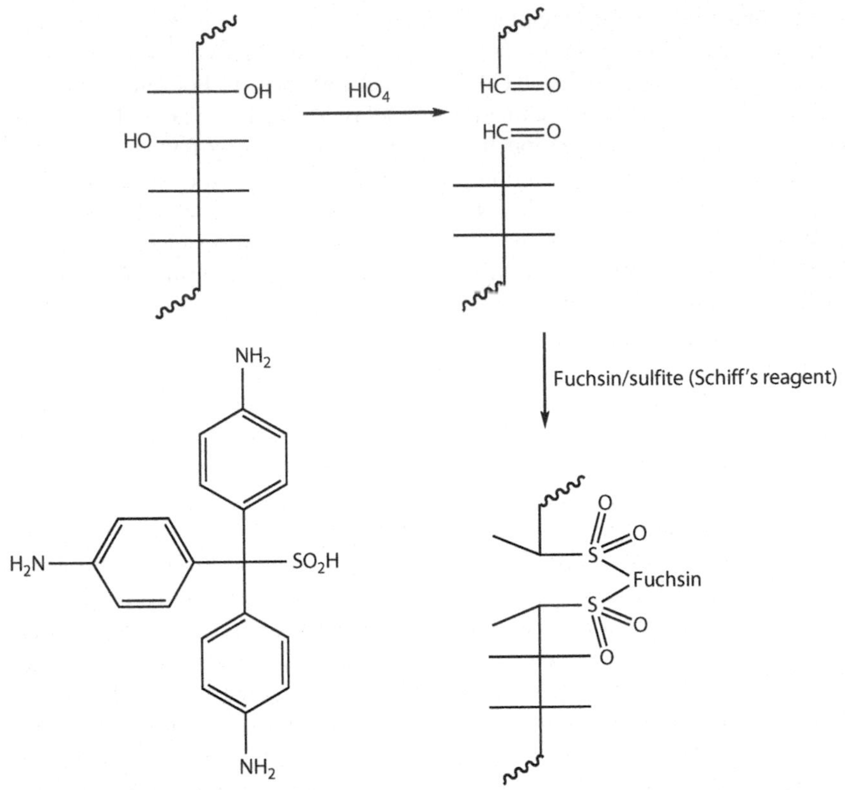

FIGURE 7.1 The PAS reaction for carbohydrate. Shown is the oxidation of a vicinal diol in a monosaccharide yielding dialdehyde, which then reacts with Schiff reagent (fuchsin-sulfite). Schiff reagent is prepared by dissolving 4-rosanilin (*p*-rosanilin) in sulfurous acid (sulfur dioxide in water). (For further information, see Jackson, E.L. and Hudson, C.S., The structure of the products of the periodic acid oxidation of starch and cellulose, *J. Am. Chem. Soc.* 60, 989–991, 1938; Hotchkiss, R.D., A microchemical reaction resulting in the staining of polysaccharide structures in fixed tissue preparations, *Arch. Biochem.* 16, 131–141, 1948; McManus, J.F.A. and Cason, J.E., Carbohydrate histochemistry studied by acetylation techniques, *J. Exp. Med.* 91, 651–654, 1950.)

matrix (Figure 7.2).[84] A related approach was used by Park and coworkers[85] who prepared hydrazide derivatives of oligosaccharides, which were then coupled to an epoxide matrix. Pochechueva and coworkers[86] prepared oligosaccharides labeled with terminal biotin, which could be bound to streptavidin-coated beads. Hsiao and coworkers[87] used a boronate matrix for binding oligosaccharides for preparation of carbohydrate microarray.

In consideration of their value in biotechnology, the chemical modification of starch, dextran, cellulose, hyaluron, and chitin is worth considering in some detail.

Starch is the most common polysaccharide and is found in green plants where it is used as energy storage. Starch is composed of glucose units joined by α-1,4-glycosidic bonds (amylose) and can contain branched structures (amylopectin),

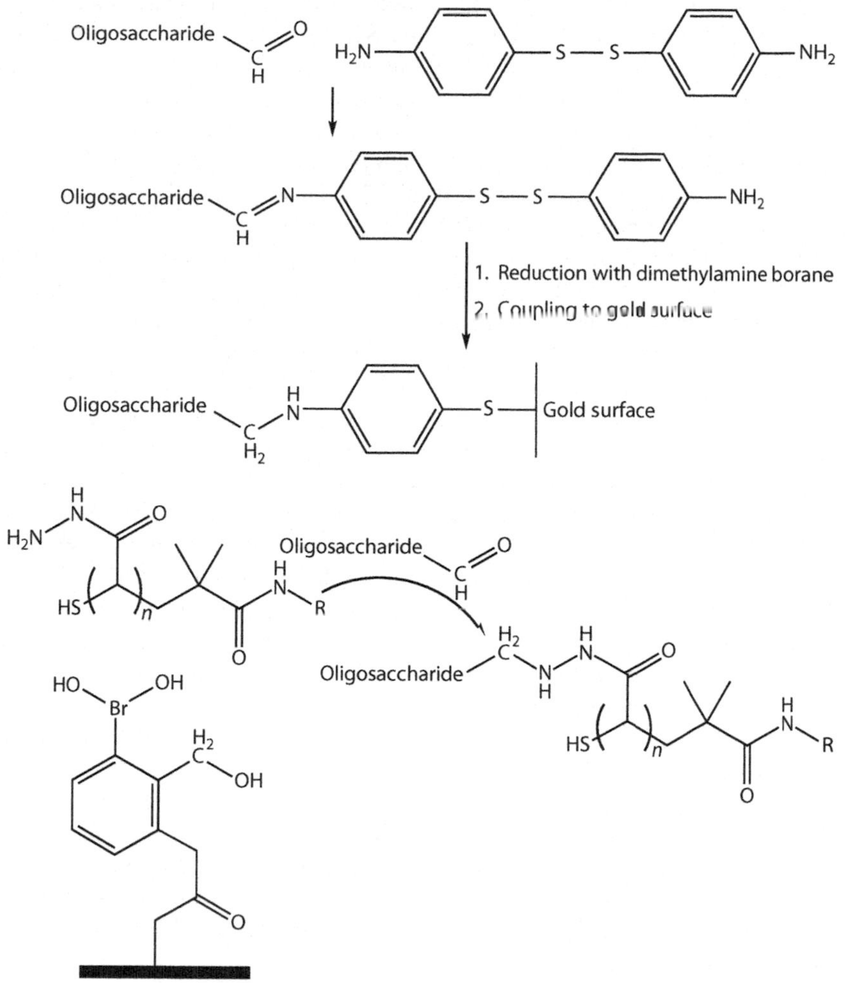

FIGURE 7.2 The reaction of terminal reducing sugars of oligosaccharides. Shown is the immobilization of an oligosaccharide to a gold surface via coupling via the terminal reducing sugars. The oligosaccharide is coupled to 4-aminodisulfide via the aldehyde function and then reduced; while the disulfide is used for the coupling, the interaction with the gold surface occurs via the sulfhydryl groups (Seo, J.H., Adachi, K., and Lee, B.K., Facile and rapid direct gold surface immobilization with controlled orientation for carbohydrates, *Bioconjug. Chem.* 18, 2197–2201, 2007). Shown below is the coupling of oligosaccharide to a hydrazide scaffold (Godula, K. and Bertozzi, C.R., Synthesis of glycopolymers for microarray applications via ligation of reducing sugars to a poly(acryloyl hydrazide) scaffold, *J. Am. Chem. Soc.* 132, 9963–9965, 2010). Also shown is 5-amino-2-hydroxyphenyl boronic acid coupled to a matrix for use as a carbohydrate microarray (Hsiao, H.Y., Chen, M.L., Wu, H.T. et al., Fabrication of carbohydrate microarrays through boronate formation, *Chem. Commun.* 47, 1187–1189, 2010).

which are cross-linked by α-1,6-glycosidic bonds. The amylose chains can be quite long (1000 monomer units), while the amylopectin molecules are smaller. The review of starch chemistry by Tomasik and Schilling[68] has been cited earlier as well as the issue of specific chemical modification of polysaccharides. The majority of reports on the chemical modification of starch have focused on developing a relatively homogeneous property, which is to be used in the general area of bioplastics. First, a short note on bioplastics; bioplastics may refer to biodegradable plastics made from precursors derived from oil, biodegradable plastics made from biologically derived precursors such as polyhydroxyalkanoic acids,[88–90] or the term might refer to "plastics" made from the processing of polysaccharides or proteins.[91] It should be recognized that polysaccharides are organic polymers that can, in the presence of plasticizers, form useful plastics.[92,93] The hydroxyl groups of starch can be modified with organic anhydrides[94] to obtain a thermoplastic product. In this study, Shin and coworkers used maleic anhydride to obtain a thermoplastic starch. Octenyl succinic anhydride (Figure 7.3) has been used to modify starch improving physical properties.[95,96] The extent of modification can be controlled by varying the amount of octenyl succinic anhydride added to the reaction mixture. The product of the reaction can be characterized by infrared spectroscopy. Wetzel and coworkers[97] used a synchrotron infrared source with confocal imaging to show heterogeneity of modification of a single starch granule by octenyl succinic anhydride. Hydroxypropylation of starch (Figure 7.3) results in a product with altered properties, which can be used for a variety of purposes[98–100] including two-phase separation of proteins.[101,102] Hydroxyethyl starch (Figure 7.3) is extensively as a plasma substitute similar to albumin.

Cyclodextrins are small, cyclic polysaccharides that are formed from amylose by enzyme action.[103] Cyclodextrins contain varying numbers of monomer units; the most common is β-cyclodextrin, which contains seven monomer units.[104] Cyclodextrins have the ability to form inclusion complexes, which has proved useful in drug delivery[105–112]; cyclodextrin derivatives have also proved useful for chiral separation.[113] The preparation of chemically modified forms such as the hydroxypropyl derivative (Figure 7.3) and the sulfobutylether (Figure 7.3) have proved to improve pharmacokinetic behavior.[114,115]

Dextran is polysaccharides produced by bacterial starch in that it is composed of glucose units predominantly linked by 1,6-glycosidic bonds, but with 1–2 glycoside bonds, 1–3 glycosidic bonds, and 1,4-glycosidic bonds providing for considerable branching and relative resistance to hydrolytic enzymes.[116–120] The stability of dextran has permitted a variety of uses including use as a matrix for size-exclusion chromatography.[121–123] Dextran has considerable use as a plasma expander[124–127] as described earlier for hydroxyethyl starch. Native dextran is subjected to processing including hydrolysis and fractionation to obtain materials used in clinical studies.[120] Dextran is subject to a variety of chemical modifications including phosphorylation, sulfation, formation of organic esters, and ethers including sulfopropyl and 2-diethylaminoethyl (DEAE) dextran.[120] These are industrial scale processes using conventional chemistry. There are several modifications of dextrans worth further comment. First, Richard and coworkers[128] coupled dextran to a hydroxylamine derivative of Brij-58P to produce an amphiphilic polysaccharide

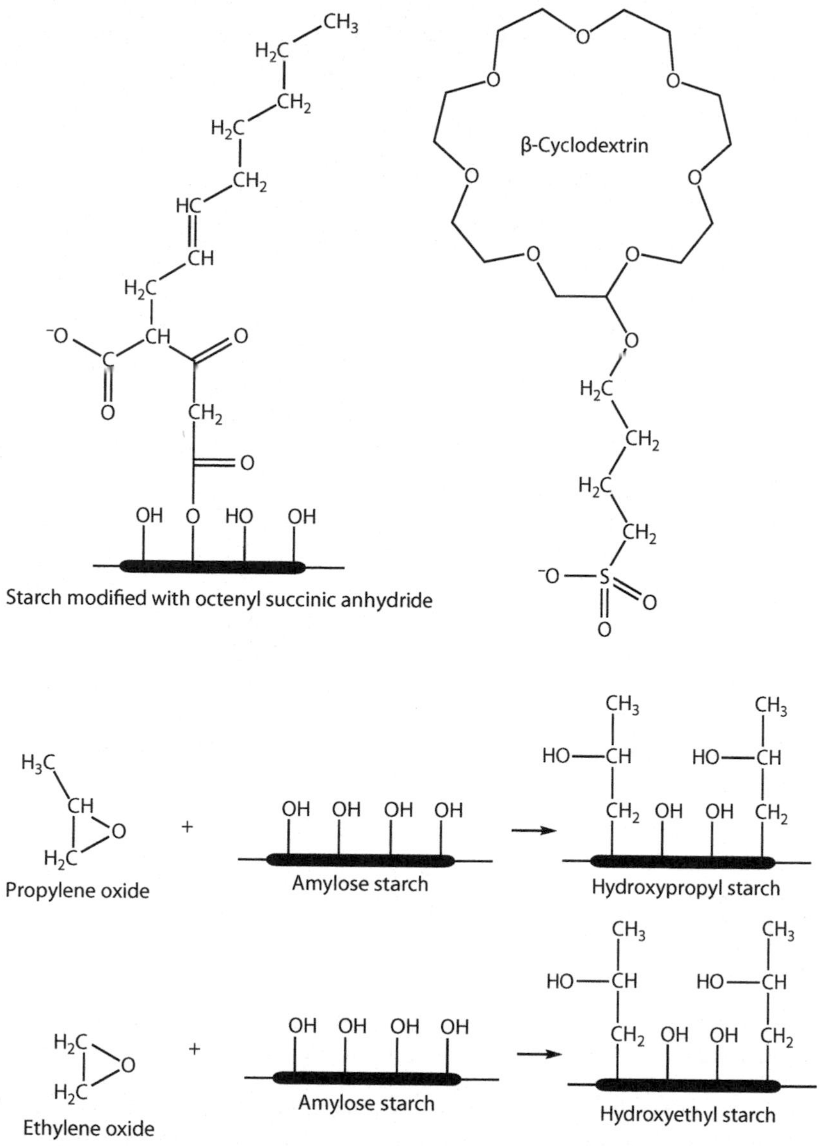

FIGURE 7.3 The chemical modification of starch. Shown is the product of the reaction of octenyl succinic anhydride with amylose starch (Wang, J., Su, L., and Wang, S., Physicochemical properties of octenyl succinic anhydride-modified potato starch with different degrees of substitution, *J. Sci. Food Agric.* 90, 424–429, 2010) and a schematic representation of sulfobutylether-β-cyclodextrin (Zin, V., Rajewski, R.A., and Stella, V.J., Effect of cyclodextrin charge on complexation of neutral and charged substrates: Comparison of $(SBE)_{7M}$-β-CD to HP-β-CD, *Pharm. Res.* 18, 667–673, 2001). Also shown is the reaction of propylene oxide with starch to form hydroxypropyl starch (Pal, J., Singhai, P.R., and Kulkarni, P.R., A comparative account of conditions of synthesis of hydroxypropyl derivative from corn and amaranth starch, *Carbohyd. Polym.* 43, 55–103, 2000) and with ethylene oxide to form hydroxyethyl starch.

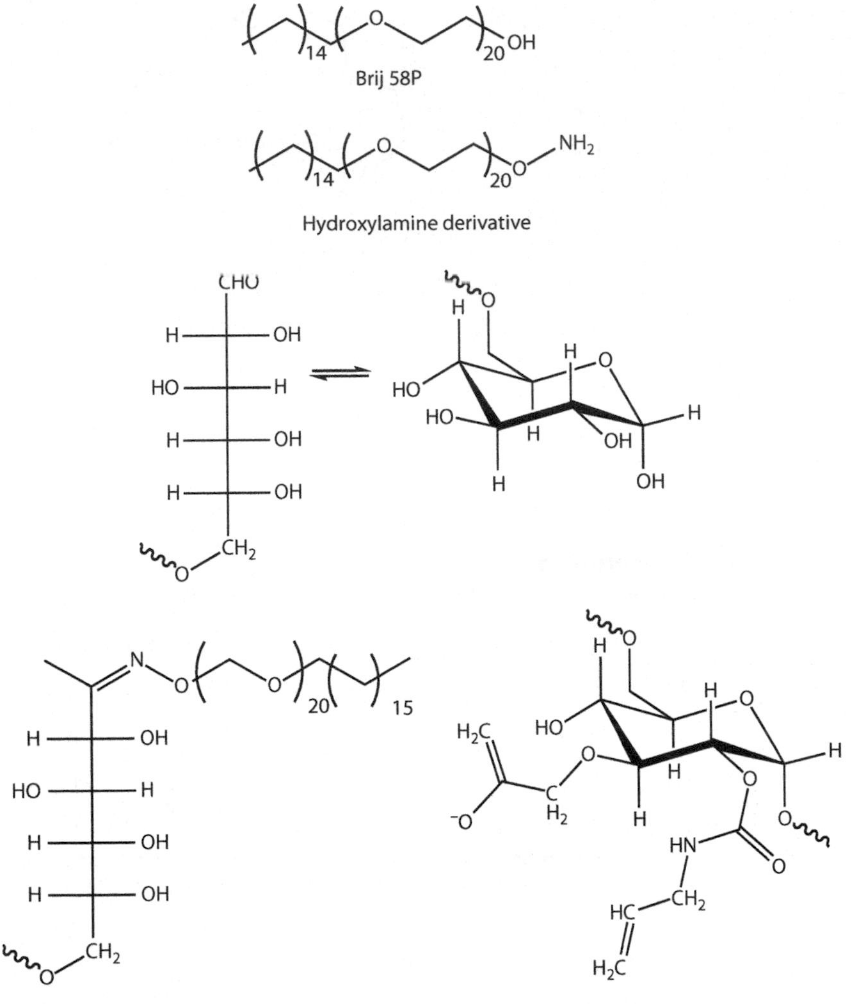

Brij 58P

Hydroxylamine derivative

FIGURE 7.4 Example of the chemical modification of dextran. Shown is the modification of dextran via the terminal reducing sugar (Richard, A., Barras, A., Younes, A.B. et al., Minimal chemical modification of reductive end of dextran to produce an amphiphilic polysaccharide able to incorporate onto lipid nanocapsules, *Bioconjug. Chem.* 19, 1491–1495, 2008). Also shown is a dextran derivative, which can be cross-linked by UV irradiation (Sun, G., Shen, Y.I., Ho, C.C. et al., Functional groups affect physical and biological properties of dextran-based hydrogels, *J. Biomed. Mater. Res.* 93A, 1080–1090, 2010).

(Figure 7.4) that could insert into a lipid nanocapsule. There has been considerable interest in the use of dextrans in hydrogels.[129–132] Sun and coworkers[132] have recently described a group of chemically modified dextrans that can be cross-linked via an acrylate derivative. Proteins have been coupled to dextran in attempts to improve functional properties. Takakura and coworkers[133] coupled soybean trypsin inhibitor to periodate-oxidized dextran; the Schiff base conjugate was stabilized by sodium borohydride. Similar chemistry was used by Sundaram and Venkatesh[134] to couple dextran to chymotrypsin. Low-molecular-weight drugs may be coupled to dextran and other polysaccharides yielding prodrugs with improved targeting capabilities.[135]

Cellulose is a structural polysaccharide found in plants. Cellulose is large with some 3000–4000 monomer units connected by β-1,4-glycosidic bonds unlike the α-1,4-glycosidic bonds found in starch. Cellulose is insoluble in neutral aqueous solvents but can be dissolved in strong acids, strong bases, and some organic solvents. The β-1,4-glycosidic bond is resistant to most glycosidases. The chemical properties of cellulose provide the basis for its structural stability. The periodate oxidation of cellulose and other polysaccharides has been recently reviewed by Kristiansen and coworkers.[136] Periodate oxidation does change the physical structure of polysaccharides. Periodate oxidation converts *cis*-diols to dialdehydes; the dialdehydes in cellulose can be further converted to diacids by mild acid sodium chlorite.[137] Cellulose that is 100% oxidized to the dicarboxylic acid form is soluble, while 70% oxidation provides a product that is still insoluble and has the capacity to bind metal ions. O'Connell and coworkers[138] have reviewed other chemical modifications of cellulose such as modification of wood pulp with citric acid anhydride, which provides adsorbents for heavy metal ions. Isogai and coworkers[139] have used electrochemical TEMPO-mediated oxidation to obtain carboxyl groups in cellulose. Other approaches include grafting and other general chemical modifications.[140–142] The process of Kraft pulping can be viewed as a chemical modification process and the generation of ethanol involves the chemical modification of cellulose.[143] It is reported that the chemical modification of cellulose does diminish enzymatic hydrolysis.[144]

Discussion of the chemical modification to this point has focused on what might be considered industrial polysaccharides, which are produced in tons. There are some polysaccharides that might be considered to be boutique polysaccharides and include hyaluronan, heparin, and chitin. All of these polymers have important medical uses and are chemically modified during *in vivo* "manufacture" and thus are more chemically heterogeneous than the homopolymers described earlier.

Hyaluronan is a linear polysaccharide consisting of a repeating disaccharide with one residue possessing an acetamido group and the other a carboxylic acid (uronic acid).[145] Hyaluronan can be quite large (6000–8000 kDa). The large size and linear nature combined with the high negative charge are responsible for the viscosity of hyaluronan solutions, which is critical for its function as a lubricating agent.[146–149] Most of the chemical modification of hyaluronan (Figure 7.5) focuses on hydrogel development.[150–153] Zhao and coworkers[150] prepared a "double-cross-linked" hyaluronan hydrogel in a two-step process using the same cross-linking reagents but different reaction conditions. The first step performed under strong alkaline conditions yields a hyaluronan derivative cross-linked by ether bonds between hydroxyl groups (Figure 7.5). The second step performed at acid pH yielded an ester derivative between carboxyl groups. Segura and coworkers[151] modified carboxyl groups with a diamine using carbodiimide chemistry as a method for attaching biotin to hyaluronan. These investigators also used a poly (ethylene glycol) diepoxide to cross-link hyaluronan

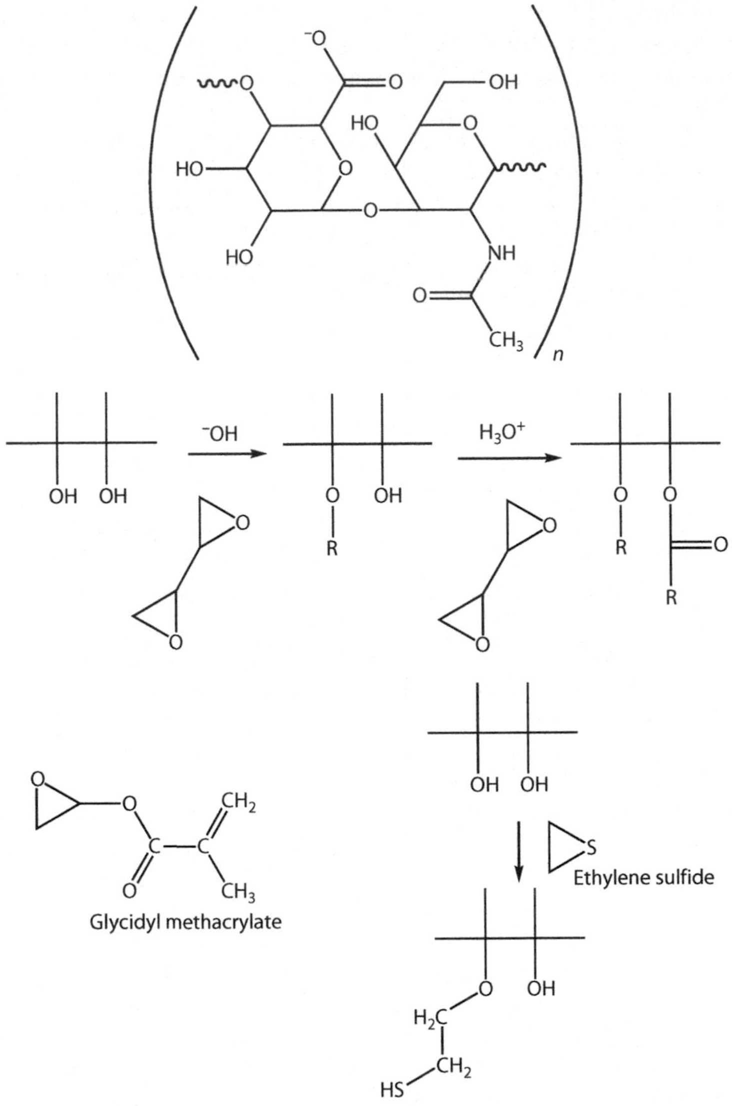

FIGURE 7.5 Chemical modification of hyaluronan. Shown is a schematic structure of the repeated glucosamine-glucuronic acid disaccharide of hyaluronan and a two-step cross-linking process that uses pH as the control for differentiating between ether formation and ester formation (Zhao, X.B., Fraser, J.E., Alexander, C. et al., Synthesis and characterization of a novel double cross-linked hyaluronan hydrogel, *J. Mater. Sci. Mater. Med.* 13, 11–16, 2002). Also shown is glycidyl methacrylate, which was used for photocross-linking of hyaluronan (Jha, A.K., Malik, M.S., Farach-Carson, M.C. et al., Hierarchically structured hyaluronic acid-based hydrogel matrices via the covalent integration of microgels into macroscopic networks, *Soft Matter* 6, 5045–5055, 2010) and the reaction of ethylene sulfide with hyaluronan to yield a thioethyl ether derivative (Serban, M.A., Yang, G., and Prestwich, G.D., Synthesis, characterization and chondroprotective properties of a hyaluronan thioethyl ether derivative, *Biomaterials* 29, 1388–1399, 2008).

via hydroxyl group on the pyranose ring. Richter and coworkers[152] modified hyaluronan with 4-vinylaniline using carbodiimide technology. The modified hyaluronan was photopolymerized to the polylactic acid membrane. Jha and coworkers[153] have modified hyaluronan with glycidyl methacrylate (Figure 7.5) and then used photo-cross-linking to prepare hydrogels. Serban and coworkers[154] prepared a thioethyl ether derivative of hyaluronan (Figure 7.5), which had unique therapeutic potential. Mlcochová and coworkers[155] used cyanogen bromide coupling to prepare carbamate-linked alkyl derivatives of hyaluronan. The reader is directed to a review[3] by Prestwich and Kuo for further discussion of the development and use of chemically modified hyaluronan derivatives.

Heparin is a sulfated glycosaminoglycan/proteoglycan derived from biological tissue[156,157] and used primarily as an acute anticoagulant drug, which also has an effect on lipid metabolism.[158] Heparin, with its various modifications (Figure 7.6) including the variable content of protein remaining from the manufacturing processes, is a heterogeneous protein with multiple sites available for modification. Iverius[159] suggested that the cyanogen bromide coupling of heparin to agarose beads occurs via a serine or peptide residue at the reducing end of the polysaccharide. Gentry and Alexander[160] also used cyanogen bromide to prepare heparin bound to agarose. Danishefsky and coworkers[161] compared cyanogen bromide coupling of heparin to agarose with heparin coupled to aminohexyl-agarose using carbodiimide technology. Some differences were observed in the performance of the two matrices. Nadkarni and coworkers[162] modified the reducing end of heparin producing several derivatives (Figure 7.6) including 2,6-diaminopyridinyl and a lactone. Fry and coworkers[163] used iminothiolane (Traut's reagent) to modify heparin with a terminal amino group to obtain a derivative with a free sulfhydryl available for coupling to a matrix. Other chemical modification studies on heparin have focused on inclusion of hydrogels via cross-linking through the carboxyl or the amino (after deacetylation) groups.[164] Tae and coworkers[165] converted the carboxyl groups into thiol functions by carbodiimide-mediated reaction with cystamine. Kim and coworkers[166] also used thiolated heparin to form a hydrogel with acrylated poly (ethylene glycol) for encapsulating cells.

Chitin is another long-chain glucose-based polymer (Figure 7.6); the monomer unit is *N*-acetylglucosamine.[167] Chitin is a major component of the exoskeletons of insects, crabs, lobsters, and other related organisms. Chitosan is derived from chitin by alkaline hydrolysis[168] and is a copolymer of *N*-acetylglucosamine (20%) and *N*-glucosamine (80%)[169] and a heterogeneous material.[170] Chitin is the second most abundant naturally occurring polysaccharide[169] and has a role in the mineralization of the exoskeleton of arthropods (phylum: Arthropoda).[169] Chitin is also suggested for clinical application for hard tissue problems.[171–173] Chitin is similar to cellulose in composition (chitin is essentially a homopolymer of *N*-acetylglucosamine while cellulose is a homopolymer of glucose) and function (both are structural components). Chitin is less soluble and less reactive than cellulose. Thus, the modification of chitin usually requires vigorous conditions and modification is not site specific (residue specific). There is a recent report[174] that selective protection/deprotection can permit regioselective modification

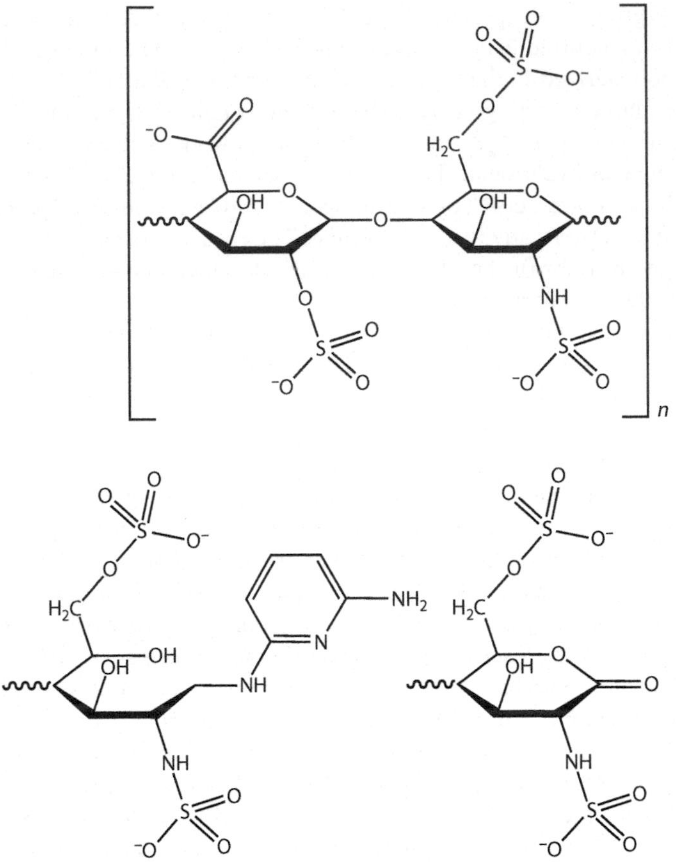

2,4-Diaminopyridinyl heparin Heparin lactone

FIGURE 7.6 Heparin structure and chemical modification. Shown is the representation of the functional disaccharide unit with carboxyl group, *O*-sulfation, and *N*-sulfation (see Petitou, M., Casu, B., and Lindahl, U., 1976–1983, a critical period in the history of heparin: The discovery of the antithrombin binding site, *Biochimie* 85, 83–89, 2003 for more accurate structural information). It is important to emphasize that the *N*-sulfated derivative is presented; the amino group may also be acetylated. Shown also are derivatives of the reducing end unit permitting orientation on a support (Nadkarni, V.D., Pervin, A., and Linhardt, R.J., Directional immobilization of heparin onto beaded supports, *Anal. Biochem.* 222, 59–67, 1994).

(regioselective in this sense refers to driving modification toward one of the several hydroxyl groups on the pyranose ring). The observation[175] that β-chitin obtained from squid is more reactive than α-chitin obtained from shrimp has proved useful in subsequent studies where trimethylsialylation was used to modify hydroxyl groups.[176] The modification of chitin can use chitin whiskers,[177,178] which are microfibrils containing protein in addition to the polysaccharide.

Chitin whiskers can be incorporated into natural rubber by chemical modification to form nanocomposites.[179] Cunha and Gandini[180] have recently reviewed progress on using chemical and/or physical modification to convert chitin and chitosan to more hydrophobic derivatives.

The most useful chemical modification of chitin is the conversion to chitosan by alkaline hydrolysis.[181] The use of chitin deacetylase[182] is attracting attention as the chitosan product is more homogeneous than the product derived from alkaline hydrolysis; however, the physical characteristics of chitin present issues with respect to process efficiency.[183] Chitin deacetylase is of critical importance in the catabolism of chitin in the marine environment.[184] The presence of a free amino group provides more opportunities for the chemical modification both in somewhat improved solubility and a reasonable nucleophile for modification. Chitosan, nonetheless, is still essentially a large homopolymer with limited solubility in neutral solvent.[185] There is considerable interest in chitosan for drug delivery including gene therapy vectors.[183–189] In addition to gene therapy applications,[190–192] Alatorre-Meda et al.[192] have evaluated the effect of chitosan size and charge density on the transfection ability. The term valence is used together with charge density and pH as attributes to DNA binding and transfection efficiency. The term valence is described by Maurstad and coworkers[193] to describe total charge per chitosan molecule. This is different from the common understanding of valence in chemistry that refers to the number of bonding electrons in an atom. The term is also used to describe number of antigen binding sites on an antibody or vice versa. A quick check of PubMed shows that the term is also used in psychology but *The Oxford English Dictionary*[194] is mute on this point but does define valency as strength. Chitosan is of value for colon drug delivery.[195–204]

Chitosan can be subjected to chemical modification using techniques as described earlier for other polysaccharides to improve characteristics for therapeutic applications.[168] This is an appropriate time to emphasize that the various biological polymers that have been described in the current work are, in fact, chemical polymers such as polyurethane, polystyrene, or polyacrylate but differ from these polymers in the variety and complexity of the monomer units. Polysaccharides, such as chitosan, can be combined via graft polymerization with classical polymer monomers such as methyl acrylate or acrylonitrile.[205–207] Jenkins and Hudson[205] prepared a graft polymer of chitosan and methyl acrylate using heterogeneous graft copolymerization (Figure 7.7). Chitosan was trifluoroacetylated and methyl acrylate polymer "built" from the trichloromethyl group upon reaction with manganese carbonyl using photoactivation. For the non-polymer chemist, the term heterogeneous graft polymerization describes a reaction occurring in two phases.[208] El-Sherbiny and coworkers[209] address the issue of chitosan solubility[185] by preparing the carboxymethyl derivative (Figure 7.7) using reaction with chloroacetic acid. This derivative was then graft copolymerized with methacrylic acid (Figure 7.7) and combined with alginate as biodegradable hydrogel for oral drug delivery. Graft polymerization of carboxymethyl chitosan with acrylic acid has also been reported.[210]

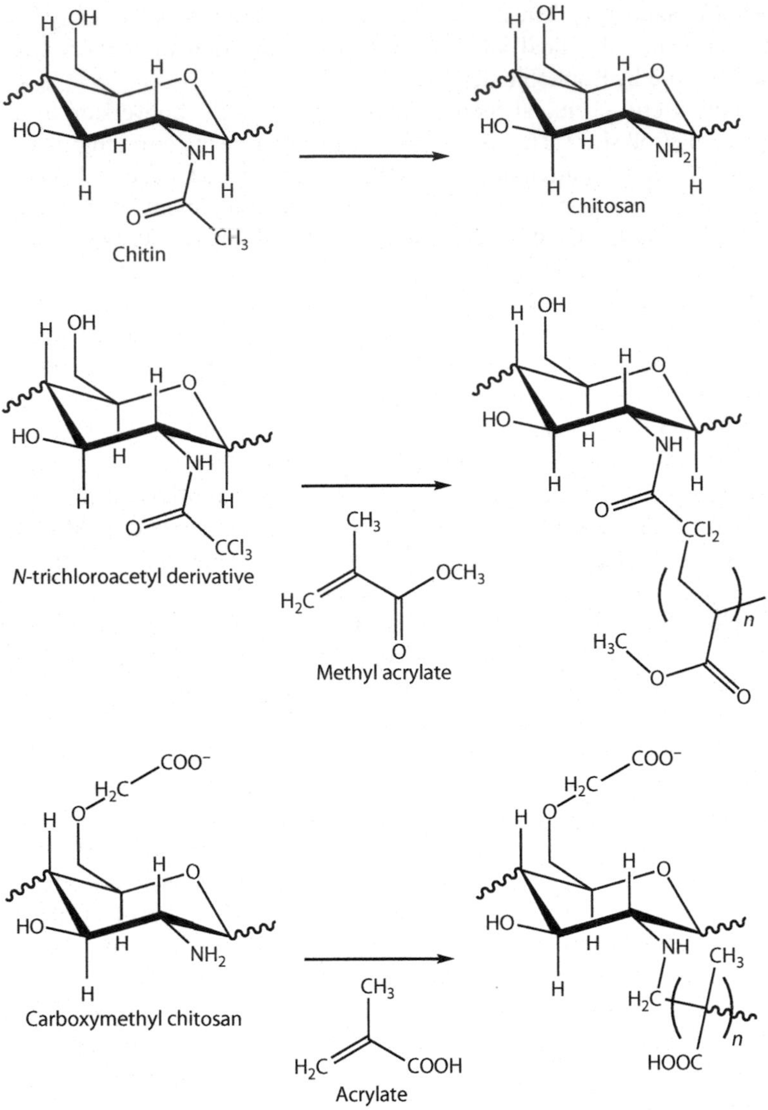

Chitin

Chitosan

N-trichloroacetyl derivative

Methyl acrylate

Carboxymethyl chitosan

Acrylate

FIGURE 7.7 Structure and reactions of chitin and chitosan. The monomer unit of chitin is *N*-acetylglucosamine (left), which can be converted to chitosan where the monomer unit is glucosamine. Chitosan can be modified by trichloroacetic anhydride to yield the *N*-trichloroacetyl derivative, which can be subjected to heterogeneous copolymerization with methyl acrylate (Jenkins, J.W. and Hudson, S.M., Heterogeneous graft copolymerization of chitosan powder with methyl acrylate using trichloroacetyl-manganese carbonyl co-initiation, *Macromolecules* 35, 3413–3419, 2002). Also shown is the formation of a graft copolymer of carboxymethyl chitosan and methyl acrylate (El-Sherbiny, I.M., Abdel-Bary, E.M., and Harding, D.R.K., Preparation and *in vitro* evaluation of new pH-sensitive hydrogel beads for oral delivery of protein drugs, *J. Appl. Polym. Sci.* 115, 2828–2837, 2010).

REFERENCES

1. O'Dell, J., The use of modified starch in the food industry, in *Polysaccharides in Food*, J.M.V. Blanshard and J.R. Mitchell (eds.), Butterworths, London, U.K., Chapter 11, pp. 171–181, 1979.
2. Hoover, R. and Sosulski, F.W., Composition, structure, functionality, and chemical modification of legume starches: A review, *Can. J. Physiol. Pharmacol.* 69, 79–92, 1991.
3. Prestwich, G.D. and Kuo, J.W., Chemically-modified HA for therapy and regenerative medicine, *Curr. Pharm. Biotechnol.* 9, 242–245, 2008.
4. Lee, K.Y., Jeong, L., Kang, Y.O. et al., Electrospinning of polysaccharides for regenerative medicine, *Adv. Drug Deliv. Rev.* 61, 1020–1032, 2009.
5. Hansen, N.M. and Plackett, D., Sustainable films and coatings from hemicelluloses: A review, *Biomacromolecules* 9, 1493–1505, 2008.
6. Coutinho, D.F., Sant, S.V., Shin, H. et al., Modified gellan gum hydrogels with tunable physical and mechanical properties, *Biomaterials* 31, 7494–7502, 2010.
7. Katzenelienbogen, E., Kubler, J., Garnian, A. et al., Structural study and serological characterization of the *O*-specific polysaccharide of *Hafnia alvei* PCM 1185, another Hafnia *O*-antigen that contains 3,6-dideoxy-3-[(R)-3-hydroxybutyramido]-D-glucose, *Carbohydr. Res.* 293, 61–70, 1996.
8. Lesur, D., Gassama, A., Moreau, V. et al., Electrospray ionization mass spectrometry: A key analytical tool for the characterization of regioselectively derivatized maltooligosaccharides obtained starting from natural β-cyclodextrin, *Rapid Commun. Mass Spectrom.* 20, 747–754, 2006.
9. Campa, C., Coslovi, A., Fiamigni, A., and Rossi, M., Overview on advances in capillary electrophoresis-mass spectrometry of carbohydrates: A tabulated review, *Electrophoresis* 27, 2027–2050, 2006.
10. Perepelov, A.V., Shevelev, S.D., Liu, B. et al., Structures of the *O*-antigens of *Escherichia coli* 013, 0129, and 0135 related to the *O*-antigens of *Shigella flexneri*, *Carbohydr. Res.* 345, 1595–1599, 2010.
11. Shah, U. and Augsburger, L., Multiple sources of sodium starch glycolate, NF: Evaluation of functional equivalence and development of standard performance tests, *Pharmaceut. Dev. Technol.* 7, 345–359, 2002.
12. Zhang, Y., Zhang, J., Mo., X. et al., Modification, characterization and structure-anticoagulant activity relationships of persimmon polysaccharides, *Carbohyd. Polym.* 82, 515–520, 2010.
13. Hind, H.G., Manufacture of heparin, *Manuf. Chemist* 34, 510–514, 1963.
14. Coyne, E., Heparin—Past, present, and future, in *Chemistry and Biology of Heparin*, R.L. Lundblad, W.V. Brown, K.G. Mann, and H.R. Roberts (eds.), Elsevier/North Holland, New York, 1981.
15. Coyne, E., From heparin to heparin fractions and derivatives, *Semin. Thromb. Hemost.* 11, 10–12, 1985.
16. Casu, B. and Lindahl, U., Structural and biological interactions of heparin and heparan sulfate, *Adv. Carbohydr. Chem. Biochem.* 57, 159–206, 2001.
17. Ciancanu, L. and Kerek, F., A simple and rapid method for permethylation of carbohydrates, *Carbohydr. Res.* 131, 209–217, 1984.
18. Bourne, E.J., Stacey, M., Tatlow, J.C., and Tedder, J.M., Studies on the trifluoroacetic anhydride as a promoter of ester formation between hydroxy compounds and carboxylic acids, *J. Chem. Soc.* 2976–2979, 1949.
19. Aduru, S. and Chait, B.T., Californium-252 plasma desorption mass spectrometry of oligosaccharides and glycoconjugates. Control of ionization and fragmentation, *Anal. Chem.* 63, 1621–1625, 1991.

20. Rauvala, H., Finne, J., Krusius, T. et al., Methylation techniques in the structural analysis of glycoproteins and glycolipids, *Adv. Carbohydr. Chem. Biochem.* 38, 389–416, 1981.
21. Azadi, P. and Heiss, C., Mass spectrometry of *N*-linked glycans, *Methods Mol. Biol.* 534, 37–51, 2009.
22. Hanisch, F.G. and Müller, S., Analysis of methylated *O*-glycan alditols by reversed-phase NanoLC coupled CAD-ESI mass spectrometry, *Methods Mol. Biol.* 534, 107–115, 2009.
23. BeMiller, J.N., Occurrence and significance, in *Glycoscience Chemistry and Biology*, B.O. Fraser-Reid, K. Tatsuta, and J. Thiem (eds.), Springer-Verlag, Berlin, Germany, Chapter 6.1, pp. 1865–1881, 2001.
24. IUPAC, Nomenclature for carbohydrates, *Adv. Carbohydr. Chem. Biochem.* 52, 43–177, 1997.
25. Chester, M.A., Nomenclature for glycolipids, *Adv. Carbohydr. Chem. Biochem.* 55, 311–326, 2000.
26. Lundblad, R.L. and MacDonald, F.M. (eds.), *Handbook of Biochemistry and Molecular Biology*, 4th edn., CRC/Taylor & Francis, Boca Raton, FL, Section 5, 2010.
27. Lindberg, B., Methylation analysis of polysaccharides, *Methods Enzymol.* 28, 178 195, 1972.
28. Lindberg, B. and Lönngren, J., Methylation analysis of complex carbohydrates: General procedure and application for sequence analysis, *Methods Enzymol.* 50, 3–33, 1978.
29. Franks, F., Structural interactions and the solution behavior of carbohydrates, in *Polysaccharides in Food*, J.M.V. Blanshard and J.R. Mitchell (eds.), Butterworths, London, U.K., Chapter 3, pp. 33–50, 1979.
30. Morris, E.R., Polysaccharides structure and conformation in solution and gels, *Polysaccharides in Food*, J.M.V. Blanshard and J.R. Mitchell (eds.), Butterworths, London, U.K., Chapter 2, pp. 15–31, 1979.
31. Mitchell, J.R., Rheology of polysaccharide solutions and gels, in *Polysaccharides in Food*, J.M.V. Blanshard and J.R. Mitchell (eds.), Butterworths, London, U.K., Chapter 4, pp. 51–72, 1979.
32. Blanshard, J.M.V., Physicochemical aspects of starch gelatinization, in *Polysaccharides in Food*, J.M.V. Blanshard and J.R. Mitchell (eds.), Butterworths, London, U.K., Chapter 9, pp. 139–152, 1979.
33. Silioc, C., Maleki, A., Zhu, K. et al., Effect of hydrophobic modification on rheological and swelling features during chemical gelation of aqueous polysaccharides, *Biomacromolecules* 8, 719–728, 2007.
34. Song, Y., Zhou, J., Li, Q. et al., Solution properties of the acrylamide-modified cellulose polyelectrolytes in aqueous solution, *Carbohydr. Res.* 344, 1332–1339, 2009.
35. Witezak, Z.J., Properties, in *Glycoscience Chemistry and Chemical Biology*, B.O. Fraser-Reid, K. Tatsuta, and J. Thiem (eds.), Springer-Verlag, Berlin, Germany, Chapter 6.2, pp. 1883–1893, 2001.
36. Allen, H.J. and Kisailus, E.C. (eds.), *Glycoconjugates. Composition, Structure, and Function*, Marcel Dekker, New York, 1992.
37. Fantini, J., Interaction of proteins with lipid rafts through glycolipid binding domains: Biochemical background and potential therapeutic applications, *Curr. Med. Chem.* 14, 2911–2917, 2007.
38. Heath, E.C., Complex polysaccharides, *Annu. Rev. Biochem.* 40, 29–56, 1971.
39. Bhattacharjya, S., *De novo* designed lipopolysaccharide binding peptides: Structure based development of antiendotoxic and antimicrobial drugs, *Curr. Med. Chem.* 17, 3080–3093, 2010.
40. Brandenburg, K., Schromm, A.B., and Gutsmann, T., Endotoxins: Relationship between structure, function, and activity, *Subcell. Biochem.* 53, 53–67, 2010.

41. Caliceti, P. and Veronese, F.M., Pharmacokinetics and biodistribution properties of poly (ethylene glycol)—Protein conjugates, *Adv. Drug Deliv. Rev.* 55, 1261–1277, 2003.
42. Chilukuri, N., Parikh, K., Sun, W. et al., Polyethylene glycosylation prolongs the circulatory stability of recombinant human butyrylcholinesterase, *Chem. Biol. Interact.* 157–158, 115–121, 2005.
43. Solá, R.J. and Griebenow, K., Glycosylation of therapeutic proteins: An effective strategy to optimize efficacy, *BioDrugs* 24, 9–21, 2010.
44. Schneider, S., Ueberberg, S., Korobeynikov, A. et al., Synthesis and evaluation of a glibenclamide glucose-conjugate: A potential new lead compound for substituted glibenclamide derivatives as islet imaging agents, *Regul. Pept.* 139, 122–127, 2007.
45. Prante, O., Einsidel, J., Hauber, R. et al., 3,4,6-Tri-*O*-acetyl-2-deoxy-2[^{18}F] fluoroglucopyranosyl phenythiosulfonate: A thiol-reactive agent for the chemoselective ^{18}F-glycosylation of peptides, *Bioconjug. Chem.* 18, 254–262, 2007.
46. Cai, G., Jiang, M., Zhang, B. et al., Preparation and biological evaluation of a glycosylated fusion interferon directed to hepatic receptors, *Biol. Pharm. Bull.* 32, 440–443, 2009.
47. Yang, W., Mou, T., Guo, W. et al., Fluorine-18 labeled galactosylated chitosan for asialoglycoprotein-receptor-mediated hepatocyte imaging, *Bioorg. Med. Chem. Lett.* 20, 4840–4844, 2010.
48. Di Stefano, G., Lanza, M., Busi, C. et al., Conjugates of nucleoside analogues with lactosaminated human albumin to selectively increase the drug levels in liver blood: Requirements for a regional chemotherapy, *J. Pharmacol. Exp. Ther.* 301, 638–642, 2002.
49. Fiume, L. and Di Stefano, G., Lactosaminated human albumin, a hepatotropic carrier of drugs, *Eur. J. Pharm. Sci.* 40, 253–262, 2010.
50. Kogelberg, H., Tolner, B., Sharma, S.K. et al., Clearance mechanism of a mannosylated antibody-enzyme fusion protein used in experimental cancer therapy, *Glycobiology* 17, 36–45, 2007.
51. Pandhal, J. and Wright, P.C., *N*-Linked glycoengineering for human therapeutic proteins in bacteria, *Biotechnol. Lett.* 32, 1189–1198, 2010.
52. Pozsgay, V., Oligosaccharide-protein conjugates as vaccine candidates against bacteria, *Adv. Carbohydr. Chem. Biochem.* 56, 153–199, 2001.
53. Hecht, M.L., Stallforth, P., Silva, D.V. et al., Recent advances in carbohydrate-based vaccines, *Curr. Opin. Chem. Biol.* 13, 354–359, 2009.
54. Guo, Z. and Wang, Q., Recent development in carbohydrate-based cancer vaccines, *Curr. Opin. Chem. Biol.* 13, 608–613, 2009.
55. Nicotra, F., Cipolla, L., Peri, R. et al., Chemoselective neoglycosylation, *Adv. Carbohydr. Chem. Biochem.* 61, 353–398, 2007.
56. Goff, R.D. and Thorson, J.S., Assessment of chemoselective neoglycosylation methods using chlorambucil as a model, *J. Med. Chem.* 53, 8129–8139, 2010.
57. Fitremann, J., Sucrose chemistry and application of sucrochemicals, *Adv. Carbohydr. Chem. Biochem.* 61, 217–292, 2008.
58. Queneau, Y., Fitreman, J., and Tronbotto, S., The chemistry of unprotected sucrose: The selectivity issue, *Compt. Rendus Chim.* 7, 177–188, 2004.
59. Queneau, Y., Chambert, S., Besset, C. et al., Recent progress in the synthesis of carbohydrate-based amphiphilic materials: The example of sucrose and isomaltulose, *Carbohydr. Res.* 343, 1999–2009, 2008.
60. Heidenreich, O., Pieken, W., and Eckstein, F., Chemically modified RNA: Approaches and applications, *FASEB J.* 7, 90–96, 1993.
61. Han, J. and Burke, J.M., Model for general acid-base catalysis by the hammerhead ribozyme: pH-activity relationships of G8 and G12 variants at the putative active site, *Biochemistry* 44, 7864–7870, 2005.

62. Lyne, P.D. and Karplus, M., Determination of the pK_a of the 2′-hydroxyl group on a phosphorylated ribose: Implications for the mechanism of hammerhead ribozyme catalysis, *J. Am. Chem. Soc.* 122, 166–167, 2000.
63. Acharya, S., Földesi, A., and Chattophadhyaya, J., The pK_a of the internucleotidic 2′-hydroxyl group in diribonucleoside (3′→5′) monophosphates, *J. Org. Chem.* 68, 1906–1910, 2003.
64. Saksmerprome, V. and Burke, D.H., Deprotonation stimulates productive folding in allosteric TRAP hammerhead ribozymes, *J. Mol. Biol.* 341, 685–694, 2004.
65. Bruice, T.C., Fife, T.H., Bruno, J.J., and Brandon, N.E., Hydroxyl group catalysis. II. The reactivity of the hydroxyl group of serine. The nucleophilicity of alcohols and the case of hydrolysis of their acetyl esters as related to their pK_a, *Biochemistry* 1, 6–12, 1962.
66. Ray, W.J., Jr., Long, J.W., and Owens, J.D., An analysis of the substrate-induced rate effect in the phosphoglucomutase system, *Biochemistry* 15, 4006–4017, 1976.
67. Silva, A.M.N., Kong, X., and Hider, R.C., Determination of the pK_a value of the hydroxyl group in the α-hydroxycarboxylates citrates, malate and lactate by 13C NMR: Implications for metal coordination in biological systems, *Biometals* 22, 771–778, 2009.
68. Tomasik, P. and Schilling, C.H., Chemical modification of starch, *Adv. Carbohydr. Chem.* 59, 175–403, 2004.
69. Malaprade, M.L., Action of polyalcohols on periodic acid or alkaline periodate, *Bull. Soc. Chim. France* 5, 833–852, 1934.
70. McManus, J.F.A., Histological demonstration of mucin after periodic acid, *Nature* 158, 202, 1946.
71. Wislocki, G.B., Rheingold, J.J., and Dempsey, E.W., The occurrence of the periodic acid-Schiff reaction in various normal cells of blood and connective tissue, *Blood* 4, 562–568, 1949.
72. Scott, H.R. and Clayton, B.P., A comparison of the staining affinities of aldehyde-fuchsin and the Schiff reagent, *J. Histochem. Cytochem.* 5, 336–352, 1952.
73. Hotchkiss, R.D., A microchemical reaction resulting in the staining of polysaccharide in fixed tissue preparations, *Arch. Biochem.* 16, 131–141, 1948.
74. McManus, J.F.A. and Cason, J.E., Carbohydrate histochemistry studied by acetylation techniques, *J. Exp. Med.* 91, 651–654, 1950.
75. Kapitany, R.A. and Zebrowski, E.J., A high resolution PAS stain for polyacrylamide gel electrophoresis, *Anal. Biochem.* 56, 361–369, 1973.
76. Jackson, E.L. and Hudson, C.S., The structure of the products of the periodic acid oxidation of starch and cellulose, *J. Am. Chem. Soc.* 60, 989–991, 1938.
77. Shoesmith, J.B., Sosson, C.E., and Hetherington, A.C., Abnormal reaction of certain aromatic aldehydes with Schiff's reagent, *J. Chem. Soc.* 2221–2230, 1927.
78. MacCallum, D.K., Positive Schiff reactivity of aortic elastin without prior periodic acid oxidation. Influence of maturity and a suggested source of the aldehyde, *Stain Technol.* 48, 117–122, 1973.
79. Kimura, Y. and Stadtman, T.C., Glycine reductase selenoprotein A is not a glycoprotein: The positive periodic acid-Schiff reagent test is the result of peptide bond cleavage and carbonyl group generation, *Proc. Natl. Acad. Sci. USA* 92, 2189–2193, 1995.
80. Rumpf, P., The color reaction of aldehydes known as the Schiff reaction, *Bull. Soc. Chim. France* 51, 503–528, 1932.
81. Bellavia, G., Cottone, G., Giuffrida, S. et al., Thermal denaturation of myoglobin in water—Disaccharide matrixes: Relation with the glass transition of the system, *J. Phys. Chem. B* 113, 11543–11549, 2009.
82. Kroh, L.W. and Schulz, A., News on the Maillard reaction of oligomeric carbohydrates: A survey, *Nahrung* 45, 160–163, 2001.
83. Seo, J.H., Adachi, K., Lee, B.K. et al., Facile and rapid direct gold surface immobilization with controlled orientation for carbohydrates, *Bioconjug. Chem.* 18, 2197–2201, 2007.

84. Godula, K. and Bertozzi, C.R., Synthesis of glycoproteins for microarray applications via ligation of reducing sugars in a poly(acryloyl hydrazide) scaffold, *J. Am. Chem. Soc.* 132, 9963–9965, 2010.
85. Park, S., Lee, M.R., and Shin, I., Chemical microarrays constructed by selective attachment of hydrazide-conjugated substances to epoxide surfaces and their applications, *Methods Mol. Biol.* 669, 195–208, 2010.
86. Pochechueva, T., Chinarev, A., Spengler, M. et al., Multiplex suspension array for human anti-carbohydrate antibody profiling, *Analyst* 136, 560–569, 2011.
87. Hsiao, H.Y., Chen, M.L., Wu, H.T. et al., Fabrication of carbohydrate microarrays through boronate formation, *Chem. Commun.* 47, 1187–1189, 2011.
88. Keshavarz, T. and Roy, I., Polyhydroxyalkanoates: Bioplastics with a green agenda, *Curr. Opin. Microbiol.* 13, 321–326, 2010.
89. Mato, T., Ben, M., Kennes, C., and Veiga, M.C., Valuable product production from wood mill effluents, *Water Sci. Technol.* 62, 2294–2300, 2010.
90. Chen, G.Q., Editorial: Sustainable bioplastics for future applications, *Biotechnol. J.* 5, 1117, 2010.
91. Gonzalez-Gutierrez, J., Partal, P., Garcia-Morales, M., and Gallegos, C., Development of highly-transparent protein/starch-based bioplastics, *Bioresour. Technol.* 101, 2007–2013, 2010.
92. Mathew, A.P. and Dufresne, A., Morphological investigation of nanocomposites from sorbitol plasticized starch and tunicin whiskers, *Biomacromolecules* 3, 609–617, 2002.
93. Lu, Y., Weng, L., and Cao, X., Biocomposition of plasticized starch reinforced with cellulose crystallites from cottonseed linter, *Macromol. Biosci.* 5, 1101–1107, 2005.
94. Shin, B.Y., Narayan, R., Lee, S.I., and Lee, T.J., Morphology and rheological properties of blends of chemically modified thermoplastic starch and polycaprolactone, *Poly. Sci. Eng.* 48, 2126–2133, 2008.
95. Bao, J., Xing, J., Phillips, D.L., and Corke, H., Physical properties of octenyl succinic anhydride modified rice, wheat, and potato starches, *J. Agric. Food Chem.* 51, 2283–2287, 2003.
96. Wang, J., Su, L., and Wang, S., Physicochemical properties of octenyl succinic anhydride-modified potato starch with different degrees of substitution, *J. Sci. Food Agric.* 90, 424–429, 2010.
97. Wetzel, D.L., Shi, Y.C., and Reffner, J.A., Synchrotron infrared confocal microspectroscopical detection of heterogeneity within chemically modified single starch granules, *Appl. Spectrosc.* 64, 282–285, 2010.
98. Plank, J., Applications of biopolymers and other biotechnological products in building materials, *Appl. Microbiol. Biotechnol.* 66, 1–9, 2004.
99. Karim, A.A., Sulfa, E.H., and Zaidul, I.S., Dual modification of starch via partial enzymatic hydrolysis in the granular state and subsequent hydroxypropylation, *J. Agric. Food Chem.* 56, 10901–10907, 2008.
100. Lan, C., Yu, L., Chen, P. et al., Design, preparation and characterization of self-reinforced starch films through chemical modification, *Macromol. Mater. Eng.* 295, 1025–1030, 2010.
101. Venâncio, A., Teixeira, J.A., and Mota, M., Evaluation of crude hydroxylpropyl starch as a bioseparation aqueous-phase-forming polymer, *Biotechnol. Prog.* 9, 635–639, 1993.
102. Farkas, T., Stålbrand, H., and Tjerneld, F., Partitioning of β-mannanase and α-galactosidase from *Aspergillus niger* in Ucon/Reppal aqueous two-phase systems and using temperature-induced phase separation, *Bioseparation* 6, 147–157, 1996.
103. Brecher, M.E., Owen, H.G., and Bandarenko, N., Alternatives to albumin: Starch replacement for plasma exchange, *J. Clin. Apher.* 12, 146–153, 1997.

104. Van der Linden, P. and Ickx, B.E., The effects of colloid solutions on hemostasis, *Can. J. Anaesth.* 53(6 Suppl), S30–S39, 2006.
105. Ertmer, C., Rehberg, S., Van Aken, H., and Westphal, M., Relevance of non-albumin colloids in intensive care medicine, *Best Pract. Res. Clin. Anaesthesiol.* 23, 193–212, 2009.
106. Boldt, J., Safety of nonblood plasma substitutes: Less frequently discussed issues, *Eur. J. Anaesthesiol.* 27, 495–500, 2010.
107. Friedman, R.B. (ed.), *Biotechnology of Amylodextrin Oligosaccharides*, American Chemical Society, Washington, DC, 1991.
108. Lindner, K. and Saenger, W., β-Cyclodextrin-dodecahydrat: Häufung von Wassermolekülen in einer hydrophoben Höhlung, *Angew. Chem.* 90, 738–740, 1978.
109. Albers, E. and Müller, B.W., Cyclodextrin derivatives in pharmaceutics, *Crit. Rev. Ther. Drug Carrier Syst.* 12, 311–337, 1995.
110. Strickley, R.G., Solubilizing excipients in oral and injectable formulations, *Pharm. Res.* 21, 201–230, 2004.
111. Stella, V.J. and He, Q., Cyclodextrins, *Toxicol. Pathol.* 36, 30–42, 2008.
112. Kaur, I.P., Chhabra, S., and Aggarwal, D., Role of cyclodextrins in ophthalmics, *Curr. Drug. Deliv.* 1, 351–360, 2010.
113. Bicchi, C., D'Amato, A., and Rubiolo, P., Cyclodextrin derivatives as chiral selectors for direct gas chromatographic separation of enantiomers in the essential oil, aroma, and flavor fields, *J. Chromatog. A* 843, 99–121, 1999.
114. Preskorn, S.H., Pharmacokinetics and therapeutics of acute intramuscular ziprasidone, *Clin. Pharmacokinet.* 44, 1117–1133, 2005.
115. Luke, D.B., Tomaszewski, K., Damle, B., and Schlamm, H.T., Review of the basic and clinical pharmacology of sulfobutyl-β-cyclodextrin (SBECD), *J. Pharm. Sci.* 99, 3291–3301, 2010.
116. Neely, W.B., Dextran: Structure and synthesis, *Adv. Carbohydr. Chem.* 15, 341–369, 1960.
117. Leonard, G.J., Uncertainties in the use of periodate oxidation for determination of dextran structure, *Carbohydr. Res.* 41, 143–152, 1975.
118. Seymour, F.R., Knapp, R.D., and Bishop, S.H., Determination of the structure of dextran by ^{13}C-nuclear magnetic resonance spectroscopy, *Carbohydr. Res.* 51, 179–194, 1976.
119. Cerning, J., Exocellular polysaccharides produced by lactic acid bacteria, *FEMS Microbiol. Rev.* 7, 113–130, 1990.
120. Heinze, T., Liebert, T., Heublein, B., and Hornig, S., Functional polymers based on dextrans, *Adv. Polym. Sci.* 205, 199–291, 2006.
121. Porath, J., Fractionation of polypeptides and proteins on dextran gels, *Clin. Chim. Acta* 4, 776–778, 1959.
122. Porath, J. and Flodin, P., Gel filtration: A method for desalting and group separation, *Nature* 183, 1657–1659, 1959.
123. Tiselius, A., Porath, J., and Albertsson, P.A., Separation and fractionation of macromolecules and particles, *Science* 141, 13–20, 1963.
124. Leonsins, A.J., Dextran; a valuable plasma volume expander, *S. Afr. Med. J.* 26, 546–549, 1952.
125. Bowman, H.W., Clinical evaluation of dextran as a plasma volume expander, *J. Am. Med. Assoc.* 153, 24–26, 1953.
126. Terry, R., Yuile, C.L., Golodetz, A. et al., Metabolism of dextran: As plasma volume expander; studies of radioactive carbon-labeled dextran in dogs, *J. Lab. Clin. Med.* 42, 6–15, 1953.
127. Dargan, E.L., Metcalf, W., Hehre, E.J., and Ohin, A., Clinical evaluation of a new dextran plasma expander, *JAMA* 179, 203–206, 1962.
128. Richard, A., Barras, A., Younes, A.B. et al., Minimal chemical modification of reductive end of dextran to produce an amphiphilic polysaccharides able to incorporate onto lipid nanocapsules, *Bioconjug. Chem.* 19, 1491–1495, 2008.

129. Lévesque, S.G. and Shoichet, M.S., Synthesis of cell-adhesive dextran hydrogels and macroporous scaffolds, *Biomaterials* 27, 5277–5285, 2006.
130. Van Tomme, S.R. and Hennink, W.E., Biodegradable dextran hydrogels for protein delivery applications, *Expert Rev. Med. Devices* 4, 147–164, 2007.
131. Klouda, L., Hacker, M.C., Kretlow, J.D., and Mikos, A.G., Cytocompatibility of evaluation of amphiphilic thermally responsive and chemically crosslinkable macromers for *in situ* forming hydrogels, *Biomaterials* 30, 4558–4566, 2009.
132. Sun, G., Shen, Y.I., Ho, C.C. et al., Functional groups affect physical and biological properties of dextran-based hydrogels, *J. Biomed. Mater. Res.* 93A, 1080–1090, 2010.
133. Takakura, Y., Kaneko, Y., Fujita, T. et al., Control of pharmaceutical properties of soybean trypsin inhibitor by conjugation with dextran. I: Synthesis and characterization, *J. Pharm. Sci.* 78, 117–121, 1989.
134. Sundaram, P.V. and Venkatesh, R., Retardation of thermal and urea induced inactivation of α-chymotrypsin by modification with carbohydrate polymers, *Protein Eng.* 11, 699–705, 1998.
135. Mehvar, R., Recent trends in the use of polysaccharides for improved delivery of therapeutic agents: Pharmacokinetic and pharmacodynamic perspectives, *Curr. Pharm. Biotechnol.* 4, 283–302, 2003.
136. Kristiansen, K.A., Potthast, A., and Christensen, B.E., Periodate oxidation of polysaccharides for modification of chemical and physical properties, *Carbohydr. Res.* 345, 1264–1271, 2010.
137. MacKewa, E. and Koshijima, T., Properties of 2,3-dicarboxylic cellulose combined with various metallic ions, *J. Appl. Polym. Sci.* 29, 2289–2297, 1984.
138. O'Connell, D.W., Birkinshaw, C., and O'Dwyer, T.F., Heavy metal adsorbents prepared from the modification of cellulose: A review, *Bioresour. Technol.* 99, 6709–6724, 2008.
139. Isogai, T., Saito, T., and Isogai, A., TEMPO electromediated oxidation of some polysaccharides including regenerated cellulose fiber, *Biomacromolecules* 11, 1593–1599, 2010.
140. Nyström, D., Lindqvist, J., Ostmark, E. et al., Superhydrophobic and self-cleaning bio-fiber surfaces via ATRP and subsequent postfunctionalization, *ACS Appl. Mater. Interfaces* 1, 816–823, 2009.
141. Siqueira, G., Bras, J., and Dufresne, A., A new process of chemical grafting of cellulose nanoparticles with a long chain isocyanates, *Langmuir* 26, 402–411, 2010.
142. Dugresne, A., Processing of polymer nanocomposites reinforced with polysaccharide nanocrystals, *Molecules* 15, 4111–4128, 2010.
143. Huang, H.J., Ramaswamy, S., Al-Dajani, W.W., and Tschirner, U., Process modeling and analysis of pulp mill-based integrated biorefinery with hemicellulose pre-extraction for ethanol production: A comparative study, *Bioresour. Technol.* 101, 624–631, 2010.
144. Mao, J.D., Holtman, K.M., and Franqui-Villanueva, D., Chemical structures of corn stover and its residue after dilute acid prehydrolysis and enzymatic hydrolysis: Insight into factors limiting enzymatic hydrolysis, *J. Agric. Food Chem.* 58, 11680–11687, 2010.
145. Cowman, M.K. and Matsuoka, S., Experimental approaches to hyaluronan structure, *Carbohydr. Res.* 340, 791–809, 2005.
146. Fam, H., Kontopoulou, M., and Bryant, J.T., Effect of concentration and molecular weight on the rheology of hyaluronic acid/bovine calf serum solutions, *Biorheology* 46, 31–43, 2009.
147. Guillaumie, F., Furrer, P., Felt-Baeyens, O. et al., Comparative studies of various hyaluronic acids produced by microbial fermentation for potential topical ophthalmic applications, *J. Biomed. Mater. Res. A* 92, 1421–1430, 2010.
148. Nyström, B., Kjøniksen, A.L., Beheshti, N. et al., Characterization of polyelectrolyte features in polysaccharide systems and mucin, *Adv. Colloid Interface Sci.* 158, 108–118, 2010.

149. James, D.F., Rick, G.M., and Baines, W.D., A mechanism to explain physiological lubrication, *J. Biomech. Eng*. 132, 071002, 2010.
150. Zhao, X.B., Fraser, J.E., Alexander, C. et al., Synthesis and characterization of a novel double crosslinked hyaluronan hydrogel, *J. Mater. Sci. Mater. Med.* 13, 11–16, 2002.
151. Segura, T., Anderson, B.C., Chung, P.H. et al., Crosslinked hyaluronic acid hydrogels: A strategy to functionalize and pattern, *Biomaterials* 26, 359–371, 2005.
152. Richter, C., Reinhardt, M., Giselbrecht, S. et al., Spatially controlled cell adhesion on three-dimensional substrates, *Biomed. Microdevices* 12, 787–795, 2010.
153. Jha, A.K., Malik, M.S., Farach-Carson, M.C. et al., Hierarchically structured, hyaluronic acid-based hydrogel matrices via the covalent integration of microgels into macroscopic networks, *Soft Matter* 6, 5045–5055, 2010.
154. Serban, M.A., Yang, G., and Prestwich, G.D., Synthesis, characterization and chondroprotective properties of a hyaluronan thioethyl ether derivative, *Biomaterials* 29, 1388–1399, 2008.
155. Mlcochová, P., Bystrický, S., Steiner, B. et al., Synthesis and characterization of new biodegradable hyaluronan alkyl derivatives, *Biopolymers* 82, 74–79, 2006.
156. Lindahl, U., Structure of the heparin-protein linkage region, *Arkiv. Kemi.* 26, 101–110, 1966.
157. Seethanathan, P. and Ehrlich, K., Anticoagulant and antilipemic activities of heparin proteoglycan from bovine intestinal mucosa, *Thromb. Res.* 19, 95–102, 1980.
158. Lundblad, R.L., Brown, W.V., Mann, K.G., and Roberts, H.R. (eds.), *Chemistry and Biology of Heparin*, Elsevier/North Holland, New York, 1981.
159. Iverius, P.H., Coupling of glycosaminoglycans to agarose beads (Sepharose 4B), *Biochem. J.* 124, 677–683, 1971.
160. Gentry, P.W. and Alexander, B., Specific coagulation adsorption to insoluble heparin, *Biochem. Biophys. Res. Commun.* 50, 500–509, 1973.
161. Danishefsky, I., Tzeng, F., Ahrens, M., and Klein, S., Synthesis of heparin-sepharoses and their binding with thrombin and antithrombin-heparin cofactor, *Thromb. Res.* 8, 131–140, 1976.
162. Nadkarni, V.D., Pervin, A., and Linhardt, R.J., Directional immobilization of heparin onto beaded supports, *Anal. Biochem.* 222, 59–67, 1994.
163. Fry, A.K., Schilke, K.F., McGuire, J., and Bird, K.E., Synthesis and anticoagulant activity of heparin immobilized "end-on" to polystyrene microspheres coated with end-group activated polyethylene oxide, *J. Biomed. Mater. Res. B* 94, 187–198, 2010.
164. Kiick, K.L., Peptide- and protein-mediated assembly of heparinized hydrogels, *Soft Matter* 4, 29–37, 2008.
165. Tae, G., Kim, Y.J., Choi, W.I. et al., Formation of a novel heparin-based hydrogel in the presence of heparin-binding biomolecules, *Biomacromolecules* 8, 1979–1986, 2007.
166. Kim, M., Lee, J.Y., Jones, C.N. et al., Heparin-based hydrogel as a matrix for encapsulation and cultivation of primary hepatocytes, *Biomaterials* 31, 3596–3603, 2010.
167. Morgulis, S., The chemical constitution of chitin, *Science* 44, 866–867, 1916.
168. Alves, N.M. and Mann, J.F., Chitosan derivatives obtained by chemical modification for biomedical and environmental applications, *Int. J. Biol. Macromol.* 43, 401–414, 2008.
169. Tharanthan, R.N. and Kittur, F.S., Chitin—The undisputed biomolecule of great potential, *Crit. Rev. Food Sci. Nutr.* 43, 61–87, 2003.
170. Muzzarelli, R.A.A., *Chitin*, Pergamon Press, New York, 1977.
171. Li, X., Liu, X., Dong, W. et al., *In vitro* evaluation of porous poly (L-lactic acid) scaffold reinforced by chitin fibers, *J. Biomed. Mater. Res. B* 90, 503–509, 2009.
172. Ge, H., Zhao, B., Lai, Y. et al., From crabshell to chitosan-hydroxyapatite composite material via a biomorphic mineralization synthesis method, *J. Mater. Sci. Mater. Med.* 21, 1781–1787, 2010.

173. Swetha, M., Sahithi, K., Moothi, A. et al., Biocomposites containing natural polymers and hydroxyapatite for bone tissue engineering, *Int. J. Biol. Macromol.* 47, 1–4, 2010.
174. Kurita, K., Yoshida, Y., and Umemura, T., Finely selective protection and deprotection of multifunctional chitin and chitosan to synthesize key intermediates for regioselective chemical modifications, *Carbohyd. Polym.* 81, 434–440, 2010.
175. Kurita, K., Ishii, S., Tomita, K. et al., Reactivity characteristics of squid β-chitin as compared to those of shrimp chitin: High potentials of squid chitin as a starting material for facile chemical modifications, *J. Polymer Sci. A* 32, 1027–1032, 1994.
176. Kurita, K., Sugita, K., Kodaira, N. et al., Preparation and evaluation of trimethylsilylated chitin as a versatile precursor for facile chemical modifications, *Biomacromolecules* 6, 1414–1418, 2005.
177. Gopalan Nair, K., Dufresne, A., Gandni, A., and Gelgacem, M.N., Crab shell chitin whiskers reinforced natural rubber nanocomposites. 1. Processing and swelling behavior, *Biomacromolecules* 4, 657–665, 2003.
178. Lertwattanaseri, T., Ichikawa, N., Mizoguchi, T. et al., Microwave technique for efficient deactylation of chitin nanowhiskers to a chitosan nanoscaffold, *Carbohydr. Res.* 344, 331–335, 2009.
179. Gopalan Nair, K., Dufresne, A., Gandni, A., and Gelgacem, M.N., Crab shell chitin whiskers reinforced natural rubber nanocomposites. 3. Effects of chemical modification of chitin whiskers, *Biomacromolecules* 4, 1835–1842, 2003.
180. Cunha, A.G. and Gandini, A., Turning polysaccharides into hydrophobic materials: A critical review. Part 2. Hemicelluloses, chitin/chitosan, starch, pectin and alginates, *Cellulose* 17, 1045–1065, 2010.
181. Hirano, S. and Usutani, A., Hydrogels of *N*-acylchitosans and their cellulose composites generated from the aqueous alkaline solutions, *Int. J. Biol. Macromol.* 20, 245–249, 1997.
182. Tsigos, I., Martinou, A., Kafetzopoulos, D. et al., Chitin deacetylases: New, versatile tools in biotechnology, *Trends Biotechnol.* 18, 305–312, 2000.
183. Win, N.N. and Stevens, W.F., Shrimp chitin as substrate for fungal chitin deacetylase, *Appl. Microbiol. Biotechnol.* 57, 334–341, 2001.
184. Li, X., Wang, L.X., Wang, X., and Roseman, S., The chitin catabolic cascade in the marine bacterium *Vibrio cholerae*: Characterization of a unique chitin oligosaccharide deacetylase, *Glycobiology* 17, 1377–1387, 2007.
185. Zhang, J., Xia, W., Liu, P. et al., Chitosan modification and pharmaceutical/biomedical applications, *Mar. Drugs* 8, 1962–1987, 2010.
186. Prabaharan, M. and Mano, J.F., Chitosan-based particles as controlled drug delivery systems, *Drug. Deliv.* 12, 41–57, 2005.
187. Park, J.H., Saravanakumar, G., Kim, K., and Kwon, I.C., Targeted delivery of low molecular drugs using chitosan and its derivatives, *Adv. Drug. Deliv. Rev.* 62, 28–41, 2010.
188. Tong, H., Qin, S., Fernandex, J.C. et al., Progress and prospects of chitosan and its derivatives as non-viral gene vectors in gene therapy, *Curr. Gene Ther.* 9, 495–502, 2009.
189. Xu, Q., Wang, C.H., and Pack, D.W., Polymeric carriers for gene delivery: Chitosan and poly(amidoamine) dendrimers, *Curr. Pharm. Des.* 16, 2350–2368, 2010.
190. Rudzinski, W.E. and Aminabhavi, T.M., Chitosan as a carrier for targeted delivery of small interfering RNA, *Int. J. Pharm.* 399, 1–11, 2010.
191. Raviña, M., Cubillo, E., Olmeda, D. et al., Hyaluronic acid/chitosan-g-poly(ethylene glycol) nanoparticles for gene therapy: An application for pDNA and siRNA delivery, *Pharm. Res.* 27, 2544–2555, 2010.
192. Alatorre-Meda, M., Taboada, P., Hartl, F. et al., The influence of chitosan valence on the complexation and transfection of DNA: The weaker the DNA-chitosan binding the higher the transfection efficiency, *Colloids Surf. B. Biointerfaces* 82, 54–62, 2011.

193. Maurstad, G., Danielsen, S., and Stokke, B.T., The influence of charge density of chitosan in the compaction of the polyanions DNA and xanthan, *Biomacromolecules* 8, 1124–1130, 2007.
194. *Oxford English Dictionary*, Oxford University Press, Oxford, U.K., 2010.
195. Hejazi, R. and Amiji, M., Chitosan-based gastrointestinal delivery systems, *J. Control. Release* 89, 151–165, 2003.
196. Chourasia, M.K. and Jain, S.K., Polysaccharides for colon targeted drug delivery, *Drug Deliv.* 11, 129–148, 2004.
197. Kosaraju, S.L., Colon targeted delivery systems: Review of polysaccharides for encapsulation and delivery, *Crit. Rev. Food Sci. Nutr.* 45, 251–258, 2005.
198. Saboktakin, M.R., Tabatabaie, R.M., Maharramov, A., and Ramzanov, M.A., Synthesis and characterization of chitosan hydrogels containing 5-aminosalicylic acid nanopendents for colon: Specific drug delivery, *J. Pharm. Sci.* 99, 4955–4961, 2010.
199. Chávarri, M., Marañón, I., Area, R. et al., Microencapsulation of a probiotic and prebiotic in alginate-chitosan capsules improves survival in simulated gastro-intestinal conditions, *Int. J. Food Microbiol.* 142, 185–189, 2010.
200. Dubey, R., Dubey, R., Omrey, P. et al., Development and characterization of colon specific drug delivery systems bearing 5-ASA and camylofine dihydrochloride for treatment of ulcerative colitis, *J. Drug Target.* 18, 589–601, 2010.
201. Thakral, N.K., Ray, A.R., and Majumdar, D.K., Eudragit S-100 entrapped chitosan microspheres of valdecoxib for colon cancer, *J. Mater. Sci. Mater. Med.* 21, 2691–2699, 2010.
202. Hiorth, M., Skøien, T., and Sande, S.A., Immersion coating of pellet cores consisting of chitosan and calcium intended for colon drug delivery, *Eur. J. Pharm. Biopharm.* 75, 245–253, 2010.
203. Kadiyala, I., Loo, Y., Roy, K. et al., Transport of chitosan-DNA nanoparticles in human intestinal M-cell model versus normal intestinal enterocytes, *Eur. J. Pharm. Sci.* 39, 103–109, 2010.
204. Laroui, H., Theiss, A.L., Yan, Y. et al., Functional TNFα gene silencing mediated by polyethyleneimine/TNFα siRNA nanocomplexes in inflamed colon, *Biomaterials* 32, 1218–1228, 2011.
205. Jenkins, D.W. and Hudson, S.M., Heterogeneous graft copolymerization of chitosan power with methyl acrylate using trichloroacetyl-manganese carbonyl co-initiation, *Macromolecules* 35, 3413–3419, 2002.
206. Prashanth, K.V.H. and Tharanathan, R.N., Studies on graft polymerization of chitosan with synthetic monomers, *Carbohyd. Polym.* 54, 343–351, 2003.
207. Lv, P., Bin, Y., Yongqiang, C. et al., Studies on graft copolymerization of chitosan with acrylonitrile by the redox system, *Polymer* 50, 5675–5680, 2009.
208. Ulbricht, M. and Riedel, M., Ultrafiltration membrane surfaces with grafted polymer 'tentacles': Preparation, characterization and application for covalent protein binding, *Biomaterials* 19, 1229–1237, 1998.
209. El-Sherbiny, J.M., Aldel-Bary, E.M., and Harding, D.R.K., Preparation and *in vitro* evaluation of new, pH-sensitive hydrogel beads for oral delivery of protein drugs, *J. Appl. Polym. Sci.* 115, 2828–2837, 2010.
210. El-Sherbiny, E.M. and Elmahdy, M.M., Preparation, characterization, structure, and dynamics of carboxymethyl chitosan grafted with acrylic acid sodium salt, *J. Appl. Polym. Sci.* 118, 2134–2143, 2010.

Index

A

I

L

M

N

P

T

V